Managing Forest Ecosystems

Volume 38

Well-managed forests and woodlands are a renewable resource, producing essential raw material with minimum waste and energy use. Rich in habitat and species diversity, forests may contribute to increased ecosystem stability. They can absorb the effects of unwanted deposition and other disturbances and protect neighbouring ecosystems by maintaining stable nutrient and energy cycles and by preventing soil degradation and erosion. They provide much-needed recreation and their continued existence contributes to stabilizing rural communities.

Forests are managed for timber production and species, habitat and process conservation. A subtle shift from multiple-use management to ecosystems management is being observed and the new ecological perspective of multi-functional forest management is based on the principles of ecosystem diversity, stability and elasticity, and the dynamic equilibrium of primary and secondary production.

Making full use of new technology is one of the challenges facing forest management today. Resource information must be obtained with a limited budget. This requires better timing of resource assessment activities and improved use of multiple data sources. Sound ecosystems management, like any other management activity, relies on effective forecasting and operational control.

More information about this series at http://www.springer.com/series/6247

Gidi Ne'eman • Yagil Osem
Editors

Pines and Their Mixed Forest Ecosystems in the Mediterranean Basin

Springer

Editors
Gidi Ne'eman
Department of Biology and the Environment
University of Haifa-Oranim
Tivon, Israel

Yagil Osem
Department of Natural Resources – Institute of Plant Sciences
Agricultural Research Organization – Volcani Center
Rishon LeZion, Israel

ISSN 1568-1319 ISSN 2352-3956 (electronic)
Managing Forest Ecosystems
ISBN 978-3-030-63624-1 ISBN 978-3-030-63625-8 (eBook)
https://doi.org/10.1007/978-3-030-63625-8

This Springer imprint is published by the registered company Springer Nature Switzerland AG
The registered company address is: Gewerbestrasse 11, 6330 Cham, Switzerland

Foreword

This book, titled *Pines and Their Mixed Forest Ecosystems in the Mediterranean Basin*, edited by Prof. Ne'eman and Dr. Osem, is a collection of 34 chapters contributed by top Mediterranean researchers; it summarizes the up-to-date knowledge that has accumulated during the last two decades. The book deals with many aspects concerning the major pine species growing in various habitats in the Mediterranean Basin with the exception of *P. canariensis*. The chapters are organized into seven parts, each dedicated to a specific subject. This solid scientific information should be the basis for constructing local and specific forest management policies and action for the development of future diverse and sustainable forests that will provide a wide range of ecological services for the benefit of local human populations.

Part I. Evolution, Genetics and Distribution includes five chapters: **Grivet and Olsson** review the phylogenetic, phylogeographic, and demographic relationships among nine Mediterranean pine species and their responses to variation in climate and fire regimes. All these are essential to understanding the past evolutionary success of Mediterranean pines, their current state, and their future response to environmental changes. **Vendramin et al.** review the population genetics and genomics of Aleppo pine within the Mediterranean Basin. Based on molecular markers, this chapter presents recent knowledge that has accumulated on population diversity and structure, demographic history, gene flow, and adaptation. It also provides an historical overview of the evolutionary history of *P. halepensis*. **Climent et al.** review the conservation and breeding of Mediterranean pines. They analyze the different evolutionary factors affecting the conservation of genetic resources in Mediterranean pines, the threats to these genetic resources, and the strategies and actions implemented for their conservation. They also review the breeding activities for Mediterranean pines and their achievements. **Benito Garzón and Vizcaino-Palomar** review the main drivers of local adaptation and phenotypic plasticity and their contribution to phenotypic variation in fitness-related traits across the distribution ranges of Mediterranean pines. They also present examples of species distribution models based on tree growth measured in common gardens. **Richardson and Nsikani** update the increasing importance of Mediterranean pines as invasive species in temperate parts of the Southern Hemisphere. They review human-mediated changes to the ranges of Mediterranean pines, the history of plantings, and the emergence of invasions by these pines in the Southern Hemisphere.

Part II. Ecophysiology of Mediterranean Pines: Resistance to Stress and Perturbation includes six chapters: **Resco de Dios** reviews the literature to understand the coordination and trade-offs between physiological mechanisms conferring resistance to drought, shade, and fire. He concludes that these trade-offs can partially explain the distribution of conifer species across productivity gradients. **Klein** reviews and synthesizes the recent literature on carbon allocation dynamics in Mediterranean pines. Allocation to stem growth is highly plastic in phenology and magnitude, whereas allocation to needle growth is sensitive to drought and competition. Relocation fluxes occur during summer drought, and these carbon fluxes might represent an important drought acclimation mechanism. **Sarris and Mazza** review the capabilities of the root systems of Mediterranean pines to adapt and function under drought and adverse soil and bedrock conditions. Pines can explore both shallow and deep-water resources under drought and are adapted to long summer drought periods. **Rubio-Cuadrado et al.** present a general review of tree mortality events in Mediterranean pinewoods. They identify the main mortality events and examine the importance of climate change, past management, land-use, pests, and diseases, as factors causing tree mortality. They also revise the main mitigation practices advocated in the academic literature. **Oliva** discusses the dieback of weakened trees and forest stands caused by fungi as a result of increasing aridity. The main threat is from invasive pathogens that enter Europe via the live plant trade. However, native pathogens may become more aggressive under global warming. Several research directions and management recommendations are provided. **Vázquez-González et al.** review how pine populations are expected to cope with the increased activity of insects. Defense production can be fine-tuned by induced defenses; thus, pine resistance to biotic threats must be understood as a general optimal resource allocation problem across life functions and biotic and abiotic environments.

Part III. Pine and Mixed Forest Ecosystems Under Global Climate Change includes six chapters: **Camarero et al.** review the effects of global change on the growth and vigor of Mediterranean pines. They analyze annual changes in tree defoliation, a proxy of vigor, mortality, and radial growth. Peaks in severe defoliation patterns reflect punctual responses to extreme drought and heat stress and often precede canopy dieback events. They discuss possible responses of some Mediterranean pine species to global change based on their observed reactions to drought stress. **Herrero et al.** assess the vulnerability and resilience of Mediterranean pines under global climate change. Using long-term or large-scale demographic data, they analyze drought-related impacts and vulnerabilities. They suggest that pine species are highly vulnerable to climate change in the Mediterranean Basin, especially small-sized trees. **Rotenberg et al.** present a case study of carbon and energy balance of two dry Mediterranean Aleppo pine forests. Yatir Forest, growing in a desert site with seven dry months and average rainfall of 285 mm, was a carbon sink during the wet season and a carbon source in summer. Birya forest, with 755 mm annual precipitation, was a carbon sink both in the wet and dry seasons. This case study shows that pine forests can adjust to conditions at the dry, semi-arid timberline and store significant amounts of carbon below ground with a long residence time. **Gallardo** reviews key indicators of carbon and nutrient cycling, such as microbial biomass, soil respiration, enzyme activity, and nutrient availability. Soil properties and biogeochemical functioning in Mediterranean pine forests are characterized by wet and cool winters in contrast to dry and warm summers, but soil dry-rewetting cycles are also frequent in winter. This restricts soil microbial activity to short periods and determines the unique properties of nutrient cycling in these ecosystems. **Grünzweig and Gliksman** review litter decomposition as affected by climatic factors. They assess the potential impact of climate change by direct and indirect impacts through changes in forest structure and composition. Global change with decreased water availability is expected to reduce rates of rain-based microbial degradation of plant litter, while increases in the openness of canopies due to defoliation or mortality may enhance photo- and thermal degradation and microbial activity, driven by air humidity, dew, or fog.

Part IV. Forest Dynamics, Biodiversity and Biotic Interactions includes four chapters: **Waitz and Sheffer** review the dynamics of mixed pine-oak forests. A long history of large-scale afforestation with pines, changes in land use and management, and conservation priorities has created a mosaic of contiguous patches of pine stands and oak woodlands. This has created opportunities for cross-colonization. Using frameworks of forest succession and gap dynamics they analyze the evidence for pine and oak regeneration within mixed pine-oak forests, and evaluate the fate of Mediterranean mixed forests. **Díaz et al.** review bird diversity in Mediterranean pine and mixed forests. Many forest birds occupy pine and mixed forests due to the biogeographic history of Mediterranean forest birds. Pine forest bird communities show clear responses to changes in forest distribution, structure, and composition, at both local and landscape scales. Bird diversity provides the regulatory ecosystem services of pest control and seed dispersal, cultural services, and even production services in the case of game birds. Forest plantation and harvesting, forest and landscape fragmentation, and climate change may have strong negative effects on bird communities. **Radea and Arianoutsou** review soil micro- and macro-fauna including species richness, spatiotemporal diversity, and surface activity of soil invertebrate communities in various Mediterranean pure and mixed pine forests. They also discuss the effects of methodological issues and knowledge gaps in systematics and ecology on the research of soil fauna. **Rincón et al.** review mycorrhizae in Mediterranean pine and mixed forests. Mycorrhizal fungi are particularly important for pines that need them for survival and development. They improve water and nutrient uptake by the host tree, in exchange for photosynthetic carbohydrates, directly affecting the productivity and response of trees to the environmental stress that is particularly important in the context of global change.

Part V. Forest Fire Ecology and Management includes 4 chapters: **Moreno et al.** review past, current, and future fire activity in Mediterranean pine forests. Fire activity has increased during the Holocene alongside human activities, leading to the disappearance of many pinewoods. Fire frequency was low until the mid-twentieth century; it then increased, reaching maximum values in the late 1980s.

Thereafter, the burnt area decreased probably due to fire suppression. Global climate change is expected to increase fire frequency and severity; thus, despite growing firefighting capacity, the risk of extremely large and devastating fires is expected to increase. **Ne'eman and Arianoutsou** focus on the adaptation of pines to fire. They describe the main fire-related traits that contribute to pine species fitness in response to fire, and discuss their syndromes and evolutionary history. They discuss the relationships between the fire-adaptive traits of different Mediterranean pine species in relation to fire-related conditions in their native habitats. **Wittenberg** reviews the issue of post-fire soil erosion. Fires affect vegetation structure and soil-forming dynamics. They increase runoff and soil erosion by removing vegetation, and change soil properties. Fire-induced erosion varies markedly across spatio-temporal scales; at the burned plot scale it is 5–7 orders of magnitude higher than at unburned sites. Ecosystem recovery may last decades, during which the soil is exposed to elevated erosion, considerably higher than natural soil formation rates. **Ungar et al.** present a case study about the effect of small-ruminant grazing on the reduction of the understory fuel load for reducing fire hazards. In order to estimate fire hazards, they implement Rothermel's surface-fire model in a spreadsheet application that connects forest survey data with the input parameters of the model. They use local data sets to examine the ranges of biomass variation in the major understory fuel types and explore the response of the model to variation within those ranges. They used GPS tracking data to estimate offtake by grazing animals, in units of animal-days per hectare. According to the model, the biomass variation of the edible fuel has a strong effect on rate of spread of a surface fire and the risk of a crown fire. The authors also discuss the role of non-consumptive processes such as trampling, and what grazing can and cannot be expected to achieve. **Castro** discusses post-fire restoration. The increase in fire severity and frequency in recent decades often makes restoration actions necessary. Recovery of burnt pine stands is currently based largely on immediate and intense management actions across the landscape. These actions include post-fire salvage logging, mulching, construction of log erosion barriers, and planting, which involve the use of heavy machinery. However, such actions may hamper natural regeneration by negatively affecting nutrient cycling, plant-animal interactions, seedlings, and the resprout (bud) bank. Castro concludes that restoration should focus on promoting the elements of the ecosystem that accelerate recovery of the desired forest while reducing management impacts. **Williams et al.** focus on post-fire regeneration in the yellow pine and mixed conifer forests of California, under a Mediterranean-type climate. In recent years, researchers have developed models and other tools to predict the extent to which natural regeneration will meet desired tree density thresholds. They describe the current state of post-fire regeneration modeling, how management and restoration efforts can benefit from these new information sources, and how such efforts can evolve to better meet management needs as the climate warms and fires become more extensive and severe. The use of these tools should also be examined for pine species in the Mediterranean Basin. **Fernández-Manso et al.** present a case study that tests remote sensing techniques for monitoring fire damage and recovery of *P. pinaster* and *P. halepensis* forests. Derived products from satellite images are becoming the most effective tool for analyzing and mapping post-fire damage, and for monitoring post-fire forest recovery. However, remote sensing-based fire damage maps should be validated with field data. They review current techniques including close-range, airborne, and satellite approaches (thermal and optical multispectral) and RADAR and LiDAR techniques.

Part VI. Policy and Ecosystem Services of Mediterranean Pine and Mixed Forest Ecosystems includes three chapters: **Picard and Garavaglia** examine Mediterranean forests in the light of the United Nations Sustainable Development Goals. Several of the sustainable development goals defined in the 2030 Agenda for sustainable development of the United Nations are relevant for Mediterranean pine and mixed forests, especially those related to the restoration of degraded forest ecosystems and to climate change mitigation and adaptation. **Torres et al.** review the ecosystem services provided by pine forests and plantations that are responsible for provisioning services such as timber and firewood, but are also rich in non-wood forest products such as resin, pine nuts and edible mushrooms. Of similar importance are the regulating services that pine woodlands provide, such as preventing soil erosion and hydrological regulation, carbon stocking, and their importance as habitats for multiple species. Pine forests also provide cultural and recreation services. Increasing temperatures and drought negatively affect pine forests and the services they supply. **Orenstein** assesses the cultural ecosystem services supplied by Mediterranean pine and mixed forests. Assessing cultural ecosystem services is challenging, because of the dynamic nature of socio-ecological systems that are highly context-specific, subject to change in social context, shifts in ecosystem structure, and function. He presents an inventory of the cultural services provided by, and relational values inspired by, pine and mixed-pine forests around the Mediterranean Basin and presents a case study of pine forests in Israel as an example of the challenges of assessing cultural services.

Part VII. Afforestation and Forest Management Policy includes four chapters: **Coll et al.** review the dynamics and management of western Mediterranean pinewoods that are dominant in many western Mediterranean forest landscapes. They review the origin of pine wood ecosystems, the main biotic and abiotic drivers of their current dynamics, and the factors modulating pine survival and growth. They highlight the importance of natural diversification processes for the resilience of these systems to natural disturbances and review the main silvicultural models and treatments that are applied to achieve different management objectives. **Osem and Moshe** presents the change from pine monocultures to mixed forest ecosystems in Israel that are driven by two key processes: pine regeneration by seedling recruitment and local reestablishment of native woodland. *Pinus halepensis* regeneration occurs in sites with annual rainfall exceeding 350-400 mm and more on soft than on hard bedrock, while it is negatively affected by overstory cover and grazing. Regeneration of *P. brutia* is limited, although overstory thinning appears effective in its promotion. Reestablishment of natural woodland occurs in more humid sites (> 450 mm) and increases sharply with rainfall. Woodland recruitment is also positively influenced by overstory cover. Management guidelines are proposed for converting old pine monocultures into complex, sustainable mixed forest ecosystems.

Osem et al. describe the changes in forest policy in Israel from the 1930s until today. The first forest generation was characterized by dense, monospecific conifer plantations mainly of the native Aleppo pine. In 1990, a new management policy was applied for enhancing the development of uneven-aged, mixed conifer-broad-leaved forests. Since the 1990s, a gradual development of forestry concepts in Israel has led to the embracing of a new forest management policy. The conceptual approach of this new policy includes four major changes: (1) from forest establishment to forest management, (2) from forest condition to defined forest goals, (3) from trees to ecosystems, and (4) from maximal control to rational intervention. Long-term forest plans have been initiated with the aim of providing a variety of ecosystem services. **Sabater et al.** review water-based forest management of Mediterranean pine forests. In a global change context, the predicted increments in aridity in the Mediterranean Basin render the need for in-depth ecohydrological knowledge. They describe the main functional characteristics by which Mediterranean pine forests cope with aridity and drought episodes, their water-related interactions with soil and atmosphere, and current and predicted water balance responses to hydrologically oriented forest management.

Gabriel Schiller

Professor Emeritus, Department of Field
Crops and Natural Resources
Agricultural Research Organization, Volcani Center,
Bet Dagan, Israel

Preface

The first compilation of available scientific and practical knowledge about Mediterranean pine trees and forests, *Ecology, Biogeography and Management of Pinus halepensis and P. brutia Forest Ecosystems in the Mediterranean Basin*, was published in 2000 following the first MEDPINE conference in Israel. Since then, the growing threats to forests due to global climate change and the growing human population and activity in the Mediterranean landscape have triggered many new research studies concerning the current and future states of Mediterranean pines and their forest ecosystems. The development of new scientific approaches, methods and instrumentations to cope with the emerging issues has resulted in an exponential increase of new knowledge published in various scientific journals. In an effort to gather and organize this vast amount of knowledge and information, this book includes a series of review chapters on major themes written by leading researchers, mainly from Mediterranean countries. This up-to-date knowledge addresses scientists, students, foresters and other land managers, the interested public, and decision makers in particular.

We thank our colleagues who served as co-editors – Gabriel Schiller, Delphine Grivet, Tamir Klein, Avi Perevolotsky, Efrat Sheffer and Asaf Karavani – for their considerable help in recruiting top researchers and reviewing the submitted manuscripts.

We thank the KKL-JNF Forestry Department for their financial support, which was essential for the scientific and language editing of the book.

Tivon, Israel — Gidi Ne'eman

Bet Dagan, Israel — Yagil Osem

Contents

Contributors

Ricardo Alía Forest Research Centre, National Institute of Agronomic Research, Madrid, Spain

Paraskevi Alizoti Aristotelian University, Thesaloniki, Greece

Aitor Ameztegui Department of Agriculture and Forest Engineering, University of Lleida, Lleida, Spain

Enrique Andivia Biodiversity, Ecology and Evolution Department, Universidad Complutense de Madrid, Madrid, Spain

Margarita Arianoutsou Department of Ecology and Systematics, National and Kapodistrian University of Athens, Athens, Greece

Mor Ashkenazi Department of Natural Resources, Institute of Plant Sciences, Agricultural Research Organization – The Volcani Center, Rishon LeZion, Israel

Camilla Avanzi Institute of Biosciences and Bioresources, National Research Council, Sesto Fiorentino, Italy

Juan Bellot Department of Ecology, University of Alicante, Alicante, Spain

The Multidisciplinary Institute for Environmental Studies (IMEM), University of Alicante, Alicante, Spain

Omri Bonneh Northern Region, Keren Kayemeth LeIsrael (KKL), Jerusalem, Israel

Rafael Calama INIA-CIFOR, Madrid, Spain

iuFOR. Sustainable Forest Management Research Institute UVa-INIA, Madrid, Spain

Leonor Calvo Area of Ecology, Faculty of Biological and Environmental Sciences, University of León, León, Spain

J. Julio Camarero Instituto Pirenaico de Ecología (IPE-CSIC), Zaragoza, Spain

Jorge Castro Department of Ecology, University of Granada, Granada, Spain

José Climent Forest Research Centre, National Institute of Agronomic Research, Madrid, Spain

Lluís Coll Department of Agriculture and Forest Engineering, University of Lleida, Lleida, Spain

Joint Research Unit CTFC-AGROTECNIO, Solsona, Spain

Michelle Coppoletta University of California, Davis, CA, USA

USDA Forest Service, Pacific Southwest Research Station, Albany, CA, USA

USDA Forest Service, Pacific Southwest Region, Vallejo, CA, USA

Víctor Resco de Dios School of Life Science and Engineering, Southwest University of Science and Technology, Mianyang, China

Joint Research Unit CTFC-Agrotecnio, Universitat de Lleida, Lleida, Spain

Sergio de Miguel Department of Crop and Forest Sciences, University de Lleida (UdL), Lleida, Spain

Joint Research Unit CTFC – AGROTECNIO – CERCA, Solsona, Spain

Antonio del Campo Re-ForeST (IIAMA), Universitat Politècnica de València, Valencia, Spain

Mario Díaz Department of Biogeography and Global Change, National Museum of Natural Sciences (BGC-MNCN-CSIC), Madrid, Spain

Fulvio Ducci CREA, Arezzo, Italy

Becky Estes University of California, Davis, CA, USA

USDA Forest Service, Pacific Southwest Research Station, Albany, CA, USA

USDA Forest Service, Pacific Southwest Region, Vallejo, CA, USA

Bruno Fady INRAE, Avignon, France

Alfonso Fernández-Manso Agrarian Science and Engineering Department, University of León, Ponferrada, Spain

Antonio Gallardo Department of Physical, Chemical and Natural Systems, University Pablo de Olavide, Sevilla, Spain

Valentina Garavaglia Food and Agriculture Organization of the United Nations, Forestry Department, Rome, Italy

Marta Benito Garzón University of Bordeaux, INRAE, BIOGECO, Pessac, France

Antonio Gazol Instituto Pirenaico de Ecología (IPE-CSIC), Zaragoza, Spain

Luis Gil Departamento de Sistemas y Recursos Naturales, Escuela Técnica Superior de Ingeniería de Montes, Forestal y del Medio Natural, Universidad Politécnica de Madrid, Madrid, Spain

Daniel Gliksman Faculty of Environmental Sciences, Institute for Hydrology and Meteorology, Technische Universität Dresden, Tharandt, Germany

Santiago C. González-Martínez University of Bordeaux, INRAE, UMR BIOGECO, Cestas, France

Delphine Grivet Department of Forest Ecology and Genetics, Forest Research Centre, INIA-CSIC, Madrid, Spain

José M. Grünzweig Faculty of Agriculture, Food and Environment, The Hebrew University of Jerusalem, Rehovot, Israel

Asier Herrero Forest Ecology and Restoration Group, Departamento de Ciencias de la Vida, Universidad de Alcalá, Alcalá de Henares, Madrid, Spain

Reuven Horn Rangeland Service, Ministry of Agriculture and Rural Development, Rishon LeZion, Israel

Tamir Klein Department of Plant and Environmental Sciences, Weizmann Institute of Science, Rehovot, Israel

Yusuf Kurt Molecular Biology and Genetics Department, Harran University, Sanliurfa, Turkey

Elisabeth Karlstad Larsen Mediterranean Center for Environmental Studies (CEAM Foundation), Joint Research Unit University of Alicante-CEAM, Alicante, Spain

Royal Horticultural Society Science and Collections, Wisley, UK

Jonathan W. Long University of California, Davis, CA, USA

USDA Forest Service, Pacific Southwest Research Station, Albany, CA, USA

USDA Forest Service, Pacific Southwest Region, Vallejo, CA, USA

Rosana López Departamento de Sistemas y Recursos Naturales, Escuela Técnica Superior de Ingeniería de Montes, Forestal y del Medio Natural, Universidad Politécnica de Madrid, Madrid, Spain

Xosé López-Goldar Misión Biológica de Galicia – Consejo Superior de Investigaciones Científicas, Pontevedra, Spain

Department of Ecology and Evolutionary Biology, Cornell University, Ithaca, NY, USA

Manuel E. Lucas-Borja Higher Technical School of Agricultural and Forestry Engineering, Castilla-La Mancha University, Albacete, Spain

Jaime Madrigal-González Climate Change Impacts and Risks in the Anthropocene (C-CIA), Institute for Environmental Sciences (ISE), University of Geneva, Geneva, Switzerland

Elena Marcos Area of Ecology, Faculty of Biological and Environmental Sciences, University of León, León, Spain

María Martínez-Jauregui National Institute for Agriculture and Food Research and Technology (INIA), Forest Research Centre (CIFOR), Madrid, Spain

Gianluigi Mazza CREA Research Centre for Forestry and Wood, Arezzo, Italy

Marc D. Meyer University of California, Davis, CA, USA

USDA Forest Service, Pacific Southwest Research Station, Albany, CA, USA

USDA Forest Service, Pacific Southwest Region, Vallejo, CA, USA

Maya Millet The Robert H Smith Faculty of Agriculture, Food and Environment, The Hebrew University of Jerusalem, Rehovot, Israel

César Morales-Molino Institute of Plant Sciences and Oeschger Centre for Climate Change Research, University of Bern, Bern, Switzerland

Luna Morcillo Mediterranean Center for Environmental Studies (CEAM Foundation), Joint Research Unit University of Alicante-CEAM, Alicante, Spain

José M. Moreno Departamento de Ciencias Ambientales, Universidad de Castilla-La Mancha, Toledo, Spain

Yossi Moshe Department of Natural Resources, Institute of Plant Sciences, Volcani Center, Israel

Hassane Moutahir Mediterranean Center for Environmental Studies (CEAM Foundation), Joint Research Unit University of Alicante-CEAM, Alicante, Spain

Department of Ecology, University of Alicante, Alicante, Spain

Gidi Ne'eman Department of Biology and the Environment, University of Haifa-Oranim, Tivon, Israel

Mlungele M. Nsikani Centre for Invasion Biology, Department of Botany and Zoology, Stellenbosch University, Stellenbosch, South Africa

Jonàs Oliva Department of Crop and Forest Sciences, University of Lleida, Lleida, Spain

Joint Research Unit CTFC – Agrotecnio, Lleida, Spain

Sanna Olsson Department of Forest Ecology and Genetics, Forest Research Centre, INIA-CSIC, Madrid, Spain

Daniel E. Orenstein Faculty of Architecture and Town Planning, Technion – Israel Institute of Technology, Haifa, Israel

Yagil Osem Department of Natural Resources – Institute of Plant Sciences, Agricultural Research Organization – Volcani Center, Rishon LeZion, Israel

José Luís Palau Mediterranean Center for Environmental Studies (CEAM Foundation), Joint Research Unit University of Alicante-CEAM, Alicante, Spain

Javier Parladé Institut de Recerca i Tecnologia Agroalimentaries (IRTA), Barcelona, Spain

Carmen Quintano Electronic Technology Department, University of Valladolid, Valladolid, Spain

Sustainable Forest Management Research Institute, University of Valladolid-INIA, Palencia, Spain

Francisco Pastor Mediterranean Center for Environmental Studies (CEAM Foundation), Joint Research Unit University of Alicante-CEAM, Alicante, Spain

Leticia Pérez-Izquierdo Swedish University of Agricultural Sciences (SLU), Uppsala, Sweden

Nicolas Picard Food and Agriculture Organization of the United Nations, Forestry Department, Rome, Italy

Yahel Porat Forestry Division, Keren Kayemeth LeIsrael (KKL), Jerusalem, Israel

Yakir Preisler Earth and Planetary Sciences, Weizmann Institute of Science, Rehovot, Israel

Rafat Qubaja Earth and Planetary Sciences, Weizmann Institute of Science, Rehovot, Israel

Canella Radea Department of Ecology and Systematics, National and Kapodistrian University of Athens, Athens, Greece

Sophia Ratcliffe National Biodiversity Network Trust, Nottingham, UK

David M. Richardson Centre for Invasion Biology, Department of Botany and Zoology, Stellenbosch University, Stellenbosch, South Africa

Ana María Rincón Institute of Agronomic Sciences – Spanish Research Council (ICA-CSIC), Madrid, Spain

Jesús Rodríguez-Calcerrada Departamento de Sistemas y Recursos Naturales, Escuela Técnica Superior de Ingeniería de Montes, Forestal y del Medio Natural, Universidad Politécnica de Madrid, Madrid, Spain

Isabel Rodriguez-Quilón Forest Research Centre, National Institute of Agronomic Research, Madrid, Spain

Eyal Rotenberg Earth and Planetary Sciences, Weizmann Institute of Science, Rehovot, Israel

Álvaro Rubio-Cuadrado Departamento de Sistemas y Recursos Naturales, Escuela Técnica Superior de Ingeniería de Montes, Forestal y del Medio Natural, Universidad Politécnica de Madrid, Madrid, Spain

Paloma Ruiz-Benito Forest Ecology and Restoration Group, Departamento de Ciencias de la Vida, Universidad de Alcalá, Alcalá de Henares, Madrid, Spain

Remote Sensing Research Group, Departamento de Geología, Geografía y Medio Ambiente, Universidad de Alcalá, Alcalá de Henares, Madrid, Spain

Ana María Sabater Mediterranean Center for Environmental Studies (CEAM Foundation), Joint Research Unit University of Alicante-CEAM, Alicante, Spain

Department of Ecology, University of Alicante, Alicante, Spain

Hugh D. Safford University of California, Davis, CA, USA

USDA Forest Service, Pacific Southwest Research Station, Albany, CA, USA

USDA Forest Service, Pacific Southwest Region, Vallejo, CA, USA

Luis Sampedro Misión Biológica de Galicia – Consejo Superior de Investigaciones Científicas, Pontevedra, Spain

Raúl Sánchez-Salguero Instituto Pirenaico de Ecología (IPE-CSIC), Zaragoza, Spain

Departamento Sistemas Físicos, Químicos y Naturales, Universidad Pablo de Olavide, Sevilla, Spain

Dimitrios Sarris Department of Biological Sciences, University of Cyprus, Nicosia, Cyprus

Faculty of Pure and Applied Sciences, Open University of Cyprus, Nicosia, Cyprus

KES Research Centre, Nicosia, Cyprus

Susana Suarez-Seoane Area of Ecology, Faculty of Biological and Environmental Sciences, University of León, León, Spain

Efrat Sheffer Institute of Plant Science and Genetics in Agriculture, The Robert H. Smith Faculty of Agriculture, Food and Environment, The Hebrew University of Jerusalem, Rehovot, Israel

Mario Soliño Department of Economic Analysis, Complutense University of Madrid, Madrid, Spain

Fyodor Tatarinov Earth and Planetary Sciences, Weizmann Institute of Science, Rehovot, Israel

Iván Torres Departamento de Ciencias Ambientales, Universidad de Castilla-La Mancha, Toledo, Spain

Eugene D. Ungar Department of Natural Resources, Institute of Plant Sciences, Agricultural Research Organization – The Volcani Center, Rishon LeZion, Israel

Carla Vázquez-González Misión Biológica de Galicia – Consejo Superior de Investigaciones Científicas, Pontevedra, Spain

Giovanni G. Vendramin Institute of Biosciences and Bioresources, National Research Council, Sesto Fiorentino, Italy

Eduardo Vicente Department of Ecology, University of Alicante, Alicante, Spain

The Multidisciplinary Institute for Environmental Studies (IMEM), University of Alicante, Alicante, Spain

Alberto Vilagrosa Mediterranean Center for Environmental Studies (CEAM Foundation), Joint Research Unit University of Alicante-CEAM, Alicante, Spain

Department of Ecology, University of Alicante, Alicante, Spain

Pedro Villar-Salvador Forest Ecology and Restoration Group, Departamento de Ciencias de la Vida, Universidad de Alcalá, Alcalá de Henares, Madrid, Spain

Natalia Vizcaíno-Palomar University of Bordeaux, INRAE, BIOGECO, Pessac, France

Yoni Waitz Institute of Plant Science and Genetics in Agriculture, The Robert H. Smith Faculty of Agriculture, Food and Environment, The Hebrew University of Jerusalem, Rehovot, Israel

John N. Williams University of California, Davis, CA, USA

USDA Forest Service, Pacific Southwest Research Station, Albany, CA, USA

USDA Forest Service, Pacific Southwest Region, Vallejo, CA, USA

Lea Wittenberg Department of Geography and Environmental Studies, University of Haifa, Haifa, Israel

Dan Yakir Earth and Planetary Sciences, Weizmann Institute of Science, Rehovot, Israel

Maya Zahavi Soil Survey, Land Evaluation and Remote Sensing, Ministry of Agriculture and Rural Development, Rishon LeZion, Israel

Rafael Zas Misión Biológica de Galicia – Consejo Superior de Investigaciones Científicas, Pontevedra, Spain

Miguel A. Zavala Forest Ecology and Restoration Group, Departamento de Ciencias de la Vida, Universidad de Alcalá, Alcalá de Henares, Madrid, Spain

Chanoch Zoref Mountain Region, Keren Kayemth LeIsrael (KKL), Jerusalem, Israel

Part I
Evolution, Genetics and Distribution

Delphine Grivet

Chapter 1
Phylogenesis and Evolution of Mediterranean Pines

Delphine Grivet and Sanna Olsson

1.1 Introduction

The Mediterranean pines represent an extremely heterogeneous assembly with distinct biogeographic and demographic histories. It is therefore of special interest to understand their adaptive evolution through time and space. Assessing population evolutionary trajectories in terms of demography, the evolution of important life-history traits, and the genomic basis of adaptation is central to understanding the past evolutionary success of Mediterranean pines, their current situation and their future response to environmental changes. The term "Mediterranean pines" is ambiguous to some extent as it may refer to different species combinations:

- Seven species from the subsection *Pinaster* (*P. brutia* Ten., *P. canariensis* C. Sm. ex DC., *P. halepensis* Mill., *P. heldreichii* Christ, *P. pinaster* Aiton, *P. pinea* L. and *P. roxburghii* Sarg.; Gernandt et al. 2005);
- Eleven species growing in the Mediterranean region (*P. brutia*, *P. canariensis*, *P. halepensis*, *P. heldreichii, P. mugo* Turra, *P. nigra* J.F.Arnold, *P. peuce* Griseb., *P. pinaster*, *P. pinea*, *P. sylvestris* L. and *P. uncinata* Mill. ex Mirbel; Price et al. 1998) but not necessarily under a Mediterranean climate (*P. peuce* grows in montane regions);
- Ten species with a native distribution range extending to the Mediterranean region (Fig. 1.1), which grow under a Mediterranean climate (the 11 species above without *P. peuce*; Klaus 1989; Pausas and Millán 2019). We will refer to this latter group as Mediterranean pines throughout this section.

D. Grivet (✉) · S. Olsson
Department of Forest Ecology and Genetics, Forest Research Centre, INIA-CSIC, Madrid, Spain
e-mail: dgrivet@inia.es

G. Ne'eman, Y. Osem (eds.), *Pines and Their Mixed Forest Ecosystems in the Mediterranean Basin*, Managing Forest Ecosystems 38,
https://doi.org/10.1007/978-3-030-63625-8_1

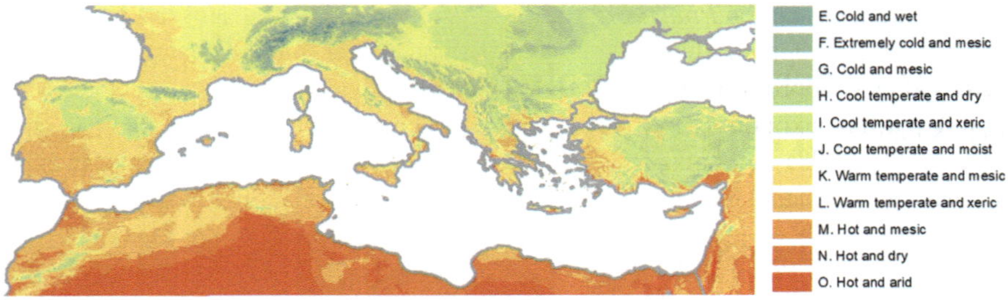

Fig. 1.1 The Mediterranean Basin with environmental stratification characteristics (adapted from Metzger et al. 2013)

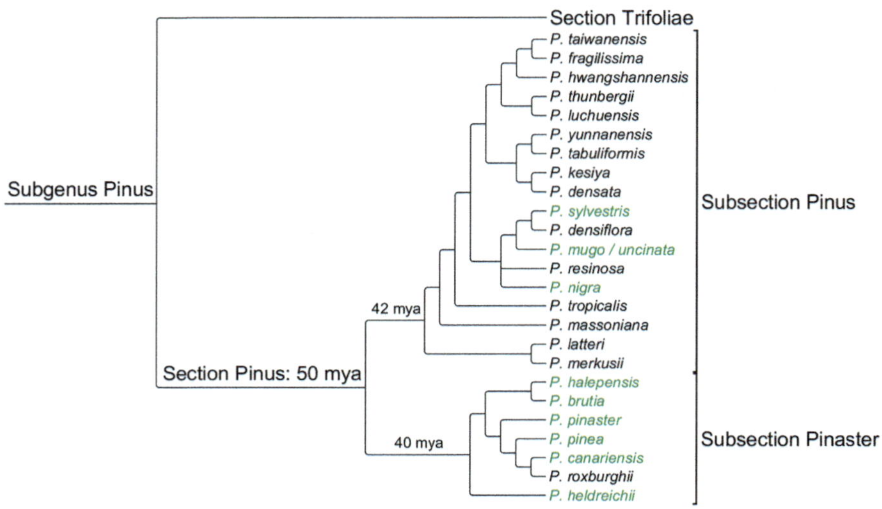

Fig. 1.2 Phylogenetic relationships of Mediterranean pines (in green) treated in this chapter. The phylogenetic relationships represent a synthesis from several studies (Gernandt et al. 2005; Eckert and Hall 2006; Grivet et al. 2013; Saladin et al. 2017); support values are therefore not displayed

1.2 Adaptive Evolution Through Time and Space for Individual Species

1.2.1 Phylogeny

The most comprehensive phylogenetic studies dealing with Mediterranean pines include the studies by Wang et al. (1999), Grotkopp et al. (2004), Gernandt et al. (2005), Eckert and Hall (2006), Grivet et al. (2013), Gallien et al. (2016) and Saladin et al. (2017) (Fig. 1.2). The most recent study by Saladin et al. (2017), based on a more extensive fossil set and several different dating approaches, estimated the divergence time of section *Pinus* to be approximately 50 million years ago (mya), while that of subsection *Pinus* approximately 42 mya and that of subsection *Pinaster* approximately 40 mya. These estimates are much earlier than estimates from previous studies – approximately 13, 6 and 10 mya, respectively (e.g. Gernandt et al. 2008). Earlier taxonomic classifications identified more subsections, for instance, Price et al. (1998) recognized *Canarienses*, *Halepenses* and *Pineae* at the level of subsection. Gernandt et al. (2005) proposed to establish subsection *Pinaster* to include these subsections together with *P. pinaster* and *P. heldreichii*. The division into subsections *Pinus* and *Pinaster* was confirmed by all successive studies. The topology of the relationships between these species has remained essentially the same but with better resolution and support in later studies based on nuclear markers (Grivet et al. 2013) and a larger number of chloroplast markers (Eckert and Hall 2006; Gallien et al. 2016; Saladin et al. 2017) (Fig. 1.2).

Due to the scarcity of data based on nuclear and mitochondrial markers for many pine species, most of the studies inferring phylogenetic relationships in Mediterranean pines have been based on a limited number of markers from the chloroplast genome, except for Grivet et al. (2013), who used nuclear data, and Grotkopp et al. (2004), who combined molecular and morphological data. The production of full chloroplast sequences for many species, including pines (Parks et al. 2012), together with their accumulation in public databases (https://www.ncbi.nlm.nih.gov/genome/organelle/) and the transferability across pine species due to the conservative nature of chloroplast genomes (Sugiura 1992), have made this molecular resource an attractive option for building multi-species phylogenies. However, the use of chloroplast markers to infer pine phylogenies is not always straightforward due to the existence of chloroplast capture (Gernandt et al. 2018). With the development of new genomic resources, most Mediterranean pines have some type of transcriptomic data nowadays (Table 1.1). These resources will allow us to gain further insights on their evolutionary history, ranging from phylogenomics to demographic histories and their evolutionary adaptation in response to abiotic and biotic factors (e.g. drought and pest resistance, wood formation, fire resistance, etc.).

1.2.2 Phylogeography and Demography

Based on the available molecular data, phylogeographic studies (Hewitt 2001) have been conducted on most of the Mediterranean pines with the aim of studying the historical processes responsible for the contemporary distribution of individuals in the light of genetics. Comparative studies revealed that, compared to other conifers in the Mediterranean and elsewhere in the world, Mediterranean pines present a high level of genetic differentiation and low genetic diversity within populations, along with an increasing level of genetic diversity from west to east and a low level at low elevation (Fady and Conord 2010; Soto et al. 2010; Fady 2012). The main factors explaining this pattern are linked to ancestral adaptation to wildfire, reduction of effective population size during the Last Glacial Maximum (LGM) and long-distance dispersal during the Holocene (Fady 2012). Although pines growing in the

Table 1.1 Genomic data available for Mediterranean pines

Species	Genomic resource	Experimental design	References
Pinus brutia	Transcriptome wide SNP data	Genetic lineages	Olsson et al. (2021)
Pinus canariensis	*De novo* transcriptome	Wood formation	Chano et al. (2017)
Pinus halepensis	*De novo* transcriptome *De novo* transcriptome *De novo* transcriptomes	Fire (serotiny) Drought resistance Genetic lineages	Pinosio et al. (2014) Fox et al. (2018) Olsson et al. (2021)
Pinus mugo	Transcriptome wide SNP data	Related species	Wachowiak et al. (2015)
Pinus nigra	*De novo* transcriptomes	Genetic lineages	Olsson et al. (2020)
Pinus pinaster	Transcriptomic database *De novo* transcriptome *De novo* transcriptome *De novo* transcriptome *De novo* transcriptome Genetic linkage map SNP assays	Several tissues General Pathogen resistance Embryogenesis Pathogen resistance General General	Fernández-Pozo et al. (2011) Canales et al. (2014) Gaspar et al. (2017) Rodrigues et al. (2018) Hernández-Escribano (2019) Chancerel et al. (2013) Lepoittevin et al. (2010); Plomion et al. (2016)
Pinus pinea	*De novo* transcriptome *De novo* transcriptome and SNP assays	Pathogen resistance Breeding	Santos et al. (2012) In preparation within the framework of the EU project B4EST
Pinus sylvestris	*De novo* transcriptome *De novo* transcriptome *De novo* transcriptome *De novo* transcriptome	Related species Developmental stages Pollen General	Wachowiak et al. (2015) Merino et al. (2016) Höllbacher et al. (2017) Ojeda Alayon et al. (2018)
Pinus uncinata	Transcriptome wide SNP data	Related species	Wachowiak et al. (2015)

Mediterranean Basin display some common patterns, each species displays a specific evolutionary history, as described below.

The Canary Island pine, ***P. canariensis,*** is endemic to the Canary Islands (Barbéro et al. 1998). It has comparatively high levels of neutral genetic diversity, similar to those found in the continental Mediterranean pine species *P. pinaster* and *P. halepensis* (Gómez et al. 2003; Vaxevanidou et al. 2006). The species probably colonized the Canary Islands from a single continental source and spread from West to East (Gómez et al. 2003; Navascués et al. 2006). The genetic characteristics of central and marginal populations differ from each other and have been affected by the complex evolutionary histories of populations on each island, including frequent extinctions and recolonizations (Navascués et al. 2006; Vaxevanidou et al. 2006).

The Aleppo pine, ***P. halepensis***, and Turkish pine, ***P. brutia***, form a group of related species that can intercross, but occupy different geographical ranges and bioclimates (Ne'eman and Trabaud 2000). They cover extensive areas in the western (*P. halepensis*) and eastern (mostly in Turkey for *P. brutia*) Mediterranean Basin (Ne'eman and Trabaud 2000). The demographic history of *P. halepensis* is complex and has not yet been fully elucidated (Vendramin et al. this volume, Chap. 2). Several sets of markers point to increasing genetic diversity from west to east, as well as a spatial genetic structure separating a western and an eastern cluster with the presence of genetically admixed populations. The impoverished western populations probably resulted from past founder effects or bottlenecks (due to climatic or/and fire events), while the eastern populations would have been less affected by the LGM and consequently would have retained more genetic diversity. *Pinus brutia* populations display globally low genetic differentiation and higher levels of genetic diversity than those of *P. halepensis* (Bucci et al. 1998; Kurt et al. 2012).

The distribution range of the Bosnian pine, ***P. heldreichii*** (including var. *leucodermis*), within southern Italy and the Balkan Peninsula, is small and isolated, resulting in reduced gene flow, high inbreeding rates and low effective population sizes (Vendramin et al. 2008a). Genetic studies have confirmed the impoverishment of some isolated populations as well as the separation of Balkan and Italian lineages (Piotti et al. 2015). Native to montane regions, *P. heldreichii* is suitable for reforestation of dry and high-altitude areas due to its ability to adapt to extreme environmental conditions and its resistance to insects.

The mountain pine, ***P. mugo,*** and the vicariant species, ***P. uncinata,*** belong to the morphologically variable species complex *P. mugo* Turra s.l. and frequently hybridize in overlapping growing zones (Christensen 1987). *Pinus mugo* grows in the east and south of the distribution range of the species complex, from the Southern and Eastern Alps to the Balkan Peninsula, while *P. uncinata* grows in the west and north of the distribution range, from the Spanish Sierra de Javalambre and the Pyrenees to the Western Alps (Barbéro et al. 1998). The morphological and taxonomic relationships within *P. mugo* s.l. were not supported by a genetic study based on a few chloroplast microsatellite markers (Heuertz et al. 2010). The same study suggested that the genetic signs of expansion resulted from demographic events that pre-dated the LGM and that core and peripheral populations of *P. mugo* have had different demographic histories.

The European black pine, ***P. nigra,*** is a morphologically and ecologically diverse species with a large, but highly fragmented distribution range, mainly in montane regions across Europe and Asia Minor, reaching its southern limits in North Africa (Vidakovic 1991). This pine shows weak spatial genetic structure in spite of phenotypic differentiation across its range (Fady and Conord 2010; Naydenov et al. 2016; Scotti-Saintagne et al. 2019). The species has undergone a complex evolutionary history where isolation, speciation and fragmentation have affected populations differently (Afzal-Rafii and Dodd 2007; Naydenov et al. 2016), leading to the coexistence of several lineages (Olsson et al. 2020). The diversification of these genetic lineages has been estimated to date back approximately 10 million years (Saladin et al. 2017; Scotti-Saintagne et al. 2019).

The maritime pine, ***P. pinaster,*** is an economically important conifer that forms large populations in Southwestern Europe, inhabiting both wet-coastal and seasonally dry continental forests, and exhibits a strong population structure associated with past climate and demographic changes (Bucci et al. 2007; Jaramillo-Correa et al. 2015). The present footprints of genetic diversity are the results of events that probably took place before the LGM as well as throughout the LGM and the Holocene, with isolated populations confined in multiple refugia from which they would have recolonized (Bucci et al. 2007; Naydenov et al. 2014; Jaramillo-Correa et al. 2020).

Pinus pinea, the Italian stone pine, occurs naturally in scattered populations throughout the Mediterranean Basin (Barbéro et al. 1998). This pine has also been widely planted for a long time, at least since the seventeenth century for reforestation (Mutke et al. 2019). It is currently becoming domesticated for the production of pine nuts, and within the last hundred years its forest area has more than doubled to 0.75 million ha due to it being used for forestation, soil protection, ecosystem restoration and sustainable production of both timber and pine nuts (Mutke et al. 2019). Its demographic history is characterized by a severe and prolonged demographic bottleneck, followed by subsequent natural and human-mediated dispersal across the Mediterranean Basin (Vendramin et al. 2008b; Jaramillo-Correa et al. 2020). Although this pine displays exceptionally low genetic variation (Vendramin et al. 2008b; Pinzauti et al. 2012; Jaramillo-Correa et al. 2020), a high degree of phenotypic plasticity (Mutke et al. 2019 and references therein) and epigenetic variation (Sáez-Laguna et al. 2014) could play a role in the adaptation of this species.

Pinus sylvestris, the Scots pine, is one of the most economically important and widely distributed pine species, with a native distribution ranging from Western Europe to Eastern Siberia, northern Scandinavia and the Mediterranean (Farjon 2010). The northern and southern populations of *P. sylvestris* have different colonization histories. On one hand an early bottleneck occurred before the last glacial period in Central and Northern European populations (Pyhäjärvi et al. 2007). The nuclear polymorphism and very wide continuous distribution of the species together indicate a lack of discrete populations; thus, the trees growing across most of Northern Europe can be considered a single panmictic population (Pyhäjärvi et al. 2019). On the other hand, the Mediterranean populations display unique genetic characteristics reflecting Mediterranean refugia during the last glacial period (Sinclair et al. 1999; Soranzo et al. 2000; Cheddadi et al. 2006; Naydenov et al. 2007).

1.2.3 Adaptation at the Molecular Level

With the development of new genomic resources for Mediterranean pines, a growing number of studies have examined the basis of pine adaptation. Studies have focused on proteins (e.g. Garcés et al. 2014; Cañas et al. 2015) or loci involved in adaptive variation. For the latter, different approaches have been used to identify loci targeted by natural selection, including: (i) genetic differentiation (e.g. Eveno et al. 2008; Ruiz Daniels et al. 2019); (ii) genetic and environmental associations (e.g. Grivet et al. 2011; Mosca et al. 2012; Ruiz Daniels et al. 2018;

Jaramillo-Correa et al. 2015); (iii) gene expression (e.g. Perdiguero et al. 2013; Cañas et al. 2015; Hernández-Escribano 2019); and (iv) genetic and phenotypic associations (e.g. Jaramillo-Correa et al. 2015; Kujala et al. 2017). These targeted proteins and genes represent valuable resources as they not only allow the identification of biotic and abiotic factors involved in pine adaptation but also their genetic basis. Contrary to neutral variation, which provides insights into common and individual demographic histories, adaptive variation may more directly inform us about the evolutionary potential of the Mediterranean pines. Both neutral and adaptive genetic information are the key to guiding genetic resources for forest conservation and management (Climent et al. this volume, Chap. 3).

1.3 Evolutionary Processes Based on Phylogeny: Traits and Genes

The evolution of pines should be interpreted in relation to climate (extreme temperatures), geology (oligotrophic soils) and fire (frequency and severity of fire regimes), all constituting environmental stressors since the Mesozoic (Keeley 2012). In particular, pine evolution has been linked to spatial and temporal variations in climate and fire regimes, as summarized by Keeley (2012), who divided pine evolution into three stages: (i) Mesozoic origin and diversification, (ii) Early Cenozoic fragmentation under the equable climates of the Eocene, and (iii) Oligocene and Miocene diversification due to expansion of fire-prone and abiotically stressful habitats. The evolutionary history of pines in comparative studies may be approached from different angles, of which historical biogeography (e.g. Eckert and Hall 2006; Badik et al. 2018) and reconstruction of ancestral ecological traits (e.g. Lamont et al. 2019) together with their underlying molecular architecture (e.g. Palmé et al. 2009; Grivet et al. 2013) are very relevant, as they can inform us about evolution across phylogenies of related or distant species.

1.3.1 Evolution of Traits

Mediterranean pines share many common life-history traits. Nonetheless, the few studies that have compared their historical biogeography and ancestral trait conditions together have focused mainly on the evolution of fire syndromes (e.g. Tapias et al. 2004; He et al. 2012; Keeley 2012; He et al. 2016; Badik et al. 2018), and less on other traits such as growth, development, reproduction (e.g. Grivet et al. 2013) or defense (e.g. Carillo-Gavilán et al. 2015). Understanding the origin and distribution of traits relevant to species adaptation has been made possible thanks to the advances made in trait-assignment techniques applied to date molecular phylogenies (e.g. Lamont et al. 2019). This approach allows us to assess the ancestral state of a trait, the time of origin of a trait, the evolution of a trait through time, or the environmental conditions associated with these times, shedding light on traits and stressors involved in species evolution.

1.3.1.1 Fire

The importance of fire as a selective agent in the subgenus *Pinus* is reflected by the vast majority of taxa being distributed in fire-prone environments (Keeley 2012; Badik et al. 2018). Pine species, even distantly related ones that have thrived in fire-prone habitats, such as those found in Mediterranean climate zones, have most likely developed traits to cope with the presence of fire (Badik et al. 2018; Ne'eman and Arianoutsou this volume, Chap. 22). Consequently, many studies have focused on understanding the evolution of fire-related traits, including height, self-pruning of dead branches, bark thickness, seedling regeneration, needle characteristics, cone serotiny or resprouting ability (Keeley 2012; Lamont et al. 2019). Badik et al. (2018) led a comparative study reconstructing the ancestral fire syndromes of pines in relation to their geographical distribution and showed that species growing in the Mediterranean Basin display a variety of fire adaptions (including avoider, evader, resister and combinations of the three), the ancestor of subgenus *Pinus* being an evader. The evolution of some well-studied fire-adapted traits is described herein:

- **Serotiny**: Serotiny, the retention of mature cones that open promptly only after exposure to high temperatures generated by fire, is a mechanism that allows the optimal germination of seeds after fire, when environmental conditions are highly favorable (He et al. 2012). In Pinaceae, molecular and fossil evidence points to serotiny as an ancestral trait emerging 89 mya during the middle Cretaceous (He et al. 2012; He et al. 2016; Mays et al. 2017); nowadays, pines growing under a Mediterranean climate in the Old World present both serotinous and non-serotinous cones.
- **Bark thickness**: Thick bark insulates the cambium from the heat generated by surface fires, allowing individual survival after fire (see references in He et al. 2012). Ancestral state reconstruction and trait evolution through time have shown that thin bark was the ancestral state, and that nowadays thick bark >15 mm is present in all 10 Mediterranean pines, except for *P. mugo* (He et al. 2012). Comparing bark thickness and bud tolerance to heat, Fernandes et al. (2008) ranked pine resistance to heat in the following order: *P. pinaster* > *P. pinea* > *P. nigra* > *P. halepensis* > *P. brutia* > *P. sylvestris* > *P. uncinata*.
- **Branch shedding**: Self-pruning or branch shedding, which creates a fuel gap between the forest floor and the crown, limits the ability of surface fires to ascend into the canopy (He et al. 2012; He et al. 2012). Branch retention is ancestral and nowadays few pines growing under the Mediterranean climate present branch shedding (He et al. 2012).

- **Resprouting capacity**: Resprouting among pines is rare and occurs on sites of low productivity and high fire frequency (in He et al. 2012). It occurs only in *P. canariensis* in association with thick bark (He et al. 2012).

These functional traits related to fire are under strong genetic control (see references in He et al. 2012) and have evolved through time, resulting in alternative fire strategies (Lamont et al. 2019) in different pine species, including: (i) extra thick bark with old-branch shedding, representing a fire strategy associated with surface fire regimes in high productivity sites (fire-avoiders), (ii) thin bark and old-branch retention coupled with serotiny, representing a fire embracing strategy associated with crown fires in low productivity sites (seeders, fire-embracers), and (iii) thick bark and branch retention coupled with epicormic resprouting and recovery (resprouters, fire-tolerators).

1.3.1.2 Other Life-History Traits

In addition to the strategies they have developed to withstand fires in the Mediterranean Basin, pines have also adapted to other extreme events (Tapias et al. 2004, and references therein) with a suite of phenotypic traits that have been postulated as central for understanding their life-history strategies. These traits are related to vegetative growth (Grotkopp et al. 2002), seedling allometry and development (e.g. Strauss and Ledig 1985), sexual reproduction (e.g. Tapias et al. 2001; Ne'eman et al. 2011), seed dispersal ability (e.g. Benkman 1995) and, particularly, tolerance to fire (Pausas et al. 2004; He et al. 2012; Pausas 2015), drought (e.g. Martínez-Vilalta et al. 2004) and frost (Climent et al. 2009). The Physical properties of their wood have also received much attention recently, since a close link with contrasting life histories has been shown in different tree species (e.g. Poorter et al. 2010). Very few studies have addressed the integration of several traits in pines (e.g. Santini et al. 2019).

A meta-analysis conducted on pines from the subsection *Pinaster,* which analyzed several life-history traits linked to growth, development and reproduction, measured in common-garden experiments, revealed no concerted trait evolution (Grivet et al. 2013), probably reflecting an heterogeneous assembly as previously reported (e.g. Klaus 1989; Tapias et al. 2004).

1.3.1.3 Biotic Resistance

Another pivotal component of pine adaptation is linked to biotic resistance. Selection imposed by biotic agents such as nematodes, insects, fungi, or pathogens, can lead to the evolution of shared chemical defenses in closely related species, leading to phylogenetic conservatism (Agrawal 2007). Studies that investigated the patterns of variation in defensive strategies across pine species revealed different abiotic and biotic selective pressures leading to marked differences in chemical defenses and defensive strategies among species (Zas et al. 2011; Carrillo-Gavillan et al. 2012). Within a comparative framework, a study comparing pines' defense strategies in response to herbivory, in section *Trifoliae* and section *Pinus* (comprising eight pines growing in the Mediterranean Basin), concluded that both phylogeny and biogeography have played an important role in the evolution of constitutive defenses of needles among pine species, with Mediterranean species showing lower constitutive concentration of total phenolics than their relatives (Carrillo-Gavillan et al. 2015). It may well be that both less persistent folivory pressure and/or past selective forces are responsible for the specific patterns revealed in Mediterranean pines. Yet, induced defenses may depend more on the particular climate, biotic interactions and resources available in each species' range (Carrillo-Gavillan et al. 2015).

1.3.2 Adaptive Molecular Evolution

Studying the genetic variation of molecular markers involved in adaptive traits (along with the traits of interest themselves, see Sect. 1.3.1) is key to understanding past evolutionary trajectories of natural populations as well as to predict their evolutionary potential. A phylogenetic approach used to detect footprints of selection of the same set of 17 nuclear genes across species from the subsection *Pinaster* identified two genes involved in defense (locus lp3.1 belonging to the ASR family – abscisic acid/water stress/ripening induction) and stress response (locus homologue to a metacaspase) that have likely played a role in the adaptation of Mediterranean pines to their environment (Grivet et al. 2013). As genomic and transcriptomic resources become more available for conifer species (see Sect. 1.2), more genes and gene networks can be examined to search for concerted evolution in a phylogenetic context, thereby providing a more complete picture of the molecular evolution of Mediterranean pines.

1.4 Conclusion and Perspectives

Because pines are important in Mediterranean landscapes, both from an ecological and economical perspective, understanding their evolution is of paramount importance for their conservation and sustainable management. The knowledge accumulated on Mediterranean pines in recent decades indicates that despite the existence of common patterns due to major past events (Fady 2012; Keeley 2012), this heterogeneous assembly displays singular and complex evolutionary and demographic trajectories. Gaining knowledge on their demographic histories, their response to environmental clues, and the evolution of some central traits and their underlying molecular basis, will be greatly aided with the omics techniques (including phenomics, metabolomics, genomics, phylogenomics and epigenomics). Identifying the drivers responsible for the trajectories of life history adaptation in Mediterranean pine evolution, together with the raw material on which they act, will be of great value for predicting their future response to environmental changes.

Acknowledgments We would like to thank José Climent for his help with the references related to "Other life-history traits" as well as Ricardo Alía and two reviewers for constructive comments on the chapter. Funding was provided by the project INIA-MAPAMA EG17-048 co-financed by FEADER (75%) according to EU Regulation 1305/2013. We acknowledge financial support from the European Union's Horizon 2020 research and innovation program under grant agreement No 676876 GenTree. SO received funding from the Spanish Ministry of Economy and Competitiveness (MINECO) under PTA2015-10836-I contract.

References

Afzal-Rafii Z, Dodd RS (2007) Chloroplast DNA supports a hypothesis of glacial refugia over postglacial recolonization in disjunct populations of black pine (*Pinus nigra*) in Western Europe. Mol Ecol 16:723–736

Agrawal AA (2007) Macroevolution of plant defense strategies. Trends Ecol Evol 22:103–109

Badik KJ, Jahner JP, Wilson JS (2018) A biogeographic perspective on the evolution of fire syndromes in pine trees (*Pinus*: Pinaceae). R Soc Open Sci 5:172412

Barbéro M, Loisel R, Quézel P et al (1998) Pines of the Mediterranean Basin. In: Richardson DM (ed) Ecology and biogeography of *Pinus*. Cambridge University Press, Cambridge, pp 219–249

Benkman CW (1995) Wind dispersal capacity of pine seeds and the evolution of different seed dispersal modes in pines. Oikos 73:221–224

Bucci G, Anzidei M, Madaghiele A et al (1998) Detection of haplotypic variation and natural hybridization in *halepensis*-complex pine species using chloroplast simple sequence repeat (SSR) markers. Mol Ecol 7:1633–1643

Bucci G, González-Martínez SC, Le Provost G et al (2007) Range-wide phylogeography and gene zones in *Pinus pinaster* Ait. revealed by chloroplast microsatellite markers. Mol Ecol 16:2137–2153

Canales J, Bautista R, Label P et al (2014) *De novo* assembly of maritime pine transcriptome: implications for forest breeding and biotechnology. Plant Biotechnol J 12:286–299

Cañas RA, Canales J, Muños-hernández C et al (2015) Understanding developmental and adaptive cues in pine through metabolite profiling and co-expression network analysis. J Exp Bot 66:3113–3127

Carrillo-Gavilán A, Moreira X, Zas R et al (2012) Early resistance of alien and native pines against two native generalist insect herbivores: no support for the natural enemy hypothesis. Funct Ecol 26:283–293

Carrillo-Gavilán A, Moreira X, Zas R et al (2015) Phylogenetic and biogeographical patterns in defensive strategies and quantitative allocation to chemical defences in Palaearctic and Nearctic pine trees. J Biogeogr 42:684–693

Chancerel E, Lamy J-B, Lesur I et al (2013) High density linkage mapping in a pine tree reveals a genomic region associated with inbreeding depression and provides clues to the extent and distribution of meiotic recombination. BMC Biol 11:50

Chano V, Lopez de Heredia U, Collada C et al (2017) Transcriptomic analysis of juvenile wood formation during the growing season in *Pinus canariensis*. Holzforschung 71:919–937

Cheddadi R, Vendramin GG, Litt T et al (2006) Imprints of glacial refugia in the modern genetic diversity of *Pinus sylvestris*. Glob Ecol Biogeogr 15:271–282

Christensen KI (1987) Taxonomic revision of the *Pinus mugo* complex and *P. X rhaetica* (*P. mugo X sylvestris*) (Pinaceae). Nord J Bot 7:838–408

Climent JM, Costa e Silva F, Chambel R et al (2009) Freezing injury in primary and secondary needles of Mediterranean pine species of contrasting ecological niches. Ann For Sci 66:407–415
Climent JM, Alizoti P, Rodriguez-Quilón I, Kurt Y, Ducci F, Fady B, Bozzano M, Alía R (this volume) Conservation and breeding of Mediterranean pines. In: Ne'eman G, Osem Y (eds) Pines and their mixed forest ecosystems in the Mediterranean Basin. Springer, Cham
Eckert AJ, Hall BD (2006) Phylogeny, historical biogeography, and patterns of diversification for *Pinus* (Pinaceae): phylogenetic tests of fossil-based hypotheses. Mol Phylogenet Evol 40:166–182
Eveno E, Collada C, Guevara MA et al (2008) Contrasting patterns of selection at *Pinus pinaster* Ait. drought stress candidate genes as revealed by genetic differentiation analyses. Mol Biol Evol 25:417–437
Fady B (2012) Biogeography of neutral genes and recent evolutionary history of pines in the Mediterranean Basin. Ann For Sci 69:421–428
Fady B, Conord C (2010) Macroecological patterns of species and genetic diversity in vascular plants of the Mediterranean basin. Divers Distrib 16:53–64
Farjon A (2010) A handbook of the world's conifers. Brill EJ, Leiden/Boston
Fernandes PM, Vega JA, Jimenez E et al (2008) Fire resistance of European pines. For Ecol Manag 256:246–255
Fernández-Pozo N, Canales J, Guerrero-Fernández D et al (2011) EuroPineDB: a high coverage web database for maritime pine transcriptome. BMC Genomics 12:366
Fox H, Doron-Faigenboim A, Kelly G et al (2018) Transcriptome analysis of *Pinus halepensis* under drought stress and during recovery. Tree Physiol 38:423–441
Gallien L, Saladin B, Boucher FC et al (2016) Does the legacy of historical biogeography shape current invasiveness in pines? New Phytol 209:1096–1105
Garcés M, Le Provost G, Lalanne C et al (2014) Proteomic analysis during ontogenesis of secondary xylem in maritime pine. Tree Physiol 34:1263–1277
Gaspar D, Trindade C, Usié A et al (2017) Expression profiling in *Pinus pinaster* in response to infection with the pine wood nematode *Bursaphelenchus xylophilus*. Forests 8:279
Gernandt DS, Geada López G, Ortiz García S et al (2005) Phylogeny and classification of *Pinus*. Taxon 54:29–42
Gernandt DS, Magallón S, López G et al (2008) Use of simultaneous analyses to guide fossil-based calibrations of Pinaceae phylogeny. Int J Plant Sci 169:1086–1099
Gernandt DS, Reséndiz Arias C, Terrazas T et al (2018) Incorporating fossils into the Pinaceae tree of life. Am J Bot 105:1329–1344
Gómez A, Gonzalez-Martinez SC, Collada C et al (2003) Complex population genetic structure in the endemic Canary Island pine revealed using chloroplast microsatellite markers. Theor Appl Genet 107:1123–1131
Grivet D, Sebastiani F, Alía R et al (2011) Molecular footprints of local adaptation in two Mediterranean conifers. Mol Biol Evol 28:101–116
Grivet D, Climent JM, Zabal-Aguirre M et al (2013) Adaptive evolution of Mediterranean pines. Mol Phylogenet Evol 68:555–566
Grotkopp E, Rejmanek M, Rost TL (2002) Towards a causal explanation of plant invasiveness: seedling growth and life-history strategies of 29 pine *Pinus* species. Am Nat 159:396–419
Grotkopp E, Rejmánek M, Sanderson MJ et al (2004) Evolution of genome size in pines (*Pinus*) and its life-history correlates: supertree analyses. Evolution 58:1705–1729
He T, Pausas JG, Belcher CM et al (2012) Fire-adapted traits of *Pinus* arose in the fiery cretaceous. New Phytol 194:751–759
He T, Belcher CM, Lamont BB et al (2016) A 350-million-year legacy of fire adaptation among conifers. J Ecol 104:352–363
Hernández-Escribano L (2019) *Fusarium circinatum*-host interaction: ecological and molecular aspects of the pathogenic and endophytic association. PhD thesis, Universidad Politécnica de Madrid

Heuertz M, Teufel J, González-Martínez SC et al (2010) Geography determines genetic relationships between species of mountain pine (*Pinus mugo* complex) in Western Europe. J Biogeogr 37:541–556

Hewitt GM (2001) Speciation, hybrid zones and phylogeography – or seing genes in space and time. Mol Ecol 10:537–549

Höllbacher B, Schmitt AO, Hofer H et al (2017) Identification of proteases and protease inhibitors in allergenic and non-allergenic pollen. Int J Mol Sci 18:1199

Jaramillo-Correa JP, Rogríguez-Quilón I, Grivet D et al (2015) Molecular proxies for climate maladaptation in a long-lived tree (*Pinus pinaster* Aiton, Pinaceae). Genetics 199:1–15

Jaramillo-Correa JP, Bagnoli F, Grivet D et al (2020) Evolutionary rate and genetic load in an emblematic Mediterranean tree following an ancient and prolonged population collapse. Mol Ecol 29:4797–4811

Keeley JE (2012) Ecology and evolution of pine life histories. Ann For Sci 69:445–453

Klaus W (1989) Mediterraneaen pines and their history. Plant Syst Evol 162:133–163

Kujala ST, Knürr T, Kärkkäinen K et al (2017) Genetic heterogeneity underlying variation in a locally adaptive clinal trait in *Pinus sylvestris* revealed by a Bayesian multipopulation analysis. Heredity 118:413–423

Kurt Y, González-Martínez SC, Alía R et al (2012) Genetic differentiation in *Pinus brutia* Ten. using molecular markers and quantitative traits: the role of altitude. Ann For Sci 69:345–351

Lamont BB, He T, Yan Z (2019) Evolutionary history of fire-stimulated resprouting, flowering, seed release and germination. Biol Rev 94:903–928

Lepoittevin C, Frigerio JM, Garnier-Géré P et al (2010) *In vitro vs. in* silico detected SNPs for the development of a genotyping array: what can we learn from a non-model species? PLoS One 5:e11034

Martínez-Vilalta J, Sala A, Piño J (2004) The hydraulic architecture of Pinaceae – a review. Plant Ecol 171:3–13

Mays C, Cantrill DJ, Bevitt JJ (2017) Polar wildfires and conifer serotiny during the cretaceous global hothouse. Geology 45:1119–1122

Merino I, Abrahamsson M, Sterck L et al (2016) Transcript profiling for early stages during embryo development in scots pine. BMC Plant Biol 16:1–18

Metzger MJ, Bunce RGH, Jongman RHG et al (2013) A high-resolution bioclimate map of the world: a unifying framework for global biodiversity research and monitoring. Glob Ecol Biogeogr 22:630–638

Mosca E, Eckert AJ, Di Pierro EA et al (2012) The geographical and environmental determinants of genetic diversity for four alpine conifers of the European Alps. Mol Ecol 21:5530–5545

Mutke S, Vendramin GG, Fady B et al (2019) Molecular and quantitative genetics of stone pine (*Pinus pinea* L.). In: Nandwani D (ed) Genetic diversity in horticultural plants. Springer International Publishing, Cham, pp 61–84

Navascués M, Vaxevanidou Z, González-Martínez SC et al (2006) Chloroplast microsatellites reveal colonisation and metapopulation dynamics in the Canary Island pine. Mol Ecol 15:2691–2698

Naydenov K, Senneville S, Beaulieu J et al (2007) Glacial vicariance in Eurasia: mitochondrial DNA evidence from scots pine for a complex heritage involving genetically distinct refugia at mid-northern latitudes and in Asia minor. BMC Evol Biol 7:1–12

Naydenov KD, Alexandrov A, Matevski V et al (2014) Range-wide genetic structure of maritime pine predates the last glacial maximum: evidence from nuclear DNA. Hereditas 151:1–13

Naydenov KD, Naydenov MK, Alexandrov A et al (2016) Ancient split of major genetic lineages of European black pine: evidence from chloroplast DNA. Tree Genet Genomes 12:1–18

Ne'eman G, Arianoutsou M (this volume) Mediterranean pines – adaptations to fire. In: Ne'eman G, Osem Y (eds) Pines and their mixed forest ecosystems in the Mediterranean Basin. Springer, Cham

Ne'eman G, Trabaud L (2000) Ecology, biogeography and management of *Pinus halepensis* and *P. brutia* forest ecosystems in the Mediterranean basin. Backhuys Publishers, Leiden

Ne'eman G, Goubitz S, Werger MJ et al (2011) Relationships between tree size, crown shape, gender segregation and sex allocation in *Pinus halepensis*, a Mediterranean pine tree. Ann Bot 108:197–206

Ojeda Alayon DI, Mattila TM, Ruttink T et al (2018) Utilization of tissue ploidy level variation in de novo transcriptome assembly of *Pinus sylvestris*. G3 9:3409–3421

Olsson S, Lorenzo Z, Zabal-Aguirre M et al (2021) Evolutionary history of the Mediterranean *Pinus halepensis-brutia* species complex using gene-resequencing and transcriptomic approaches. Plant Mol Biol. https://doi.org/10.1007/s11103-021-01155-7 (in press)

Olsson S, Grivet D, Cattonaro F et al (2020) Evolutionary relevance of lineages in the European black pine (*Pinus nigra*) in the transcriptomic era. Tree Genet Genomes 16:30

Palmé AE, Pyhäjärvi T, Wachowiak W et al (2009) Selection on nuclear genes in a *Pinus* phylogeny. Mol Biol Evol 26:893–905

Parks M, Cronn R, Liston A (2012) Separating the wheat from the chaff: mitigating the effects of noise in a plastome phylogenomic data set from *Pinus* L. (Pinaceae). BMC Evol Biol 12:100

Pausas JG (2015) Evolutionary fire ecology: lessons learned from pines. Trends Plant Sci 20:318–324

Pausas JG, Millán MM (2019) Greening and browning in a climate change hotspot: the Mediterranean Basin. Bioscience 69:143–151

Pausas J, Bradstock R, Keith D et al (2004) Plant functional traits in relation to fire in crown-fire ecosystems. Ecology 85:1085–1100

Perdiguero P, Barbero M d C, Cervera MT et al (2013) Molecular response to water stress in two contrasting Mediterranean pines (*Pinus pinaster* and *Pinus pinea*). Plant Physiol Biochem 67:199e208

Pinosio S, González-Martínez SC, Bagnoli F et al (2014) First insights into the transcriptome and development of new genomic tools of a widespread circum-Mediterranean tree species, *Pinus halepensis* Mill. Mol Ecol Resour 14:846–856

Pinzauti F, Sebastiana F, Budde KB et al (2012) Nuclear microsatellites for *Pinus pinea* (Pinaceae), a genetically depauperate tree, and their transferability to *P. halepensis*. Am J Bot 99:e362–e365

Piotti A, Pini S, Bagnoli F et al (2015) Conservation genetics of *Pinus heldreichii* var. *leucodermis* disjunct populations in the Pollino National Park (southern Italy). Sostenere il pianeta, boschi per la vita, Ricerca e innovazione per la tutela e la valorizzazione delle risorse forestali10 Congresso Nazionale SISEF, Firenze, abstract book Posters, p 22

Plomion C, Bartholomé J, Lesur I et al (2016) High-density SNP assay development for genetic analysis in maritime pine (*Pinus pinaster*). Mol Ecol Resour 16:574–587

Poorter L, McDonald I, Alarcón A et al (2010) The importance of wood traits and hydraulic conductance for the performance and life history strategies of 42 rainforest tree species. New Phytol 185:481–492

Price RA, Liston A, Strauss SH (1998) Phylogeny and systematics of *Pinus*. In: Richardson DM (ed) Ecology and biogeography of *Pinus*. Cambridge University Press, Cambridge, pp 49–68

Pyhäjärvi T, García-Gil MR, Knürr T et al (2007) Demographic history has influenced nucleotide diversity in European *Pinus sylvestris* populations. Genetics 177:1713–1724

Pyhäjärvi T, Kujala S, Savolainen O (2019) 275 years of forestry meets genomics in *Pinus sylvestris*. Evol Appl 13(1):11

Rodrigues AS, De Vega JJ, Miguel CM (2018) Comprehensive assembly and analysis of the transcriptome of maritime pine developing embryos. BMC Plant Biol 18:379

Ruiz Daniels R, Taylor RS, Serra-Varela MJ et al (2018) Inferring selection in instances of long-range colonization: the Aleppo pine (*Pinus halepensis*) in the Mediterranean Basin. Mol Ecol 2:3331–3345

Ruiz Daniels R, Taylor RS, González-Martínez SC et al (2019) Looking for local adaptation: convergent microevolution in Aleppo pine (*Pinus halepensis*). Genes 10:673

Sáez-Laguna E, Guevara M-A, Díaz L-M et al (2014) Epigenetic variability in the genetically uniform forest tree species *Pinus pinea* L. PLoS One 9(8):e103145

Saladin B, Leslie AB, Wüest RO et al (2017) Fossils matter: improved estimates of divergence times in *Pinus* reveal older diversification. BMC Evol Biol 17:95

Santini F, Climent JM, Voltas J (2019) Phenotypic integration and life history strategies among populations of *Pinus halepensis*: an insight through structural equation modelling. Ann Bot 124:1161–1171

Santos CS, Pinheiro M, Silva AI et al (2012) Searching for resistance genes to *Bursaphelenchus xylophilus* using high throughput screening. BMC Genomics 13:599

Scotti-Saintagne C, Giovannelli G, Scotti I et al (2019) Recent, late Pleistocene fragmentation shaped the phylogeographic structure of the European black pine (*Pinus nigra* Arnold). Tree Genet Genomes 15:76

Sinclair WT, Morman JD, Ennos RA (1999) The postglacial history of Scots pine (*Pinus sylvestris* L.) in Western Europe: evidence from mitochondrial DNA variation. Mol Ecol 8:83–88

Soranzo N, Alia R, Provan J et al (2000) Patterns of variation at a mitochondrial sequence-tagged-site locus provides new insights into the postglacial history of European *Pinus sylvestris* populations. Mol Ecol 9:1205–1211

Soto A, Robledo-Arnuncio JJ, González-Martínez SC et al (2010) Climatic niche and neutral genetic diversity of the six Iberian pine species: a retrospective and prospective view. Mol Ecol 19:1396–1409

Strauss SH, Ledig FT (1985) Seedling architecture and life-history evolution in pines. Am Nat 125:702–715

Sugiura M (1992) The chloroplast genome. Plant Mol Biol 19:149–168

Tapias R, Gil L, Fuentes-Utrilla P et al (2001) Canopy seed banks in Mediterranean pines of South-Eastern Spain: a comparison between *Pinus halepensis* Mill., *P. pinaster* Ait., *P. nigra* Arn. and *P. pinea* L. J Ecol 89:629–638

Tapias R, Climent J, Pardos JA et al (2004) Life histories of Mediterranean pines. Plant Ecol 171:53–68

Vaxevanidou Z, González-Martínez S, Climent J et al (2006) Tree populations bordering on extinction: a study case in the endemic Canary Islands pine. Biol Conserv 129:451–460

Vendramin GG, Fineschi S, Fady B (2008a) EUFORGEN technical guidelines for genetic conservation and use of Bosnian pine (*Pinus heldreichii*). Bioversity International, Rome

Vendramin GG, Fady B, González-Martínez SC et al (2008b) Genetically depauperate but widespread: the case of an emblematic Mediterranean pine. Evolution 62:680–688

Vendramin GG, Avanzi C, González-Martínez SC et al (this volume) Population genetics and genomics of Aleppo pine (*Pinus halepensis*). In: Ne'eman G, Osem Y (eds) Pines and their mixed forest ecosystems in the Mediterranean Basin. Springer, Cham

Vidakovic M (1991) Conifers: morphology and variation)Revised English). Graficki Zavod Hrvatske, Zagreb

Wachowiak W, Trivedi U, Perry A et al (2015) Comparative transcriptomics of a complex of four European pine species. BMC Genomics 16:234

Wang X-R, Tsumura Y, Yoshimaru H et al (1999) Phylogenetic relationships of Eurasian pines (*Pinus*, Pinaceae) based on chloroplast *rbcL*, *MATK*, *RPL20-RPS18* spacer, and *TRNV* intron sequences. Am J Bot 86:1742–1753

Zas R, Moreira X, Sampedro L (2011) Tolerance and induced resistance in a native and an exotic pine species: relevant traits for invasion ecology. J Ecol 99:1316–1326

Chapter 2
Population Genetics and Genomics of Aleppo Pine (*Pinus halepensis*)

Giovanni G. Vendramin, Camilla Avanzi, Santiago C. González-Martínez, and Delphine Grivet

2.1 Patterns at Full Distribution Scale

Aleppo pine (*Pinus halepensis* Mill.) is an emblematic conifer native to the Mediterranean region (Fig. 2.1); its ecological and economic relevance (Ne′eman and Trabaud 2000) has spurred great interest in understanding the evolutionary history of its populations at the full distribution scale. The history of *P. halepensis* in the Mediterranean Basin was first reconstructed by Morgante et al. (1998) in a range-wide study based on paternally-inherited, chloroplast microsatellite markers (cpSSRs). The results of their study grouped *P. halepensis* populations into two main clusters (Greece vs. rest of the range), characterized by very different levels of haplotypic diversity. In particular, Greek populations harbored many ancient lineages, exhibiting extraordinarily high within-population genetic diversity. In contrast, populations from the rest of species range were characterized by only a limited subset of lineages, showing low genetic diversity and differentiation. This picture suggests that Greek populations experienced a clearly distinct history from that of all other populations. The most plausible hypothesis to describe the level and distribution of haplotypic diversity found by Morgante et al. (1998) is that the vast majority of *P. halepensis* populations experienced dramatic bottlenecks and drastic reductions of population size throughout their history. Thus, their extant chloroplast gene pool composition resulted from a severe founder effect. Similarly, low levels

G. G. Vendramin (✉) · C. Avanzi
Institute of Biosciences and Bioresources, National Research Council, Sesto Fiorentino, Italy
e-mail: giovanni.vendramin@ibbr.cnr.it

S. C. González-Martínez
University of Bordeaux, INRAE, UMR BIOGECO, Cestas, France

D. Grivet
Department of Forest Ecology and Genetics, Forest Research Centre, INIA-CSIC, Madrid, Spain

G. Ne'eman, Y. Osem (eds.), *Pines and Their Mixed Forest Ecosystems in the Mediterranean Basin*, Managing Forest Ecosystems 38,
https://doi.org/10.1007/978-3-030-63625-8_2

Fig. 2.1 Distribution range of *P. halepensis* (green shading) based on EUFORGEN 2009 (www.euforgen.org), with the background in different colors representing the environmental zoning adapted from Metzger et al. (2013)

of diversity within *P. halepensis* populations were detected by Schiller et al. (1986) using isozyme markers and, in this case too, they were interpreted as the consequence of one or several bottlenecks. Morgante et al. (1998) proposed a straightforward hypothesis to reconstruct the biogeographic history of the species: Greece likely represented a glacial refugium where *P. halepensis* populations survived during glacial periods, without experiencing significant genetic bottlenecks. Thus, extant Greek populations represented the ancestral gene pool of the species, while all the other populations were successively established by migration of a restricted number of individuals.

In the same year of Morgante's landmark study, Bucci et al. (1998) used cpSSRs to further describe genetic variation of three closely related species belonging to the *halepensis* complex (i.e. *P. halepensis*, *P. brutia* Ten. and *P. eldarica* Medw.). In *P. halepensis*, the authors found a similar pattern to that described by Morgante et al. (1998), with large among-population differentiation and low within-population genetic diversity, except for two Greek and one southern Italian population showing an outstanding amount of genetic variation, as well as a large proportion of private haplotypes. Overall, these findings support the hypothesis proposed by Morgante et al. (1998) of a post-glacial expansion of the species, shaped by the migration of a few individuals (founder effect), as well as by population dynamics regulated by genetic bottlenecks (likely prompted by wildfires). Three highly informative cpSSRs were successfully used to distinguish *P. halepensis* from the other two species of the *halepensis* complex, and to monitor the occurrence of natural hybridization in a Turkish sympatric population of *P. halepensis* and *P. brutia* (Bucci et al. 1998). Signatures of introgression of *P. halepensis* haplotypes into *P. brutia* seeds were clearly identified, but not in the opposite direction. The existence of unidirectional gene flow in sympatric populations of *P. halepensis* and *P. brutia* reinforced previous evidence about partial reproductive barriers between the two species (Panetsos 1975; Moulalis et al. 1976; Korol et al. 1995). In particular, previous results from controlled crossings showed that inter-specific matings were successful only when *P. halepensis* was the pollen donor and *P. brutia* was the female parent (Panetsos 1981). Although the two species can intercross, they occupy distinct geographical ranges and bioclimates (Ne′eman and Trabaud 2000), with both *P. brutia* and *P. halepensis* showing large overall haplotypic variation and genetic divergence (Bucci et al. 1998).

The widespread application of Sanger sequencing to forest trees paved the way for a new generation of studies aimed at disentangling the genomic signatures left by neutral processes from those caused by selection and adaptation. Grivet et al. (2009) used cpSSRs and coalescent-based modeling on nuclear genes to dig deeper into the demographic history of *P. halepensis*. Additionally, as these nuclear genes were selected because of their putative involvement in tree responses to drought, the authors examined their patterns of polymorphism to assess whether selective pressures for drought tolerance existed along the species range, considering plausible demographic scenarios. To this end, numerous demographic scenarios were tested to identify the one that best fitted the empirical data. Although the results should be taken with caution, as they were based on a relatively low number of genes, the most likely demographic scenario for *P. halepensis* indicated an ancient and stable relict Greek population, together with genetically distinct western Mediterranean populations showing molecular signatures of intense bottlenecks. This scenario was consistent with previous genetic analyses performed at the species level (Morgante et al. 1998; Gómez et al. 2001) as well as with paleoecological evidence suggesting a recent westwards expansion of the species in the last 10,000 years (Nahal 1962). Additional support for this scenario comes from the range-wide study by Ruiz Daniels et al. (2018): first, the authors found a strong pattern of Isolation-by-Distance (IBD); and second, the directionality index ψ indicated an eastern origin of the expansion range (Fig. 2.2). An alternative hypothesis was suggested by Fady

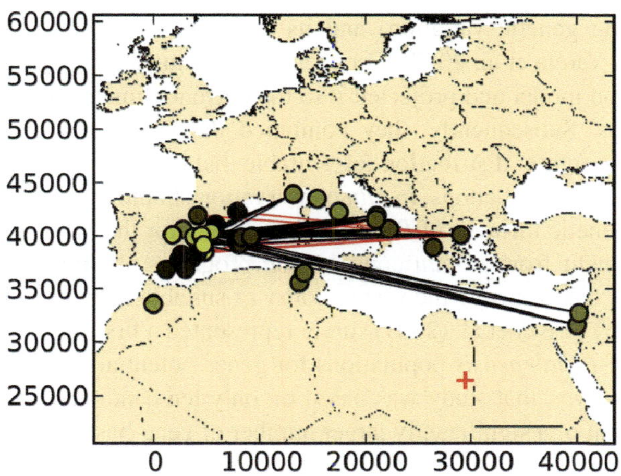

Fig. 2.2 Detection of range expansion using PooGui.py (Peter and Slatkin 2013) from Ruiz Daniels et al. (2018). Circles correspond to populations of *P. halepensis* genotyped with 294 SNPs. Each line corresponds to an estimate of the pairwise ψ statistic (directionality index), with thicker and brighter lines corresponding to higher values. ψ is sensitive to patterns created by range expansions, as it detects the allele frequency clines created by successive founder events. The expectation of ψ is zero in an equilibrium Isolation-by-Distance model, and its expectation is positive in the direction of the expansion. Black and red lines denote eastward and westward migration, respectively. The red cross indicates the approximate estimated single origin of range expansion

and Conord (2010), who postulated that western populations of *P. halepensis* suffered strong demographic bottlenecks during the Last Glacial Maximum (LGM), while eastern populations did not; thereby the latter retained more and unique genetic diversity. This scenario did not involve a long-range westward colonization and relied on the existence of local populations throughout the Mediterranean Basin during the LGM. More extensive demographic modeling and additional genomic sampling would be needed to distinguish between these alternative scenarios.

Interestingly, contrasting patterns of polymorphism between eastern and western populations were observed at both neutral and functional loci in Grivet et al. (2009), likely indicating the combined action of genetic drift and natural selection on front populations during the colonization of new environments – a process that has been named 'allele or gene surfing' (Edmonds et al. 2004; see review in Excoffier and Ray 2008). Altogether, the results from Grivet et al. (2009) pointed to distinct evolutionary units for *P. halepensis*, among which the Greek populations represented a particularly relevant target for conservation purposes, as they harbored most of the neutral and functional variation found in the species. Interestingly, Greek provenances were often outliers for various adaptive features, exhibiting particularly fast-growing behaviors (Matziris 2000; Chambel et al. 2007; Climent et al. 2008; Voltas et al. 2008). Greek forests of *P. halepensis* could indeed serve as sources of genetic variants potentially pre-adapted to future climate conditions, as they already grow in more arid conditions than those found in western Mediterranean populations (Arianoutsou and Ne'eman 2000). Indications for the *in-situ* and *ex-situ* conservation of *P. halepensis* have also come from the evaluation of species vulnerability both in terms of its adaptive potential (measured as adaptive trait differentiation and/or standing genetic variation) and its exposure to habitat loss and climate change (Serra-Varela et al. 2017). For each genetic cluster, the authors fitted a species distribution model and projected it to the current climate and to 42 future climate scenarios. Subsequently, they combined the projections to obtain a map describing the future distribution of suitable habitats for each genetic group. According to these predictions, *in-situ* conservation should guarantee the conservation of most genetic lineages of *P. halepensis*, except for the Moroccan populations that would benefit from *ex-situ* conservation programs, as those would likely be exposed to the risk of losing the vast majority of suitable habitats in the future.

The study by Grivet et al. (2009) surely represented a first step towards the characterization of *P. halepensis* populations for genes potentially involved in drought tolerance. However, that study was based on only ten candidate genes, calling for the development of a significantly larger number of gene-based molecular markers. To this end, Pinosio et al. (2014) characterized the transcriptome of two *P. halepensis* individuals that were phenotypically divergent for fire-related traits using Illumina next-generation sequencing (NGS). The authors annotated the transcriptome and developed a new set of gene-based markers for population genomics and genetic association studies. The assembled *P. halepensis* transcriptome comprised 48,629 contigs and covered about 54.6 Mbp. It showed similar characteristics to those of other conifers (e.g. *P. contorta* Douglas, Parchman et al. 2010; *Taxus mairei* (Lemée & Lév.) S.Y. Hu *ex* T.S. Liu, Hao et al. 2011; *Picea abies* (L.) H. Karst,

Chen et al. 2012), suggesting that patterns of transcription were conserved across species. By comparing *P. halepensis* transcripts with those of *Picea sitchensis* (Bong.) Carr., 34,014 annotated SNPs were detected. Several genes were differentially expressed across the two phenotypically divergent *P. halepensis* individuals, including a glutathione-transferase, a cellulose synthase and a cobra-like protein (the last two expressed only in the fire-adapted tree). Finally, a large number of new amplifiable markers (3,334 nuclear SSRs and 28,236 SNPs) were identified. For microsatellites, di- and tri-nucleotide repeats comprised the most abundant SSR motif classes, while tetra-and penta-nucleotide repeats were detected at much lower frequencies. A 384-SNP oligo pool assay for genotyping with Illumina Vera Code technology was finally designed, characterized by a high overall SNP conversion rate (76.6%) that was comparable (or even superior) to those of other conifer species (e.g. 66.9% in *P. taeda* L., Eckert et al. 2009; 51% in *P. pinaster* Ait., Lepoittevin et al. 2010; 72.5% in *Pseudotsuga menziesii* (Mirbel) Franco, Howe et al. 2013).

The 384-SNP array developed by Pinosio et al. (2014) was used to further distinguish between the confounding effects of drift and selection acting within *P. halepensis* populations (Ruiz Daniels et al. 2018, 2019). During long-range expansions, rare alleles could spread over wide areas, significantly increasing their frequencies, simply due to the stochastic effect of strong genetic drift (i.e. 'allele surfing', Edmonds et al. 2004). Similarly, a beneficial mutation could be largely propagated by selection and increase its allelic frequency (Excoffier and Ray 2008). Hence, it is crucial to account for the effect of allele surfing when searching for loci potentially involved in adaptive responses (Lotterhos and Whitlock 2014, 2015), especially in species that may have undergone a range expansion followed by severe bottlenecks, such as *P. halepensis*. Despite its relevance, the impact of allele surfing has been rarely investigated in empirical studies of forest trees, and even more rarely in the context of adaptation to new environments. The novelty of the methodological framework used by Ruiz Daniels et al. (2018) to infer selection lies precisely in the opportunity of simultaneously accounting for two important sources of false positives: shared history and allele surfing. The authors first characterized *P. halepensis* populations across the Mediterranean Basin with both nuclear SSR and SNP markers. Subsequently, they simulated a SNP dataset that mimicked the effects of genetic drift and range expansions, but assumed no selection. Finally, they searched for SNP loci potentially involved in local adaptation processes using environmental association analysis and F_{ST}-based methods, while correcting for the rate of false positives caused by allele surfing. Taking advantage of a larger number of molecular markers and populations, Ruiz Daniels et al. (2018) revealed a previously unsuspected complex genetic structure across the range of *P. halepensis*. Both nuclear SSR and SNP data indeed indicated the existence of multiple gene pools in the species, particularly in the western part of the range. The data also revealed the presence of admixture zones between different gene pools (see also Serra-Varela et al. 2017).

Combining the results of their study with those of previous genetic investigations (Bucci et al. 1998; Morgante et al. 1998; Gómez et al. 2005; Grivet et al. 2009, 2011; Jaramillo-Correa et al. 2010), as well as with paleoecological data (Nahal

1962; Pons 1992), Ruiz Daniels et al. (2018) proposed a comprehensive demographic history for *P. halepensis*. The species probably reached higher latitudes during the Tertiary, occupying an overall larger area. When climatic conditions changed at the end of the Tertiary, the species was probably confined to southern Europe (Nahal 1962). Then, during the last interglacial (125,000 years BP) a range expansion likely occurred, as suggested by the separation between African and Iberian populations, dated to 110,000 years BP (Jaramillo-Correa et al. 2010). During the LGM (25,000–18,000 years BP) climatic conditions became harsher, especially in the western part of the Mediterranean Basin (Van Andel 2002), leading to bottlenecks (e.g. in Italy; Grivet et al. 2011) and range contractions. The species persisted in a few local glacial refugia, probably located across the Mediterranean Basin, but mainly in its eastern part. When climatic conditions gradually became more favorable at the end of the last glaciation (Van Andel 2002), a new range expansion occurred (Nahal 1962; Schiller et al. 1986; Pons 1992), leading to the current distribution of the species and to the genetically impoverished western Mediterranean populations (Ruiz Daniels et al. 2018). Overall, this last range expansion was probably accompanied by gene flow, resulting in admixture zones throughout the entire distribution range of the species and in a complex population genetic structure.

Besides these new insights on the evolutionary history of *P. halepensis*, Ruiz Daniels et al. (2018) showed evidence for selection acting on SNP loci involved in responses to drought. In fact, seven SNP loci were found to be significantly associated with climatic variables related to aridity, which is known to be a driver of growth and reproduction in the species (Climent et al. 2008; Voltas et al. 2008, 2015; Girard et al. 2011, 2012; De Luis et al. 2013; Santos-del-Blanco et al. 2013; Hernández-Serrano et al. 2014). Interestingly, five of these seven SNP loci were located within genes that have been putatively involved in adaptive responses to a broad variety of selective drivers in other trees species (Li et al. 2009; Verne et al. 2011; Perdiguero et al. 2013; Wei et al. 2013; Fan et al. 2014; Zhang et al. 2014). In particular, one SNP was located in a gene that was expressed in response to drought in *P. pinea* L. and *P. pinaster* (Perdiguero et al. 2013). This locus showed a certain degree of homology with a peroxisomal membrane protein that might be involved in the establishment of a ROS-mediated scavenging mechanism in drought responses (Cruz de Carvalho 2008).

2.2 Patterns at the Local Scale

The history of *P. halepensis* has also been investigated at a more local geographical scale, in relation to other pine species representing marked differences in ecological requirements and demographic histories. For instance, cpSSRs have been employed to reconstruct the history of the species, along with other pines, in the Iberian Peninsula (Gómez et al. 2005; Soto et al. 2010). Overall, *P. halepensis* (together with *P. pinea*) displayed lower genetic diversity than other pines (*P. pinaster*, *P. uncinata* Turra, *P. sylvestris* L. and *P. nigra* J.F.Arnold) and lower genetic differentiation than *P. uncinata* and *P. pinaster*, but similar to that of *P. sylvestris* and *P. nigra* (*P. pinea* showed no divergence; Soto et al. 2010). The low genetic diversity observed for *P. halepensis* in the Iberian Peninsula supports the hypothesis of the western-range populations resulting from both bottlenecks and founder effects associated with range expansion. The different genetic patterns observed reflect the role of historical events and ecological processes in strongly shaping the level and distribution of genetic diversity of Iberian pine species. Such a substantial body of evidence should be carefully taken into account for defining species-specific conservation strategies, such as the establishment of *in situ* genetic reserves.

Besides reconstructing the evolutionary history of the species, SSR and SNP markers have also been successfully used to investigate spatial genetic structure (SGS) and gene flow processes at the population level. Troupin et al. (2006) analyzed SGS changes over time in an expanding population of *P. halepensis* in Israel. Most of the factors that shape SGS (i.e. seed and pollen dispersal, mating system, unequal reproductive success and micro-environmental selection) could indeed vary over time, as the density of reproductive individuals increases or decreases. Jointly analyzing aerial photos, tree-ring time series and neutral genetic data, Troupin et al. (2006) reconstructed the 30-year expansion of a *P. halepensis* population from five initial individuals. In the early stages of the expansion, no SGS was detected, but fine-scale SGS (<20 m) developed at a certain point, as the density of adults started to increase considerably. SGS likely derived from the spatial arrangement and genetic variation of these few founding individuals, while limitations to seed dispersal apparently did not play a significant role in structuring genotypes in space. Alternatively, the development of SGS might be caused by fine-scale selection acting on different microhabitats that hosted *P. halepensis* regeneration (Heywood 1991; Epperson 1992). If certain genotypes were preferred in certain micro-environmental conditions, then the increased density of these genotypes in distinct patches would result in SGS.

Molecular markers have also been used in *P. halepensis* to explore historical and/or contemporary gene flow patterns, as well as among- and within-population differences in mating systems and reproductive success. Steinitz et al. (2011) applied a parentage-based approach to elucidate seed dispersal dynamics in an Israeli population of *P. halepensis*. Within the same expanding population already studied in detail by Troupin et al. (2006), the authors sampled the five ancestor trees from which the recolonization began in the 1940s (between 1944 and 1957, as emerged from both tree-ring data and aerial photos) and another 122 adult trees that were their potential offspring. All sampled trees were genotyped with nine highly polymorphic nuclear SSRs. After having assigned each (now mature) offspring to its most likely parent pair or single parent, the authors used parentage assignments to characterize the realized distribution of the effective seed dispersal distances. Subsequently, the effective dispersal kernel was compared to the potential dispersal kernel resulting from a mechanistic dispersal model that mimicked the wind regime experienced by the five ancestor trees. Steinitz et al. (2011) found that the effective seed dispersal had strong directionality, corresponding to the prevailing wind directions during the dispersal season. Moreover, the authors detected genetic evidence for a Janzen-Connell (JC) effect: the mode of the effective dispersal kernel was shifted farther away (20 m) from mother trees. This reflected a higher mortality rate close to the mother trees. In *P. halepensis*, such low survival of seeds close to the mother trees was likely due to high seed predation rates close to adults (Nathan and Ne'eman 2004) and/or to lower germination and growth rate under unfavorable light conditions below the mothers' canopy (Schiller 1979). Interestingly, the mode of the effective seed dispersal kernel increased with age, shifting from 5–10 m when considering 3-year-old saplings to 10–15 m when considering 4- and 5-year-old saplings (Nathan et al. 2000), and further increasing to 20 m when considering (mature) adult trees (Steinitz et al. 2011). The strengthening of the JC effect with age might be due not only to a higher incidence of predators, parasites or pathogens near mother trees, but also to increased competition with conspecific adult trees. Overall, this body of evidence highlighted the role of post-dispersal processes in shaping the reproductive dynamics of forest trees throughout their life cycle.

In the Mediterranean Basin, wildfires strongly affect the genetic layout of tree species, causing severe bottlenecks, altering seed dispersal and microsite conditions, and also acting as a selective driver on adaptive traits. *P. halepensis* has indeed evolved an obligate seeder strategy, regenerating after fire only from seed banks stored in the canopy, within serotinous cones (Fig. 2.3) that open under hot and dry weather or after fire (Ne'eman et al. 2004). The effect of fire on pollen flow was investigated in an Israeli natural stand experiencing a drastic fire that killed 96% of the pine trees in the stand (Shohami and Nathan 2014). Seven of the 13 surviving trees, grouped in two distinct patches 80 m apart from each other, exhibited pre-fire serotinous cones that did not open despite the fire. Genotyping such closed cones allowed characterization of the pre-fire pollen pool that was compared to the post-fire pollen pool from newly produced cones. A substantial fraction (30–60%) of the sampled pollen originated from the 13 surviving trees (i.e. *identified* pollen). The *identified* pollen was highly genetic differentiated between the two patches, just as the adult trees were, and such differentiation was detected both before and after the fire, pointing to very local pollen dispersal dynamics. Interestingly, the pollen that was not produced by the 13 surviving trees (i.e. *unidentified* pollen) showed significant SGS and high kinship before the fire, but it became spatially homogeneous after the fire. Assuming that the surviving trees could be a reliable sample of the pre-fire population, these results suggest that the landscape opening caused by disturbances could increase gene flow, rather than reduce it, by favoring wind-mediated pollen dispersal. This might help to alleviate the negative effects of a strong reduction in population size on the population genetic layout. Another study explored the neutral and adaptive effects of wildfires within *P. halepensis* populations in the Iberian Peninsula (Budde et al. 2017). The authors compared the genetic diversity and SGS of stands that suffer contrasting fire regimes (high vs low frequency and intensity of wildfires: *HiFi* vs *LoFi*), using both nuclear SSR and SNP loci from candidate genes. The SNP array was enriched in genes potentially involved in the architecture of fire-related traits, as many SNPs came from those selected by Pinosio et al. (2014) in their comparative transcriptomic study. Differences in the frequency and intensity of wildfires did not have any effect on genetic diversity parameters.

Fig. 2.3 Different aerial parts of *P. halepensis*. From top-left to bottom-right: male strobili, female strobili, natural population from southern Spain, clustered female cones, serotinous cones, post-fire natural population from southern Spain. (Photos: Climent J; De Santos L)

HiFi stands exhibited stronger SGS than *LoFi* stands, especially at SNP loci. This pattern might confirm the simultaneous post-fire recruitment of co-dispersed related seeds occurring after fire events (Ne'eman and Izhaki 1998). In contrast, *LoFi* stands exhibited weak SGS, as coherently expected in a wind-pollinated and wind-dispersed species where within-population genetic variation is efficiently reshuffled. Finally, three SNP loci were found to be highly differentiated between *HiFi* and *LoFi* stands, and they were identified as potential targets of selection. Such SNP loci have been located within genes playing a role in the regulation of wood formation (Oakley et al. 2007) and circadian rhythms (Mathews 2010). These candidate genes might be involved in the expression of fire-related traits, such as increased production of serotinous cones observed in *HiFi* stands (Hernández-Serrano et al. 2014). Altogether, the findings of Budde et al. (2017) suggested that the future increase in the frequency and intensity of wildfires due to climate change will likely promote stronger SGS and selective pressures in *P. halepensis* populations.

Finally, an additional study performed by Ruiz Daniels et al. (2019) aimed at detecting genomic signatures of local adaptation in *P. halepensis* using the 384-SNP array developed by Pinosio et al. (2014). The authors used a nested experimental design, sampling pairs/triplets of populations along three altitudinal gradients (in Spain, France and Italy) belonging to different genetic clusters (see Ruiz Daniels et al. 2018). This experimental design provided the opportunity of testing whether patterns of local adaptation held across different phylogeographic groups with independent evolutionary history. Three alternative methods were applied to identify SNP loci potentially under selection (i.e. PCAdapt, Bayenv2, Baypass), all accounting for population structure. No SNP locus was found to be potentially under selection by all three methods, while eight SNP loci were found to be potentially under selection by two methods. Only one of these eight loci exhibited the same pattern of changes in allele frequencies across the three altitudinal gradients, pointing to a scenario of convergent adaptation to similar selective pressures related to increasing altitude. Interestingly, this SNP was located in the non-coding region of a PIN-like protein (PIN2 in *P. tabuliformis*; GenBank: AJP06341.1) that acts as an auxin transporter and could be involved in the regulation of both plant development and responses to environmental stimuli (Křeček et al. 2009). This pattern of convergent evolution could be due to parallel selective pressures, but also to shared ancestral standing variation and/or to the spreading of favored alleles by gene flow. A more thorough understanding of the causes of such a scenario of convergent evolution will be crucial for forecasting whether *P. halepensis* still has enough adaptive potential to cope with future changes. With the climatic changes that are now occurring in the Mediterranean Basin and that will dramatically worsen in future decades (Intergovernmental Panel on Climate Change 2014), *P. halepensis* might track optimal climatic conditions by moving uphill, rather than through latitudinal or longitudinal migrations (Ruiz-Labourdette et al. 2012; Benito-Garzón and Fernández-Manjarrés 2015).

2.3 Conclusions – Perspectives

Altogether, the genetic information accumulated over the last 20 years has revealed a more complex evolutionary history of *P. halepensis* in the Mediterranean Basin than previously thought. This knowledge has contributed to improving our understanding of phylogeographic patterns for Mediterranean forest trees. Although genetic markers have clearly shown an east–west decreasing trend in genetic diversity, and point to distinct population demographic histories, the exact demographic events and their timing are yet to be fully deciphered – demographic modeling using newly-generated SNP datasets is a promising approach to tackle this challenge. Moreover, given the environmental constraints of the Mediterranean Basin, particularly drought and fire, expanding our knowledge on the adaptive potential of the species is essential. Understanding how gene flow, genetic drift, migration and selection interact will undoubtedly shed some light on the capacity of this pine to withstand the ongoing selective pressures. Finally, the genetic knowledge accumulated over the years will be of great assistance to forest management, conservation and breeding programs (Climent et al. this volume, Chap. 3).

Acknowledgments We acknowledge financial support from the European Union's Horizon 2020 research and innovation program under grant agreement No 676876 GenTree.

References

Arianoutsou M, Ne'eman G (2000) Post-fire regeneration of natural *Pinus halepensis* forests in the East Mediterranean Basin. In: Ne'eman G, Trabaud L (eds) Ecology, biogeography and management of *Pinus halepensis* and *P. brutia* forest ecosystems in the Mediterranean Basin. Backhuys Publishers, Leiden, pp 269–289

Benito-Garzón M, Fernández-Manjarrés JF (2015) Testing scenarios for assisted migration of forest trees in Europe. New For 46:979–994

Bucci G, Anzidei M, Madaghiele A et al (1998) Detection of haplotypic variation and natural hybridization in halepensis-complex pine species using chloroplast simple sequence repeat (SSR) markers. Mol Ecol 7:1633–1643

Budde KB, González-Martínez SC, Navascués M et al (2017) Increased fire frequency promotes stronger spatial genetic structure and natural selection at regional and local scales in *Pinus halepensis* mill. Ann Bot 119:1061–1072

Chambel MR, Climent J, Alía R (2007) Divergence among species and populations of Mediterranean pines in biomass allocation of seedlings grown under two watering regimes. Ann For Sci 64:87–97

Chen J, Uebbing S, Gyllenstrand N et al (2012) Sequencing of the needle transcriptome from Norway spruce (*Picea abies* Karst L.) reveals lower substitution rates, but similar selective constraints in gymnosperms and angiosperms. BMC Genomics 13:589

Climent J, Prada MA, Calama R et al (2008) To grow or to seed: ecotypic variation in reproductive allocation and cone production by young female Aleppo pine (*Pinus halepensis*, Pinaceae). Am J Bot 95:833–842

Climent JM, Alizoti P, Rodriguez-Quilón I et al (this volume) Conservation and breeding of Mediterranean pines. In: Ne'eman G, Osem Y (eds) Pines and their mixed forest ecosystems in the Mediterranean Basin. Springer, pp 33–70

Cruz de Carvalho MH (2008) Drought stress and reactive oxygen species: production, scavenging and signalling. Plant Signal Behav 3:156e165

De Luis M, Čufar K, Di Filippo A et al (2013) Plasticity in dendroclimatic response across the distribution range of Aleppo pine (*Pinus halepensis*). PLoS One 8:e83550

Eckert AJ, Pande B, Ersoz ES et al (2009) High-throughput genotyping and mapping of single nucleotide polymorphisms in loblolly pine (*Pinus taeda* L.). Tree Genet Genomes 5:225–234

Edmonds CA, Lillie AS, Cavalli-Sforza LL (2004) Mutations arising in the wave front of an expanding population. Proc Natl Acad Sci U S A 101:975–979

Epperson BK (1992) Spatial structure of genetic variation within populations of forest trees. New For 6:257–278

Excoffier L, Ray N (2008) Surfing during population expansions promotes genetic revolutions and structuration. Trends Ecol Evol 23:347–351

Fady B, Conord C (2010) Macroecological patterns of species and genetic diversity in vascular plants of the Mediterranean Basin. Divers Distrib 16:53–64

Fan F, Cui B, Zhang T et al (2014) The temporal transcriptomic response of *Pinus massoniana* seedlings to phosphorus deficiency. PLoS One 9:e105068

Girard F, Vennetier M, Ouarmim S et al (2011) Polycyclism, a fundamental tree growth process, decline with recent climate change: the example of *Pinus halepensis* Mill. in Mediterranean France. Trees 25:311–322

Girard F, Vennetier M, Guibal F et al (2012) *Pinus halepensis* Mill. crown development and fruiting declined with repeated drought in Mediterranean France. Eur J For Res 131:919–931

Gómez A, Alía R, Bueno MA (2001) Genetic diversity of *Pinus halepensis* Mill. populations detected by RAPD loci. Ann For Sci 58:869–875

Gómez A, Vendramin GG, González-Martínez SC et al (2005) Genetic diversity and differentiation of two Mediterranean pines (*Pinus halepensis* Mill. and *Pinus pinaster* Ait.) along a latitudinal cline using chloroplast microsatellite markers. Divers Distrib 11:257–263

Grivet D, Sebastiani F, González-Martínez SC et al (2009) Patterns of polymorphism resulting from long-range colonization in the Mediterranean conifer Aleppo pine. New Phytol 184:1016–1028

Grivet D, Sebastiani F, Alia R et al (2011) Molecular footprints of local adaptation in two Mediterranean conifers. Mol Biol Evol 28:101–116

Hao DC, Ge GB, Xiao PG et al (2011) The first insight into the tissue specific Taxus transcriptome via Illumina second generation sequencing. PLoS One 6:e21220

Hernández-Serrano A, Verdu M, Santos-del-Blanco L et al (2014) Heritability and quantitative genetic divergence of serotiny, a fire persistence plant trait. Ann Bot 114:571–577

Heywood JS (1991) Spatial analysis of genetic variation in plant populations. Ann Rev Ecol Syst 22:335–355

Howe GT, Yu J, Knaus B et al (2013) A SNP resource for Douglas-fir: *de novo* transcriptome assembly and SNP detection and validation. BMC Genomics 14:137

Intergovernmental Panel on Climate Change (IPCC) (2014) Climate change 2014: synthesis report. Contribution of Working Groups I, II and III to the Fifth Assessment Report of the Intergovernmental Panel on Climate Change. Core Writing Team: Pachauri RK, Meyer LA, Eds. IPCC, Geneva. Available online: http://www.ipcc.ch/report/ar5/syr/

Jaramillo-Correa JP, Grivet D, Terrab A et al (2010) The Strait of Gibraltar as a major biogeographic barrier in Mediterranean conifers: a comparative phylogeographic survey. Mol Ecol 19:5452–5468

Korol L, Madmony A, Riov Y et al (1995) *Pinus halepensis* x *Pinus brutia* subsp. *brutia* hybrids? Identification using morphological and biochemical traits. Silvae Genet 44:186–190

Křeček P, Skůpa P, Libus J et al (2009) The PIN-FORMED (PIN) protein family of auxin transporters. Genome Biol 10:249

Lepoittevin C, Frigerio JM, Garnier-Géré P et al (2010) In vitro vs in silico detected SNPs for the development of a genotyping array: what can we learn from a non-model species? PLoS One 5:e11034

Li X, Wu HX, Dillon SK et al (2009) Generation and analysis of expressed sequence tags from six developing xylem libraries in *Pinus radiata*. BMC Genomics 10:41

Lotterhos KE, Whitlock MC (2014) Evaluation of demographic history and neutral parameterization on the performance of F_{ST} outlier tests. Mol Ecol 23:2178–2192

Lotterhos KE, Whitlock MC (2015) The relative power of genome scans to detect local adaptation depends on sampling design and statistical method. Mol Ecol 24:1031–1046

Mathews S (2010) Evolutionary studies illuminate the structural–functional model of plant phytochromes. Plant Cell 22:4–16

Matziris D (2000) Genetic variation and realized genetic gain from Aleppo pine tree improvement. Silvae Genet 49:5–10

Metzger MJ, Bunce RGH, Jongman RHG et al (2013) A high-resolution bioclimate map of the world: a unifying framework for global biodiversity research and monitoring. Glob Ecol Biogeogr 22:630–638

Morgante M, Felice N, Vendramin GG (1998) Analysis of hypervariable chloroplast microsatellites in *Pinus halepensis* reveals a dramatic genetic bottleneck. In: Karp A, Issac PG, Ingram DS (eds) Molecular tools for screening biodiversity. Springer, Dordrecht, pp 407–412

Moulalis D, Bassiotis C, Mitsopoulos L (1976) Controlled pollinations among pine species in Greece. Silvae Genet 25:95–107

Nahal I (1962) Le pin d'Alep (*Pinus halepensis* Mill.). Etude taxonomique, phytogéographique, écologique et sylvicole. Ann de l'Ecole Nationale Des Eaux et Forêst et de la Station de Recherches et Expériences Forestières 19:7–207

Nathan R, Ne'eman G (2004) Spatiotemporal dynamics of recruitment in Aleppo pine (*Pinus halepensis* Miller). Plant Ecol 171:123–137

Nathan R, Safriel UN, Noy-Meir I et al (2000) Spatiotemporal variation in seed dispersal and recruitment near and far from *Pinus halepensis* trees. Ecology 81:2156–2169

Ne'eman G, Izhaki I (1998) Stability of pre- and post-fire spatial structure of pine trees in Aleppo pine forest. Ecography 21:535–542

Ne'eman G, Trabaud L (2000) Ecology, biogeography and management of *Pinus halepensis* and *P. brutia* forest ecosystems in the Mediterranean basin. Backhuys Publishers, Leiden

Ne'eman G, Goubitz S, Nathan R (2004) Reproductive traits of *Pinus halepensis* in the light of fire – a critical review. Plant Ecol 17:69–79

Oakley RV, Wang YS, Ramakrishna W et al (2007) Differential expansion and expression of alpha- and beta-tubulin gene families in *Populus*. Plant Physiol 145:961–973

Panetsos CP (1975) Natural hybridization between *Pinus halepensis* and *Pinus brutia* in Greece. Silvae Genet 24:163–168

Panetsos CP (1981) Monograph of *Pinus halepensis* (Mill.) and *Pinus brutia* (Ten.). Ann Forestales 9:39–77

Parchman TL, Geist KS, Grahnen JA et al (2010) Transcriptome sequencing in an ecologically important tree species: assembly, annotation, and marker discovery. BMC Genomics 11:180

Perdiguero P, Barbero MC, Cervera MT et al (2013) Molecular response to water stress in two contrasting Mediterranean pines (*Pinus pinaster* and *Pinus pinea*). Plant Physiol Biochem 67:199–208

Peter BM, Slatkin M (2013) Detecting range expansions from genetic data. Evolution 67:3274–3289

Pinosio S, González-Martínez SC, Bagnoli F et al (2014) First insights into the transcriptome and development of new genomic tools of a widespread circum-Mediterranean tree species, *Pinus halepensis* Mill. Mol Ecol Res 14:846–856

Pons A (1992) Les enseignements des données historiques concernant le Pin d'Alep. Forêt Méditerranéenne 12:155–157

Ruiz Daniels R, Taylor RS, Serra-Varela MJ et al (2018) Inferring selection in instances of long-range colonization: the Aleppo pine (*Pinus halepensis*) in the Mediterranean Basin. Mol Ecol 27:3331–3345

Ruiz Daniels R, Taylor RS, González-Martínez SC et al (2019) Looking for local adaptation: convergent microevolution in Aleppo pine (*Pinus halepensis*). Genes 10:673

Ruiz-Labourdette D, Nogués-Bravo D, Ollero HS et al (2012) Forest composition in Mediterranean mountains is projected to shift along the entire elevational gradient under climate change. J Biogeogr 39:162–176

Santos-del-Blanco L, Bonser SP, Valladares F et al (2013) Plasticity in reproduction and growth among 52 range-wide populations of a Mediterranean conifer: adaptive responses to environmental stress. J Evol Biol 26:1912–1924

Schiller G (1979) Factors involved in natural regeneration of Aleppo pine. PhD thesis, Tel Aviv University, Tel Aviv

Schiller G, Conkle MT, Grunwald C (1986) Local differentiation among Mediterranean populations of Aleppo pine in their isoenzymes. Silvae Genet 35:11–18

Serra-Varela MJ, Alía R, Ruiz Daniels R et al (2017) Assessing vulnerability of two Mediterranean conifers to support genetic conservation management in the face of climate change. Divers Distrib 23:507–516

Shohami D, Nathan R (2014) Fire-induced population reduction and landscape opening increases gene flow via pollen dispersal in *Pinus halepensis*. Mol Ecol 23:70–81

Soto A, Robledo-Arnuncio JJ, González-Martínez SC et al (2010) Climatic niche and neutral genetic diversity of the six Iberian pine species: a retrospective and prospective view. Mol Ecol 19:1396–1409

Steinitz O, Troupin D, Vendramin GG et al (2011) Genetic evidence for a Janzen-Connell recruitment pattern in reproductive offspring of *Pinus halepensis* trees. Mol Ecol 20:4152–4164

Troupin D, Nathan R, Vendramin GG (2006) Analysis of spatial genetic structure in an expanding *Pinus halepensis* population reveals development of fine-scale genetic clustering over time. Mol Ecol 15:3617–3630

Van Andel TH (2002) The climate and landscape of middle part of Weichselian glaciation in Europe: the stage 3 project. Quat Res 57:2–8

Verne S, Jaquish B, White R et al (2011) Global transcriptome analysis of constitutive resistance to the white pine weevil in spruce. Genome Biol Evol 3:851–867

Voltas J, Chambel MR, Prada MA et al (2008) Climate-related variability in carbon and oxygen stable isotopes among populations of Aleppo pine grown in common garden tests. Trees 22:759–769

Voltas J, Lucabaugh D, Chambel MR et al (2015) Intraspecific variation in the use of water sources by the circum Mediterranean conifer *Pinus halepensis*. New Phytol 208:1031–1041

Wei H, Yordanov YS, Georgieva T et al (2013) Nitrogen deprivation promotes *Populus* root growth through global transcriptome reprogramming and activation of hierarchical genetic networks. New Phytol 200:483–497

Zhang J, Jiang D, Liu B et al (2014) Transcriptome dynamics of a desert poplar (*Populus pruinosa*) in response to continuous salinity stress. Plant Cell Rep 33:1565–1579

Chapter 3
Conservation and Breeding of Mediterranean Pines

José Climent, Paraskevi Alizoti, Isabel Rodriguez-Quilón, Yusuf Kurt, Fulvio Ducci, Bruno Fady, and Ricardo Alía

3.1 Introduction

Mediterranean forests result from the long-term interplay between natural processes and silvopastoral practices (Blondel 2006) and are of high economic value for the whole Mediterranean Basin. Differences though occur, depending on the part of the Mediterranean Basin (i.e. North vs South, East vs West) and the forest species considered, stemming from large differences among areas in terms of forest use, abandonment of agricultural and pastoral lands, and forest recolonization *vs.* degradation that lead to fragmentation or deforestation (Gauquelin et al. 2018). When accounting for the ecosystem services, products, incomes, and environmental assets provided from these forests, their total economic value should be much more higher (e.g. up to 4.5 times higher in southern Spanish forests) than that estimated by using the traditional economic evaluation methods applied in the national or regional statistics, that are based on the products and incomes generated (Campos et al. 2019). Besides their vital contribution to rural economies, Mediterranean forests are also extremely important from a cultural, historic and biodiversity point of view.

J. Climent (✉) · I. Rodriguez-Quilón · R. Alía
Forest Research Centre, National Institute of Agronomic Research, Madrid, Spain
e-mail: climent@inia.es

P. Alizoti
Aristotelian University, Thesaloniki, Greece

Y. Kurt
Molecular Biology and Genetics Department, Harran University, Sanliurfa, Turkey

F. Ducci
CREA, Arezzo, Italy

B. Fady
INRAE, Avignon, France

G. Ne'eman, Y. Osem (eds.), *Pines and Their Mixed Forest Ecosystems in the Mediterranean Basin*, Managing Forest Ecosystems 38,
https://doi.org/10.1007/978-3-030-63625-8_3

Mediterranean forests are among the most endangered areas globally, due to an array of threats, according to different predictions (Gauquelin et al. 2018). We expect dramatic changes in rainfall and temperature during the twenty-first century, but also in the intensity and frequency of forest fires – a main driver for adaptation in the area – while socio-economic changes will affect the provision of goods and services in the Mediterranean (Fargeon et al. 2020).

Mediterranean conifers have played (and continue to play) an important role in the provision of products (e.g. resin, pine seeds, timber, firewood, honey production) and other ecosystem services essential for the local communities. Activies related to the management of genetic resources, genetic improvement and production of forest reproductive material (FRM) of Mediterranean conifers can differ among species. The different management practices applied in Mediterranean pine forests do not seem to have significantly affected the diversity of the populations (Ratnam et al. 2014). From a genetic perspective, Mediterranean conifers are characterized both by higher within and among populations genetic diversity, when compared to the average genetic diversity reported for conifers worldwide (Fady-Welterlen 2005), with a clear longitudinal gradient where eastern populations show higher diversity than western populations reported for 17 Mediterranean conifer species (Conord et al. 2012). Moreover, there are contrasting patterns in terms of diversity and differentiation from the center to the rear edge of the distributions, attributed to different evolutionary factors (Hampe and Petit 2005). Therefore, Mediterranean pines, as keystone species in these ecosystems, play an important role in the conservation of genetic resources of many other species of flora and fauna, a fact that poses important questions for the implementation of programs for the conservation and use of genetic resources in the Mediterranean region.

Due to the long tradition of management of Mediterranean conifers, particularly Mediterranean pines, conservation and sustainable use of their genetic resources are strongly interlinked, and in most cases the appropriate use of the genetic resources provides an opportunity to favor conservation. These activities can be considered at different scales: entire range of distribution (the Mediterranean Basin), landscape or stand levels. Usually the different scales provide or require different pieces of information, as the genetic and socioeconomic factors determining the conservation strategies will differ. For instance, the influence of genetically improved material on the natural genetic resources can only be assessed when analyzing specific populations subjected to gene introgression by such genetic material, while the conservation of breeding populations or the effective implementation of conservation networks could be of interest in genetic conservation at the national and the entire range of distribution scales.

Mediterranean pines cannot be considered one homogeneous group in terms of conservation and use of forest genetic resources, as they differ in key aspects (i.e. socio-economic context, importance for forest management, provision of goods and services) even at the population level within each species.

This chapter focuses on the conservation and breeding of Mediterranean pines (see Chap. 1 for a discussion on the species considered). We review important aspects of current and future use of their genetic resources in the Mediterranean

Basin under the forest management and conservation perspectives: conservation of genetic resources, breeding activities, procurement and deployment of FRM. Furthermore, we review the different constraints existing among the above mentioned aspects that could influence the adaptation of Mediterranean pine forests and their production, together with the main gaps in managing the genetic diversity of Mediterranean pine species under uncertain scenarios. We review the knowledge derived from existing national and European programs, and we make an attempt to fill in some of the existing knowledge gaps on the subject, to highlight others, and finally to discuss future research needs.

We focus in pine species that have a circum-Mediterranean native distribution range, including all the islands, with a Mediterranean climate: *Pinus brutia* Ten., *P. canariensis* C.Sm., *P. halepensis* Mill., *P. heldreichii* H.Christ, *P. mugo* Turra, *P. nigra* J.F.Arn, *P. pinaster* Ait., *P. pinea* L., *P. sylvestris* L. and *P. uncinata* Mill. ex Mirbel (Fady 2012; Pausas and Millán 2019). For species distributing in the Mediterranean Basin and beyond (eurasian pines), we focus only on the Mediterranean part of their range.

3.2 Genetics of Mediterranean Pines

Pines are mainly outcrossing species, although they have a mixed mating system (selfing and outcrossing). Gene flow through pollen is highly efficient, resulting in large effective population sizes in many species, and usually low population structure (Neale and Savolainen 2004). An important genetic load in pines can prevent inbreeding, but a recessive load can also explain a high rate of seed abortion (Savolainen et al. 1992; Kärkkäinen et al. 1999). However, these characteristics differ among Mediterranean pines, as these species and their populations show different demographic and geographical characteristics.

Different evolutionary factors affect differentiation among populations, as well as within-population variation. **Mutation and recombination** increase both and we cannot expect many differences among pine species (Neale and Savolainen 2004) or populations in this respect.

Mediterranean pine species differ greatly in their levels of diversity and fragmentation, and also in their levels of quantitative differentiation (Table 3.1). Mediterranean pines have been extensively studied in recent years using different sets of markers and quantitative traits to analyze the geographical patterns of distribution of the genetic diversity and other evolutionary aspects.

Table 3.1 Intensity of evolutionary factors prevailing in different Mediterranean pines. The intensity scale ranges from very low (+) to very high (++++), or unknown (?). The values are based on the references included in this chapter

Species	Fragmentation	Quantitative differentiation	Genetic drift	Diversity	Inbreeding
P. brutia	+	++	+	++++	+
P. canariensis	++++	++++	++++	++++	+
P. halepensis	+	++	+	+++	+
P. heldreichii	++++	?	++++	++	+++
P. mugo	++++	++++	++	++++	+
P. nigra	++	++	+++	++++	+
P. pinaster	++++	++++	++++	++++	+
P. pinea	+	+	+	+	++++
P. sylvestris	++++	++++	+++	++++	+/++++
P. uncinata	++++	++++	++	++++	+

3.2.1 Gene Flow

Gene flow is an important evolutionary factor, as it tends to reduce the levels of differentiation among populations and increase the level of diversity within the populations if excessive, or increase the differentiation among populations if limited or lacking (Kremer et al. 2012). In pines, particularly Mediterranean pines, we need to distinguish gene flow occurring via seeds from that occurring via pollen, – most importantly, long-distance gene flow – as their estimates are quite different. Also important is the distinction between short-distance and long-distance pollen gene flow, as they have different implications for adaptation (Kremer et al. 2012). Finally, we need to consider both ancient and contemporary gene flow, as the former has shaped the history of the species while the latter shapes the actual mating system of the populations. Dispersion by seeds has been reported at short distances (20–50 m, Kellomäki et al. 1987; González-Martínez et al. 2002b, 2006; Mukassabi et al. 2012). This is coherent with strong genetic structure at short distances reported for some natural populations (González-Martínez et al. 2002b).

3.2.2 Natural Selection

The comparison of neutral vs. quantitative differentiation (Q_{ST} vs F_{ST}) has allowed elucidation of the effects of divergent selection for different growth related traits in most cases analyzed (Alía et al. 2001; González-Martínez et al. 2002a; Kurt et al. 2012; Rodriguez-Quilon et al. 2016), although stabilizing selection has also been reported for cavitation resistance in maritime pine (*P. pinaster*) (Lamy et al. 2011). Alternatively, clear signals of selection have been reported, judging from the existence of significant climate-trait correlations in some of the Mediterranean pines (Alía et al. 1997; Climent et al. 2008; Ruiz Daniels et al. 2018).

3.2.3 Phenotypic Plasticity

Significant effects of both phenotypic plasticity and genetic differences in plasticity among populations have been reported for different species and traits (eg. *P. pinaster*, *P. pinea*, *P. halepensis*, *P. nigra*, *P. canariensis*) (Alía et al. 1995; Chambel et al. 2004; Corcuera et al. 2010). Theoretical models have emphasized the importance of these two factors for species adaptability (Chevin et al. 2010, 2012), but information is still lacking for most of the Mediterranean pines in traits related to adaptation to future climatic conditions, and also for endangered populations, preventing the effective assessment of their evolutionary potential.

3.2.4 Population Differentiation and Local Adaptation

The distribution of the Mediterranean pines is highly fragmented, but the level of genetic differentiation among populations (F_{ST}) varies widely among species. Analyses using different neutral genetic markers (differing in the ability to separate recent vs ancient fragmentation) have shown different patterns depending on the species, with *P. brutia*, *P. halepensis*, and *P. pinea* showing comparatively low levels of differentiation among populations, with clear geographical structuration of the diversity (Fallour et al. 1997; Bucci et al. 1998; Vendramin et al. 2008a; Grivet et al. 2009; Kurt et al. 2012). Similarly, for *P. canariensis*, *P. pinaster*, *P. nigra*, *P. sylvestris*, *P. mugo* and *P. uncinata*, there is clear range wide structure of the diversity, that increases as species are more cold tolerant (Provan et al. 1998; Soranzo et al. 2000; Gomez et al. 2003; Wahid et al. 2004; Robledo-Arnuncio et al. 2005; Bucci et al. 2007; Heuertz et al. 2009). One main barrier for the populations in the Western Mediterranean is the Straits of Gibraltar, which establishes a genetic disjunction across conifers; however, most of the observed genetic breaks seem to reflect the individual biological history of each species (Jaramillo-Correa et al. 2010).

In *P. canariensis*, this differentiation is related to the geological age of the island on which the species grows, while for the other species, it is mostly related to ancient fragmentation in different environments that resulted in distinct genetic groups.

The populations can also differ in size and ecological conditions. Many different marginal populations have been identified, and the use of genetic markers allows us to determine whether this fragmentation is recent and mainly caused by anthropogenic effects. Therefore, the value of these populations for conservation depends on the genetic effects of fragmentation: differentiation by local adaptation or genetic drift (e.g. in *P. pinaster* populations; Derory et al. 2002), or recent fragmentation without a detectable effect in the population.

Local adaptation may be more commonly found in large populations (Leimu and Fischer 2008), as usual in some Mediterranean pines (Grivet et al. 2011), while genetic drift may play an important role in the differentiation of some small populations of Mediterranean pines, without evidences of local adaptation (Boscherini et al. 1994; Naydenov et al. 2005; Heuertz et al. 2009).

3.2.5 Within-Population Diversity

It has been observed that low-elevation Mediterranean pines display significantly less genetic diversity within populations than mesophilous and montane Mediterranean pines (Fady and Conord 2010). All studies have shown high within-population diversity of Mediterranean pines and populations, except for *P. pinea* (Fallour et al. 1997; Vendramin et al. 2008a). *Pinus pinea* has a relatively large distribution range and is resistant to different pests and diseases, as well as being well adapted to drought and low-intensity forest fires. The causes of these low levels of genetic diversity and the exceptional ability of this species to cope with unfavorable environments are still unknown, but this fact has important implications for conservation and breeding activities for *P. pinea*. The variation found for important traits (cone production, growth) may indicate that epigenetic effects can play an important role in this species to favor adaptation (Sáez-Laguna et al. 2014), or that the low genetic variation has not reduced variability in important genomic regions.

For the other species, it is usual to find high levels of within-population diversity with high gene flow due to seed and pollen dispersal within populations. In populations with more than 200 m diameter, as much as 80% of pollen and seeds have been shown to come from individuals outside the area (González-Martínez et al. 2006). Correlated paternity, therefore, in this type of population is quite low, maintaining a robust mating system.

3.2.6 Inbreeding

Most pines are almost entirely outbreeders; however, *P. heldreichii* presents a low outcrossing rate determined not only by mating among relatives, but also reflecting significant levels of selfing (Morgante et al. 1991). This could also be the case for *P. pinea* (Vendramin et al. 2008a). An important increase in inbreeding and correlated paternity has been reported in small and isolated populations, when the population size is clearly reduced (< 30 individuals) (Robledo-Arnuncio et al. 2004a; Restoux et al. 2008). Marginal populations raise important questions for the implementation of conservation of genetic resources practices, as we cannot wait for long-term assessment of the effects of inbreeding in these populations before defining a conservation strategy. The high within-population variability in the majority of the populations allow selection to occur (Alía et al. 2014), but the evaluation of such changes is difficult in natural populations.

3.2.7 *Other Aspects*

We should also consider **outbreeding** as an important factor in small populations of Mediterranean pines. Important gene flow via pollen from non-native sources has been observed in small population of Scots pine (*P. sylvestris*) and maritime pine (>16%), but this seed is more intensively selected against during the first phases of recruitment (Ramirez-Valiente and Robledo-Arnuncio 2014; Unger et al. 2016; Robledo-Arnuncio and Unger 2018). However, seeds from non-native material will still survive in this new location, and will eventually change the structure of small populations.

There are some special characteristics of the Mediterranean region that should also be considered. One major issue that is still not well understood is the role of different **reproductive strategies** in the adaptation of Mediterranean pines. Serotiny is a paramount adaptive trait related to forest fires, which has important implications for long-term local adaptation among populations of some species (Tapias et al. 2004; Budde et al. 2014). Since different seed cohorts coexist in the same mother tree, differing in the year of pollination (with possible different paternal contribution and different epigenetic effects), we could expect high genetic diversity in one single recruitment event after fire in contrast to non-serotinous species or populations. This cumulative diversity could reduce the effect of the low population size in small populations, but the contribution to the next generation it is still unknown, especially in the highly variable (in both space and time) Mediterranean environments. We note that serotiny and the aerial seed bank in *P. halepensis* have been shown to be negatively affected by drought in a possibly maladaptive way (Martín-Sanz et al. 2016).

3.3 Conservation of Forest Genetic Resources

Forest tree conservation programs are mainly concerned with maintaining the evolutionary potential of the populations, aiming to construct a dynamic conservation system where evolutionary concepts are taken into consideration (Eriksson et al. 1993). For most of the forest tree species, including Mediterranean pines, the populations of a given species, not the species, are the objective of genetic resource conservation. However, implementation of such a conservation framework must consider some specific aspects of the species. The populations are usually subjected to forest management and it would be necessary to integrate their management (silvicultural systems, regeneration schemes, thinning and harvesting selection criteria) into the rules of the sustainable management of forest genetic resources, to favor adaptation in the long term. Therefore, we will need to analyze the genetic and evolutionary aspects of Mediterranean pines in order to formulate and implement cost-efficient genetic resource conservation actions.

Another relevant point to consider is the different scales (in time and space) for effective conservation programs. In forest management, we make decisions at the generation scale, but the effects of many evolutionary factors (e.g., gene flow, adaptation, differentiation, and inbreeding) are minor and can only be properly observed after several generations or by refining the methods of analysis (Krutovsky et al. 2012). A major current concern is to develop effective methods to assess selection intensity in the wild, contemporary gene flow, and the effect of inbreeding and outbreeding depression among others, for the implementation of an effective conservation strategy for forest genetic resources.

Conservation of forest genetic resources is a national responsibility, according to the CBD. However there are some coordination efforts in the European Mediterranean region related to the activities of EUFORGEN (European Forest Genetic Resource Conservation Program www.euforgen.eu). This program was established in 1994 following a resolution adopted in 1990 by the first Ministerial Conference of the Forest Europe process to promote the conservation and sustainable use of forest genetic resources in Europe as an integral part of sustainable forest management, and to serve as a platform for pan-European collaboration in this field. Mediterranean pines are included in the conservation strategy (Lefevre et al. 2012; Koskela et al. 2013; de Vries et al. 2015). EUFORGEN provides guidelines and the minimum requirements to save the conserved population data in the EUFGIS information system (European Information System on Forest Genetic Resources. www.eufigis.eu). We need to consider that some of the species extend beyond Europe, where the program operates and where coordinated activities are implemented at a pan-European scale, leading to potentially different levels of conservation in different parts of their distribution. Mediterranean pines are a heterogeneous group of species and populations, but they are a good example of conservation of outcrossing trees, as we can describe different situations; therefore, we should apply different methodologies for conservation of different species and populations.

Conservation of forest genetic resources, within the EUFORGEN program, aims to conserve the evolutionary potential of populations (dynamic conservation) both *in situ* and *ex situ* (Lefevre et al. 2012; de Vries et al. 2015). Different risks may be associated with some populations, due to genetic or evolutionary factors (e.g. inbreeding, fragmentation, genetic drift) but mostly due to threats from other sources (e.g. land-use changes, climate change, social changes); therefore, the conservation of genetic resources must consider both aspects (genetic and non-genetic) when implementing conservation activities on those populations.

3.3.1 Threats to the Genetic Resources of Mediterranean Pines

The main threats to the genetic resources of Mediterranean pines come from outside the forestry sector (e.g. land use changes, global change, forest fires) and result in the fragmentation of populations, reduction of population sizes and difficulties in forest regeneration (Table 3.2). Only in some cases (e.g *Pinus nigra* and the

Table 3.2 Threats to forest genetic resources of Mediterranean pines, and activities for conservation and breeding

Species	Major threats to forest genetic resources	In situ Conservation Units and ex situ collections[a]	Basic material[b]
P. brutia	Habitat loss, forest fires	TR, GR	SO
P. canariensis	Climate change, wildfire	ES	
P. halepensis	Wildfire	BiH, ES, FR, IT, GR, TR, MO	SO
P. heldreichii	Habitat loss, climate change	BiH, IT, CR, SL	
P. mugo (P. uncinata)	Habitat loss, climate change	ES	SO
P. nigra	Habitat loss, hybridization with other subspecies, wildfire	BiH, ES, FR, GR, TR	SO
P. pinaster	Wildfire, pests and diseases	ES, FR, IT, GR, MO	SO, PF
P. pinea	Habitat loss, wildfire	ES, PT, IT, TR	CL, CLM
P. sylvestris	Habitat loss, climate change	BiH, ES, FR, AT, IT, TR, GR	SO

[a]Country codes: Croatia (CR), Bosnia and Herzegovina (BiH), France (FR), Greece (GR), Italy (IT), Morocco (MO), Slovenia (SL), Spain (ES), Turkey (TR)

[b]Basic material for producing FRM derived of individual selection:, seed orchard (SO), Parents of family (PF), clone (CL) and Clonal Mixture (CM)

hybridization of native subspecies with planted exotic black pines, and some populations in other Mediterranena pines) the main threat come from the forestry sector. In some cases it is only possible to mitigate the threats in order to conserve the genetic resources of the species and populations, as they do not depend on actions that can be easily corrected by the managers of the genetic resources.

Global change (i.e. climate change (Schueler et al. 2012) and the associated socio-economic changes in the Mediterranean) is the most important threat to the conservation of forest genetic resources. The different species and populations differ in terms of susceptibility to climate change, or the possibility of escaping to more favorable areas in the future by migration. For instance, *P. uncinata* and *P. mugo* populations are now at the upper limit of the mountain ranges (Benito-Garzón et al. 2008). In contrast, some of the *P. sylvestris* or *P. nigra* populations can still migrate within the same area to upper regions, and the risk is lower at shorter time scales. However, the lowland pine forests of central Spain are quite endangered, as extinction of species such as *P. pinea* or *P. pinaster* is expected due to lack of appropriate environmental niches in the future (Benito-Garzón et al. 2011; Serra-Varela et al. 2017). The future predictions for *P. halepensis* and *P. pinaster* indicate that most of the conservation units defined, covering a broad range of distribution of the species (Fig. 3.1), will likely be unsuitable or uncertain for their future. These predictions also reflect the predictions for the entire Mediterranean region (Thuiller et al. 2005).

These models do not take into consideration the local adaptation and the adaptability of the populations, and therefore the predictions should only be used to prioritize the populations for *ex situ* conservation, and to implement some actions in the species or populations concerned.

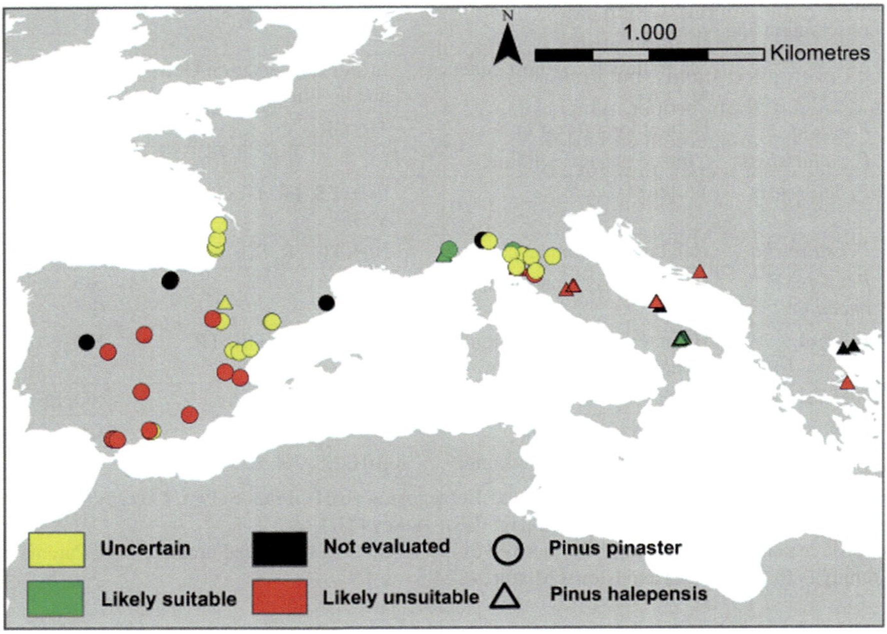

Fig. 3.1 Exposure assessment for the EUFGIS (European Information System on Forest Genetic Resources) dynamic conservation units of *Pinus pinaster* and *Pinus halepensis* along their natural distribution (Modified from Serra-Varela et al. 2017)

Another issue related to climate change is the outbreak of pests and diseases (e.g. *Fusarium*, *Matsococcus*, nematodes, forest decline, but also already existing ones, such as *Diplodia*, *Armillaria*, processionary moth). Mediterranean pines differ in susceptibility and also in the associated risks to their distribution range (Mendel 1984; Schvester and Ughetto 1986; Elvira-Recuenco et al. 2014; Menéndez-Gutiérrez et al. 2017).

Another risk is the lack of regeneration for some populations. Under climate change we may observe changes in the mating system of the species: crop production changes during mast years, a reduction of produced cones and/or conelets, and reduced seedling survival during the early phases of recruitment. There is strong selective pressure during the regeneration phase (Petit and Hampe 2006); therefore, it is the most critical stage for predicting the future adaptation of the forest trees. Moreover, a lack of regeneration will result in a loss of the evolutionary potential of the populations/species of interest.

The Mediterranean region experienced an intensive afforestation program during the last century (Vallejo et al. 2004; Sheffer 2012). These afforestations included genetic material of both native and non-native origins.

Socio-economic changes also pose a threat in some of these populations. Often, use of such populations has been extensive, leading to overexploitation for different purposes, but other instances include the abandonment of rural areas, or a lack of forest management, which can lead to deterioration of these populations (Gauquelin et al. 2018). Some populations have been traditionally managed for producing pure and even-aged stands, but changes that occur in the structure of the populations, as they age without proper management, can result in a low number of young trees.

Forest management can affect the genetic resources (Ratnam et al. 2014), mainly through the regeneration system, or the selection of trees (thinning, mother trees) for the future generation. However, we have scant documentation of these factors with respect to Mediterranean pines. In large Scots pine populations, the silvicultural systems did not have any impact on their genetic diversity, when considering shelterwood systems *vs.* clear cutting in central Spain (Robledo-Arnuncio et al. 2004b). In *P. brutia*, improved material was not found to pose any threat to the genetic diversity of the species (Içgen et al. 2006). However some effects of inbreeding have been observed in overexploited forests (Lise et al. 2007). This can be understood to have general application for Mediterranean pines, due to their mating system, when the population size is large enough.

3.3.2 Conservation Strategies and Actions

There are two main conservation strategies for forest trees, the *in situ*, and *ex situ* conservation of the genetic resources strategies. Due to the low level of domestication of forest tree species, the high value of their genetic resources for the ecosystems, and the cost of the different conservation options, the *in situ* conservation strategy has been traditionally prioritized, while the *ex situ* conservation one has been applied in cases where *in situ* conservation was not feasible.

Dynamic conservation has been defined as the main objective of the EUFORGEN conservation program, to preserve the adaptive and evolutionary potential of the populations (Lefevre et al. 2012). The implementation of the conservation programs must rely on our knowledge of the evolutionary and social factors affecting the conservation of the genetic resources of Mediterranean pines.

A reasonable starting point is to consider that a conservation network should span a species' entire standing genetic variation (Crandall et al. 2000). Following this, two main approaches have been used to define tree conservation units: (1) methods based on the contribution of particular populations to the total neutral genetic diversity or differentiation of the species (Petit et al. 1998), and (2) methods based on ecological and geographical information (e.g., Hamann et al. 2005; Koskela et al. 2013). The existing European network of dynamic conservation units is largely based on this second approach (Koskela et al. 2013; de Vries et al. 2015), assuming that the main climate types in Europe are closely linked to genetic differentiation, indirectly including adaptive processes under the hypothesis of local adaptation and clinal or ecotypic variation in tree species (Alberto et al. 2013), but also because for most of the Mediterranean populations, we still lack the genetic information and knowledge to fine-tune conservation efforts and strategies.

The pan-European strategy is based on defining a core network of dynamic conservation units. These units are not interconnected by gene flow, but together they might potentially capture the current genetic diversity across the European continent. The program is based on the conservation units defined for each species. As this is a pan-European program, the core network of conservation units has been already defined for a limited number of species, and the strategy has not been implemented yet for most Mediterranean pines. A second limiting factor is that for some more valuable, and even threatened, species, genetic resources can be found outside of their European distribution range (e.g. *P. halepensis*, *P. nigra*, *P. pinaster*, *P. pinea*). Taking these factors into consideration, we can analyze the actual conservation status of the species.

In the EUFGIS (European Information System on Forest Genetic Resources) portal (Fig. 3.2) 362 conservation units have been designated for Mediterranean pines (*P. canariensis* is not included in the list, being of interest to only one country). The distribution of genetic conservation units by country and species is not even, indicating some gaps for conservation across and within species and countries.

Pinus brutia: This species shows signs of local adaptation to environmental gradients related to altitude (Kurt et al. 2012). *Pinus brutia* as such is not endangered; the main threats to its survival include factors such as pests (*Matsucoccus josephi* Bodenheimer and Harpaz in the Eastern Mediterranean) and forest fires (Fady et al. 2003). Three gene conservation units (GCUs) have been designated in Greece, two in Italy (non-native *ex situ* stands) and 57 in Turkey, covering most of the

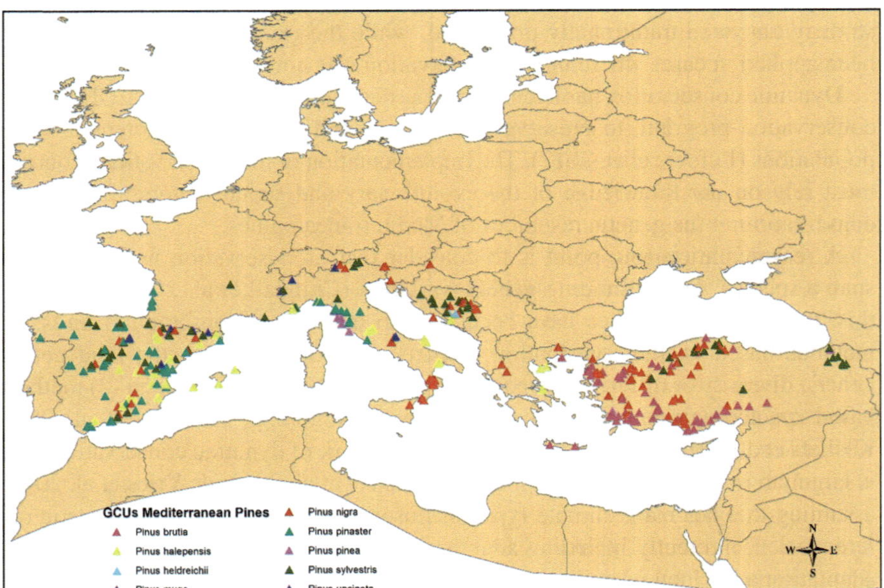

Fig. 3.2 Dynamic conservation units of Mediterranean pines included in EUFGIS (http://portal.eufgis.org/)

distribution range of the species; out of the total of 62 GCUs, six were included in the core *in situ* EUFORGEN collection. However, in areas where the species grows in Lebanon, where it is not considered endangered (Talhouk et al. 2001), and in Cyprus, there is a lack of GCUs.

Pinus canariensis: This species is not included in the EUFORGEN program. The distribution of *P. canariensis* is restricted to the Canary Islands, where it exhibits large genetic differences among populations for conservation (Gomez et al. 2003), as well as large differences in phenotypic plasticity, indicating signs of local adaptation in this species (Climent et al. 2006; López et al. 2007a). No GCUs have been defined for *P. canariensis*.

Pinus halepensis: A comprehensive analysis of the distribution range of this species has identified a minimum number of two genetic groups (western/eastern); when this number increases (up to seven), a spatial substructure appears at a finer spatial scale (Serra-Varela et al. 2017). Conservation units have been defined in many countries: one in Bosnia and Herzegovina, 19 in Spain, three in Greece and 12 in Italy, covering the distribution range in Europe and the main genetic regions defined for the species. It has been included in seven GCUs in the European core collection for *in situ* conservation. However, in North Africa, marginal populations of importance for conservation of genetic resources exist in Morocco, Algeria, Tunisia and Lybia. Human impacts may hamper the conservation of genetic resources of some of these marginal populations, and *ex situ* activities should be considered (Fady et al. 2003).

Pinus heldreichii: Bosnian pine is included in the IUCN Red List of threatened species (even though it appears as a low risk entry) because its natural distribution is currently very limited and fragmented, mostly due to forest fires (Vendramin et al. 2008b). This species' distribution range can be divided into distinct groups based on the locations of the populations (Boscherini et al. 1994; Naydenov et al. 2005). Two GCUs have been designated in Bosnia and Herzegovina, one of which is in the EUFORGEN core collection. High levels of selfing were found to occur in both natural populations and planted stands, where tree density is higher and neighborhood structures less pronounced (Puglisi 1995; Vendramin et al. 2008b). Despite the high fragmentation and inbreeding, conservation units are lacking in many areas across the distribution range of the species (Montenegro, Albania).

Pinus mugo: In the Mediterranean range of this species (excluding *P. uncinata* populations) all populations can be included in one genetic group (Heuertz et al. 2009). There are two GCUs in Italy and one in Bulgaria. The isolated populations of the species in its southern range – Abruzzo (Italy), and the western Balkan Range (Serbia and Bulgaria) – can be considered to be at risk, due to climate change and the human impact resulting from the construction of ski resorts and accompanying facilities (Alexandrov et al. 2019), as well as the use (for other purposes) of the area that could be potentially suitable for the species in the face of the climate change. *Pinus uncinata* (considered as a subspecies of *Pinus mugo*) present similar threats to those for *P. mugo*; some predictions reflect the likely reduction of the distribution range of *P. uncinata* (Benito-Garzón et al. 2008), due to the limited area for migration above its current distribution. This species has some marginal populations in

Central Spain (Alexandrov et al. 2019), and genetic differences occur among the populations (Soto et al. 2010) that belong to one genetic group within the *P. mugo* complex (Heuertz et al. 2009). Seven GCUs have been defined in Spain and three in Italy, covering most of this species' range in the Mediterranean region.

Pinus nigra: Black pine is not recognized as a threatened species although some of its sub-Mediterranean endemic populations constitute priority habitats under the EU Natura 2000 Directive, and the reduction of population size in marginal populations is the main genetic risk associated with these populations (Fady et al. 2003). Twelve GCUs have been designated in Bosnia and Herzegovina, 19 in Spain, three in Greece, one in Croatia, 28 in Italy, one in Serbia, one in Slovenia, and 41 in Turkey. In the Mediterranean region, the core EUFORGEN network comprises 14 GCUs.

Pinus pinaster: This species presents clear genetic structuration of the populations into six distinct genetic groups that have been analyzed for the genetic conservation of the species (Rodriguez-Quilon et al. 2016); within the Moroccan group, three additional main groups of populations can be distinguished based on morphological and molecular markers: (i) Mediterranean Coastal, (ii) Occidental Rif and Middle Atlas and (iii) High Atlas (Wahid et al. 2006, 2009). Forty-four GCUs have been designated in Spain, four in France, and 14 in Italy; six populations have been included in the core collection. There are, however, many important areas that are not represented in the GCU network. In Morocco, Destremau et al. (1976) described the different populations and their main characteristics, and identified overexploitation of the forest by intensive cattle grazing and land speculation as active mechanisms that still are depleting the genetic resources of *P. pinaster* (Wahid et al. 2009). Populations in Algeria and Tunisia have been less characterized, but they are assumed to harbor valuable genetic resources, based on the limited knowledge of their performance in provenance tests and genetic studies.

Pinus pinea: This species is genetically very uniform, and provenance trials do not indicate any strong geographic structuration in adaptive traits, such as vigor (Fady et al. 2008). An extensive range-wide chloroplast DNA diversity study revealed that all but a few populations located in Lebanon had the same haplotype. *Pinus pinea* is not considered a threatened species. Although its genetic diversity is low, it is comparatively rarely attacked by pests and diseases. Eight GCUs have been designated in Italy and two in Turkey, while conservation gaps across the range of the species can be recognized.

Pinus sylvestris: The distribution range of this species in the Mediterranean region is characterized by isolated and fragmented populations in comparison to its northern range. In extreme circumstances, the remnant populations harbor less than 100 trees (Matyas et al. 2004). Fifteen GCUs have been defined in Bosnia and Herzegovina, 25 in Spain, two in France, one in Croatia, seven in Italy, one in Portugal, one in Serbia, and 21 in Turkey. The core network in the Mediterranean region comprises 14 GCUs, covering the entire distribution range, although some gaps have been identified (Lefevre et al. 2012; de Vries et al. 2015).

Ex situ conservation of all the Mediterranean pines can be implemented at two levels. By plantation under different climatic conditions (e.g. favorable ones as predicted by the climatic models, such as in Serra-Varela et al. 2017), or under extreme conditions to favor adaptation in the early stages. Some other traditional plantations and breeding trials (e.g. plantations, provenance tests, and seed orchards) can also be included in this category even if their main objective is not gene conservation. Some of the EUFGIS gene conservation may fall into this category. More static methods, such as seed banks, can be easily implemented for these species, as it is easy and cheap to maintain them for 10–20 years. Only a few *ex situ* conservation measures have been implemented to date in the different countries (e.g., Tranque et al. 2018 in *P. nigra*), and they lack such a general framework as that existing for *in situ* conservation.

3.4 Breeding in Mediterranean Pines

Considering the new threats to Mediterranean forests (e.g., global change including climate change, new pests and diseases from both pre-existing and new invading organisms, need for carbon-neutral products), sustainable forestry requires tree improvement that integrates adaptation. In this sub-chapter we review the state of the art in the breeding of Mediterranean pines and propose practical recommendations that forest managers and policy makers can adopt for promoting forest resilience.

3.4.1 Breeding Activities and Actions

3.4.1.1 *Pinus pinaster*

Pinus pinaster (maritime pine) is indeed the model species for breeding among Mediterranean pines. Breeding programs have a long history in France, and more recently in Spain, Portugal, Italy and Morocco, all of them reviewed by Bouffier et al. (2013).

Breeding Program in France (Landes of Aquitaine)

Early provenance testing showed the superiority of the local Landes provenance for growth and cold resistance, despite form defects for stem straightness and branching. Three hundred and eighty plus trees were phenotypically selected from the local gene pool based on height, diameter and stem form. A progeny test demonstrated that this phenotypic selection was efficient for significantly improving growth and stem straightness (Danjon 1995). In addition, genetic variation among provenances and the performance of inter-provenance crosses were explored (Harfouche and Kremer 2000). Among all tested combinations, Landes × Corsica families proved to be the best material for growth and form (stem straightness and branch quality) in Aquitaine conditions. The breeding program followed an intra-population recurrent selection approach, with a main population composed of the Landes plus trees. Following this strategy, the main population has cycled through three generations, with more than 4500 individuals selected, and 5000 families tested over 500 ha of trials. For the next generation, the focus is on a reduction of population census size and better management of pedigrees, to optimize selection efficiency while producing regularly renewed varieties with increasing genetic gains. Eight unrelated sublines have been assembled within the breeding population based on pedigrees and breeding values, allowing the deployment of unrelated selections in clonal seed orchards. Double-pair mating designs are used to produce material for progeny tests and the base of the next generation, while polycross progeny testing is performed for parental ranking. Replicated progeny trials are established in contrasting sites, usually with single-tree plots and a large number of replications per site.

Three generations of open-pollinated seed orchards have been produced to date. The first-generation orchards were seedling seed orchards based on a very large number of full-sib families, corresponding to the progeny tests of plus trees, and were rogued after genetic assessment. These orchards yielded genetic gains of 10–15% in volume and stem straightness at about 15 years of age. Second-generation orchards were established with a reduced genetic basis and greater expected genetic gains. They were based on a few tens (usually around 30) of backward-selected genotypes, either as a classical grafted clonal orchard or as a randomized plantation of polycross families obtained by controlled pollination between selected clones (Baradat et al. 1992). The expected genetic gain over unimproved material was estimated from progeny trials at 13 years of age to be 30% for both volume and stem straightness. Since hurricane "Martin" in 1999, the area reforested with maritime pine in Aquitaine increased annually from 15,000 to 23,000 ha, out of which 70% has been planted with materials originated from second-generation orchards. Third-generation clonal and polycross-family seed orchards have been established over an area of 180 ha in total, and will enter production by 2022. This third generation also includes a Landes × Corsica variety, to be produced by controlled pollination.

Seed orchards are expected to be renewed more rapidly in the near future, to respond adequately to climate change and new market demands. Recent adaptations in the maritime pine breeding program, such as the optimization of population management through sublining and selection efficiency with BLUP (Best Linear Unbiased Prediction) evaluations, are expected to be augmented with marker-assisted selection for complex traits such as wood quality and drought resistance.

Breeding Programs in Other Countries Within the Natural Range

Provenance studies began in Spain in the 1960s (Molina 1965) to select provenances to be deployed in the high productivity areas of north-western Spain. Breeding was then followed by recurrent selection of plus trees in the most interesting regions of provenance, and clonal seed orchards and progeny tests from open-pollinated progenies were established (Pardos and Gil 1986). Different breeding zones corresponded to major geographical regions (Galicia, Central Iberian Planes, Iberian Range and Central System). In Galicia, the breeding program was based on seed orchards of the autochthonous provenance, with two breeding zones: coastal and interior, distinguished by altitude, and therefore, by frost frequency and intensity. A similar approach was followed in Portugal (Aguiar et al. 2003), where the local provenance displayed superiority over other materials (Correia et al. 2010). However, this division into different breeding zones has been demonstrated to be ineffective (De La Mata et al. 2012). It has been suggested that breeding zones should be redefined to a larger extent, and provide material suitable for the different conditions by managing only one breeding population. This approach has been implemented by establishing a single breeding program over the entire area (Majada and Alía 2012). Initially, the main objective was to increase volume production and improve stem form (especially in some of the populations); currently, wood quality characteristics are also considered essential for inclusion, along with tolerance to biotic and abiotic risks (drought and cold, diseases, pests). A set of progeny tests and provenance-progeny tests has been established for testing the material from the different selections. Usually plus trees are tested, in combination with materials from other more advanced programs or regions. At present, there is a network of genetic tests with selected populations (e.g., Portugal, Galicia, Spanish Central Range). Recently, different genetic trials have been established with more than 30 populations covering the entire distribution range of *P. pinaster*, and about 450 families across those populations. These genetic trials are essential for defining the future breeding population of maritime pine in the Iberian Peninsula. Spatial analysis has been implemented as a routine analysis of experiments to increase the efficiency of estimates (e.g., Zas and Merlo 2008; Correia et al. 2010; Saldaña et al. 2012). Different traits have been evaluated in these populations, including tree height, carbon-isotope discrimination, stem diameter, stem form, nutrient efficiency, and fecundity (Correia et al. 2008; Zas and Merlo 2008; Martins et al. 2009; Santos-del-Blanco et al. 2012). In most cases, high genetic diversity has been found, securing a basis for long-term breeding. Despite these breeding efforts, most plantations are still based on local populations and, to a lesser extent, on compatible provenances following published seed transfer recommendations (Alía et al. 2009b). Some approaches using vegetative material obtained from mass propagation of young seedlings have been tested, in order to produce material from parents of families (Majada et al. 2011).

Additionally, some selection and breeding activities have also been implemented in Spain for resin production, a highly heritable trait, in the highly productive areas of the Castilian Plateau (Tadesse et al. 2001). Clonal banks and progeny tests were established and evaluated, but currently breeding for resin production is on standby due to lack of commercial interest.

A network of *P. pinaster* provenance tests was established in Greece in 1971 (Matziris 1982). Four provenances of the species (three French and one Portuguese) were tested in seven field trials established in central and southern Greece. Seven traits were evaluated, namely survival, height growth, diameter at breast height, stem straightness, crown form, fructification and number of cones per tree. Provenances originating from the Atlantic race (Portuguese and Landes) outperformed the Cevennes and Corsican provenances in growth traits, while the most prolific provenances were the Portuguese and the Landes (French) ones. Despite the large differences in growth traits recorded among the seven sites, the provenance × environment interaction was not significant, indicating genetic stability of provenance performance.

3.4.1.2 *Pinus brutia*

Besides provenance research in Greece (Alizoti et al. 2018) and Italy, breeding of *P. brutia* (Turkish red pine, Calabrian pine) is particularly well developed in Turkey, due to the high interest for timber production of this species in Anatolia (Chambel et al. 2013).

In Turkey, in parallel with wide provenance research, a wide program of seed orchard plantation has been carried out (the only seed orchard program among the three mentioned countries). The first clonal seed orchard of *P. brutia* was established in 1976; 67 seed orchards covering 472 ha were established by 1993, with progeny testing regularly undertaken (Kaya and Isik 1997; Isik et al. 2002). Seeds orchards established during that period generally contained 30 clones. An important change was due to implementation of the National Tree Breeding and Seed Production Program (NTBSP) starting in 1994 (Koski and Antola 2013). By considering seedling number per hectare and seed amount needed for seedling production, seed requirements were determined for each species. Seed sources (seed stands and seed orchards) were planned according to species' seed demand. Breeding studies have been accelerated by the progeny trials established in the framework of the above program. In addition, seed orchards have been established with a higher number of clones (41–152 clones). Three breeding zones were delimited and plus tree selections for each breeding zone were carried out. By 2006, two new seed orchards were established in two breeding zones according to the first results (4th year) of the progeny tests. These two seed orchards will be converted to genotypic seed orchards by genetic purging after the results of the progeny tests. Currently, all seed used for plantations of *P. brutia* in Turkey comes from seed stands and orchards. Seventy-six clonal seed orchards have been established for the species with the number of clones ranging from 10 to 147. The seed orchards range in size from 1.0 ha to 18.6 ha, with an average size of 7.07 ha and cover a total area of 649.3 ha (ORTOHUM; Bilgen et al. 2013). In addition to the seed orchards, 72 seed stands have been delimited for the species, covering 10423.3 ha, 61 gene conservation forests covering 9504.1 ha and 15 clone parks covering 33.2 ha. All the above breeding efforts materialized so far support the argument that *P. brutia* is the main conifer species of interest in Turkey.

3.4.1.3 *Pinus halepensis*

Different unrelated breeding programs have been carried out for *P. halepensis* (Aleppo pine) in Greece, Spain, and Turkey, with additional provenance and eco-physiological studies in Israel (all reviewed in Chambel et al. 2013).

In Greece, breeding of Aleppo pine began in 1962, mainly involving hybridization with *P. brutia*. Standard breeding programs were initiated in 1970, with selection and collection of material on a provenance and mother-tree basis. *Pinus halepensis* exhibited significant genetic variation for all the adaptive traits studied, while their performances across sites indicated the high potential for effective across-site selection (Matziris 1997; Alizoti et al. 2000). The outstanding performance of *P. brutia* × *P. halepensis* hybrids is worth mentioning, as their superiority in growth over the pure species varied from 5 to 190%, depending on the planting site, and the specific genotypes of the two parental species that were intercrossed (Panetsos 1986). According to this author, hybrid vigor was manifested in intermediate and disturbed habitats, or outside the natural distribution of the pure species, but always within their ecological range. F1 hybrids tolerated low temperatures much better than *P. halepens*is, while *P. brutia* is the parental species tolerating low temperatures and thus conferring this trait to the hybrid progenies. F2 and back crosses were always inferior in growth, compared to F1 hybrids, at least in the testing environments.The results suggested the existence of ample genetic variation for adaptive traits among and within the natural populations of both species (Matziris 2000), the potential for selection and breeding and the need for conservation of their unique genetic resources in the face of global change. A first-generation seed orchard exists for *P. halepensis*, covering 20 ha and including 76 genotypes (2630 grafted individuals in total) from plus trees selected in the Euboea region (Alizoti et al. 2018). There are two progeny trials associated with this seed orchard, established in 1987, with 70 open-pollinated families corresponding to 70 out of the 76 seed orchard clones (Matziris 2000; Alizoti et al. 2018).

In Spain, in parallel with provenance testing, a seed orchard program was carried out in the 1980s and 1990s, starting from a selection of plus trees in three breeding zones of Eastern continental Spain (Pardos and Gil 1986). Progeny trials were established, both with open-pollinated seeds from the selected mother trees and, more recently, with seeds collected from seed orchards. A more recent set of comparative trials was planted to promote the FRM of the most successful seed orchard in the tested category (Nieto et al. 2009), even though the demand for improved seed of this species in Spain (justifying a higher price) is very low. Thanks to the precocious and prolific cone-bearing in seedlings, progeny tests can easily be converted into seedling seed orchards of 1.5 generations.

In Turkey there are currently two seed orchards of *P. halepensis* covering a total of 9.6 ha. One seed orchard comprises 10 clones and 280 ramets, while the second one 20 clones and 498 ramets (ORTOHUM; Bilgen et al. 2013). In addition, two seed stands have been delimited for the species covering an area of 199.2 ha.

3.4.1.4 *Pinus pinea*

Umbrella or Mediterranean stone pine (*P. pinea*) is a separate case for two main reasons: firstly, because of wide domestication and planting both for seed production and recreational purposes and, secondly, because of its extraordinarily low genetic variation (Vendramin et al. 2008a). Today, the most advanced breeding program for stone pine is carried out in Spain, but Portugal, Tunisia, Turkey and Italy have also developed different provenance research and breeding programs for this species.

In Spain, breeding of stone pine began with the delimitation of seed sources and definition of selected stands for reforestation and artificial regeneration for multifunctional forestry (Mutke et al. 2000). This was followed by the selection of plus trees for cone production, including their monitoring of annual cone yield for several years (to avoid bias due to masting (Catalan et al. 1995; Mutke et al. 2000, 2011; Gordo et al. 2001; Gordo 2004). Subsequently, more than five hundred plus trees were propagated by grafting in clonal banks, aiming at the evaluation of breeding values and constituting an *ex situ* reserve of breeding material for further needs of the program. These clonal banks were monitored for a number of years for growth and cone yield. A bottleneck for the registration of commercial varieties for the production of controlled genetic material was the low morphological and molecular differentiation (cpSSR – chloroplast microsatellites). Thanks to the use of high-resolution genetic markers (nSSR – nuclear microsatellites) this problem was recently overcome; ten clones have been approved as qualified basic material and five as controlled basic material. Three more clones were submitted in 2016, but their approval is still pending (Guadaño et al. 2016).

Additionally, somatic embryogenesis for selected rootstock production is being investigated (Celestino et al. 2013). Although progeny trials have also been established from plus-tree seeds, it is not yet known when they will reach reproductive maturity in order to evaluate the F1 generation for cone production to assess narrow-sense heritability.

In Portugal, 64 plus trees have been phenotypically selected for cone yield in natural stands. Actually, Portugal is the only country, apart from Spain, where progress has been made in the selection of basic material in the last 15 years, but with a different method. The first phases were the same: prospecting of more than 300 plus trees in the forest as genetic starting material and installation of two clonal banks in 2004 (within the framework of the AGRO451 project "Improvement of pine stands and optimization of cone yield management"). Each bank has 50 replicates of the 64 plus trees covering an area of 8 ha (3200 plants at 5 × 5 m). These plots, located in Coruche and Alcácer do Sal, are managed by two associations of local forest owners; namely APFC and ANSUB (Carrasquinho et al. 2010). However, in Portugal the set of these 64 selected genotypes in the mountains has been registered as a 'mixture of clones' in the National Catalogue of Basic Materials for the production of 'qualified' FRM. Since 2009, tens of thousands of scions have been traded for grafting.

In Tunisa, interest in grafted plantations for cone production has increased thanks to different cooperation programs (Piqué et al. 2013); this will pave the way for rapid deployment of improved varieties.

In Turkey, in addition to provenance testing (Acar et al. 2013), there are currently seven seed orchards with 30–64 clones. The number of ramets in the seed orchards ranges between 270 and 3068. The average seed orchard size is 11.72 ha (1.7–19.5 ha) and the total covered is 87 ha (ORTOHUM; Bilgen et al. 2013). There are also nine seed stands of the species covering 1709.9 ha, two genetic conservation forests covering 337.6 ha and two clone parks covering 9 ha. In this country, unlike Spain or Portugal, the target has so far been the use of seedling plantations in multi-purpose, community-owned forests, instead of deploying grafts from improved varieties (Sülüsoglu 2004).

3.4.1.5 *Pinus canariensis*

Together with *P. pinea,* the Canary Islands pine (*P. canariensis*) can be considered an outlier amongst Mediterrenean pines for many reasons. The fact that it is endemic in its only natural area in the Canary Islands has put the emphasis on conservation (both ecological and genetic). Provenance research has been carried out thanks to a cooperation between Spain and the Israeli Forest Service (KKL), with approximately replicated trials in four locations in the Canary Islands and three in Israel (Schiller et al. 1999; López et al. 2007b). The results of this research have been taken into account for new plantations, both within and outside the natural area of distribution. A selection of plus trees based on growth and form with subsequent grafting was carried out in Spain during the 1980s and 1990s, but both low grafting success (due to rootstock sprouting ability, although heteroblastic grafting onto *P. pinea* rootstocks was promising) and scarce interest for improved seed led to the abandonment of this breeding program. However, there is strong interest in this species as an exotic tree in countries as diverse as Australia, Ethiopia and Morocco. Its drought and fire tolerance have led this species to be recurrently identified for low input forestry in moderately warm drylands.

The current needs of reproductive materials for local afforestation are fulfilled with selected stands (Auñón et al. 2005) identified throughout all the regions and sub-regions of provenance to ensure highly diverse, adapted materials. However, there has been recent interest in developing highly drought-tolerant varieties through breeding in Gran Canaria Island, where natural stands are severely threatened by long-lasting droughts and fires.

3.4.1.6 *Pinus nigra*

Breeding of *P. nigra* has been strongly conditioned by the existence of subspecific taxa, classified as subspecies or varieties, depending on the taxonomies (Isajev et al. 2004). Some subspecies have been widely used in forestry for timber production (since the wood of the species is one of the finest due to its good mechanical properties, natural durability and excellent quality due to the favorable tree form). These have been mostly the subspecies *laricio* (from Corsica, in some classifications *P. nigra* subsp. *corsicana*) and *nigra*, from central Europe. Other Mediterranean subspecies are subsp. *salzmannii*, from Spain and southern France, and *laricio* var. *callabrica* (subsp. *callabrica* in other classifications). Subsp. *pallasiana*, from Greece and Turkey also occupies areas with Mediterranean influence. Finally, the subspecies *mauretanica*, from the Atlas Mountains in Morocco is under high extinction risk due to climate change. A conflict among productive forestry and genetic conservation arose from the historical use of allochthonous subspecies that may introgress via gene flow into local gene pools of autochthonous provenances. This is the case across large areas of *P. nigra* subsp. *salzmannii* in southern France and Spain. Specific breeding programs for *P. nigra* have been carried out in Greece and Turkey (subsp. *pallasiana*), Spain (subsp. *nigra* and *salzmannii*) and France (subsp. *laricio*).

In Greece, in addition to a network of provenance trials, four clonal seed orchards of *P. nigra* were established in different locations between 1978 and 1987 (Alizoti et al. 2010). In Northern Greece (Chalkidiki Peninsula), a clonal seed orchard was established in 1980 with 60 clones of *P. nigra* subsp. *pallasiana* from plus trees selected in five natural populations (Mt. Olympus, Mt. Pieria, Mt. Vermion, Chalkidiki Peninsula and Thasos Island (Alizoti et al. 2010, 2018)). Two clonal seed orchards were established in 1981 in Central Greece, the first one with 91 clones of subsp. *pallasiana* originating from plus trees selected in the Pindos Mountains, and another one with 49 clones of the introduced subsp. *nigra* (or var. *austriaca*). Finally, another seed orchard was established in Southern Greece (Elea) with 52 clones derived from plus trees selected in the autochthonous populations of the Peloponnese (Matziris 1993, 2005; Ioannidis 2017, Alizoti et al. 2018). This seed orchard has been progeny tested with three open pollinated progeny trials. The realized genetic gain estimated from the performance of progenies at 9 years of age was 8% for height, 11% for diameter at breast height and 32% for volume (Matziris 2005). In Turkey, *P. nigra* (subsp. *pallasiana*) was identified as one of the target species within the National Tree Breeding and Seed Production Program (NTBSP) which began in 1994 (Çengel et al. 2012). Since then, 55 seed orchards have been established in different breeding zones with a variable number of clones (10–120), and ramets (75–3225) covering 467.4 ha (ORTOHUM; Bilgen et al. 2013). In addition, this species has 71 seed stands covering 9090.8 ha and 42 gene conservation forests covering 6589.2 ha.

In Spain, three separate breeding programs have been carried out for *P. nigra*, one with a highly demanded local provenance from the introduced subsp. *nigra* in North-eastern Spain (49 clones, pre-Pyrenean Mountains from Navarra, Huesca and Zaragoza provinces), and two others from autochthonous provenances of *P. nigra* subsp. *salzmannii*, both from the Iberian Range (72 clones) and the Baetic Mountains (50 clones). In these last two cases, the established seed orchards were aimed more towards preservation of genetically diverse basic material than towards genetic improvement. Progeny tests have been planted with open pollinated progenies of the Iberian Range seed orchard. Recently, a conservation seed orchard with material from scattered endangered populations of the Central Range has been established, including 94 clones from nine populations (Tranque et al. 2018).

In France, while traditional breeding has been based solely on subsp. *laricio* (one seed orchard of Calabrian pine and two seed orchards of Corsican pine, (Isajev et al. 2004) there have also been recent efforts to ensure the genetic conservation of subsp. *salzmannii* from Southern France (Fady et al. 2010), with *ex situ* collections from several populations clonally propagated by grafting.

3.4.1.7 *Pinus sylvestris*

While breeding activities for *P. sylvestris* are widespread in central Europe, Scotland and Fennoscandia, in the Mediterranean-climate areas, breeding is restricted to three countries: Turkey, Greece and Spain. In France, breeding of *P. sylvestris* is limited to the continental northeast regions.

Pinus sylvestris was also included as a target species in Turkey's National Tree Breeding and Seed Production Program (NTBSP) initiated in 1994 (Alan et al. 2007). To date, 21 seed orchards covering 109.6 ha have been planted, with corresponding progeny testing. The number of clones ranges from 8 to 152, and the number of ramets ranges from 94 to 2287 (Bilgen et al. 2013). The species also has 35 seed stands covering 4642.5 ha, 22 gene conservation forests covering 2291.6 ha and three clone parks covering 5.9 ha (ORTOHUM).

In Spain, multi-site provenance testing in most natural regions of provenance showed high differentiation among edaphic ecotypes (calcareous vs acidic soils, Chambel 2006)). Seed orchard programs were initiated in the 1980s (Pardos and Gil 1986) with plus trees selected for growth and form focusing on three provenances: Central Range (1990, 72 clones, 5 ha), North Iberian Range (2000, 64 clones, 1.3 ha) and Aragon Pyrenees (1990, 49 clones). Multi-site progeny testing is underway.

In Greece, provenance testing has been carried out with a broad range of Greek provenances as well as provenances from Bulgaria and North Macedonia.

3.4.1.8 High Mountain Pines (*P. heldreichii, P. uncinata, P. mugo*)

The emphasis in the management of genetic resources of these species in Mediterranean countries is conservation (e.g., Vendramin et al. 2008b); therefore, these narrow-niche species have not been the subject of breeding programs to date. A seed orchard of *P. uncinata* was established in 2000 in Spain with 49 clones from the Pyrenees (Climent et al. 1997). The objective, however, was to ensure the supply of qualified seed, easily collectable at low cost, rather than genetic improvement. In fact, this can be considered more correctly to be an *ex situ* genetic conservation initiative.

3.5 Transfer and Use of Forest Reproductive Material

The FRM of most of the Mediterranean pine species considered here is under different levels of regulation. In Europe, these species (except *P. mugo* and *P. uncinata*) are under the directive EC105/99 (with national additions to Annex I) for marketing of FRM in forestry (Council EU 2000). This regulation is similar to the one established by the OECD for the international market (Nanson 2001). However in Southern Mediterranean countries (e.g. Morocco, Tunisia, Algeria, Turkey), the regulations are established at the national level, and can be considered similar to those in Northern Mediterranean countries.

One main aspect that we should consider is that the EC105/99 directive regulates the production of FRM, and its marketing in EU countries. Therefore, this regulation defines the type of basic material (seed sources, stands, parents of families, seed orchards, clones and clonal mixtures) and their requisites to be included in the EU list of basic material to produce reproductive material of the different categories (source identified, selected, qualified and tested). One major concern is that the EU regulations impose some constraints on the marketing of FRM from non EU countries. The description of these regulations and the comparison with the OECD scheme can be found elsewhere (Nanson 2001; Alía et al. 2009a), but in this section we will focus on some general aspects related to the characteristics of the forest reproductive material, and the implications for its transfer and use.

3.5.1 Regions of Provenance and Basic Material

One main aspect, not always well understood by the end user of FRM, is the difference between the procurement zone (where the FRM is obtained) and the deployment zone (where the FRM is going to be used) (van Buijtenen 1992). The regions of provenance are the basic units for the marketing of source identified and selected FRM, and they are the basic procurement zones (according to the EU regulations the material from a region of provenance can be mixed).

The regions are established as units in which the populations of a given species are phenotypically or genetically similar or experience similar environmental factors (e.g Alía et al. 1992; de Dato et al. 2017). The main implication of this definition is that materials from the same region should share some common characteristics in terms of diversity, performance, and local adaptation (if it exists) given that they could have evolved under common environmental conditions over long time periods, or have resulted from other evolutionary factors (e.g. genetic drift, inbreeding).

Therefore, source identified material from areas without any selection, can be (if properly recollected) similarly diverse than the corresponding region of provenance. Selected material (from stands subjected to mass selection) cannot provide any significant improvement of the FRM beyond that derived from differences among populations in the same region, that can reach 5% for some growth traits (Gülcü and

Çelik 2009). Therefore, these materials can have an influence on the genetic structure of the populations in the long term (several generations), as they represent a first step in the domestication process.

Qualified (under individual phenotypic selection) or tested (under genetic evaluation and estimation of the improved value) materials are those delivered by non-intensive and intensive breeding programs, as described previously. They can have an impact on the genetic structure of local genetic resources, if used in their vicinity, by genetic introgression via pollen or seeds when they reach reproductive maturity (Cagelli and Lefevre 1995).

However, in Mediterranean pines, most of the basic material available is from source identified and selected material (representing over 95% of the total basic material in Spain, France and Italy). Source identified and selected reproductive materials are the most important basic material for most Mediterranean pines (Table 3.3), while qualified and tested materials are important only for some species and populations within these species (see Sect. 3.3). We do not have precise information from other countries, but we can expect similar patterns to those described for the below mentioned Northern Mediterranean countries.

3.5.2 Seed Sourcing in Mediterranean Pines

To select the reproductive material to be used in a given deployment zone, we can define two main approaches: local provenancing (selection of the local provenance) and predictive provenancing (selection based on the response under different environments) (Parker 1992; Lindgren and Ying 2000; Wang et al. 2006, 2010; Farjat et al. 2017). Other methods, such as composite provenancing, admixture provenancing and assisted migration can be implemented to increase resilience, with the use of decision charts for choosing the best provenance (Breed et al. 2013); however,

Table 3.3 Basic material in the EU list by category

Species	Source identified	Selected	Qualified	Countries[a]
P. brutia	4	2	1	CR, **CY**, **IT**
P. canariensis	22	11		**ES**
P. halepensis	381	37	1	**FR**, SL, **SP**, IT
P. heldreichii	30	1		**IT**
P. mugo (P. uncinata)	na[b]	na	na	ES
P. nigra	11	283	38	CR, **FR, IT, SL**
P. pinaster	318	171	21	**ES**, IT, **FR**
P. pinea	103	45	10	**ES**, **FR**, IT, CR, CY
P. sylvestris	739	194	6	**ES**, SL, IT, CR, **FR**

Source: Forematis, https://ec.europa.eu/forematis/index.xhtml. Access, 12/2019
[a]Country codes: Croatia (CR), Cyprus (CY), France (FR), Italy (IT), Slovenia (SL), Spain (ES)
[b]*na* not available

some previous information is usually required for effective application of these methods..

The strict use of local populations is widespread (Breed et al. 2013) based on the expectation that populations are locally adapted (Kawecki and Ebert 2004; Savolainen et al. 2007; Holliday et al. 2010; Alberto et al. 2013). However, local is not always the best, due to lack of adaptation or low performance of the local material (Namkoong 1969; Leimu and Fischer 2008; Jones 2013), or a reduced population size resulting in a high degree of inbreeding (Robledo-Arnuncio et al. 2004a). Nevertheless, the method of using the local provenance is widely used in Mediterranean pines.

Alternatively, we can select the most suitable FRM based on predictions of future performance, from results obtained in comparative genetic tests. These models, and results of provenance tests, are limited by the existing provenances and sites available for many of these species (e.g for *P. pinaster* (Alía et al. 1997; Benito-Garzón et al. 2013), *P. halepensis* (Voltas et al. 2018) or *P. brutia* (Kandemir et al. 2010). One limitation of the existing information is that only in special cases can we select the best provenances for very unfavorable environments such as the desert transition zone for *P. halepensis* provenances (Schiller and Atzmon 2009).

There are also some other considerations when transferring genetic resources in the landscape. To avoid any risk to local genetic resources it is necessary to check for the existence of endangered populations when deploying FRM in a given area. In some of these situations, it is mandatory to use local material or seedlots obtained for the objective of conservation of genetic resources. Also, there are restrictions on the use of FRM in and around protected areas.

However, any transfer can produce undesirable side effects on biodiversity, and therefore should be carefully examined (Maestre and Cortina 2004), as the Mediterranean ecosystems are complex and comprise many aspects (functioning, dynamics, etc.) that are still not well known (Méndez et al. 2008). This is especially important when the main objective is the restoration of Mediterranean areas, in which active management of old plantations may be required due to the abiotic characteristics of the plantations and the composition of the stands (Vallejo et al. 2004; Gómez-Aparicio et al. 2009).

3.6 Perspectives and Concluding Remarks

Genetic resources of Mediterranean pines are not endangered as such at the species level, but across the Mediterranean Basin these species present fragmented distributions with marginal populations. Moreover, there are contrasting situations among the group of countries included in the EUFORGEN program and other countries, mostly North African ones. Therefore, to advance the conservation of Mediterranean pines, we need to implement some actions, as summarized below, which coincide with the objectives of EUFORGEN, but should be extended to the entire

Mediterranean Basin region, by coordinating these activities with other regional initiatives, including *Silva Mediterranea*.

Collate, maintain and disseminate reliable information on forest genetic resources. One of the main problems when addressing conservation activities in this region is the lack of reliable and comparable information among countries on the distribution of the species, their demographic situation, and other aspects related to their *in situ* conservation. EUFORGEN maps provide a rough depiction of their distribution, but cannot be used for identification of conservation gaps, as many of the populations are not included or their information is not reliable. These maps are the first step for planning and implementing coordinated activities, and we need to improve the level of information provided, by including the activities of the national forest inventories or other national initiatives that can be coordinated among countries. Some actions have been initiated, but they lack the capacity of updating this information on a regular (10–20 year) basis. Furthermore, for the implementation of the programs we need to obtain information on the plantations established in the region and the origin of the material. The assessment of threat levels to the local genetic resources depends on the efficient use of detailed information that will enable the identification of risks related to introgression or reduction of population sizes, and for these reasons more informative studies based on molecular and quantitative traits are needed.

Coordinate and monitor the conservation of forest genetic resources. One main question in the implementation of the programs is the extent to which the activities are coordinated among countries or regions, for instance, with respect to the criteria for selecting a conservation unit, monitoring of the conservation status, etc. These actions should rely on scientific information provided by different studies. For Mediterranean pines, we can expect that many of the studies can be more easily generalized to other species and situations due to the homogeneous mating system of the species. However, the socio-economic context is quite variable, and has many implications when trying to maintain a successful conservation program in the region. Therefore, we should focus additionally in understanding well the socio-economic context and plan well the implementation of activities in accordance with the conditions.

Develop guidelines and analyses on topics and issues relevant for the use of forest genetic resources. A focus of this work effort will be to improve the way genetic aspects are incorporated into the production and use of FRM. Some Mediterranean pines are important for afforestation and plantations. Therefore, we need to develop actions to minimize the impact of domestication in the natural ranges of Mediterranean pines, or at least in the GCUs. The valuable resources of these species should be maintained for future generations, as a basis for subsequent activities related to future demands on Mediterranean forests.

Despite the risk of many populations, *in situ* conservation must play a more important role in genetic conservation, since we cannot allow our valuable genetic resources to be disconnected from the ecosystems and social values of the stands in which they currently grow.

Investigate whether climate and global changes are threatening the adaptation of improved varieties. Probably the most urgent problem is the spread of new pests and diseases. While the most recent breeding materials include some tolerance to endemic herbivores, such as *Hilobyus abietis* L., *Diorictria mendacella* (Staudinger) and *Thaumatopoea*, trade-offs among high growth rate and defense ability have been reported in *P. pinaster* (Sampedro et al. 2011). More frequent and intense drought episodes could also affect naturally induced defense mechanisms, as reported for Aleppo pine (Suárez-Vidal et al. 2019). This implies a need to regularly test current breeding materials for potential performance changes, due to their interactions with abiotic-biotic stresses, driven by climate change. Furthermore, new emerging threats are currently demanding rapid testing for genotypes tolerant to pests, such as *Fusarium circinatum* Nirenberg & O'Donnell and the wood pine nematode (*Bursaphelenchus xylophilus* (Steiner & Buhrer) Nickle). The recent widespread infestation of *P. pinea* by the weevil, *Leptoglossus occidentalis* Heidermann, affecting the highly valuable edible seeds, is causing severe economic losses. In this case, it is unlikely that the few catalogued improved clones include any tolerance to this new threat.

Investigate intrinsic trade-offs between economically relevant and other key life-history traits to illustrate the selection process, such that a broad general defense ability is ensured even if this implies some loss of potential gain (see, for example, Sampedro et al. 2014; Santini et al. 2019). For example, a single selection event for growth and stem form was shown to affect importantly early reproductive allocation in maritime pine (Santos-del-Blanco et al. 2015), a highly predictable outcome in the light of life-history theory as part of a domestication process.

Both water use efficiency and product quality (mostly wood) are progressively being included in the most developed breeding programs, such as *P. pinaster* ones in France and to a lesser extent, Spain and Portugal. Platforms for phenotyping of different traits are being developed with the assistance of international projects (for example, through transnational access within the recent EU projects, Noveltree, Trees4Future and B4EST). Near infra-red (NIR) technologies are increasingly being used to assess wood density and different chemical components (e.g., lignin and extractives). Micro-densitometry analysis offers a precise view of short-term reactions to water availability, and can be combined with δ13C (carbon isotope discrimination) as a proxy for water use efficiency. All these phenotyping platforms could significantly improve the genetic evaluation and efficiency of selection methods in the very near future.

The discovery of genes related to wood quality and water stress resistance by quantitative trait loci (QTL) detection has facilitated important progress with maritime pine as a model species (Brendel et al. 2002), but approaches based on association studies have also been fruitful (Rodriguez-Quilon 2017). Moreover, the development of large sets of single-nucleotide polymorphisms (SNPs) has provided high quality fingerprinting tools and allowed characterization of the geographical origins of this species (Jaramillo-Correa et al. 2015).

In the very near future, genome-wide selection would help the progress of new breeding lines (Isik et al. 2016); it can also help to develop new approaches to low-input breeding for some Mediterranean pine species (El-Kassaby and Lstibůrek 2009; Lstibůrek et al. 2015). Genome-wide selection will help in these new approaches and new challenges stemming out from global change (while traditional breeding programs are probably too rigid and costly) and provide sound deployment recommendations for forest managers in the near future. Genome-wide selection could contribute to faster selection procedures and speed up breeding cycles especially for traits controlled by major genes, but still need to be tested for adaptive traits controlled by numerous genes whose role may change in different ontogenetic phases.

Acknowledgments We would like to thank P. Grant for language improvements and D. Grivet, and reviewers for useful comments on the manuscript. Funding was supported by Futurpin Project RTI2018-094691-B-C32, RTA2017-00063-C04-01, EG17-047 and the European Union's Horizon 2020 research and innovation program under grant agreement No 676876 GenTree.

References

Acar FC, Altun ZG, Boza A, Bilgin F (2013) Provenance trials of stone pine (Pinus pinea L.) in the Aegean region: tenth year evaluation. Options Méditérr 105:67–72

Aguiar A, Almeida MA, Borralho N (2003) Genetic control of growth, wood density and stem characteristics of Pinus pinaster in Portugal. Silva Lusit 11:131–139

Alan M, Ozturk H, Siklar S (2007) Seed orchard planning and management in Turkey Seed Orchard planning and management in Turkey. In: Seed Orchard Conference. Umeå, Sweden, 26–28 September 2007, pp 11–12

Alberto FJ, Aitken SN, Alía R et al (2013) Potential for evolutionary responses to climate change – evidence from tree populations. Glob Chang Biol 19:1645–1661. https://doi.org/10.1111/gcb.12181

Alexandrov A, von Wühlisch G, Vendramini GG (2019) EUFORGEN technical guidelines for genetic conservation and use for mountain pine (Pinus mugo). European Forest Genetic Resources Programme (EUFORGEN), European Forest Institute, Barcelona, Spain

Alía R, Martín S, Gil P (1992) Delineation of pine provenance regions and select stand selection in Spain. FBVA-Ber 65:155–162

Alía R, Gil LA, Pardos JA (1995) Performance of 43 Pinus pinaster Ait. provenances on 5 locations in Central Spain. Silvae Genet 44:75–81

Alía R, Moro J, Denis JB, Moro-Serrano J (1997) Performance of Pinus pinaster provenances in Spain: interpretation of the genotype by environment interaction. Can J For Res 27:1548–1559. https://doi.org/10.1139/cjfr-27-10-1548

Alía R, Gomez A, Agúndez D et al (2001) Levels of genetic differentiation in Pinus halepensis Mill. in Spain using quantitative traits, isozymes, RAPDs and cp-microsatellites. Genet Response Syst Chang Environ Cond 70:151–160

Alía R, Alba N, Chambel MR et al (2009a) Genetic quality of forest reproductive materials in land restoration programmes. In: Vallejo VR, Bautista S, Aaronson J (eds) Innovative approaches in forest restoration. CEAM, Valencia, pp 89–103

Alía R, García-del-Barrio JM, Iglesias S et al (2009b) Regiones de procedencia de especies forestales en España. O.A. Parques Nacionales. MMARM, Madrid

Alía R, Chambel R, Notivol E et al (2014) Environment-dependent microevolution in a Mediterranean pine (Pinus pinaster Aiton). BMC Evol Biol 14:200. https://doi.org/10.1186/s12862-014-0200-5

Alizoti PG, Panetsos KP, Moulalis DC (2000) Genetic parameters and performance of Pinus halepensis Mill. In: Proceedings of the 8th conference of the Hellenic Scientific Society of Plant Breeding. Arta, Greece, pp 37–49

Alizoti PG, Kilimis K, Gallios P (2010) Temporal and spatial variation of flowering among *Pinus nigra* Arn. clones under changing climatic conditions. For Ecol Manag 259:786–797. https://doi.org/10.1016/j.foreco.2009.06.029

Alizoti PG, Aravanopoulos FA, Ioannidis K (2018) Ex situ conservation of forest genetic resources in Greece. In: Šijačić-Nikolić MMN (ed) Forests of Southeast Europe under a changing climate: conservation of genetic resources. Springer, Cham, pp 292–230

Auñón FJ, Dopazo M, Alía R, Iglesias S (2005) Caracterización fenotípica de material de base para la producción de materiales forestales de reproducción seleccionado del género Pinus, con objetivo producción de madera. In: IV Congreso Forestal Español. SECF, Zaragoza (España)

Baradat P, Durel CE, Pastuszka P (1992) The polycross seed orchard: an original concept. In: Afocel E (ed) Proceedings of the IUFRO–AFOCEL symposium, mass production technology for genetically improved fast growing forest tree species, Tome II. Bordeaux, pp 53–62

Benito Garzón M, Sánchez De Dios R, Sainz Ollero H (2008) Effects of climate change on the distribution of Iberian tree species. Appl Veg Sci 11:169–178. https://doi.org/10.3170/2008-7-18348

Benito-Garzón M, Alía R, Robson TM, Zavala MA (2011) Intra-specific variability and plasticity influence potential tree species. Glob Ecol Biogeogr 20:766–788. https://doi.org/10.1111/j.1466-8238.2010.00646.x

Benito-Garzón M, Ruiz-Benito P, Zavala MA (2013) Interspecific differences in tree growth and mortality responses to environmental drivers determine potential species distributional limits in Iberian forests. Glob Ecol Biogeogr n/a–n/a. https://doi.org/10.1111/geb.12075

Bilgen BB, Alan M, Kurt Y (2013) Importance of effective clone number in seed orchards: a comparative study on seven conifer species in Turkey. Sumar List 137:297–305

Blondel J (2006) The “design” of Mediterranean landscapes: a millennial story of humans and ecological systems during the historic period. Hum Ecol 34:713–729. https://doi.org/10.1007/s10745-006-9030-4

Boscherini G, Morgante M, Rossi P, Vendramin GG (1994) Allozyme and chloroplast DNA variation in Italian and Greek populations of Pinus leucodermis. Heredity (Edinb) 73:284–290. https://doi.org/10.1038/hdy.1994.135

Bouffier L, Raffin A, Alia R (2013) Maritime pine – Pinus pinaster Ait. In: Mullin TJ, Lee SJ (eds) Best practice for tree breeding in Europe. Skogforsk, Gävle, pp 65–76

Breed MF, Stead MG, Ottewell KM et al (2013) Which provenance and where? Seed sourcing strategies for revegetation in a changing environment. Conserv Genet 14:1–10. https://doi.org/10.1007/s10592-012-0425-z

Brendel O, Pot D, Plomion C et al (2002) Genetic parameters and QTL analysis of δ 13 C and ring width in maritime pine. Plant Cell Environ 25:945–953

Bucci G, Anzidei M, Madaghiele A, Vendramin GG (1998) Detection of haplotypic variation and natural hybridization in halepensis -complex pine species using chloroplast simple sequence repeat (SSR) markers. Mol Ecol 7:1633–1643

Bucci G, González-Martínez SC, Le Provost G et al (2007) Range-wide phylogeography and gene zones in Pinus pinaster Ait. revealed by chloroplast microsatellite markers. Mol Ecol 16:2137–2153. https://doi.org/10.1111/j.1365-294X.2007.03275.x

Budde KB, Heuertz M, Hernández-Serrano A et al (2014) In situ genetic association for serotiny, a fire-related trait, in Mediterranean maritime pine (*Pinus pinaster*). New Phytol 201:230–241. https://doi.org/10.1111/nph.12483

Cagelli L, Lefevre F (1995) The conservation of Populus nigra L. and gene flow with cultivated poplars in Europe. For Genet 2:135–144

Campos P, Caparrós A, Oviedo JL et al (2019) Bridging the gap between national and ecosystem accounting application in Andalusian Forests, Spain. Ecol Econ 157:218–236. https://doi.org/10.1016/j.ecolecon.2018.11.017

Carrasquinho I, Freire J, Rodrigues A et al (2010) Selection of Pinus pinea L. plus tree candidates for cone production. Ann For Res 67:41. https://doi.org/10.1051/forest/2010050

Catalan G, Dietl T, Enciso E (1995) Selección de árboles sobresalientes, en cuanto a producción de fruto, en los pinares de pino piñonero en la región "Valles del Tiétar y del Alberche". Parámetros de selección. In: El pino piñonero como árbol productor de frutos secos en los países mediterráneos. I Reunión de la Red de Frutos Secos de la FAO sobre el Pino Piñonero. FAO/INIA. Madrid, Spain

Celestino C, Carneros E, Ruiz-Galea M et al (2013) Cloning stone pine (Pinus pinea L.) by somatic embryogenesis. Options Méditérr 105:89–98

Çengel B, Tayanç Y, Kandemir G et al (2012) Magnitude and efficiency of genetic diversity captured from seed stands of Pinus nigra (Arnold) subsp. pallasiana in established seed orchards and plantations. New For 43:303–317

Chambel MR (2006) Variabilidad adaptativa y plasticidad fenotípica en procedencias de pinos ibéricos. Dep. Silvopascicultura, Madrid, p 118

Chambel MR, Climent J, Alía R (2004) Intra-specific variation of phenotypic plasticity for biomass allocation in Mediterranean pines. In: Arianoutsou M, Thanos C (eds) Proceedings 10th MEDECOS Conference, Rhodes, Greece, April. Millpress, Rhodas

Chambel MR, Climent JM, Ducci F, Pichot C (2013) Species breeding monograph: Aleppo pine (*Pinus halepensis* Mill.) and Brutia pine (*Pinus brutia* Ten.). In: Pâques L (ed) Forest tree breeding across Europe. Springer, Dordrecht

Chevin L-M, Lande R, Mace GM (2010) Adaptation, plasticity, and extinction in a changing environment: towards a predictive theory. PLoS Biol 8:e1000357. https://doi.org/10.1371/journal.pbio.1000357

Chevin L-M, Collins S, Lefèvre F (2012) Phenotypic plasticity and evolutionary demographic responses to climate change: taking theory out to the field. Funct Ecol 27:966. https://doi.org/10.1111/j.1365-2435.2012.02043.x

Climent JM, Gil A, Pardos JA (1997) Estado actual y perspectivas de la red de huertos semilleros del género Pinus en España. Cuad la SECF 5:73–82

Climent J, Chambel MR, Lopez R et al (2006) Population divergence for heteroblasty in the Canary Island pine (Pinus canariensis, Pinaceae). Am J Bot 93:840–848

Climent J, Prada MA, Calama R et al (2008) To grow or to seed: ecotypic variation in reproductive allocation and cone production by young female Aleppo pine (*Pinus halepensis*, Pinaceae). Am J Bot 95:833–842. https://doi.org/10.3732/ajb.2007354

Conord C, Gurevitch J, Fady B (2012) Large-scale longitudinal gradients of genetic diversity: a meta-analysis across six phyla in the mediterranean basin. Ecol Evol 2:2600–2614. https://doi.org/10.1002/ece3.350

Corcuera L, Gil-Pelegrin E, Notivol E (2010) Phenotypic plasticity in Pinus pinaster δ13C: environment modulates genetic variation. Ann For Sci 67:812–812. https://doi.org/10.1051/forest/2010048

Correia I, Almeida MH, Aguiar A et al (2008) Variations in growth, survival and carbon isotope composition (δ13C) among Pinus pinaster populations of different geographic origins. Tree Physiol 28:1545–1552. https://doi.org/10.1093/treephys/28.10.1545

Correia I, Alía R, Yan W et al (2010) Genotype × environment interactions in Pinus pinaster at age 10 in a multienvironment trial in Portugal: a maximum likelihood approach. Ann For Sci 67:612–612. https://doi.org/10.1051/forest/2010025

Council EU (2000) COUNCIL DIRECTIVE 1999/105/EC of 22 December 1999 on the marketing of forest reproductive material. Off J Eur Communities 11:17–40

Crandall KA, Bininda-emonds ORP, Mace GM, Wayne RK (2000) Considering evolutionary processes in conservation biology. TREE 15:290–295

Danjon F (1995) Observed selection effects on height growth, diameter and stem form in maritime pine. Silvae Genet 44:10–19

de Dato G, Tenai A, Mattioni C et al (2017) Delineation of seed collection zones based on environmental and genetic characteristics for Quercus suber L. in Sardinia, Italy. iForest 11:651–659. https://doi.org/10.3832/ifor2572-011
De La Mata R, Voltas J, Zas R (2012) Phenotypic plasticity and climatic adaptation in an Atlantic maritime pine breeding population. Ann For Sci 69:477–487. https://doi.org/10.1007/s13595-011-0173-0
de Vries SMG, Alan M, Bozzano M et al (2015) Pan-European strategy for genetic conservation of forest trees and establishment of a core network of dynamic conservation units. Bioversity International, Rome (Italy)
Derory J, Mariette S, González-Martínez SC et al (2002) What can nuclear microsatellites tell us about maritime pine genetic resources conservation and provenance certification strategies? Ann For Sci 59:699–708
Destremau D, Jolly H, Tafri T (1976) Contribution à la connaissance des provenances de Pinus pinaster. Ann Rech For Maroc 16:101–153
El-Kassaby YA, Lstibůrek M (2009) Breeding without breeding. Genet Res 91:111–120
Elvira-Recuenco M, Iturritxa E, Majada J et al (2014) Adaptive potential of maritime pine (Pinus pinaster) populations to the emerging pitch canker pathogen, Fusarium circinatum. PLoS One 9:e114971. https://doi.org/10.1371/journal.pone.0114971
Eriksson G, Namkoong G, Roberds JH (1993) Dynamic gene conservation for uncertain futures. For Ecol Manag 62:15–37. https://doi.org/10.1016/0378-1127(93)90039-P
Fady B (2012) Biogeography of neutral genes and recent evolutionary history of pines in the Mediterranean Basin. Ann For Sci 69:421–428. https://doi.org/10.1007/s13595-012-0219-y
Fady B, Conord C (2010) Macroecological patterns of species and genetic diversity in vascular plants of the Mediterranean basin. Divers Distrib 16:53–64. https://doi.org/10.1111/j.1472-4642.2009.00621.x
Fady B, Semerci H, Vendramin GG (2003) EUFORGEN technical guidelines for genetic conservation and use for Aleppo pine (Pinus halepensis) and Brutia pine (Pinus brutia). International Plant Genetic Resources Institute, Rome. 6 p
Fady B, Fineschi S, Vendramin G (2008) EUFORGEN Technical guidelines for genetic conservation and use for Italian stone pine (Pinus pinea). European Forest Genetic Resources Programme (EUFORGEN), European Forest Institute, Barcelona, Spain
Fady B, Brahic P, Cambon D et al (2010) Valoriser et conserver le Pin de Salzmann en France. Forêt Méditerranéenne XXXI:3–14
Fady-Welterlen B (2005) Is there really more biodiversity in Mediterranean forest ecosystems? Taxon 54:905–910. https://doi.org/10.2307/25065477
Fallour D, Fady B, Lefèvre F (1997) Study on isozyme variation in Pinus pinea L.: evidence for low polymorphism. Silvae Genet 46:201–207
Fargeon H, Pimont F, Martin-StPaul N et al (2020) Projections of fire danger under climate change over France: where do the greatest uncertainties lie? Clim Chang 160:479. https://doi.org/10.1007/s10584-019-02629-w
Farjat A, Reich BJ, Guinness J et al (2017) Optimal seed deployment under climate change using spatial models: application to Loblolly Pine in the Southeastern US. J Am Stat Assoc 112:909–920. https://doi.org/10.1080/01621459.2017.1292179
Gauquelin T, Michon G, Joffre R et al (2018) Mediterranean forests, land use and climate change: a social-ecological perspective. Reg Environ Chang 18:623–636. https://doi.org/10.1007/s10113-016-0994-3
Gomez A, González-Martínez SC, Collada C et al (2003) Complex population genetic structure in the endemic Canary Island pine revealed using chloroplast microsatellite markers. Theor Appl Genet 107:1123–1131
Gómez-Aparicio L, Zavala MA, Bonet FJ, Zamora R (2009) Are pine plantations valid tools for restoring Mediterranean forests? An assessment along abiotic and biotic gradients. Ecol Appl 19:2124–2141. https://doi.org/10.1890/08-1656.1

González-Martínez SC, Alía R, Gil LA (2002a) Population genetic structure in a Mediterranean pine (Pinus pinaster Ait.): a comparison of allozyme markers and quantitative traits. Heredity (Edinb) 89:199–206. https://doi.org/10.1038/sj.hdy.6800114

González-Martínez SC, Gerber S, Cervera MT et al (2002b) Seed gene flow and fine-scale structure in a Mediterranean pine (Pinus pinaster Ait.) using nuclear microsatellite markers. Theor Appl Genet 104:1290–1297. https://doi.org/10.1007/s00122-002-0894-4

González-Martínez SCSC, Burczyk J, Nathan R et al (2006) Effective gene dispersal and female reproductive success in Mediterranean maritime pine (Pinus pinaster Aiton). Mol Ecol 15:4577–4588. https://doi.org/10.1111/j.1365-294X.2006.03118.x

Gordo J (2004) Selección de grandes productores de fruto de Pinus pinea L. en la Meseta Norte. Universidad Politécnica de Madrid, Madrid

Gordo FJ, Mutke S, Gil L (2001) Modelo individual de producción de piñón de Pinus pinea L. como criterio de selección fenotípica. In: III Congreso Forestal Español, Granada, Spain. Sociedad Española de Ciencias Forestales

Grivet D, Sebastiani F, González-Martínez SC, Vendramin GG (2009) Patterns of polymorphism resulting from long-range colonization in the Mediterranean conifer Aleppo pine. New Phytol 184:1016–1028. https://doi.org/10.1111/j.1469-8137.2009.03015.x

Grivet D, Sebastiani F, Alía R et al (2011) Molecular footprints of local adaptation in two Mediterranean conifers. Mol Biol Evol 28:101–116. https://doi.org/10.1093/molbev/msq190

Guadaño C, Iglesias S, León D et al (2016) Establecimiento de plantaciones clonales de Pinus pinea para la producción de piñón mediterráneo. Instituto Nacional de Investigación y Tecnología Agraria y Alimentaria (INIA), Madrid

Gülcü S, Çelik S (2009) Genetic variation in Pinus brutia Ten. seed stands and seed orchards for growth, stem form and crown characteristics. Afr J Biotechnol 8:4387–4394. https://doi.org/10.5897/AJB09.947

Hamann A, Smets P, Yanchuk AD, Aitken SN (2005) An ecogeographic framework for in situ conservation of forest trees in British Columbia. Can J For Res 35:2553–2561. https://doi.org/10.1139/x05-181

Hampe A, Petit RJ (2005) Conserving biodiversity under climate change: the rear edge matters. Ecol Lett 8:461–467. https://doi.org/10.1111/j.1461-0248.2005.00739.x

Harfouche A, Kremer A (2000) Provenance hybridization in a diallel mating scheme of maritime pine (Pinus pinaster). I. Means and variance components. Can J For Res 30:1–9. https://doi.org/10.1139/x99-178

Heuertz M, Teufel J, González-Martínez SC et al (2009) Geography determines genetic relationships between species of mountain pine (Pinus mugo complex) in western Europe. J Biogeogr 37:1–16. https://doi.org/10.1111/j.1365-2699.2009.02223.x

Holliday JA, Ritland K, Aitken SN (2010) Widespread, ecologically relevant genetic markers developed from association mapping of climate-related traits in Sitka spruce (Picea sitchensis). New Phytol 188:501–514. https://doi.org/10.1111/j.1469-8137.2010.03380.x

Içgen Y, Kaya Z, Çengel B et al (2006) Potential impact of forest management and tree improvement on genetic diversity of Turkish red pine (Pinus brutia Ten.) plantations in Turkey. For Ecol Manag 225:328–336. https://doi.org/10.1016/j.foreco.2006.01.009

Ioannidis K (2017) Identification of black pine (Pinus nigra Arn.) heartwood as a rich source of bioactive stilbenes by qNMR. J Sci Food Agric 97:1708–1716

Isajev V, Fady B, Semerci H, Andonovski V (2004) EUFORGEN technical guidelines for genetic conservation and use for European black pine (Pinus nigra). European Forest Genetic Resources Programme (EUFORGEN), European Forest Institute, Barcelona, Spain

Isik F, Isik K, Yildrim T, Li B (2002) Annual shoot growth components related to growth of Pinus brutia. Tree Physiol 22:51–58

Isik F, Bartholomé J, Farjat A et al (2016) Genomic selection in maritime pine. Plant Sci 242:108–119. https://doi.org/10.1016/j.plantsci.2015.08.006

Jaramillo-Correa JP, Grivet D, Terrab A et al (2010) The strait of Gibraltar as a major biogeographic barrier in Mediterranean conifers: a comparative phylogeographic survey. Mol Ecol 19:5452–5468. https://doi.org/10.1111/j.1365-294X.2010.04912.x
Jaramillo-correa JP, Prunier J, Vázquez-Lobo A et al (2015) Molecular signatures of adaptation and selection in forest trees. Adv Bot Res 74. https://doi.org/10.1016/bs.abr.2015.04.003
Jones TA (2013) When local isn't best. Evol Appl 6:1109–1118. https://doi.org/10.1111/eva.12090
Kandemir GE, Kaya Z, Temel F, Önde S (2010) Genetic variation in cold hardiness and phenology between and within Turkish red pine (Pinus brutia Ten.) populations: implications for seed transfer. Silvae Genet 59:49–57. https://doi.org/10.1515/sg-2010-0006
Kärkkäinen K, Savolainen O, Koski V (1999) Why do plants abort so many developing seeds : bad offspring or bad maternal genotypes? Evol Ecol 13:305–317
Kawecki TJ, Ebert D (2004) Conceptual issues in local adaptation. Ecol Lett 7:1225–1241. https://doi.org/10.1111/j.1461-0248.2004.00684.x
Kaya Z, Isik F (1997) The pattern of genetic variation in shoot growth of Pinus brutia populations sampled from the Taurus Mountains. Silvae Genet 46:73–81
Kellomäki S, Hänninen H, Kolström T, Kotisaari A, Pukkala T (1987) A tentative model for describing the effects of some regenerative process on the properties of natural seedling stands. Silva Fennica 21(1):1–10
Koskela J, Lefèvre F, Schueler S et al (2013) Translating conservation genetics into management: Pan-European minimum requirements for dynamic conservation units of forest tree genetic diversity. Biol Conserv 157:39–49
Koski V, Antola J (2013) National tree breeding and seed production programme for Turkey 1994-2003. Directorate of Forest Tree Seeds and Tree Breeding. Ankara, Turkey, 49 pp
Kremer A, Ronce O, Robledo-Arnuncio JJ et al (2012) Long-distance gene flow and adaptation of forest trees to rapid climate change. Ecol Lett 15:378–392. https://doi.org/10.1111/j.1461-0248.2012.01746.x
Krutovsky KV, Burczyk J, Chybicki I et al (2012) Gene flow, spatial structure, local adaptation, and assisted migration in trees. In: Genomics of tree crops. Springer, New York
Kurt Y, González-Martínez SC, Alia R, Isik K (2012) Genetic differentiation in Pinus brutia ten. using molecular markers and quantitative traits: the role of altitude. Ann For Sci 69:345–351
Lamy J-B, Bouffier L, Burlett R et al (2011) Uniform selection as a primary force reducing population genetic differentiation of cavitation resistance across a species range. PLoS One 6:e23476. https://doi.org/10.1371/journal.pone.0023476
Lefevre F, Koskela J, Hubert J et al (2012) Dynamic conservation of forest genetic resources in 33 European countries. Conserv Biol 27:1–12. https://doi.org/10.1111/j.1523-1739.2012.01961.x
Leimu R, Fischer M (2008) A meta-analysis of local adaptation in plants. PLoS One 3:1–8. https://doi.org/10.1371/journal.pone.0004010
Lindgren D, Ying CC (2000) A model integrating seed source adaptation and seed use. New For 20:87–104
Lise Y, Kaya Z, Isık F et al (2007) The impact of over-exploitation on the genetic structure of Turkish Red Pine (Pinus brutia Ten.) populations determined by RAPD markers. Silva Fenn 41:211–220
López R, Zehavi A, Climent J et al (2007a) Contrasting ecotypic differentiation for growth and survival in Pinus canariensis. Aust J Bot 55:759. https://doi.org/10.1071/BT07016
López R, Zehavi A, Climent JM, Gil L (2007b) Contrasting ecotypic differentiation for growth and survival in Pinus canariensis (Pinaceae). Aust J Bot 55:1–11
Lstibůrek M, Hodge GR, Lachout P (2015) Uncovering genetic information from commercial forest plantations—making up for lost time using "breeding without breeding". Tree Genet Genomes 11:55. https://doi.org/10.1007/s11295-015-0881-y
Maestre FT, Cortina J (2004) Are Pinus halepensis plantations useful as a restoration tool in semiarid Mediterranean areas? For Ecol Manag 198:303–317. https://doi.org/10.1016/j.foreco.2004.05.040
Majada J, Alía R (2012) Plataforma de mejora de Pinus pinaster en España. 12 pp

Majada J, Martínez-Alonso C, Feito I et al (2011) Mini-cuttings: an effective technique for the propagation of Pinus pinaster Ait. New For 41:399–412. https://doi.org/10.1007/s11056-010-9232-x
Martins P, Sampedro L, Moreira X, Zas R (2009) Nutritional status and genetic variation in the response to nutrient availability in Pinus pinaster. A multisite field study in Northwest Spain. For Ecol Manag 258:1429–1436. https://doi.org/10.1016/j.foreco.2009.06.041
Martín-Sanz RC, Santos-Del-Blanco L, Notivol E et al (2016) Disentangling plasticity of serotiny, a key adaptive trait in a mediterranean conifer. Am J Bot 103:1582–1591. https://doi.org/10.3732/ajb.1600199
Matyas C, Ackzell L, Samuel CJA (2004) EUFORGEN technical guidelines for genetic conservation and use for Scots pine (Pinus sylvestris). International Plant Genetic Resources Institute, Rome
Matziris D (1982) Variation in growth and quality characters in Pinus pinaster provenances grown at seven sites in Greece. Silvae Genet 31:168–173
Matziris DI (1993) Variation in cone production in a clonal seed orchard of black pine. Silvae Genet 42:136–141
Matziris DI (1997) Variation in growth, flowering and cone production in a clonal seed orchard of Aleppo pine grown in Greece. Silvae Genet 46:224–228
Matziris DI (2000) Genetic variation and realized genetic gain from Aleppo pine tree improvement. Silvae Genet 49:5–10
Matziris DI (2005) Genetic variation and realized genetic gain from black pine tree improvement. Silvae Genet 54:96–104
Méndez M, García D, Maestre FT, Escudero A (2008) More ecology is needed to restore mediterranean ecosystems: a reply to Valladares and Gianoli. Restor Ecol 16:216. https://doi.org/10.1111/j.1526-100X.2008.00390.x
Menéndez-Gutiérrez M, Alonso M, Toval G, Díaz R (2017) Variation in pinewood nematode susceptibility among Pinus pinaster Ait. provenances from the Iberian Peninsula and France. Ann For Sci 74:1. https://doi.org/10.1007/s13595-017-0677-3
Mendel Z (1984) Provenance as a factor in susceptibility of Pinus halepensis to Matsucoccus joseph (homoptera: Margarodidae). For Ecol Manag 9:259–266
Molina F (1965) Comportamiento racial de Pinus pinaster en el noroeste de España. An del IFIE 2:221–238
Morgante M, Vendramin GG, Olivieri AM (1991) Mating system analysis in pinus leucodermis ant: detection of self-fertilization in natural populations. Heredity (Edinb) 67:197–203. https://doi.org/10.1038/hdy.1991.80
Mukassabi TA, Polwart A, Coleshaw T, Thomas PA (2012) Scots pine seed dynamics on a waterlogged site. Trees Struct Funct 26:1305–1315
Mutke S, Gordo J, Gil L (2000) The stone pine (Pinus pinea L.) breeding programme in Castile-Leon(Central Spain). FAO-CIHEAM NUCIS-Newsl 9:50–55
Mutke S, Gordo J, Prada MA et al (2011) Mediterranean stone pine provenances' performance in inner Spain – low differentiation in height growth and shoot phenology. In: Besacier C, Ducci F, Malagnoux M, Souvannavong O (eds) Status of the experimental network of Mediterranean forest genetic resources. CRA SEL & FAO, Arezzo, Rome, pp 145–149
Namkoong G (1969) Nonoptimality of local races. In: Proceedings of the tenth southern Conference on forest tree improvement. School of Forestry, North Carolina State College; Southern Forest Tree Improvement Committee, Houston
Nanson A (2001) The new OECD scheme for the certification of forest reproductive materials. Silvae Genet 50:181–187
Naydenov KD, Tremblay FM, Bergeron Y et al (2005) Dissimilar patterns of Pinus heldreichii Christ. populations in Bulgaria revealed by chloroplast microsatellites and terpenes analysis. Biochem Syst Ecol 33:133–148
Neale DB, Savolainen O (2004) Association genetics of complex traits in conifers. Trends Plant Sci 9:325–330

Nieto P, Aguado A, Ruano JR et al (2009) Ensayo de evaluación genética para la admisión de la Unidad HS-24/46/001 como material de base para la producción de material forestal de reproducción de la categoría 'CONTROLADO'. In: V Congreso Forestal Español, Ávila, Spain. Sociedad Española de Ciencias Forestales.
Panetsos KP (1986) Genetics and breeding in the group halepensis. Fôret Méditerranéene VIII:5–12
Pardos JA, Gil L (1986) Los huertos semilleros. Estudios básicos para su establecimiento en España. ICONA, Madrid
Parker WH (1992) Focal point seed zones: site-specific seed zone delineation using geographic information systems. Can J For Res 22:267–271
Pausas JG, Millán MM (2019) Greening and browning in a climate change hotspot: the Mediterranean basin. Bioscience 69:143–151. https://doi.org/10.1093/biosci/biy157
Petit RJ, Hampe A (2006) Some evolutionary consequences of being a tree. Annu Rev Ecol Evol Syst 37:187–214
Petit RJ, Mousadik AEL, Pons O (1998) Identifying populations for conservation on the basis of genetic markers. Conserv Biol 12:844–855. https://doi.org/10.1046/j.1523-1739.1998.96489.x
Piqué M, Ammari Y, Solano D et al (2013) Production and management of stone pine (Pinus pinea) for early nut production: grafted plantations as an alternative for restoring degraded areas and generating income in rural communities of Tunisia. Options Méditérr A:43–47
Provan J, Soranzo N, Wilson NJ et al (1998) Gene-pool variation in caledonian and European Scots pine (Pinus sylvestris L.) revealed by chloroplast simple-sequence repeats. Proc Biol Sci 265:1697–1705. https://doi.org/10.1098/rspb.1998.0491
Puglisi S (1995) Gene conservation in pine species. For Genet 2:145–153
Ramirez-Valiente JA, Robledo-Arnuncio JJJ (2014) Adaptive consequences of human-mediated introgression for indigenous tree species: the case of a relict Pinus pinaster population. Tree Physiol 34:1376–1387. https://doi.org/10.1093/treephys/tpu097
Ratnam W, Rajora OP, Finkeldey R et al (2014) Genetic effects of forest management practices: global synthesis and perspectives. For Ecol Manag 333:52–65. https://doi.org/10.1016/j.foreco.2014.06.008
Restoux G, Silva DE, Sagnard F et al (2008) Life at the margin: the mating system of Mediterranean conifers. Web Ecol 8:94–102. https://doi.org/10.5194/we-8-94-2008
Robledo-Arnuncio JJ, Unger GM (2018) Measuring viability selection from prospective cohort mortality studies: a case study in Maritime Pine. Evol Appl 12:0–2. https://doi.org/10.1111/eva.12729
Robledo-Arnuncio JJ, Alía R, Gil LA (2004a) Increased selfing and correlated paternity in a small population of a predominantly outcrossing conifer, Pinus sylvestris. Mol Ecol 13:2567–2577. https://doi.org/10.1111/j.1365-294X.2004.02251.x
Robledo-Arnuncio JJ, Smouse PE, Gil LA, Alía R (2004b) Pollen movement under alternative silvicultural practices in native populations of Scots pine (Pinus sylvestris L.) in Central Spain. For Ecol Manag 197:245–255. https://doi.org/10.1016/j.foreco.2004.05.016
Robledo-Arnuncio JJJ, Collada C, Alía R, Gil LA (2005) Genetic structure of montane isolates of Pinus sylvestris L. in a Mediterranean refugial area. J Biogeogr 32:595–605. https://doi.org/10.1111/j.1365-2699.2004.01196.x
Rodriguez-Quilon I (2017) Ecological and association genetics in two mediterranean pine species. UPM
Rodriguez-Quilon I, Santos-del-Blanco L, Serra-Varela MJ et al (2016) Capturing neutral and adaptive genetic diversity for conservation in a highly structured tree species. Ecol Appl 26:2254–2266. https://doi.org/10.1002/eap.1361
Ruiz Daniels R, Taylor RS, Serra-Varela MJ et al (2018) Inferring selection in instances of long-range colonization: the Aleppo pine (Pinus halepensis) in the Mediterranean Basin. Mol Ecol 27:3331–3345. https://doi.org/10.1111/mec.14786
Sáez-Laguna E, Guevara M-Á, Díaz L-M et al (2014) Epigenetic variability in the genetically uniform forest tree species Pinus pinea L. PLoS One 9:e103145. https://doi.org/10.1371/journal.pone.0103145

Saldaña A, Borralho N, Majada JRA (2012) Evaluation of early growth of provenance-progeny tests of maritime pine across sites and implications for breeding in Spain. In: Final Conference Noveltree. Tree breeding, genomics and evolutionary biology, Helsinki. METLA

Sampedro L, Moreira X, Zas R (2011) Costs of constitutive and herbivore-induced chemical defences in pine trees emerge only under low nutrient availability. J Ecol 99:818–827. https://doi.org/10.1111/j.1365-2745.2011.01814.x

Sampedro L, Voltas J, Martín-Sanz R et al (2014) Trade-offs among life history traits underlying population differentiation in Aleppo pine. In: 5th International Conference on Mediterranean Pines (MEDPINE 5). Instituto Nacional de Investigaciones Agrarias, Solsona

Santini F, Climent JM, Voltas J (2019) Phenotypic integration and life history strategies among populations of Pinus halepensis: an insight through structural equation modelling. Ann Bot in press 124:1161. https://doi.org/10.1093/aob/mcz088

Santos-del-Blanco L, Climent JM, González-Martínez SC, Pannell JR (2012) Genetic differentiation for size at first reproduction through male versus female functions in the widespread Mediterranean tree Pinus pinaster. Ann Bot 110:1449–1460

Santos-del-Blanco L, Alía R, González-Martínez SC et al (2015) Correlated genetic effects on reproduction define a domestication syndrome in a forest tree. Evol Appl 8:403–410. https://doi.org/10.1111/eva.12252

Savolainen O, Karkkainen K, Kuittinen H (1992) Estimating numbers of embryonic lethals in conifers. Heredity 69:308–314

Savolainen O, Pyhajarvi T, Knurr T et al (2007) Gene flow and local adaptation in tees. Annu Rev Ecol Evol Syst 38:595–619. https://doi.org/10.1146/annurev.ecolsys.38.091206.095646

Schiller G, Atzmon N (2009) Performance of Aleppo pine (Pinus halepensis) provenances grown at the edge of the Negev desert: a review. J Arid Environ 73:1051–1057. https://doi.org/10.1016/j.jaridenv.2009.06.003

Schiller G, Korol L, Ungar ED et al (1999) Canary Islands pine (Pinus canariensis Chr. Sm. ex DC.). 1. Differentiation among native populations in their isoenzymes. For Genet 6:257–276

Schueler S, Kapeller S, Konrad H et al (2012) Adaptive genetic diversity of trees for forest conservation in a future climate: a case study on Norway spruce in Austria. Biodivers Conserv 22:1151. https://doi.org/10.1007/s10531-012-0313-3

Schvester D, Ughetto F (1986) (Homoptera: Margarodidae) selon les provenances de pin maritime (Pinus pinaster AIT). Ann Sci For 43:459–474

Serra-Varela MJ, Alía R, Daniels RR et al (2017) Assessing vulnerability of two Mediterranean conifers to support genetic conservation management in the face of climate change. Divers Distrib 23:507–516. https://doi.org/10.1111/ddi.12544

Sheffer E (2012) A review of the development of Mediterranean pine-oak ecosystems after land abandonment and afforestation: are they novel ecosystems? Ann For Sci 69:429–443. https://doi.org/10.1007/s13595-011-0181-0

Soranzo N, Alía R, Provan J, Powell W (2000) Patterns of variation at a mitochondrial sequence-tagged- site locus provides new insights into the postglacial history of European Pinus sylvestris populations. Mol Ecol 9:1205–1211

Soto A, Robledo-Arnuncio JJ, González-Martínez SC et al (2010) Climatic niche and neutral genetic diversity of the six Iberian pine species: a retrospective and prospective view. Mol Ecol 19:1396–1409. https://doi.org/10.1111/j.1365-294X.2010.04571.x

Suárez-Vidal E, Sampedro L, Voltas J et al (2019) Drought stress modifies early effective resistance and induced chemical defences of Aleppo pine against a chewing insect herbivore. Environ Exp Bot 162:550–559. https://doi.org/10.1016/j.envexpbot.2019.04.002

Sülüsoglu M (2004) The management of villagers owned stone pine (Pinus pinea L.) plantations in Kozak Region, Turkey, a case study, p 48

Tadesse W, Nanos N, Auñon FJ et al (2001) Genetic improvement of resin yield from maritime pine in Spain. For Chem Rev 111:10–16

Talhouk SN, Zurayk R, Khuri S (2001) Conservation of the coniferous forests of Lebanon: past, present and future prospects. Oryx 35:206–215. https://doi.org/10.1046/j.1365-3008.2001.00180.x

Tapias R, Climent J, Pardos JA, Gil LA (2004) Life histories of Mediterranean pines. Plant Ecol 171:53–68

Thuiller W, Lavorel S, Araujo MB et al (2005) Climate change threats to plant diversity in Europe. Proc Natl Acad Sci 102:8245–8250. https://doi.org/10.1073/pnas.0409902102

Tranque FJ, Iglesias A, Climent J (2018) Avanzando en la conservación ex situ de los recursos genéticos de Pinus nigra subsp. salzmannii de la Sierra de Gredos. Montes 133:23–29

Unger GM, Heuertz M, Vendramin GG, Robledo-Arnuncio JJ (2016) Assessing early fitness consequences of exotic gene flow in the wild: a field study with Iberian pine relicts. Evol Appl 9:367. https://doi.org/10.1111/eva.12333

Vallejo R, Pausas JG, Blade C et al (2004) Pines and oaks in the restoration of Mediterranean landscapes of Spain: new perspectives for an old practice – a review. Plant Ecol 171:209–220

van Buijtenen JP (1992) Fundamental genetic principles. In: Fins AT, Brotschol JV, Friedman L (eds) Handbook of quantitative forest genetics. Kluwer Academic Publishers, Dordrecht, pp 29–68

Vendramin GG, Fady B, González-Martínez SC et al (2008a) Genetically depauperate but widespread: the case of an emblematic Mediterranean pine. Evolution (NY) 62:680–688. https://doi.org/10.1111/j.1558-5646.2007.00294.x

Vendramin GG, Fineschi B, Fady B (2008b) EUFORGEN technical guidelines for genetic conservation and use for Bosnian pine (Pinus heldreichii). European Forest Genetic Resources Programme (EUFORGEN), European Forest Institute, Barcelona, Spain

Voltas J, Shestakova TA, Patsiou T et al (2018) Ecotypic variation and stability in growth performance of the thermophilic conifer Pinus halepensis across the Mediterranean basin. For Ecol Manag 424:205–215. https://doi.org/10.1016/j.foreco.2018.04.058

Wahid N, González-Martínez SC, El Hadrami I, Boulli A (2004) Genetic structure and variability of natural populations of maritime pine (Pinus pinaster Aiton) in Morocco. Silvae Genet 53:93–99

Wahid N, González-Martínez SC, El Hadrami I, Boulli A (2006) Variation of morphological traits in natural populations of maritime pine (Pinus pinaster Ait.) in Morocco. Ann For Sci 63:83–92. https://doi.org/10.1051/forest

Wahid N, González-Martínez SC, Alía R, Boulli A (2009) Exploration et conservation des ressources génétiques du pin maritime au Maroc. Fôret Méditerranéene 1970:245–256

Wang T, Hamann A, Yanchuk A et al (2006) Use of response functions in selecting lodgepole pine populations for future climates. Glob Chang Biol 12:2404–2416. https://doi.org/10.1111/j.1365-2486.2006.01271.x

Wang T, O'Neill GA, Aitken SN (2010) Integrating environmental and genetic effects to predict responses of tree populations to climate. Ecol Appl 20:153–163

Zas R, Merlo E (2008) El programa de mejora genética de Pinus pinaster en Galicia. Boletín CIDEU 6–7:5–24

Chapter 4
Biogeographical Patterns of Local Adaptation and Plasticity of Mediterranean Pines and Their Implications Under Climate Change

Marta Benito Garzón and Natalia Vizcaíno-Palomar

4.1 Introduction

Climate change is reshuffling species distribution ranges (Lenoir et al. 2008; Poloczanska et al. 2013). In order to survive under changing climates, trees can move towards more favorable conditions (Chen et al. 2011; Sunday et al. 2011), persist *in-situ* by evolutionary processes, implying adaption to new climates, or adjust by acclimation, i.e. without changes in their genetic structure (West-Eberhard 2003; Pulido and Berthold 2004). Evolutionary responses occur over several generations that, in the case of long-lived organisms such as trees, may be too slow to track climate change (Pedlar and McKenney 2017; Vizcaíno-Palomar et al. 2020; Fréjaville et al. 2020). In addition, some populations may present maladaptation to current climate, which will likely increase for those populations at the trailing edge of the distribution ranges under climate change (Pedlar and McKenney 2017; Fréjaville et al. 2020). In contrast, plasticity implies a rapid response to environmental changes which can help populations to persist, at least in the short term. Hence, it is important to understand how local adaptation and phenotypic plasticity contribute to phenotypic variation in adaptive traits across tree species' distribution ranges.

Mediterranean pines are economically important species having generally scattered distribution ranges (including the Mediterranean distribution range of Scots pine) with populations typically adapted to local conditions (Table 4.1). Their current distributions, initially shaped by their recent demographic history after the Last Glacial Maximum (Médail and Diadema 2009), are the consequence of their low competitive capacity in comparison with broadleaved species (Gómez-Aparicio et al. 2011) and the intensive, long-term forest management that could have favored

M. Benito Garzón (✉) · N. Vizcaíno-Palomar
University of Bordeaux, INRAE, BIOGECO, Pessac, France
e-mail: marta.benito-garzon@inrae.fr

G. Ne'eman, Y. Osem (eds.), *Pines and Their Mixed Forest Ecosystems in the Mediterranean Basin*, Managing Forest Ecosystems 38,
https://doi.org/10.1007/978-3-030-63625-8_4

Table 4.1 Main drivers of genetic adaptation on functional and demographic traits of some Mediterranean *Pinus* species. Drivers are associated with traits when information existed. Please note that the review is not exhaustive

Species	Traits	Drivers	References
P. pinaster	Candidate genes	Drought	Grivet et al. (2011) and Jaramillo-Correa et al. (2015)
	Growth	Winter precipitation	Vizcaíno-Palomar et al. (2016), Aranda et al. (2009), Grivet et al. (2011), Sánchez-Salguero et al. (2018), and Vizcaíno-Palomar et al. (2020)
	Height–diameter allometry		
	Height growth		
	Thick bark, early cones	Fire	Budde et al. (2014)
P. halepensis	Single Nucleotide Polymorphisms (SNPs)	Drought	Grivet et al. (2011), Atzmon et al. (2004), de Luis et al. (2013), Vizcaíno-Palomar et al. (2016), and Ruiz Daniels et al. 2018)
	Height–diameter allometry	Temperature	Atzmon et al. (2004), Grivet et al. (2011), de Luis et al. (2013), Vizcaíno-Palomar et al. (2016), and Ruiz Daniels et al. (2018)
	Survival		
	Radial growth		
	Predawn needle water potential, sap flow in the xylem, photosynthesis, water-use efficiency		Atzmon et al. (2004)
	Onset of cone production, cone serotiny.		Hernández-Serrano et al. (2013)
P. pinea	Photosynthetic rate		Pardos and Calama (2018) and Vizcaíno-Palomar et al. (2020)
	Shoot growth		
	Plant morphology		
	Radial growth		
	Height growth	Summer precipitation	Pardos and Calama (2018) and Vizcaíno-Palomar et al. (2020)
P. nigra	Height–diameter allometry	Photoperiod (latitude) and annual water availability	Vizcaíno-Palomar et al. (2016)
	Height growth		Vizcaíno-Palomar et al. (2020)

broadleaved species over conifers. Therefore, Mediterranean pines are relegated to live under harsh conditions, such as sites with poor soils, with high frequency and intensity of fires and/or high intensity of drought events, etc., to which they are generally adapted (Tapias et al. 2004; Hernández-Serrano et al. 2013; Budde et al. 2014). Generally, broadleaved species present higher plasticity and less genetic adaptation than Mediterranean pines (Benito Garzón et al. 2019), which would favor broadleaved species' persistence under climate change at the expense of conifers (Vayreda et al. 2016), although this is likely to be related to land abandonment during recent decades in the Mediterranean Basin that favored broadleaved species' expansion. Nevertheless, we could expect that Mediterranean pines, with overall high tolerance to harsh conditions, would have an advantage over the less drought-resistant broadleaved species under climate change (DeLucia et al. 2000; Maherali et al. 2004; Choat et al. 2012).

The main goal of this chapter is to present some insights on the vulnerability of Mediterranean pines to climate change through the understanding of local adaptation and phenotypic plasticity of fitness-related traits that may vary across species distribution ranges. We first describe the climatic niche of the Mediterranean pines (Sect. 4.2) and the main climatic drivers of local adaptation (Sect. 4.3), and explain the relative contribution of phenotypic plasticity and local adaptation to the phenotypic variation across species distribution ranges (Sect. 4.4). Finally, we focus on the importance of phenotypic plasticity and local adaptation for explaining the distribution ranges of Mediterranean pine species under climate change (Sect. 4.5), and end with some perspectives and an outline of required further work (Sect. 4.6).

4.2 The Climatic Niche of Mediterranean Pines

Mediterranean pines are the most widespread conifers in the Mediterranean region. They include circum-Mediterranean species such as *Pinus halepensis* Mill. and *P. nigra* Arn.; *P. pinea* L. occurring mainly in the western Mediterranean; and other pines with restricted distributions, such as *P. pinaster* Ait., which is confined mainly to the western Mediterranean, *P. brutia* Ten., *P. heldreichii* H. Christ and *P. peuce* Griseb. with scattered distributions in the eastern Mediterranean and the southern populations of the widely distributed *P. sylvestris* L.

The distribution ranges of Mediterranean pines share some characteristics: they present highly scattered distribution ranges with populations adapted to local conditions, which is reflected by generally high among-population differentiation in fitness-related traits but lower genetic diversity within populations than other conifers (Fady 2012); and they share part of their realized climatic niches, which reflects the climatic conditions where species occur. Here we estimated the realized climatic niches of the Mediterranean pines with a principal component analysis (PCA, prcomp function) built on a random sample of raster points comprising the 3% of the climate (10,000 points approximately) of each species' distribution range. The two first axes of the PCA, that explained 80% of the climatic variation, were used to outline the climatic niche of each species by calculating convex hull polygons (Fig. 4.1).

The realized climatic niches of Mediterranean pines are largely defined by temperature: from those with cold requirements (*P. heldreichii* and *P. peuce*) to those requiring warmer conditions (*P. brutia*, *P. halepensis*, *P. pinaster* and *P. pinea*) (Fig. 4.1). More specifically, *P. halepensis* and *P. pinaster* can inhabit the warmest sites of the Mediterranean Basin, followed by *P. brutia* and *P. pinea,* whilst

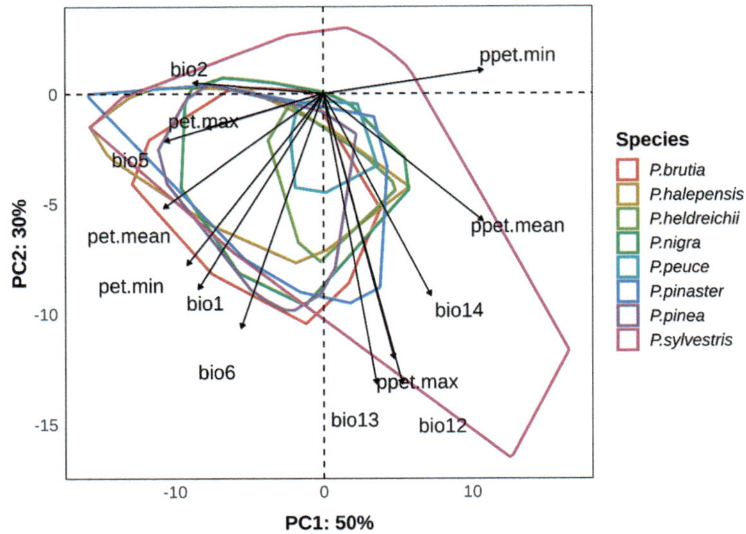

Fig. 4.1 The realized climatic niches of Mediterranean pines from a PCA of *Pinus brutia*, *P. halepensis*, *P. heldreichii*, *P. nigra*, *P. peuce*, *P. pinaster*, *P. pinea* and *P. sylvestris* (the species distribution ranges were taken from EUFORGEN). Climate-related variables used to perform the PCA: bio1: annual daily mean temperature (°C); bio2: mean diurnal temperature range (max–min, °C); bio5: maximum temperature of the warmest month, (°C); bio6: minimum temperature of the coldest month (°C); bio12: annual precipitation (mm); bio13: precipitation of the wettest month (mm); bio14: precipitation of the driest month (mm); pet.mean: annual potential evapotranspiration (mm); ppet.mean: annual water availability (mm); pet.min: potential evapotranspiration of the wettest month (mm); pet.max: potential evapotranspiration of the driest month (mm); ppet.min: water availability of the driest month (mm); ppet.max: water availability of the wettest month (mm)

P. heildreichii and *P. peuce* require colder and relatively wetter environments. The climatic niche of *P. nigra* is restricted between warm and cool climates, not reaching sites with very hot climates. Finally, the climatic niche of *P. sylvestris* is the largest among the Mediterranean pines and it inhabits a large range of temperatures. The realized climatic niche of the Mediterranean pines reflects the ecological tolerances to the climatic conditions considered but may also reflect their low capacity to compete with broadleaved species, their intensive management over the years (Ruiz-Benito et al. 2012) and their evolutionary histories (Fady 2012).

4.3 Environmental Drivers of Genetic Adaptation in Mediterranean Pines

Mediterranean pines have evolved under similar environments, characterized by high climate variability, intense and recurrent periods of drought, heat waves and, in many cases, intense fires (Tapias et al. 2004; Hernández-Serrano et al. 2013; Budde et al. 2014). Therefore, we could expect to find footprints of these drivers shaping the adaptive trait variation that the Mediterranean pines currently present. Many adaptive traits may be driven by the same factors (Table 4.1), although they can present trade-offs, as for example between growth and reproduction (Santos-del-Blanco et al. 2013). Overall, Mediterranean conifers have similar drivers of local adaptation (Table 4.1), being often adapted to extreme climatic conditions (Grivet et al. 2013; Jaramillo-Correa et al. 2015). Adaptive selection is generally related to temperature, evapotranspiration and drought (Grivet et al. 2011; Vizcaíno-Palomar et al. 2020). Some Mediterranean pines show adaptation to fire, which is a common characteristic of Mediterranean environments. For example *P. pinaster* and *P. halepensis* can present serotinous cones (Tapias et al. 2004).

4.4 Phenotypic Plasticity and Genetic Adaptation in Phenotypic Traits at Large Geographical Scales

The capacity of populations to respond rapidly to changes in the environment through plasticity can be advantageous in the short-term but it may delay evolutionary processes in the long-term, highlighting the importance of considering both processes to analyse species' capacity to persist under rapid climate change (Chevin et al. 2010). Over the generations, Mediterranean pines have evolved through genetic adaptation to local conditions, which might confer advantages to certain populations under climate change, but may also lead to maladaptation to new climates in isolated, southern populations (Fréjaville et al. 2020; Vizcaíno-Palomar et al. 2019a).

One way to estimate range-wide local adaptation and phenotypic plasticity at large geographical scales is based on population-level reaction norms measured from phenotypic traits gathered in common gardens. For example, in the case of *P. pinea* almost all the contribution to trait variation comes from phenotypic plasticity, with a negligible genetic effect (Vizcaíno-Palomar et al. 2020). Plant morphology and tree height are usually highly plastic (Chambel et al. 2007; Pardos and Calama 2018; Vizcaíno-Palomar et al. 2020). In contrast, *P. halepensis* shows high plasticity for growth whereas reproduction is more genetically determined (Santos-del-Blanco et al. 2013).

Tree height of *P. nigra, P. pinaster, P. pinea and P. sylvestris*, measured in extensive common garden networks (Vizcaíno-Palomar et al. 2019b) is used here as an example to illustrate population reaction norms (Fig. 4.2). The relative contributions of the effects of the population (local adaptation) and the trial site (phenotypic plasticity) differ among the four pines, but phenotypic plasticity always makes the greatest contribution to tree height variation (Fig. 4.2). *Pinus nigra, P. pinaster* and *P. sylvestris* present moderate-to-high levels of phenotypic plasticity and among-population differentiation (Fig. 4.2a, b and d, concave curves along the provenance and trial axes), which might allow these three species to respond either to rapid

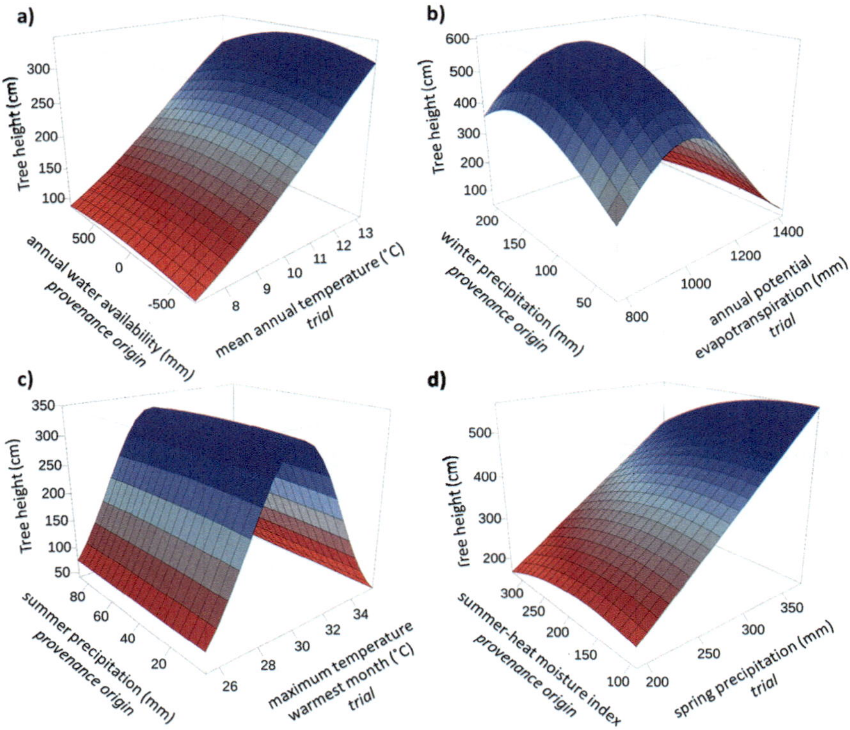

Fig. 4.2 Provenance (genetic adaptation) and trial (phenotypic plasticity) effects on tree height for (**a**) *Pinus nigra*, (**b**) *P. pinaster*, (**c**) *P. pinea* and (**d**) *P. sylvestris*, shown by three-dimensional graphs of population reaction norms. (Adapted from (Vizcaíno-Palomar et al. 2019a, 2020)). See Table 4.2 for further explanation of the models

changes through phenotypic plasticity or to slowly evolve through natural selection. However, most of the tree height variation presented in *P. pinea* is associated with phenotypic plasticity combined with very low differentiation among populations (Fig. 4.2c, flat line along the provenance axis and concave curve along the trial axis), indicating that plasticity is the main mechanism for this species' response to changes in the environment.

4.5 Phenotypic Plasticity, Local Adaptation and Distribution Ranges of Mediterranean Pines

Understanding species distribution ranges has always attracted the attention of ecologists, particularly in recent times, given our awareness that the persistence of species distribution ranges depends on the tolerance and adaptive capacity of populations to novel conditions created by a changing climate. New emerging approaches based

Table 4.2 Models that spatially predict fitness-related traits or species distribution ranges using information related to phenotypic variation in Mediterranean *Pinus* species

Species	Model	Geographical region	Model calibration	Main results	Reference
P. halepensis	Ecotypic stability models (linear-mixed effects models)	Species distribution range	Tree height and survival from common gardens	Higher genetic adaption than phenotypic plasticity for the ecotypes analysed	Voltas et al. (2018)
P. nigra	ΔTraitSDM (linear-mixed effect models)	Species distribution range	Tree height growth from common gardens	Moderate tree height increments across the species distribution range; comparing present and future conditions for 2050 under RCP 8.5 (Fig. 4.3a)	Fig. 4.3 – Adapted from Vizcaíno-Palomar et al. (2020)
P. pinaster	ΔTraitSDM (random forests)	Spain	Tree height growth and survival measured in common gardens	Suitable area predicted for tree growth and survival for 2050 is reduced (30%) in the future predictions with respect to current predictions in Spain.	Benito Garzón et al. (2011)
	ΔTraitSDM (linear-mixed effects models)	Species distribution range	Tree height growth measured in common gardens	Moderate-to-high tree height decrease, except in the north-western, centre-inland and east coast of Spain, north part of west France, and the Mediterranean coast of France; comparing present and future conditions for 2050 under RCP 8.5 (Fig. 4.3b)	Fig. 4.3 – Adapted from Vizcaíno-Palomar et al. (2020)
	SDM	Species distribution range	Population genetic structure based on mitochondrial, chloroplast and nuclear molecular markers	Projected habitat suitability is enlarged in SDM performed at the population level as defined by genetic structure	Serra-Varela et al. (2015)

(continued)

Table 4.2 (continued)

Species	Model	Geographical region	Model calibration	Main results	Reference
P. pinea	ΔTraitSDM (linear-mixed effects models)	Species distribution range	Tree height growth from common gardens	Moderate-to-high tree height increase, except for moderate decreases in the Guadalquivir Basin in Spain and in some populations located in the eastern part of the distribution; comparing present and future conditions for 2050 under RCP 8.5 (Fig. 4.3c)	Fig. 4.3 – Adapted from Vizcaíno-Palomar et al. (2020)
P. sylvestris	ΔTraitSDM (random forests)	Spain	Tree growth and survival from common gardens	Suitable area predicted for tree growth and survival for 2050 is only slightly reduced (1%) in future predictions with respect to current predictions in Spain.	Benito Garzón et al. (2011)
	ΔTraitSDM (linear-mixed effects models)			Tree height decrease, mainly in low altitude areas, and tree height increase in high altitude areas; comparing present and future conditions for 2050 under RCP 8.5 (Fig. 4.3d).	Fig. 4.3 – Adapted from Vizcaíno-Palomar et al. (2019a)

on phenotypic plasticity and local adaptation to delimit species distribution models based on adaptive traits are a promising tool towards more realistic predictions of fitness-related traits (ΔTraitSDM; Benito Garzón et al. 2019). Among them, a few analyses have been done for Mediterranean pines (Table 4.2) showing that accounting for the plasticity and genetic adaptation of populations would give a more realistic prediction of species distribution ranges and a less pessimistic future than that predicted by species distribution models based on species occurrence.

Based on the population-level reaction norms estimated from the wide-range networks of provenance tests (Fig. 4.2), ΔTraitSDM of tree height (Fig. 4.3) shows an overall expected increase in tree height by the year 2050 under the RCP 8.5 scenario for *P. nigra* and *P. pinea* with some exceptions (see Table 4.2 and Fig. 4.3), while for *P. pinaster* and *P. sylvestris* (here, referring to the southern part of the species distribution) tree height is expected to decrease over a large proportion of the species distribution range (Fig. 4.3).

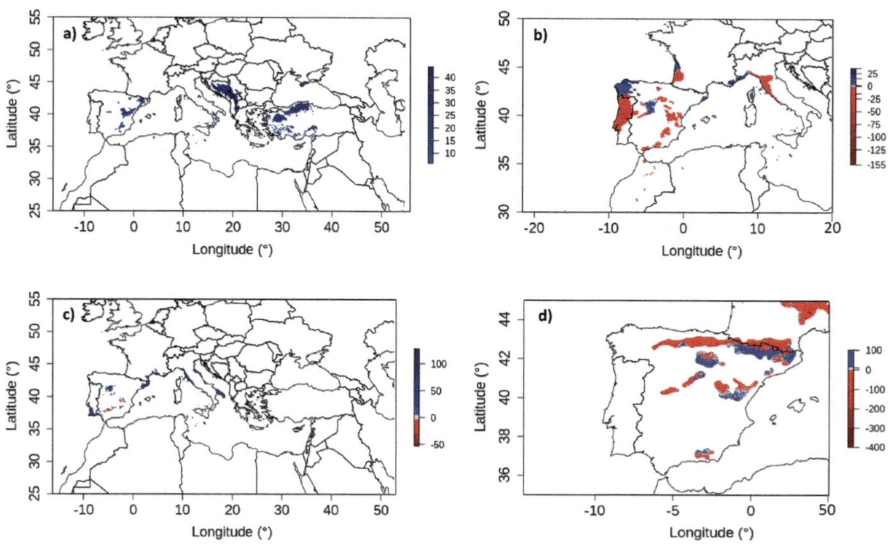

Fig. 4.3 Spatial predictions of the difference in tree height (cm) between future (2050, RCP 8.5) and current conditions using ΔTraitSDM of (**a**) *Pinus nigra*, (**b**) *P. pinaster*, (**c**) *P. pinea* and (**d**) *P. sylvestris*. (Adapted from (Vizcaíno-Palomar et al. 2019a, 2020)). Blue shades indicate an increase in predicted tree height, grey indicates no change and red shades indicate a decrease in tree height preditions

4.6 Mediterranean Conifers Under Climate Change: Perspectives and Further Research

In this chapter, we showed that Mediterranean pines are adapted to drought and to a broad range of temperatures, which may confer on Mediterranean pines the ability to persist in the Mediterranean Basin with the increase in frequency and intensity of drought and heat-wave events expected under climate change. However, Mediterranean pines show lower genetic diversity than the widely-distributed European pines (Alberto et al. 2013), which would compromise the evolutionary process of Mediterranean pines in the long-term, due to the effects of genetic drift (Le Corre and Kremer 2012) and low gene flow among populations in fragmented distributions. In the short-term, it is difficult to predict to what extent Mediterranean pines will be able to handle rapid climate change through phenotypic plasticity, and we still need more experimental studies to measure the limits of phenotypic plasticity of Mediterranean pines.

To date, only tree height has been explored range-wide, thus limiting our knowledge on the tolerance to new climates of other fitness-related and functional traits that may have evolved differently from tree growth. The most important traits that need to be examined to determine the vulnerability of Mediterranean pines to climate change include: (1) the timing of growth and reproduction, to explore potential trade-offs (Rohde and Bhalerao 2007; Hänninen and Tanino 2011); (2) early fitness components of germination and establishment to explore recruitment, and cold and drought tolerance (Laanisto and Niinemets 2015), and (3) resistance to emergent pathogens (Perry et al. 2016).

References

Alberto FJ, Aitken SN, Alía R, González-Martínez SC, Hanninen H, Kremer A, Lefevre F, Lenormand T, Yeaman S, Whetten R, Savolainen O (2013) Potential for evolutionary responses to climate change - evidence from tree populations. Glob Chang Biol 19:1645–1661

Aranda I, Alía R, Ortega U, Dantas ÂK, Majada J (2009) Intra-specific variability in biomass partitioning and carbon isotopic discrimination under moderate drought stress in seedlings from four *Pinus pinaster* populations. Tree Genet Genomes 6:169–178

Atzmon N, Moshe Y, Schiller G (2004) Ecophysiological response to severe drought in *Pinus halepensis* Mill. trees of two provenances. Plant Ecol 171:15–22

Benito Garzón M, Alía R, Robson TM, Zavala MA (2011) Intra-specific variability and plasticity influence potential tree species distributions under climate change. Glob Ecol Biogeogr 20:766–778

Benito Garzón M, Robson TM, Hampe A (2019) ΔTrait SDMs: species distribution models that account for local adaptation and phenotypic plasticity. New Phytol 222:1757–1765

Budde KB, Heuertz M, Hernández-Serrano A, Pausas JG, Vendramin GG, Verdú M, González-Martínez SC (2014) In situ genetic association for serotiny, a fire-related trait, in Mediterranean maritime pine (*Pinus pinaster*). New Phytol 201:230–241

Chambel MR, Climent J, Alia R (2007) Divergence among species and populations of Mediterranean pines in biomass allocation of seedlings grown under two watering regimes. Ann For Sci 64:87–97

Chen IC, Hill JK, Ohlemüller R, Roy DB, Thomas CD (2011) Rapid range shifts of species associated with high levels of climate warming. Science 333:1024–1026

Chevin L-M, Lande R, Mace GM (2010) Adaptation, plasticity, and extinction in a changing environment: towards a predictive theory. PLoS Biol 8:e1000357

Choat B, Jansen S, Brodribb T, Cochard H, Delzon S, Bhaskar R, Bucci S, Feild T, Gleason S, Hacke U, Jacobsen A, Lens F, Maherali H, Martinez-Vilalta J, Mayr S, Mencuccini M, Mitchell A (2012) Global convergence in the vulnerability of forests to drought. Nature 491:752–755

de Luis M, Čufar K, Di Filippo A, Novak K, Papadopoulos A, Piovesan G, Rathgeber CBK, Raventós J, Saz MA, Smith KT (2013) Plasticity in dendroclimatic response across the distribution range of Aleppo pine (Pinus halepensis). PLoS One 8:e83550–e83550

DeLucia EH, Maherali H, Carey EV (2000) Climate-driven changes in biomass allocation in pines. Glob Chang Biol 6:587–593

Fady B (2012) Biogeography of neutral genes and recent evolutionary history of pines in the Mediterranean Basin. Ann For Sci 69:421–428

Fréjaville T, Vizcaíno-Palomar N, Fady B, Kremer A, Benito Garzón M (2020) Range margin populations show high climate adaptation lags in European trees. Glob Chang Biol 26:484–495

Gómez-Aparicio L, García-Valdés R, Ruiz-Benito P, Zavala MA (2011) Disentangling the relative importance of climate, size and competition on tree growth in Iberian forests: implications for forest management under global change. Glob Chang Biol 17:2400–2414

Grivet D, Sebastiani F, Alía R, Bataillon T, Torre S, Zabal-Aguirre M, Vendramin GG, González-Martínez SC (2011) Molecular footprints of local adaptation in two Mediterranean conifers. Mol Biol Evol 28:101–116

Grivet D, Climent JM, Zabal-Aguirre M, Neale DB, Vendramin GG, González-Martínez SC (2013) Adaptive evolution of Mediterranean pines. Mol Phylogenet Evol 68:555–566

Hänninen H, Tanino K (2011) Tree seasonality in a warming climate. Trends Plant Sci 16:412–416

Hernández-Serrano A, Verdú M, González-Martínez SC, Pausas JG (2013) Fire structures pine serotiny at different scales. Am J Bot 100:2349–2356

Jaramillo-Correa J-P, Rodríguez-Quilón I, Grivet D, Lepoittevin C, Sebastiani F, Heuertz M, Garnier-Géré PH, Alía R, Plomion C, Vendramin GG, González-Martínez SC (2015) Molecular proxies for climate maladaptation in a long-lived tree (Pinus pinaster Aiton, Pinaceae). Genetics 199:793–807

Laanisto L, Niinemets Ü (2015) Polytolerance to abiotic stresses: how universal is the shade-drought tolerance trade-off in woody species? Glob Ecol Biogeogr 24:571–580

Le Corre V, Kremer A (2012) The genetic differentiation at quantitative trait loci under local adaptation. Mol Ecol 21:1548–1566

Lenoir J, Gegout JC, Marquet PA, de Ruffray P, Brisse H (2008) A significant upward shift in plant species optimum elevation during the 20th century. Science 320:1768–1771

Maherali H, Pockman WT, Jackson RB (2004) Adaptive variation in the vulnerability of woody plants to xylem cavitation. Ecology 85:2184–2199

Médail F, Diadema K (2009) Glacial refugia influence plant diversity patterns in the Mediterranean Basin. J Biogeogr 36:1333–1345

Pardos M, Calama R (2018) Responses of *Pinus pinea* seedlings to moderate drought and shade: is the provenance a differential factor? Photosynthetica 56:786–798

Pedlar JH, McKenney DW (2017) Assessing the anticipated growth response of northern conifer populations to a warming climate. Sci Rep **7**:srep43881

Perry A, Brown AV, Cavers S, Cottrell JE, Ennos RA (2016) Has Scots pine (*Pinus sylvestris*) co-evolved with *Dothistroma septosporum* in Scotland? Evidence for spatial heterogeneity in the susceptibility of native provenances. Evol Appl 9:982–993

Poloczanska ES, Brown CJ, Sydeman WJ, Kiessling W, Schoeman DS, Moore PJ, Brander K, Bruno JF, Buckley LB, Burrows MT, Duarte CM, Halpern BS, Holding J, Kappel CV, O'Connor MI, Pandolfi JM, Parmesan C, Schwing F, Thompson SA, Richardson AJ (2013) Global imprint of climate change on marine life. Nat Clim Chang 3:919–925

Pulido F, Berthold P (2004) Microevolutionary response to climatic change. Adv Ecol Res 35:151–183

Rohde A, Bhalerao RP (2007) Plant dormancy in the perennial context. Trends Plant Sci 12:217–223

Ruiz Daniels R, Taylor RS, Serra-Varela MJ, Vendramin GG, González-Martínez SC, Grivet D (2018) Inferring selection in instances of long-range colonization: the Aleppo pine (*Pinus halepensis*) in the Mediterranean Basin. Mol Ecol 27:3331–3345

Ruiz-Benito P, Gómez-Aparicio L, Zavala MA (2012) Large-scale assessment of regeneration and diversity in Mediterranean planted pine forests along ecological gradients. Divers Distrib 18:1092–1106

Sánchez-Salguero R, Camarero JJ, Rozas V, Génova M, Olano JM, Arzac A, Gazol A, Caminero L, Tejedor E, de Luis M, Linares JC (2018) Resist, recover or both? Growth plasticity in response to drought is geographically structured and linked to intraspecific variability in *Pinus pinaster*. J Biogeogr 45:1126–1139

Santos-del-Blanco L, Bonser SP, Valladares F, Chambel MR, Climent J (2013) Plasticity in reproduction and growth among 52 range-wide populations of a Mediterranean conifer: adaptive responses to environmental stress. J Evol Biol 26:1912–1924

Serra-Varela MJ, Grivet D, Vincenot L, Broennimann O, Gonzalo-Jiménez J, Zimmermann NE (2015) Does phylogeographical structure relate to climatic niche divergence? A test using maritime pine (*Pinus pinaster* Ait.). Glob Ecol Biogeogr 24:1302–1313

Sunday JM, Bates A, Dulvy NK (2011) Global analysis of thermal tolerance and latitude in ectotherms. Proc R Soc B Biol Sci 278:1823–1830

Tapias R, Pardos JA, Gil L, Climent J (2004) Life histories of Mediterranean pines. Plant Ecol 171:53–68

Vayreda J, Martinez-Vilalta J, Gracia M, Canadell JG, Retana J (2016) Anthropogenic-driven rapid shifts in tree distribution lead to increased dominance of broadleaf species. Glob Chang Biol 22:3984–3995

Vizcaíno-Palomar N, Ibáñez I, González-Martínez SC, Zavala MA, Alía R (2016) Adaptation and plasticity in aboveground allometry variation of four pine species along environmental gradients. Ecol Evol 6

Vizcaíno-Palomar N, González-Muñoz N, González-Martínez SC, Alía R, Benito Garzón M (2019a) Most southern scots pine populations are locally adapted to drought for tree height growth. Forests 10:555

Vizcaíno-Palomar N, Benito Garzón M, Alia R, Giovanelli G, Huber G, Mutke S, Pastuszka P, Raffin A, Sbay H, Seho M, Vauthier D, Fady B (2019b) Geographic variation of tree height of three pine species (Pinus nigra Arn., P. pinaster Aiton and P. pinea L.) gathered from common gardens in Europe and North-Africa. Ann For Sci 76:77

Vizcaíno-Palomar N, Fady B, Alia R, Raffin A, Mutke S, Benito Garzón M (2020) The legacy of climate variability over the last century on populations' phenotypic variation in tree height. Sci Total Environ 749:141454

Voltas J, Shestakova TA, Patsiou T, di Matteo G, Klein T (2018) Ecotypic variation and stability in growth performance of the thermophilic conifer *Pinus halepensis* across the Mediterranean basin. For Ecol Manag 424:205–215

West-Eberhard MJ (2003) Developmental plasticity and evolution. Oxford University Press, incl, New York

Chapter 5
Mediterranean Pines as Invasive Species in the Southern Hemisphere

David M. Richardson and Mlungele M. Nsikani

5.1 Introduction

Ten *Pinus* species have natural ranges in the Mediterranean Basin (including the Canary Islands): *P. brutia* Tenore, *P. canariensis* C.Sm., *P. halepensis* Miller, *P. heldreichii* H.Christ, *P. mugo* Turra, *P. nigra* J.F.Arnold, *P. pinaster* Aiton, *P. pinea* L., *P. sylvestris* L. and *P. uncinata* Mill. Ex Mirb., although the ranges of four of these species (*P. mugo*, *P. nigra*, *P. sylvestris* and *P. uncinata*) extend significantly beyond the Mediterranean Basin.

This chapter first reviews how human activities have induced range changes within the natural range of pines in the Mediterranean Basin. It then addresses the history of plantings of Mediterranean pines in other parts of the world. The history of naturalization and invasion of Mediterranean pines is then reviewed. Subsequent sections deal with the ecology of Mediterranean pine invasions in the Southern Hemisphere, the impacts of these invasions, and efforts to deal with them.

5.2 Human-Induced Range Changes Within the Natural Range of Mediterranean Pines

Human activities over almost a million years within the range of Mediterranean pines have radically changed the natural distributions of pines, including the genetic diversity of taxa. The late-Holocene "pine-rise" in Europe has been attributed to range expansions of pines into human-modified habitats (Huntley and Birks 1983).

D. M. Richardson (✉) · M. M. Nsikani
Centre for Invasion Biology, Department of Botany and Zoology, Stellenbosch University, Stellenbosch, South Africa
e-mail: rich@sun.ac.za

G. Ne'eman, Y. Osem (eds.), *Pines and Their Mixed Forest Ecosystems in the Mediterranean Basin*, Managing Forest Ecosystems 38,
https://doi.org/10.1007/978-3-030-63625-8_5

Intentional translocations and complex combinations of fire, grazing, browsing, and diverse practices that expose the soil continue to change the abundance and/or geographic range of several Mediterranean pine species. Human activities that have led to range changes in the Mediterranean Basin include: (1) clearing of forests for timber; (2) large-scale planting of pines over many centuries, leading to profound range extensions of some species; (3) the movement of seeds (especially those of *P. pinea*) along old trade routes; (4) planting of pines (especially *P. pinaster*) to stabilize dunes in Portugal since the fourteenth century; (5) changing patterns of land use, including urbanization and land abandonment that create conditions favorable for the expansion of some species; (6) altering fire regimes and various diseases and insect pests (and interactions between these factors) that have ravaged pine forests in some areas (Barbéro et al. 1998; Le Maitre 1998; Richardson et al. 2007). These and other factors have blurred the distinction between the natural and adventive ranges for all native pines in the Mediterranean Basin. At least one Mediterranean pine taxon (*P. nigra* subsp. *nigra*) is considered invasive outside its natural range in the region (Chauchard et al. 2006). The fact that pines still cover vast areas of the region attests to their ability to persist and even flourish under a wide range of intense human pressures (Richardson et al. 2007; Singh et al. 2018).

Many studies have described the features of Mediterranean pines and the environmental conditions associated with range expansions (e.g., Abbas et al. 1984; Achérar et al. 1984; Salim et al. 2010; Selvi et al. 2016). Preadaptation to the wide range of environmental conditions in the Mediterranean Basin and especially to the pervasive influences of human-induced disturbance has equipped these species for dealing with diverse conditions in foreign environments.

5.3 The History of Planting Mediterranean Pines Outside Their Natural Range

Afforestation has been practiced within the Mediterranean Basin for centuries, but only following European colonialism in the fifteenth century were trees native to the Mediterranean widely planted in other parts of the world. Because of their many uses and their ability to survive and flourish in a wide range of harsh environments, pines were widely planted in Dutch, English, French, Portuguese and Spanish colonies. Pines first arrived in South Africa in the second half of the seventeenth century, in Australia soon after 1770, and in New Zealand in the early 1800s. These were the three centers of pine planting in the Southern Hemisphere in the nineteenth century and for most of the twentieth century. Large-scale afforestation with pines began much later in South America (Richardson et al. 2008). Early introductions generally followed a "shotgun" approach, with little consideration given to species-site matching or selection of the best species or provenances for particular purposes and sites. Some species were widely planted from the beginning (notably *P. pinaster*), but most other species were planted sporadically, many only in small experimental stands. Large-scale commercial forestry was underway in Australia, New Zealand and South Africa by the 1880s. Poynton's (1979) list of 80 *Pinus* species grown in South Africa included all the Mediterranean pines. Booth and Saunders (1984) list 132 pine taxa (species, subspecies and varieties – some no longer recognized), including all the Mediterranean pines, that have been planted in field trials in Australia. All 10 Mediterranean pines have been planted in New Zealand (McGregor et al. 2012).

Pines have many traits that make them attractive for plantation forestry (Richardson 2011). The extent and dynamics of the planting of pines in different regions has been influenced, among other things, by: (1) the shortage or absence of certain products and/or the need for certain services that are potentially better provided by pines than any native tree species or other non-native trees; (2) the potential of different pine taxa to fulfill various roles (determined by their growth form and architecture, establishment and reproduction requirements, climatic and edaphic requirements, growth rates, quality of timber, ability to produce large seeds ("nuts") etc.).

Mediterranean pines are much less important for commercial forestry in all parts of the Southern Hemisphere than some other *Pinus* species, notably *P. caribaea* Morelet, *P. elliottii* Engelm., *P. kesiya* Royle ex Gordon, *P. oocarpa* Schiede ex Schltdl., *P. patula* Schiede ex Schltdl. & Cham., *P. taeda* L., and especially *P. radiata* D. Don (Le Maitre 1998). In many areas, these species and others are replacing Mediterranean pines that were initially planted. Nonetheless, Mediterranean pines have many uses that will ensure their continued importance in many parts of the world outside their natural range. Important uses of Mediterranean pines include: seeds as food for humans (*P. pinea*); foliage (*P. nigra*); resin (*P. halepensis*, *P. nigra*, *P. pinaster*, *P. sylvestris*); solid wood (most species); wood fiber (mainly *P. pinaster*); wood charcoal (mainly *P. halepensis*); "protection" [including hedges, shelter belts, erosion control, mine revegetation and driftsand stabilization] (*P. halepensis*, *P. nigra*, *P. pinaster*, *P. pinea*, *P. sylvestris*), ornamental (especially *P. canariensis* and *P. pinea*); Christmas trees (most species are used where they are grown). Planting for these uses (and for other uses for which Mediterranean pines are not particularly suitable but for which they were initially used) has ensured the widespread presence of these species across a wide range of Southern Hemisphere ecosystems. A good sample of genotypes of Mediterranean pines has thus sampled a wide range of environments in the Southern Hemisphere, especially in temperate regions that have similar climatic conditions to those in the Mediterranean Basin.

5.4 The History of Mediterranean Pine Invasions

Orians (1986, p. 135) noted that "almost no conifers introduced into Europe or into North America from elsewhere have established feral populations even though most of them have been extensively planted and most do not require the services of animals for either pollen or seed dispersal". Pines are much less widespread as non-native invaders in the Northern than in the Southern Hemisphere. Mortenson and Mack (2006) reviewed the fate of non-native pines in the USA and concluded that the failure of any species to become a major invader could be ascribed to a combination of several factors: short residence time and the small size of founder populations; pests; and the absence of requisite mycorrhizae. Some pines, including Mediterranean species, are, however, naturalized or invasive outside their natural ranges in the Northern Hemisphere. Examples in North America include *P. halepensis* that is naturalized near San Francisco, California (McClintock et al. 1990). The lack of success of *P. halepensis* as an invader in California is striking, given the scale of planting of the species in the state. *Pinus nigra* has spread from plantings on the sand dunes around Lake Michigan (Leege and Murphy 2000) and *P. sylvestris* has invaded peatlands in Ontario, Canada (Marinich and Powell 2017). *Pinus pinea* is one of the "worst weeds" on Santa Cruz Island off the coast of California (National Park Service 2019). *Pinus pinaster* is invasive on Maui, Hawaii (Oppenheimer 2002). In Wales, *P. nigra* has invaded dunes on the south-west coast of Anglesey (Hodgkin 1984) and *P. nigra* is invasive in the French Pyrenees (Chauchard et al. 2006). Despite these examples, nowhere in the Northern Hemisphere do invasions of Mediterranean pines (or any other non-native pines) match those in parts of the Southern Hemisphere in terms of extent or impact. One obvious reason for the relatively poor performance of these species as invaders in the Northern Hemisphere compared with the Southern Hemisphere is that there has been much less planting of these species in Northern Hemisphere regions (hence fewer propagules and fewer foci to launch invasions) (Richardson and Higgins 1998). Another reason is the presence of native congeners for pines in the Northern Hemisphere, which are lacking in the Southern Hemisphere. The presence of closely related species may increase the colonization rates of natural enemies and lead to pines in the Northern Hemisphere recruiting pathogens and phytophages that decrease their ability to spread and become invasive (Essl et al. 2010). Further work to elucidate the "invasion windows" for the species mentioned above in the Northern Hemisphere would be profitable. The rest of this chapter deals with invasions of Mediterranean pines in the Southern Hemisphere where invasions are much more widespread and damaging, and thus have been studied more thoroughly.

Seven of the 10 Mediterranean pines are known to be invasive in the Southern Hemisphere where they have invaded a wide range of forest, grassland and shrubland vegetation types in at least the following countries: Australia (6 species), Chile (2 species), New Zealand (6 species), South Africa (4 species) and Uruguay (1 species) (Table 5.1; Fig. 5.1). Another 12 species of *Pinus* from other parts of the world are also invasive in the Southern Hemisphere (Richardson and Higgins 1998; Richardson and Rejmánek 2004; Simberloff et al. 2010; Rejmánek and Richardson 2013).

The first record of prolific natural regeneration (and possible spread) of pines in the Southern Hemisphere was for *P. halepensis*, which was noted to be spreading (or at least regenerating prolifically) near Cape Town, South Africa, as early as 1855, about 25 years after its introduction in the country (Richardson and Higgins 1998). As the time since introduction and the extent of planting of many species increased,

Table 5.1 List of Mediterranean *Pinus* taxa known to be "naturalized" or "**invasive**" (invasive in bold) in the Southern Hemisphere

Pinus taxon	Z-score (Rejmánek et al. 2005)	Country/region where the species is invasive or naturalized
P. brutia	7.49	Australia (WA; Vic#)
P. canariensis	−12.25	Australia (WA; SA); **South Africa**
P. heldreichii	12.34	No evidence of naturalization/invasion
P. halepensis	11.76	**Argentina** (Zalba and Villamil 2002); **Australia** (SA; Vic; NSW (Hosking et al. 2007); WA#; TAS#); **New Zealand**; **South Africa**
P. mugo	11.63	**New Zealand**
P. nigra	4.24	**Australia** (NSW; Vic; SA); **New Zealand**
P. pinaster	7.89	**Australia** (SA; Vic; NSW; TAS; WA#; Qld#); **Chile**; **New Zealand**; Reunion (Baret et al. 2006); **South Africa**; **Uruguay**
P. pinea	−8.37	Australia (NSW); **South Africa**
P. sylvestris	7.75	Argentina; **Chile**; **New Zealand**
P. uncinata	9.75	New Zealand (Bellingham et al. 2004; Froude 2011)

Updated from Richardson and Rejmánek (2004), Richardson and Petit (2005), and Rejmánek and Richardson (2013). Definitions for "naturalized" and "invasive" follow Richardson et al. (2000) and Pyšek et al. (2004). Z-scores are derived from the discriminant function of Rejmánek et al. (2005), based on traits of 12 invasive and 11 non-invasive pine species. Positive Z-scores suggest high levels of inherent invasiveness, negative Z scores are more likely non-invasive. Key citations not included in the above references are provided. #: Kate Blood, pers comm December 2019

more species became naturalized and then invasive. Self-sown stands of *P. pinaster* were also widespread throughout the Cape region by the end of the nineteenth century. Pines were also spreading in New Zealand at about this time. The first few decades of the twentieth century saw rapid increases in the extent of pine invasions in New Zealand and South Africa. Pine invasions in Australia began several decades later (Richardson and Higgins 1998), but invasions in South America only began in the late 1980s (Richardson et al. 2008; Simberloff et al. 2010).

The importance of pines as "environmental weeds" (non-native plant taxa that invade natural vegetation, usually adversely affecting native biodiversity and/or ecosystem functioning) has increased at about the same rate as the overall increase in the problem of biological invasions worldwide, i.e., a massive, exponential increase in abundance and the magnitude of associated problems since about the mid-twentieth century. Pines now feature prominently on regional, national and global lists of important or potentially important weeds (references in Richardson and Higgins 1998). Studies of pine invasions in several widely-scattered parts of the Southern Hemisphere, and general or anecdotal accounts from other areas, show that the extent of the phenomenon has increased rapidly in recent decades and that the nuisance value of pines outside plantations is escalating rapidly [see Richardson and Higgins (1998) for details on the extent of invasion by different pine species].

Fig. 5.1 Several Mediterranean pines have invaded very large areas of natural vegetation in the Southern Hemisphere. The most widespread invaders are *Pinus pinaster* in South African fynbos vegetation (top; photo: A. Turner) and *P. nigra* in montane grasslands and shrublands in New Zealand (bottom; photo: P. Raal). Different management strategies have evolved for controlling pine invasions in different regions. Mechanical clearing combined with fire (top insert; photo: D.M. Richardson) and application of herbicides from helicopters (bottom insert; photo: P. Raal) are the main methods applied in South Africa and New Zealand respectively

5.5 The Ecology of Pine Invasions in the Southern Hemisphere

Many publications have documented the distribution and anecdotal aspects of the ecology of Mediterranean pines as invaders in the Southern Hemisphere, but there have been few detailed studies of factors mediating these invasions (see Richardson et al. 1994; Richardson and Higgins 1998 for detailed reviews of this literature). Only five of the 10 Mediterranean pine species have been studied in any detail. It is beyond the scope of this chapter to review these studies and readers seeking detailed information should consult the sources listed in Table 5.2.

A general understanding of the dynamics of pine invasions in the Southern Hemisphere has emerged from studies of the natural experiment provided by the widespread planting of many species in many localities where a wide range of factors have interacted to determine to what extent pines have invaded in different

Table 5.2 Detailed studies of Mediterranean pines as invaders in the Southern Hemisphere

Species	Study location	Study subject	Studies
P. canariensis	South Africa	Invasion ecology	Rouget et al. (2001)
P. halepensis	South Africa	History of planting and early invasion	Shaughnessy (1986)
		Invasion ecology	Richardson (1988), Richardson and Cowling (1994) and Rouget et al. (2001)
P. nigra	New Zealand	Invasion ecology	Hunter and Douglas (1984), Ledgard and Belton (1985), Ledgard (1988), Belton and Ledgard (1991), Buckley et al. (2005) and Coutts et al. (2012)
P. pinaster	Australia	Invasion ecology	Van Etten et al. (2020)
	South Africa	History of planting and early invasion	Shaughnessy (1986)
		Invasion ecology	Kruger (1977), van Wilgen and Siegfried (1986), Richardson and Cowling (1994), Higgins et al. (2001), Rouget et al. (2001) and Rouget and Richardson (2003)
		The role of plantings in driving invasions	McConnachie et al. (2015)
P. pinea	South Africa	History of planting and early invasion	Shaughnessy (1986)
		Invasion ecology	van Wilgen and Siegfried (1986), Richardson et al. (1990) and Richardson and Cowling (1994)
P. sylvestris	New Zealand	Invasion ecology	Belton and Ledgard (1991)

areas (Richardson and Bond 1991; Richardson et al. 1994; Richardson and Higgins 1998; Richardson 2006). These studies utilized invasion events involving 19 pine species to build a general model of pine invasions in the Southern Hemisphere; they provide a correlative basis for predicting pine invasions. Factors that emerged as being most important are: species attributes; residence time and the extent of planting; elements of the disturbance regime; and special features of the native biota (see Richardson and Higgins 1998, p. 465 for details). Pine invasions in the Southern Hemisphere can be explained, and future changes predicted, using this scheme (Richardson and Higgins 1998).

In an attempt to "operationalize" the correlates of invasive success for pines, Higgins and Richardson (1998) constructed a spatially-explicit, individual-based model of pine invasions. The model simulates invasions of two "pine types" (a "pioneer-pine" and a "late-seral-pine") using suites of life-history traits closely associated with two of Keeley and Zedler's (1998) life-history types that include most of the invasive pines in the Southern Hemisphere. The first type was parameterized using *P. halepensis* as a model species (serotinous cones, short juvenile period, low probability of surviving intense fires, shade-intolerant seedlings, annual cone production). The second type is non-serotinous, has a slightly longer juvenile period, thicker bark (making it more likely to survive fires), seedlings that are more shade tolerant, and less regular cone production. The second type conforms in some respects with *P. sylvestris*. This model provides a useful means for summarizing key aspects of the ecology of invasions of Mediterranean pines in the Southern Hemisphere.

Invasions were simulated in the three vegetation types most frequently invaded by pines in the Southern Hemisphere: fire-prone shrublands, montane grasslands, and temperate broad-leaved forests. Changing disturbance regimes were parameterized in terms of increased fire frequency (shrublands), increased grazing and interactions with fire regimes (grasslands), and changed patterns of treefall/logging (forests). The simulation study showed (as predicted from correlative evidence) that: (1) invasion rates increase when disturbance levels increase; (2) grasslands and fire-prone shrublands are more invasible than forest; and (3) the pioneer-pine is more invasive than the late-seral-pine. These generalizations do not hold for all situations. The two pine types differ substantially in their responses. For example, the pioneer-type invaded grassland and shrubland rapidly, and forest slowly; its invasiveness increased linearly with increasing disturbance in shrubland, asymptotically in grassland and unimodally in forest. The late-seral-pine shows a less spectacular, but more consistent, ability to invade under all conditions. Forest was most resistant to invasion by the late-seral pine. Invasiveness of this pine increased linearly with disturbance in forest and responded unimodally with disturbance in grassland. Disturbance of shrubland did not influence the invasion rate of the late-seral pine. These results emphasize the importance of interactions between life-history traits of invading pines and environmental features in shaping invasion dynamics. This model also has value for generating management recommendations. For example: pine invasions in temperate forests can be prevented by limiting tree felling levels; the invasion of montane grasslands can be controlled by promoting frequent and intense fires; invasions in shrublands can be slowed (to some extent) by maintaining infrequent and hence intense fires. Franzese and Raffaele (2017) reviewed the pivotal role of fire as a driver of pine invasions in the Southern Hemisphere.

Pine invasions have been particularly useful in shedding light on a fundamental question in invasion ecology, namely: what attributes make some plant species more invasive? Using a sample of 24 pine species (including four Mediterranean pines) that are widely cultivated outside their natural ranges, Rejmánek and Richardson (1996) derived a robust discriminant function that was consistently reliable in separating invasive from non-invasive pines in a separate sample. The three traits that were most important in separating invasive species were: mean seed mass, minimum juvenile period and the mean interval between large seed crops (see also Richardson et al. 1990; Rejmánek 1999 for discussion). In several cases, seed dispersal by vertebrates confounds predictions on invasiveness. For example, *P. pinea* is more invasive than expected in South Africa because introduced squirrels disperse its seeds. Similarly, invasion of *P. pinaster* in Western Australia is augmented by cockatoos, which carry seeds into fragmented natural woodlands (Stock et al. 2013). Further insights on the superior inherent capacity of species to invade are emerging from studies of the genome size of woody species, including pines. Small genome size (the result of selection for short generation time) has been shown to be a strong determinant of the invasiveness of *Pinus* species (Grotkopp et al. 2004).

5.6 Impacts of Invasive Pines in the Southern Hemisphere

Impacts of invading pines on natural and semi-natural systems are generated by the transformation of grassland or shrublands into pine woodlands or forests, involving a major shift in life-form dominance, reduced structural diversity, a massive increase in biomass (with implications for hydrology and fire behavior), and altered nutrient cycling (Richardson and Higgins 1998; Richardson and van Wilgen 2004). Impacts have been studied in the greatest detail in South African fynbos; invasions of non-native trees (including *P. pinaster*) in this region threaten many plant taxa with extinction (e.g., on the Cape Peninsula; Richardson et al. 1996), have reduced run-off from catchments by up to 70%, and cause major problems when intense fires in self-sown pine stands result in massive erosion due to heat-induced water repellency near the soil surface. Cost-benefit analyses show that the substantial financial investment required to clear invasive pines is warranted (van Wilgen et al. 1996).

In Argentinean Pampas grasslands, *P. halepensis* invasions have been shown to reduce native plant diversity by displacing endemic species and promote invasions by other non-native species (Zalba and Villamil 2002). Similarly, *P. sylvestris* has replaced native species in Andean *Nothofagus* forests in Argentina and led to a reduction in species richness of understory vascular plants, birds and epigeal beetles (Paritsis and Aizen 2008). Invasion by *P. nigra* in New Zealand grasslands has altered native invertebrate assemblages, with the impacts more pronounced in high-density stands, i.e. canopy cover > 50% (Pawson et al. 2010).

5.7 Management of Pine Invasions in the Southern Hemisphere

The control of invading pines has received considerable attention in South Africa and New Zealand. A systematic control strategy for pine invasion has evolved over the past 20 years in South African fynbos where major constraints are the rugged terrain which hampers access, and the need to integrate control plans with schedules for prescribed burning in the fire-prone vegetation (Roura-Pascual et al. 2009; van Wilgen et al. 2016). The slow fuel-accumulation rates mean that fynbos can seldom support fires at intervals of less than 5 years. Also, fire cycles of less than 10 years are usually detrimental to native shrubs. Most current management plans prescribe burns at intervals of between 12 and 15 years. Management of invasive pines involves felling trees 12–18 months before scheduled burns, allowing release of seeds from the serotinous cones (Fig. 5.1). This "cut, burn and follow-up" method is very expensive but is the only practical way of clearing stands of non-native pines and reducing impacts (van Wilgen et al. 1992). A landscape-scale implementation of the spatially explicit modeling approach described above has been used to compare the costs of different clearing strategies, the time to remove all the invading plants, and the impacts of the invaders on native biodiversity. This model highlighted the value of prioritizing low-density sites dominated by juvenile non-native plants (Higgins et al. 2000). Restoration of native plant communities is often feasible after pine control because abiotic variables remain relatively unchanged (Gaertner et al. 2012). Mechanical clearing of the pines is often the only action required to kick-start natural recovery of native plant communities. However, if the structural threshold has been crossed during invasion (changes in growth form and species composition) additional measures may be needed to ensure recovery of native plant communities, such as reintroduction of key native species (Fig. 5.2).

Control of invasive pines ("wildings") in New Zealand has focused on the unimproved grasslands in the high country of the South Island and on conservation land on the Central Volcanic Plateau. Given the huge extent and rapid expansion of the invasions, remote detection methods using manned aircraft and unmanned aerial vehicles have been developed and applied to identify and monitor pine invasions. These methods assist managers in prioritizing areas where pines are to be controlled and allocating appropriate resources for control (Dash et al. 2019; Sprague et al. 2019). Intensive grazing (sometimes combined with herbicide application) is effectively applied to control wildings in some areas (see Richardson et al. 1994 for details). Fire is seldom used because of the risks of soil erosion. Mechanical control involving cutting and hand-pulling was the standard for many years and is still widely used in easily accessible areas. However, rugged and inaccessible terrain is a significant barrier to wilding management in many areas. A novel method for aerial application of herbicide to the crowns of trees from a helicopter ("aerial spot application") has been developed and applied to control *P. nigra* and *P. sylvestris* (Fig. 5.1). This method overcomes challenges presented by rugged and inaccessible terrain, allows isolated trees to be easily targeted and large areas to be treated (Gous

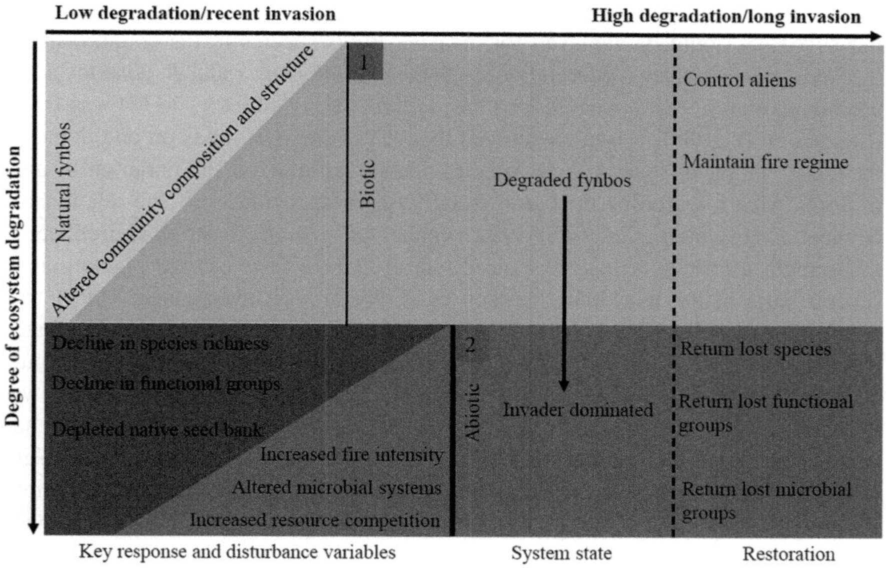

Fig. 5.2 Impacts of *Pinus* invasion on community dynamics and implications for restoration in South African fynbos vegetation, illustrating different ecosystem states and thresholds as invasion intensifies. Triangles represent invasion impacts by non-native *Pinus* and the disturbance variables that lead to each system state; the matrix represents the fynbos ecosystem and its response variables. Thresholds are indicated by the thick black lines. Appropriate management actions to restore the ecosystem are listed in the right-hand column. (Modified from Gaertner et al. (2012))

et al. 2015). This approach is currently under trial in South Africa with the hope of integrating its use with current approaches for managing the vast areas invaded by invasive pines (especially *P. pinaster*) in the Western and Eastern Cape provinces.

Physical removal, sometimes complemented by herbicide application, is the major control measure applied in Australia, although fire is used to control *P. pinaster* at French Island and at the Wonthaggi Heathland and Coastal Reserve in Victoria (Kermode 1993).

Efforts to control pines in South America are very limited and have mainly used mechanical methods. Cutting and hand pulling have been used to control *P. halepensis* in Argentinean Pampas grasslands (Zalba et al. 2008; Cuevas and Zalba 2009, 2010, 2013). Native vegetation recovery has been observed following mechanical control, suggesting that complementary restoration interventions are not needed. This is most likely because pine control was conducted early in the invasion process and native vegetation recovery was not seed-limited. These results demonstrate the capacity for natural recovery following pine control and the importance of early intervention to prevent changes that could be difficult to overcome later in the invasion process (Cuevas and Zalba 2010, 2013).

Biological control has not yet been implemented against invasive pines anywhere in the world. Following successful implementation of biological control using a range of agents on other invasive non-native tree species in South Africa (notably

Australian *Acacia* species; Moran et al. 2005), researchers were optimistic that similar successes could be achieved for invasive *Pinus* species in the country. Initial research focused on two Mediterranean pines: *P. halepensis* and *P. pinaster* and a promising cone-feeding weevil, *Pissodes validirostris* (Sahlberg, 1834) was found (Lennox et al. 2009). Research showed that *P. validirostris* feeds on and damages the leader shoots of several *Pinus* species that are naturalized and under cultivation in South Africa, including the commercially important *P. radiata*. The relationship between *P. validirostris* adults and pitch canker (*Fusarium circinatum* Nirenberg & O'Donnell, a pathogen that has caused major damage to *P. radiata* plantations in several parts of the world; Richardson et al. 2007) was investigated. Under trial conditions, weevils did not transmit the fungus but their feeding damage facilitated ingress of the fungus into *P. radiata* plants. This finding suggested that there could be unacceptable consequences for commercial forestry if *P. validirostris* was released for the biocontrol of other pines in South Africa. This risk was weighed against the potential benefits to be gained from potential reductions in damages caused by invasive pines and a decision was made to discontinue research on options for biological control of pines in South Africa for the foreseeable future (Hoffmann et al. 2011).

5.8 Conclusions

Non-native tree invasions are growing in importance as an environmental problem in many parts of the world and research on the phenomenon has escalated in the last two decades (Richardson et al. 2014). Pines are among the most widespread and damaging invasive trees globally (Richardson and Rejmánek 2011), and the *Pinus* genus has featured prominently in multi-faceted research on tree invasions (e.g. Richardson 2006; Essl et al. 2010; Procheş et al. 2012; Nuñez et al. 2017).

Several species of Mediterranean pines have become major weeds in Southern Hemisphere regions where they have been widely planted for over a century (and much longer in some cases). Some of these invasions have had severe impacts that diminish the net value of these pines and justify expensive interventions. Information on the types and magnitude of impacts is sketchy and more work is needed to provide objective assessments to inform management decisions in different regions and contexts. Problems with invasive pines are increasing as the total area and number of foci of pine plantings, the time since introduction, and the magnitude of human-induced disturbance to natural and semi-natural systems increase. Further invasions are inevitable in regions with recent widespread plantings of Mediterranean pines. Therefore, best practices should be used to reduce negative impacts of Mediterranean pines and their risk of escaping from plantation sites in such areas. Another factor that is exacerbating the problem is the increasing genetic diversity of most non-native pines in these areas as a result of adaptation (the establishment of landraces), intentional addition of new genotypes (new provenances introduced for specific purposes) and, in some cases, human-mediated manipulation of genotypes (notably through hybridization). These factors (separately and in concert) will result in more genotypes that are suitable for more Southern Hemisphere habitats, contributing to the substantial invasion debt for Mediterranean pines (sensu Rouget et al. 2016). Global changes (including climate change and altered disturbance regimes) are enhancing conditions for invading pines. For example, increasing ignition events due to human activities in South Africa's Cape Floristic Region (Slingsby et al. 2020) is enhancing opportunities for recruitment, persistence and spread of non-native trees, including *P. halepensis* and *P. pinaster* in fynbos shrubs. Native shrubs are unable to recruit and are being eliminated by repeated short intervals between fires, while non-native trees such as pines can persist and flourish due to their short juvenile periods and capacity to recolonize via seed dispersal by wind (Richardson and Cowling 1992).

Increasing emphasis is being placed on developing means for the objective assessment of the costs associated with non-native species that also have clear benefits; such species generate complex conflicts of interest that complicate management (van Wilgen and Richardson 2012, 2014). Invasive Mediterranean pines provide good case studies for the development of such methodologies (Le Maitre et al. 1996; van Wilgen et al. 1996; van Wilgen and Richardson 2012, 2014).

References

Abbas H, Barbéro M, Loisel R (1984) Réflexions sur le dynamisme actuel de la régéneration naturelle du pin d' Alep (*Pinus halepensis* Mill.) dans les pinèdes incendiéés en Provence Calcaire (de 1973 à 1979). Ecol Mediterr 10:85–104

Achérar M, Lepart J, Debusschc M (1984) La colonisation des friches par le pin d' Alep (*Pinus halepensis* Miller) en Languedoc méditerranéen. Acta Oecol Oecologia Plant 5:179–189

Barbéro M, Loisel R, Quézel P, Richardson DM, Romane F (1998) Pines of the Mediterranean Basin. In: Richardson DM (ed) Ecology and biogeography of Pinus. Cambridge University Press, Cambridge, pp 153–170

Baret S, Rouget M, Richardson DM et al (2006) Current distribution and potential extent of the most invasive alien plant species on La Réunion (Indian Ocean, Mascarene Islands). Austral Ecol 31:747–758

Bellingham PJ, Duncan RP, Lee WG et al (2004) Seedling growth rate and survival do not predict invasiveness in naturalised woody plants in New Zealand. Oikos 106:308–316

Belton MC, Ledgard NJ (1991) A study of the spread of exotic trees in Canterbury high country. Review 13, New Zealand Mountainlands Institute, Lincoln University, Canterbury

Booth TH, Saunders JC (1984) A catalogue of tree species evaluated in field trials in Australia. CSIRO Division of Water and Land Resource, Divisional Report 84/2, CSIRO, Canberra

Buckley YM, Brockerhoff E, Langer L et al (2005) Slowing down a pine invasion despite uncertainty in demography and dispersal. J Appl Ecol 42:1020–1030

Chauchard S, Pille G, Carcaillet C (2006) Large herbivores control the invasive potential of nonnative Austrian black pine in a mixed deciduous Mediterranean forest. Can J For Res 36:1047–1053

Coutts SR, Caplat P, Cousins K et al (2012) Reproductive ecology of *Pinus nigra* in an invasive population: individual- and population-level variation in seed production and timing of seed release. Ann For Sci 69:467–476

Cuevas YA, Zalba SM (2009) Control de pinos invasores en el Parque Provincial Ernesto Tornquist (Buenos Aires): áreas prioritarias y análisis de costos. BioScriba 2:76–89

Cuevas YA, Zalba SM (2010) Recovery of native grasslands after removing invasive pines. Restor Ecol 18:711–719

Cuevas YA, Zalba SM (2013) Efecto del tipo de corte y de tratamientos en el mantillo para la restauraci6n de pastizales naturales invadidos por Pinus halepensis. Bol Soc Argent Bot 48:315–329

Dash JP, Watt MS, Paul TS et al (2019) Early detection of invasive exotic trees using UAV and manned aircraft multispectral and LiDAR Data. Remote Sens 11:1812

Essl F, Moser D, Dullinger S et al (2010) Selection for commercial forestry determines global patterns of alien conifer invasions. Divers Distrib 16:911–921

Franzese J, Raffaele E (2017) Fire as a driver of pine invasions in the southern hemisphere: a review. Biol Invasions 19:2237–2246

Froude VA (2011) Wilding conifers in New Zealand: status report. Ministry of Agriculture and Forestry, Auckland, p 206

Gaertner M, Holmes PM, Richardson DM (2012) Biological invasions, resilience and restoration. In: Andel J, Aronson J (eds) Restoration ecology: the new frontier. Wiley-Blackwell, Oxford, pp 265–280

Gous S, Raal P, Kimberley MO, Watt MS (2015) Chemical control of isolated invasive conifers using a novel aerial spot application method. Weed Res 55:380–386

Grotkopp E, Rejmánek SMJ, Rost TL (2004) Evolution of genome size in pines (*Pinus*) and its life history correlates: supertree analyses. Evolution 58:1705–1729

Higgins SI, Richardson DM (1998) Pine invasions in the Southern hemisphere: modelling interactions between organism, environment and disturbance. Plant Ecol 135:79–93

Higgins SI, Richardson DM, Cowling RM (2000) Using a dynamic landscape model for planning the management of alien plant invasions. Ecol Appl 10:1833–1848

Higgins SI, Richardson DM, Cowling RM (2001) Validation of a spatial simulation model of a spreading alien plant population. J Appl Ecol 38:571–584

Hodgkin SE (1984) Scrub encroachment and its effects on soil fertility on Newborough Warren, Anglesey, Wales. Biol Conserv 29:99–119

Hoffmann JH, Moran VC, van Wilgen BW (2011) Prospects for the biological control of invasive Pinus species (Pinaceae) in South Africa. Afr Entomol 19:393–401

Hosking JR, Conn BJ, Lepschi BJ et al (2007) Plant species first recognised as naturalised for New South Wales in 2002 and 2003, with additional comments on species recognised as naturalised in 2000–2001. Cunninghamia 10:139–166

Hunter GS, Douglas MH (1984) Spread of exotic conifers on South Island rangelands. N Z J For 29:78–96

Huntley B, Birks HJB (1983) An atlas of past and present pollen maps for Europe: 0–13000 years ago. Cambridge University Press, Cambridge

Keeley JE, Zedler PH (1998) Evolution of life histories in *Pinus*. In: Richardson DM (ed) Ecology and biogeography of *Pinus*. Cambridge University Press, Cambridge, pp 219–251

Kermode L (1993) The invasion of heath vegetation by *Pinus pinaster* at French Island and Wonthaggi, Victoria. Dissertation, La Trobe University

Kruger FJ (1977) Invasive woody plants in the Cape fynbos with special reference to the biology and control of *Pinus pinaster*. Proceedings of the Second National Weeds Conference of South Africa. Balkema, Cape Town, pp 57–74

Le Maitre DC (1998) Pines in cultivation: a global view. In: Richardson DM (ed) Ecology and biogeography of Pinus. Cambridge University Press, Cambridge, pp 407–431

Le Maitre DC, van Wilgen BW, Chapman RA et al (1996) Invasive plants and water resources in the Western Cape Province, South Africa: modelling the consequences of a lack of management. J Appl Ecol 33:161–172

Ledgard NJ (1988) The spread of introduced trees in New Zealand's rangelands – South Island High Country Experience. Rev Tussock Grassl Mt Lands Inst 44:1–7

Ledgard NJ, Belton MC (1985) Exotic trees in the Canterbury High Country. N Z J For Sci 15:298–323

Leege LM, Murphy PG (2000) Growth of the non-native *Pinus nigra* in four habitats on the sand dunes of Lake Michigan. For Ecol Manag 126:191–200

Lennox CL, Hoffmann JH, Coutinho TA et al (2009) A threat of exacerbating the spread of pitch canker precludes further consideration of a cone weevil, *Pissodes validirostris*, for biological control of invasive pines in South Africa. Biol Control 50:179–184

Marinich A, Powell K (2017) Scots pine (*Pinus sylvestris* L.) best management practices in Ontario. Ontario Invasive Plant Council, Peterborough

McClintock E, Reeberg P, Knight W (1990) A Flora of the San Bruno Mountains. California Native Plant Society, Sacramento

McConnachie MM, Richardson DM, van Wilgen BW et al (2015) Estimating the effect of plantations on pine invasions in protected areas: a case study from South Africa. J Appl Ecol 52:110–118

McGregor KF, Watt MS, Hulme PE et al (2012) How robust is the Australian weed risk assessment protocol ? A test using pine invasions in the Northern and Southern hemispheres. Biol Invasions 14:987–998

Moran VC, Hoffmann JH, Zimmermann HG (2005) Biological control of invasive alien plants in South Africa: necessity, circumspection, and success. Front Ecol Environ 3:71–77

Mortenson SG, Mack RN (2006) The fate of alien conifers in long-term plantings in the USA. Divers Distrib 12:456–466

National Park Service (2019) Terrestrial invasive plants. https://www.nps.gov/chis/learn/nature/terrestrial-invasive.htm. Accessed 3 Dec 2019

Nuñez MA, Chiuffo M, Torres A et al (2017) Ecology and management of invasive Pinaceae around the world: progress and challenges. Biol Invasions 19:3099–3120

Oppenheimer HL (2002) The spread of gymnosperms on Maui: a neglected element of the Modern Hawaiian Flora. Bishop Mus Occas Pap 68:19–23

Orians GH (1986) Site characteristics favoring invasions. In: Mooney HA, Drake JA (eds) Ecology of biological invasions of North America and Hawaii. Springer, New York, pp 133–148

Paritsis J, Aizen MA (2008) Effects of exotic conifer plantations on the biodiversity of understory plants, epigeal beetles and birds in *Nothofagus dombeyi* forests. For Ecol Manag 255:1575–1583

Pawson SM, McCarthy JK, Ledgard NJ et al (2010) Density-dependent impacts of exotic conifer invasion on grassland invertebrate assemblages. J Appl Ecol 47:1053–1062

Poynton RJ (1979) Tree planting in Southern Africa, volume 1, The pines. Department of Forestry, Pretoria

Procheş Ş, Wilson JRU, Richardson DM et al (2012) Native and naturalised range size in *Pinus*: relative importance of biogeography, introduction effort and species traits. Glob Ecol Biogeogr 21:513–523

Pyšek P, Richardson DM, Rejmánek M, Webster GL, Williamson M, Kirschner J (2004) Alien plants in checklists and floras: towards better communication between taxonomists and ecologists. Taxon 53:131–143

Rejmánek M (1999) Invasive plant species and invasible ecosystems. In: Sandlund QT, Schei PJ, Viken A (eds) Invasive species and biodiversity management. Kluwer, Dordrecht, pp 79–102

Rejmánek M, Richardson DM (1996) What attributes make some plant species more invasive? Ecology 77:1655–1661

Rejmánek M, Richardson DM (2013) Trees and shrubs as invasive alien species–2013 update of the global database. Divers Distrib 19:1093–1094

Rejmánek M, Richardson DM, Higgins SI et al (2005) Ecology of invasive plants: state of the art. In: Mooney HA, McNeely JA, Nevile N, Schei PJ, Waage JP (eds) Invasive alien species: a new synthesis. Island Press, Washington, DC, pp 104–161

Richardson DM (1988) Age structure and regeneration after fire in a self-sown *Pinus halepensis* forest on the Cape Peninsula, South Africa. S Afr J Bot 54:140–144

Richardson DM (2006) *Pinus*: a model group for unlocking the secrets of alien plant invasions? Preslia 78:375–388
Richardson DM (2011) Forestry and agroforestry. In: Simberloff D, Rejmánek M (eds) Encyclopedia of biological invasions. University of California Press, Berkeley, pp 241–248
Richardson DM, Bond WJ (1991) Determinants of plant distribution: evidence from pine invasions. Am Nat 137:639–668
Richardson DM, Cowling RM (1992) Why is mountain fynbos invasible and which species invade? In: Van Wilgen BW, Richardson DM, Kruger FJ et al (eds) Fire in South African mountain fynbos. Springer, Berlin, pp 161–181
Richardson DM, Cowling RM (1994) The ecology of invasive pines (*Pinus* spp.) in the Jonkershoek Valley, Stellenbosch, South Africa. Bontebok 9:1–14
Richardson DM, Higgins SI (1998) Pines as invaders in the southern hemisphere. In: Richardson DM (ed) Ecology and biogeography of *Pinus*. Cambridge University Press, Cambridge, pp 450–473
Richardson DM, Petit R (2005) Pines as invasive aliens: outlook on transgenic pine plantations in the Southern Hemisphere. In: Williams CG (ed) Landscapes, genomics and transgenic conifers. Springer, Dordrecht, pp 169–188
Richardson DM, Rejmánek M (2004) Invasive conifers: a global survey and predictive framework. Divers Distrib 10:321–331
Richardson DM, Rejmánek M (2011) Trees and shrubs as invasive alien species–a global review. Divers Distrib 17:788–809
Richardson DM, van Wilgen BW (2004) Invasive alien plants in South Africa: how well do we understand the ecological impacts? S Afr J Sci 100:45–52
Richardson DM, Cowling RM, Le Maitre DC (1990) Assessing the risk of invasive success in *Pinus* and *Banksia* in south African Mountain fynbos. J Veg Sci 1:629–642
Richardson DM, Williams PA, Hobbs RJ (1994) Pine invasions in the Southern Hemisphere: determinants of spread and invadability. J Biogeogr 21:511–527
Richardson DM, van Wilgen BW, Higgins SI et al (1996) Current and future threats to biodiversity on the Cape Peninsula. Biodivers Conserv 5:607–647
Richardson DM, Pyšek P, Rejmánek M et al (2000) Naturalization and invasion of alien plants: concepts and definitions. Divers Distrib 6:93–107
Richardson DM, Rundel PW, Jackson ST et al (2007) Human impacts in pine forests: past, present and future. Annu Rev Ecol Evol Syst 38:275–297
Richardson DM, van Wilgen BW, Nuñez MA (2008) Alien conifer invasions in South America: short fuse burning? Biol Invasions 10:573–577
Richardson DM, Hui C, Nuñez M et al (2014) Tree invasions: patterns, processes, challenges and opportunities. Biol Invasions 16:473–481
Rouget M, Richardson DM (2003) Inferring process from pattern in alien plant invasions: a semimechanistic model incorporating propagule pressure and environmental factors. Am Nat 162:713–724
Rouget M, Richardson DM, Milton SJ et al (2001) Invasion dynamics of four alien *Pinus* species in a highly fragmented semi-arid shrubland in South Africa. Plant Ecol 152:79–92
Rouget M, Robertson MP, Wilson JRU et al (2016) Invasion debt—quantifying future biological invasions. Divers Distrib 22:445–456
Roura-Pascual N, Richardson DM, Krug RM et al (2009) Ecology and management of alien plant invasions in South African fynbos: accommodating key complexities in objective decision making. Biol Conserv 142:1595–1604
Salim K, Naydenov KD, Benyounes H et al (2010) Genetic signals of ancient decline in Aleppo pine populations at the species' southwestern margins in the Mediterranean Basin. Hereditas 147:165–175
Selvi F, Carrari E, Coppi A (2016) Impact of pine invasion on the taxonomic and phylogenetic diversity of a relict Mediterranean forest ecosystem. For Ecol Manag 367:1–11

Shaughnessy G (1986) A case study of some woody plant introductions to the Cape Town area. In: Macdonald IAW, Kruger FJ, Ferrar AA (eds) Ecology and Management of Biological Invasions in Southern Africa. Oxford University Press, Cape Town, pp 37–43
Simberloff D, Nuñez M, Ledgard NJ et al (2010) Spread and impact of introduced conifers in South America: lessons from other southern hemisphere regions. Austral Ecol 35:489–504
Singh SP, Inderjit SJS et al (2018) Insights on the persistence of pines (*Pinus* species) in the late Cretaceous and their increasing dominance in the Anthropocene. Ecol Evol 8:10345–10359
Slingsby JA, Moncrieff GR, Rogers AJ et al (2020) Altered ignition catchments threaten a hyperdiverse fire-dependent ecosystem. Glob Change Biol 26:616–628
Sprague R, Godsoe W, Hulme PE (2019) Assessing the utility of aerial imagery to quantify the density, age structure and spatial pattern of alien conifer invasions. Biol Invasions 21:2095–2106
Stock WD, Finn H, Parker J et al (2013) Pine as fast food: foraging ecology of an endangered cockatoo in a forestry landscape. PLoS One 8:e61145
Van Etten EJB, Belen CA, Calvino-Cancela M (2020) Invasion patterns of *Pinus pinaster* in south-west Australia in relation to fire, vegetation type and plantation management. For Ecol Manag 463:118042
van Wilgen BW, Richardson DM (2012) Three centuries of managing introduced conifers in South Africa: benefits, impacts, changing perceptions and conflict resolution. J Environ Manag 106:56–68
van Wilgen BW, Richardson DM (2014) Managing invasive alien trees: challenges and trade-offs. Biol Invasions 16:721–734
van Wilgen BW, Siegfried WR (1986) Seed dispersal properties of three pine species as a determinant of invasive potential. S Afr J Bot 52:546–548
van Wilgen BW, Bond WJ, Richardson DM (1992) Ecosystem management. In: Cowling RM (ed) The ecology of fynbos: nutrients, fire and diversity. Oxford University Press, Cape Town, pp 345–371
van Wilgen BW, Cowling RM, Burgers CJ (1996) Valuation of ecosystem services. A case study from South African fynbos ecosystems. BioScience 46:184–189
van Wilgen BW, Carruthers J, Cowling RM et al (2016) Ecological research and conservation management in the Cape Floristic Region between 1945 and 2015: history, current understanding and future challenges. Trans R Soc S Afr 71:207–303
Zalba SM, Villamil CB (2002) Woody plant invasion in relictual grasslands. Biol Invasions 4:55–72
Zalba SM, Cuevas YA, Boo RM (2008) Invasion of *Pinus halepensis* Mill. following a wildfire in an Argentine grassland nature reserve. J Environ Manag 88:539–546

Part II
Ecophysiology of Mediterranean Pines: Resistance to Stress and Perturbation

Tamir Klein

Chapter 6
What Is a Mediterranean Pine? Physiological Trade-Offs Under Stress and Perturbation

Víctor Resco de Dios

6.1 Introduction

In this chapter we will try to understand the physiological traits that define a Mediterranean pine and how they differ from those present in other conifers, as well as in other co-occurring Mediterranean trees (oaks). We will focus particularly on traits related to shade, drought and fire survival and test to what extent they can help us understand the distribution of Mediterranean conifers across productivity gradients. We will begin by exploring how Mediterranean conifers vary in terms of shade, drought and fire tolerance. We will then discuss in greater depth the key attributes explaining drought tolerance, photosynthesis and growth rates, as well as some key fire-related response parameters. Finally, we will integrate and examine whether physiological responses to stress (shade, drought) and perturbation (fire) are coordinated or whether some trade-offs occur. We conclude that the distribution of Mediterranean conifers across productivity gradients in the Mediterranean Basin may be, at least partly, driven by such trade-offs. We hope that this chapter serves as a general but integrated introduction to the ecophysiology of Mediterranean pines, which will be developed in more detail in subsequent book chapters.

V. Resco de Dios (✉)
School of Life Science and Engineering, Southwest University of Science and Technology, Mianyang, China

Joint Research Unit CTFC-Agrotecnio, Universitat de Lleida, Lleida, Spain
e-mail: victor.resco@udl.cat

G. Ne'eman, Y. Osem (eds.), *Pines and Their Mixed Forest Ecosystems in the Mediterranean Basin*, Managing Forest Ecosystems 38,
https://doi.org/10.1007/978-3-030-63625-8_6

6.2 Shade and Drought and Fire Tolerances

Pine species are prevalent across much of the Mediterranean Basin, from coastal to mountain environments. They thus demonstrate a variety of adaptations to survive under such a wide array of environmental conditions. The first visually apparent difference is canopy architecture (Fig. 6.1). We find that the degree of epinastic control diminishes with aridity, such that species from drier environments show more open and round crowns than those from wetter and colder environments. Another important canopy trait that varies with habitat productivity is the degree of self-thinning. Species from drier environments show a lower degree of self-thinning than those from colder and wetter environments.

These differences arise from the different constraints on productivity and, potentially, also on reproduction that exist among these environments. Water limitation obviously prevails in drylands, whereas light limitation is more prevalent in mesic environments (Stephenson 1990). Photosynthetic optimization under water scarcity thus relies on having more open canopies that enhance light penetration, whereas the opposite (tight epinastic control that favors overtopping the neighbors) is true for mesic or montane regions.

In addition to light and water, perturbation, particularly fire, is another factor that interacts with canopy architecture. A counter-intuitive feature is that species from drier areas, where fire activity is more common, often show a lower degree of

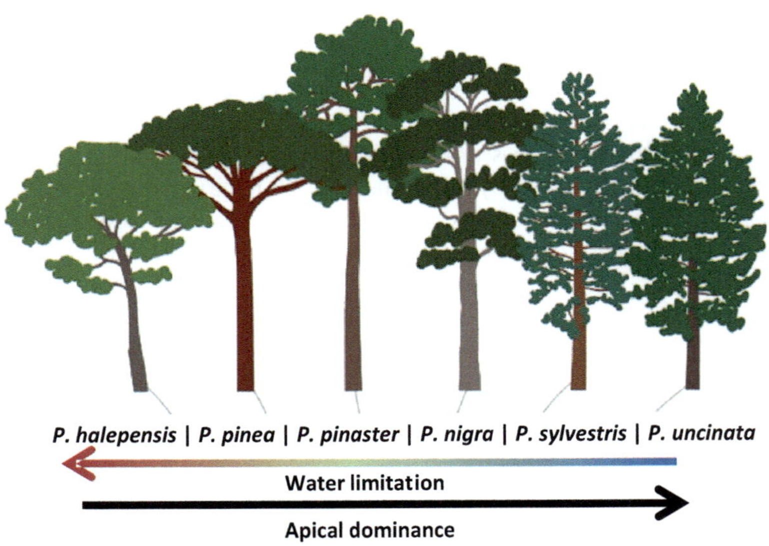

Fig. 6.1 The degree of apical dominance in Western European pines increases as the degree of water limitation decreases. Note that *Pinus pinea* L. may have lower apical dominance than *P. halepensis* Mill. although the latter lives in environments with a higher degree of water limitation. (Modified from J. Luis Ordóñez (2019, CREAF))

self-pruning than those from wetter areas. This is counter-intuitive because lower self-pruning increases vertical fuel continuity (fuel ladder) and, consequently, increases the risk of canopy fire. Self-pruning results from branch shedding, which leads to a negative carbon balance (higher respiratory costs than photosynthetic gains) and eventually to leaf and branch senescence. The fact that species with a lower degree of self-pruning occur in drier areas likely results from having more open canopies, which allows higher light penetration to lower branches, and, consequently, a positive carbon balance in the lower branches.

Despite the common trade-off between shade and drought tolerance, we note that Mediterranean pines are primarily light-demanding species, with a limited degree of shade tolerance. A study assessing shade tolerance as a trait varying from 0 (low) to 5 (high) based on expert judgment observed that gymnosperms are predominantly shade-intolerant species (Niinemets and Valladares 2006) (Fig. 6.2) and that Mediterranean pines are no exception, as their shade tolerances ranges from 1.3 to 2.1 (with the exception of *Pinus cembra*[1] L. which is a more shade-tolerant species with a value of 2.87). This study also assessed drought tolerance and contrastingly observed that while gymnosperms generally show limited drought resistance, Mediterranean pines tend towards drought tolerance, with values between 3.9 and 5, again with the exception of the subalpine *P. cembra*, which had a value of 3.

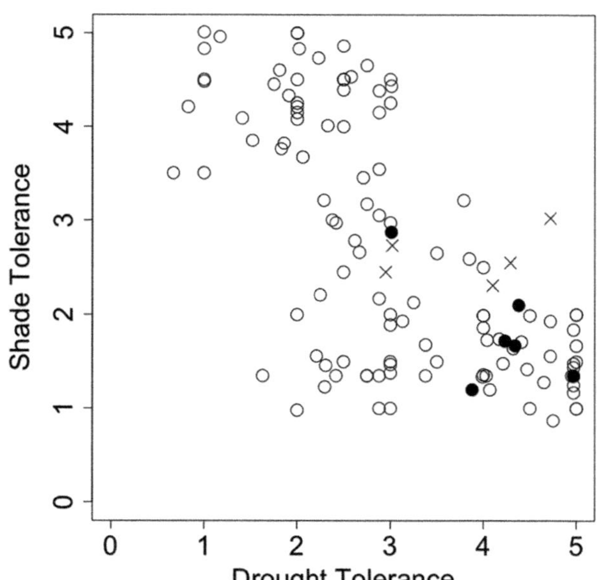

Fig. 6.2 Variation in shade and drought tolerances in conifers (open dots), Mediterranean conifers (filled dots) and Mediterranean oaks (crosses). (Data from Niinemets and Valladares (2006))

[1] Note that some authors would not consider *P. cembra* as a Mediterranean species. It is included as it occurs within the Mediterranean basin

If we compare these values with those from co-occurring oak species, we can observe that the latter are more shade tolerant (values between 2.3 and 3.0) but similarly drought tolerant (values between 2.95 and 4.72). The inference that one may draw from this simple exercise is of course limited, but one may raise some hypotheses to be tested: pines are more frugal species that occur earlier in succession (lower shade tolerance) and both species will suffer climate change-induced drought to a similar extent (similar drought tolerance). However, one needs to examine more fully the mechanisms underlying these interactions before making such statements.

6.3 Drought Resistance in Mediterranean Pines

6.3.1 Cavitation Resistance

One of the most common indicators of drought resistance is the capacity to resist cavitation. As drought advances, a phase change in xylem sap, from liquid to vapor, occurs. This phenomenon, termed cavitation, results in a decrease in the plant's capacity to transport water. For conifers, a particularly important threshold is the point at which 50–80% of the hydraulic conductivity is lost (Brodribb and Cochard 2009; Hammond et al. 2019). This is because conifers are usually unable to recover their cavitated tracheids, they often do not resprout, and mortality driven by dehydration starts to occur from that point onwards.

We can compare the resistance to cavitation in Mediterranean pines, relative to other species, by examining variation in global datasets. Figure 6.3 indicates the relationship between the minimum pressure potential observed in the field (Ψ_{min}) and the pressure potential where 50% of the hydraulic conductivity is lost (Ψ_{50}). Plants usually operate above the 1:1 line in this relationship and the difference between Ψ_{min} and Ψ_{50} is known as the safety margin. That is, if a plant is operating in the field at a value of water potential (Ψ) lower than Ψ_{50}, it would lose more than 50% of its capacity to transport water and, consequently, its probability for dehydration mortality increases.

There are a few lessons that can be learnt from this exercise. The first is that Mediterranean pines are not particularly resistant to cavitation, at least relative to other conifers. That is, the range of Ψ_{50} values measured for conifers ranges from −14.1 MPa to −1.4 MPa, and the median value is −4.4 MPa. However, Ψ_{50} in Mediterranean pines only ranges from −5.0 to −2.8 MPa, with a median value of −3.6 MPa. This observation is surprising at first because conifers are widely distributed throughout the world, from cold and wet ecosystems such as boreal or montane tropical forests, to dry and hot environments. Mediterranean environments are often regarded as drought-prone ecosystems and one could thus expect a higher resistance to cavitation for trees in these environments. However, Ψ_{50} of Mediterranean pines shows similar values to those of species growing in colder environments, such as

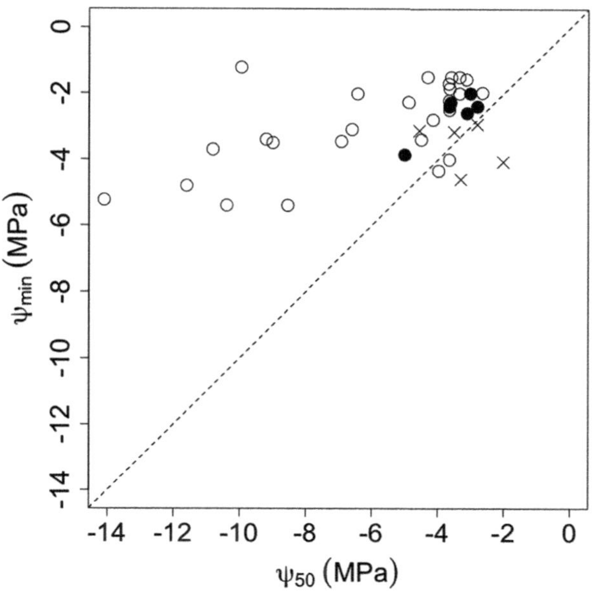

Fig. 6.3 Variation in minimum pressure potential and the pressure potential creating a 50% loss of hydraulic conductivity in conifers (open dots), Mediterranean conifers (filled dots) and Mediterranean oaks (crosses). The dashed line indicates the 1:1 relationship. (Data from (Choat et al. 2012))

spruce or fir, but higher than desert species such as cypress and juniper, which have the highest cavitation resistance.

Another interesting observation is that Mediterranean pines operate at smaller safety margins than other conifers. That is, the slope of the relationship between Ψ_{min} and Ψ_{50} in Mediterranean pines (0.71 ± 0.2, mean ± SE) is steeper than that in the global dataset (0.25 ± 0.05). This indicates that Mediterranean pines have a less conservative water use strategy than other conifers. This could result from differences in stomatal behavior between Mediterranean pines and the junipers that show higher resistance to cavitation and underlie the larger safety margin in the global dataset. Pines are isohydric species, meaning that they regulate their stomata to maintain relatively constant water potential. However, junipers are anisohydric species that do not seek homeostasis in water potential and, consequently, they would need to operate under wider safety margins.

A comparison between Mediterranean pines and coexisting oaks also reveals interesting patterns. The first is that there are no significant differences in the ranges of cavitation that they can tolerate. The range of Ψ_{50} in oaks, from −2.0 MPa to −7.0 MPa with a median value of −3.4 MPa, is similar to that observed in Mediterranean pines. There is however a difference in the safety margin under which they operate, which is larger in conifers than in hardwoods. The latter operate near the value of Ψ_{50}. This difference might be explained by the fact that while critical cavitation thresholds for conifers occur at 50% loss of hydraulic conductivity, the critical mortality threshold in hardwoods occurs at 80% loss of hydraulic conductivity (Resco et al. 2009). This implies that, although there are no differences in Ψ_{50}, hardwoods are effectively more resistant to cavitation. However, it should be noted that oaks are ring porous species. That is, functional xylem will often be limited to the outer ring, whereas several years will be functional in pines.

6.3.2 Water Use and Rooting Patterns

In addition to xylem structural properties, morphological features, such as root length, also affect drought resistance. In fact, studies comparing Mediterranean pine species (*P. halepensis*) with oaks (*Quercus ilex* L.) have found that growth in the former is more sensitive to precipitation than in the latter (Ferrio et al. 2003). This could be expected because Mediterranean pine species often lack deep tap roots, which are characteristic of Mediterranean oaks and which confer them access to groundwater (del Castillo et al. 2016).

The pattern of water availability at least in the Western Mediterranean Basin is characterized by a peak in late spring and early summer, a decline as the summer advances, and a small recovery in early fall. There is a certain degree of plasticity in water uptake in pine roots to accommodate such a temporal pattern in water availability. Water uptake shifts to deeper water sources as the summer drought advances (del Castillo et al. 2016), and biomass increases during drought are restricted to root growth (Klein et al. 2011). However, transpiration is often inhibited during the drier part of the summer season, consistent with an isohydric response strategy whereby leaf water potential is maintained constant (Klein et al. 2011). Furthermore, Mediterranean pines also have the capacity for osmotic adjustments (Calamassi et al. 2001; López et al. 2009), indicating that leaves will accumulate more solutes as water stress increases to avoid plasmolysis.

6.4 Photosynthesis and Growth Rates

The relative growth rate of Mediterranean pines is lower than that of coexisting oak species (Fig. 6.4a). This can be explained by the higher photosynthetic rates and nitrogen use efficiencies of Mediterranean oaks compared to pines (Fig. 6.4b). However, Mediterranean pines tend to have higher photosynthetic rates than other conifers existing worldwide, and they also demonstrate a higher slope of the photosynthesis vs nitrogen relationship, indicating higher nitrogen use efficiency.

The lower photosynthetic rates and, consequently, growth rates of Mediterranean pines may result from differences in leaf longevity. Higher leaf life spans often show lower photosynthetic rates. Pine needles last for at least three growing seasons. Leaf longevity in the evergreen *Q. ilex* is limited to 2 years, and it is obviously shorter in the deciduous oaks.

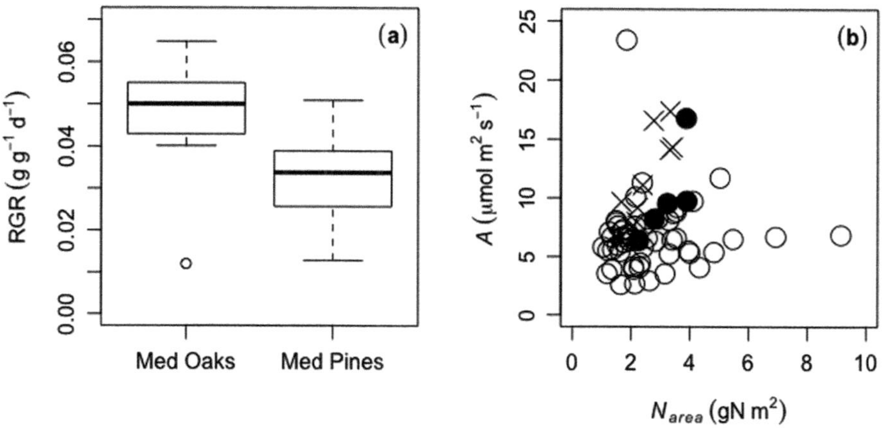

Fig. 6.4 (**a**) Differences in relative growth rate (RGR) between Mediterranean oaks and pines. (**b**) Relationship between photosynthesis (A) and nitrogen (N_{area}) for conifers (open dots), Mediterranean pines (filled dots) and Mediterranean oaks (crosses). (Data from the TRY database (Kattge et al. 2020))

Growth in Mediterranean (i.e. xeric) species is often sensitive to precipitation and characterized by a bimodal pattern, with peaks in spring and fall (Camarero et al. 2010). However, the growth pattern in mesic species, where water is not limiting to such a large degree, is driven more by photoperiod limitations (Camarero et al. 2010).

Understanding the processes underlying photoperiod limitation is currently the focus of a very active field of research. Photoperiod has an effect on both leaf phenology and leaf gas exchange (Way and Montgomery 2015; Granda et al. 2020). Furthermore, photoperiod could also control the timing of cambial expansion, but the underlying mechanisms are still not fully resolved (Cuny et al. 2015).

6.5 The Physiology of Pine-Fire Interactions

Pine-fire interactions are a two-way street. On the one hand, differences in pine traits may affect fire behavior and, on the other, fire may also shape pine traits. The topic of pine adaptations to fire will be covered by Ne'eman and Arianoutsou (this volume, Chap. 22); however, some physiological responses will be addressed here. In particular we will explore the strategies linking growth, drought and fire.

One of the major traits affecting fire behavior, which results from the interactions between growth and drought, is leaf moisture. Leaf moisture, expressed on a dry matter basis such as in fuel moisture calculations, averages 106% in Mediterranean pine species (Yebra et al. 2019). If we examine the temporal variation in this trait we observe that it shows very minor changes over a season (Fig. 6.5). That is, leaf

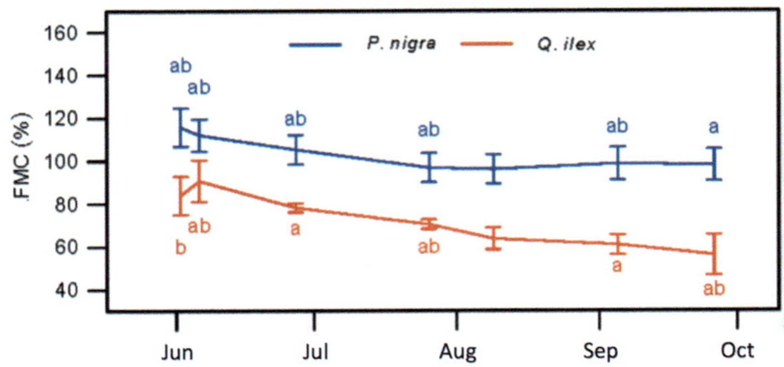

Fig. 6.5 Patterns of foliage moisture content (FMC) in *Pinus nigra* J.F.Arnold and *Quercus ilex* remain relatively stable throughout a fire season. (Redrawn with permission from (Nolan et al. 2018))

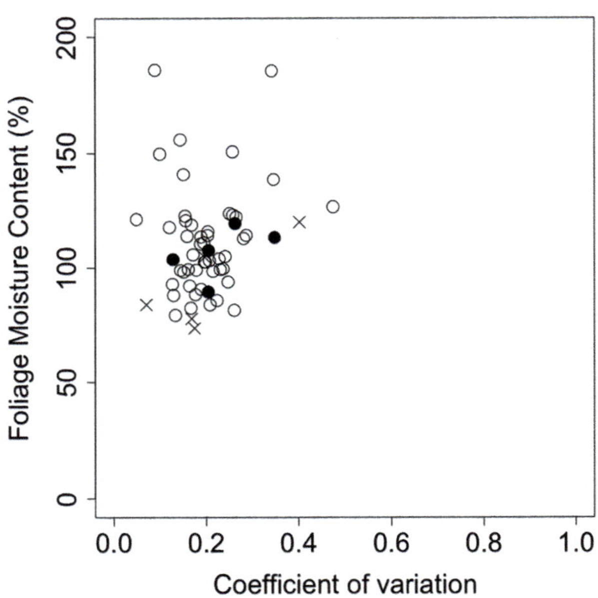

Fig. 6.6 Variation in foliage moisture content in conifers (open dots), Mediterranean pines (filled dots) and Mediterranean oaks (crosses). (Data from (Yebra et al. 2019))

moisture content remains stable as water deficit increases during the growing season because of the drought adaptations previously discussed (Sect. 6.3).

Leaf moisture in Mediterranean pines ranges from 89% to 119% (Yebra et al. 2019), which is towards the lower end of the leaf moisture spectrum in conifers (79–186%) (Fig. 6.6). The highest moisture values recorded in conifers are in *Abies grandis* (Douglas ex D.Don) Lindley (185%) and *Larix occidentalis* Nutt. (186%). Other pine species with high leaf moisture values include *P. elliotti* Engelm. (150%)

and *P. virginiana* Mill. (150%). The lower leaf moisture exhibited by Mediterranean pines indicates that they are more flammable (at a leaf scale) than other conifers.

It is interesting to note that leaf moisture in Mediterranean oaks (92% on average) is lower than in Mediterranean pines. Leaf moisture can be as low as 73% for *Q. ilex*, and increase up to 120% in *Q. faginea* Lam. (Yebra et al. 2019). Lower leaf moisture values in oaks indicate that they will have higher flammability at the leaf scale.

Leaf moisture varies with age, and young leaves can have much higher water contents than older leaves. Furthermore, a common strategy in Mediterranean pines for coping with increasing drought is to shed older leaves. In *P. halepensis*, for instance, needles can live for up to 3 years. As drought increases in the summer, some populations will drop their leaves. This often happens towards late June and early July. Interestingly, the first half of July is the period of the year with the highest wildfire activity in the Western Mediterranean Basin. It is thus tempting to hypothesize that increased fire activity in early summer results from drought-induced leaf shedding, as it increases the likelihood of crown fire (Resco de Dios 2020). However, this hypothesis is yet to be tested.

6.6 Physiological Adaptations to Stress and Disturbance

There are long-standing ecological questions on whether adaptations to stress and perturbation are antagonistic or complementary (Resco de Dios 2020; Grime 1977). In some instances, we may discover that physiological mechanisms protecting against stress also confer protection against disturbance.

This is for instance the case regarding resistance to cavitation. Fire may act as a very quick and intense drought, along with a heat wave, with vapor pressure deficit reaching values of 270 kPa or more (Kavanagh et al. 2010). The primary cause underlying fire-induced mortality in non-resprouting species, such as Mediterranean pines, is defoliation and bud consumption. Additional mechanisms include cambium charring and xylem cavitation (Karavani et al. 2018; Rundel 1973; Midgley et al. 2011) as well as root damage (Hood et al. 2018). Mechanisms enhancing resistance to hydraulic failure should thus also favor fire survival.

Hydraulic failure during a fire may occur via either embolism or cell wall deformation (Bar et al. 2019; Resco de Dios 2020). Such hydraulic failure may create long-lasting legacies on tree performance. That is, conifers that have survived a fire may become more vulnerable to future stresses if their hydraulic system was severely damaged during the fire. Under this scenario, adaptations to drought and fire are complementary.

6.7 Trade-Offs in Adaptations to Stress and Disturbance

In many instances, however, we observe that there are trade-offs in the adaptation to stress and disturbance. Perhaps the most clear case is the allocation trade-off that occurs between the bark and the stem (Resco de Dios et al. 2018; Karavani et al. 2018). It has indeed been documented that Mediterranean pines may exhibit either thick bark, which protects against surface fires, or xylem resistant to cavitation, which protects against drought. However, it is rare to encounter species with both thick bark and xylem resistant to cavitation, a response that has been attributed to an allocation trade-off (Resco de Dios et al. 2018; Karavani et al. 2018).

Another example of a trade-off in drought and fire resistances, additionally involving shade tolerance, lies in the degree of self-pruning and fire risk. Mediterranean pines with a low degree of self-pruning are shade intolerant species from xeric areas. These species often seek to maximize carbon uptake by increasing leaf area, and a lack of light limitation inhibits the development of self-pruning. However, a low degree of self-pruning enhances fuel ladders, which increases the likelihood of crown fire and, consequently, lowers fire resistance.

Drought-induced leaf shedding is yet another example of the trade-offs that may occur under stress and perturbation. As water scarcity increases in the summer, Mediterranean pines seek to diminish leaf area in order to reduce transpirational area and water loss. This is a strategy that enhances drought survival. However, it occurs at the expense of increasing fire risk. Leaf senescence predates leaf shedding and, during that time, leaf moisture is very low. That is, once older years senescence, a substantial portion of the canopy has the moisture content of dead fuel, always below 30% and typically below 15% (Resco de Dios 2020). Such low water content makes the canopy much more flammable as live fuel moisture is much higher (Fig. 6.7).

6.8 Drought, Fire and Shade Tolerances as Major Drivers of Pine Species Distributions across Mediterranean Productivity Gradients

A result of the physiological responses to fire and drought is the differential and coordinated response of plant traits across productivity gradients (Keeley 2012; Resco de Dios et al. 2018). Conifers may be considered to be fire-tolerant, fire-embracing or fire-avoiding. Fire-tolerant species are those occurring in mesic, high productivity environments. These species have higher shade tolerance and a higher degree of epinastic control. They generally lack drought adaptations but exhibit thick bark and are capable of withstanding surface fires. Typical fire-tolerant species are *P. pinaster* Aiton, *P. sylvestris* L. and *P. nigra*.

Fire-embracing species are those with a low degree of self-pruning (which consequently enhances crown fires), thin bark and high resistance to cavitation, as well

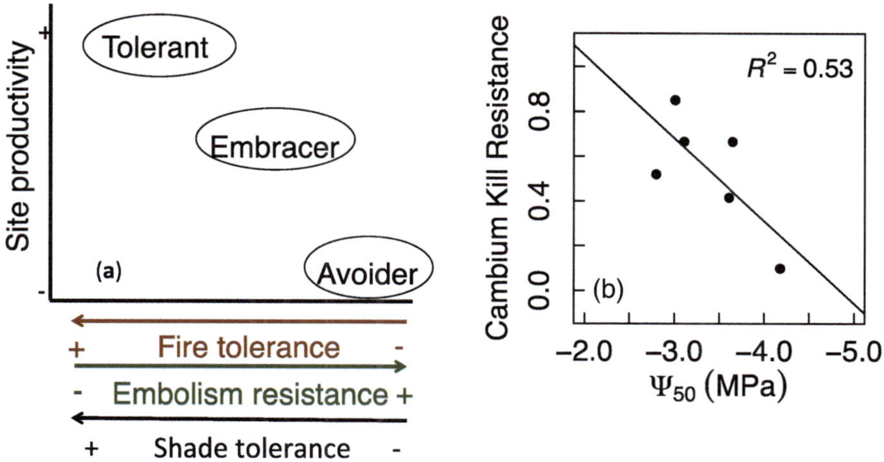

Fig. 6.7 (**a**) Trade-offs between fire tolerance, embolism resistance and shade tolerance offer a partial explanation for the distribution of conifer species. An example of these trade-offs is presented in (**b**), where cambium kill resistance, a function of bark thickness, correlates negatively with resistance to embolism indicated by Ψ_{50}. (Figure modified from Resco de Dios et al. (2018) and reproduced with permission)

as a lower degree of shade tolerance and low epinastic control. *Pinus halepensis*, which occurs at more xeric sites, is a typical fire-embracing species. Fire-embracing species depend on canopy fires for regeneration as they exhibit serotinous cones.

Fire-avoiding species are those occurring where fires are rare, such as deserts or subalpine environments; they exhibit thin bark and they succumb and cannot regenerate after fire. *Pinus uncinata* Mill. ex Mirbel and *P. cembra* would fall within this category.

It is worth noting that such classification is ambiguous, and some species, such as *P. pinea*, exhibit traits belonging to more than one category. However, understanding trade-offs among traits conferring shade, drought and fire tolerances is a useful approach for explaining the distribution of Mediterranean pine species.

6.9 Conclusions

Returning to our original question we can thus conclude that Mediterranean pines show a similar degree of drought resistance to conifers growing in wetter habitats and that their response to fire, as well as their low shade tolerance, are some of the main differentiating features. Similarly, Mediterranean pines are not differentiated from coexisting oaks by their resistance to cavitation; rather, other traits such as shade tolerance or fire response play a more important role. Finally, it is important to note that fire and drought tolerances often involve multiple trade-offs, such that species highly tolerant to both stress and disturbance are lacking. If drought stress and fire intensity increase during climate warming, as predicted, we could expect major niche contractions for Mediterranean pines.

Acknowledgments I acknowledge support from the talent funds of Southwest University of Science and Technology (No.18ZX7131) and the Spanish MICINN (RTI2018-094691-B-C31)

References

Bar A, Michaletz ST, Mayr S (2019) Fire effects on tree physiology. New Phytol. https://doi.org/10.1111/nph.15871

Brodribb TJ, Cochard H (2009) Hydraulic failure defines the recovery and point of death in water-stressed conifers. Plant Physiol 149(1):575–584. https://doi.org/10.1104/pp.108.129783

Calamassi R, Gianni DR, Falusi M, Paoletti E, Strati S (2001) Resistance to water stress in seedlings of eight European provenances of *Pinus halepensis* Mill. Ann For Sci 58:663–672. https://doi.org/10.1051/forest:2001153

Camarero JJ, Olano JM, Parras A (2010) Plastic bimodal xylogenesis in conifers from continental Mediterranean climates. New Phytol 185(2):471–480. https://doi.org/10.1111/j.1469-8137.2009.03073.x

Choat B, Jansen S, Brodribb TJ, Cochard H, Delzon S, Bhaskar R, Bucci SJ, Feild TS, Gleason SM, Hacke UG, Jacobsen AL, Lens F, Maherali H, Martinez-Vilalta J, Mayr S, Mencuccini M, Mitchell PJ, Nardini A, Pittermann J, Pratt RB, Sperry JS, Westoby M, Wright IJ, Zanne AE (2012) Global convergence in the vulnerability of forests to drought. Nature 491(7426):752–755. https://doi.org/10.1038/nature11688

Cuny HE, Rathgeber CBK, Frank D, Fonti P, Mäkinen H, Prislan P, Rossi S, EMD C, Campelo F, Vavrčík H, Camarero JJ, Bryukhanova MV, Jyske T, Ji G, Vr G, Luis MD, Vieira J, Čufar K, Kirdyanov AV, Oberhuber W, Treml V, Huang J-G, Li X, Swidrak I, Deslauriers A, Liang E, Nöjd P, Gruber A, Nabais C, Morin H, Krause C, King G, Fournier M (2015) Woody biomass production lags stem-girth increase by over one month in coniferous forests. Nat Plants 1:15160. https://doi.org/10.1038/nplants.2015.160

del Castillo J, Comas C, Voltas J, Ferrio JP (2016) Dynamics of competition over water in a mixed oak-pine Mediterranean forest: Spatio-temporal and physiological components. For Ecol Manag 382:214–224. https://doi.org/10.1016/j.foreco.2016.10.025

Ferrio JP, Florit A, Vega A, Serrano L, Voltas J (2003) Δ13C and tree-ring width reflect different drought responses in Quercus ilex and Pinus halepensis. Oecologia 137(4):512–518

Granda E, Baumgarten F, Gessler A, Gil-Pelegrin E, Peguero-Pina JJ, Sancho-Knapik D, Zimmerman NE, Resco de Dios V (2020) Day length regulates seasonal patterns of stomatal conductance in Quercus species. Plant Cell Environ 43(n/a):28–39. https://doi.org/10.1111/pce.13665

Grime JP (1977) Evidence for the existence of three primary strategies in plants and its relevance to ecological and evolutionary theory. Am Nat 111:1169–1194

Hammond WM, Yu K, Wilson LA, Will RE, Anderegg WRL, Adams HD (2019) Dead or dying? Quantifying the point of no return from hydraulic failure in drought-induced tree mortality. New Phytol. https://doi.org/10.1111/nph.15922

Hood SM, Varner JM, van Mantgem P, Cansler CA (2018) Fire and tree death: understanding and improving modeling of fire-induced tree mortality. Environ Res Lett 13(11):113004. https://doi.org/10.1088/1748-9326/aae934

Karavani A, Boer MM, Baudena M, Colinas C, Díaz-Sierra R, Pemán J, de Luís M, Enríquez-de-Salamanca Á, Resco de Dios V (2018) Fire-induced deforestation in drought-prone Mediterranean forests: drivers and unknowns from leaves to communities. Ecol Monogr 88:141–169

Kattge J et al (2020) TRY plant trait database – enhanced coverage and open access. Glob Chang Biol 26:119–188

Kavanagh KL, Dickinson MB, Bova AS (2010) A way forward for fire-caused tree mortality prediction: modeling a physiological consequence of fire. Fire Ecol 6:80–94

Keeley JE (2012) Ecology and evolution of pine life histories. Ann For Sci 69:445–453. https://doi.org/10.1007/s13595-012-0201-8

Klein T, Cohen S, Yakir D (2011) Hydraulic adjustments underlying drought resistance of Pinus halepensis. Tree Physiol 31(6):637–648. https://doi.org/10.1093/treephys/tpr047

López R, Aranda I, Gil L (2009) Osmotic adjustment is a significant mechanism of drought resistance in "Pinus pinaster" and "Pinus canariensis". Invest Agraria Sist Recursos Forestales 18:159–166. https://doi.org/10.5424/fs/2009182-01059

Midgley JJ, Kruger LM, Skelton R (2011) How do fires kill plants? The hydraulic death hypothesis and Cape Proteaceae "fire-resisters". S Afr J Bot 77:381–386. https://doi.org/10.1016/j.sajb.2010.10.001

Ne'eman G, Arianoutsou M (this volume) Mediterranean pines – adaptations to fire. In: Ne'eman G, Osem Y (eds) Pines and their mixed forest ecosystems in the Mediterranean Basin. Springer, pp 457–480

Niinemets Ü, Valladares F (2006) Tolerance to shade, drought, and waterlogging of temperate northern hemisphere trees and shrubs. Ecol Monogr 76:521–547. https://doi.org/10.1890/0012-9615(2006)076[0521:TTSDAW]2.0.CO;2

Nolan RH, Hedo J, Arteaga C, Sugai T, Resco de Dios V (2018) Physiological drought responses improve predictions of live fuel moisture dynamics in a Mediterranean forest. Agric For Meteorol 263:417–427. https://doi.org/10.1016/j.agrformet.2018.09.011

Ordóñez JL (2019) Cómo diferenciar los pinos autóctonos peninsulares. In Justamante, A. (27th May 2019). Claves para identificar las diferentes especies de pinos, ¡infografía disponible [Blog post]. Retrieved from http://blog.creaf.cat/es/noticias/claves-para-identificar-las-diferentes-especies-de-pinos-infografia-disponible/

Resco de Dios V (2020) Plant fire interactions. Applying Ecophysiology to Wildfire Management. Springer

Resco de Dios V, Arteaga C, Hedo J, Gil-Pelegrín E, Voltas J (2018) A trade-off between embolism resistance and bark thickness in conifers: are drought and fire adaptations antagonistic? Plant Ecol Divers 11(3):253–258. https://doi.org/10.1080/17550874.2018.1504238

Resco V, Ewers BE, Sun W, Huxman TE, Weltzin JF, Williams DG (2009) Drought-induced hydraulic limitations constrain leaf gas exchange recovery after precipitation pulses in the C_3 woody legume, *Prosopis velutina*. New Phytol 181:672–682

Rundel PW (1973) The relationship between basal fire scars and crown damage in Giant Sequoia. Ecology 54:210–213. https://doi.org/10.2307/1934393

Stephenson NL (1990) Climatic control of vegetation distribution: the role of the water balance. Am Nat 135:649–670

Way DA, Montgomery RA (2015) Photoperiod constraints on tree phenology, performance and migration in a warming world. Plant Cell Environ 38(9):1725–1736. https://doi.org/10.1111/pce.12431

Yebra M, Scortechini G, Badi A, Beget ME, Boer MM, Bradstock R, Chuvieco E, Danson FM, Dennison P, Resco de Dios V, Di Bella CM, Forsyth G, Frost P, Garcia M, Hamdi A, He B, Jolly M, Kraaij T, Martín MP, Mouillot F, Newnham G, Nolan RH, Pellizzaro G, Qi Y, Quan X, Riaño D, Roberts D, Sow M, Ustin S (2019) Globe-LFMC, a global plant water status database for vegetation ecophysiology and wildfire applications. Sci Data 6:155. https://doi.org/10.1038/s41597-019-0164-9

Chapter 7
Carbon Allocation Dynamics in Mediterranean Pines Under Stress

Tamir Klein

7.1 Introduction

Carbon (C) allocation is a central biological process for any plant. In trees, and particularly in large trees, such as pines, this process is of even greater importance. This is partly because trees are long-lived organisms, and partly because of their size and compartmentalization. As a result, C moves in multiple ways in trees: it forms the wood, it can be stored in short- and long-term reserves, and it feeds back to the environment in multiple forms, from litter production, through CO_2 and volatile emissions, to root exudation, and more (Körner 2003). When considering tree C allocation dynamics, we actually ask three different questions about the fate of assimilated C (Klein and Hoch 2015). The first question is simply *where* does the C go? By this we mean the tree compartment, i.e. foliage, stem and branches, and the root system. Of particular importance is the partitioning between aboveground and belowground allocation. The second question is *what is the form* of C? Most C in trees is structural, i.e. in long carbohydrate chains of cellulose, lignin, and other compounds. As in other living organisms, some C is embedded in primary metabolites such as proteins and fatty acids; a smaller part resides in secondary metabolites such as terpenoids and alkaloids; and another in nonstructural carbohydrates, i.e. soluble sugars and starch (de Simón et al. 2018). The third question is *how* does the C flow? This means the physiological process, e.g. respiration and growth. Here we term these processes "fluxes" of C, either internally within the tree (i.e. from C sources to sinks), but mostly externally, between the tree and its soil and atmospheric environments. Within this question is also the quantitative information, i.e. *how much* C, which is the flux magnitude. Taken together, we can say that any C

T. Klein (✉)
Department of Plant and Environmental Sciences, Weizmann Institute of Science, Rehovot, Israel
e-mail: tamir.klein@weizmann.ac.il

G. Ne'eman, Y. Osem (eds.), *Pines and Their Mixed Forest Ecosystems in the Mediterranean Basin*, Managing Forest Ecosystems 38,
https://doi.org/10.1007/978-3-030-63625-8_7

atom in a tree is embedded in a particular molecule, which resides in a particular tree compartment, where it takes part in a particular tree C flux. Since C comprises anywhere between 40% and 50% of the biomass of any tree compartment, deciphering C allocation dynamics means understanding tree function. And this is all the more important, if we remember that about 90% of the biological C on Earth is stored in trees, many of which are pines.

In this chapter, we will try to provide an up-to-date view of C allocation dynamics in Mediterranean pines. We will consider the three aspects of C allocation, beginning with the tree compartment, continuing with the C compound, and ending with the C fluxes. Our work is focused on "true" Mediterranean pines, i.e., those pine species that grow either exclusively, or mostly, in the Mediterranean Basin. This includes the pan-Mediterranean *Pinus halepensis* Mill., its north-east parallel, *P. brutia* Ten., and the other important species, *P. pinea* L., *P. nigra* J.F.Arnold, and *P. pinaster* Ait.. As in many other aspects of Mediterranean pines, the largest share of scientific research has been devoted to those species with a larger distribution range, namely, *P. halepensis* and *P. pinaster*. Information on the more local mountain species, *P. heldreichii* H.Christ and *P. peuce* Griseb., which are found in the Balkan and Italian peninsulas, and *P. uncinata* from Iberia and S. France, is still scarce, and will hopefully be complemented by future research. While C allocation is a fundamental plant process, much can be learned about the tree response to change when it is measured under stress. Here too, the Mediterranean has its own unique characteristics. Unlike in other regions, the term "stress" in the Mediterranean seldom means stress created by freezing or flooding, for example. Instead, most stress situations relate to fire, drought, heat, and biotic attacks (Klein et al. 2019). Cold-induced stress in Mediterranean pines is typically restricted to inland forests. On top of these stressors, competition between trees can cause stress to any individual tree in a forest stand, exacerbating other stresses (Bottero et al. 2017). Research conducted in the last decade has significantly improved our knowledge of tree C allocation dynamics across different biomes (McDowell and Sevanto 2010; Sala et al. 2012; Palacio et al. 2014; Martínez-Vilalta et al. 2016). The significance of this process for major questions such as C sink–source relationships, tree drought resistance, and the roles of C reserves in tress, has enhanced research in this direction. Fortunately, this trend has included Mediterranean pines; hence, the majority of the studies used in this review were published between 2010 and 2019. Together with older studies, they form what we currently know about C allocation in Mediterranean pines, under benign and stress conditions alike.

7.2 Carbon Allocation Across Tree Compartments in Mediterranean Pines

Among the three questions presented in the introduction (*Where*; *Which form*; and *How*), the most well-studied is the first. The distribution of C across tree compartments is a fundamental question related to tree morphology, stemming from long traditions of forestry in the Mediterranean region. In modern scientific research, some of the first important experiments were conducted on *P. halepensis* by INRA in France (Gerant et al. 1996; Kytöviita et al. 1999). Together with drought and elevated CO_2, elevated atmospheric ozone was of major concern during the 1990s. In 3-year-old *P. halepensis* trees, exposure to high ozone (100 ppb) created disequilibrium of C allocation between shoots and roots, reducing transfer to roots (Gerant et al. 1996). When combined with mild drought, ozone significantly reduced C fixation and C allocation belowground, significantly reducing root growth. This ozone effect was confirmed in another experiment, this time combined with elevated CO_2, and inoculation with mycorrhizal fungi in the soil (Kytöviita et al. 1999). While ozone decreased the overall biomass of *P. halepensis* seedlings, and the partitioning to roots in particular, exposure to elevated CO_2 (700 ppm) and the mycorrhizal inoculation (*Paxillus involutus* (Batsch) Fr.) partly reversed these effects. On the other side of the Mediterranean, C allocation research began on *P. halepensis* at the very dry edge of forest existence (Klein et al. 2005; Maseyk et al. 2008b). In a semi-arid afforestation in Israel, drought is the norm. Thus, an irrigation experiment was set up in the forest, to test eco-physiological tree responses in this unique setting, with <300 mm annual precipitation. In spite of the prevailing hot and dry air, *in-situ* irrigation during the long dry season doubled and quadrupled C allocation to stem wood (Klein et al. 2005). In needles, a 25% increase was observed during the first year of irrigation, but not in the second year. Competition between individual trees at the same site decreased needle length for many years (Tsamir et al. 2019). The higher plasticity of C allocation to the stem compared with the needles was also expressed by the irrigation effect on their respective growth phenologies. While the rate of C allocation to needles was stable, C allocation to the stem was sensitive to irrigation, with increased allocation in the dry season (Klein et al. 2005). In other words, while C allocation to needles is genetically programed, C allocation to the stem is adaptable, and even opportunistic. This trend can probably be generalized to other pines (e.g. Camarero et al. 2010), as observed by a comparison of the growth season phenology across a south–north European transect through pine forests. The peak C fixation season moves from summer to spring as latitude decreases, and ultimately to winter, in the case of the dry *P. halepensis* forest in Israel (Rotenberg and Yakir 2010).

Indeed, research on other Mediterranean pine species was not late in catching up. A detailed C partitioning study was conducted on *P. pinaster* in INRA, considering its major role in the French timber industry. While the C content in wood biomass was traditionally assumed to be 50%, measured values yielded 54% and 52% in shoots and roots, respectively (Bert and Danjon 2006). Laying the foundations for

C partitioning across tree compartments in *P. pinaster*, the study also provided precious data on the root system structure. In particular, it demonstrated the proliferation of horizontal roots, with a low number of sinker roots, and a relatively short tap root (Bert and Danjon 2006).

Is C partitioning to tree compartments conserved across Mediterranean pines? A comparison between the two most important species shows large differences, with high allocation to crown compartments in *P. halepensis*, and high allocation to the stem and to belowground compartments in *P. pinaster* (Fig. 7.1). Considering the similar stand density and age of the studied trees, these differences probably reflect genetic differences, which are also expressed in height differences (20 m in *P. pinaster* species vs. 10 m *P. halepensis*). While some of these C allocation patterns are representative at the species level, others change from one provenance to the next (Correia et al. 2008). Specifically in Iberian *P. pinaster*, major differences were found between Mediterranean and Atlantic provenances, with higher growth in the latter. How do provenance and drought stress interact in shaping C partitioning? In a greenhouse drought experiment on *P. pinaster* seedlings, C partitioning to roots did not change with drought, but was consistently higher in a southern, montane provenance (from Morocco, relative to provenances from Spain and France; Aranda et al. 2010).

At the forest scale, C allocation patterns translate into C stocks, which are particularly dynamic in our region, due to recurring wildfire and forest management activities. In *P. pinaster* forests in Portugal, the effects of recurring fires on C sequestration were found to depend completely on pine regeneration (Santana et al. 2016). Fire risk was a major motivation of another study, focused on aboveground C partitioning of small-diameter *P. brutia* trees in Turkey, taking into account their

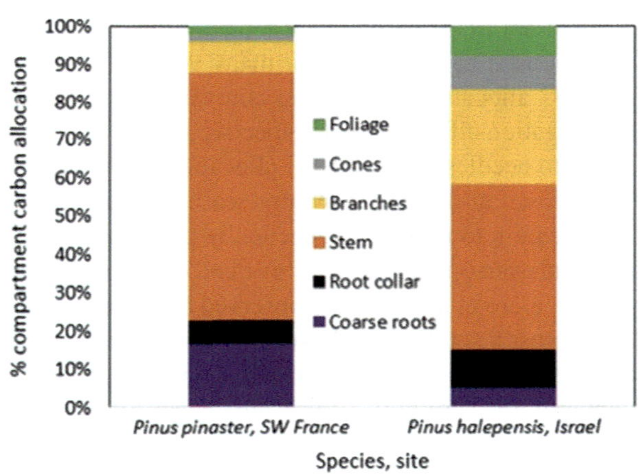

Fig. 7.1 Carbon allocation to tree compartments in 50-year-old *Pinus pinaster* (left; SW France) and 40-year-old *Pinus halepensis* (right; Israel). Stand densities in France and Israel are 225 and 300 trees ha^{-1}, respectively; Mean annual precipitation and temperature are 12.5 and 18.0 °C and 930 and 280 mm, respectively. (Adapted from Bert and Danjon 2006; Grünzweig et al. 2007; Klein and Hoch 2015)

high flammability and the role of small trees in forest fire dynamics (Sakici et al. 2018). On average, C allocation to stem, branches, and needles, was 23%, 52%, and 25%, respectively. However, these ratios changed rapidly with tree age and size, increasing in the stem compartment to >50%, and decreasing in the branch compartment, and more so in the foliage (Sakici et al. 2018).

A major endeavor in tree C research, and Mediterranean pines are no exception, is the study of stable C isotope fractionation, namely the dynamics of ^{13}C composition in tree organs (Ferrio et al. 2003; Voltas et al. 2008). The link between ^{13}C fractionation and water-use efficiency, via the discrimination of ^{13}C in photosynthesis, means that a large part of this research has been directed towards tree water relations and tree drought resistance (Klein et al. 2005, 2012; Sarris et al. 2013; Camarero et al. 2015a). Still, important insights have been gained in cases where ^{13}C composition has been studied in detail within and among tree organs. In a study of *P. nigra* needles from both Croatia and Spain, the base of the needle was always ^{13}C-enriched compared to its tip (Barszczowska and Jedrysek 2005). The authors interpreted this trend as evidence for the lack of re-translocation of C during needle growth. It was suggested that the needle base, formed early in the season, contained C from internal reserves, whereas the later needle growth relied on concurrent photosynthates, also reflecting the seasonal variation in the ^{13}C content of atmospheric CO_2. Interestingly, an opposite trend was reported for *P. halepensis* needles in Israel during the very same year (Klein et al. 2005). Here, the base of the needle was always ^{13}C-depleted compared to its tip. These seemingly contradictory observations are settled by considering the mismatch between peak C assimilation, in early spring, and needle growth, in summer, in the semi-arid site. As conditions become harsher during summer, stomatal closure means that more ^{13}C is assimilated and allocated to needle tissue. In mature *P. pinaster* needles, ^{13}C composition was studied across seven provenances, rather than along the season (Correia et al. 2008). Here, ^{13}C correlated with growth and survival, albeit in contrasting ways: high growth and survival were related to ^{13}C-enrichment in Atlantic provenances, and to ^{13}C-depletion in Mediterranean provenances. Again, these trends reflected climatic differences and their effect on growth patterns.

7.3 Carbon Reserves and Their Use in Mediterranean Pines Under Stress

The second question in tree C allocation is *the form* of C, i.e. its chemical compound. In stress-related research, most of the attention has been given to C storage, i.e. the accumulation of C reserves, mostly starch and soluble sugars, and their use in situations limiting photosynthetic C uptake. This line of research is part of a larger endeavor to understand the role of C reserves in tree survival and mortality under stress, particularly under drought (McDowell and Sevanto 2010; Sala et al. 2012; Camarero et al. 2015a; Martínez-Vilalta et al. 2016). Mediterranean pines provide a good testing ground, since they accumulate nonstructural C (NSC) and are regularly exposed to water limitation. Indeed, in mature *P. halepensis* growing in a semi-arid forest, seasonal NSC dynamics in stem and roots were shown to provide the required C missing from the asynchrony between C source and C sink activities (Klein and Hoch 2015). The authors calculated that starch degradation was sufficient to buffer a transient C imbalance, at the whole tree scale, of up to 57 g per day. Comparing trees at different drought stress levels at that same site showed uniform starch degradation in the stem in late spring, with only sugars evidencing the effect of drought (Klein et al. 2014). The major, up to threefold, difference in NSC concentrations between healthy and stressed trees, was in the roots. This observation was recently confirmed in *P. halepensis* seedlings under induced drought (Suárez-Vidal et al. 2019) and in heat and drought treatment combinations (Birami et al. 2018). Constituting up to 10% of the root biomass (Klein et al. 2014), root NSC might represent an important, long-term C storage reserve in Mediterranean pines. In another semi-arid *P. halepensis* forest, in Spain, NSC concentrations in wood decreased under irrigation (Pacheco et al. 2018). Irrigated trees grew at significantly higher rates than control trees; hence, their lower C reserves could reflect a dilution effect. Other interpretations include higher demand for C for growth (Pacheco et al. 2018), or lower necessity of C reserves in the absence of water limitation. A link between drought stress, NSC concentration, and tree mortality was made in a *P. nigra* population following the dry summer of 2012 (Savi et al. 2019). Declining trees had significantly lower levels of starch and sucrose compared to their healthy neighbors. In another study, drought-induced dieback and growth decline lead to an increase of sapwood NSC concentrations in Aleppo pine (Camarero et al. 2015a). Nevertheless, the study of C reserves has not been limited to drought stress. Interruptions in C uptake can result from needle defoliation due to insect herbivores, for example. In a well-designed experiment on young *P. pinaster*, Puri et al. (2015) showed that severe defoliation reduced growth, but not NSC concentrations. In another study on *P. nigra* subjected to defoliation by the pine processionary moth, no C depletion was observed after a severe outbreak which caused a drastic reduction of wood formation (Palacio et al. 2012). These observations could reflect the prioritization of C storage over growth, as part of a safety mechanism. Indeed, activation of the molecular machinery allowing for rapid starch synthesis during recovery from drought was shown in *P. halepensis* needles (Fox et al. 2017). Still, it could be related to a critical loss of N and P reserves, or to hormonal changes affecting cambial activity (Puri et al. 2015). Here too, NSC were considerably higher in roots than in the stem.

A few recent studies have made an important step forward and included measurements of compounds other than NSC. A unique greenhouse experiment simulating heat-waves on *P. halepensis* seedlings reported interesting dynamics of additional metabolites. Carboxylic acids decreased in shoots under drought, and more so in roots under a combination of heat and drought (Birami et al. 2018). Cyclitols, i.e. inositol and pinitol, significantly increased in shoots, but decreased in roots, under heat and drought. These observations have exposed some of the less-studied tree C allocation responses to stress, including the synthesis, degradation, and mobilization of specific compounds. Secondary metabolites such as terpenoids and alkaloids have also been studied. Many of these C compounds are important players in tree defense against herbivores. For example, *P. halepensis* seedlings were shown to respond to a chewing insect herbivore by decreasing production of polyphenols and condensed tannins, increasing production of diterpenes, and modifying the profiles of major terpenes (Suárez-Vidal et al. 2019). When combined with drought, the responses of polyphenols were reduced, whereas the increase in diterpenes was highest under moderate drought. Finally, the first metabolomic study on a Mediterranean pine species has recently been reported for needles of *P. pinaster* under elevated CO_2 (de Simón et al. 2018). Among the dozens of C compounds included in the analysis, the effect of elevated CO_2 was more pronounced on secondary, than primary, metabolites, and in juvenile, rather than adult, needles. There were metabolic differences among genotypes, in terpenoids and free fatty acids, but not in other C compounds (de Simón et al. 2018). All in all, the fate of C in the cellular metabolic networks of a tree organ depends very much on the genetic and environmental situation, in turn affecting the entire physiological response. This topic is yet very little studied in adult trees.

7.4 Carbon Fluxes in Mediterranean Pines

The third question in tree C allocation is *the flux* of C, i.e. its fate and physiological context within the tree. Assimilated C is partitioned into respiration, growth, export, and litter processes in each of the different tree compartments (Klein and Hoch 2015). While C assimilation is considered the C source, each of these fluxes is a tree C sink. For many years, stem growth was the only flux being studied. This is expected, since forest science developed from forestry, where a commercial viewpoint was applied. For example, a comparative study of climate change effects on tree species of the north Mediterranean reported various growth responses (Sabaté et al. 2002). In both *P. halepensis* and *P. pinaster*, the increasing level of atmospheric CO_2 was predicted to increase needle turnover, while warming would induce higher stem growth, especially in *P. pinaster*. This prediction was given for rainfall-abundant sites, where in drought years, both species were expected to suffer growth reductions and even tree mortality (Sabaté et al. 2002). This was partially supported by tree-ring width data in field research showing no CO_2-induced fertilization effect on growth due to rising water-use efficiency, and an overriding role of drought as constraint of pine radial growth (Camarero et al. 2015b; Shestakova et al. 2017). A major step forward in looking at tree fluxes beyond stem growth was the research on *P. halepensis* at a semi-arid site in Israel, which was also important for its drought stress response characterization. Here, a temporal separation between stem and needle growth was described, with rather conserved timing of relative growth rate of each compartment between years (Maseyk et al. 2008a, b; Rotenberg and Yakir 2010). Carbon allocation to needle growth was restricted to the summer months, while the timing of stem growth shifted to the winter and spring months. For the first time, C respiration rates were measured in a Mediterranean pine species, showing the seasonal dynamics in needles and stem CO_2 exchange (Maseyk et al. 2008a). Respiration rates were higher in developing needles than in mature foliage, and stem respiration peaked in May, possibly due to carbohydrate re-mobilization and cambial activity. These measurements, together with additional flux measurements and estimates from the same trees at the site, yielded the detailed flux partitioning which followed Klein and Hoch (2015). It turned out that a ~ 20-cm-diameter, ~10-m-high *P. halepensis* tree assimilated about 24 kg of C per year; a value which was later confirmed by down-scaling canopy CO_2 exchange, estimated by eddy covariance, at the same site (Klein et al. 2016). Unexpectedly, 70% of this C partitioned into respiratory fluxes, whereas only 17% and 13% partitioned to growth and litter, respectively (plus export to soil). While C respiration was dominated by roots, growth and litter production were dominated by the foliage (Klein and Hoch 2015). There was minor C allocation to reproductive tissue, i.e. cone production, which was detected in litter during March and April. The mass balance approach applied in that study enabled the identification of relocation fluxes during summer, i.e. from stem to foliage, and from roots to stem. Although minor in magnitude, these C fluxes might represent an important drought acclimation mechanism for pines in the Mediterranean.

The study of C flux partitioning in *P. halepensis* has intensified in the last decade, involving both modeling (Gea-Izquierdo et al. 2015; Santini et al. 2019) and new measurements (Birami et al. 2018). In a study on the effects of heat-waves and drought, seedlings which were exposed to two 4-day heat periods responded immediately by decreasing photosynthesis, while dark respiration remained high (Birami et al. 2018). Furthermore, in a combined heat and drought treatment, most seedlings perished, and survivors used most of their C reserves. The authors concluded that heat-waves in combination with moderate drought can result in increased mortality or delayed recovery of *P. halepensis* in the region. Looking beyond the species level, a comparison among 51 *P. halepensis* populations across the Mediterranean revealed important variation in life history strategies, including C allocation patterns (Santini et al. 2019). It was found that slow-growing populations invested significantly more in reproduction and C reserves than fast-growing populations. Such a large divergence was explained by local adaptation of ecotypes to contrasting climates, controlled by drought as the main stressor, and by fire as the main ecological disturbance (Santini et al. 2019). Studies of C fluxes in other Mediterranean pines are also emerging. Focusing on respiratory fluxes, a stand-scale study of *P. pinaster* was conducted in central Italy (Matteucci et al. 2015). Root respiration accounted for 26% of soil respiration, with a one-week time lag between rainfall pulses and root respiration peaks. This delay was probably due to the translocation of recently assimilated C from the canopy to the root system. Interestingly, total tree respiration (considering both above- and belowground compartments) accounted for 60% of total canopy respiration (Matteucci et al. 2015), supporting the view that respiration is the major C sink in Mediterranean pines, and probably in trees in general. Studies focusing on tree growth continue to appear, however at much finer detail: one study found that stem growth in *P. pinea* is bimodal, i.e. separated between spring and autumn (Camarero et al. 2010; Pacheco et al. 2017; Castagneri et al. 2018). It was found that the lumen diameter of the first earlywood tracheids was related to winter precipitation, whereas later tracheids were shaped by mid-spring precipitation, and latewood tracheids by mid-autumn precipitation. Carbon isotope measurements showed that earlywood was likely formed using both recently and formerly assimilated C, while latewood relied mostly on C assimilated many months prior to its formation (Castagneri et al. 2018). Such detailed investigations become important when considering the sensitivity of C allocation to climate, particularly, in view of ongoing climate change.

7.5 Conclusions

In this chapter I have tried to expose, and to a lesser extent, synthesize, some of the key recent findings on C allocation in Mediterranean pines under stress. Overall, the wealth of information was surprising, and in some aspects, research in our region represents the cutting edge of tree C allocation research everywhere. This is noteworthy, especially considering the fact that research centers in North America, Central and Northern Europe, Japan, and China, are far larger than those in the Mediterranean. Among the many interesting discoveries on patterns of C allocation in Mediterranean pines, those listed below stand out. Some might be unique to the studied species, while others may represent patterns general to pines, or trees in general.

1. Carbon allocation to stem growth is highly plastic in phenology and magnitude, whereas C allocation to needle growth is sensitive to growing conditions (e.g. drought, competition), but is rather genetically programmed in timing and magnitude.
2. Divergence in C partitioning among tree compartments is linked to interspecific differences in tree morphology, with high allocation to crown compartments in *P. halepensis*, and high allocation to stem and belowground compartments in *P. pinaster*.
3. C partitioning to roots can diverge among provenances, and isotopic composition of C correlates with growth and survival, albeit in contrasting ways: High growth and survival are related to ^{13}C-enrichment in some provenances, and to ^{13}C-depletion in others. Similarly, slow-growing populations invest significantly more in reproduction and C reserves than fast-growing populations. Such divergences might be explained by local adaptation of ecotypes to contrasting climates, controlled by drought as the main climatic stressor, and by fire as the main ecological disturbance.
4. Starch is a major C reserve in Mediterranean pines; it provides C under limited photosynthesis for weeks and months, is replenished during recovery before growth is resumed, and is stored at high concentrations in the root system.

5. Carbon respiration is dominated by roots, while growth and litter production are dominated by the foliage. Relocation fluxes occur under summer drought, i.e. from stem to foliage, and from roots to stem. Although minor in magnitude, these C fluxes might represent an important drought acclimation mechanism for pines in the Mediterranean
6. Heat-waves induce decreasing photosynthesis, while dark respiration remains high.
7. Herbivore stress can induce decreased production of polyphenols and condensed tannins, increased production of diterpenes, and modification of the profiles of major terpenes. When combined with drought, the responses of polyphenols are reduced, whereas diterpenes increase the most under moderate drought. Overall, herbivory did not lead to C depletion.
8. Elevated CO_2 affects mostly secondary, rather than primary, metabolites. The increasing level of atmospheric CO_2 is predicted to increase needle turnover, while warming would induce higher stem growth in rain-abundant sites.

To what degree are these findings relevant? As stated above, understanding tree C allocation means understanding tree physiology. The Mediterranean is a cradle of human culture, and as a result, Mediterranean forests have been under constant change for thousands of years. Nevertheless, the current change is faster than any of those which preceded it. For this reason, forest research in the Mediterranean is in a race to the unknown (Klein 2019). The ongoing change includes elevated CO_2, warming, drying, and higher frequency of extreme events such as drought, fire, storms, and pest attacks. Therefore, observations from the hot and dry edges of the Mediterranean, together with well-designed experiments on adult individuals, are becoming increasingly important. The results of these studies, some of which are documented above, constitute the levels and limits to which Mediterranean pines can acclimate through C allocation responses. In addition, they attract European forest researchers, as climate conditions become habitable for Mediterranean pine species.

Acknowledgment The author would like to thank JJ Camarero of the Spanish National Research Council for his useful comments on a previous version of this chapter.

References

Aranda I, Alía R, Ortega U et al (2010) Intra-specific variability in biomass partitioning and carbon isotopic discrimination under moderate drought stress in seedlings from four *Pinus pinaster* populations. Tree Genet Genomes 6(2):169–178

Barszczowska L, Jedrysek MO (2005) Carbon isotope distribution along pine needles [*Pinus nigra* Arnold]. Acta Soc Bot Pol 74(2):93–98

Bert D, Danjon F (2006) Carbon concentration variations in the roots, stem and crown of mature *Pinus pinaster* (Ait.). For Ecol Manag 222(1–3):279–295

Birami B, Gattmann M, Heyer AG et al (2018) Heat waves alter carbon allocation and increase mortality of Aleppo pine under dry conditions. Front For Glob Chang 1:8. https://doi.org/10.3389/ffgc.2018.00008

Bottero A, D'Amato AW, Palik BJ et al (2017) Density-dependent vulnerability of forest ecosystems to drought. J Appl Ecol 54(6):1605–1614

Camarero JJ, Olano JM, Parras A (2010) Plastic bimodal xylogenesis in conifers from continental Mediterranean climates. New Phytol 185:471–480

Camarero JJ, Gazol A, Sangüesa-Barreda G, Oliva J, Vicente-Serrano SM (2015a) To die or not to die: early-warning signals of dieback in response to a severe drought. J Ecol 103:44–57

Camarero JJ, Gazol A, Tardif JC, Conciatori F (2015b) Attributing forest responses to global-change drivers: limited evidence of a CO2-fertilization effect in Iberian pine growth. J Biogeogr 42:2220–2233

Castagneri D, Battipaglia G, Von Arx G et al (2018) Tree-ring anatomy and carbon isotope ratio show both direct and legacy effects of climate on bimodal xylem formation in *Pinus pinea*. Tree Physiol 38(8):1098–1109

Correia I, Almeida MH, Aguiar A et al (2008) Variations in growth, survival and carbon isotope composition (δ13C) among *Pinus pinaster* populations of different geographic origins. Tree Physiol 28(10):1545–1552

de Simón BF, Cadahía E, Aranda I (2018) Metabolic response to elevated CO_2 levels in *Pinus pinaster* Aiton needles in an ontogenetic and genotypic-dependent way. Plant Physiol Biochem 132:202–212

Ferrio JP, Florit A, Vega A et al (2003) Δ 13 C and tree-ring width reflect different drought responses in *Quercus ilex* and *Pinus halepensis*. Oecologia 137(4):512–518

Fox H, Doron-Faigenboim A, Kelly G et al (2017) Transcriptome analysis of *Pinus halepensis* under drought stress and during recovery. Tree Physiol 38(3):423–441

Gea-Izquierdo G, Guibal F, Joffre R et al (2015) Modelling the climatic drivers determining photosynthesis and carbon allocation in evergreen Mediterranean forests using multiproxy long time series. Biogeosciences 12:3695–3712

Gerant D, Podor M, Grieu P et al (1996) Carbon metabolism enzyme activities and carbon partitioning in *Pinus halepensis* Mill, exposed to mild drought and ozone. J Plant Physiol 148(1–2):142–147

Grünzweig JM, Gelfand I, Yakir D (2007) Biogeochemical factors contributing to enhanced carbon storage following afforestation of a semi-arid shrubland. Biogeosciences 4(5):891–904

Klein T (2019) A race to the unknown: contemporary research on tree and forest drought resistance, an Israeli perspective. J Arid Environ 172:104045

Klein T, Hoch G (2015) Tree carbon allocation dynamics determined using a carbon mass balance approach. New Phytol 205(1):147–159

Klein T, Hemming D, Lin T et al (2005) Association between tree-ring and needle δ 13 C and leaf gas exchange in *Pinus halepensis* under semi-arid conditions. Oecologia 144(1):45–54

Klein T, Di Matteo G, Rotenberg E et al (2012) Differential ecophysiological response of a major Mediterranean pine species across a climatic gradient. Tree Physiol 33(1):26–36

Klein T, Hoch G, Yakir D, Körner C (2014) Drought stress, growth and nonstructural carbohydrate dynamics of pine trees in a semi-arid forest. Tree Physiol 34(9):981–992

Klein T, Rotenberg E, Tatarinov F et al (2016) Association between sap flow-derived and eddy covariance-derived measurements of forest canopy CO 2 uptake. New Phytol 209(1):436–446

Klein T, Cahanovitc R, Sprintsin M et al (2019) A nation-wide analysis of tree mortality under climate change: Forest loss and its causes in Israel 1948–2017. For Ecol Manag 432:840–849

Körner C (2003) Carbon limitation in trees. J Ecol 91(1):4–17

Kytöviita MM, Pelloux J, Fontaine V et al (1999) Elevated CO_2 does not ameliorate effects of ozone on carbon allocation in *Pinus halepensis* and *Betula pendula* in symbiosis with *Paxillus involutus*. Physiol Plant 106(4):370–377

Martínez-Vilalta J, Sala A, Asensio D et al (2016) Dynamics of non-structural carbohydrates in terrestrial plants: a global synthesis. Ecol Monogr 86(4):495–516

Maseyk K, Grünzweig JM, Rotenberg E et al (2008a) Respiration acclimation contributes to high carbon-use efficiency in a seasonally dry pine forest. Glob Chang Biol 14(7):1553–1567
Maseyk KS, Lin T, Rotenberg E et al (2008b) Physiology–phenology interactions in a productive semi-arid pine forest. New Phytol 178(3):603–616
Matteucci M, Gruening C, Ballarin IG et al (2015) Components, drivers and temporal dynamics of ecosystem respiration in a Mediterranean pine forest. Soil Biol Biochem 88:224–235
McDowell NG, Sevanto S (2010) The mechanisms of carbon starvation: how, when, or does it even occur at all? New Phytol 186(2):264–266
Pacheco A, Camarero JJ, Ribas M, Gazol A, Gutiérrez E, Carrer M (2017) Disentangling the climate-driven bimodal growth pattern in coastal and continental Mediterranean pine stands. Sci Total Environ
Pacheco A, Camarero JJ, Carrer M (2018) Shifts of irrigation in Aleppo pine under semi-arid conditions reveal uncoupled growth and carbon storage and legacy effects on wood anatomy. Agric For Meteorol 253:225–232
Palacio S, Hernández R, Maestro-Martínez M, Camarero JJ (2012) Fast replenishment of initial carbon stores after defoliation by the pine processionary moth and its relationship to the re-growth ability of trees. Trees 26:1627–1640
Palacio S, Hoch G, Sala A et al (2014) Does carbon storage limit tree growth? New Phytol 201(4):1096–1100
Puri E, Hoch G, Körner C (2015) Defoliation reduces growth but not carbon reserves in Mediterranean *Pinus pinaster* trees. Trees 29(4):1187–1196
Rotenberg E, Yakir D (2010) Contribution of semi-arid forests to the climate system. Science 327(5964):451–454
Sabaté S, Gracia CA, Sánchez A (2002) Likely effects of climate change on growth of *Quercus ilex*, *Pinus halepensis*, *Pinus pinaster*, *Pinus sylvestris* and *Fagus sylvatica* forests in the Mediterranean region. For Ecol Manag 162(1):23–37
Sakici OE, Kucuk O, Ashraf MI (2018) Compatible above-ground biomass equations and carbon stock estimation for small diameter Turkish pine (*Pinus brutia* Ten.). Environ Monit Assess 190(5):285
Sala A, Woodruff DR, Meinzer FC (2012) Carbon dynamics in trees: feast or famine? Tree Physiol 32(6):764–775
Santana VM, González-Pelayo O, Maia PA et al (2016) Effects of fire recurrence and different salvage logging techniques on carbon storage in *Pinus pinaster* forests from northern Portugal. Eur. J For Res 135(6):1107–1117
Santini F, Climent JM, Voltas J (2019) Phenotypic integration and life history strategies among populations of *Pinus halepensis*: an insight through structural equation modelling. Ann Bot 124(7):1161–1172
Sarris D, Siegwolf R, Körner C (2013) Inter- and intra-annual stable carbon and oxygen isotope signals in response to drought in Mediterranean pines. Agric For Meteorol 168:59–68
Savi T, Casolo V, Dal Borgo A et al (2019) Drought-induced dieback of *Pinus nigra*: a tale of hydraulic failure and carbon starvation. Conserv Physiol 7(1):coz012
Shestakova TA, Camarero JJ, Ferrio JP, Knorre AA, Gutiérrez E, Voltas J (2017) Increasing drought effects on five European pines modulate Δ13C-growth coupling along a Mediterranean altitudinal gradient. Func Ecol. https://doi.org/10.1111/1365-2435.12857
Suárez-Vidal E, Sampedro L, Voltas J et al (2019) Drought stress modifies early effective resistance and induced chemical defences of Aleppo pine against a chewing insect herbivore. Environ Exp Bot 162:550–559
Tsamir M, Gottlieb S, Preisler Y et al (2019) Stand density effects on carbon and water fluxes in a semi-arid forest, from leaf to stand-scale. For Ecol Manag 453:117573
Voltas J, Chambel MR, Prada MA et al (2008) Climate-related variability in carbon and oxygen stable isotopes among populations of Aleppo pine grown in common-garden tests. Trees 22(6):759–769

Chapter 8
Mediterranean Pine Root Systems Under Drought

Dimitrios Sarris and Gianluigi Mazza

8.1 Mediterranean Pines´ Root Structure and Function

Plant survival during drought requires not only aboveground water conservation strategies, but also utilization of all possible moisture sources available underground to sustain plant cell turgor. Roots are the vital organ permitting such functioning.

The genus *Pinus* in the Mediterranean includes drought-adapted species such as *P. halepensis* Mill., *P. brutia* Ten., *P. pinaster* Aiton., *P. pinea* L. and *P. nigra* J.F.Arnold (Barbéro et al. 1998), although drought-resistance varies from one species to the other, with *P. halepensis and P. brutia* capable of colonizing the most arid of the habitats that the genus inhabits in the Mediterranean (Dorman et al. 2015a; Mauri et al. 2016). Aboveground adjustments to drought in conifers include flexibility in their seasonality of cambial activity (Liphschitz and Lev-Yadun 1984; Sarris et al. 2013), whereas after severe and frequent droughts, conifers can produce shorter needles and shoots (Vennetier et al. 2013) and reduce green biomass through shedding needles, resulting in changes in tree architecture. Nonetheless, this mechanism has its limitations since not all needles can be shed to prevent confer desiccation (Körner et al. 2005; Sarris et al. 2007; Allen et al. 2010). Drought-adapted tree species tend to invest more biomass in longer-lasting root organs, as well as minimizing water loss from transpiration (Brunner et al. 2015). These patterns have contributed to the hypotheses that trees respond to water deficit by increasing

D. Sarris (✉)
Department of Biological Sciences, University of Cyprus, Nicosia, Cyprus

Faculty of Pure and Applied Sciences, Open University of Cyprus, Nicosia, Cyprus

KES Research Centre, Nicosia, Cyprus
e-mail: desarris@ucy.ac.cy

G. Mazza
CREA Research Centre for Forestry and Wood, Arezzo, Italy

G. Ne'eman, Y. Osem (eds.), *Pines and Their Mixed Forest Ecosystems in the Mediterranean Basin*, Managing Forest Ecosystems 38,
https://doi.org/10.1007/978-3-030-63625-8_8

root-to-shoot ratios and rooting depth. Indeed, species adapted to dry climatic regimes generally have higher root-to-shoot ratios and deeper root systems than species that are more suited to mesic climatic conditions (see Brunner et al. 2015 and references therein).

If deep roots are indeed the most vital organs that pines possess to deal with drought, understanding the responses of pines under moisture stress presents a serious challenge. Deep roots are their less-studied organs, as is the general case for tree roots (Brunner et al. 2015; Fan et al. 2017). This is because the biomass of root systems is difficult to measure in any forest ecosystem, in contrast to the several established methods for estimating the biomass of aboveground tree components (Sanford and Cuevas 1996). Roots remain hidden belowground, giving little access for direct scientific measurements, particularly under field conditions (Brunner et al. 2015). This difficulty is even greater when it comes to deep roots.

According to Guo et al. (2013), direct measurement of coarse roots is conducted using destructive methods (e.g., excavation, uprooting, soil block, and profile wall technique), which are laborious and time consuming, thus restricting the manageability of sampling numbers and the repeatability of measurements (Oliveira et al. 2000; Van Noordwijk et al. 2000; Polomski and Kuhn 2002; Reubens et al. 2007). Guo et al. (2013) agree that such methods are not always appropriate for studying roots as they introduce external disturbance to the rhizosphere environment and/or destroy root systems if not done appropriately, making long-term repeated measurements inaccurate or impossible (Van Noordwijk et al. 2000; Nadezhdina and Čermák 2003; Reubens et al. 2007; Danjon and Reubens 2008). Consequently, significant efforts (e.g., Berntson et al. 1995; Hruška et al. 1999; Čermák et al. 2008; Zenone et al. 2008; Leucci 2010) have been made to develop non-destructive coarse root analysis methods in recent decades, including labeling methods (e.g., radioisotope and stable isotope labeling), sap flow approaches, and geophysical imaging techniques (e.g., electrical resistivity tomography, seismic refraction tomography, and ground penetrating radar; Guo et al. 2013). Among these techniques, stable isotopes (Rose et al. 2003; Sarris et al. 2013) can provide considerable information about how deep pine roots function under drought. Other methods include tree-ring stem analysis. By correlating tree-ring width with precipitation variables that determine groundwater availability it is possible to indirectly assess whether pines are deep rooted and what is the origin of the moisture sources their roots have access to. This can provide an inexpensive and easily reproduced, non-destructive method for assessing *in situ* groundwater utilization variability over the course of a pine's entire life span (Sarris et al. 2007; Dorman et al. 2015b; Mazza et al. 2018). Information offered using the above methods suggests that deep pine roots and their rapid growth indeed hold some of the key traits that have enabled the genus to successfully colonize the semiarid habitats of the Mediterranean (Andivia et al. 2018) and beyond.

8.1.1 Pine Roots and Genetic Adaptations to Drought

Xerophytic plants have roots characterized by their superficial and horizontal dispersal starting from 5–10 cm. A part of the roots penetrates to the depth to which rainwater penetrates (Fahn 1964; Gindel 1973). In phreatophytes this depth can reach tens of meters belowground as an evolutionary adaptation of tree and shrub species for surviving throughout the dry season in arid and semiarid habitats (Thomas 2014; Fan et al. 2017). Pine species of the semi-arid Mediterranean environment, having developed the genetic potential for lateral and tap roots (Sarris et al. 2013), should be considered as "facultative" or "temporary" phreatophytes. Their tap roots have the potential to access the water table, although this is not necessary for their successful growth. Nonetheless, they are usually not considered to be phreatophytes (Thomas 2014).

Most pine populations around the world grow in low-temperature biomes; thus, the mean rooting depth of this genus has been considered to be ca. 2.5–6.5 m. Compared to phreatophytic genera present mostly in warm climates this may not be too impressive. For example, *Eucalyptus* is considered to have a mean rooting depth of ca. 9–18 m, while *Acacia* appears to be even more deep-rooted (ca. 13–31 m mean rooting depth; Fan et al. 2017). Nevertheless, pine species in warm, semiarid climates, unlike their cold-adapted relatives, can develop a deep taproot which can explore water sources many meters belowground reaching water tables up to 20 m deep (Fan et al. 2017) or even deeper (24 m taproot for *P. ponderosa*; Cannon 1960 *in* Stone and Kalisz 1991). This may very well also be the case for some typical lower-elevation Mediterranean pine species, such as *P. halepensis* or *P. brutia,* although deep rooting is a trait that higher-elevation pine species such as *P. nigra* may also possess (Fig. 8.1).

Long lateral roots may also assist in expanding the range of topsoil moisture acquisition under drought before it evaporates or before it is utilized by competing species. Some Mediterranean pines that are not known to produce very deep roots, such as *P. pinea,* can grow lateral roots up to 38 m long (Frattegiani et al. 1994) or even longer, as recorded for *P. pinaster* (44 m; Sudmeyer et al. 2004).

8.1.2 Pine Root Functioning and Its Environmental Drivers

In addition to genetics, topography and hydrology regulate plant rooting depth (Fan et al. 2017). Therefore, root architecture in pines is determined by the combined effect of species´ genetics and environmental factors such as moisture availability, water infiltration and oxygen availability.

Pine species such as *P. halepensis* or *P. brutia,* which can access deep moisture sources, appear to be able to grow deep roots only where wet season rainfall is sufficient to penetrate deep into the soil. In such cases, these drought-tolerant pine species can utilize moisture from deeper soil/bedrock layers that has accumulated

Fig. 8.1 The dimorphic root structure of *Pinus brutia* (left) and *P. nigra* subsp. *pallasiana* (right), with surface (S) and deeper roots (D) uncovered after land/rockslides on Mt. Ambelos (Samos Island), Greece. The deeper roots are visible down to 4 m for *P. brutia* and to 10 m for *P. nigra*. Note that they exceed black pine's height. Photos Sarris & Christodoulakis

from rainfall events that have occurred over the course of up to 5–6 years (Sarris et al. 2007; Dorman et al. 2015b). Such root functioning permits drought-adapted pines not only to overcome the lack of topsoil moisture during the dry season, but also to deal with the effects of several consecutive drought years (Sarris et al. 2013). Other pine species such as *P. nigra* (Mazza et al. 2018) or even *P. pinea* (Mazza and Manetti 2013) have been shown to demonstrate similar performance. Pines, however, do not appear to grow deep roots when rainfall is not sufficient to penetrate deep into the soil, and/or evapotranspiration rates are too strong to permit sufficient rainfall to infiltrate deep into the soil (Dorman et al. 2015b). This is not unexpected, as not only the deepest, but also the shallowest roots in the world can be found in biomes adapted to arid regions (Fahn 1964; Kleidon and Heimann 1998; Fan et al. 2017).

Where deep pine roots exist in semiarid habitats with sufficient rainwater infiltration or with access to the water table, species such as *P. halepensis* are considered to perform hydraulic lift (Filella and Peñuelas 2003). This occurs when surface soils dry up during the dry season, but deep water is still available to roots. Hydraulic lift can support water and nutrient uptake during prolonged dry periods (Breda et al. 2006) in a similar way that hydraulic redistribution may permit fine roots to retain their life span under drought at levels similar to roots under conditions without water stress (Bauerle et al. 2008; Prieto et al. 2012).

Plants growing close to hydraulic-lifting plants may use such water (Prieto et al. 2010) to improve their moisture content and growth rates compared to those that do not have access to this source of water (Filella and Peñuelas 2003; Zou et al. 2005), especially during long periods of drought (Caldwell and Richards 1989; Dawson 1993; Horton and Hart 1998; Filella and Peñuelas 2003). Pine roots are known for their symbiosis with mycorrhizal fungi (Gardes and Bruns 1996; Ashkannejhad and Horton 2006). The uplifted water that reaches surface pine roots in dry season soil may also be valuable for sustaining the vitality of such microbial communities. It is believed that when surface soils are very dry, water is redirected, by water potential gradients, from tree shoots to fine roots, and then into mycorrhizal fungal hyphae, whereas hydraulic lift may buffer rhizosphere hyphae against the negative effects of severe soil drying (Querejeta et al. 2003, 2007). In contrast, common mycorrhizal networks provide a potential pathway for the transfer of hydraulically lifted water between plants (Egerton-Warburton et al. 2007). Thus, pines' deep-rooting potential may assist ecosystem productivity under drought by sustaining the vitality of soil microbiota as well as understory vegetation, although such an ecological trait may not only be limited to pines, but rather, it may be typical of many tree species adapted to semiarid conditions.

Apart from mycorrhizal networks trees may transport water, minerals and organic compounds by natural grafting of surface roots, a phenomenon that has been observed in many pine species and received various explanations for its occurrence (Graham and Bormann 1966; Liphschitz and Bonneh 1987; Lev-Yadun 2011; Tarroux and DesRochers 2011). The extent of this phenomenon in Mediterranean pine forests and its ecological impact is difficult to assess as it has not well been investigated. Nonetheless, its potentials effects may be stronger the closer pines grow together and the longer their lateral roots. Thus, its existence may be less frequent in open dry habitat pine stands or in pines that invest in deep rather than surface roots (Loehle and Jones 1990).

Time is an additional factor to consider. Water utilization by pine roots may considerably change as aridity varies within the growing season, but also as climate changes over longer climatic periods.

In semiarid conditions pines absorb moisture from the topsoil during the wet season. As these reservoirs dry up, roots of some, but not all plants, track water depletion down the soil/bedrock moisture profile (Bledsoe et al. 2014). Pines in Mediterranean climates have been known to track water (Fig. 8.2). As the seasonal climate becomes drier and the overlying substrate is depleted of moisture, progressively deeper water sources, including weathered bedrock can be exploited to a depth of several meters, enabling trees to escape desiccation (Rose et al. 2003 for *P. jeffreyi*; Sarris et al. 2013 for *P. brutia*). Water uptake by roots shifts back to topsoil moisture once the wet season resumes (Sarris et al. 2013). Not only roots, but mycorrhizal fungi also extend into the bedrock and/or search out suspended pockets of water (Hubbert et al. 2001; Querejeta et al. 2007; Estrada-Medina et al. 2013; Bledsoe et al. 2014). Mycorrhizal fungal hyphae can also track roots through fractures in the rock and then extend from root tips penetrating the bedrock for water extraction (Allen 2006; Bledsoe et al. 2014). Thus, we must ask whether shifts from top to deeper moisture utilization, under increasing drought stress, involve or are driven by changes in the growth characteristics and/or the community structure of

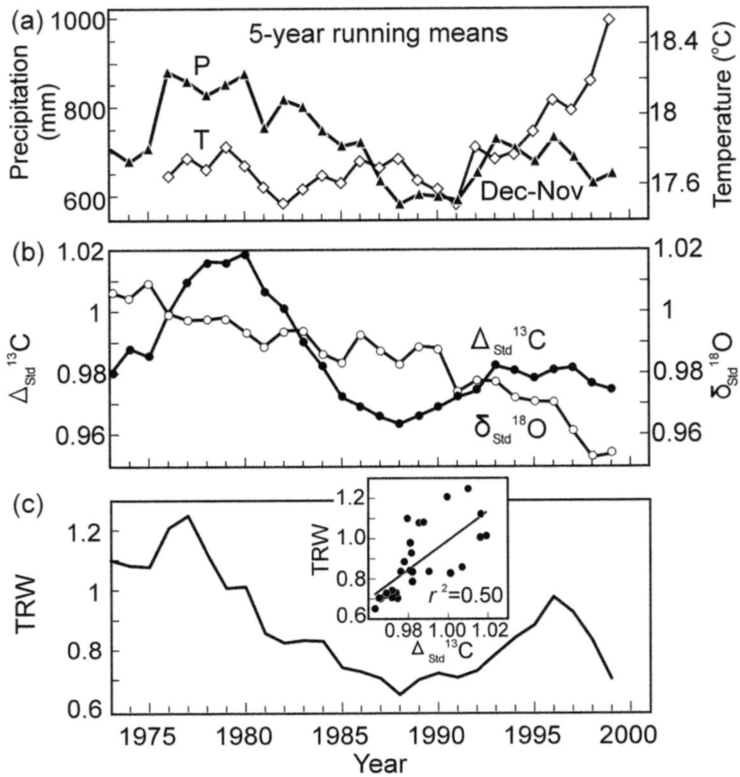

Fig. 8.2 (**a**) Full-year (Dec–Nov) precipitation (*P*) and temperature (*T*) for Samos Island, Greece; (**b**) $\Delta_{std}{}^{13}C$ and $\delta_{std}{}^{18}O$ tree-ring signatures from *Pinus brutia* ($S_{n=3}$ trees); and (**c**) corresponding tree-ring width (TRW). The insert shows the linear regression ($r^2 = 0.50$, $p < 0.001$) between the $\Delta_{std}{}^{13}C$ tree-ring signals and corresponding TRW ($S_{n=3}$ trees) appearing in (**b**) and (**c**). As long-term drought intensifies (**a**), drought stress increases as displayed by a decline in $\Delta_{std}{}^{13}C$ (**b**), and growth declines as displayed by a decline in TRW (**c**). Moreover, deep roots in *P. brutia* become more important, as indicated by the gradual utilization of deeper moisture pools, displayed by the decline in $\delta_{std}{}^{18}O$ (**b**). (See Sarris et al. 2013 for details)

pine root mycorrhiza, since changes in soil moisture have been found to have a strong impact on mycorrhiza (Jarvis et al. 2013 for *P. sylvestris*).

A shift in the utilization of water from different soil horizons is typical of plants with dimorphic roots in Mediterranean-type ecosystems (Dawson and Pate 1996). The phenomenon can also be observed from one decade to another, with a series of wet years permitting pines to rely more on topsoil moisture, whereas a series of dry years can force pines such as *P. brutia* and *P. halepensis* into lower productivity driven by deeper moisture uptake (Sarris et al. 2007; Dorman et al. 2015b). Nevertheless, when a series of dry years causes deeper moisture pools to become depleted, these pines shift their water uptake back to surface moisture inputs (Dorman et al. 2015b; Fig. 8.3). In such a case, soil depth, by affecting rooting depth, together with the degree of exposure to the sun's radiation, are among the

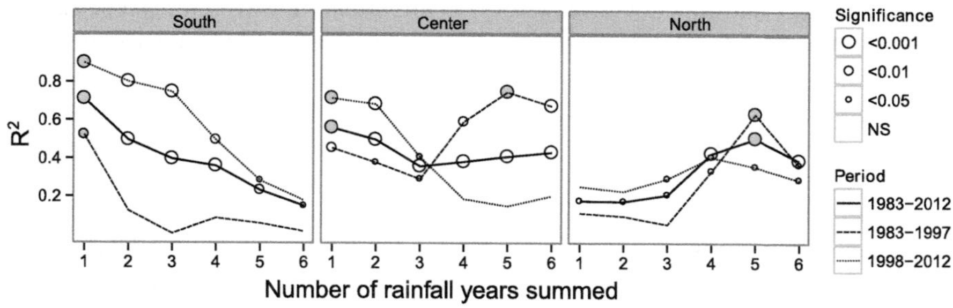

Fig. 8.3 Linear regression of *Pinus halepensis* growth (BAI) and different rainfall integration timescales (1–6 years) for three periods (1983–1997 wet climate, 1998–2012 dry climate, and 1983–2012) in three regions of Israel. The integration period that produced the highest r^2 in each region and period is marked with a *filled circle*. Under a wetter climate, deep-rooted *P. halepensis* trees were identified, utilizing moisture from deeper soil supplied by the last 5 years of rainfall (North). Under an extremely dry climate, coinciding with pine desiccation, shallow-rooted pines that depended on current year's rainfall were identified (South). Under intermediate moisture, pines depended on deeper moisture supplies in rainy years (Central). However, after severe drought, deeper moisture pools most likely became depleted, and pines in Central Israel shifted their water uptake to surface moisture inputs. (See Dorman et al. 2015b for details)

prime factors that determine the fate of pine desiccation or survival (Dorman et al. 2015c). Deep soils may affect tree architecture, prohibiting deep rooting into the bedrock, which in turn can lead to elevated tree vulnerability in times of drought on south-facing slopes with sparse pines, particularly when the entire soil profile dries out, while water stored in rock layers is beyond the reach of tree roots (Fensham and Fairfax 2007; Dorman et al. 2015c).

A similar response can be observed for pine species adapted to sandy habitats with a shallow water table, such as *P. pinea*. *Pinus pinea* roots reach toward the capillary rise of the water table under normal climatic conditions. Under drought, when the water table declines below 4 m, the deepest rooting depth observed for the species (Stone and Kalisz 1991), *P. pinea* can only rely on surface moisture inputs for survival (Mazza and Sarris, pers comm). Losing contact with the water table likely reduces these pines' productivity and makes them more vulnerable to drought stress, which may further intensify due to competition from better drought-adapted species with deeper root system capacity (e.g., *P. halepensis*).

8.2 Pine Roots and Forest Management Under Climatic Change

Management of Mediterranean pine forests for resistance to drought intensification should aim at increasing moisture availability within tree stands. Thinning has been widely applied and is still considered an important tool for achieving this goal, although it has its limitations (Dorman et al. 2015b, c). The effect of such practices on the structure and function of pine roots is important; however, it has been given little attention to date. Thus, successful management practices to increase pine stand resilience to drought should aim at both increasing root depth, where deeper moisture pools do exist, and removing pines that have low productivity and shallow root systems. However, the chances of increasing erosion and heat stress on soil due to thinning (Dorman et al. 2015b, c) should also be considered.

After thinning, trees may develop more extensive individual root systems over time, hence, increasing their capacity to extract water from the soil during and after drought periods, compared to trees in unthinned stands (Whitehead et al. 1984; Aussenac and Granier 1988; Misson et al. 2003). Comparing two *P. pinaster* stands with different thinning intensity under decreasing precipitation revealed significant differences in groundwater utilization. The stand with the lower tree density, unlike its denser counterpart, utilized deeper moisture supplies from past rainy years, especially for earlywood formation (Mazza et al. 2014). As the upper soil layers dry up faster in open woodland than in dense forested stands (Raz-Yaseef et al. 2010), pines in low-density stands created by thinning may explore deeper soil horizons, thus reaching water resources stored from past years' precipitation, particularly during periods of drought.

In turn, understanding rooting functioning can assist forest thinning. Based on an intra-stand dendroecological approach using tree rings, Mazza et al. (2018) discovered that in wet years most *P. nigra* subsp. *laricio* trees utilized not only surface but also deeper moisture pools at a Calabrian site within the pines' mesic to xeric distribution range. In contrast, under a drier climate, approximately one-third of the trees within the stand significantly reduced their capacity to utilize deeper soil/bedrock moisture (Fig. 8.4). The survival of these trees, apparently reliant on surface roots during drought, depended mostly on any surface rainfall input provided by summer rainfall. This suggests that they are the most vulnerable to summer drought within the stand.

The intra-stand dendroecological approach could also prove useful in supporting the management of mixed stands (e.g., pine-oak forests). The management of mixed stands is one of the great challenges for increasing forest resilience to climate change; understanding root structure and functioning may hold one of the keys for their successful management. In the Mediterranean, most of the pine stands planted for the purpose of reforesting degraded lands are at the end of their pioneering role, as the chain of ecological succession continues towards more resilient ecosystems characterized by native broadleaved species and higher levels of biodiversity (Pausas et al. 2004). The root systems of oaks, such as *Quercus cerris*, may allow them to explore a more extensive surface soil layer than pines, especially in mixed stands with *P. nigra* trees. In turn, *P. nigra* trees appear to have deeper roots that utilize deeper moisture pools to avoid competition with *Q. cerris* trees in the upper soil layers (Mazza et al. 2021).

Therefore, where deep rooting permits higher resilience to drought, identifying and quantifying the shallow-rooted pines more vulnerable to drought stress is vital for applying appropriate site-specific silvicultural treatments (e.g., selective thinning or modulating thinning intensity; Mazza et al. 2018). Such an approach may

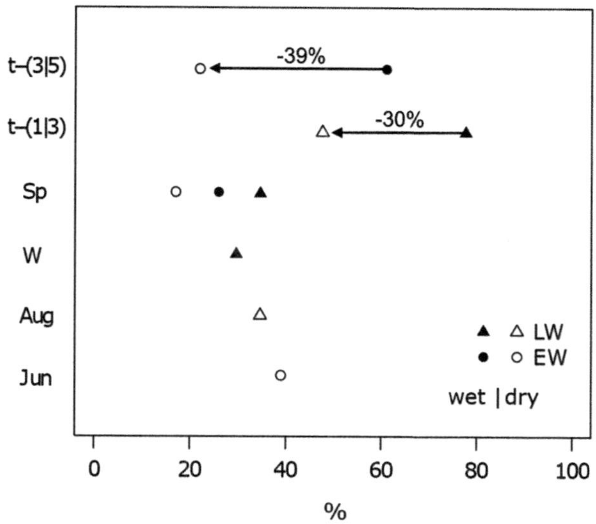

Fig. 8.4 Percentage of *Pinus. nigra* subsp. *laricio* trees from Southern Italy producing significant correlations with three precipitation integration time-scales (monthly, seasonal 1–3 and 3–5 past years) for earlywood (EW) and latewood (LW) formation during wet and dry climatic periods. Only the best climate drivers are shown. Arrows indicate the percentage of trees that depleted deeper moisture pools during dry climate and shifted their water uptake from multiple year precipitation to summer rainfall using surface roots. These trees were considered to exhibit higher vulnerability to future drought events. (See Mazza et al. 2018 for details)

increase the resilience of single stands or mixed pine stands to drought, which is expected to affect large areas of the Mediterranean region under climate change.

References

Allen MF (2006) Water dynamics of mycorrhizas in arid soils. In: Gadd GM (ed) Fungi in biogeochemical cycles. Cambridge University Press, Cambridge, pp 74–97

Allen CD, Macalady AK, Chenchouni H et al (2010) A global overview of drought- and heat-induced tree mortality reveals emerging climate change risks for forests. For Ecol Manag 259:660–684

Andivia E, Zuccarini P, Grau B et al (2018) Rooting big and deep rapidly: the ecological roots of pine species distribution in southern Europe. Trees. https://doi.org/10.1007/s00468-018-1777-x

Ashkannejhad S, Horton TR (2006) Ectomycorrhizal ecology under primary succession on coastal sand dunes: interactions involving *Pinus contorta*, suilloid fungi and deer. New Phytol 169(2):345–354

Aussenac G, Granier A (1988) Effects of thinning on water-stress and growth in Douglas-fir. Can J For Res 18:100–105

Barbéro M, Loisel R, Quézel P et al (1998) Pines of the Mediterranean Basin. In: Richardson DM (ed) Ecology and biogeography of *Pinus*. Cambridge University Press, Cambridge

Bauerle TL, Richards JH, Smart DR et al (2008) Importance of internal hydraulic redistribution for prolonging the lifespan of roots in dry soil. Plant Cell Environ 31:177–186. https://doi.org/10.1111/j.1365-3040.2007.01749.x

Berntson GM, Farnsworth EJ, Bazzaz FA (1995) Allocation, within and between organs, and the dynamics of root length changes in two birch species. Oecologia 101:439–447. https://doi.org/10.1007/BF00329422

Bledsoe CS, Allen MF, Southworth D (2014) Beyond mutualism: complex mycorrhizal interactions. In: Lüttge U, Beyschlag W, Cushman J (eds) Progress in botany 75. Springer, Heidelberg/New York/Dordrecht/London, pp 311–334

Breda N, Huc R, Granier A et al (2006) Temperate forest trees and stands under severe drought: a review of ecophysiological responses, adaptation processes and long-term consequences. Ann For Sci 63:625–644. https://doi.org/10.1051/forest:2006042

Brunner I, Herzog C, Dawes MA et al (2015) How tree roots respond to drought. Front Plant Sci. https://doi.org/10.3389/fpls.2015.00547

Caldwell MM, Richards JH (1989) Hydraulic lift: water efflux from upper roots improves effectiveness of water uptake by deep roots. Oecologia 79(1):1–5. https://doi.org/10.1007/s11104-010-0638-1

Cannon HL (1960) The development of botanical methods of prospecting for Uranium on the Colorado Plateau. US Geol Surv Bull 1085-A:1–50

Čermák J, Nadezhdina N, Meiresonne L et al (2008) Scots pine root distribution derived from radial sap flow patterns in stems of large leaning trees. Plant Soil 305:61–75. https://doi.org/10.1007/s11104-007-9433-z

Danjon F, Reubens B (2008) Assessing and analyzing 3D architecture of woody root systems, a review of methods and applications in tree and soil stability, resource acquisition and allocation. Plant Soil 303:1–34. https://doi.org/10.1007/s11104-007-9470-7

Dawson TE (1993) Hydraulic lift and water use by plants: implications for water balance, performance and plant-plant interactions. Oecologia 95(4):554–565

Dawson TE, Pate JS (1996) Seasonal water uptake and movement in root systems of Australian phraeatophytic plants of dimorphic root morphology: a stable isotope investigation. Oecologia 107:13–20

Dorman M, Perevolotsky A, Sarris D et al (2015a) Amount vs. temporal pattern: on the importance of intra-annual climatic conditions on tree growth in a dry environment. J Arid Environ 118:65–68

Dorman M, Perevolotsky A, Sarris D et al (2015b) The effect of rainfall and competition intensity on forest response to drought: lessons learned from a dry extreme. Oecologia. https://doi.org/10.1007/s00442-015-3229-2

Dorman M, Svoray T, Perevolotsky A et al (2015c) What determines tree mortality in dry environments? A multi-perspective approach. Ecol Appl 25:1054–1071

Egerton-Warburton LM, Querejeta JI, Allen MF (2007) Common mycorrhizal networks provide a potential pathway for the transfer of hydraulically lifted water between plants. J Exp Bot 58:1473–1483

Estrada-Medina H, Graham RC, Allen MF et al (2013) The importance of limestone bedrock and dissolution karst features on tree root distribution in northern Yucatan, Mexico. Plant Soil 362:37–50. https://doi.org/10.1007/s11104-012-175-x

Fahn A (1964) Some anatomical adaptations of desert plants. Phytomorphology 14:93–102

Fan Y, Miguez-Machob G, Jobbágyc EG et al (2017) Hydrologic regulation of plant rooting depth. PNAS. https://doi.org/10.1073/pnas.1712381114

Fensham RJ, Fairfax RJ (2007) Drought-related tree death of savanna eucalypts: species susceptibility, soil conditions and root architecture. J Veg Sci 18:71–80

Filella I, Peñuelas J (2003) Indications of hydraulic lift by *Pinus halepensis* and its effects on the water relations of neighbor shrubs. Biol Plant 47(2):209–214. https://doi.org/10.1023/B:BIOP.0000022253.08474.fd

Frattegiani M, Mencuccini M, Mercurio R et al (1994) Quantitative analysis of Stone pine (*Pinus pinea* L.) root systems morphology and its relationships with water table and soil characters. Invest Agraria, Fuera de Serie 3:405–416

Gardes M, Bruns TD (1996) Community structure of ectomycorrhizal fungi in a *Pinus muricata* forest: above- and below-ground views. Can J Bot 74:1572–1583

Gindel I (1973) Xerophytism. In: A new ecophysiological approach to forest-water relationships in arid climates. Springer, Dordrecht

Graham BF, Bormann FH (1966) Natural root grafts. Bot Rev 32:255–292

Guo L, Chen J, Cui X et al (2013) Application of ground penetrating radar for coarse root detection and quantification: a review. Plant Soil 362:1–23. https://doi.org/10.1007/s11104-012-1455-5

Horton JL, Hart SC (1998) Hydraulic lift: a potentially important ecosystem process. Trends Ecol Evol 13(6):232–235. https://doi.org/10.1016/S0169-5347(98)01328-7

Hruška J, Čermák J, Sustek S (1999) Mapping tree root systems with ground-penetrating radar. Tree Physiol 19:125–130. https://doi.org/10:1093/treephs/19.2.125

Hubbert KR, Beyers JL, Graham RC (2001) Roles of weathered bedrock and soil in seasonal water relations of *Pinus jeffreyi* and *Arctostaphylos patula*. Can J For Res 31:1947–1957

Jarvis S, Woodward S, Alexander IJ (2013) Regional scale gradients of climate and nitrogen deposition drive variation in ectomycorrhizal fungal communities associated with native scots pine. Glob Chang Biol 19:1688–1696. https://doi.org/10.1111/gcb.12178

Kleidon A, Heimann M (1998) A method of determining rooting depth from a terrestrial biosphere model and its impacts on the global water and carbon cycle. Glob Chang Biol 4:275–286

Körner C, Sarris D, Christodoulakis D (2005) Long-term increase in climatic dryness in the East-Mediterranean evidenced for the island of Samos. Reg Environ Chang 5:27–36

Leucci G (2010) The use of three geophysical methods for 3D images of total root volume of soil in urban environments. Explor Geophys 41:268–278. https://doi.org/10.1071/eg09034

Lev-Yadun S (2011) Why should trees have natural root grafts? Tree Physiol 31:575–578

Liphschitz N, Bonneh O (1987) Living stumps-circumstantial evidence for root grafting in *Pinus halepensis* and *P. brutia* plantations in Israel. Isr J Bot 36:41–43

Liphschitz N, Lev-Yadun S, Rosen E, Waisel Y (1984) The annual rhythm of activity of the lateral meristems (cambium and phellogen) in Pinus halepensis mill. and Pinus pinea L. IAWA Bulletin 5:263–274

Loehle C, Jones RH (1990) Adaptive significance of root grafting in trees. Funct Ecol 4:268–271

Mauri A, Di Leo M, de Rigo D et al (2016) *Pinus halepensis* and *Pinus brutia* in Europe: distribution, habitat, usage and threats. In: San-Miguel-Ayanz J, de Rigo D, Caudullo G et al (eds) European atlas of forest tree species. Publ Off EU, Luxembourg, p e0166b8+

Mazza G, Manetti MC (2013) Growth rate and climate responses of *Pinus pinea* L. in Italian coastal stands over the last century. Clim Chang 121(4):713–725

Mazza G, Cutini A, Manetti MC (2014) Influence of tree density on climate-growth relationships in a *Pinus pinaster* Ait. forest in the northern mountains of Sardinia (Italy). iForest 8(4):456–463

Mazza G, Sarris D, Chiavetta U (2018) An intra-stand approach to identify intra-annual growth responses to climate in *Pinus nigra* subsp. *laricio* Poiret trees from southern Italy. For Ecol Manag 425:9–20

Mazza G, Markou L, Sarris D (2021) Species-specific growth dynamics and vulnerability to drought at the single tree level in a Mediterranean reforestation. Trees (in press)

Misson L, Antoine N, Joel G (2003) Effects of different thinning intensities on drought response in Norway spruce (*Picea abies* (L.) Karst.). For Ecol Manag 183:47–60

Nadezhdina N, Čermák J (2003) Instrumental methods for studies of structure of and function of root systems of large trees. J Exp Bot 54:1511–1521. https://doi.org/10.1093/jxb/erg154

Oliveira MRG, van Noordwijk M, Gaze SR et al (2000) Auger sampling, ingrowth cores and pinboard methods. In: Smit AL, Bengough AG, van Noordwijk M et al (eds) Root methods: a hand book. Springer, Berlin, pp 175–210

Pausas JG, Bladé C, Valdecantos A, Seva JP, Fuentes D, Alloza JA, Vilagrosa A, Bautista S, Cortina J, Vallejo R (2004) Pines and oaks in the restoration of Mediterranean landscapes of Spain: new perspectives for an old practice – a review. Plant Ecol 171:209–220
Polomski J, Kuhn N (2002) Root research methods. In: Weisel Y, Eshel A, Kafkafi U (eds) Plant roots: the hidden half, 3rd edn. Marcel Dekker, New York, pp 295–322
Prieto I, Kikvidze Z, Pugnaire FI (2010) Hydraulic lift: soil processes and transpiration in the Mediterranean leguminous shrub *Retama sphaerocarpa* (L) Boiss. Plant Soil 329(1–2):447–456. https://doi.org/10.1007/s11104-009-0170-3
Prieto I, Armas C, Pugnaire FI (2012) Water release through plant roots: new insights into its consequences at the plant and ecosystem level. New Phytol 193:830–841. https://doi.org/10.1111/j.1469-8137.2011.04039.x
Querejeta JI, Egerton-Warburton LM, Allen MF (2003) Direct nocturnal water transfer from oaks to their mycorrhizal symbionts during severe soil drying. Oecologia 134:55–64
Querejeta JI, Egerton-Warburton LM, Allen MF (2007) Hydraulic lift may buffer rhizosphere hyphae against the negative effects of severe soil drying in a California oak savanna. Soil Biol Biochem 39:409–417
Raz-Yaseef N, Rotenberg E, Yakir D (2010) Effects of spatial variations in soil evaporation caused by tree shading on water flux partitioning in a semi-arid pine forest. Agric For Meteorol 150:454–462. https://doi.org/10.1016/j.agrformet.2010.01.010
Reubens B, Poesen J, Danjon F et al (2007) The role of fine and coarse roots in shallow slope stability and soil erosion control with a focus on root system architecture: a review. Tree Struct Funct 21:385–402. https://doi.org/10.1007/s00468-007-0132-4
Rose KL, Graham RC, Parker DR (2003) Water source utilization by *Pinus jeffreyi* and *Arctostaphylos patula* on thin soils over bedrock. Oecologia 134(1):46–54
Sanford RL, Cuevas E (1996) Root growth and rhizosphere interactions in tropical forest. In: Mulkey SS, Chazdon RL, Smith AP (eds) Tropical forest plant ecophysiology. Chapman Hall, New York, pp 269–300
Sarris D, Christodoulakis D, Körner C (2007) Recent decline in precipitation and tree growth in the eastern Mediterranean. Glob Chang Biol 13:1187–1200
Sarris D, Siegwolf R, Körner C (2013) Inter- and intra-annual stable carbon and oxygen isotope signals in response to drought in Mediterranean pines. Agric For Meteorol 168:59–68
Stone EL, Kalisz PJ (1991) On the maximum extent of tree roots. For Ecol Manag 46:59–102
Sudmeyer R, Speijers J, Nicholas BD (2004) Root distribution of *Pinus pinaster, P. radiata*, *Eucalyptus globulus* and *E. kochii* and associated soil chemistry in agricultural land adjacent to tree lines. Tree Physiol 24(12):1333–1346
Tarroux E, DesRochers A (2011) Effect of natural root grafting on growth response of Jack pine (*Pinus banksiana*; Pinaceae). Am J Bot 98(6):967–974
Thomas FM (2014) Ecology of phreatophytes. In: Lüttge U, Beyschlag W, Cushman J (eds) Progress in botany, vol 75. Springer, Berlin/Heidelberg
Van Noordwijk M, Brouwer G, Meijboom F et al (2000) Trench profile techniques and core BreakMethods. In: Smit AL, Bengough AG, van Noordwijk M et al (eds) Root methods: a hand book. Springer, Berlin, pp 211–234
Vennetier M, Thabeet A, Didier C et al (2013) Climate change impact on tree architectural development and leaf area. INTECH Open Access Publisher. https://doi.org/10.5772/51510
Whitehead D, Jarvis PG, Waring RH (1984) Stomatal conductance, transpiration, and resistance to water-uptake in a *Pinus sylvestris* spacing experiment. Can J For Res 14:692–700
Zenone T, Morelli G, Teobaldelli M et al (2008) Preliminary use of ground-penetrating radar and electrical resistivity tomography to study tree roots in pine forests and poplar plantations. Funct Plant Biol 35:1047–1058. https://doi.org/10.1071/fp08062
Zou CB, Barnes PW, Archer S et al (2005) Soil moisture redistribution as a mechanism of facilitation in savanna tree-shrub clusters. Oecologia 145(1):32–40. https://doi.org/10.1007/s00442-005-0110-8

Chapter 9
Stress and Tree Mortality in Mediterranean Pine Forests: Anthropogenic Influences

Álvaro Rubio-Cuadrado, Rosana López, Jesús Rodríguez-Calcerrada, and Luis Gil

9.1 Introduction

In recent years, there has been a proliferation of studies related to forest decline and tree mortality in many ecosystems throughout the world (Allen et al. 2010; Caudullo and Barredo 2019). Most of these studies describe the impact on forests of increasing temperatures and drought resulting from climate change. However other factors, often overlooked, can predispose forests to suffer an excess of stress, leading to an increase in mortality.

Mediterranean ecosystems have been significantly transformed by man for thousands of years. In addition, land-use changes in recent decades have had a considerable impact upon Mediterranean forests; they include abandonment of non-productive lands, abandonment of traditional uses, a trend towards non-management of forests, spread of insects beyond their natural habitat, etc. All these anthropogenic factors affect tree health and resilience and may influence tree mortality events. In this study, we perform: (i) a review of the use and management history of pine forests, focusing on Spain as one of the countries where pine afforestation was commonplace and where the suitability of its use has given rise to heated debates; (ii) a bibliographic review of all mortality events occurring in Mediterranean pine forests, discussing the geographical and temporal trends encountered; (iii) an analysis of the factors conducive to tree decline and mortality and a description of mortality mechanisms in pine forests; and (iv) a review of management techniques designed to mitigate stress and reduce mortality in Mediterranean pine forests.

Á. Rubio-Cuadrado (✉) · R. López · J. Rodríguez-Calcerrada · L. Gil
Departamento de Sistemas y Recursos Naturales, Escuela Técnica Superior de Ingeniería de Montes, Forestal y del Medio Natural, Universidad Politécnica de Madrid, Madrid, Spain
e-mail: alvaro.rubio.cuadrado@upm.es

G. Ne'eman, Y. Osem (eds.), *Pines and Their Mixed Forest Ecosystems in the Mediterranean Basin*, Managing Forest Ecosystems 38,
https://doi.org/10.1007/978-3-030-63625-8_9

9.2 History of Use and Management

Forest history is linked to both extensive deforestation and extensive afforestation; the *Pinus* genus has played a key role in both phenomena. For example, eroded lands were unable to sustain afforestation with *Quercus* spp. (Caro 1908) while pine afforestation was successful, which justified the use of the genus. Paleobotanical studies have demonstrated that forest degradation is related to the local and regional extinction of pinewoods (Carrión 2012). However the erroneous concept of "natural potential vegetation" (Carrión and Fernández 2009), promoted by phytosociologists, excludes pine woodlands as the last stage of plant succession, ignoring old historical sources which report large areas occupied by pines in the past. For example, the first forestry registry in Spain indicated a total of 2,171,120 ha of pinewoods (Engineers Corps 1859).

Except for the oro-Mediterranean area, phytosociologists have assumed that pines were a secondary and unnatural species (e.g. Rivas-Martínez 1987; and see Grove and Rackam's opinion, 2001, regarding *Pinus halepensis*). This opinion has had a major influence upon the philosophies that determine the policies and management of the Forestry Administration and they predominate in society. For example, Rivas-Martínez et al., who were commissioned to conduct the Cartography and Inventory of Habitat of Directive 92/43/EEC in Spain (Rivas-Martínez et al. 1993), did not consider the presence of pinewoods in most of their natural habitats; this conception was also assimilated into the Natura 2000 Network. For instance in Spain, in areas where *P. uncinata* has become extinct, such as in the Cantabrian Mountains (Venturas et al. 2013), the Central Range or Sierra Nevada (Fig. 9.1), heaths are the only habitat recognized as natural. Another example is the disappearance of the Azaba pine forest (Salamanca, Spain), the existence of which is not recognized, although this forest was described by Nieulant in 1770 (Fig. 9.2). Currently, the locations in which this forest of *P. pinaster* once thrived are occupied by meadows with poorly developed centennial holm oak afforestations.

State-sponsored afforestation began in Spain in the nineteenth century with a view to protecting the soil and restoring deforested areas. However, until 1940 the forested area increased by less than 50,000 ha (Ortuño and Ceballos Jiménez 1977). In 1940, a policy of large-scale afforestation was initiated (Table 9.1) with the fundamental social goal of creating rural employment (Ximénez de Embún Oseñalde and Ceballos 1939). At that time, the importance of seed origin, i.e. provenance,

Fig. 9.1 Granada Mountains in 1930. (Photo: Kindel)

Fig. 9.2 Map of the border between Spain and Portugal in 1643. It highlights the importance of Azaba pines. (Simancas Archive)

Table 9.1 Evolution of the area occupied by pine trees in Spain from 1940 to 2007 (×1000 ha)

Species	Forested area 1940	Afforested area 1940–1993	Forested area 2007	1993–2007 increase	1993–2007 increase (%)
P. halepensis Mill.	806	550	2121.2	765.2	56.43
P. pinaster Aiton	652	840	1364.0	−128.0	−8.58
P. nigra J.F.Arnold	386	404	921.7	131.7	16.67
P. sylvestris L.	436	611	1196.1	149.1	14.24
P. pinea L.	179	263	455.0	13.0	2.94
P. uncinata Mill. ex Mirbel	58	17	20.0	−55.0	−73.33
P. canariensis C. Sm ex DC.	52	29	79.4	−1.6	−1.98
Total	**2569**	**2714**	**6157.4**	**874.4**	**16.55**

was unknown. By 1993, the land area occupied by pines had doubled (106%) and that of *P. pinaster* had increased by 129%; *Pinus sylvestris* had expanded its area by 171% in relation to 1940. For *P. nigra*, the difficulty of obtaining local seeds prompted the choice of the non-native *P. nigra* subsp. *nigra*. The reduction in the land area used for agriculture and grazing since 1970 has not been accompanied by the expected increase in the area of pine forests. According to the Spanish Forest Map (SFM 2007), pinewood area in relation to 1993 has increased only 16.55%.

Similarly, throughout the Mediterranean Basin, there has been an important afforestation effort during the nineteenth and, particularly, twentieth century

Table 9.2 Total area covered by forest (×1000 ha), afforested area (×1000 ha and %) and increase of the afforested area between 1990 and 2015 of the Mediterranean countries (FAO 2015)

Country	Total forest area	Afforested area	Afforested area (%)	1990–2015 increase (%)
Turkey	11,715	3386	28.9	126.5
Spain	18,418	2909	15.8	42.8
France	16,989	1967	11.6	28.7
Bosnia and Herzegovina	2185	999	45.7	−4.6
Portugal	3182	891	28	11.1
Tunisia	1041	725	69.6	147.4
Morocco	5632	706	12.5	47.7
Italy	9297	639	6.9	16.8
Algeria	1956	556	28.4	67.0
Syria	491	294	59.8	68.1
Libya	217	217	100	0.0
Serbia	2720	215	7.9	451.3
Greece	4054	140	3.5	18.6
Macedonia	998	105	10.5	0.0
Albania	772	90	11.7	−12.9
Israel	165	89	53.9	34.8
Croatia	1922	75	3.9	−18.5
Egypt	73	73	100	65.9
Jordan	98	47	48.1	–
Slovenia	1248	34	2.7	0.0
Cyprus	173	31	17.8	26.3
Lebanon	137	11	7.9	–
Montenegro	827	8	1	–

The data for Portugal and Syria correspond to 2000–2015 and 1990–2010, respectively, due to the lack of data for the whole period. Note that due to the use of different data sources, the data for Spain do not match those in Table 9.1

(Table 9.2). Only between 1995 and 2015 the afforested area has increase a 47.6%, and nowadays, the Mediterranean Basin is, after Asia, the region of the world with the highest percentage of planted forests (16.8%; FAO 2015).

The failure of some afforestations due to the use of poorly adapted non-local provenances (Alia et al. 1995) has led to the false belief that pine mortality is an inherent feature of pine afforestations. Criticism of forestry activities caused the pine afforestations to remain unmanaged. Afforestations, which generally involved the planting of 1600 pines/ha, when direct sowing was not performed, have not received silvicultural treatment and can easily be destroyed by fire. Rural depopulation, abandonment of forest traditional uses, lack of budget for silvicultural interventions and low value of extracted wood in the first silvicultural interventions limit the application of an adequate management, more so when the administration is increasingly imposing the principle of "non-management" as the "natural" activity in forestry practice. The presence of non-local provenances, together with intraspecific competition, defines the recent history of an important area of our pinewoods.

9.3 The Magnitude of the Decline in Mediterranean Pine Forests

To evaluate the current status of the pine forests we performed an exhaustive search for published papers and consulted regional experts in order to find published or unpublished information related with pine decline (Fig. 9.3 and Table 9.3 within Annex). From all the information obtained, we have only considered those cases with significant adult tree mortality at the stand or population level, due to biotic or abiotic causes, but excluding fires and occasional atmospheric events (windstorms, snowstorms, etc.). Here *P. brutia* was considered to be a subspecies of *P. halepensis* (i.e., *P. halepensis* subsp. *brutia*).

The species most affected by mortality events are the temperate species, *P. sylvestris*, mainly in Spain, Slovenia and Italy, and *P. nigra*, mainly in Slovenia and to a lesser extent in Spain (Fig. 9.3). However, other temperate species, such as *P. uncinata* and *P. mugo*, hardly presented any mortality events, possibly because the forests of these species tend to have lower densities and have been less transformed by man. Conversely, the more typically Mediterranean species *P. halepensis*, and to a lesser extent, *P. pinaster*, also suffered abundant mortality events, throughout the Mediterranean, in the case of *P. halepensis* (Fig. 9.4), and mainly in Spain and France, in the case of *P. pinaster*. Finally, for *P. pinea* only a few mortality events have been reported, possibly due to the low densities usually maintained for the benefit of pine nut production.

Drought-induced tree mortality was reported in 45% of the events (Table 9.3 within Annex) while pest and/or disease were reported in 65%. In 15% of the previous percentages, the combined effects of drought and pest or disease were involved in mortality. Only in 5% of the events was not due to any of the above causes. In

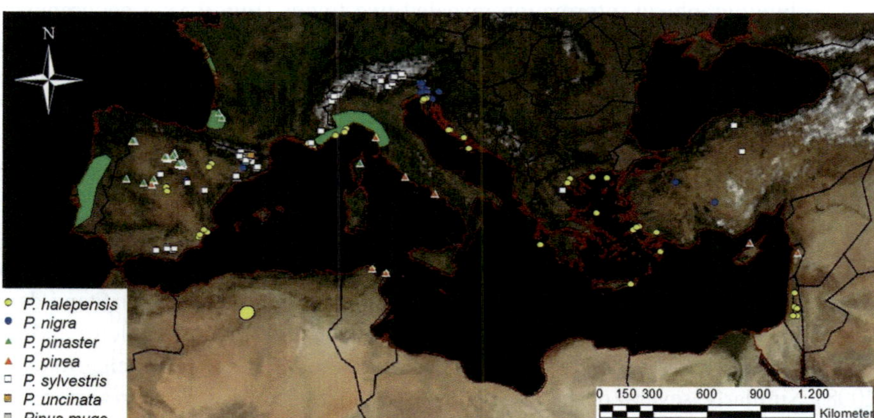

Fig. 9.3 Localized events of and large areas affected by mortality in recent decades. The color of the areas indicates the affected species according to the legend. Mortality data are presented in Table 9.3 within Annex. (Satellite image retrieved from ASF)

Fig. 9.4 *Pinus halepensis* forest located in Spain with an abundance of dying trees in 2014. (Photo: Alberto Vilagrosa)

79% of the events in which drought was considered the main cause of mortality, the climate anomalies (CA in Table 9.3) had negative values, that is, these mortality events ocurred in years that were drier than average, whereas, in 21% of these events, climatic anomalies had positive values, i.e. they correspond to humid years. However, we should consider that frequently drought does not kill the trees immediately but weakens them and mortality happens a few years after (Bigler et al. 2007).

Our bibliographical information revealed two trends (Fig. 9.5): a geographical trend, with an increase in mortality from South to North and form East to West; and a temporal trend, with a considerable increase over time in the number of papers focusing on the study of forest mortality events. Concerning the spatial trend, some regions may have site and historical characteristics that make pine forests in these regions more sensitive to climate change. Spain, where documented cases of tree mortality have been especially abundant, is the second Mediterranean country with the largest artificially afforested area (Table 9.2), with these afforestations being generally older than those carried out in other countries, such as Turkey, the country with the largest afforested area. These afforestations often present high stand density, a low coefficient of variation of tree diameter, high monospecificity and, in some cases, are established beyond their natural climatic range (Gil-Pelegrín et al. 2008). Similarly, several of the affected forests in Israel are *P. halepensis* afforestations growing in regions with an annual rainfall of less than 450 mm (Dorman et al. 2013), a value below which there are no natural *P. halepensis* forests (Schiller 2000). These afforestations performed beyond the species' natural range were often conducted during a particularly rainy period and based on climatic records

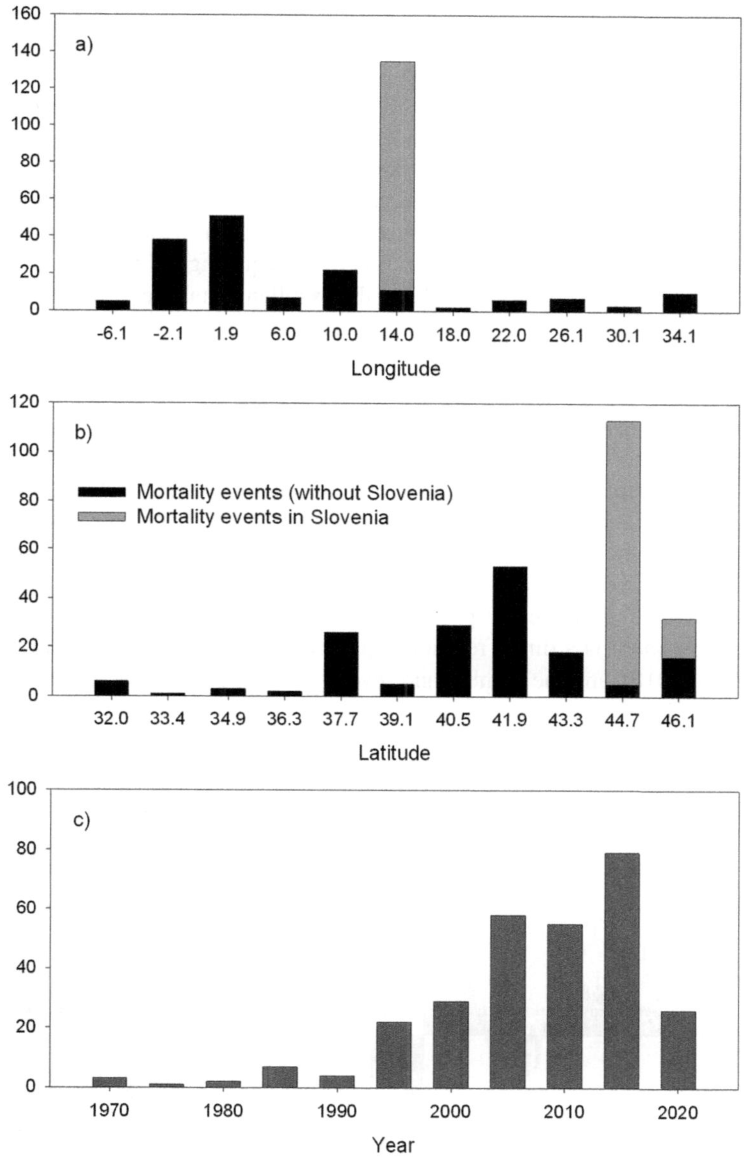

Fig. 9.5 Number of recorded mortality events (see Table 9.3 within Annex) according to longitude (**a**), latitude (**b**) and time (**c**). In graphs (**a**) and (**b**) data provided by Slovenian Forest Service were differentiated. For mortality events prolonged over several consecutive years, the mean date has been considered. Likewise, for those mortality events studied over large areas, the mean geographic coordinates have been used. The width of the intervals for longitude is 4.022 (being −8.125/−4.103 the first interval), 1.408 for latitude (being 31.333/32.741 the first interval) and 5 years for time (being 1968/1972 the first interval). Note that the last time interval in (**c**) only includes mortality events occurring in 2018 and 2019

overestimating the real humidity of these areas (Gil-Pelegrín et al. 2008). In France also, large areas of forests affected by decay and mortality are afforestations undertaken with an inappropriate seed source (Guyon and Magnin 1991; Le Tacon et al. 1994). Meanwhile, the low number of mortality events in the Southern and part of the Eastern Mediterranean may be related to the higher anthropogenic pressure on forests in these regions (Talhouk et al. 2003; FAO 2010; Rifai et al. 2018), reducing their density and thus enabling them to cope better with droughts. In any case, the geographical trend is distorted by the differences in the research and technical resources in each country, which prevent us from acquiring unbiased knowledge of regional differences. An example of this is the excellent database provided by Marija Kolšek of the Slovenian Forest Service, which could lead one to mistakenly think that their forests are particularly affected by mortality events.

Regarding the temporal trend (Fig. 9.5c), climate change and its possible impact on forests has considerably increased the volume of academic literature devoted to this type of phenomenon, in some cases creating alarm (Mátyás 2010). However, decline and mortality events have occurred naturally throughout history (Allen et al. 2010), and are just two of the wide range of factors influencing ecosystem dynamics. In addition, there have been past instances of scientific and social alarm over supposedly massive forest declines, such as many of the studies linking declines to pollution, which subsequently proved to be erroneous (Skelly and Innes 1994). Nevertheless, there is a direct relationship between droughts and mortality events (Allen et al. 2010), and the number and severity of droughts in the Mediterranean is increasing (Fig. 9.6). In addition, the increase in stand density and competition in

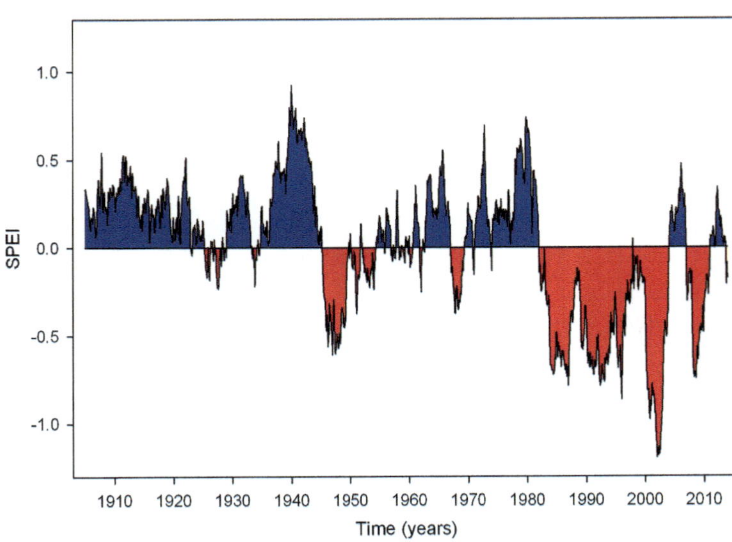

Fig. 9.6 Temporal evolution of the SPEI drought index (Vicente-Serrano et al. 2010) averaged over the entire Mediterranean Basin. This index was calculated using 48-month-long scales with negative values indicating dry conditions

many European forests in recent decades is increasing tree stress during periods of drought and reducing the forests' defenses against insect or fungal attacks. For example, some previous studies have found that a significant increase in defoliation rates has occurred during recent decades in Southern Europe and in arid Mediterranean areas, in contrast to the stable state observed in more humid locations (Dorman et al. 2013), which may be related to an increase in stress and mortality in these dry areas.

In 80 out of 284 mortality events the forest origin is known (Table 9.3 within Annex). Of these, half of the forests had a natural origin and the other half were afforestations. However, stands of artificial origin account for 16.8% of all the forest area in the Mediterranean Basin (FAO 2015), pointing to higher incidence of mortality in afforestations. Natural stands are not exempt from suffering these events, but apparently to a lower extent than artificial ones.

Most of the studies presented in Fig. 9.3 mention mortality indirectly, without a direct calculation of either the number of trees or the area affected. In addition, almost all are highly localized studies. However, if climate change is the cause of most of these tree mortality events, it must be observed not only in specific locations but also across large areas, and it must be demonstrated that this mortality is significantly higher than other mortality events that occurred in the past. Thus, studies must be conducted at large spatial and temporal scales (for example, combining information obtained from national forest inventories, carried out with permanent plots, and satellite images), trying to identify the effects of different factors such as climate, density, pests and diseases in mortality.

9.4 Decline Predisposing Factors and Tree Mortality Mechanisms

An extensive body of research has demonstrated the complex mechanisms driving forest decline and associated tree mortality (McDowell et al. 2008; Sala et al. 2010; Choat et al. 2018). Indeed, from particular sequences of climate stress to stand life histories, via insect pests and diseases, the issue of tree mortality remains something of a mystery and our ability to predict when and how a tree will die is still very limited (McDowell et al. 2008; Choat et al. 2018). In order to deal with this complexity, Manion's decline spiral model (1981) has been frequently used as a framework (Fig. 9.7). This model outlines three factors leading to mortality. The first is a long-term stressor or predisposing factor, such as aging, competition or poor site conditions. The second is a more severe and short-term stressor or inciting factor, such as drought or heat wave, which causes tree death. The third contributing factor is represented by opportunistic biotic drivers that amplify or are amplified by the negative effects of the stress.

Frequently, long and severe droughts are the most extended driver of tree mortality in the Mediterranean Basin (Allen et al. 2010). McDowell et al. (2008) suggested

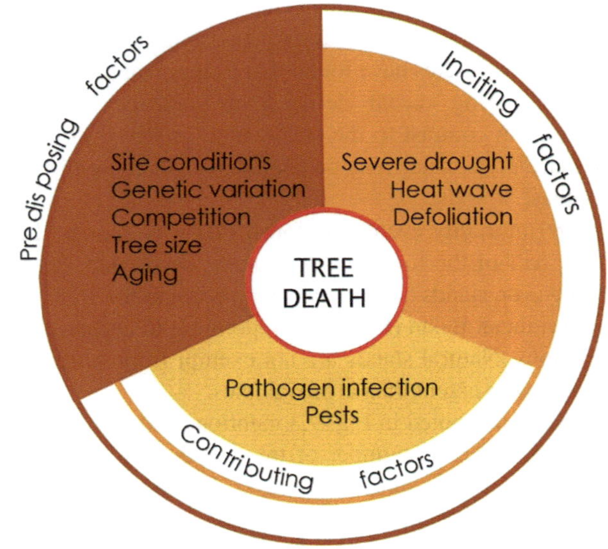

Fig. 9.7 Adaptation of Manion's spiral of tree decline with factors that predispose, incite and contribute to tree mortality

the presence of two non-exclusive physiological mechanisms behind this drought-related mortality: (i) hydraulic failure due to massive xylem embolism occurring when soil water content does not meet transpiration demand and (ii) carbon starvation caused by depletion of non-structural carbohydrate pools associated to plant respiration under drought-induced stomatal closure. Biotic agents, such as pathogens and insects, would amplify both mechanisms, overcoming the defenses of their previously stressed tree hosts. The prevalence of one factor or another would depend on the duration and intensity of drought and the water strategy of the species, such that more anisohydric species would be more likely to die of hydraulic failure, while isohydric species would be more prone to carbon starvation. A third mechanism predisposing plants to mortality, related to constraints in the production and translocation of sugars, resins, and other secondary metabolites at low water potentials, was later added to the equation (Sala et al. 2010).

Although hydraulic failure has been extensively reported as the main mechanism involved in extensive crown death and tree mortality during drought (Choat et al. 2018), pines have some characteristics that can make them particularly vulnerable to drought-induced restrictions in carbon uptake. Pines have classically been considered to be drought avoidant species, with tight regulation of leaf water potential and stomatal control, and thus, prone to suffer from carbon starvation (but see Martínez-Vilalta and Garcia-Forner (2017) for a reconsideration of the isohydric/anisohydric classification). Moreover, trees that succumb to insect and pathogen attacks show weak resin production, consistent with limitations on carbon uptake, cellular metabolism, and/or tree water relations (Allen et al. 2010).

Mediterranean pines are more vulnerable to embolism than other coexistent conifers such as cypresses, cedars or junipers (Bouche et al. 2014). Values of P_{50}, water potential causing 50% loss of hydraulic conductivity, range from −3 MPa in *P. sylvestris* to −5 MPa in *P. canariensis* and *P. halepensis* (López et al. 2013; Rosas et al. 2019), with little evidence of adaptive variation in embolism resistance in some populations of these species (Martínez-Vilalta et al. 2009; Lamy et al. 2014) but not in others (see López et al. 2013 and David-Schwartz et al. 2016). Even lower is the variability of turgor loss point among species, meaning that stomatal closure occurs at similar water potentials (Rosas et al. 2019). Pines mostly rely on changes in the allocation pattern between below-ground and aboveground organs and reduction of the leaf to sapwood area ratio in order to guarantee water supply to leaves and maintain the water potential within the functional limits of the xylem (Martínez-Vilalta et al. 2009; López et al. 2016). If drought is so severe that it forces trees beyond their physiological safety margins, they will succumb, as occurred for example with southern populations of *P. sylvestris* and *P. nigra* in northeast and southeast Spain after the droughts of the 1990s and 2000s (Martínez-Vilalta and Piñol 2002; Sánchez-Salguero et al. 2012).

Although heat and drought impact the carbon economy of pines (Galiano et al. 2011; Garcia-Forner et al. 2016), the carbon starvation mechanism as such remains controversial (Sala et al. 2010; Hartmann 2015), since depletion of carbon reserves is not routinely found in declining trees but just in some decline episodes (Salmon et al. 2015; Savi et al. 2019). What is frequently observed, though, is a strong reduction of growth for several years or decades prior to death and changes in the morphology of the xylem in declining trees compared with non-declining trees growing in the same stand (Hereş et al. 2014; Cailleret et al. 2017; Savi et al. 2019), suggesting limitations in carbohydrate availability (Hereş et al. 2014; Savi et al. 2019). The attempt of declining trees to maintain stomata open at low water potentials can be interpreted as a compensatory mechanism to maintain a positive carbon balance, but at the risk of systemic hydraulic failure (Hereş et al. 2014; Savi et al. 2019), again highlighting the intricate relationship between carbon and water fluxes in trees.

9.4.1 Global Change-Related Threats

As mentioned above, the landscapes of the Mediterranean Basin have been transformed over thousands of years by expanding human populations, resulting in profoundly modified anthropogenic systems. Less evident is the impact of ongoing climate change on forests. Widespread forest mortality, associated with climate-induced physiological stress and interactions with other climate-mediated processes, such as insect outbreaks and wildfires, raises the possibility that pine forests are increasingly responding to ongoing warming and changes in the hydrological cycle.

9.4.1.1 Land-Use change

Land use has changed drastically in most Mediterranean countries over the last 40–60 years. The rural exodus that accompanied the socioeconomic changes of the mid-twentieth century put an end to an age-old need for arable and grazing lands, and a long history of forest exploitation for timber and firewood. Pine species benefited from these changes by recruiting in abandoned agricultural lands, naturally, and with the assistance of public administrations and private owners through afforestations in subsequent decades (Vadell et al. 2016). At the same time, the reduced interest in forest products and services resulted in a decline in forest management activities. Both factors have had long-term consequences for vegetation dynamics and stand resilience (Puerta-Piñero et al. 2012). On the one hand, in some sites, increasing pine canopy closure has facilitated the establishment of later-successional, more shade-tolerant trees, which can eventually replace or co-dominate the pine canopy in the absence of fire, or even in the occurrence of fire if shade-tolerant species colonizing the understory have the ability to resprout (e.g. *Quercus pyrenaica* Willd. in *P. pinaster* forests; Torres et al. 2016). On the other hand, increasing tree density and fuel load, together with ongoing climate changes, have increased fire recurrence (Pausas and Fernández-Muñoz 2012). This has had a negative impact on stands of pines less adapted to fire (e.g., *P. sylvestris* and *P. nigra*), but also on those of pines with thicker bark (e.g., *P. pinaster*), and with shorter juvenile periods and serotinous cones (e.g., *P. halepensis*) better adapted to fire. Entire stands have disappeared (in favor of shrublands) where fire-damaged canopy and soil seed banks have not been replaced by nearby (usually <100 m) seed sources (Pausas et al. 2004; Karavani et al. 2018).

These problems, jeopardizing the persistence of many pine stands, are probably exacerbated in pine afforestations. Lack of management has resulted in the abandonment of large-scale pine afforestations established in the late nineteenth and early to mid-twentieth centuries to recover forest cover. Compared to natural stands, unmanaged pine afforestations have higher tree densities and fuel load (Fig. 9.8). High intraspecific competition weakens the trees and makes them more vulnerable to abiotic and biotic stresses such as drought, pest outbreaks (e.g., *Thaumetopoea pityocampa* Denis & Schiffermüller) and infection by pathogens (e.g., *Bursaphelenchus xylophilus* (Steiner & Buhrer) Nickle) and parasitic plant species (e.g., *Viscum album* L.). This could be particularly so in afforestations in inappropriate sites, or made using poor plant material or non-local provenances, given the evidence of drought stress tolerance varying among populations, e.g. of *P. pinaster* (Fernández et al. 2000) or *P. canariensis* (López et al. 2009). Land-use and climate change also affect the depth of the water table. In central Spain, warmer temperatures, longer droughts and aquifer overuse for irrigation are contributing to the decline of *P. pinaster* populations in sandy soils with low soil water availability and likely reliance on groundwater to survive dry periods (Gómez Sanz and Garcia-Viñas 2011; Gea-Izquierdo et al. 2019). Interactions between disturbance and stress factors can also occur in the current context of global change. Post-fire drought might hinder stand recovery by limiting pine seedling survival, but pre-fire drought

Fig. 9.8 Excessively dense afforestation of *Pinus pinea* located in Tunisia. (Photo: Sven Mutke)

might also increase fire occurrence and severity, as well as reduce post-fire recovery by affecting cone production, seed production and viability, or resprouting capacity in *P. canariensis* (Karavani et al. 2018).

9.4.1.2 Climate Change

Many episodes of mortality in pine forests throughout the Mediterranean have occurred after major drought and heat spells. These are likely to increase in the future, as climate models predict an increase in aridity linked to warmer temperatures (up to 6 °C by the end of the century in the Mediterranean basin), higher atmospheric evaporative demand, a decrease in annual precipitation of up to 20%, and more frequent droughts and heat waves in a region already vulnerable to climate variability (Stocker et al. 2013).

Ongoing responses of pine forests to climate change include reduced radial growth in southern populations, extended phenology, reduced seedling recruitment, long-term shifts in community composition and increased fire, insect and pathogen damage. Long-term growth decline and defoliation observed in southern populations of *P. sylvestris*, *P. nigra* or *P. halepensis* (Sarris et al. 2007; Martínez-Vilalta et al. 2008; Carnicer et al. 2011; Sánchez-Salguero et al. 2012) will challenge the resilience and recovery capacity of these forests following major disturbances (Camarero et al. 2018). The growing season will be extended due to warmer temperatures in winter and autumn, but the decreasing trend in spring precipitation and longer summer droughts could counteract this positive effect on growth (Camarero et al. 2013, 2015b). The substitution of *Pinaceae* by *Fagaceae* is a global trend in temperate forests (Alfaro Reyna et al. 2018) and will have a profound impact on plant and animal communities (Martínez-Vilalta and Lloret 2016). Fire frequency and burn extension have increased in the Western Mediterranean Basin since the 1970s as a result of fuel accumulation, drier conditions, and extreme weather events (Pausas and Fernández-Muñoz 2012) and future wildfire risk is projected to increase. Finally, a higher incidence of disease and insect outbreaks has been observed in many Mediterranean forests (Desprez-Loustau et al. 2006). Rising temperatures and prolonged water deficits impact both host physiology and insect population dynamics. Tree vigor and production of defense compounds are affected, increasing susceptibility to pests and pathogens (Desprez-Loustau et al. 2006). Outbreaks of some aggressive bark beetles, such as *Matsucoccus* or Curcullionidae, mistletoe infections and fungi spreading in roots and phloem, are closely tied to warming and drought (Mendel 1987; Gea-Izquierdo et al. 2019; Morcillo et al. 2019).

Pine forests will eventually adjust to the new climate, but it is also true that some populations inhabiting the xeric range edges of the species are particularly threatened, and that the legacy effects of forest decline will affect forest services and functions for many years to come (Navarro-Cerrillo et al. 2018). The outcome will depend on how fast the local environment changes, the frequency of extreme meteorological events, the extent of local adaptation and plasticity of key adaptive traits (Valladares et al. 2014), and our ability to take early measures to adapt pine forests to climate changes.

9.5 Mitigation Practices Through Forest Management

In summary, many of the episodes of pine forest decline that are currently taking place are probably related to land-use and climate changes that have occurred in recent decades. The study of silvicultural measures that may help to adapt Mediterranean pine forests to such changes must be promoted (Fig. 9.9). Researchers recommend managing forests to favor mixed pinewoods and pine-broadleaved

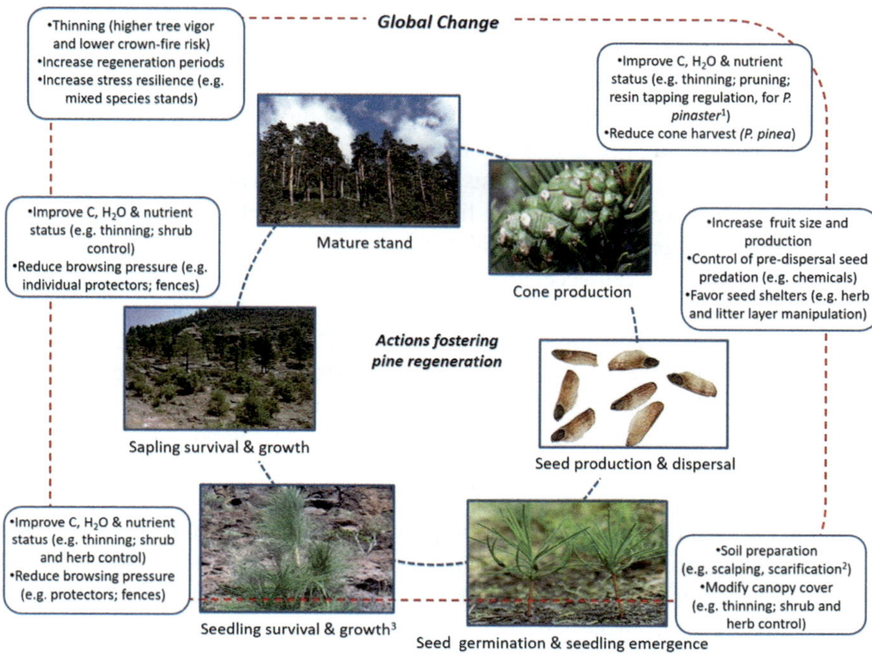

Fig. 9.9 Silvicultural actions that may foster pine regeneration in the context of climate change. [1]Based on a hypothetical trade-off between wound healing and reproduction; [2]litter effects on germination and seedling emergence vary from negative to positive across species and populations (e.g. of *P. sylvestris*); [3]a trade-off is often reported between conditions favoring survival and growth

stands; the associated increase in structural, functional and species diversity would benefit productivity and stress resistance and resilience at the stand level (Pausas et al. 2004; Puerta-Piñero et al. 2012). Actions should also aim to increase the vigor of trees, e.g., via pruning or competition removal. Thinning alleviates competition for light, water and nutrients and, in general, improves the water status of remaining adult trees, which grow faster and produce more cones and viable seeds (Moya et al. 2008; del Río et al. 2017). Thus, thinning can reduce drought susceptibility and plague risks in natural pine stands and particularly in abandoned pine afforestations. In relation to fire risk, although it can be reduced by thinning (e.g. due to elimination of suppressed trees and improved tree water status), care is also needed to avoid understory colonization by shrubs, which can increase fuel load and the risk of crown fires in more open understories with higher wind penetration (Molina et al. 2011).

Another key action to adapt pine forests to global changes is to foster regeneration. Canopy closure is a major determinant of plant survival at the seedling and early-sapling stages (Fyllas et al. 2008). Canopy-seedling interactions change from competition to facilitation as abiotic stress increases. Thus, a moderate degree of canopy cover usually provides optimal growth conditions (in terms of air vapor pressure deficit and soil water availability), and enhances the survival of seedlings of Mediterranean pines, e.g., *P. pinea* and *P. pinaster* (Calama et al. 2017), and boreal pines in Mediterranean populations (e.g. *P. sylvestris*; Castro et al. 2004), which tend to recruit in large clearings in boreal and temperate populations. Other factors add to or interact with canopy closure across further stages of regeneration. In the absence of fire, drought is the most important factor governing vegetation dynamics in Mediterranean pine forests due to its impact on all regeneration phases, accelerating adult tree mortality and limiting cone production, seed germination, and seedling emergence and survival (Fyllas et al. 2008; Calama et al. 2017). Regeneration cuttings can accelerate regeneration. However, their type, intensity, and timing should be based on the previous identification of the main regeneration bottlenecks for each population.

Compared to the twentieth century, there are currently very few large-scale pine afforestations in Mediterranean countries. However, assisting natural pine regeneration with afforestations would initiate forest recovery in degraded soils (e.g. after fire or massive mortality episodes) and accelerate the expansion of small pine populations, improving forest adaptation to climate change. For this, it is essential to avoid the errors of the past. We need to improve tree breeding practices, consider the provenance and progeny of the plant material, and maintain the efforts to ensure afforestation success and resilience over time. Extensive afforestations with pines of unknown genetic provenances have decreased resilience and regeneration after disturbances such as fire (Gil et al. 2009).

Possible response approaches to the impact of climate change on forestry include short- and long-term strategies that focus on enhancing ecosystem resistance and resilience and tackling potential limits to carbon accumulation (Millar et al. 2007; Nabuurs et al. 2013). Fragmented small-scale forest ownership can constrain adaptive capacity (Lindner et al. 2010). Landscape planning and fuel load management may reduce the risk of wildfires, but these actions may be constrained by higher flammability owing to warmer and drier conditions (Vallejo et al. 2012). Moreover, assisted migration and afforestations with species or populations better adapted to warmer and drier environmental conditions could be an option for managers. Strategies to reduce forest mortality include preference for species better adapted to relatively warm environmental conditions (Resco De Dios et al. 2007). The selection of tolerant or resistant families and clones may also reduce the risk of damage by pests and diseases in pure stands (Jactel et al. 2009).

Acknowledgments We are grateful to Matej Kravanja, Boštjan Košiček and Zoran Zavrtanik of the Slovenian Forest Service and in particular to Marija Kolšek for providing detailed information about mortality events in Slovenia; to Dimitrios N. Avtzis for the mortality event data from Greece; Andrea Nardini, Giovanna Battipaglia, Roberto Tognetti, Giorgio Vacchiano and Alfredo Di Filippo for data and photographs from Italy; and to Alberto Vilagrosa and Sven Mutke for providing data, and the photographs shown in Figs. 9.4 and 9.8, respectively.

Annex

Table 9.3 Mortality events in recent decades

Species (I)	Country	Longitude	Latitude	Cause of mortality	Year of the mortality	CA	Forest origin	References
P. halepensis	Algeria	2.00–3.08	34.57–34.75	Drought	1988–2016		Natural origin	Guit et al. (2016)
P. halepensis	Croatia	16.409	43.514	Drought and pest	2018	3.34		Pernek et al. (2019)
P. halepensis	Croatia	16.679	42.961	Drought and pest	2018	1.71		Pernek et al. (2019)
P. halepensis	Croatia	15.480	43.931	Drought and pest	2018	1.20		Pernek et al. (2019)
P. halepensis	Croatia	15.349	43.954	Drought and pest	2018	1.10		Pernek et al. (2019)
P. halepensis	France	5.37–6.25	43.01–43.24	Polluted sea spray	Recent			Garrec (1994)
P. halepensis	Greece	26.634	37.700	Drought	2001	−5.46	Natural origin	Sarris et al. (2007)
P. halepensis	Greece	27.033	37.767	Drought	2001	−4.85	Natural origin	Sarris et al. (2007)
P. halepensis	Greece	26.850	37.783	Drought	2001	−4.93	Natural origin	Sarris et al. (2007)
P. halepensis	Greece	23.184	40.786	Pest and excessive density	2018	2.16		DN Avtzis, pers comm
P. halepensis	Greece	22.982	40.632	Pest, drought and density	2019		Afforestation	DN Avtzis, pers comm
P. halepensis	Greece	25.607	40.487	Pest and excessive density	2019			DN Avtzis, pers comm
P. halepensis	Greece	20.684	37.768	Drought	2000–2001	−3.20	Natural origin	Sarris et al. (2011)
P. halepensis	Greece	24.511	38.869	Drought	2000–2001	−4.39	Natural origin	Sarris et al. (2011)
P. halepensis	Greece	25.873	35.057	Drought	2000–2001	−2.36	Natural origin	Sarris et al. (2011)
P. halepensis	Greece	28.029	36.307	Pest and excessive density	2016–2017	−3.25		DN Avtzis, pers comm
P. halepensis	Greece	24.664	40.686	Pest and excessive density	2017–2018	2.40		DN Avtzis, pers comm
P. halepensis	Israel	35.049	31.793	Drought	2011	−3.49	Afforestation	Calev et al. (2016)

(continued)

Table 9.3 (continued)

Species (I)	Country	Longitude	Latitude	Cause of mortality	Year of the mortality	CA	Forest origin	References
P. halepensis	Israel	35.050	31.333	Drought and soil	1996–2011		Afforestation	Preisler et al. (2019)
P. halepensis	Israel	34.851	31.372	Drought	1998–2015		Afforestation	Dorman et al. (2015b)
P. halepensis	Israel	34.860	31.380	Drought	Recent		Afforestation	Dorman et al. (2015a)
P. halepensis	Israel	34.850	31.370	Drought	Recent		Afforestation	Dorman et al. (2015a)
P. halepensis	Israel	35.160	31.670	Pest			Afforestation	Mendel (1987)
P. halepensis	Israel	35.310	32.450	Pest			Afforestation	Mendel (1987)
P. halepensis	Italy	7.554	43.782	Polluted sea spray	Recent			Nicolotti et al. (2005)
P. halepensis	Italy	8.236	44.102	Polluted sea spray	Recent			Nicolotti et al. (2005)
P. halepensis	Italy	8.164	43.952	Polluted sea spray	Recent			Nicolotti et al. (2005)
P. halepensis	Slovenia	13.977	45.673	Fungal disease	2000	−0.04		Data provided by Slovenian Forest Service
P. halepensis	Slovenia	13.977	45.673	Fungal disease	2001	−0.04		Data provided by Slovenian Forest Service
P. halepensis	Slovenia	13.911	45.585	Fungal disease	2001	0.26		Data provided by Slovenian Forest Service
P. halepensis	Slovenia	13.911	45.585	Fungal disease	2002	0.26		Data provided by Slovenian Forest Service
P. halepensis	Slovenia	13.735	45.500	Fungal disease	2007	−1.54		Data provided by Slovenian Forest Service
P. halepensis	Slovenia	13.964	45.610	Fungal disease	2007	−1.79		Data provided by Slovenian Forest Service
P. halepensis	Slovenia	13.685	45.451	Fungal disease	2019			Data provided by Slovenian Forest Service
P. halepensis	Slovenia	13.710	45.540	Fungal disease	2019			Data provided by Slovenian Forest Service

Species (II)	Country	Longitude	Latitude	Cause of mortality	Year of the mortality	CA	Forest origin	References
P. halepensis	Spain	−0.470	38.387	Drought	2014	−2.88	Afforestation	García de la Serrana et al. (2015)
P. halepensis	Spain	−0.950	38.101	Drought	2014	−2.60	Afforestation	García de la Serrana et al. (2015)
P. halepensis	Spain	−0.883	38.238	Drought	2014	−2.77	Afforestation	García de la Serrana et al. (2015)
P. halepensis	Spain	−0.693	38.518	Drought	2014	−2.85	Afforestation	García de la Serrana et al. (2015)
P. halepensis	Spain	−0.730	41.789	Drought	1995 and 2005	−2.10	Natural origin	Camarero et al. (2015a)
P. halepensis	Spain	−3.482	40.280	Drought and disease	Recent		Afforestation	Morcillo et al. (2019)
P. halepensis	Spain	−3.377	40.083	Drought and disease	Recent		Afforestation	Morcillo et al. (2019)
P. halepensis	Spain	−0.887	41.615	Pest	Recent		Afforestation	Sangüesa-Barreda et al. (2015)
P. halepensis	Tunisia	10.067	37.183	Drought and sea spray	Recent		Afforestation	Bachtobji Bouachir et al., (2017)
P. halepensis	Tunisia	10.950	36.950	Drought and sea spray	Recent		Afforestation	Bachtobji Bouachir et al. (2017)
P. halepensis	Turkey	28.000	37.300	Air pollution	Recent			Tolunay (1997)
P. mugo	Slovenia	13.713	46.365	Fungal disease	2014	2.99		Data provided by Slovenian Forest Service
P. nigra	Italy	13.862	45.63	Drought	2003–2016		Afforestation	Savi et al. (2019)
P. nigra	Slovenia	13.837	45.834	Fungal disease	1984	0.38		Data provided by Slovenian Forest Service
P. nigra	Slovenia	13.770	45.791	Fungal disease	1984	1.29		Data provided by Slovenian Forest Service
P. nigra	Slovenia	13.837	45.834	Fungal disease	1986	0.38		Data provided by Slovenian Forest Service

(continued)

Table 9.3 (continued)

Species (II)	Country	Longitude	Latitude	Cause of mortality	Year of the mortality	CA	Forest origin	References
P. nigra	Slovenia	13.836	45.827	Fungal disease	1989	−0.28		Data provided by Slovenian Forest Service
P. nigra	Slovenia	13.977	45.673	Fungal disease	2000	−0.04		Data provided by Slovenian Forest Service
P. nigra	Slovenia	13.911	45.585	Fungal disease	2000	−0.21		Data provided by Slovenian Forest Service
P. nigra	Slovenia	13.865	45.712	Fungal disease	2000	−1.01		Data provided by Slovenian Forest Service
P. nigra	Slovenia	13.951	45.687	Pest	2000	−0.02		Data provided by Slovenian Forest Service
P. nigra	Slovenia	13.938	45.598	Pest	2000	−0.82		Data provided by Slovenian Forest Service
P. nigra	Slovenia	13.977	45.673	Fungal disease	2001	−0.04		Data provided by Slovenian Forest Service
P. nigra	Slovenia	13.911	45.585	Fungal disease	2001	−0.21		Data provided by Slovenian Forest Service
P. nigra	Slovenia	13.656	45.855	Pest	2001	−0.81		Data provided by Slovenian Forest Service
P. nigra	Slovenia	13.951	45.687	Pest	2001	−0.02		Data provided by Slovenian Forest Service
P. nigra	Slovenia	13.969	45.614	Pest	2001	0.63		Data provided by Slovenian Forest Service
P. nigra	Slovenia	13.991	45.602	Fungal disease	2002	−0.22		Data provided by Slovenian Forest Service
P. nigra	Slovenia	13.911	45.585	Fungal disease	2002	−0.21		Data provided by Slovenian Forest Service
P. nigra	Slovenia	13.786	45.797	Fungal disease	2002	−0.70		Data provided by Slovenian Forest Service

P. nigra	Slovenia	13.656	45.855	Pest	2002	−0.81		Data provided by Slovenian Forest Service
P. nigra	Slovenia	14.122	45.676	Pest	2002	−0.28		Data provided by Slovenian Forest Service
P. nigra	Slovenia	13.780	45.790	Fungal disease	2003	−2.91		Data provided by Slovenian Forest Service
P. nigra	Slovenia	13.952	45.658	Fungal disease	2003	−2.78		Data provided by Slovenian Forest Service
P. nigra	Slovenia	13.656	45.855	Pest	2003	−0.81		Data provided by Slovenian Forest Service
P. nigra	Slovenia	13.935	45.540	Pest	2003	−2.77		Data provided by Slovenian Forest Service
P. nigra	Slovenia	13.812	45.790	Fungal disease	2004	−0.78		Data provided by Slovenian Forest Service
P. nigra	Slovenia	13.964	45.610	Fungal disease	2004	−0.83		Data provided by Slovenian Forest Service
P. nigra	Slovenia	13.898	45.553	Pest	2004	−1.02		Data provided by Slovenian Forest Service

Species (III)	Country	Longitude	Latitude	Cause of mortality	Year of the mortality	CA	Forest origin	References
P. nigra	Slovenia	13.786	45.799	Fungal disease	2005	−0.02		Data provided by Slovenian Forest Service
P. nigra	Slovenia	13.964	45.610	Fungal disease	2005	−0.83		Data provided by Slovenian Forest Service
P. nigra	Slovenia	13.955	45.704	Pest	2005	−0.12		Data provided by Slovenian Forest Service
P. nigra	Slovenia	13.964	45.610	Fungal disease	2006	−0.83		Data provided by Slovenian Forest Service

(continued)

Table 9.3 (continued)

Species (III)	Country	Longitude	Latitude	Cause of mortality	Year of the mortality	CA	Forest origin	References
P. nigra	Slovenia	14.896	45.881	Fungal disease	2006	0.58		Data provided by Slovenian Forest Service
P. nigra	Slovenia	13.898	45.553	Pest	2006	−1.02		Data provided by Slovenian Forest Service
P. nigra	Slovenia	13.964	45.610	Fungal disease	2007	−0.83		Data provided by Slovenian Forest Service
P. nigra	Slovenia	13.898	45.553	Pest	2007	−1.02		Data provided by Slovenian Forest Service
P. nigra	Slovenia	13.788	45.804	Fungal disease	2008	−1.11		Data provided by Slovenian Forest Service
P. nigra	Slovenia	14.234	45.618	Fungal disease	2008	−1.56		Data provided by Slovenian Forest Service
P. nigra	Slovenia	13.709	46.066	Fungal disease	2008	1.07		Data provided by Slovenian Forest Service
P. nigra	Slovenia	13.956	45.732	Pest	2008	−1.20		Data provided by Slovenian Forest Service
P. nigra	Slovenia	13.692	45.930	Pest	2008	−1.06		Data provided by Slovenian Forest Service
P. nigra	Slovenia	13.916	45.885	Fungal disease	2009	0.34		Data provided by Slovenian Forest Service
P. nigra	Slovenia	13.926	45.602	Pest	2009	−0.08		Data provided by Slovenian Forest Service
P. nigra	Slovenia	14.130	45.679	Fungal disease	2010	−1.05		Data provided by Slovenian Forest Service
P. nigra	Slovenia	13.682	46.340	Pest	2010	−0.56		Data provided by Slovenian Forest Service
P. nigra	Slovenia	13.843	45.735	Pest	2010	1.16		Data provided by Slovenian Forest Service

P. nigra	Slovenia	14.116	45.681	Fungal disease	2011	−1.68		Data provided by Slovenian Forest Service
P. nigra	Slovenia	13.936	45.667	Fungal disease	2011	−1.68		Data provided by Slovenian Forest Service
P. nigra	Slovenia	13.958	46.053	Pest	2011	−1.39		Data provided by Slovenian Forest Service
P. nigra	Slovenia	13.721	45.829	Pest	2011	−1.52		Data provided by Slovenian Forest Service
P. nigra	Slovenia	14.130	45.679	Fungal disease	2012	−1.05		Data provided by Slovenian Forest Service
P. nigra	Slovenia	13.682	46.340	Pest	2012	−0.56		Data provided by Slovenian Forest Service
P. nigra	Slovenia	13.900	45.554	Pest	2012	−3.59		Data provided by Slovenian Forest Service
P. nigra	Slovenia	13.902	45.861	Fungal disease	2013	2.13		Data provided by Slovenian Forest Service
P. nigra	Slovenia	13.996	45.696	Fungal disease	2013	1.02		Data provided by Slovenian Forest Service
P. nigra	Slovenia	13.709	46.066	Fungal disease	2013	1.07		Data provided by Slovenian Forest Service
P. nigra	Slovenia	13.595	45.881	Pest	2013	1.23		Data provided by Slovenian Forest Service
P. nigra	Slovenia	13.992	45.655	Fungal disease	2014	3.04		Data provided by Slovenian Forest Service
P. nigra	Slovenia	13.822	45.837	Pest	2014	3.15		Data provided by Slovenian Forest Service
P. nigra	Slovenia	13.709	46.066	Fungal disease	2014	1.07		Data provided by Slovenian Forest Service

(continued)

Table 9.3 (continued)

Species (III)	Country	Longitude	Latitude	Cause of mortality	Year of the mortality	CA	Forest origin	References
P. nigra	Slovenia	13.902	45.861	Pest	2014	2.13		Data provided by Slovenian Forest Service
P. nigra	Slovenia	13.727	46.177	Fungal disease	2014	1.48		Data provided by Slovenian Forest Service
P. nigra	Slovenia	13.922	45.639	Pest	2014	3.04		Data provided by Slovenian Forest Service
P. nigra	Slovenia	14.034	45.848	Pest	2014	1.04		Data provided by Slovenian Forest Service
P. nigra	Slovenia	13.995	45.603	Fungal disease	2015	−0.09		Data provided by Slovenian Forest Service
P. nigra	Slovenia	14.114	45.691	Fungal disease	2015	−0.19		Data provided by Slovenian Forest Service
P. nigra	Slovenia	14.008	45.702	Pest	2015	−0.16		Data provided by Slovenian Forest Service

Species (IV)	Country	Longitude	Latitude	Cause of mortality	Year of the mortality	CA	Forest origin	References
P. nigra	Slovenia	13.880	45.584	Pest	2015	0.00		Data provided by Slovenian Forest Service
P. nigra	Slovenia	13.527	46.327	Fungal disease	2015	0.97		Data provided by Slovenian Forest Service
P. nigra	Slovenia	13.925	45.614	Pest	2015	−0.01		Data provided by Slovenian Forest Service
P. nigra	Slovenia	13.875	45.548	Fungal disease	2016	0.53		Data provided by Slovenian Forest Service
P. nigra	Slovenia	14.367	45.532	Pest	2016	0.31		Data provided by Slovenian Forest Service

P. nigra	Slovenia	14.034	45.848	Pest	2016	1.04		Data provided by Slovenian Forest Service
P. nigra	Slovenia	13.733	46.175	Fungal disease	2016	0.95		Data provided by Slovenian Forest Service
P. nigra	Slovenia	13.925	45.604	Pest	2016	0.45		Data provided by Slovenian Forest Service
P. nigra	Slovenia	13.954	45.911	Pest	2016	0.64		Data provided by Slovenian Forest Service
P. nigra	Slovenia	13.895	45.562	Fungal disease	2017	−0.77		Data provided by Slovenian Forest Service
P. nigra	Slovenia	14.147	45.679	Fungal disease	2017	−0.61		Data provided by Slovenian Forest Service
P. nigra	Slovenia	14.355	45.536	Pest	2017	−0.73		Data provided by Slovenian Forest Service
P. nigra	Slovenia	13.727	46.177	Fungal disease	2017	1.48		Data provided by Slovenian Forest Service
P. nigra	Slovenia	14.034	45.848	Pest	2017	1.04		Data provided by Slovenian Forest Service
P. nigra	Slovenia	14.149	45.604	Pest	2017	−0.63		Data provided by Slovenian Forest Service
P. nigra	Slovenia	13.866	45.542	Fungal disease	2018	4.26		Data provided by Slovenian Forest Service
P. nigra	Slovenia	13.992	45.698	Fungal disease	2018	3.69		Data provided by Slovenian Forest Service
P. nigra	Slovenia	13.968	45.561	Pest	2018	3.58		Data provided by Slovenian Forest Service
P. nigra	Slovenia	13.862	45.713	Fungal disease	2019			Data provided by Slovenian Forest Service

(continued)

Table 9.3 (continued)

Species (IV)	Country	Longitude	Latitude	Cause of mortality	Year of the mortality	CA	Forest origin	References
P. nigra	Slovenia	13.907	45.642	Fungal disease	2019			Data provided by Slovenian Forest Service
P. nigra	Slovenia	13.840	45.703	Pest	2019			Data provided by Slovenian Forest Service
P. nigra	Slovenia	14.009	45.680	Pest	2019			Data provided by Slovenian Forest Service
P. nigra	Slovenia	13.860	45.713	Pest	2019			Data provided by Slovenian Forest Service
P. nigra	Slovenia	13.847	45.697	Pest	2019			Data provided by Slovenian Forest Service
P. nigra	Slovenia	13.838	45.702	Pest	2019			Data provided by Slovenian Forest Service
P. nigra	Slovenia	13.862	45.670	Pest	2019			Data provided by Slovenian Forest Service
P. nigra	Slovenia	13.903	45.726	Pest	2019			Data provided by Slovenian Forest Service
P. nigra	Slovenia	13.862	45.526	Pest	1982–1983	−0.88		Data provided by Slovenian Forest Service
P. nigra	Slovenia	13.862	45.526	Pest	1992–1993	−0.88		Data provided by Slovenian Forest Service
P. nigra	Spain	1.049	41.311	Drought	1994–1998	−0.34	Natural origin	Martínez-Vilalta and Piñol (2002)
P. nigra	Spain	−2.612	37.244	Drought	2002–2005	0.14	Afforestation	Navarro Cerrillo et al. (2007)
P. nigra	Spain	−2.440	37.227	Drought	2002–2005	0.30	Afforestation	Navarro Cerrillo et al. (2007)
P. nigra	Spain	−2.380	37.315	Drought	2002–2005	−0.32	Afforestation	Navarro Cerrillo et al. (2007)
P. nigra	Spain	−2.841	37.273	Drought	2002–2005	0.53	Afforestation	Navarro Cerrillo et al. (2007)

P. nigra	Slovenia	13.73–14.20	45.48–45.82	Fungal disease	1986	−0.35		Jurc et al. (2000)
P. nigra	Spain	−3.457	37.097	Drought	2005	−2.16	Afforestation	Herrero et al. (2013)
P. nigra	Spain	−2.211	40.815	Drought	2005	−1.90		Herguido et al. (2016)
P. nigra	Spain	1.454	41.821	Drought	2012	−2.06		Chaparro et al. (2017)
P. nigra	Spain	1.440	41.797	Drought	2012	−2.06		Chaparro et al. (2017)

Species (V)	Country	Longitude	Latitude	Cause of mortality	Year of the mortality	CA	Forest origin	References
P. nigra	Spain	1.026	41.344	Drought	1994–1998	−0.62	Natural origin	Martínez-Vilalta and Piñol (2002)
P. nigra	Turkey	29.950	39.450	Air pollution	2002	3.41	Natural origin	Makineci and Sevgi (2006)
P. nigra	Turkey	32.038	37.978	Drought	2002–2007	−0.28		Semerci et al. (2008)
P. pinaster	France	−3.13-1.28	45.95–47.48	Wrong provenance and soil	1983	3.28	Afforestation	Guyon and Magnin (1991)
P. pinaster	France	−1.32-0.15	43.53–44.50	Wrong provenance	1985	0.86	Afforestation	Le Tacon et al. (1994)
P. pinaster	France	9.234	42.431	Non-native insect	1997	−1.42	Natural and afforestation	Jactel et al. (1998)
P. pinaster	France	6.32–6.82	43.26–43.51	Non-native insect	1957–1978		Natural origin	Schvester and Fabre (2001)
P. pinaster	France	−0.418	44.138	Non-native nematode	Recent			Baujard et al. (1979)
P. pinaster	France	−0.693	44.370	Non-native nematode	Recent			Baujard et al. (1979)
P. pinaster	Italy	7.12–11.29	43.62–44.73	Non-native insect	1973–2004			APAT (2005)
P. pinaster	Italy	10.305	43.717	Polluted sea spray	1960–1981		Afforestation	Gellini et al. (1983)
P. pinaster	Portugal	−8.94-7.31	37.16–41.08	Non-native nematode	1999–2019			Calvão et al. (2019)
P. pinaster	Spain	−2.809	37.414	Drought	2012	−0.89	Natural and afforestation	Navarro-Cerrillo et al. (2018)

(continued)

Table 9.3 (continued)

Species (V)	Country	Longitude	Latitude	Cause of mortality	Year of the mortality	CA	Forest origin	References
P. pinaster	Spain	−2.761	37.343	Drought	2012	−1.17	Natural and afforestation	Navarro-Cerrillo et al. (2018)
P. pinaster	Spain	1.049	41.311	Drought	1994–1998	−0.34	Afforestation	Martínez-Vilalta and Piñol (2002)
P. pinaster	Spain	1.026	41.344	Drought	1994–1998	−0.62	Afforestation	Martínez-Vilalta and Piñol (2002)
P. pinaster	Spain	−4.300	40.500	Drought	Recent		Natural origin	Gea-Izquierdo et al. (2019)
P. pinaster	Spain	−3.747	41.772	Drought	Recent		Natural origin	Prieto-Recio et al. (2015)
P. pinaster	Spain	−3.867	41.763	Drought	Recent		Natural origin	Prieto-Recio et al. (2015)
P. pinaster	Spain	−3.769	41.701	Drought	Recent		Natural origin	Prieto-Recio et al. (2015)
P. pinaster	Spain	−3.855	41.746	Drought	Recent		Natural origin	Prieto-Recio et al. (2015)
P. pinaster	Spain	−3.678	41.728	Drought	Recent		Natural origin	Prieto-Recio et al. (2015)
P. pinaster	Spain	−3.724	41.700	Drought	Recent		Natural origin	Prieto-Recio et al. (2015)
P. pinaster	Spain	−2.640	41.633	Drought	Recent		Natural origin	Prieto-Recio et al. (2015)
P. pinaster	Spain	−2.621	41.584	Drought	Recent		Natural origin	Prieto-Recio et al. (2015)
P. pinaster	Spain	−2.614	41.502	Drought	Recent		Natural origin	Prieto-Recio et al. (2015)
P. pinaster	Spain	−2.849	41.499	Drought	Recent		Natural origin	Prieto-Recio et al. (2015)
P. pinaster	Spain	−2.828	41.548	Drought	Recent		Natural origin	Prieto-Recio et al. (2015)
P. pinaster	Spain	−2.923	41.561	Drought	Recent		Natural origin	Prieto-Recio et al. (2015)
P. pinaster	Spain	−3.310	42.043	Drought	Recent		Natural origin	Prieto-Recio et al. (2015)
P. pinaster	Spain	−3.291	41.947	Drought	Recent		Natural origin	Prieto-Recio et al. (2015)
P. pinaster	Spain	−3.360	41.839	Drought	Recent		Natural origin	Prieto-Recio et al. (2015)
P. pinaster	Spain	−6.093	42.219	Drought	Recent		Natural origin	Prieto-Recio et al. (2015)
P. pinaster	Spain	−6.216	42.204	Drought	Recent		Natural origin	Prieto-Recio et al. (2015)
P. pinaster	Spain	−6.165	42.204	Drought	Recent		Natural origin	Prieto-Recio et al. (2015)

P. pinaster	Spain	−3.883	41.767	Drought and fungal disease	Recent			Álvarez et al. (2015)
P. pinaster	Spain	−4.963	40.275	Drought and fungal disease	Recent			Álvarez et al. (2015)
P. pinaster	Spain	−6.100	40.300	Drought and fungal disease	Recent			Zarco-Tejada et al. (2019)
P. pinea	Italy	10.305	43.717	Polluted sea spray	1960–1981		Afforestation	Gellini et al. (1983)

Species (VI)	Country	Longitude	Latitude	Cause of mortality	Year of the mortality	CA	Forest origin	References
P. pinea	Italy	12.314	41.720	Polluted sea spray	1970s		Afforestation	Gisotti (1979)
P. pinea	Italy	14.187	40.827	Pest	2014–2017	−2.27	Afforestation	Garonna et al. (2018)
P. pinea	Lebanon	36.117	34.300	Pest	2005–2015		Natural origin	Nemer (2015)
P. pinea	Lebanon	33.544	35.551	Pest	2005–2015		Natural origin	Nemer (2015)
P. pinea	Spain	−4.400	40.300	Drought	2005–2016			Calama (2017)
P. pinea	Tunisia	10.067	37.183	Drought and sea spray	Recent		Afforestation	Bachtobji Bouachir et al. (2017)
P. pinea	Tunisia	10.950	36.950	Drought and sea spray	Recent		Afforestation	Bachtobji Bouachir et al. (2017)
P. sylvestris	France	5.839	43.345	Drought	2003–2005	−1.32		Thabeet et al. (2009)
P. sylvestris	France	5.839	43.339	Drought	2003–2005	−1.32		Thabeet et al. (2009)
P. sylvestris	France	6.499	43.922	Drought	2003–2005	−1.68		Thabeet et al. (2009)
P. sylvestris	Greece	22.580	40.270	Pest and fungal disease	2005–2010	0.17	Natural origin	Chrysopolitou et al. (2013)
P. sylvestris	Italy	7.476	45.731	Drought	1992	−0.74		Vertui and Tagliaferro (1998)
P. sylvestris	Italy	10.493	46.740	Drought and pest	1993	0.00	Natural and afforestation	Minerbi (1993)
P. sylvestris	Italy	10.656	46.608	Drought and pest	1993	0.82	Natural and afforestation	Minerbi (1993)

(continued)

Table 9.3 (continued)

Species (VI)	Country	Longitude	Latitude	Cause of mortality	Year of the mortality	CA	Forest origin	References
P. sylvestris	Italy	10.724	46.513	Drought and pest	1993	0.35	Natural and afforestation	Minerbi (1993)
P. sylvestris	Italy	11.233	46.237	Drought and pest	1993	−0.11	Natural and afforestation	Minerbi (1993)
P. sylvestris	Italy	11.521	46.404	Drought and pest	1993	0.18	Natural and afforestation	Minerbi (1993)
P. sylvestris	Italy	11.903	46.638	Drought and pest	1993	−0.27	Natural and afforestation	Minerbi (1993)
P. sylvestris	Italy	11.985	46.679	Drought and pest	1993	−0.55	Natural and afforestation	Minerbi (1993)
P. sylvestris	Italy	12.211	46.654	Drought and pest	1993	−1.44	Natural and afforestation	Minerbi (1993)
P. sylvestris	Italy	12.323	46.701	Drought and pest	1993	−1.54	Natural and afforestation	Minerbi (1993)
P. sylvestris	Italy	12.209	46.817	Drought and pest	1993	−1.09	Natural and afforestation	Minerbi (1993)
P. sylvestris	Italy	11.538	46.607	Drought	2003–2004	−2.45	Natural and afforestation	Minerbi et al. (2006)
P. sylvestris	Italy	11.612	46.660	Drought	2003–2004	−2.37	Natural and afforestation	Minerbi et al. (2006)
P. sylvestris	Italy	11.618	46.678	Drought	2003–2004	−2.46	Natural and afforestation	Minerbi et al. (2006)
P. sylvestris	Italy	11.647	46.703	Drought	2003–2004	−2.28	Natural and afforestation	Minerbi et al. (2006)
P. sylvestris	Italy	11.640	46.725	Drought	2003–2004	−2.32	Natural and afforestation	Minerbi et al. (2006)
P. sylvestris	Italy	7.183	45.433	Drought	2003–2006	−1.87		Vacchiano et al. (2012)

P. sylvestris	Italy	7.307	45.716	Drought	2005–2006	−1.44		Gonthier et al. (2007)
P. sylvestris	Slovenia	13.812	45.790	Fungal disease	2004	−0.78		Data provided by Slovenian Forest Service
P. sylvestris	Slovenia	13.786	45.799	Fungal disease	2005	−0.02		Data provided by Slovenian Forest Service
P. sylvestris	Slovenia	13.964	45.610	Fungal disease	2006	−0.06		Data provided by Slovenian Forest Service
P. sylvestris	Slovenia	13.964	45.610	Fungal disease	2006	−0.06		Data provided by Slovenian Forest Service
P. sylvestris	Slovenia	14.234	45.618	Fungal disease	2008	−1.56		Data provided by Slovenian Forest Service
P. sylvestris	Slovenia	13.936	45.667	Fungal disease	2011	−1.68		Data provided by Slovenian Forest Service
P. sylvestris	Slovenia	14.130	45.679	Fungal disease	2012	−3.25		Data provided by Slovenian Forest Service
P. sylvestris	Slovenia	13.527	46.327	Fungal disease	2013	1.08		Data provided by Slovenian Forest Service
P. sylvestris	Slovenia	14.109	45.683	Pest	2013	1.02		Data provided by Slovenian Forest Service
P. sylvestris	Slovenia	13.992	45.655	Fungal disease	2014	3.04		Data provided by Slovenian Forest Service

Species (VII)	Country	Longitude	Latitude	Cause of mortality	Year of the mortality	CA	Forest origin	References
P. sylvestris	Slovenia	13.822	45.837	Pest	2014	3.15		Data provided by Slovenian Forest Service
P. sylvestris	Slovenia	13.709	46.066	Fungal disease	2014	2.89		Data provided by Slovenian Forest Service
P. sylvestris	Slovenia	13.692	45.930	Fungal disease	2015	0.21		Data provided by Slovenian Forest Service

(continued)

Table 9.3 (continued)

Species (VII)	Country	Longitude	Latitude	Cause of mortality	Year of the mortality	CA	Forest origin	References
P. sylvestris	Slovenia	13.995	45.603	Fungal disease	2015	−0.09		Data provided by Slovenian Forest Service
P. sylvestris	Slovenia	14.008	45.702	Pest	2015	−0.16		Data provided by Slovenian Forest Service
P. sylvestris	Slovenia	13.527	46.327	Fungal disease	2015	1.08		Data provided by Slovenian Forest Service
P. sylvestris	Slovenia	13.902	45.861	Pest	2015	−0.04		Data provided by Slovenian Forest Service
P. sylvestris	Slovenia	14.005	45.991	Pest	2016	0.73		Data provided by Slovenian Forest Service
P. sylvestris	Slovenia	13.733	46.175	Fungal disease	2016	0.95		Data provided by Slovenian Forest Service
P. sylvestris	Slovenia	14.147	45.679	Fungal disease	2017	−0.61		Data provided by Slovenian Forest Service
P. sylvestris	Slovenia	13.595	46.249	Pest	2018	4.39		Data provided by Slovenian Forest Service
P. sylvestris	Spain	−2.120	40.662	Drought	2005	−1.92		Herguido et al. (2016)
P. sylvestris	Spain	−3.457	37.097	Drought	2005	−2.16	Afforestation	Herrero et al. (2013)
P. sylvestris	Spain	1.927	42.119	Drought	2012	−2.34		Chaparro et al. (2017)
P. sylvestris	Spain	−4.010	40.860	Drought	1970–2014			Gea-Izquierdo et al. (2014)
P. sylvestris	Spain	1.049	41.311	Drought	1994–1998	−0.34	Natural origin	Martínez-Vilalta and Piñol (2002)
P. sylvestris	Spain	1.026	41.344	Drought	1994–1998	−0.62	Natural origin	Martínez-Vilalta and Piñol (2002)
P. sylvestris	Spain	1.000	41.317	Drought	1994–2014			Aguadé et al. (2015)
P. sylvestris	Spain	−0.955	40.436	Drought	1995 and 2005	−2.77	Natural origin	Camarero et al. (2015a)
P. sylvestris	Spain	−2.533	37.217	Drought	1995–2012		Afforestation	Sánchez-Salguero et al. (2012)

P. sylvestris	Spain	−2.833	37.367	Drought	1995–2012		Afforestation	Sánchez-Salguero et al. (2012)
P. sylvestris	Spain	1.158	42.561	Drought	2001 and 2005–2008	−1.74		Hereş et al. (2012)
P. sylvestris	Spain	1.012	41.308	Drought	2001–2003 and 2006–2008	−0.41		Hereş et al. (2012)
P. sylvestris	Spain	−2.612	37.244	Drought	2002–2005	0.14	Afforestation	Navarro Cerrillo et al. (2007)
P. sylvestris	Spain	−2.440	37.227	Drought	2002–2005	0.30	Afforestation	Navarro Cerrillo et al. (2007)
P. sylvestris	Spain	−2.380	37.315	Drought	2002–2005	−0.32	Afforestation	Navarro Cerrillo et al. (2007)
P. sylvestris	Spain	−2.841	37.273	Drought	2002–2005	0.53	Afforestation	Navarro Cerrillo et al. (2007)
P. sylvestris	Spain	1.108	42.379	Drought	2005–2007	−2.22	Natural and afforestation	Galiano et al. (2010)
P. sylvestris	Spain	1.577	42.165	Drought and pest	2005–2014			Jaime et al. (2019)
P. sylvestris	Spain	1.543	42.105	Drought and pest	2005–2014			Jaime et al. (2019)
P. sylvestris	Spain	1.573	42.220	Drought and pest	2005–2014			Jaime et al. (2019)
P. sylvestris	Spain	1.642	42.318	Drought and pest	2005–2014			Jaime et al. (2019)
P. sylvestris	Spain	1.653	42.313	Drought and pest	2005–2014			Jaime et al. (2019)
P. sylvestris	Spain	1.625	42.337	Drought and pest	2005–2014			Jaime et al. (2019)
P. sylvestris	Spain	1.567	42.383	Drought and pest	2005–2014			Jaime et al. (2019)
P. sylvestris	Spain	1.710	42.332	Drought and pest	2005–2014			Jaime et al. (2019)
P. sylvestris	Spain	1.323	42.243	Drought and pest	2005–2014			Jaime et al. (2019)
P. sylvestris	Spain	1.317	42.265	Drought and pest	2005–2014			Jaime et al. (2019)
P. sylvestris	Spain	1.282	42.308	Drought and pest	2005–2014			Jaime et al. (2019)

(continued)

Table 9.3 (continued)

Species (VIII)	Country	Longitude	Latitude	Cause of mortality	Year of the mortality	CA	Forest origin	References
P. sylvestris	Spain	1.530	42.092	Drought and pest	2005–2014			Jaime et al. (2019)
P. sylvestris	Spain	1.560	42.185	Drought and pest	2005–2014			Jaime et al. (2019)
P. sylvestris	Spain	2.383	41.922	Drought and pest	2005–2014			Jaime et al. (2019)
P. sylvestris	Spain	2.325	41.848	Drought and pest	2005–2014			Jaime et al. (2019)
P. sylvestris	Spain	2.318	41.833	Drought and pest	2005–2014			Jaime et al. (2019)
P. sylvestris	Spain	2.358	41.927	Drought and pest	2005–2014			Jaime et al. (2019)
P. sylvestris	Spain	1.088	42.348	Drought and pest	2005–2014			Jaime et al. (2019)
P. sylvestris	Spain	1.088	42.363	Drought and pest	2005–2014			Jaime et al. (2019)
P. sylvestris	Spain	1.050	42.353	Drought and pest	2005–2014			Jaime et al. (2019)
P. sylvestris	Spain	1.052	42.365	Drought and pest	2005–2014			Jaime et al. (2019)
P. sylvestris	Spain	1.008	41.338	Drought and pest	2005–2014			Jaime et al. (2019)
P. sylvestris	Turkey	34.583	41.317	Fungal disease	2011	3.95	Natural origin	Lehtijärvi et al. (2012)
P. sylvestris	Turkey	34.616	39.972	Drought	2002–2007	−0.74		Semerci et al. (2008)
P. uncinata	Spain	1.809	42.463	Air pollution	Recent		Natural origin	Diaz-de-Quijano et al. (2016)
P. uncinata	Spain	1.850	42.460	Air pollution	Recent		Natural origin	Diaz-de-Quijano et al. (2016)

When the geographic coordinates are not specified in the text of the reference, approximate coordinates are given. A "recent" year of mortality indicates that the mortality event occurred in the years prior to the publication of the reference paper. CA refers to the climate anomaly, calculated as the difference between the average Palmer Drought Severity Index (PDSI; Alley 1984) of the mortality year and the average PDSI during the period 1980–2018, with negative values meaning dry conditions. The PDSI data were extracted from the TerraClimate database (Abatzoglou et al. 2018). CA has not been calculated for long mortality periods, over which dry and wet years occur, or for 2019, for which no drought data were available at the time of writing

References

Abatzoglou JT, Dobrowski SZ, Parks SA, Hegewisch KC (2018) Terraclimate, a high-resolution global dataset of monthly climate and climatic water balance from 1958–2015. Sci Data 5:170191. https://doi.org/10.1038/sdata.2017.191

Aguadé D, Poyatos R, Gómez M et al (2015) The role of defoliation and root rot pathogen infection in driving the mode of drought-related physiological decline in Scots pine (Pinus sylvestris L.). Tree Physiol 35:229–242. https://doi.org/10.1093/treephys/tpv005

Alfaro Reyna T, Retana J, Martínez-Vilalta J (2018) Is there a substitution of Pinaceae by Fagaceae in temperate forests at the global scale? Glob Planet Change 166:41–47. https://doi.org/10.1016/j.gloplacha.2018.04.001

Alia R, Gil L, Pardos JA (1995) Performance of 43 Pinus pinaster Ait. provenances on 5 locations in central Spain. Silvae Genet 44:75–81

Allen CD, Macalady AK, Chenchouni H et al (2010) A global overview of drought and heat-induced tree mortality reveals emerging climate change risks for forests. For Ecol Manag 259:660–684. https://doi.org/10.1016/j.foreco.2009.09.001

Alley WM (1984) The Palmer Drought Severity Index: Limitations and assumptions. J Clim Appl Meteorol 23:1000–1009. https://doi.org/10.1175/1520-0450(1984)023<1100:TPDSIL>2.0.CO;2

Álvarez G, Fernández M, Diez JJ (2015) Ophiostomatoid fungi associated with declined Pinus pinaster stands in Spain. For Syst 24:e-006. https://doi.org/10.5424/fs/2015241-05707

APAT (2005) La cocciniglia del pino marittimo in Italia. Strategie di contenimento del Matsucoccus feytaudi Ducasse e orientamenti per gli interventi di recupero ambientale delle pinete di Pinus pinaster Aiton. Agenzia per la protezione dell'ambiente e per i servizi tecnici, Roma

Bachtobji Bouachir B, Khorchani A, Guibal F et al (2017) Dendroecological study of Pinus halepensis and Pinus pinea in northeast coastal dunes in Tunisia according to distance from the shoreline and dieback intensity. Dendrochronologia 45:62–72. https://doi.org/10.1016/j.dendro.2017.06.008

Baujard P, Boulbria A, Ham R et al (1979) Premières données sur la nématofaune associée aux dépérissements du pin maritime dans l'Ouest de la France. Ann des Sci For 36:331–339. https://doi.org/10.1051/forest/19790405

Bigler C, Gavin DG, Gunning C, Veblen TT (2007) Drought induces lagged tree mortality in a subalpine forest in the Rocky Mountains. Oikos 116:1983–1994. https://doi.org/10.1111/j.2007.0030-1299.16034.x

Bouche PS, Larter M, Domec JC et al (2014) A broad survey of hydraulic and mechanical safety in the xylem of conifers. J Exp Bot. https://doi.org/10.1093/jxb/eru218

Cailleret M, Jansen S, Robert EMR (2017) A synthesis of radial growth patterns preceding tree mortality. Glob Chang Biol 23:1675–1690. https://doi.org/10.1111/gcb.13535

Calama R (2017) La gestión forestal como herramienta para la adaptación al cambio climático: ¿realidad o ficción científica? Cuad Soc Esp Cienc For 43:59–90. https://doi.org/10.31167/csef.v0i43.17527

Calama R, Manso R, Lucas-Borja ME et al (2017) Natural regeneration in Iberian pines: a review of dynamic processes and proposals for management. For Syst 26:eR02S. https://doi.org/10.5424/fs/2017262-11255

Calev A, Zoref C, Tzukerman M (2016) High-intensity thinning treatments in mature Pinus halepensis plantations experiencing prolonged drought. Eur J For Res 135:551–563. https://doi.org/10.1007/s10342-016-0954-y

Calvão T, Duarte CM, Pimentel CS (2019) Climate and landscape patterns of pine forest decline after invasion by the pinewood nematode. For Ecol Manag 433:43–51. https://doi.org/10.1016/j.foreco.2018.10.039

Camarero JJ, Manzanedo RD, Sanchez-Salguero R et al (2013) Growth response to climate and drought change along an aridity gradient in the southernmost Pinus nigra relict forests. Ann For Sci 70:769–780. https://doi.org/10.1007/s13595-013-0321-9

Camarero JJ, Gazol A, Sangüesa-Barreda G et al (2015a) To die or not to die: early warnings of tree dieback in response to a severe drought. J Ecol 103:44–57. https://doi.org/10.1111/1365-2745.12295

Camarero JJ, Gazol A, Tardif JC, Conciatori F (2015b) Attributing forest responses to global-change drivers: limited evidence of a CO_2-fertilization effect in Iberian pine growth. J Biogeogr 42:2220–2233. https://doi.org/10.1111/jbi.12590

Camarero JJ, Gazol A, Sangüesa-Barreda G et al (2018) Forest growth responses to drought at short- and long-term scales in Spain: squeezing the stress memory from tree rings. Front Ecol Evol 6:9. https://doi.org/10.3389/fevo.2018.00009

Carnicer J, Coll M, Ninyerola M et al (2011) Widespread crown condition decline, food web disruption, and amplified tree mortality with increased climate change-type drought. Proc Natl Acad Sci U S A 108:1474–1478. https://doi.org/10.1073/pnas.1010070108

Caro E (1908) Plantaciones de pino rodeno (Pinus pinaster Sol.). Tipografía e Hijos de Manuel Durán, Ronda

Carrión JS (2012) Paleoflora y paleovegetación de la Península Ibérica e Islas Baleares. Ministerio de Economía y Competitividad and Universidad de Murcia, Murcia

Carrión JS, Fernández S (2009) The survival of the "natural potential vegetation" concept (or the power of tradition). J Biogeogr 36:2202–2203. https://doi.org/10.1111/j.1365-2699.2009.02209.x

Castro J, Zamora R, Hódar JA et al (2004) Seedling establishment of a boreal tree species (Pinus sylvestris) at its southernmost distribution limit: consequences of being in a marginal Mediterranean habitat. J Ecol 92:266–277. https://doi.org/10.1111/J.0022-0477.2004.00870.X

Caudullo G, Barredo JI (2019) A georeferenced dataset of drought and heat-induced tree mortality in Europe. One Ecosyst 4:e37753. https://doi.org/10.3897/oneeco.4.e37753

Chaparro D, Vayreda J, Vall-Llossera M et al (2017) The role of climatic anomalies and soil moisture in the decline of drought-prone forests. IEEE J Sel Top Appl Earth Obs Remote Sens 10:503–514. https://doi.org/10.1109/JSTARS.2016.2585505

Choat B, Brodribb TJ, Brodersen CR et al (2018) Triggers of tree mortality under drought. Nature. https://doi.org/10.1038/s41586-018-0240-x

Chrysopolitou V, Apostolakis A, Avtzis D et al (2013) Studies on forest health and vegetation changes in Greece under the effects of climate changes. Biodivers Conserv 22:1133–1150. https://doi.org/10.1007/s10531-013-0451-2

David-Schwartz R, Paudel I, Mizrachi M et al (2016) Indirect evidence for genetic differentiation in vulnerability to embolism in Pinus halepensis. Front Plant Sci 7:768. https://doi.org/10.3389/fpls.2016.00768

del Río M, Barbeito I, Bravo-Oviedo A et al (2017) Mediterranean pine forests: management effects on carbon stocks. In: Bravo F, LeMay V, Jandl R (eds) Managing forest ecosystems: the challenge of climate change. Springer, Cham, pp 301–327. https://doi.org/10.1007/978-3-319-28250-3_15

Desprez-Loustau M-L, Marçais B, Nageleisen L-M et al (2006) Interactive effects of drought and pathogens in forest trees. Ann For Sci 63:597–612. https://doi.org/10.1051/forest:2006040

Diaz-de-Quijano M, Kefauver S, Ogaya R et al (2016) Visible ozone-like injury, defoliation, and mortality in two Pinus uncinata stands in the Catalan Pyrenees (NE Spain). Eur J For Res 135:687–696. https://doi.org/10.1007/s10342-016-0964-9

Dorman M, Svoray T, Perevolotsky A et al (2013) Forest performance during two consecutive drought periods: diverging long-term trends and short-term responses along a climatic gradient. For Ecol Manag 310:1–9. https://doi.org/10.1016/J.FORECO.2013.08.009

Dorman M, Perevolotsky A, Sarris D et al (2015a) The effect of rainfall and competition intensity on forest response to drought: lessons learned from a dry extreme. Oecologia 177:1025–1038. https://doi.org/10.1007/s00442-015-3229-2

Dorman M, Svoray T, Perevolotsky A et al (2015b) What determines tree mortality in dry environments? a multi-perspective approach. Ecol Appl 25:1054–1071. https://doi.org/10.1890/14-0698.1

Engineers Corps (1859) Clasificación General de los Montes Públicos. Imprenta nacional, Madrid

FAO (2010) Global forest resources assessment 2010. FAO Forestry Paper No. 163. Rome, Italy

FAO (2015) Global forest resources assessment 2015. Food and Agriculture Organization of the United Nations, Rome

Fernández M, Gil L, Pardos JA (2000) Effects of water supply on gas exchange in Pinus pinaster Ait. provenances during their first growing season. Ann For Sci 57:9–16. https://doi.org/10.1051/forest:2000107

Fyllas NM, Dimitrakopoulos PG, Troumbis AY (2008) Regeneration dynamics of a mixed Mediterranean pine forest in the absence of fire. For Ecol Manag 256:1552–1559. https://doi.org/10.1016/J.FORECO.2008.06.046

Galiano L, Martínez-Vilalta J, Lloret F (2010) Drought-induced multifactor decline of Scots pine in the Pyrenees and potential vegetation change by the expansion of co-occurring oak species. Ecosystems 13:978–991. https://doi.org/10.1007/s10021-010-9368-8

Galiano L, Martínez-Vilalta J, Lloret F (2011) Carbon reserves and canopy defoliation determine the recovery of Scots pine 4 yr after a drought episode. New Phytol 190:750–759. https://doi.org/10.1111/j.1469-8137.2010.03628.x

García de la Serrana R, Vilagrosa A, Alloza JA (2015) Pine mortality in southeast Spain after an extreme dry and warm year: interactions among drought stress, carbohydrates and bark beetle attack. Trees 29:1791–1804. https://doi.org/10.1007/s00468-015-1261-9

Garcia-Forner N, Sala A, Biel C et al (2016) Individual traits as determinants of time to death under extreme drought in Pinus sylvestris L. Tree Physiol 36:1196–1209. https://doi.org/10.1093/treephys/tpw040

Garonna AP, Foscari A, Russo E et al (2018) The spread of the non-native pine tortoise scale Toumeyella parvicornis (Hemiptera: Coccidae) in Europe: a major threat to Pinus pinea in southern Italy. IForest 11:628–634. https://doi.org/10.3832/ifor2864-011

Garrec J-P (1994) Les dépérissements littoraux d'arbres forestiers. Rev For Fr 5:454–457. https://doi.org/10.4267/2042/26571

Gea-Izquierdo G, Viguera B, Cabrera M et al (2014) Drought induced decline could portend widespread pine mortality at the xeric ecotone in managed Mediterranean pine-oak woodlands. For Ecol Manag 320:70–82. https://doi.org/10.1016/J.FORECO.2014.02.025

Gea-Izquierdo G, Férriz M, García-Garrido S et al (2019) Synergistic abiotic and biotic stressors explain widespread decline of Pinus pinaster in a mixed forest. Sci Total Environ 685:963–975. https://doi.org/10.1016/j.scitotenv.2019.05.378

Gellini R, Pantani F, Grossoni P et al (1983) Survey of the deterioration of the coastal vegetation in the park of San Rossore in central Italy. For Pathol 13:296–304. https://doi.org/10.1111/j.1439-0329.1983.tb00130.x

Gil L, López R, García-Mateos Á et al (2009) Seed provenance and fire-related reproductive traits of Pinus pinaster in central Spain. Int J Wildl Fire 18:1003–1009. https://doi.org/10.1071/WF08101

Gil-Pelegrín E, Peguero Pina JJ, Camarero JJ et al (2008) Drought and forest decline in the Iberian Peninsula: a simple explanation for a complex phenomenon? In: Sánchez JM (ed) Droughts: causes, effects and predictions. Nova Science Publishers, New York

Gisotti G (1979) Il deperimento della pineta di Castel Fusano. Agric Ambient 1:68–81

Gómez Sanz V, Garcia-Viñas JI (2011) Soil moisture spatio-temporal behavior of Pinus pinaster stands on sandy flatlands of central Spain. For Syst 20:293. https://doi.org/10.5424/fs/2011202-11186

Gonthier P, Giordano L, Nicolotti G (2007) Sui disseccamenti acuti e generalizzati del pino silvestre nell'Envers della media Valle d'Aosta. Inf Agric 23:41–45

Grove AT, Rackham O (2001) The nature of Mediterranean Europe: an ecological history. Yale University Press, New Haven. https://doi.org/10.1002/esp.247

Guit B, Nedjimi B, Chakali G et al (2016) État sanitaire des peuplements de Pin d'Alep (Pinus halepensis Mill.) dans le massif forestier de Senalba (Djelfa, Algérie). Rev d'Ecologie (Terre Vie) 71:156–167

Guyon J, Magnin H (1991) Dépérissement du pin maritime (Pinus pinaster Aït) en Vendée. Les causes écologiques. Ann des Sci For 48:333–346. https://doi.org/10.1051/forest:19910308
Hartmann H (2015) Carbon starvation during drought-induced tree mortality – are we chasing a myth? J Plant Hydraul 2:005. https://doi.org/10.20870/jph.2015.e005
Hereş AM, Martínez-Vilalta J, Claramunt López B (2012) Growth patterns in relation to drought-induced mortality at two Scots pine (Pinus sylvestris L.) sites in NE Iberian Peninsula. Trees 26:621–630. https://doi.org/10.1007/s00468-011-0628-9
Hereş AM, Camarero JJ, López BC et al (2014) Declining hydraulic performances and low carbon investments in tree rings predate Scots pine drought-induced mortality. Trees Struct Funct 28:1737–1750. https://doi.org/10.1007/s00468-014-1081-3
Herguido E, Granda E, Benavides R et al (2016) Contrasting growth and mortality responses to climate warming of two pine species in a continental Mediterranean ecosystem. For Ecol Manag 363:149–158. https://doi.org/10.1016/j.foreco.2015.12.038
Herrero A, Castro J, Zamora R et al (2013) Growth and stable isotope signals associated with drought-related mortality in saplings of two coexisting pine species. Oecologia 173:1613–1624. https://doi.org/10.1007/s00442-013-2707-7
Jactel H, Menassieu P, Ceria A et al (1998) Une pullulation de la cochenille Matsucoccus feytaudi provoque un début de dépérissement du Pin maritime en Corse. Rev For Fr 1:33–45. https://doi.org/10.4267/2042/5510
Jactel H, Nicoll BC, Branco M et al (2009) The influences of forest stand management on biotic and abiotic risks of damage. Ann For Sci 66:701. https://doi.org/10.1051/forest/2009054
Jaime L, Batllori E, Margalef-Marrase J et al (2019) Scots pine (Pinus sylvestris L.) mortality is explained by the climatic suitability of both host tree and bark beetle populations. For Ecol Manag 448:119–129. https://doi.org/10.1016/j.foreco.2019.05.070
Jurc D, Jurc M, Sleber TN et al (2000) Endophytic Cenangium ferruginosum (Ascomycotd) as a reservoir for an epidemic of Cenangium dieback in Austrian pine. Phyton (B Aires) 40:103–108
Karavani A, Boer MM, Baudena M et al (2018) Fire-induced deforestation in drought-prone Mediterranean forests: drivers and unknowns from leaves to communities. Ecol Monogr 88:141–169. https://doi.org/10.1002/ecm.1285
Lamy JB, Delzon S, Bouche PS et al (2014) Limited genetic variability and phenotypic plasticity detected for cavitation resistance in a Mediterranean pine. New Phytol 201:874–886. https://doi.org/10.1111/nph.12556
Le Tacon F, Bonneau M, Gelpe J et al (1994) Le dépérissement du pin maritime dans les landes de gascogne à la suite des introductions de graines d'origine ibérique et des grands froids des années 1962–1963 et 1985. Rev For Fr 5:474–484. https://doi.org/10.4267/2042/26575
Lehtijärvi A, Doğmuş-Lehtijärvi HT, Aday AG (2012) Armillaria ostoyae associated with dying 60-year-old Scots pines in northern Turkey. For Pathol 42:267–269. https://doi.org/10.1111/j.1439-0329.2011.00756.x
Lindner M, Maroschek M, Netherer S et al (2010) Climate change impacts, adaptive capacity, and vulnerability of European forest ecosystems. For Ecol Manag 259:698–709. https://doi.org/10.1016/j.foreco.2009.09.023
López R, Rodríguez-Calcerrada J, Gil L (2009) Physiological and morphological response to water deficit in seedlings of five provenances of Pinus canariensis: potential to detect variation in drought-tolerance. Trees 23:509–519. https://doi.org/10.1007/s00468-008-0297-5
López R, López De Heredia U, Collada C et al (2013) Vulnerability to cavitation, hydraulic efficiency, growth and survival in an insular pine (Pinus canariensis). Ann Bot 111:1167–1179. https://doi.org/10.1093/aob/mct084
López R, Cano FJ, Choat B et al (2016) Plasticity in vulnerability to cavitation of Pinus canariensis occurs only at the driest end of an aridity gradient. Front Plant Sci 7:769. https://doi.org/10.3389/fpls.2016.00769
Makineci E, Sevgi O (2006) Investigation of some needle characteristics between dead and healthy Austrian pine (Pinus nigra Arnold.) trees in forest decline sites around the air-polluted Kütahya province. Fresenius Environ Bull 15:470–476
Manion PD (1981) Tree disease concepts, 2nd edn. Prentice-Hall, Englewood Cliffs

Martínez-Vilalta J, Garcia-Forner N (2017) Water potential regulation, stomatal behaviour and hydraulic transport under drought: deconstructing the iso/anisohydric concept. Plant Cell Environ 40:962–976. https://doi.org/10.1111/pce.12846

Martínez-Vilalta J, Lloret F (2016) Drought-induced vegetation shifts in terrestrial ecosystems: The key role of regeneration dynamics. Glob Planet Change 144:94–108. https://doi.org/10.1016/j.gloplacha.2016.07.009

Martínez-Vilalta J, Piñol J (2002) Drought-induced mortality and hydraulic architecture in pine populations of the NE Iberian Peninsula. For Ecol Manag 161:247–256. https://doi.org/10.1016/S0378-1127(01)00495-9

Martínez-Vilalta J, López BC, Adell N et al (2008) Twentieth century increase of Scots pine radial growth in NE Spain shows strong climate interactions. Glob Chang Biol 14:2868–2881. https://doi.org/10.1111/j.1365-2486.2008.01685.x

Martínez-Vilalta J, Cochard H, Mencuccini M et al (2009) Hydraulic adjustment of Scots pine across Europe. New Phytol 184:353–364. https://doi.org/10.1111/j.1469-8137.2009.02954.x

Mátyás C (2010) Forecasts needed for retreating forests. Nature 464:1271. https://doi.org/10.1038/4641271a

McDowell N, Pockman WT, Allen CD et al (2008) Mechanisms of plant survival and mortality during drought: why do some plants survive while others succumb to drought? New Phytol 178:719–739. https://doi.org/10.1111/j.1469-8137.2008.02436.x

Mendel Z (1987) Major pests of man-made forests in Israel: origin, biology, damage and control. Phytoparasitica 15:131–137. https://doi.org/10.1007/BF02980912

Millar CI, Stephenson NL, Stephens SL (2007) Climate change and forests of the future: managing in the face of uncertainty. Ecol Appl 17:2145–2151. https://doi.org/10.1890/06-1715.1

Minerbi S (1993) Wie gesund sind unsere Wälder? Bericht über den Zustand der Wälder in Südtirol. Autonome Provinz Bozen Südtirol, Bolzano

Minerbi S, Cescatti A, Cherubini P et al (2006) Scots pine dieback in the Isarco Valley due to severe drought in the summer of 2003. For Obs 2(3):89–144

Molina JR, Rodriguez y Silva F, Herrera MA (2011) Potential crown fire behavior in Pinus pinea stands following different fuel treatments. For Syst 20:266. https://doi.org/10.5424/fs/2011202-10923

Morcillo L, Gallego D, González E et al (2019) Forest decline triggered by phloem parasitism-related biotic factors in Aleppo pine (Pinus halepensis). Forests 10:608. https://doi.org/10.3390/f10080608

Moya D, De las Heras J, López-Serrano FR et al (2008) Optimal intensity and age of management in young Aleppo pine stands for post-fire resilience. For Ecol Manag 255:3270–3280. https://doi.org/10.1016/J.FORECO.2008.01.067

Nabuurs GJ, Lindner M, Verkerk PJ et al (2013) First signs of carbon sink saturation in European forest biomass. Nat Clim Chang 3:792–796. https://doi.org/10.1038/nclimate1853

Navarro Cerrillo RM, Varo MA, Lanjeri S et al (2007) Cartografía de defoliación en los pinares de pino silvestre (Pinus sylvestris L.) y pino salgareño (Pinus nigra Arnold.) en la Sierra de los Filabres. Rev Ecosistemas 16:163–171. https://doi.org/10.7818/RE.2014.16-3.00

Navarro-Cerrillo R, Rodriguez-Vallejo C, Silveiro E et al (2018) Cumulative drought stress leads to a loss of growth resilience and explains higher mortality in planted than in naturally regenerated Pinus pinaster stands. Forests 9:358. https://doi.org/10.3390/f9060358

Nemer N (2015) Report on insect pests associated with conelet losses and their management in Pinus pinea forests in Lebanon. Rome, Italy

Nicolotti G, Rettori A, Paoletti E et al (2005) Morphological and physiological damage by surfactant-polluted seaspray on Pinus pinea and Pinus halepensis. Environ Monit Assess 105:175–191. https://doi.org/10.1007/s10661-005-3497-0

Nieulant L (1770) Departamento de EL BASTÓN de la Muy Noble y Muy Leal ciudad de Ciudad Rodrigo. Publicaciones de la Junta de Colonización Interior, Madrid

Ortuño F, Ceballos Jiménez A (1977) Los bosques españoles. Incafo, Madrid

Pausas JG, Fernández-Muñoz S (2012) Fire regime changes in the Western Mediterranean Basin: from fuel-limited to drought-driven fire regime. Clim Chang 110:215–226. https://doi.org/10.1007/s10584-011-0060-6

Pausas JG, Bladé C, Valdecantos A et al (2004) Pines and oaks in the restoration of Mediterranean landscapes of Spain: new perspectives for an old practice – a review. Plant Ecol 171:209–220. https://doi.org/10.1023/B:VEGE.0000029381.63336.20

Pernek M, Lacković N, Lukić I et al (2019) Outbreak of Orthotomicus erosus (Coleoptera, Curculionidae) on Aleppo pine in the Mediterranean region in Croatia. South-East Eur For 10:19–27. https://doi.org/10.15177/seefor.19-05

Preisler Y, Tatarinov F, Grünzweig JM et al (2019) Mortality versus survival in drought-affected Aleppo pine forest depends on the extent of rock cover and soil stoniness. Funct Ecol 33:901–912. https://doi.org/10.1111/1365-2435.13302

Prieto-Recio C, Martín-García J, Bravo F et al (2015) Unravelling the associations between climate, soil properties and forest management in Pinus pinaster decline in the Iberian Peninsula. For Ecol Manag 356:74–83. https://doi.org/10.1016/j.foreco.2015.07.033

Puerta-Piñero C, Espelta JM, Sánchez-Humanes B et al (2012) History matters: previous land use changes determine post-fire vegetation recovery in forested Mediterranean landscapes. For Ecol Manag 279:121–127. https://doi.org/10.1016/J.FORECO.2012.05.020

Resco De Dios V, Fischer C, Colinas C (2007) Climate change effects on Mediterranean forests and preventive measures. New For 33:29–40. https://doi.org/10.1007/s11056-006-9011-x

Rifai N, Khattabi A, Moukrim S et al (2018) Assessment of land cover dynamics in the Tahaddart Ramsar wetland (north-western Morocco). Rev d Ecol terre la vie 73:142–152

Rivas-Martínez S (1987) Memoria del mapa de series de vegetación de España. ICONA, Madrid

Rivas-Martínez S, Asensi A, Costa M et al (1993) Proyecto de Cartografía e Inventariación de los tipos de Hábitat de la Directiva 92/43/CEE en España. In: Colloques Phytosociologiques XXII. Baylleul, France, pp 611–661

Rosas T, Mencuccini M, Barba J et al (2019) Adjustments and coordination of hydraulic, leaf and stem traits along a water availability gradient. New Phytol 223:632–646. https://doi.org/10.1111/nph.15684

Sala A, Piper F, Hoch G (2010) Physiological mechanisms of drought-induced tree mortality are far from being resolved. New Phytol. https://doi.org/10.1111/j.1469-8137.2009.03167.x

Salmon Y, Torres-Ruiz JM, Poyatos R et al (2015) Balancing the risks of hydraulic failure and carbon starvation: a twig scale analysis in declining Scots pine. Plant Cell Environ 38:2575–2588. https://doi.org/10.1111/pce.12572

Sánchez-Salguero R, Navarro-Cerrillo RM, Camarero JJ et al (2012) Selective drought-induced decline of pine species in southeastern Spain. Clim Chang 113:767–785. https://doi.org/10.1007/s10584-011-0372-6

Sangüesa-Barreda G, Linares JC, Camarero JJ (2015) Reduced growth sensitivity to climate in bark-beetle infested Aleppo pines: connecting climatic and biotic drivers of forest dieback. For Ecol Manag 357:126–137. https://doi.org/10.1016/j.foreco.2015.08.017

Sarris D, Christodoulakis D, Körner C (2007) Recent decline in precipitation and tree growth in the eastern Mediterranean. Glob Chang Biol 13:1187–1200. https://doi.org/10.1111/j.1365-2486.2007.01348.x

Sarris D, Christodoulakis D, Körner C (2011) Impact of recent climatic change on growth of low elevation eastern Mediterranean forest trees. Clim Chang 106:203–223. https://doi.org/10.1007/s10584-010-9901-y

Savi T, Casolo V, Dal Borgo A et al (2019) Drought-induced dieback of Pinus nigra: a tale of hydraulic failure and carbon starvation. Conserv Physiol 7:coz012. https://doi.org/10.1093/conphys/coz012

Schiller G (2000) Ecophysiology of Pinus halepensis Mill. and P. brutia Ten. In: Ne'eman G, Trabaud L (eds) Ecology, biogeography and management of Pinus Halepensis and P. Brutia forest ecosystems in the Mediterranean Basin. Backhuys Publishers, Leiden, p 407

Schvester D, Fabre JP (2001) Le dépérissement du Pin maritime des Maures et de l'Estérel provoqué par Matsucoccus feytaudi Ducasse (Homoptera, Matsucoccidae), connaissances acquises de 1964 à 1990 (1ère partie: historique et rôle des insectes). Forêt Méditerranéenne 22:213–234

Semerci A, Sanlı BN, Sahin O et al (2008) Examination of tree mortalities in semi-arid central Anatolian region of Turkey during last six-year period (2002–2007). In: International conference on adaptation of forests and forest management to changing climate with emphasis on forest health: a review of science, policies and practices, Umeå, Sweden

SFM (2007) Mapa Forestal de España. Ministerio de Agricultura, Alimentación y Medio Ambiente, Madrid

Skelly JM, Innes JL (1994) Waldsterben in the forests of central Europe and eastern North America: fantasy or reality? Plant Dis 78:1021. https://doi.org/10.1094/PD-78-1021

Stocker TF, Qin D, Plattner GK et al (2013) Climate change 2013 the physical science basis: Working Group I contribution to the fifth assessment report of the intergovernmental panel on climate change. Cambridge University Press, Cambridge, UK. https://doi.org/10.1017/CBO9781107415324

Talhouk SN, Zurayk R, Khuri S (2003) Conifer conservation in Lebanon. Acta Hortic:411–414. https://doi.org/10.17660/ActaHortic.2003.615.46

Thabeet A, Vennetier M, Gadbin-Henry C et al (2009) Response of Pinus sylvestris L. to recent climatic events in the French Mediterranean region. Trees Struct Funct 23:843–853. https://doi.org/10.1007/s00468-009-0326-z

Tolunay D (1997) The effects of waste gases from Yatağan thermal power plant to the diameter increment of the Pinus brutia ten. forest. In: Air quality management at urban regional and global scales, Istanbul, Turkey

Torres I, Pérez B, Quesada J et al (2016) Forest shifts induced by fire and management legacies in a Pinus pinaster woodland. For Ecol Manag 361:309–317. https://doi.org/10.1016/J.FORECO.2015.11.027

Vacchiano G, Garbarino M, Borgogno Mondino E et al (2012) Evidences of drought stress as a predisposing factor to Scots pine decline in Valle d'Aosta (Italy). Eur J For Res 131:989–1000. https://doi.org/10.1007/s10342-011-0570-9

Vadell E, de-Miguel S, Pemán J (2016) Large-scale reforestation and afforestation policy in Spain: a historical review of its underlying ecological, socioeconomic and political dynamics. Land Use Policy 55:37–48. https://doi.org/10.1016/J.LANDUSEPOL.2016.03.017

Valladares F, Matesanz S, Guilhaumon F et al (2014) The effects of phenotypic plasticity and local adaptation on forecasts of species range shifts under climate change. Ecol Lett 17:1351–1364. https://doi.org/10.1111/ele.12348

Vallejo VR, Arianoutsou M, Moreira F (2012) Post-fire management and restoration of southern European forests. Post-fire For. Manag. South. Eur. a COST action Gather. disseminating. Sci Knowl 24:93–119. https://doi.org/10.1007/978-94-007-2208-8

Venturas M, García Álvarez S, Fajardo Alcántara M et al (2013) Species selection for reforestations: what happens with historic local extinctions and habitat protection zones? A case study in the Cantabrian Range. Eur J For Res 132:107–120. https://doi.org/10.1007/s10342-012-0660-3

Vertui F, Tagliaferro F (1998) Scots pine (Pinus sylverstris L.) die-back by unknown causes in the Aosta Valley, Italy. In: Chemosphere, pp 1061–1065. https://doi.org/10.1016/S0045-6535(97)10172-2

Vicente-Serrano SM, Beguería S, López-Moreno JI et al (2010) A new global 0.5° gridded dataset (1901–2006) of a multiscalar drought index: comparison with current drought index datasets based on the Palmer Drought Severity Index. J Hydrometeorol 11:1033–1043. https://doi.org/10.1175/2010JHM1224.1

Ximénez de Embún Oseñalde J, Ceballos L (1939) Plan General para la Repoblación Forestal de España. In: Organismo Autónomo Parques Nacionales (ed) Tres Trabajos Forestales. Ministerio de Medio Ambiente, Madrid, pp 8–388

Zarco-Tejada PJ, Hornero A, Beck PSA et al (2019) Chlorophyll content estimation in an open-canopy conifer forest with Sentinel-2A and hyperspectral imagery in the context of forest decline. Remote Sens Environ 223:320–335. https://doi.org/10.1016/j.rse.2019.01.031

Chapter 10
Forest Disease Affecting Pines in the Mediterranean Basin

Jonàs Oliva

10.1 Introduction

Patterns of forest disease are changing across the world. Pine woodlands in the Mediterranean Basin deliver important ecosystem functions that may be affected by pathogen damage. This chapter will provide a brief theoretical background for the main drivers of disease and how those can affect pine forests. The role of native diseases in natural ecosystems will be introduced, as opposed to the impacts caused by alien pathogens or by emerging diseases caused by native microbes that have increased their pathogenicity. Subsequently, the one-pathogen one-disease model will be compared with a multi-pathogen disease approach, and their application to pine diseases will be briefly discussed. The involvement of society in disease prevention and how this may need to be developed in the Mediterranean context will also be discussed. The role of environmental factors in disease will be explained together with some examples of the most prevalent stressors in pines, with a special focus on the role of forest management. Some examples of modeling approaches for improving management decisions such as whether or not to treat stumps in pure and mixed stands will be provided, among other suggestions of future research directions and management strategies.

J. Oliva (✉)
Department of Crop and Forest Sciences, University of Lleida, Lleida, Spain

Joint Research Unit CTFC – Agrotecnio, Lleida, Spain
e-mail: jonas.oliva@udl.cat

G. Ne'eman, Y. Osem (eds.), *Pines and Their Mixed Forest Ecosystems in the Mediterranean Basin*, Managing Forest Ecosystems 38,
https://doi.org/10.1007/978-3-030-63625-8_10

10.2 Drivers of Disease in the Mediterranean Region

Current drivers of forest disease are global, and as such, they also affect Mediterranean pinewoods. Three main processes are responsible for high mortality episodes in forests: globalization, climate change and land-use change (Stenlid and Oliva 2016). Globalization has favored the international trade of many goods, including plant material, and has brought exotic pathogens in contact with hosts that lack effective recognition and defense systems (Liebhold et al. 2012). Trade networks are very complex, so there are multiple possibilities by which forests in a given area may be exposed to exotic pathogens. Burgess and Wingfield (2017) studied disease records on planted *Eucalyptus* trees around the world and found evidence of attacks by native/exotic pathogens on both native/exotic hosts, including non-commercial eucalypts, planted within or outside their native range. One rather typical case of trade has been the movement of seeds and seedlings in plantation forestry. Exotic pathogens have been commonly introduced when host trees have been planted outside their natural distribution range. These plantations have acted themselves as sources of propagating material, which in some cases have turned to be a venue for new pathogens to the native range of the host. Classical diseases such as *Cronartium ribicola* J.C. Fisch., affecting white pine (*Pinus strobus* L.) in northern USA, follow this pattern (Geils et al. 2010). This is also likely to be the case for *Fusarium circinatum* Nirenberg & O'Donnell, once a major pest of *P. radiata* D. Don (native of California), which is now affecting native *P. pinaster* Ait. in northern Spain (Landeras et al. 2005) and posing a risk for other susceptible Mediterranean pines. The origin of the pathogen is assumed to be somewhere in Mexico (Wikler and Gordon 2000).

Shifts in the climatic regime can enable a pathogen to expand its range of its capacity to cause disease (Redondo et al. 2015; Brodde et al. 2019). Climate can increase disease by either favoring the pathogen e.g. new conditions are better for pathogen multiplication, or by affecting the host e.g. altered defensive system. The host can become more susceptible as new conditions may bring more stress, but may also modify the host's capacity to escape the pathogen by shifting phenological synchrony or by affecting the host phenotype and making it more susceptible through trade-offs e.g. growth vs. defense (Stenlid and Oliva 2016).

Finally, it is likely that land-use change has played a role too. In the Mediterranean region, the forest landscape has changed dramatically following the abandonment of traditional agricultural and forest practices in the latter part of the twentieth century. This has yielded a landscape of dense, young forests with epidemiological consequences for disease development.

10.3 Different Types of Disease Affecting Pines

10.3.1 Disease as Part of the Natural Dynamics of Forests

Pathogens play an important role in the functioning of forest ecosystems (Oliva et al. 2020). At the ecosystem level, for example, pathogens contribute to nutrient cycling by affecting decomposition of dead tissues and by affecting primary producers (Hicke et al. 2012), and at the needle level pathogens can impair gas exchange and photosynthetic capacity (Manter et al. 2000). In the case of conifers, mortality centers caused by root rots represent one of the most significant impacts on ecosystem functioning (Fig. 10.1a, b). The wave-like pattern of mortality increases nutrient turnover, and structures forests, creating variation of tree age. Root rots can selectively affect different tree species and vertebrate forest composition. Mortality centers in pine forests have been typically studied in wet, mountainous areas (Bendel et al. 2006), but little is known about their occurrence and dynamics in drier areas. In the Mediterranean area, *Heterobasidion annosum* has been isolated in plantations and natural forests on *P. nigra* J.F.Arnold and *P. pinea* L. (Oliva et al. 2008a; Gonthier et al. 2014). Further studies such as the one by Lung-Escarmant and Guyon (2004) on *P. pinaster* in the Landes forest in France should be carried out, where mortality rates as high as of 35% over a 17-year period were observed, possibly causing a significant ecological and economic impact.

Host specificity is a powerful mechanism by which pathogens are able to maintain forest tree diversity, at least in tropical areas (Mangan et al. 2010). The Janzen Connell hypothesis predicts that rare tree species can survive in a mixture by taking advantage of the accumulation of pathogens specific to the more dominant tree species (Gilbert 2002). In dry areas, plant-soil feedbacks are not only driven by pathogens but also by mycorrhiza. In fact, an extensive study on Australia showed that ectomycorrhizal species (such as pines) showed positive rather than negative feedbacks, i.e. species would regenerate better under their own canopy than under other species (Teste et al. 2017). Plugging into the mycorrhizal network brings not only benefits in terms of nutrient acquisition but also in terms of pathogen protection (Bennett et al. 2017). In the Mediterranean Basin, pines are often found mixed with other pine species or with other ectomycorrhizal species such as oaks, indicating the possibility of stabilizing forces against pure stands involving pathogens. For example, little is known on how mycorrhiza and pathogens interact with plants under water stress conditions. Also, to date, plant soil feedbacks have focused on soil pathogens, and little is known on the role of the local feedback provided by species-specific pathogens fruiting in needles and branches in the canopy. Some shoot, cone or needle pathogens such as *Diplodia sapinea* or *Lophodermium* sp. may play a suppressive role in pine regeneration (Munck et al. 2009), and produce negative feedbacks similar to those of soil-associated pathogens.

Fig. 10.1 Root rots on pine stands. (**a**) Mortality gap caused by *Armillaria* sp. on mixed *Pinus sylvestris*/*P. uncinata* stand, (**b**) detail of stunted growth caused by *Armillaria* sp. prior to death of *P. sylvestris*, (**c**) simultaneous thinning and stump treatment with *Phlebiopsis gigantea* (colored in blue) performed by a harvesting machine, (**d**) stump protection increases resistance to wind storms, shown are the experimental plots of *Picea abies* from Oliva et al. (2008b) where windfelling was widepread 15 years after artificial and natural stump infection, while no wind daamages ocurred in plots where thinning stumps had been treated

10.3.2 Novel Diseases Caused by Invasive Pathogens

In contrast to native pathogens, exotics can cause mortality rates higher than what would be expected under natural conditions (Manion 2003; Díaz-Yáñez et al. 2020). Historically, Mediterranean forests have been attacked by invasive pathogens (Santini et al. 2013). Regardless of whether this is due to some intrinsic characteristics of this area, such as mild climates, tree diversity or a high human presence, they have been the arena of some of the most dramatic cases of exotic tree diseases such as ink disease, chestnut blights or sudden oak death (Garbelotto 2009). Pines have also been affected. The most alarming disease is the pine wood nematode (*Bursaphelenchus xylophylus*), currently affecting pine forests in Portugal and supposedly eradicated in Spain. Another threat is the one posed by *Heterobasidion irregulare* in Italy (Gonthier et al. 2014). This pathogen was first reported in *P. pinea* in early 2000 (Gonthier et al. 2004); since then it has spread and interbreed with native *Heterobasidion* species (Gonthier and Garbelotto 2011) and has been found affecting the much more widespread *P. halepensis* (Scirè et al. 2008). Other pine species in the Mediterranean Basin are also likely to be susceptible.

Other invasive species with the potential to cause damage are *Dothistroma septosporum/pini* and *Lecanosticta acicola*, both present in several countries around the Mediterranean Basin, and *Fusarium circinatum*, currently widespread in northwest Spain (Ortíz de Urbina et al. 2017). Efforts to control these diseases focus on understanding the conditions that favor the disease, finding their modes of spread i.e. the role of insect vectoring (Bezos et al. 2015), or finding methods of controlling them either by biological means (Romeralo et al. 2018) or by planting resistant material (Elvira-Recuenco et al. 2014).

The behavior of invasive pathogens in Mediterranean conditions should also be a focus of study, possibly because an imperfect match between the invader and the new environment. Invasive pathogens remain cryptic for years to then cause explosive epidemics when the right conditions occur – the so-called "boom and bust" patterns (Blackburn et al. 2011). This could be the case of *Dothistroma* spp. causing dramatic defoliations on *P. nigra* spp. *salzmannii* in the Pyrenees following a very humid, cold year (Fig. 10.2). The delay between arrival and invasion makes it difficult to act on eradicating pathogens that are able to spread without causing major epidemics.

10.3.3 Disease Caused by Multiple Biotic Agents Including Pests

Pathogens often interact with other micro- and macro-organisms and stressors (Oliva et al. 2013). Interactions can occur at different times and with variable intensity. We know that trees continue the legacy of previous stressing factors that prime their defense systems (Sampedro et al. 2011), but such factors can also weaken trees

Fig. 10.2 Red band caused by *Dothistroma* spp. on pine: (**a**) Damage to stands with dense understory and high moisture conditions. Strong defoliation occurs on (**b**) regenerating and (**c**) adult trees. (**d**) Needles are girdled and dead, dry parts remain on the tree

and increase their likelihood of dying. Based on the latter process, Paul D. Manion developed the concept of forest decline (Manion 1981), according to which pathogens can act as predisposing, inciting or contributing factors to tree mortality. This may be a useful mindset for predicting death under Mediterranean conditions, in which variable weather conditions can favor different types of pests or pathogens at different times. Legacy effects of a pathogen attack have been shown for *Gremmeniella abietina* on Scots pine trees that suffered one (predisposing) or two (predisposing and inciting) consecutive attacks. Trees that were recovering from the first defoliation event were three times more likely to be attacked by *Tomicus piniperda* (contributing factor) and die. The ones that survived both the predisposing and the inciting event, displayed stagnated growth and were unable to restore their crown 10 years after the second *Gremmeniella* outbreak (Oliva et al. 2016b). From the perspective of the tree, the variable conditions typical of the Mediterranean climate probably cause the trees to be predisposed in one way or another. A study of drought-induced mortality in *P. sylvestris* at its drier distribution limit showed that the root rot *Onnia* sp. was associated with a worsening of hydraulic conductivity and sapwood width and therefore acted as a predisposing factor (Aguadé et al. 2015). In general, there is a lack of studies on the interactive causes of death in the driest conditions. In many cases, the main challenge is to identify the primary pathogen, as it often appears overshadowed by the appearance of secondary pests and pathogens together at the time of death.

10.4 Interactive Factors Leading to Disease

10.4.1 Interaction of Disease with Abiotic Factors

Abiotic factors contribute to disease by affecting both pathogen and host performance (Oliva et al. 2013). In the Mediterranean context, it is not rare for forests to be affected by drought, hail, fire and wind-storm events. Drought can cause tree mortality by itself although pathogens often actively contribute to the tree's death (Desprez-Loustau et al. 2006). Drought-induced mortality implies a disruption of the water and carbon dynamics in the tree. The contribution of pathogens can be predicted by the type of trophic interactions established with the host (Oliva et al. 2014). Pathogens are often classified as biotrophs, necrotrophs or vascular wilts. These types have marked differences in how they interact with the carbon and water system of the tree. From a carbon perspective, theory shows that necrotrophs will benefit from carbon starvation under drought stress, while biotrophs will suffer and their impact will be reduced. This framework allows for general predictions for pines too, so that necrotrophs such as root rots, canker pathogens or shoot blights are likely to benefit from drought, while needle diseases or rusts with initial biotrophic phases are likely to pose a lesser risk. These predictions are well supported by cases in which drought has been associated with greater damage by root rots (Linares et al. 2010), cankers (Munck et al. 2016), or shoot blights (Stanosz et al. 2001), although further testing is required, especially considering the different water economy strategies of the host.

Much less is known about other abiotic factors, such as hail or wind storms. Hail is known to be associated with tip blight caused by *D. sapinea* (Fig. 10.3). Defoliations and shoot death can be severe, and produce almost complete loss of latewood growth (Brodde et al. 2019). The mechanisms by which the pathogen benefits from mechanical damage are currently unknown. Under drought, *D. sapinea* seems to benefit from proline as a nitrogen source and as protection against reactive oxygen species produced by the tree (Sherwood et al. 2015). It is possible

Fig. 10.3 *Diplodia sapinea* attacks following hail storms on (**a**) *Pinus radiata*, (**b**) *P. nigra*/*P. halepensis*, (**c**) *P. pinaster* and (**d**) *P. nigra*/*P. sylvestris*. Detail of shoot blight on *P. sylvestris* (**e**) the same year of the attack, and (**f**) 1 year after the attack. Note that the needles remain connected to the trunk and fruiting bodies protrude at the base of the needles in (**f**)

that hail increases proline or other N-rich metabolites following mechanical damage, and provides similar benefits to the pathogen as drought stress (Oliva et al. 2021). Hail and rain storms usually bring high winds. Root rots are well known for their capacity to weaken root systems and increase the probability of mass windthrow (Fig. 10.1c) (Oliva et al. 2008b). Windthrown trees can remain alive for some time and become a breeding ground for bark beetles.

Fire is probably the most devastating disturbance affecting pines in the Mediterranean region. In northern latitudes, pine regeneration after fire can be compromised by the growth of xerophytic pathogens such as *Rhizina undulata* on burned slash (Lygis et al. 2005). Spores of these pathogens remain dormant in the soil and require a heat-shock of 35–45 °C in order to germinate. No records of the impact of such pathogens on tree regeneration following forest fires are available for drier areas. Pathogens may also affect fire severity. Many of the needle pathogens kill pine needles before the tree is able to shed them. In the case of *D. sapinea*, necrosis beings on the shoot, so the needles dry out and remain attached (Fig. 10.2e, f). In the case of *Dothistroma* sp., needles are girdled at the infection point, which makes them look half dry and half alive (Fig. 10.3d). Both mechanisms can increase the amount of flammable material in the forest canopy. Crown transparency after defoliation may also reduce moisture in the understory vegetation. Overall, information on how pathogens may affect fire behavior is largely lacking.

10.4.2 Interaction of Disease with Anthropogenic Activities, Such as Forest Management and Forest Plantations (Native and Exotic)

Management (including forest abandonment) is a major cause of disease in pine forests. Historically, pine forests have been favored by plantation forestry or by silviculture, either because of their economic value or because of their role in restoration and land protection. Moreover, pines as early colonizers have benefitted from land abandonment. Pines are thus subjected to a constellation of management situations with pathological consequences. For plantation forestry, the main risk has been the introduction of pathogens via plantation of infected nursery stock (already discussed in Sect. 10.2). Silvicultural management of both natural and planted forests has also benefited pathogens. The most conspicuous case has been the increase of root rots, particularly *H. annosum* following thinning operations (Garbelotto and Gonthier 2013). Pines are affected by *Heterobasidion* species of the so-called P (pine) group, such as *H. annosum* (native to Europe) or *H. irregulare* (native to North America). Airborne *Heterobasidion* spores infect freshly cut stumps and spread to neighboring living trees via root-to-root contacts. *Heterobasidion* causes growth losses and mortality and increases susceptibility to other pests (Wang et al. 2014). Because of the inconspicuous presence of fruiting bodies, the importance of this pathogen has been largely underestimated in southern areas (but see Mesanza and Iturritxa 2012). The large areas covered with young pine forests in the Mediterranean Basin represent the potential for this pathogen to become a problem in future, as has already begun in some parts of Spain and Turkey (Prieto-Recio et al. 2012; Doğmuş-Lehtijärvi et al. 2016).

Other problems may arise from the lack of management. Dense stands resulting from encroachment may provide favourable, moist and warm conditions for *Dothistroma* needle cast (Fabre et al. 2012) or/and may increase water stress. The type of vegetation surrounding the pine stands may also influence pathogens, for example, mistletoes. An interesting study on *P. halepensis* showed that the presence of olive trees attracted birds vectoring *Turdus viscivorus* to hibernate and thus feed and spread mistletoe seeds (Roura i Pascual et al. 2012). Another clear example of the effect of surrounding vegetation concerns alternative hosts for rust species. In a study on *P. sylvestris*, Mattila (2005) found that the risk of twisting rust caused by *Melampsora pinitorqua* was nine times higher when located near aspen trees.

10.5 Modeling of Forest Disease for Management and Policy

Disease management needs to be considered within a framework of losses and gains. This is especially difficult when the affected species has little economic value and the impact of the disease is on ecosystem functioning. Models can be used to predict losses and to calculate costs and benefits from a certain treatment. The real challenge comes from diseases that interact directly with management, such as root rots. Mechanistic models can be developed when the biology of the pathogen and epidemiological factors involved in the disease development are known. One such example is Rostand (Pukkala et al. 2005), a mechanistic model that can simulate root rot caused by *H. annosum* under different forest management regimes, considering alternative methods to control disease, such as stump treatment, winter thinning or stumping (see Fig. 10.4 as an example). Once correctly parameterized, it can predict the disease severity expected from normal forest operations (Oliva and Stenlid 2011).

Because of the growing role of forests as a source of recreation and landscape, the general public has a growing role in preventing diseases either by changing their recreational and consumption habits, alerting about new outbreaks or by demanding stricter policies in terms of plant health (Stenlid et al. 2011). A general survey across Europe showed that awareness of exotic pine diseases was different across Mediterranean countries such as Italy, France, Portugal, Spain and Turkey, however, it was clear that most people had not experienced diseases directly, and had obtained information from the media or other indirect sources (Eriksson et al. 2018). Pine forests need to be healthy in order to deliver goods and services; however, it is important to conduct open dialog with stakeholders and society to design effective policies. Understanding the needs of forest managers is also crucial. The presence of a market for timber products was found to be the deal breaker for forest managers

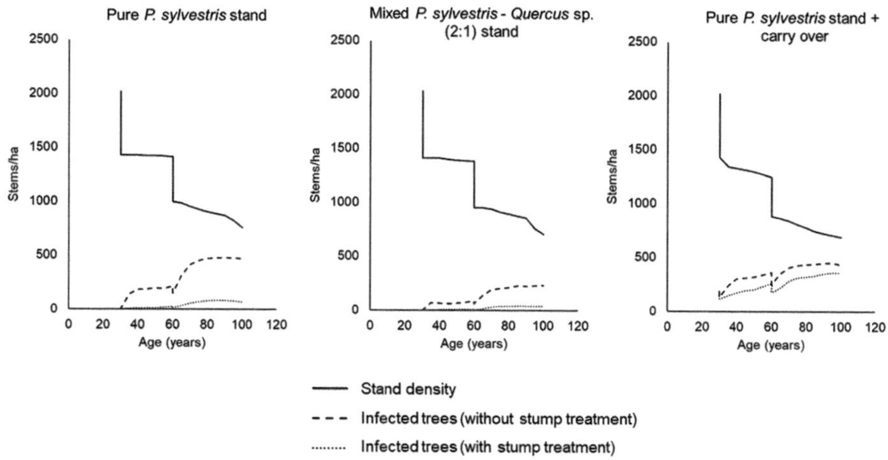

Fig. 10.4 Simulating *Heterobasidion annosum* root rot development in pine. Simulations have been carried out with Rotstand on first-generation pure pine stands or mixed stands (2:1 ratio between pine and oak), or second-generation pine stands established on a forest site infected in the previous rotation. Simulations include two thinnings to remove 30% of the basal area, with or without stump treatment. In pure stands, stump treatment reduces the proportion of infected trees from 62% to 10%. In mixed stands with oaks, disease development is slower on pines than in pure stands. At the end of the rotation, 33% of trees are infected, generating an individual risk for pine trees of 50%. With stump treatment, the risk of infection is rather similar to that in pure stands (9%). Preventing tree infection is critical for the following rotation. When including 300 stumps/ha of which 30% are infected, infection occurs earlier and the effect of stump treatment during thinning is smaller (63% vs 47%). These simulations advocate for preventive measures against *H. annosum* on healthy, first-generation pine stands

to apply biological treatment rather than any concern for the health of the particular tree species (Oliva et al. 2016a).

10.6 Research Directions: Particular Issues of Forest Disease Under Dry Mediterranean Conditions

We lack basic information on the role of pathogens in dry Mediterranean areas, particularly for pines, where disease symptoms are often masked by abiotic stressors. Some pine forests may lack economic value but serve important ecological functions. Understanding how pathogens modify those functions is a priority (see Bjelke et al. 2016). A good example of this approach could be to focus on understanding the ecological role of pathogens in plant-soil feedbacks. This process has consequences ranging from predicting the stability of mixed forests to helping forest managers ensure good regeneration at the end of the rotation or after a fire.

Invasive pathogens pose a real threat to pine woodlands, and thanks to advances in genomic technology, it will be soon possible to understand the genetic basis for tolerance/resistance in tree hosts (Santos et al. 2012), as well as the origins of virulence in the pathogen (Santos et al. 2012). Pathogen functional traits have been shown to be informative of invasion risk (Redondo et al. 2018a, b). Further study and characterization of pathogen communities as well as development of databases for susceptibility will improve our capacity to protect pines from exotic pathogens (Oliva et al. 2020).

At the theoretical level, we need to develop our understanding of the interactions between pathogens and abiotic factors. More efforts should be spent on bringing pathogens into our examination of drought-induced mortality from a physiological angle. Particularly important for such processes is the study of facultative endophytic pathogens that inhabit living plant tissues but can switch to the pathogenic state under host stress. Little is known about how these microbes establish in different tissues of the tree and what function they provide to the tree.

10.7 Management Recommendations for Pines in Semiarid Regions

Disease management should be based on a comprehensive assessment of the impacts we are trying to mitigate. When disease impacts are clearly monetary, costs should not outweigh benefits. In that sense, mechanistic models become useful in assisting forest managers. Simulations carried out on pure and mixed stands support the role of mixed stands in suppressing disease by promoting pine mixtures with non-host species such as oaks (Fig. 10.4). Mixing pine with 30% oaks decreases the probability of becoming infected from 62% to 50%. However, even stronger than the effect of forest mixture is the effect of stump treatment with the biocontrol fungus *Phlebiopsis gigantea* (62–10%). Keeping *Heterobasidion*-free stands is critical as stumps left in the final felling can carry over the inoculum to the next rotation. If pine stands establish on a previous site inhabited by infected pines, the impact of stump treatment will be largely reduced (Fig. 10.4). In those cases, only stumping may reduce inoculum in the ground, albeit at a much higher economic cost (Cleary et al. 2013). Stump treatment has been shown to reduce root rot and increase resistance to wind felling in other conifer species (Fig. 10.1d) (Oliva et al. 2008b, 2010). Treating thinning stumps as part of regular forest operations should be seriously considered across the Mediterranean Basin (Fig. 10.1b), not only due to the presence of native *Heterobasidion* species, but because of the current increase of *H. irregulare* and hybrid genotypes.

We should strive to find strategies to communicate with society about the importance of pine forests in the Mediterranean region and the sanitary risk posed by the introduction of novel species. Many native pathogens may become more prevalent if forests are abandoned and intra- and interspecific tree competition increases. We could potentially mitigate these impacts by managing tree competition; however, care should be taken to understand the key triggering factor in each case, and whether other pathogens may benefit from our actions. This is only possible when based on a thorough understanding of pathogen biology and disease. For instance, if we want to decrease the impact of *D. sapinea* in a dense stand assuming that drought was causing stress and promoting tip blight, thinning the stand would be a good option. However, opening the stand may also raise the overall temperature of the forest, allowing further sporulation of the pathogen. Also, the timing of the thinning should be adjusted so minimal infection on the stumps by *Heterobasidion* occurs. By thinning the stand, we may decrease the risk of a red band disease outbreak if a wet summer is expected. Models integrating several pathogens and their interaction with biotic and abiotic factors would be particularly useful in the Mediterranean context.

References

Aguadé D, Poyatos R, Gómez M et al (2015) The role of defoliation and root rot pathogen infection in driving the mode of drought-related physiological decline in Scots pine (*Pinus sylvestris* L.). Tree Physiol 35:229–242

Bendel M, Kienast F, Bugmann H et al (2006) Incidence and distribution of *Heterobasidion* and *Armillaria* and their influence on canopy gap formation in unmanaged mountain pine forests in the Swiss Alps. Eur J Plant Pathol 116:85–93

Bennett JA, Maherali H, Reinhart KO et al (2017) Plant-soil feedbacks and mycorrhizal type influence temperate forest population dynamics. Science 355:181–184

Bezos D, Martínez-Álvarez P, Diez JJ et al (2015) The pine shoot beetle *Tomicus piniperda* as a plausible vector of *Fusarium circinatum* in northern Spain. Ann For Sci 72:1079–1088

Bjelke U, Boberg J, Oliva J et al (2016) Dieback of riparian alder caused by the *Phytophthora alni* complex: projected consequences for stream ecosystems. Freshw Biol 61:565–579

Blackburn TM, Pysek P, Bacher S et al (2011) A proposed unified framework for biological invasions. Trends Ecol Evol 26:333–339

Brodde L, Adamson K, Julio Camarero J et al (2019) Diplodia tip blight on its way to the North: drivers of disease emergence in Northern Europe. Front Plant Sci. https://doi.org/10.3389/fpls.2018.01818

Burgess TI, Wingfield MJ (2017) Pathogens on the move: a 100-year global experiment with planted eucalypts. Bioscience 67:14–25

Cleary MR, Arhipova N, Morrison DJ et al (2013) Stump removal to control root disease in Canada and Scandinavia: a synthesis of results from long-term trials. For Ecol Manag 290:5–14

Desprez-Loustau M-L, Marçais B, Nageleisen L-M et al (2006) Interactive effects of drought and pathogens in forest trees. Ann For Sci 63:597–612

Díaz-Yáñez O, Mola-Yudego B, Timmermann V et al (2020) The invasive forest pathogen *Hymenoscyphus fraxineus* boosts mortality and triggers niche replacement of European ash (*Fraxinus excelsior*). Sci Rep 10:5310

Doğmuş-Lehtijärvi HT, Erdoğan RC, Lehtijärvi A et al (2016) Pathogenicity of *Heterobasidion annosum* (Fr.) Bref. *sensu stricto* on coniferous tree species in Turkey. For Pathol 46:22–28

Elvira-Recuenco M, Iturritxa E, Majada J et al (2014) Adaptive potential of maritime pine (*Pinus pinaster*) populations to the emerging pitch canker pathogen, *Fusarium circinatum*. PLoS One 9:e114971

Eriksson L, Boberg J, Cech TL et al (2018) Invasive forest pathogens in Europe: cross-country variation in public awareness but consistency in policy acceptability. Ambio 48:1–12

Fabre B, Ioos R, Piou D et al (2012) Is the emergence of Dothistroma needle blight of pine in France caused by the cryptic species *Dothistroma pini*? Phytopathology 102:47–54

Garbelotto M (2009) Molecular analyses to study invasions by forest pathogens: examples from Mediterranean ecosystems. Phytopathol Mediterr 47:183–203

Garbelotto M, Gonthier P (2013) Biology, epidemiology, and control of *Heterobasidion* species worldwide. Annu Rev Phytopathol 51:39–59

Geils BW, Hummer KE, Hunt RS (2010) White pines, *Ribes*, and blister rust: a review and synthesis. For Pathol 40:147–185

Gilbert GS (2002) Evolutionary ecology of plant diseases in natural ecosystems. Annu Rev Phytopathol 40:13–43

Gonthier P, Garbelotto M (2011) Amplified fragment length polymorphism and sequence analyses reveal massive gene introgression from the European fungal pathogen *Heterobasidion annosum* into its introduced congener *H. irregulare*. Mol Ecol 20:2756–2770

Gonthier P, Warner R, Nicolotti G et al (2004) Pathogen introduction as a collateral effect of military activity. Mycol Res 108:468–470

Gonthier P, Anselmi N, Capretti P et al (2014) An integrated approach to control the introduced forest pathogen *Heterobasidion irregulare* in Europe. Forestry 87:471–481

Hicke JA, Allen CD, Desai AR et al (2012) Effects of biotic disturbances on forest carbon cycling in the United States and Canada. Glob Chang Biol 18:7–34

Landeras E, García P, Fernández Y et al (2005) Outbreak of pitch canker caused by *Fusarium circinatum* on *Pinus* spp. in Northern Spain. Plant Dis 89:1015

Liebhold AM, Brockerhoff EG, Garrett LJ et al (2012) Live plant imports: the major pathway for forest insect and pathogen invasions of the US. Front Ecol Environ 10:135–143

Linares JC, Camarero JJ, Bowker MA et al (2010) Stand-structural effects on *Heterobasidion abietinum*-related mortality following drought events in *Abies pinsapo*. Oecologia 164:1107–1119

Lung-Escarmant B, Guyon D (2004) Temporal and spatial dynamics of primary and secondary infection by *Armillaria ostoyae* in a *Pinus pinaster* plantation. Phytopathology 94:125–131

Lygis V, Vasiliauskas R, Stenlid J (2005) Clonality in the postfire root rot ascomycete *Rhizina undulata*. Mycologia 97:788–792

Mangan SA, Schnitzer SA, Herre EA et al (2010) Negative plant–soil feedback predicts tree-species relative abundance in a tropical forest. Nature 466:752

Manion PD (1981) Tree disease concepts. Prentice Hall, Englewood Cliffs

Manion PD (2003) Evolution of concepts in forest pathology. Phytopathology 93:1052–1055

Manter DK, Bond BJ, Kavanagh KL et al (2000) Pseudothecia of Swiss needle cast fungus, *Phaeocryptopus gaeumannii*, physically block stomata of Douglas fir, reducing CO2 assimilation. New Phytol 148:481–491

Mattila U (2005) Probability models for pine twisting rust (*Melampsora pinitorqua*) damage in Scots pine (*Pinus sylvestris*) stands in Finland. For Pathol 35:9–21

Mesanza N, Iturritxa E (2012) Root and butt rot caused by *Heterobasidion annosum* in Atlantic coniferous ecosystems of Spain. For Pathol 42:514–520

Munck IA, Smith DR, Sickley T et al (2009) Site-related influences on cone-borne inoculum and asymptomatic persistence of Diplodia shoot blight fungi on or in mature red pines. For Ecol Manag 257:812–819

Munck IA, Luther T, Wyka S et al (2016) Soil and stocking effects on Caliciopsis canker of *Pinus strobus* L. Forests 7:269

Oliva J, Stenlid J (2011) Validation of the Rotstand model for predicting *Heterobasidion annosum* root rot in *Picea abies* stands. For Ecol Manag 261:1841–1851

Oliva J, Ibarra N, Colinas C et al (2008a) Molecular detection of *Heterobasidion annosum s.s.* after thinning operations in a *Pinus nigra* plantation in Pyrenees [Spain] (in Spanish). Bol San Veg Plagas 34:415–416

Oliva J, Samils N, Johansson U et al (2008b) Urea treatment reduced *Heterobasidion annosum s.l.* root rot in *Picea abies* after 15 years. For Ecol Manag 255:2876–2882

Oliva J, Thor M, Stenlid J (2010) Long term effects of mechanized stump treatment against *Heterobasidion annosum s.l.* root rot in *Picea abies*. Can J For Res 40:1020–1033

Oliva J, Boberg JB, Hopkins AJ et al (2013) Concepts of epidemiology of forest diseases. In: Gonthier P, Niccolotti G (eds) Infectious forest diseases. CAB International, Wallingford, pp 1–28

Oliva J, Stenlid J, Martínez-Vilalta J (2014) The effect of fungal pathogens on the water and carbon economy of trees: implications for drought-induced mortality. New Phytol 203:1028–1035

Oliva J, Castaño C, Baulenas E et al (2016a) The impact of the socioeconomic environment on the implementation of control measures against an invasive forest pathogen. For Ecol Manag 380:118–127

Oliva J, Stenlid J, Grönkvist-Wichmann L et al (2016b) Pathogen-induced defoliation of *Pinus sylvestris* leads to tree decline and death from secondary biotic factors. For Ecol Manag 379:273–280

Oliva J, Redondo MA, Stenlid J (2020) Functional ecology of forest disease. Annu Rev Phytopathol 58:343-361

Oliva J, Ridley M, Redondo MA, Caballol M (2021) Competitive exclusion amongst endophytes determines shoot blight severity on pine. Funct Ecol 35:239–254

Ortíz De Urbina E, Mesanza N, Aragonés A et al (2017) Emerging needle blight diseases in Atlantic pinus ecosystems of Spain. Forests 8:18

Prieto-Recio C, Romeralo C, Bezos D et al (2012) First report of *Heterobasidion annosum* on *Pinus pinaster* in Spain. Plant Dis 96:770

Pukkala T, Möykkynen T, Thor M et al (2005) Modeling infection and spread of *Heterobasidion annosum* in even-aged Fennoscandian conifer stands. Can J For Res 35:74–84

Redondo MA, Boberg J, Olsson CH et al (2015) Winter conditions correlate with *Phytophthora alni* subspecies distribution in Southern Sweden. Phytopathology 105:1191–1197

Redondo MA, Boberg J, Stenlid J et al (2018a) Contrasting distribution patterns between aquatic and terrestrial *Phytophthora* species along a climatic gradient are linked to functional traits. ISME J 12:2967–2980

Redondo MÁ, Boberg J, Stenlid J et al (2018b) Functional traits associated with the establishment of introduced *Phytophthora* spp. in Swedish forests. J Appl Ecol 55:1538–1552

Romeralo C, Bezos D, Martínez-Álvarez P et al (2018) Vertical transmission of *Fusarium circinatum* mitoviruses FcMV1 and FcMV2-2 via microconidia. Forests 9:356

Roura i Pascual N, Brotons L, García D, Zamora R, De Cáceres M (2012) Local and landscape-scale biotic correlates of mistletoe distribution in Mediterranean pine forests. For Syst 21:179–188

Sampedro L, Moreira X, Zas R (2011) Resistance and response of *Pinus pinaster* seedlings to *Hylobius abietis* after induction with methyl jasmonate. Plant Ecol 212:397–401

Santini A, Ghelardini L, De Pace C et al (2013) Biogeographical patterns and determinants of invasion by forest pathogens in Europe. New Phytol 197:238–250

Santos CS, Pinheiro M, Silva AI et al (2012) Searching for resistance genes to *Bursaphelenchus xylophilus* using high throughput screening. BMC Genomics 13:599

Scirè M, D'amico L, Motta E et al (2008) North American P type of *Heterobasidion annosum* shows pathogenicity towards *Pinus halepensis* in Italy. For Pathol 38:299–301

Sherwood P, Villari C, Capretti P et al (2015) Mechanisms of induced susceptibility to Diplodia tip blight in drought-stressed Austrian pine. Tree Physiol 35:549–562

Stanosz GR, Blodgett JT, Smith DR et al (2001) Water stress and *Sphaeropsis sapinea* as a latent pathogen of red pine seedlings. New Phytol 149:531–538

Stenlid J, Oliva J (2016) Phenotypic interactions between tree hosts and invasive forest pathogens in the light of globalization and climate change. Philos Trans R Soc B 371. https://doi.org/10.1098/rstb.2015.0455

Stenlid J, Oliva J, Boberg JB et al (2011) Emerging diseases in European forest ecosystems and responses in society. Forests 2:486–504

Teste FP, Kardol P, Turner BL et al (2017) Plant-soil feedback and the maintenance of diversity in Mediterranean-climate shrublands. Science 355:173–176

Wang L, Zhang J, Drobyshev I et al (2014) Incidence and impact of root infection by *Heterobasidion* spp., and the justification for preventative silvicultural measures on Scots pine trees: a case study in southern Sweden. For Ecol Manag 315:153–159
Wikler K, Gordon TR (2000) An initial assessment of genetic relationships among populations of *Fusarium circinatum* in different parts of the world. Can J Bot 78:709–717

Chapter 11
Genetic and Ecological Basis of Resistance to Herbivorous Insects in Mediterranean Pines

Carla Vázquez-González, Luis Sampedro, Xosé López-Goldar, and Rafael Zas

11.1 Introduction

Pines represent a successful genus of trees that have occupied vast areas around the Mediterranean Basin for the last million years (Grivet et al. 2013). As massive, sessile, long-lived and prominent organisms potentially providing a huge amount of nutritional resources (Schulman 1984), Mediterranean pines are constantly exposed to a wide plethora of enemies that feed on their tissues. Insect herbivores stand out as one of the most important biotic threats for pine trees, causing important impacts on growth, reproduction, survival and ultimately on pine fitness (Paine and Lieutier 2016).

During a long co-evolutionary history, pines and herbivores have evolved multiple and complex mechanisms aimed at maintaining their fitness in the context of their interactions. Despite the potential evolutionary disadvantage imposed by the asymmetry between the generation times of pines and herbivores (Petit and Hampe 2006), pine trees have been able to persist and dominate extensive tracts of land around the Mediterranean Sea until current times (Carrión et al. 2000). Part of this success must rely on efficient physiological and ecological mechanisms that protect pine trees (i.e. resistance) or minimize the impact of insect damage on pine fitness (i.e. tolerance) (Mumm and Hilker 2006).

C. Vázquez-González (✉) · L. Sampedro · R. Zas
Misión Biológica de Galicia – Consejo Superior de Investigaciones Científicas, Pontevedra, Spain
e-mail: cvazquez@mbg.csic.es

X. López-Goldar
Misión Biológica de Galicia – Consejo Superior de Investigaciones Científicas, Pontevedra, Spain

Department of Ecology and Evolutionary Biology, Cornell University, Ithaca, NY, USA

G. Ne'eman, Y. Osem (eds.), *Pines and Their Mixed Forest Ecosystems in the Mediterranean Basin*, Managing Forest Ecosystems 38,
https://doi.org/10.1007/978-3-030-63625-8_11

A diverse community of enemies exerts complex and heterogeneous selection pressures to which pine defenses are expected to adaptively respond (Poelman and Kessler 2016). However, rather than by directional responses to specific pine-enemy interactions, pine defenses are expected to be the result of diffuse selective pressures exerted not only by the complex community of herbivore species (Wise and Rausher 2013), but also by the interaction with the abiotic environment (Vogan and Schoettle 2015). Pine defenses against insect herbivory are highly costly to produce (Sampedro et al. 2011a), and thus, resources invested in defense are no longer available for other plant functions such as growth or reproduction. Consequently, pine trees must thus optimize resource allocation among different functions according to the particular biotic and abiotic environment (Sampedro 2014).

Altogether, both the diffuse selection pressures exerted by the highly diverse enemy community and the integrated compromises for maximizing the different life functions have contributed to maintain the huge variation among and within pine species in defensive mechanisms and strategies against insect herbivory (Petit and Hampe 2006). Aiming to provide some insights into the huge variation of defensive traits and strategies within Mediterranean pines, in the present chapter we first introduce the main insect pests affecting Mediterranean pines, with special emphasis on their negative effects on the host tree. After that, we describe the defensive mechanisms implicated in biotic resistance to such pests in pine species, including direct (chemical and anatomical) and indirect defenses. We then discuss compromises between allocation to growth and defenses in Mediterranean pines. Finally, we review current knowledge on the sources of variation of pine resistance to insect herbivory, paying special attention to the genetic, plastic and ecological factors modulating investment in chemical and physical defenses and/or effective resistance against particular enemies. We focus on the four most abundant Mediterranean pine species (*Pinus pinaster* Ait., *P. halepensis* Mill., *P. pinea* L. and *P. brutia* Ten.) on which most of the available literature is focused.

11.2 Main Insect Herbivores of Mediterranean Pines

The number of phytophagous insect species that are associated with Mediterranean pines (*P. pinaster, P. halepensis, P. pinea* and *P. brutia*) is enormous. Based on a literature search and their own data, Mendel (2000) reported more than 110 insect species feeding on *P. halepensis* and *P. brutia* in southeast Europe, most of them (62%) belonging to five families: Scolytidae, Buprestidae, Cerambycidae, Lachnidae, and the superfamily Coccoidea. In Southwestern Europe, Muñoz-López et al. (2007) listed up to 66 species producing significant damage to native Pinaceae species. Insects feeding on Mediterranean pines include piercing-suckers, borers and chewers, producing damage in all pine tissues, including needles, buds, stems and cones. Despite the large diversity of insects feeding on Mediterranean pines, only a few cause damage of ecological and economic relevance with relatively high frequency (Fig. 11.1).

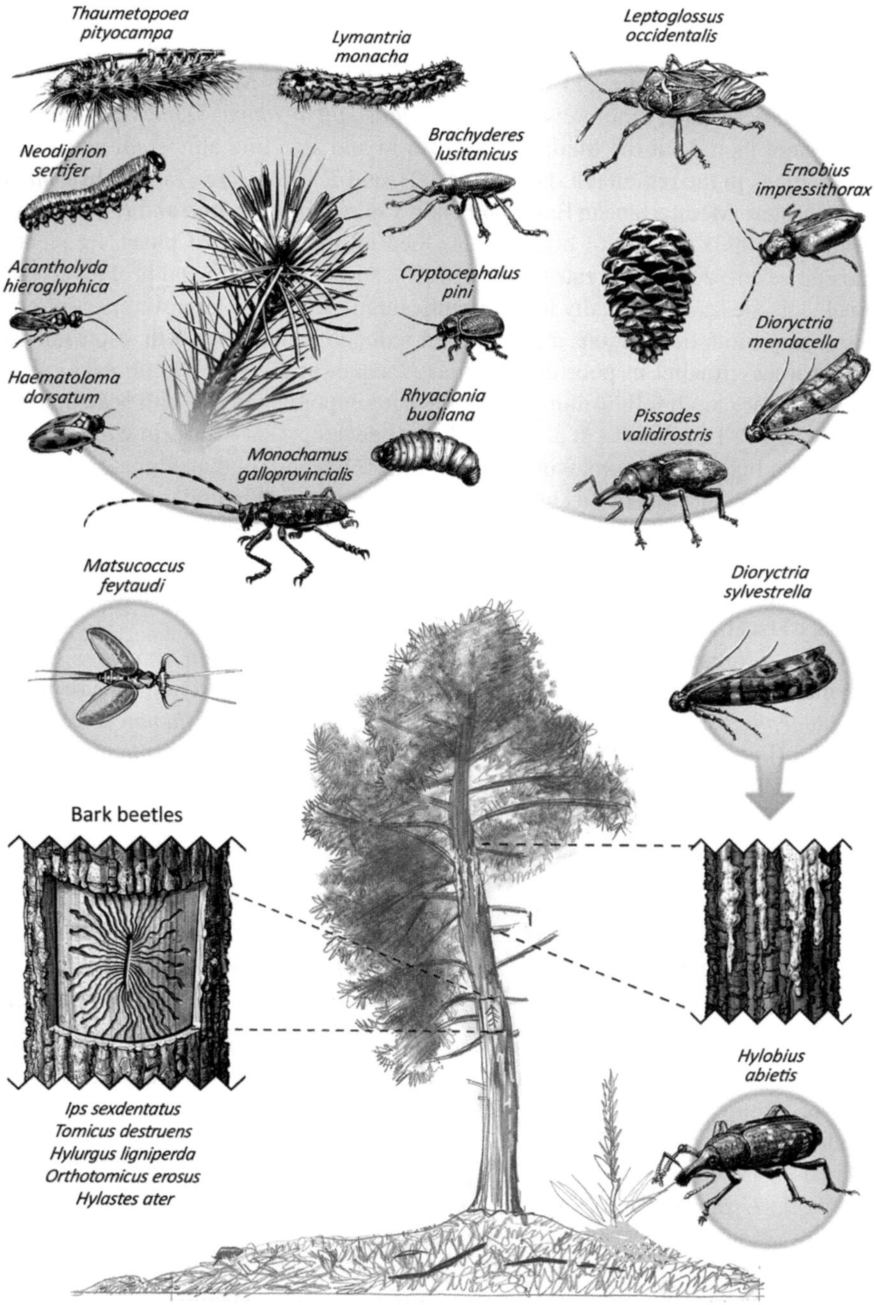

Fig. 11.1 Main insects feeding on living tissues of Mediterranean pines

11.2.1 Defoliators

Probably the most iconic and well-known insect pests of Mediterranean pines are the pine processionary moths, with *Thaumetopoea pityocampa* (Dennis and Schiff.), distributed thoughout the Mediterranean Basin and affecting almost all native and exotic pines in the region (Hodar et al. 2002), and *T. wilkinsoni* (Tams), distributed in the eastern Mediterranean Basin and mainly damaging *P. brutia* and *P. halepensis* (Halperin 1990). Larvae of both species feed on the needles of pines, frequently causing severe defoliation rates (up to 100%). Although severe and prolonged outbreak episodes can eventually led to significant mortality, the largest impact of processionary moth defoliations are important reductions in tree growth, especially in young trees growing in poor quality sites (Jacquet et al. 2012). Outbreaks of the processionary moth follow more or less regular temporal patterns with peak defoliations every 7–11 years (Li et al. 2015). Larval development occurs in winter and is mainly limited by minimum winter temperatures. Because of that, climate change is favoring the expansion of this insect to higher altitudes and latitudes (Pimentel et al. 2011; Battisti et al. 2016a). In the Mediterreanean Basin, the processionary moth is probably the forest pest for which investment in control measures – mainly aerial insecticide applications – is the highest. However, the efficacy of such management practices has been questioned (Cayuela et al. 2011).

Other defoliators of Mediterranean pines include *Neodiprion sertifer* Geoff. (Hymenoptera, Diprionidae), *Lymantria monacha* L. (Lepidoptera, Lymantriidae), *Cryptocephalus pini* L. (Coleoptera, Curculionidae), *Panolis flammea* Denis & Schiff (Lepidoptera, Noctuidae) and *Acantholyda hieroglyphica* Christ. (Hymenoptera, Pamphiliidae) to mention a few, but in all cases both their incidence on Mediterranean pines and the information available about them are much more limited (Battisti et al. 2016b).

11.2.2 Stem Borers, Phloem Feeders and Bud Miners

An important group of insects that causes severe damage in Mediterranean pine forests are bark beetles (Coleoptera, Scolytidae). Bark beetles are an extremely diverse group of insects, with more than 40 species associated with typical Mediterranean conifers, many of them highly specialized to one or a few highly-related host species (Lieutier et al. 2016). *Ips sexdentatus* Börner, *Tomicus destruens* Woll., *Hylurgus ligniperda* Fab., *Orthotomicus erosus* Woll., and *Hylastes ater* Pay. are among the most frequent bark beetles infesting Mediterranean pines. Bark beetles oviposit in galleries beneath the bark of branches, trunks or roots, where developing larvae feed on the innermost part of the bark, causing tree mortality due to the breakdown of the vascular system. Bark beetles typically attack weakened trees, but epidemic populations during outbreaks can also kill healthy trees. As their counterparts in boreal pine forest, bark beetles of Mediterranean pines can act on host trees in symbiotic association with an array of microorganisms, including pathogenic fungi such as *Ophistoma* sp. (Kirisits 2004). Pine host resistance to bark beetles has been reviewed elsewhere (Krokene 2015).

Another important stem borer of Mediterranean pines is the European stem borer *Dioryctria sylvestrella* (Ratz.) (Lepidoptera, Pyralidae), which causes extensive damage in pine forest plantations, through trunk malformations and resin pockets within the stem wood, and an increase in the risk of stem breakage (Budde et al. 2017).

One insect that is gaining increasing attention in the Mediterranean Basin is the pine sawyer *Monochamus galloprovincialis* Olivier (Coleoptera, Cerambycidae), which feeds in the inner stem of pine branches. Although the direct feeding impact of these longhorn beetles is not considered to be very significant (Evans et al. 2004), adults of this species vector the exotic nematode, *Bursaphelenchus xylophilus* (Steiner and Buhrer) Nickle, responsible for pine wilt disease (Sousa et al. 2001). The devastating effects of this exotic invasive organism have motivated vast research efforts aimed at controlling its expansion throughout Europe; an important part of this effort has been to study the biology, ecology and control of the insect vector (Naves et al. 2016).

Insect damage on Mediterranean pines can occur at all tree ages; the early stages of development are assumed to be more critical, as insect damage can easily lead to seedling mortality (Barton and Koricheva 2010). One insect that can cause great damage at early ontogenetic stages of Mediterranean pines is the pine weevil, *Hylobius abietis* L. (Coleoptera, Curculionidae). Adults of this insect feed on the phloem and bark of young seedlings, easily leading to stem girdling and seedling death, with damage being especially problematic in natural or artificial regenerations after clear-cuts (Day et al. 2004). This insect has an enormous economic impact for the regeneration of conifer forests in northern and central Europe, but it can also attack Mediterranean pines such as *P. pinaster* (Sampedro et al. 2009), *P. halepensis* (Suarez-Vidal et al. 2019), or *P. brutia* (Semiz et al. 2017).

Bud miners of Mediterranean pines include lepidopteran species such as *Rhyacionia buoliana* (Dennis and Schiff.) and *R. duplana* Hübner (Lepidoptera, Tortricidade), two moths that can cause important economic damage on young plantations producing fancy stem deformations such as forked or crooked stems.

11.2.3 Sap Suckers

Several sap-sucker species of aphids, coccids and bugs are also known to affect Mediterranean pines, reducing tree vigor and favoring the entrance of secondary pests (Sallé and Battisti 2016). Among them, *Haematoloma dorsatum* Ahrens (Hemiptera: Cercopidae), *Leucaspis pusilla* Loew, *L. pini* Hartig, and *L. lowi* Colvée (Hemiptera: Diaspididae) are considered potential pests (Sallé and Battisti 2016). Two pine bast scales (*Matsucoccus* sp., Hemiptera, Matsucoccidae) are also potentially highly destructive for Mediterranean pines (Mendel et al. 2016). *Matsucoccus feytaudi* Duc. is native to the western Mediterranean Basin and is specific to maritime pine, while *M. josephi* Bodenhiemer and Harpaz appears in the eastern Mediterranean and feeds mainly on *P. brutia* and *P. halepensis*. In their native range, both species cohabit with their host in equilibrium. However, the two species have recently spread out of their natural range, causing extensive damage and great mortality (Mendel et al. 2016; Roversi et al. 2013). The absence of efficient defensive mechanisms in the native pine populations of the invaded range due to absence of a co-evolutionary history with the insect (Schvester and Ughetto 1986) and lack of natural enemies in the new range (Jactel et al. 2006) seem to underlie these contrasting patterns of susceptibility.

11.2.4 Cone and Seed Feeders

Cone and seed feeders are another important group of phytophagous insects associated with Mediterranean pines (Boivin and Auger-Rozenberg 2016). Most of the knowledge about them comes from the study of the damage and control of insects affecting the valued cones of stone pine (*P. pinea*) (Bracalini et al. 2013). Two cone moths, *Dioryctria pineae* Staudinger and *D. mendacella* Staudinger (Lepidoptera: Pyralidae), one cone weevil, *Pissodes validirostris* Gyll (Coleoptera: Curculionidae), and the anobiid beetle *Ernobius impressithorax* Pic (Coleoptera: Anobiidae) are probably the most abundant cone-damaging agents (Bracalini et al. 2013; Innocenti and Tiberi 2002). The larvae of these insects cause severe visual and internal damage to the cones and ultimately lead to high seed loss rates, and a notable decrease in the weight and viability of the remaining unattacked seeds (Bracalini et al. 2013; Calama et al. 2017). Another exotic insect, *Leptoglossus occidentalis* Heidemann (Hemiptera: Coreidae), native to North America and recently introduced into Europe (Bernardinelli and Zandigiacomo 2001), is also causing large seed yield losses in *P. pinea* orchards, but signs of its damage on the cones are not so evident (Oliveira Farinha et al. 2018).

11.3 Mechanisms and Strategies Providing Resistance to Insect Herbivores in Mediterranean Pine Trees

11.3.1 Direct Resistance Mechanisms

Pine trees have evolved a complex defensive system with multiple layers of chemical and anatomical defences (see Box 11.1). Pine defenses are always present at a baseline level, and are referred to as constitutive defenses, providing direct resistance against a vast array of herbivores and pathogens (Table 11.1). Direct resistance traits are toxic, repellent, anti-digestive chemical compounds or physical barriers that directly reduce the performance or efficiency of herbivores feeding on plant tissues, affecting their fitness at diverse stages of their life cycles.

Box 11.1: Mechanisms of Resistance to Insect Herbivores in Mediterranean Pines

Direct resistance mechanisms are those mechanisms that act directly killing or repelling the insect herbivores, and include:

- *Mechanical defenses* are displayed as concentric layers acting as defensive barriers aimed at avoiding the entrance of invader organisms. In Mediterranean pines, these include the outer bark (periderm), a highly suberized or lignified cell types, and a highly developed system of resin ducts.
- *Chemical defenses* are stored in different anatomical or cellular structures, in all pine tissues (needles, stem, roots and cones). In Mediterranean pines, chemical compounds, including at least phenolic compounds and terpenes, are released into the intercellular space upon insect damage, causing deterrence, digestive or feeding dysfunction, direct toxicity and might ultimately lead to insect mortality.

Indirect resistance mechanisms: Indirect defenses involve third partners, usually predators of insect herbivores, in mutualistic association with plant hosts that positively influence plant resistance throughout top-down regulation of herbivore populations. Indirect defenses may involve a 'call for help' carried out by plants often mediated by the emission of volatile organic compounds (VOCs) which the predator can recognize.

Induced resistance: The production of direct and indirect defenses is not constant throughout time. Conversely, their production is increased when the biotic damage is perceived, producing the so called 'induced defenses', which are specially relevant to deter the negative impacts of an insect attack. These include:

- *Induced direct defenses* in Mediterranean pines involve processes leading to increased production of physical and chemical defenses, including cell differentiation, gene expression, protein biosynthesis and hormonal regulation.
- *Induced emission of volatile organic compounds (VOCs)* lead to changes in the tree chemical airborne signature that might lead to faster within-plant and between-plant communication and to a strong 'call for help'.
- *Induced tolerance responses* refer to phenotypic changes in the host aimed at minimizing the impact caused by the herbivore on plant fitness, with no effects on the herbivore survival, growth or reproduction.

Priming of resistance: Priming of defensive mechanisms act as a 'vaccination effect' by which plants are able to perceive the risk of an attack and activate the cellular and molecular machinery to carry out a faster induced response once the damage finally occurs. Up to date priming has been described in *Picea abies*, but not yet in Mediterranean pines.

Table 11.1 Defensive traits providing resistance to specific insect herbivores in Mediterranean pines

Host	Insect herbivore	Trait	Type[a]	Pattern[b]	Observations	References
Pinus spp.	*Thaumetopoea pityocampa*	Terpenes	C	R	Oviposition deterrence	Paiva et al. (2011)
		Total phenols	C	R	Larval mortality	Schopf and Avtzis (1987)
		Silica	P	R	Larval mortality	Schopf and Avtzis (1987)
	Matsucoccus josephi	Wound periderm	P	R	Faster response in non-host species	Liphschitz and Mendel (1989)
		Tissue hypertrophy	P	S	Abnormal tissue formation in host species	Liphschitz and Mendel (1989)
	Leptoglossus occidentalis	Cone and seed size	O	S	*P. pinea* > *P. pinaster* = *P. halepensis*	Farinha et al. (2018)
P. pinea	*Tomicus destruens*	Terpenes	C	R	Limonene, β-caryophyllene	Faccoli et al. (2011)
		Terpenes	C	S	α-pinene, β-myrcene, α-terpinolene	Faccoli et al. (2008, 2011)
	Thaumetopoea pityocampa	Terpenes	C	R	(1S)-(-)-β-pinene and (R)-(+)-limonene reduced egg laying. Limonene reduce oviposition preference	Panzavolta et al. (2015) and Tiberi et al. (1999)
	Neodiprion sertifer	Terpenes	C	R	Low limonene concentration is attractive, whereas high concentration is repellent	Martini et al. (2010)
	Marchalina hellenica	Terpenes	C	R	Limonene increases resistance	Mita et al. (2002)
P. pinaster	*Dioryctria sylvestrella*	Terpenes	C	R	Terpinolene as a candidate compound providing resistance for this insect	Jactel et al. (1999)
		Terpenes	C	S	Limonene, copaene, linalool and longipinene contribute to susceptibility	Jactel et al. (1996) and Kleinhentz et al. (1999)

(continued)

Table 11.1 (continued)

Host	Insect herbivore	Trait	Type[a]	Pattern[b]	Observations	References
		Bark thickness	P	R	Bark thickness reduces infestation	Carisey et al. (1994)
	Hylobius abietis	Terpenes	C	R	Limonene provides constitutive resistance	López-Goldar et al. (2018)
			C		Induced resin acids increase resistance	
		Phenolics	C	R	Induced lignans increase resistance	López-Goldar et al. (2018)
		Non-volatile resin	C	R	MJ induction of non-volatile resin increases resistance	Sampedro et al. (2011b)
	Diprion pini	Terpenes	C	R	Resin acids reduce oviposition	Auger et al. (1994)
	Thaumetopoea pityocampa	Terpenes	C	R	Sprayed limonene reduces female oviposition on needles	Tiberi et al. (1999)
P. halepensis	*Thaumetopoea pityocampa*	Needle length	P	R	Shorter needles reduce survival of pine processionary eggs	Hezil et al. (2018)
		Needle asymmetry	P	S	Pine processionary females select asymmetric needles for oviposition	Pérez-Contreras et al. (2008)
	Tomicus piniperda	Terpenes	C	R	High levels of (+)-α-pinene, +-3-carene; low levels of (-)-α-pinene are deterrent	Almquist et al. (2006)
	Tomicus destruens	Ethanol	C	S	Increased ethanol in drought-stressed pines, which is attractive for the insect	Kelsey et al. (2014)
	Marchalina hellenica	Terpenes	C	S	High levels of α-pinene and low levels of limonene and α-terpinyl acetate	Mita et al. (2002)
P. brutia	*Tomicus destruens*	Bark thickness	P	R	Attacked trees are damaged only in thin-bark parts	Ciesla et al. (2011)

(continued)

Table 11.1 (continued)

Host	Insect herbivore	Trait	Type[a]	Pattern[b]	Observations	References
	Tomicus minor	Bark thickness	P	R	Attacked trees are damaged only in thin-bark parts of the mid and upper bole	Ciesla et al. (2011)
	Hylobius abietis	Terpenes	C	R	γ-terpinene, sabinene are negatively correlated to pine weevil damage.	Semiz et al. (2017)
					Limonene and α-terpinolene showed negative trends with weevil damage.	

[a]Type of defenses: *C* Chemical, *P* Physical, *O* Other
[b]Resistance pattern analyzed: *R* Resistance, *S* Susceptibility

11.3.1.1 Mechanical Defenses

Pine trees display a model mechanical defensive system in their stem and branches, comprising several concentric layers (Franceschi et al. 2005). The outer bark (periderm), among other functions, provides a first barrier of suberized, lignified, dead cells that hinder the entrance of insects into the stem (Krokene 2016). In the bark, the main defensive mechanisms are the polyphenolic parenchyma, with living cells containing large vacuoles filled with phenolic compounds (Nagy et al. 2004), and stored starch and lipids that could be readily released in case of attack (Franceschi et al. 2000). Vacuoles with proteinase inhibitors, phenol oxidases and other enzymes reducing the digestibility of ingested host tissues may also be present, as found in other plant taxa (Huffaker et al. 2013). Cell walls may be reinforced with compounds such as suberin, lignin and callose. Other layers of defensive cells with calcium oxalate crystals or stone cells with a defensive function have been described in other conifer species (Moreira et al. 2012a; Whitehill et al. 2016) but not yet in Mediterranean pines.

A major defensive mechanism in pine trees are resin-based defenses (Franceschi et al. 2005). Particularly, pine trees have developed a complex three-dimensional network of interconnected axial and radial resin ducts within the xylem and phloem. Resin ducts are the structures where oleoresin (a complex mixture of diterpenes and resin acids solubilized by a volatile fraction of monoterpenes and sesquiterpenes) is synthesized and stored (Celedon and Bohlmann 2019). Pine oleoresin is a sticky, toxic fluid that flows out when the resin ducts are injured, thus preventing the entrance of invading organisms.

Resin synthesis also takes place in needles (Turner et al. 2018), where axial resin ducts are also present and connected to the tree resin duct network. Cuticular waxes may prevent oviposition or reduce the mobility of defoliating insects (Hilker et al. 2002). Trichomes and spines have also been described as mechanical barriers against small insects in other species. Small spines are present in primary needles of pine seedlings but they have been not yet been associated with insect resistance.

11.3.1.2 Chemical Defenses

Among direct chemical defenses, mono-, sesqui- and diterpene acids are highly toxic compounds for many invading organisms (Whitehill et al. 2019). Pine phenolic compounds are very diverse in nature and contribute in different forms to resistance, such as the tannins binding digestive proteins or the stilbenes with direct toxicity (Nagy et al. 2004). Both terpenes and phenolics are carbon-based compounds and have been traditionally considered to be the most relevant chemical defenses for pine resistance. Up to 93 different terpenoid chemical species and 35 different phenolic compounds have been found in young *P. pinaster* seedlings, reaching a concentration up to 27 mg g^{-1} and 15 mg g^{-1}, respectively, in some tissues (López-Goldar et al. 2019). Piperidine alkaloids have been described in other conifers and could provide direct toxicity to insect herbivores or their microbial symbionts (Gerson and Kelsey 2002). Similarly, acetophenones have been reported as chemicals providing resistance against spruce budworm (Parent et al. 2020). However, alkaloids and acetophenones remain understudied in Mediterranean pines.

A particular feature of secondary metabolism in pine trees is that the large concentration of mono- and sesquiterpenes in pine tissues is actively stored in the internal network of resin ducts, rather than in cell vacuoles, external glandular trichomes or directly solved in the cell content as found for other plant groups such as oaks and most aromatic herbs. As mono- and sesquiterpenes are volatile at atmospheric temperatures, they can diffuse or be more actively released to the atmosphere from the stomata and exposed resin ducts. The emission rate and profile of such terpene mixtures have been found to vary across populations and families of maritime pine (Blanch et al. 2012; Sampedro et al. 2010). Emission of volatile organic compounds (VOCs) also depends on light conditions, temperature, and nutrient availability and thus informs about plant performance and nutritional status (Blanch et al. 2012). The emission of such an amount of VOCs confers an airborne signature variable among individuals or lineages that may have great ecological and evolutionary implications for herbivore deterrence, host selection by herbivores, host selection by herbivore predators and plant–plant interactions.

11.3.2 Indirect Resistance Mechanisms

Besides mechanisms providing direct resistance, plants have also evolved mechanisms of indirect resistance (i.e. indirect defenses) (see Box 11.1). They are based on boosting mutualistic relationships with other organisms that are potentially antagonistic to plant enemies (Heil 2008). Most of those third partners are predators and parasitoids of herbivore insects, and in most cases indirect resistance involves an active, inducible, and fairly specific 'call for help' from the plant counterpart in response to herbivore attack (Heil 2014). Many particular cases of indirect resistance have been discovered in annual plants and some have currently been exploited in agricultural crops. In pines and other trees such mechanisms remain understudied. However, it has been reported that some insectivorous birds may play such a role in scots pine (Mäntylä et al. 2018). Furthermore, there is strong evidence that wasp parasitoids provide indirect defenses against sawflies in some pine species (Hilker et al. 2002; Mumm and Hilker 2006). A particular case is the protective mutualism potentially provided by aphid-tending ant species on pine trees. Aphids feed on the pine phloem while ants feed on the honeydew produced by tended aphids. Ants clean their aphid colonies and surrounding pine tissues of any other insects and debris. Some studies have showed effective protection of ants against other herbivores (Maňák et al. 2013) and a lack of costs in terms of growth in young pines of several species (Moreira et al. 2012b). More research is needed to obtain a clear understanding of the costs and benefits of these particular aphid-mediated potential plant mutualisms along the ontogeny of the tree host and across environments.

11.3.3 Induced Resistance

Provided that plant species have the ability to perceive, identify and signal changes in the environment, they are also able to adjust their defensive phenotype to maximize survival. Plant defence theory predicts that this kind of plastic response must be favored in long-lived slow-growing species such as pine trees. For instance, when plants perceive an herbivore attack, the allocation priorities change and plants can express an "induced response" to protect their tissues (Eyles et al. 2010) (Fig. 11.2). The constitutive levels of defenses may be raised to obtain an extra-defended phenotype. Expressing induced responses is supposed to be a cost-saving strategy, as associated costs are only materialized when needed, after the attack (Cipollini and Heil 2010). Induction may be systemic, when phenotypic change affects the whole organism, or local, when changes are detectable just in localized plant parts or specific tissues (Moreira et al. 2009). In pine trees, induced responses can be classified into the following types of responses, described below.

11.3.3.1 Induced Direct Resistance

This includes those phenotypic changes in response to damage that aim to repel the attack and weaken insect performance. Producing an induced defensive phenotype implies that plants are able to identify the challenge, signal the alarm message to target tissues, and activate gene expression, protein biosynthesis and gene and

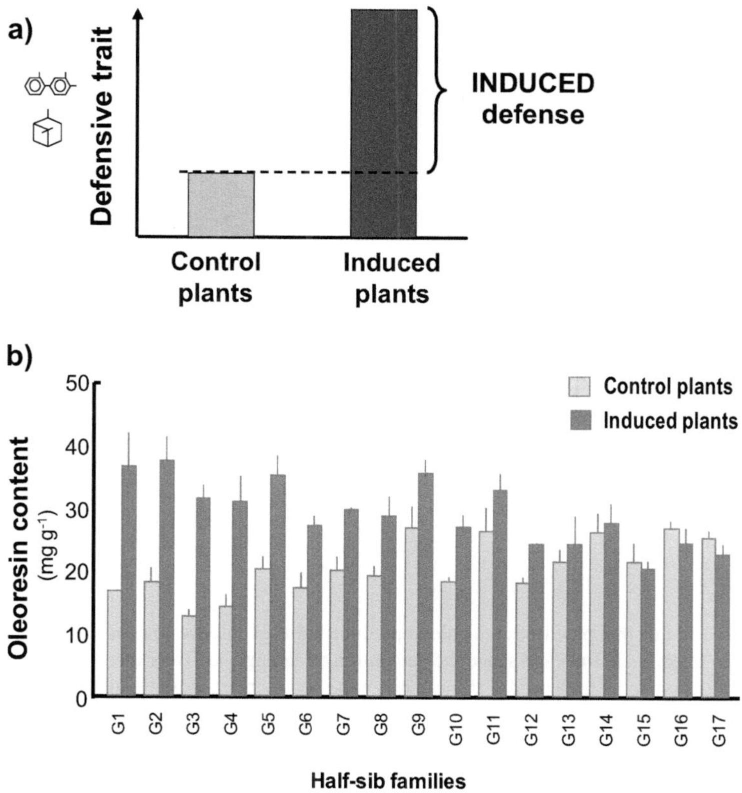

Fig. 11.2 Pannel **a** presents three common concepts regarding induced responses in trees: (i) the concentration of a given defensive trait providing resistance in control plants (light grey bar), which is the baseline level, already present before any attack (constitutive defenses); (ii) the concentration of the same defensive trait in plants exposed to herbivory, induced in response to a biotic challenge (dark grey bar, induced level); and (iii) inducibility – the difference between the two, i.e. the ability to express a plastic, induced, response in a given defensive trait, sometimes reversible after a given time. Pannel **b** illustrates how constitutive and induced defenses in pine trees, as well as the inducibility of a given defense, may be genetically variable across genetic entries, as is the case of oleoresin in the stem of *P. pinaster* half-sib families. (Modified from Sampedro et al. 2011b)

enzyme regulation to produce the proper phenotypic changes leading to improved resistance (Fig. 11.3). Upregulation of biosynthesis of plant hormones associated with herbivore signaling (such as jasmonic acid, ethylene and salicylic acid) and chemical defenses (terpenoids and phenolics) have been reported in several conifer species in response to herbivore attack (Verne et al. 2011). Reconfiguration of the entire primary metabolism may be required, with a complex integration of potential cross-talks with other plant hormonal signaling pathways (auxins, abscisic acid, brassinosteroids and gibberellins) to balance plant growth, abiotic stress tolerance and defense. Changes in physical and chemical defenses may contribute to improve effective resistance and plant fitness in time and space, thus reducing insect

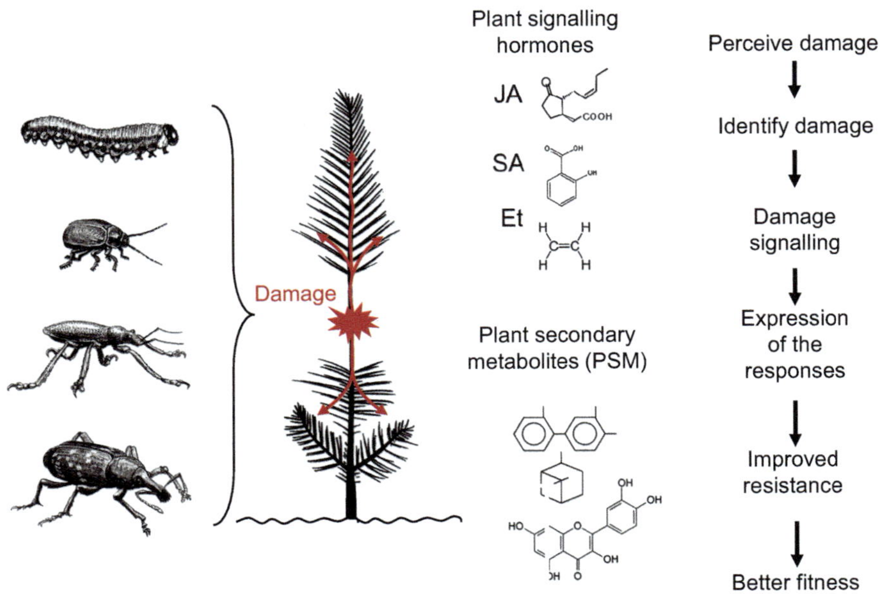

Fig. 11.3 The fact that pine trees are able to produce induced defenses means that they are able to adequately integrate in time and space the perception of damage into the expression of an adequate response, leading to an improved resistant phenotype and increasing fitness under herbivore damage. Plant hormones such as jasmonic acid, salicylic acid, ethylene and others are essential in this process, leading to the induction of defensive proteins, chemicals and anatomical defenses

performance. Most of the knowledge in this field comes from model plants and a strong translational effort to pine trees would be required.

11.3.3.2 Induced Emission of Organic Volatile Compounds (VOCs)

Changes in the emission of VOCs may contribute to faster within-plant damage-signaling among plant parts, separated anatomically but physically close (Heil and Bueno 2007; Heil 2014). Changes in the airborne message from infested pine trees may also modulate other biotic interactions leading to induced indirect resistance, with a strong "call for help" from attacked plants attracting insect predators and parasitoids (Heil and Bueno 2007). Furthermore, changes in airborne scents in the tree's neighborhood could be perceived by relatives of the focal host plant, leading to early perception of the risk of herbivore attack (Heil and Karban 2010). Again, knowledge about this kind of induced indirect resistance mechanism in pine trees is still very limited.

11.3.3.3 Induced Tolerance Responses

These include those phenotypic changes aimed at minimizing the impact of the attack on plant fitness, irrespective of the performance of the attacker. Tolerance is a plant defense strategy that seeks to maintain plant fitness with no effect on herbivore growth, survival and/or reproduction. This last point is essential for understanding this strategy. No effects on herbivore fitness will result in no selective pressure and thus no room to further co-evolutionary arms race. Induced direct tolerance responses in pine trees may include changes in growth priorities among tissues, for instance favoring apical leader recovery and diminishing branching differentiation, or stimulating fine root differentiation for water uptake. Such changes in growth patterns could be favored by active mobilization of carbon and nutrients to sink tissues, or, alternatively, by moving resources away from tissues targeted by the herbivores (roots, stems, needles). A particular case of responses aiming to achieve induced indirect tolerance, could be those boosting carbon exudation through the roots to favor decomposition and nutrient mineralization by soil microbes and/or more intense ectomycorrhizal (ECM) colonization in fine roots, which may favour water and nutrient uptake and thus quicker recovery after attack. Although pine trees are obligate mutualists with ECM, this kind of response has been understudied in pines. However, there is evidence of drastic changes in carbon allocation to fine roots and nitrogen and phosphorus remobilization in young pines in response to biotic stimuli (Moreira et al. 2012c).

11.3.4 Priming of Resistance

Although expressing induced defenses is an energy-saving strategy, the benefits of induced defenses rely on effective damage recognition and timely expression of the improved defensive phenotype. A good match between damage recognition and activation time – the time required for expressing an effective defensive phenotype – is essential for a efficient, rapid response to improve resistance. Similarly, when herbivore risk diminishes, the lag time and decay time – the time for returning to basal defensive investment – is essential for balancing the costs. Thus, time is an essential concept for understanding the costs and benefits of expressing induced defenses in each environmental circumstance.

Research on model species has shown that plants, besides recognizing herbivore damage, can perceive the risk of damage. Perception of the risk of damage prior to herbivore arrival may be crucial for activating induced defenses in time, for instance, recognizing the herbivore loading in the neighborhood, probably via changes in the airborne messages of VOC from the surroundings. Upon perception of such risk, pine trees may express an intermediate state of immune activation before herbivore attack. This state is known as the 'priming state' (see reviews by Hilker et al. 2016; Martinez-Medina et al. 2016; Hilker and Schmülling 2019). The ability to be primed after recognition of the risk of damage is an evolutionary step of adaptive induced resistance. Knowledge about priming in long-lived plants such as trees is only in its early stages; future work will lead to great advances in basic and applied research (Mumm et al. 2003; Hilker and Meiners 2006; Hilker and Fatouros 2016; Bittner et al. 2019; Hilker and Schmülling 2019; Mageroy et al. 2019).

Discovering that plants may react with a plastic response to pests and pathogens has signaled a brand new era in our understanding of plant science and led to a new discipline of plant behavior and plant immunity. The first reports of plant-induced defenses were published in the early 1950s, whereas induced defenses in conifers were reported some 30 years later (Lewinsohn et al. 1991). A literature search to date in the ISI-Web engine showed that published articles in the last 5 years with the words 'tree + induced + responses' and 'pine + induced + responses' accounted just for 15% and 2%, respectively, of those published with the words 'plant + induced + responses', the latter with a yearly average of 3050 items. Thanks to the last 20 years of research on model plants, we now know that plant resistance does not rely only on the presence and abundance of secondary chemicals. Plant resistance depends on a complex and integrated immune system that is able to recognize specific molecular patterns of plant self-damage and herbivore associated molecular patterns (HAMPs), among others (Heil and Land 2014). Moreover, herbivore offense commonly involves effector proteins and RNAs that confound and disturb the plant immune system, molecular pattern recognition or plant hormonal damage signaling. In turn, plants have evolved resistance proteins that block pathogen and herbivore effectors triggering plant triggered immunity. Effector-triggered immunity, where plants detect phytopathogenic factors that trigger a response, has been identified in other genera of trees such as *Malus*, *Salix*, *Vitis*, *Prunus*, *Castanea* and *Eucalyptus* in response to insect gallers, but not yet in conifers (Yamaguchi et al. 2012; Tooker and Helms 2014; Oates et al. 2016).

Herbivore-plant interactions are more complex than we could imagine a few years ago. The induced defensive phenotype is the outcome of many molecular steps, pathways and interactions, including recognition of insect-associated molecular patterns, insect effectors, modulators, plant defensive proteins, etc. Furthermore, induced responses against one organism can lead to susceptibility to others, as specific responses and defensive mechanisms to each organism might be affected by hormonal crosstalks. The effects of the combined exposure to multiple organisms are, however, mostly still unknown.

11.4 Trade-Offs Associated with Mediterranean Pine Defenses

Since defense production is costly and resources are limited, optimal resource allocation to defenses is expected to come at the expense of growth and reproduction, leading to the expression of trade-offs among life functions (Herms and Mattson 1991, 1992) (Fig. 11.4a). Particularly, growth–defense trade-offs have been of

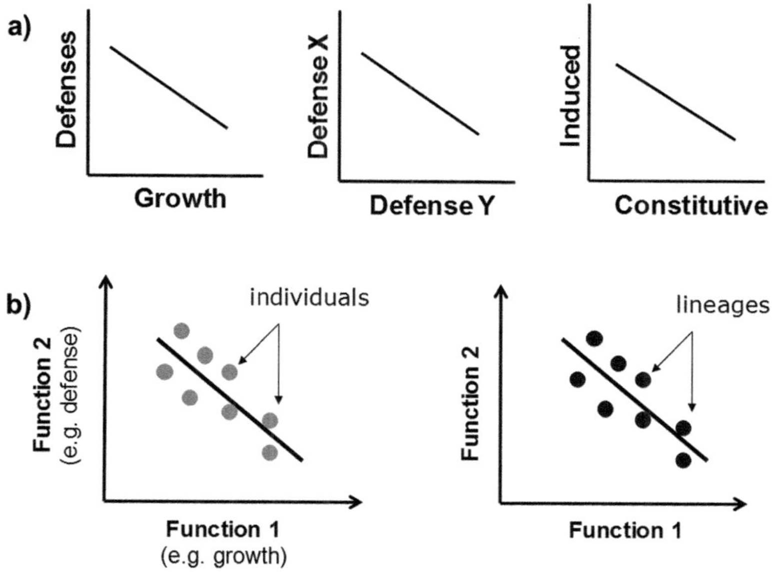

Fig. 11.4 Some of the trade-offs potentially associated with pine defenses. (**a**) As defensive investment is energetically costly and the required carbon resources are shared with growth, constraints between growth and defenses have sometimes been reported. As defenses of different nature (x and y in (**a**)) may share the resources required for their synthesis, a trade-off in the expression of both types of defenses could be expected. The same pattern could be observed if both types of defenses are to some extent functionally redundant. Moreover, a trade-off between constitutive defenses and inducibility of defenses could be also expected. The balance of costs and benefits of both strategies depends on environmental factors such as resource availability, herbivore loading and risk and predictability of herbivore attacks. Furthermore, evolutionary conflicts between both strategies may arise because the benefits of induced defenses in terms of fitness could differ for plant lineages with high and low constitutive defenses. (**b**) Conflicts in the allocation to two traits of functions sharing a common resource usually emerge as negative phenotypic correlations at the individual level (grey points in (**b**)-left), where points in the negative correlation are subjects or individuals, indicating a physiological trade-off among source-related traits. Negative genetic associations between pairs of traits may emerge as negative correlations, where points in the plot are entities with a genetic basis (black points in (**b**)-right; families, lineages, populations, species) delimiting a surface of all the possible combinations of values for both traits leading to similar fitness. Such negative genetic correlations have heritable and evolutionary relevance, and denote potential constraints to the simultaneous selection on both functions

special consideration in plant defense research since their expression is thought to maintain patterns of variation in defensive allocation among and within species (Stamp 2003; Agrawal 2011, 2020). Several ecological and evolutionary hypotheses have been postulated to provide a mechanistic basis that explains constraints between growth and defenses (Stamp 2003). For instance, the growth–differentiation balance hypothesis (GDBH – Herms and Mattson 1992) states that when resource shortage (mainly in water or key nutrients such as nitrogen and phosphorus) limits growth but not phothosistensis, a physiological trade-off between growth and defenses might be expressed, as the exceed of photoassimilates can be reallocated to defense production. Likewise, the resource availability hypothesis (RAH – Coley et al. 1985; Endara and Coley 2011) addresses the existence of an evolutionary trade-off among species. According to the RAH, habitats with high resource availability select for species with enhanced growth rates and low defensive capability. Conversely, habitats characterized by poor resource conditions select for slow-growing but highly defended plants, given the costs associated with the replacement of tissue lost by herbivore damage.

The defensive system of pine species is costly to produce, particularly in relation to resin-based defenses, as it involves the development of a system of interconnected resin ducts and the production of abundant oleoresin comprising carbon-enriched terpenoid molecules (Gershenzon 1994; Moreira et al. 2015). Furthermore, environmental conditions in the Mediterranean Basin are often strongly limiting (e.g. pronounced, long-lasting drought periods), and may therefore exert strong constraints on resource allocation among life-functions. Growth–defense trade-offs (i.e., negative correlations between both functions) are therefore expected to be expressed in Mediterranean pines, although empirical evidence is scarce and suggests that their expression is strongly context-dependent. For instance, glasshouse studies with *P. pinaster* saplings have revealed that trade-offs between chemical or anatomical defenses and growth emerged only under nutrient-limited conditions, particularly in phosphorous availability (Sampedro et al. 2011a; Moreira et al. 2015). Likewise, an experimental study with *P. halepensis* saplings also showed that growth–defense constraints in this species only emerged in response to biotic competition with other tree species (Fernandez et al. 2016). More importantly, finding evidence of negative genetic correlation (i.e. across families or populations within species; Fig. 11.4b) between growth and defenses might indicate the heritable and therefore evolutionary basis of such trade-offs (Agrawal 2020) (Fig. 11.4b). For instance, Moreira et al. (2015) reported negative intraspecific genetic correlations between growth and anatomical defenses among *P. pinaster* families, indicating that fast growing genotypes were less defended and vice versa. Likewise, Di Matteo and Voltas (2016) found that *P. pinaster* populations adapted to more favorable conditions and with higher growth potential were more susceptible to the insect *M. feytaudi*. Conversely, slow-growing populations adapted to harsher environmental conditions showed reduced insect damage (i.e., higher resistance). This latter study provides evidence of a trade-off between growth and resistance at the population level and suggests that *P. pinaster* populations evolved divergently towards differential defensive strategies.

Mediterranean pine species occupy wide distribution ranges and are therefore subject to heterogeneous environmental conditions leading to genetic adaptation and different life-history strategies (Tapias et al. 2004; Grivet et al. 2010, 2013). If growth–defense constraints are genetically determined, as seems to be the case (Moreira et al. 2015; Di Matteo and Voltas 2016), populations with higher intrinsic growth rates (i.e. growth potential) might have compromised their defensive capability. This would have immediate implications for the future management of forest genetic resources, since characterization of optimal population performance is often based on growth potential or abiotic resistance (Alía et al. 1995; Taïbi et al. 2014).

However, biotic resistance markers are less frequently considered, despite the evidence of increased biotic stresses associated with global change (Peñuelas and Filella 2001; Bale et al. 2002; Benito-Garzón and Fernández-Manjarrés 2015). Given the current evidence of genetic growth–defense trade-offs, there is an urgent need to understand to what extent and under what conditions (water or nutrient availability) such constraints are expressed between and within Mediterranean pine populations. This knowledge will help us anticipate population responses to predicted environmental shifts in the Mediterranean Basin and facilitate decision making in adaptive forest management under current scenarios of global change.

11.5 Variation in Defensive Investment

Anatomical and chemical defenses in Mediterranean pine species vary widely both among and within species (Table 11.2). Such variation is likely the result of genetic adaptation to environmental variation, both biotic (i.e. differential selective pressures imposed by phytophagous organisms) and abiotic (i.e., environmental gradients in abiotic factors along the species distribution range). This variation is translated into notable inter- and intraspecific differences in the susceptibility to many particular insect herbivores (Table 11.3). Investment in defensive traits and resistance to specific herbivores is also largely determined by plastic responses to the abiotic and biotic environmental conditions (e.g. Lombardero et al. 2000), the ecological context in which the host trees are immersed (e.g. Castagneyrol et al. 2014) and complex interactions with other symbiotic, mutualistic and higher-trophic level organisms (e.g. Giffard et al. 2012; Kanekar et al. 2018). Understanding how genetic and plastic responses modulate defensive investment in long-lived species such as pine trees is required to forecast their future performance, particularly in the current context of environmental change. In the following sections we review current knowledge on the sources of variation of defensive traits and effective resistance against particular enemies in Mediterranean pines.

11.5.1 Genetic Variation

Mediterranean pines show several evolutionary particularities (Grivet et al. 2013; see also Grivet and Olsson this volume, Chap. 1). For instance, since pine species, as other trees, display lower mutation rates per unit of time than other plants, low intraspecific genetic variation within populations might be expected (Petit and Hampe 2006). However, within-population variation is high for many quantitative traits (Zas et al. 2005; Hernandez-Serrano et al. 2014) probably favored by significant among-population gene flow in these outcrossing species (De-Lucas et al. 2008). Such intra-population variation ultimately allows for rapid local adaptation processes to environmental change (Grivet et al. 2017). Indeed, Mediterranean pine

Table 11.2 Studies reporting inter and intraspecific genetic variation in physical and chemical defenses of Mediterranean pine species

Species	Genetic variation	Trait	Reported pattern	References
Pinus spp.	Among species	Terpenes	Wide genetic variation in chemical composition and concentration	Rodrigues et al. (2017)
		Phenolics	Wide genetic variation in chemical composition and concentration	Kaundun et al. (1997) and Pimentel et al. (2017)
P. pinaster	Among populations	Terpenes	Constitutive monoterpenes and diterpenes: $Q_{ST} = F_{ST}$	López-Goldar et al. (2019)
			Constitutive sesquiterpenes: $Q_{ST} > F_{ST}$	
		Phenolics	Constitutive phenolic groups: $Q_{ST} = F_{ST}$	López-Goldar et al. (2019)
			Constitutive hydroxicinnamic acids: $Q_{ST} < F_{ST}$	
			Inducibility of eugenols	
		Resin canals	Wide genetic variation among populations for constitutive resin duct features, but not for inducibility	Zas et al. (2015) and Vázquez-González et al. (2019)
	Among families	Terpenes	Constitutive diterpenes	López-Goldar et al. (2019)
		Phenolics	Constitutive hydroxycinnamic acids and lignans	López-Goldar et al. (2019)
		Non-volatile resin	Genetic variation in constitutive and MJ-induced defenses	Sampedro et al. (2011a)
		Total phenolics	Genetic variation in constitutive and MJ-induced defenses	Sampedro et al. (2011a)
		Lignin	Additive genetic variation ($h^2 = 0.34$)	Gaspar et al. (2011)
		Resin canals	Wide genetic variation among families in inducibility, but not in constitutive resin duct features.	Moreira et al. (2015)
		Growth (tolerance)	Strong compensatory growth and large genetic variation in tolerance after herbivory by a stem-chewer insect.	Sampedro et al. (2009) and Zas et al. (2011)

(continued)

Table 11.2 (continued)

Species	Genetic variation	Trait	Reported pattern	References
P. halepensis	Among populations	Terpenes	Three differentiated provenance clusters based on chemotypic variation	Djerrad et al. (2015)
		Flavonoids	Wide genetic variation, grouped into three distinct clusters.	Kaundun et al. (1998b)
	Among and within families	Resin flow	Large inter- and intrafamilial variation	Moulalis (1991)
			Among families: $h^2 = 0.65$	
			Within families $h^2 = 0.84$	
P. brutia	Among populations	Terpenes	Wide genetic variation showing geographic clustering	Schiller and Grunwald (1987) and Schiller and Genezi (1993)
		Flavonoids	Wide genetic variation showing geographic clustering	Kaundun et al. (1998a)
P. pinea	Among populations	Terpenes	Wide genetic variation in volatile terpenes. Limonene and α-pinene as dominant metabolites	Nasri et al. (2011)

Q_{ST} = Genetic differentiation among populations displayed in quantitative traits. F_{ST} = Diferentiation among populations measured on neutral loci. h^2 = narrow sense heritability, or additive variance contributing to total phenotypic variance. MJ = Methil Jasmonate, hormone implicated in damage signaling and the production of induced defenses

species are strongly genetically differentiated among populations (Gómez et al. 2005) including large quantitative differences in almost every life-history trait (e.g. Santos-Del-Blanco et al. 2012; Voltas et al. 2015), resulting from both neutral and adaptive evolutionary processes.

Defensive traits are not an exception and are also highly variable among and within Mediterranean pine populations (Table 11.2). Intraspecific variation in defensive traits is reflected in significant differences among populations and families within populations in effective resistance to specific herbivores, although the available information about this is scarce (Table 11.3). Population differentiation in plant secondary metabolites, for example, is particularly acute in species such as maritime pine, with highly fragmented and isolated populations (Meijón et al. 2016; López-Goldar et al. 2019). The singular demographic history of Mediterranean pine species with several independent postglacial refugia suffering relevant bottle-necks, and posterior migration routes from such refugia (Petit et al. 2003) has surely contributed to this marked population differentiation (López-Goldar et al. 2019; Vázquez-González et al. 2019). Furthermore, genetic adaptation to differential selection pressures imposed by the great environmental heterogeneity within the natural distribution range of Mediterranean pines is also assumed to explain the

Table 11.3 Studies reporting among or within species variation in susceptibility of Mediterranean pines to diverse insect herbivores

Insect herbivore/ pine species	Genetic level	Outcome	References
Hylobius abietis			
P. pinaster	Families	Additive variation among 40 half-sibs in susceptibility	Zas et al. (2005)
P. pinaster	Populations	Significant variation in susceptibility among 10 pine populations	López-Goldar et al. (2018)
P. pinaster	Populations	Significant variation in susceptibility among 3 pine populations	Suarez-Vidal et al. (2017)
P. halepensis	Populations	No significant variation in susceptibility among 3 pine populations	Suarez-Vidal et al. (2019)
Matsucoccus feytaudi			
P. pinaster	Populations	Significant variation in density of nymphs among 8 pine populations	Di Matteo and Voltas (2016)
P. pinaster	Populations	Significant variation in insect density and symptoms among 8 pine populations	Schvester and Ughetto (1986)
P. pinaster	Populations	Significant variation in symptoms among 25 pine populations	Harfouche et al. (1995)
Thaumatopoea pytiocampa			
P. halepensis	Populations	Variation in susceptibility among 57 pine populations	Sbay and Zas (2018)
	Species	*P. halepensis* more attacked than *P. brutia*	Sbay and Zas (2018)
P. halepensis	Populations	Variation in susceptibility among 57 pine populations	Bariteau and Pommery (1992)
	Species	Differences in oviposition and larvae development on different hosts	Hodar et al. (2002)
	Species	Oviposition deterrence: *P. pinea* > *P. halepensis* > *P. pinaster* > *P. brutia*	Paiva et al. (2011)
	Species	Larval mortality: *P. pinea* > *P. brutia* > *P. halepensis* > *P. pinaster*	Schopf and Avtzis (1987)
Thaumatopoea wilkinsoni			
	Species	Susceptibility variation: *P. eldarica* > *P. brutia* > *P. halepensis*	Mendel (1988)
Monochamus galloprovincialis			
	Species	Differences among species in insect feeding and oviposition	Sanchez-Husillos et al. (2013)
	Species	Differences among species in insect feeding and oviposition	Naves et al. (2006)
Dyorictria sylvestrella			
P. pinaster	Families	Significant differences in resistance among full-sibs	Jactel et al. (1999)
P. pinaster	Families	Significant differences in resistance among full-sibs	Kleinhentz et al. (1998)

observed genetic differentiation among populations in defensive traits (Elvira-Recuenco et al. 2014). Both the biotic environment (e. g., variation in the herbivorous insect communities and the frequency of insect outbreaks) and the pronounced heterogeneity in the abiotic environment across the range of pine species are expected to have shaped defensive strategies and traits across populations (Serra-Varela et al. 2015; Di Matteo and Voltas 2016).

An outstanding example of how variation in defensive traits may be associated with adaptive responses to differences in the biotic environment is the variation among maritime pine populations in their susceptibility to the bast scale *M. feytaudi*. Although factors related to the invasion process such as founder effects or lack of natural enemies in the new range (Jactel et al. 2006) can also explain variation patterns, common garden tests have demonstrated that pine populations genetically differ in their susceptibility to the insect (Schvester and Ughetto 1986; Harfouche et al. 1995; Di Matteo and Voltas 2016). Western populations that have coexisted with the insect are much more resistant than eastern populations that lack efficient resistance mechanisms due to the absence of a co-evolutionary history with the insect.

Intraspecific variation in defensive traits may also result from adaptations to abiotic environmental conditions and trade-offs with other life history traits. For example, some studies have reported that population variation in both chemical and anatomical pine defenses follows climatic clines, suggesting adaptive processes related to the abiotic environment (López-Goldar et al. 2019; Vázquez-González et al. 2019). However, rather than acting through direct evolutionary responses to climate in defensive-related traits, these adaptive processes likely act indirectly through links with other life functions such as growth, reproduction or tolerance to abiotic factors (see Sect. 11.4).

Population differentiation processes are typically associated with a reduction of variation within populations. However, intra-population genetic variation in defensive traits in Mediterranean pines is commonly very high, even higher than that among populations (Tables 11.2 and 11.3). High rates of gene flow among populations, high environmental heterogeneity within population ranges, and genetic correlations with other characters are likely contributing to maintaining such wide variation. Intra-population variation constitutes the fuel of evolutionary change where natural selection operates (Petit and Hampe 2006) and the basis for breeding programs aimed at improving resistance to biotic threats (Woodcock et al. 2018). Although heritability of some defensive traits may be notably high (Rosner and Hannrup 2004), resistance to insect herbivores typically shows low heritability estimates (Kleinhentz et al. 1998; Zas et al. 2005). However, the large phenotypic variation in defensive traits results in high additive coefficients of variation, allowing for rapid and strong evolutionary responses to natural or artificial selection (Sniezko and Koch 2017).

Evidence suggests that variation in defense production may be, at least in part, genetically determined in Mediterranean pine species. Future research should, however, deepen into the genetic and evolutionary basis of effective resistance and tolerance against key pests and associated defensive mechanisms. Such knowledge would allow us to forecast the evolutionary responses of pine tree populations and create adaptive forest management programs for tackling the negative impacts of increasing biotic stresses associated with global change in the Mediterranean region.

11.5.2 Environmental Plasticity

Pine trees, like other plants, are exposed to constant environmental fluctuations in both abiotic and biotic factors. Particularly, abiotic factors such as soil nutrients, quality and amount of light, water availability and pollutants drive primary production but may also directly and indirectly affect defensive allocation. At the same time, pine trees host a large and diverse community of antagonistic and mutualistic organisms. Importantly, both biotic and abiotic factors vary in space and time, and thus pine trees must finely integrate all the environmental signals to enhance their fitness. Energy allocation must be therefore readjusted with respect to the balance between the costs and benefits of defensive allocation under a specific environmental context (Herms and Mattson 1992). Accordingly, plastic responses to both biotic and abiotic conditions are expected to play a key role in the optimal performance of pine populations. Plastic responses to biotic factors are covered in Sect. 11.3.3.

Studies exploring plastic responses to abiotic factors in Mediterranean pines often assess the effects of nutrient availability and drought under experimental conditions. For instance, higher nutrient availability has been shown to negatively affect the production of both chemical and anatomical defenses in *P. pinaster*, probably due to changes in resource allocation priorities for growth (Moreira et al. 2008, 2015; Sampedro et al. 2011a). Furthermore, reduced water availability treatments have been reported to constraint the inducibility of chemical defenses in *P. halepensis* (Suarez-Vidal et al. 2019). Similarly, light deprivation has also been shown to reduce the inducibility of chemical defenses in saplings of *P. pinaster* (Suárez-Vidal et al. 2017). It is also known that evapotranspiration rates and other climate-related variables affect resin yield and production of resin ducts (i.e. anatomical defenses) in *P. pinaster* (Rodriguez-Garcia et al. 2015). There is a considerable knowledge gap, however, regarding the effect of environmental conditions, particularly climate, on defense production in mature trees growing under natural conditions.

Climate change scenarios anticipate considerably harsh conditions for Mediterranean regions in the near future, including increased atmospheric temperature and more extreme drought periods (IPCC 2014). Plasticity in functional traits may facilitate the persistence of locally adapted forest tree populations by buffering the negative impacts of shifting climate conditions and allowing range expansions (Valladares et al. 2014). Knowledge on how climate modulates defensive allocation in Mediterranean pine species is therefore needed in order to anticipate population responses to both increased biotic and abiotic stresses. As referenced by the literature cited above, there is some experimental evidence to expect that drought may negatively affect defensive investment in Mediterranean pine species (Suarez-Vidal et al. 2019). Further studies should therefore focus on disentangling the effect of different climate factors on anatomical and chemical defensive traits in mature trees growing under natural environmental conditions. For instance, climatic response functions widely applied in dendrochronological studies have facilitated the accumulation of an extensive body of knowledge on how climatic constraints affect growth patterns in conifer species (Andreu et al. 2007; Tardif et al. 2003). In a similar way, tree-ring analysis could be applied to resin-based defenses in pine trees, particularly to axial resin ducts recorded in annual growth rings, to clarify the role of particular climate variables in defensive investment. This approach has already been applied to different Northern-American pine species demonstrating the sensitivity of axial resin ducts to temperature and precipitation (Saracino et al. 2017; Slack et al. 2017). Improving our understanding of which climate variables have the most significant effect on defensive investment will improve our ability to forecast the future responses of pine populations and the potential impacts of global warming on Mediterranean forests.

11.5.3 Other Sources of Variation

It is known that the environmental conditions experienced by parent plants may determine, at least in part, the phenotype of their progeny (Holeski et al. 2012). Such transgenerational plastic responses to the maternal environment are transmitted to the offspring phenotype without any change in the DNA sequence, and thus constitute a layer of phenotypic variation over the genomic variation. This kind of maternal environmental effect can influence the evolutionary processes and population dynamics of plant species. It has been showed that the abiotic environment experienced by mother maritime pine trees may modify germination rate, early growth patterns and resistance to root rot fungi in their progeny (Vivas et al. 2013; Zas et al. 2013; Zas and Sampedro 2015). Whether the biotic environment experienced by mother trees may modify the resistance of the progeny against insect herbivores remains unexplored in pine trees. This kind of transgenerational priming could generate powerful and relevant applications in forest management.

As discussed above (Sect. 11.5.2), differences in resource availability (light, nutrients, water) can modulate allocation to defenses, and interact with the expression of induced defenses and their costs. Heterogeneity in resource availability usually follows non-random spatial patterns in nature. Moreover, biotic interactions are usually spatially aggregated, and individuals in close proximity are more likely to share the same herbivore loadings, mycorrhizal partners, pathogen risks and exposure to VOCs. In summary, it is more likely that neighboring trees share the same biotic interactions and resource availability. Even when genetic relatedness could be expected to also show some degree of spatial aggregation, neighboring trees may express a more similar defensive phenotype than expected from their genetic relatedness. In general, we can say that biotic interactions have been overlooked as a source of spatial autocorrelation, and thus biotic interactions and their effects on the plant phenotype should be considered in a spatially explicit context.

Similarly, the composition of a given plant community may alter the relative likelihood of a focal plant species to be attacked by a given herbivore (Barbosa et al. 2009). Factors determining the associational benefits include plant community diversity, the presence or abundance of a given species, patch size, structural composition of the plant community determining habitat type, and the nature of plant antagonists such as specialist/generalist, insect/mammal, feeding guild, diet breadth, etc. Associational resistance and associational susceptibility effects could spatially alter the pressure exerted by herbivores and thus the spatial distribution of plant phenotypes.

11.6 Concluding Remarks

The ecological and genetic bases of resistance to herbivore insects in Mediterranean pines generally match those of more studied model plants. However, there are some particularities. Mediterranean pines are natural or semi-natural populations of long-lived plants; they show large genetic diversity in functional and molecular traits, even when their particular demographic history of regression, recolonization and isolation has been greatly influenced by evolution; they live in a hotspot of diversity, but also in a hotspot of global change, on one hand subjected to range expansion of alien and native pests, pathogens and other plants, and on the other hand challenged by rapid climate change in the Mediterranean region. Combining the advances in model plants in the last decade in a translational research effort with the particular life history characteristics of Mediterranean pines will allow us to unravel the patterns of resistance and tolerance to multiple pests and pathogens in these species. This knowledge will ultimately help to efficiently manage forest genetic diversity in resistance traits under a changing environment and to forecast how species and populations will face the current challenges.

Acknowledgements For the preparation of this chapter, authors received founds from the Spanish Government (grant 'RESILPINE' RTI2018-094691-B-C33 (MCIU/AEI/FEDER, UE), and 'GENMAC' 2018ES06RDEI12479V1 (MAPA/AEI-Agri/FEADER, UE)). CVG received financial support from the MINECO FPI Grant programme (MINECO-BES-2016-076624). Authors would like to thank Isaac Sucasas for his superb job preparing the hand drawings.

References

Agrawal AA (2011) Current trends in the evolutionary ecology of plant defence. Funct Ecol 25:420–432

Agrawal AA (2020) A scale-dependent framework for trade-offs, syndromes, and specialization in organismal biology. Ecology 101:e02924

Alía R, Gil L, Pardos J (1995) Performance of 43 *Pinus pinaster* Ait. provenances on 5 locations in central Spain. Silvae Genet 44:75–80

Almquist AC, Faldt J, Yart A et al (2006) Host selection in *Tomicus piniperda* L.: composition of monoterpene hydrocarbons in relation to attack frequency in the shoot feeding phase. Z Naturforsch C 61:439–444. https://doi.org/10.1515/znc-2006-5-622

Andreu L, Gutiérrez E, Macias M et al (2007) Climate increases regional tree-growth variability in Iberian pine forests. Glob Chang Biol 13:804–815

Auger MA, Geri C, Allais JP (1994) Effect of the foliage of different pine species on the development and on the oviposition of the pine sawfly *Diprion pini* L. (Hym., Diprionidae). J Appl Entomol 117:165–181. https://doi.org/10.1111/j.1439-0418.1994.tb00722.x

Bale JS, Masters GJ, Hodkinson ID et al (2002) Herbivory in global climate change research: direct effects of rising temperature on insect herbivores. Glob Chang Biol 8:1–16

Barbosa P, Hines J, Kaplan I et al (2009) Associational resistance and associational susceptibility: having right or wrong neighbors. Annu Rev Ecol Evol Syst 40:1–20. https://doi.org/10.1146/annurev.ecolsys.110308.120242

Bariteau M, Pommery J (1992) Variabilite geographique et adaptation aux contraintes du milieu mediterraneen des pins de la section halepensis: resultats (provisoires) d'un essai en plantations comparatives en France. Ann Sci For 49:261–276

Barton KE, Koricheva J (2010) The ontogeny of plant defense and herbivory: characterizing general patterns using meta-analysis. Am Nat 175:481–493

Battisti A, Avci M, Avtzis DN et al (2016a) Natural history of the processionary moths (Taumetopoea spp.): new insights in relation to climate change. In: Roques A (ed) Processionary moths and climate change: an update. Springer, Dordrecht, pp 15–79. https://doi.org/10.1007/978-94-017-9340-7_2

Battisti A, Branco M, Mendel Z (2016b) Defoliators in native insect systems of the Mediterranean Basin. In: Paine T, Lieutier F (eds) Insects and diseases of Mediterranean forest systems. Springer, pp 29–45. https://doi.org/10.1007/978-3-319-24744-1_3

Benito-Garzón M, Fernández-Manjarrés JF (2015) Testing scenarios for assisted migration of forest trees in Europe. New For 46:979–994

Bernardinelli I, Zandigiacomo P (2001) *Leptoglossus occidentalis* Heidemann (Heteroptera: Coreidae): a conifer seedbug recently found in northern Italy. J For Sci 47:56–58

Bittner N, Hundacker J, Achotegui-Castells A et al (2019) Defense of Scots pine against sawfly eggs (*Diprion pini*) is primed by exposure to sawfly sex pheromones. PNAS 116:24668–24675. https://doi.org/10.1073/pnas.1910991116

Blanch JS, Sampedro L, Llusia J et al (2012) Effects of phosphorus availability and genetic variation of leaf terpene content and emission rate in *Pinus pinaster* seedlings susceptible and resistant to the pine weevil, *Hylobius abietis*. Plant Biol 14:66–72. https://doi.org/10.1111/j.1438-8677.2011.00492.x

Boivin T, Auger-Rozenberg M-A (2016) Native fruit, cone and seed insects in the Mediterranean Basin. In: Paine TD, Lieutier F (eds) Insects and diseases of Mediterranean forest systems. Springer, pp 47–88. https://doi.org/10.1007/978-3-319-24744-1_6

Bracalini M, Benedettelli S, Croci F et al (2013) Cone and seed pests of *Pinus pinea*: assessment and characterization of damage. J Econ Entomol 106:229–234. https://doi.org/10.1603/ec12293

Budde KB, Gonzalez-Martinez SC, Navascues M et al (2017) Increased fire frequency promotes stronger spatial genetic structure and natural selection at regional and local scales in *Pinus halepensis* Mill. Ann Bot 119:1061–1072. https://doi.org/10.1093/aob/mcw286

Calama R, Fortin M, Pardos M et al (2017) Modelling spatiotemporal dynamics of *Pinus pinea* cone infestation by *Dioryctria mendacella*. For Ecol Manag 389:136–148

Carisey N, Ménassieu P, Baradat P et al (1994) Sensibilité du pin maritime *Pinus pinaster* Ait, installé dans des conditions stationnelles différentes, aux attaques de la pyrale du tronc, *Dioryctria sylvestrella* Ratz (Lepidoptera, Pyralidae). Relations avec quelques caractéristiques de croissance. Ann Sci For 51:67–75. https://doi.org/10.1051/forest:19940105

Carrión JS, Navarro C, Navarro J et al (2000) The distribution of cluster pine (*Pinus pinaster*) in Spain as derived from palaeoecological data: relationships with phytosociological classificatio. The Holocene 10:243–252

Castagneyrol B, Regolini M, Jactel H (2014) Tree species composition rather than diversity triggers associational resistance to the pine processionary moth. Basic Appl Ecol 15:516–523. https://doi.org/10.1016/j.baae.2014.06.008

Cayuela L, Hódar JA, Zamora R (2011) Is insecticide spraying a viable and cost-efficient management practice to control pine processionary moth in Mediterranean woodlands? For Ecol Manag 261:1732–1737. https://doi.org/10.1016/j.foreco.2011.01.022

Celedon JM, Bohlmann J (2019) Oleoresin defenses in conifers: chemical diversity, terpene synthases and limitations of oleoresin defense under climate change. New Phytol 224:1444–1463. https://doi.org/10.1111/nph.15984

Ciesla WM, Erkanat M, Akeson C (2011) Bark beetle associated with *Pinus brutia* and *Cupressus sempervirens* in Northern Cyprus. Silva Lusit 19:117–126

Cipollini D, Heil M (2010) Costs and benefits of induced resistance to herbivores and pathogens in plants. CAB Rev 5:1–25

Coley PD, Bryant JP, Chapin FS (1985) Resource availability and plant antiherbivore defense. Science 230:895–899

Day KR, Nordlander G, Kenis M et al (2004) General biology and life cycles of bark weevils. In: Lieutier F, Day KR, Battisti A et al (eds) Bark and wood boring insects in living trees in Europe, a synthesis. Kluwer Academic Publishers, Dordrecht, pp 331–349

De-Lucas AI, Robledo-Arnuncio JJ, Hidalgo E et al (2008) Mating system and pollen gene flow in Mediterranean maritime pine. Heredity 100:390–399

Di Matteo G, Voltas J (2016) Multienvironment evaluation of *Pinus pinaster* provenances: evidence of genetic trade-offs between adaptation to optimal conditions and resistance to the maritime pine bast scale (*Matsucoccus feytaudi*). For Sci 62:553–563. https://doi.org/10.5849/forsci.15-109

Djerrad Z, Kadik L, Djouahri A (2015) Chemical variability and antioxidant activities among *Pinus halepensis* Mill. essential oils provenances, depending on geographic variation and environmental conditions. Ind Crop Prod 74:440–449. https://doi.org/10.1016/j.indcrop.2015.05.049

Elvira-Recuenco M, Iturritxa E, Majada J et al (2014) Adaptive potential of maritime pine (*Pinus pinaster*) populations to the emerging pitch canker pathogen, *Fusarium circinatum*. PLoS One 9:e114971. https://doi.org/10.1371/journal.pone.0114971

Endara MJ, Coley PD (2011) The resource availability hypothesis revisited: a meta-analysis. Funct Ecol 25:389–398

Evans HF, Moraal LG, Pajares JA (2004) Biology, ecology and economic importance of Buprestidae and Cerambycidae. In: Lieutier F, Day KR, Battisti A et al (eds) Bark and wood boring insects in living trees in Europe, a synthesis. Springer, Berlin, pp 447–474

Eyles A, Bonello P, Ganley R et al (2010) Induced resistance to pests and pathogens in trees. New Phytol 185:893–908

Faccoli M, Anfora G, Tasin M (2008) Responses of the Mediterranean pine shoot beetle *Tomicus destruens* (Wollaston) to pine shoot and bark volatiles. J Chem Ecol 34:1162–1169. https://doi.org/10.1007/s10886-008-9503-6

Faccoli M, Anfora G, Tasin M (2011) Stone pine volatiles and host selection by *Tomicus destruens*. Silva Lusit 19:61–73

Farinha AO, Durpoix C, Valente S et al (2018) The stone pine, *Pinus pinea* L., a new highly rewarding host for the invasive *Leptoglossus occidentalis*. NeoBiota 41:1–18. https://doi.org/10.3897/neobiota.41.30041

Fernandez C, Monnier Y, Santonja M et al (2016) The impact of competition and allelopathy on the trade-off between plant defense and growth in two contrasting tree species. Front Plant Sci 7:594. https://doi.org/10.3389/fpls.2016.00594

Franceschi VR, Krokene P, Krekling T et al (2000) Phloem parenchyma cells are involved in local and distant defense responses to fungal inoculation or bark-beetle attack in Norway spruce (Pinaceae). Am J Bot 87:314–326. https://doi.org/10.2307/2656627

Franceschi VR, Krokene P, Christiansen E et al (2005) Anatomical and chemical defenses of conifer bark against bark beetles and other pests. New Phytol 167:353–376. https://doi.org/10.1111/j.1469-8137.2005.01436.x

Gaspar MJ, Alves A, Louzada JL et al (2011) Genetic variation of chemical and mechanical traits of maritime pine (*Pinus pinaster* Aiton). Correlations with wood density components. Ann For Sci 68:255–265. https://doi.org/10.1007/s13595-011-0034-x

Gershenzon J (1994) The cost of plant chemical defense against herbivory: a biochemeical perspective. In: Bernays EA (ed) Insect-plant interactions. CRC Press, Boca Raton, pp 105–173

Gerson EA, Kelsey RG (2002) Piperidine alkaloids in sitka spruce with varying levels of resistance to white pine weevil (Coleoptera: Curculionidae). J Econ Entomol 95:608–613. https://doi.org/10.1603/0022-0493-95.3.608

Giffard B, Corcket E, Barbaro L et al (2012) Bird predation enhances tree seedling resistance to insect herbivores in contrasting forest habitats. Oecologia 168:415–424

Gómez A, Vendramin GG, González-Martínez SC et al (2005) Genetic diversity and differentiation of two Mediterranean pines *Pinus halepensis* Mill. and *Pinus pinaster* Ait. along a latitudinal cline using chloropl microsatellite markers. Divers Distrib 11:257–263

Grivet D, Olsson S (this volume) Phylogenesis and evolution of Mediterranean pines. In: Ne'eman G, Osem Y (eds) Pines and their mixed forest ecosystems in the Mediterranean Basin. Springer, Cham

Grivet D, Sebastiani F, Alía R et al (2010) Molecular footprints of local adaptation in two Mediterranean conifers. Mol Biol Evol 28:101–116

Grivet D, Climent J, Zabal-Aguirre M et al (2013) Adaptive evolution of Mediterranean pines. Mol Phylogenet Evol 68:555–566. https://doi.org/10.1016/j.ympev.2013.03.032

Grivet D, Avia K, Vaattovaara A et al (2017) High rate of adaptive evolution in two widespread European pines. Mol Ecol 26:6857–6870

Halperin J (1990) Life history of *Thaumetopoea* spp. (Lep., Thaumetopoeidae) in Israel. J Appl Entomol 110:1–6. https://doi.org/10.1111/j.1439-0418.1990.tb00088.x

Harfouche A, Baradat P, Durel CE (1995) Intraspecific variability in maritime pine (*Pinus pinaster* Ait.) in the south-east of France. Variability in autochthonous populations and in the whole range of the species. Ann Sci For 52:307–328. https://doi.org/10.1051/forest:19950402

Heil M (2008) Indirect defence via tritrophic interactions. New Phytol 178:41–61. https://doi.org/10.1111/j.1469-8137.2007.02330.x

Heil M (2014) Herbivore-induced plant volatiles: targets, perception and unanswered questions. New Phytol 204:297–306. https://doi.org/10.1111/nph.12977

Heil M, Bueno JCS (2007) Within-plant signaling by volatiles leads to induction and priming of an indirect plant defense in nature. PNAS 104:5467–5472. https://doi.org/10.1073/pnas.0610266104

Heil M, Karban R (2010) Explaining evolution of plant communication by airborne signals. Trends Ecol Evol 25:137–144. https://doi.org/10.1016/j.tree.2009.09.010

Heil M, Land WG (2014) Danger signals – damaged-self recognition across the tree of life. Front Plant Sci 5:578

Herms DA, Mattson WJ (1991) Does reproduction compromise defense in woody plants? In: Baranchikov YN, Mattson WJ, Hain FP et al (eds) Forest insect guilds: patterns of interaction with host trees, vol 153. Gen. Tech. Rep. NE, Abakan, pp 35–46

Herms DA, Mattson WJ (1992) The dilemma of plants: to grow or defend. Q Rev Biol 67:283–335

Hernandez-Serrano A, Verdu M, Santos-Del-Blanco L et al (2014) Heritability and quantitative genetic divergence of serotiny, a fire-persistence plant trait. Ann Bot 114:571–577. https://doi.org/10.1093/aob/mcu142

Hezil S, Chakali G, Battisti A (2018) Plant phenotype affects oviposition behaviour of pine processionary moth and egg survival at the southern edge of its range. iForest 11:572–576. https://doi.org/10.3832/ifor2675-011

Hilker M, Fatouros NE (2016) Resisting the onset of herbivore attack: plants perceive and respond to insect eggs. Curr Opin Plant Biol 32:9–16. https://doi.org/10.1016/j.pbi.2016.05.003

Hilker M, Meiners T (2006) Early herbivore alert: insect eggs induce plant defense. J Chem Ecol 32:1379–1397. https://doi.org/10.1007/s10886-006-9057-4
Hilker M, Schmülling T (2019) Stress priming, memory, and signalling in plants. Plant Cell Environ 42:753–761. https://doi.org/10.1111/pce.13526
Hilker M, Kobs C, Varama M et al (2002) Insect egg deposition induces *Pinus sylvestris* to attract egg parasitoids. J Exp Biol 205:455–461
Hilker M, Schwachtje J, Baier M et al (2016) Priming and memory of stress responses in organisms lacking a nervous system. Biol Rev 91:1118–1133. https://doi.org/10.1111/brv.12215
Hodar JA, Zamora R, Castro J (2002) Host utilisation by moth and larval survival of pine processionary caterpillar *Thaumetopoea pityocampa* in relation to food quality in three *Pinus* species. Ecol Entomol 27:292–301
Holeski LM, Jander G, Agrawal AA (2012) Transgenerational defense induction and epigenetic inheritance in plants. Trends Ecol Evol 27:618–626. https://doi.org/10.1016/j.tree.2012.07.011
Huffaker A, Pearce G, Veyrat N et al (2013) Plant elicitor peptides are conserved signals regulating direct and indirect antiherbivore defense. PNAS 110:5707–5712. https://doi.org/10.1073/pnas.1214668110
Innocenti M, Tiberi R (2002) Cone and seed pests of *Pinus pinea* L. in Central Italy. Redia 85:21–28
IPCC (2014) Climate change 2014: synthesis report. Contribution of Working Groups I, II and III to the fifth assessment report of the Intergovernmental Panel on Climate Change. IPCC, Geneva
Jacquet J-S, Orazio C, Jactel H (2012) Defoliation by processionary moth significantly reduces tree growth: a quantitative review. Ann For Sci 69:857–866. https://doi.org/10.1007/s13595-012-0209-0
Jactel H, Kleinhentz M, Marpeau-Bezard A et al (1996) Terpene variations in maritime pine constitutive oleoresin related to host tree selection by *Dioryctria sylvestrella* Ratz. (Lepidoptera: Pyralidae). J Chem Ecol 22:1037–1050. https://doi.org/10.1007/BF02029953
Jactel H, Kleinhentz M, Raffin A et al (1999) Comparison of different selection methods for the resistance to *Dioryctria sylvestrella* Ratz. (Lepidoptera: Pyralidae) in *Pinus pinaster* Ait. In: Lieutier F, Mattson WJ, Wagner MR (eds) Physiology and genetics of tree–phytophage interactions – international symposium. Colloques De l'INRA 90:137–149
Jactel H, Menassieu F, Vetillard A et al (2006) Tree species diversity reduces the invasibility of maritime pine stands by the bast scale, *Matsucoccus feytaudi* (Homoptera: Margarodidae). Can J For Res 36:314–326
Kanekar SS, Cale JA, Erbilgin N (2018) Ectomycorrhizal fungal species differentially affect the induced defensive chemistry of lodgepole pine. Oecologia 188:395–404
Kaundun SS, Fady B, Lebreton P (1997) Genetic differences between *Pinus halepensis*, *Pinus brutia* and *Pinus eldarica* based on needle flavonoids. Biochem Syst Ecol 25:553–562. https://doi.org/10.1016/s0305-1978(97)00049-5
Kaundun SS, Lebreton P, Fady B (1998a) Genetic variation in the needle flavonoid composition of *Pinus brutia* var. *brutia* populations. Biochem Syst Ecol 26:485–494. https://doi.org/10.1016/s0305-1978(98)00019-2
Kaundun SS, Lebreton P, Fady B (1998b) Geographical variability of *Pinus halepensis* Mill. as revealed by foliar flavonoids. Biochem Syst Ecol 26:83–96. https://doi.org/10.1016/s0305-1978(97)00092-6
Kelsey RG, Gallego D, Sánchez-García FJ et al (2014) Ethanol accumulation during severe drought may signal tree vulnerability to detection and attack by bark beetles. Can J For Res 44:554–561. https://doi.org/10.1139/cjfr-2013-0428
Kirisits T (2004) Fungal associates of European bark beetles with special emphasis on the ophiostomatoid fungi. In: Lieutier F, Day KR, Battisti A et al (eds) Bark and wood boring insects in living trees in Europe, a synthesis. Kluwer, Dordrecht, pp 181–235
Kleinhentz M, Raffin A, Jactel H (1998) Genetic parameters and gain expected from direct selection for resistance to *Dioryctria sylvestrella* Ratz. (*Lepidoptera: Pyralidae*) in *Pinus pinaster* Ait., using a full diallel mating design. For Genet 5:147–154

Kleinhentz M, Jactel H, Menassieu P (1999) Terpene attractant candidates of *Dioryctria sylvestrella* in maritime pine oleoresin, needles, liber and headspace samples. J Chem Ecol 25:2741–2756. https://doi.org/10.1023/a:1020803608406

Krokene P (2015) Conifer defenses and resistance to bark beetles. In: Vega FE, Hofstetter RW (eds) Bark beetles: biology and ecology of native and invasive species. Elsevier, Oxford, pp 177–207

Krokene P (2016) Carbon castles and insect invaders: dissecting physical defences in conifer stems. Plant Cell Environ 39:1643–1645. https://doi.org/10.1111/pce.12687

Lewinsohn E, Gijzen M, Croteau R (1991) Defense mechanisms of conifers. Differences in constitutive and wound-induced monoterpene biosynthesis among species. Plant Physiol 96:44–49. https://doi.org/10.1104/pp.96.1.44

Li S, Daudin JJ, Piou D et al (2015) Periodicity and synchrony of pine processionary moth outbreaks in France. For Ecol Manag 354:309–317. https://doi.org/10.1016/j.foreco.2015.05.023

Lieutier F, Mendel Z, Faccoli M (2016) Bark beetles of Mediterranean conifers. In: Paine TD, Lieutier F (eds) Insects and diseases of Mediterranean forest systems. Springer, pp 105–197. https://doi.org/10.1007/978-3-319-24744-1_6

Liphschitz N, Mendel ZVI (1989) Interactions between hosts and non-hosts of *Pinus* spp. and *Matsucoccus josephi*: anatomical responses of stem to infestation. New Phytol 113:135–142. https://doi.org/10.1111/j.1469-8137.1989.tb04699.x

Lombardero MJ, Ayres MP, Lorio PL et al (2000) Environmental effects on constitutive and inducible resin defences of *Pinus taeda*. Ecol Lett 3:329–339

López-Goldar X, Villari C, Bonello P et al (2018) Inducibility of plant secondary metabolites in the stem predicts genetic variation in resistance against a key insect herbivore in maritime pine. Front Plant Sci 9:1651. https://doi.org/10.3389/fpls.2018.01651

López-Goldar X, Villari C, Bonello P et al (2019) Genetic variation in the constitutive defensive metabolome and its inducibility are geographically structured and largely determined by demographic processes in maritime pine. J Ecol 107:2464–2477. https://doi.org/10.1111/1365-2745.13159

Mageroy MH, Christiansen E, Långström B et al (2019) Priming of inducible defenses protects Norway spruce against tree-killing bark beetles. Plant Cell Environ 43:420–430. https://doi.org/10.1111/pce.13661

Maňák V, Nordenhem H, Björklund N et al (2013) Ants protect conifer seedlings from feeding damage by the pine weevil *Hylobius abietis*. Agric For Entomol 15:98–105. https://doi.org/10.1111/j.1461-9563.2012.00597.x

Mäntylä E, Kleier S, Lindstedt C et al (2018) Insectivorous birds are attracted by plant traits induced by insect egg deposition. J Chem Ecol 44:1127–1138. https://doi.org/10.1007/s10886-018-1034-1

Martinez-Medina A, Flors V, Heil M et al (2016) Recognizing plant defense priming. Trends Plant Sci 21:818–822. https://doi.org/10.1016/j.tplants.2016.07.009

Martini A, Botti F, Galletti G et al (2010) The influence of pine volatile compounds on the olfactory response by *Neodiprion sertifer* (Geoffroy) females. J Chem Ecol 36:1114–1121. https://doi.org/10.1007/s10886-010-9851-x

Meijón M, Feito I, Oravec M et al (2016) Exploring natural variation of *Pinus pinaster* Aiton using metabolomics: is it possible to identify the region of origin of a pine from its metabolites? Mol Ecol 25:959–976. https://doi.org/10.1111/mec.13525

Mendel Z (1988) Host selection by the pine processionary caterpillar *Thaumetopoea wilkinsoni*. Phytoparasitica 16:101–108

Mendel Z (2000) The phytophagous insect fauna of *Pinus halepensis* and *P. brutia* forests in the Mediterranean. In: Ne'eman GTL (ed) Ecology, biogeography and management of *Pinus halepensisi* and *P. brutia* forest ecosystems in the Mediterranean Basin. Backhuys Publishers, Leiden

Mendel Z, Branco M, Battisti A (2016) Invasive sap-sucker insects in the Mediterranean Basin. In: Paine T, Lieutier F (eds) Insects and diseases of Mediterranean forest systems. Springer, pp 105–197. https://doi.org/10.1007/978-3-319-24744-1_6

Mita E, Tsitsimpikou C, Tsiveleka L et al (2002) Seasonal variation of oleoresin terpenoids from *Pinus halepensis* and *Pinus pinea* and host selection of the scale insect *Marchalina hellenica* (Homoptera, Coccoidea, Margarodidae, Coelostonidiinae). Holzforschung 56:572–578. https://doi.org/10.1515/hf.2002.087

Moreira X, Sampedro L, Zas R et al (2008) Alterations of the resin canal system of *Pinus pinaster* seedlings after fertilization of a healthy and of a *Hylobius abietis* attacked stand. Trees 22:771–777. https://doi.org/10.1007/s00468-008-0237-4

Moreira X, Sampedro L, Zas R (2009) Defensive responses of *Pinus pinaster* seedlings to exogenous application of methyl jasmonate: concentration effect and systemic response. Environ Exp Bot 67:94–100. https://doi.org/10.1016/j.envexpbot.2009.05.015

Moreira X, Alfaro RI, King JN (2012a) Constitutive defenses and damage in Sitka spruce progeny obtained from crosses between white pine weevil resistant and susceptible parents. Forestry 85:87–97. https://doi.org/10.1093/forestry/cpr060

Moreira X, Mooney KA, Zas R et al (2012b) Bottom-up effects of host-plant species diversity and top-down effects of ants interactively increase plant performance. Proc R Soc B Biol Sci 279:4464–4472. https://doi.org/10.1098/rspb.2012.0893

Moreira X, Zas R, Sampedro L (2012c) Genetic variation and phenotypic plasticity of nutrient re-allocation and increased fine root production as putative tolerance mechanisms inducible by methyl jasmonate in pine trees. J Ecol 100:810–820. https://doi.org/10.1111/j.1365-2745.2011.01938.x

Moreira X, Zas R, Solla A et al (2015) Differentiation of persistent anatomical defensive structures is costly and determined by nutrient availability and genetic growth-defence constraints. Tree Physiol 35:112–123. https://doi.org/10.1093/treephys/tpu106

Moulalis D (1991) Heritabilitätsuntersuchungen über die Harzertragsleistung bei der Aleppokiefer (*Pinus halepensis* Mill.). Forstwiss Centralbl 110:381–392. https://doi.org/10.1007/bf02741271

Mumm R, Hilker M (2006) Direct and indirect chemical defence of pine against folivorous insects. Trends Plant Sci 11:351–358

Mumm R, Schrank K, Wegener R et al (2003) Chemical analysis of volatiles emitted by *Pinus sylvestris* after induction by insect oviposition. J Chem Ecol 29:1235–1252. https://doi.org/10.1023/A:1023841909199

Muñoz-López C, Pérez-Fortea V, Cobos-Suárez R et al (2007) Sanidad forestal. Ediciones Mundi-Prensa, Madrid

Nagy NE, Fossdal CG, Krokene P et al (2004) Induced responses to pathogen infection in Norway spruce phloem: changes in polyphenolic parenchyma cells, chalcone synthase transcript levels and peroxidase activity. Tree Physiol 24:505–515. https://doi.org/10.1093/treephys/24.5.505

Nasri N, Tlili N, Triki S et al (2011) Volatile constituents of *Pinus pinea* L. needles. J Essent Oil Res 23:15–19. https://doi.org/10.1080/10412905.2011.9700441

Naves PM, de Sousa EM, Quartau JA (2006) Feeding and oviposition preferences of *Monochamus galloprovincialis* for certain conifers under laboratory conditions. Entomol Exp Appl 120:99–104. https://doi.org/10.1111/j.1570-7458.2006.00430.x

Naves P, Bonifácio L, Ed S (2016) The pine wood nematode and its local vectors in the Mediterranean Basin. In: Paine TD, Lieutier F (eds) Insects and diseases of Mediterranean forest systems. Springer, pp 329–378. https://doi.org/10.1007/978-3-319-24744-1_6

Oates CN, Denby KJ, Myburg AA et al (2016) Insect gallers and their plant hosts: from omics data to systems biology. Int J Mol Sci 17:1891

Oliveira Farinha AC, Pinheiro Silva JE, Correia AC et al (2018) Is *Leptoglossus occidentalis* entirely responsible for the high damage observed on cones and seeds of *Pinus pinea*? Results from a fertirrigation trial in Portugal. For Ecol Manag 429:198–206. https://doi.org/10.1016/j.foreco.2018.07.014

Paine TD, Lieutier F (2016) Insects and diseases of Mediterranean forest systems. Springer, Dordrecht

Paiva MR, Mateus E, Santos MH (2011) Pine volatiles mediate host selection for oviposition by *Thaumetopoea pityocampa* (Lep., Notodontidae). J Appl Entomol 135:195–203

Panzavolta T, Tellini Florenzano G, Tiberi R (2015) Pine monoterpene deterrence of pine processionary moth oviposition. Urban For Urban Green 14:538–543. https://doi.org/10.1016/j.ufug.2015.05.004

Parent GJ, Méndez-Espinoza C, Giguère I et al (2020) Hydroxyacetophenone defenses in white spruce against spruce budworm. Evol Appl 13:62–75. https://doi.org/10.1111/eva.12885

Peñuelas J, Filella I (2001) Responses to a warming world. Science 294:793–795

Pérez-Contreras T, Soler JJ, Soler M (2008) Needle asymmetry, pine vigour and pine selection by the processionary moth *Thaumetopoea pityocampa*. Acta Oecol 33:213–221. https://doi.org/10.1016/j.actao.2007.11.004

Petit RJ, Hampe A (2006) Some evolutionary consequences of being a tree. Annu Rev Ecol Evol Syst 37:187–214. https://doi.org/10.1146/annurev.ecolsys.37.091305.110215

Petit R, Aguinagalde I, de Beaulieu J-L et al (2003) Glacial refugia: hotspots but not melting pots of genetic diversity. Science 300:1563–1565

Pimentel C, Calvão T, Ayres MP (2011) Impact of climatic variation on populations of pine processionary moth *Taumetopoea pityocampa* in a core area of its distribution. Agric For Entomol 13:273–281. https://doi.org/10.1111/j.1461-9563.2011.00520.x

Pimentel CS, Firmino PN, Calvão T et al (2017) Pinewood nematode population growth in relation to pine phloem chemical composition. Plant Pathol 66:856–864. https://doi.org/10.1111/ppa.12638

Poelman EH, Kessler A (2016) Keystone herbivores and the evolution of plant defenses. Trends Plant Sci 21:477–485

Rodrigues AM, Mendes MD, Lima AS et al (2017) *Pinus halepensis*, *Pinus pinaster*, *Pinus pinea* and *Pinus sylvestris* essential oil chemotypes and monoterpene hydrocarbon enantiomers, before and after inoculation with the pinewood nematode *Bursaphelenchus xylophilus*. Chem Biodivers 14. https://doi.org/10.1002/cbdv.201600153

Rodriguez-Garcia A, Martin JA, Lopez R (2015) Influence of climate variables on resin yield and secretory structures in tapped *Pinus pinaster* Ait. in central Spain. Agric For Meteorol 202:83–93. https://doi.org/10.1016/j.agrformet.2014.11.023

Rosner S, Hannrup B (2004) Resin canal traits relevant for constitutive resistance of Norway spruce against bark beetles: environmental and genetic variability. For Ecol Manag 200:77–87

Roversi PF, Sciarretta A, Marziali L (2013) A GIS-based cost distance approach to analyse the spread of *Matsucoccus feytaudi* in Tuscany, Italy (Coccoidea Matsucoccidae). Redia 96:9–16

Sallé A, Battisti A (2016) Native sap-sucker insects in the Mediterranean Basin. In: Paine T, Lieutier F (eds) Insects and diseases of Mediterranean forest systems. Springer, pp 89–103. https://doi.org/10.1007/978-3-319-24744-1_5

Sampedro L (2014) Physiological trade-offs in the complexity of pine tree defensive chemistry. Tree Physiol 34:915–918. https://doi.org/10.1093/treephys/tpu082

Sampedro L, Moreira X, Martíns P (2009) Growth and nutritional response of *Pinus pinaster* after a large pine weevil (*Hylobius abietis*) attack. Trees 23:1189–1197. https://doi.org/10.1007/s00468-009-0358-4

Sampedro L, Moreira X, Llusia J et al (2010) Genetics, phosphorus availability, and herbivore-derived induction as sources of phenotypic variation of leaf volatile terpenes in a pine species. J Exp Bot 61:4437–4447. https://doi.org/10.1093/jxb/erq246

Sampedro L, Moreira X, Zas R (2011a) Costs of constitutive and herbivore-induced chemical defences in pine trees emerge only under low nutrient availability. J Ecol 99:818–827. https://doi.org/10.1111/j.1365-2745.2011.01814.x

Sampedro L, Moreira X, Zas R (2011b) Resistance and response of *Pinus pinaster* seedlings to *Hylobius abietis* after induction with methyl jasmonate. https://doi.org/10.1007/s11258-010-9830-x

Sanchez-Husillos E, Alvarez-Baz G, Etxebeste I (2013) Shoot feeding, oviposition, and development of *Monochamus galloprovincialis* on *Pinus pinea* relative to other pine species. Entomol Exp Appl 149:1–10

Santos-Del-Blanco L, Climent J, González-Martínez SC (2012) Genetic differentiation for size at first reproduction through male versus female functions in the widespread Mediterranean tree *Pinus pinaster*. Ann Bot 110:1449–1460

Saracino A, Rita A, Rossi S (2017) Climatic signal from *Pinus leucodermis* axial resin ducts: a tree-ring time series approach. Eur J For Res 136:27–36. https://doi.org/10.1007/s10342-016-1005-4

Sbay H, Zas R (2018) Geographic variation in growth, survival, and susceptibility to the processionary moth (*Thaumetopoea pityocampa* Dennis & Schiff.) of *Pinus halepensis* Mill. and *P. brutia* Ten.: results from common gardens in Morocco. Ann For Sci 75:69. https://doi.org/10.1007/s13595-018-0746-2

Schiller G, Genizi A (1993) An attempt to identify the origin of *Pinus Brutia* plantations in Israel by needle resin composition. Silvae Genet 42:2–3

Schiller G, Grunwald C (1987) Cortex resin monoterpene composition in *Pinus brutia* provenances grown in Israel. Biochem Syst Ecol 15:389–394. https://doi.org/10.1016/0305-1978(87)90051-2

Schopf R, Avtzis N (1987) Die Bedeutung von Nadelinhaltsstoffen für die Disposition von fünf Kiefernarten gegenüber *Thaumetopoea pityocampa* (Schiff.). J Appl Entomol 103:340–350. https://doi.org/10.1111/j.1439-0418.1987.tb00993.x

Schulman E (1984) Longevity under adversity in conifers. Science 119:396–399

Schvester D, Ughetto F (1986) Differences in susceptibility to *Matsucoccus feytaudi* (Homoptera: Margarodidae) in maritime pine (*Pinus pinaster*) according to provenance. Ann For Sci 43:459–474. https://doi.org/10.1051/forest:19860403

Semiz G, Erbilgin N, Holopainen JK (2017) *Hylobius abietis* L. feeding on the novel host *Pinus brutia* Ten. increases emission of volatile organic compounds. J Appl Entomol 141:133–140

Serra-Varela MJ, Grivet D, Vincenot L (2015) Does phylogeographical structure relate to climatic niche divergence? A test using maritime pine (*Pinus pinaster* Ait.). Glob Ecol Biogeogr 24:1302–1313

Slack A, Kane J, Knapp E (2017) Contrasting impacts of climate and competition on large sugar pine growth and defense in a fire-excluded forest of the central Sierra Nevada. Forests 8:244

Sniezko RA, Koch J (2017) Breeding trees resistant to insects and diseases: putting theory into application. Biol Invasions 19:3377–3400. https://doi.org/10.1007/s10530-017-1482-5

Sousa E, Bravo MA, Pires J (2001) *Bursaphelenchus xylophilus* (Nematoda; aphelenchoididae) associated with *Monochamus galloprovincialis* (Coleoptera; Cerambycidae) in Portugal. Nematology 3:89–91

Stamp N (2003) Out of the quagmire of plant defense hypotheses. Q Rev Biol 78:23–55

Suarez-Vidal E, Lopez-Goldar X, Sampedro L et al (2017) Effect of light availability on the interaction between maritime pine and the pine weevil: light drives insect feeding behavior but also the defensive capabilities of the host. Front Plant Sci 8:1452. https://doi.org/10.3389/fpls.2017.01452

Suarez-Vidal E, Sampedro L, Voltas J et al (2019) Drought stress modifies early effective resistance and induced chemical defences of Aleppo pine against a chewing insect herbivore. Environ Exp Bot 162:550–559. https://doi.org/10.1016/j.envexpbot.2019.04.002

Taïbi K, Del Campo A, Mulet J et al (2014) Testing Aleppo pine seed sources response to climate change by using trial sites reflecting future conditions. New For 45:603–624

Tapias R, Climent J, Pardos JA et al (2004) Life histories of Mediterranean pines. Plant Ecol 171:53–68

Tardif J, Camarero JJ, Ribas M (2003) Spatiotemporal variability in tree growth in the Central Pyrenees: climatic and site influences. Ecol Monogr 73:241–257

Tiberi R, Niccoli A, Curini M et al (1999) The role of the monoterpene composition in *Pinus* spp needles, in host selection by the pine processionary caterpillar, *Thaumetopoea pityocampa*. Phytoparasitica 27:263–272

Tooker JF, Helms AM (2014) Phytohormone dynamics associated with gall insects, and their potential role in the evolution of the gall-inducing habit. J Chem Ecol 40:742–753. https://doi.org/10.1007/s10886-014-0457-6

Turner GW, Parrish AN, Zager JJ et al (2018) Assessment of flux through oleoresin biosynthesis in epithelial cells of loblolly pine resin ducts. J Exp Bot 70:217–230. https://doi.org/10.1093/jxb/ery338
Valladares F, Matesanz S, Guilhaumon F et al (2014) The effects of phenotypic plasticity and local adaptation on forecasts of species range shifts under climate change. Ecol Lett 17:1351–1364
Vázquez-González C, López-Goldar X, Zas R et al (2019) Neutral and climate-driven adaptive processes contribute to explain population variation in resin duct traits in a Mediterranean pine species. Front Plant Sci 13:1613
Verne S, Jaquish B, White R et al (2011) Global transcriptome analysis of constitutive resistance to the white pine weevil in spruce. Genome Biol Evol 3:851–867. https://doi.org/10.1093/gbe/evr069
Vivas M, Zas R, Sampedro L et al (2013) Environmental maternal effects mediate the resistance of maritime pine to biotic stress. PLoS One 8:e70148. https://doi.org/10.1371/journal.pone.0070148
Vogan PJ, Schoettle AW (2015) Selection for resistance to white pine blister rust affects the abiotic stress tolerances of limber pine. For Ecol Manag 344:110–119
Voltas J, Lucabaugh D, Chambel MR et al (2015) Intraspecific variation in the use of water sources by the circum-Mediterranean conifer *Pinus halepensis*. New Phytol 208:1031–1041. https://doi.org/10.1111/nph.13569
Whitehill JG, Henderson H, Schuetz M et al (2016) Histology and cell wall biochemistry of stone cells in the physical defence of conifers against insects. Plant Cell Environ 39:1646–1661. https://doi.org/10.1111/pce.12654
Whitehill JGA, Yuen MMS, Henderson H et al (2019) Functions of stone cells and oleoresin terpenes in the conifer defense syndrome. New Phytol 221:1503–1517. https://doi.org/10.1111/nph.15477
Wise MJ, Rausher MD (2013) Evolution of resistance to a multiple-herbivore community: genetic correlations, diffuse coevolution, and constraints on the plant's response to selection. Evolution 67:1767–1779. https://doi.org/10.1111/evo.12061
Woodcock P, Cottrell JE, Buggs RJA et al (2018) Mitigating pest and pathogen impacts using resistant trees: a framework and overview to inform development and deployment in Europe and North America. Forestry 91:1–16. https://doi.org/10.1093/forestry/cpx031
Yamaguchi H, Tanaka H, Hasegawa M et al (2012) Phytohormones and willow gall induction by a gall-inducing sawfly. New Phytol 196:586–595. https://doi.org/10.1111/j.1469-8137.2012.04264.x
Zas R, Sampedro L (2015) Heritability of seed weight in Maritime pine, a relevant trait in the transmission of environmental maternal effects. Heredity 114:116–124. https://doi.org/10.1038/hdy.2014.76
Zas R, Sampedro L, Prada E et al (2005) Genetic variation of *Pinus pinaster* Ait. seedlings in susceptibility to the pine weevil *Hylobius abietis* L. Ann For Sci 62:681–688. https://doi.org/10.1051/forest:2005064
Zas R, Moreira X, Sampedro L (2011) Tolerance and induced resistance in a native and an exotic pine species: relevant traits for invasion ecology. J Ecol 99:1316–1326. https://doi.org/10.1111/j.1365-2745.2011.01872.x
Zas R, Cendán C, Sampedro L (2013) Mediation of seed provisioning in the transmission of environmental maternal effects in Maritime pine (*Pinus pinaster* Aiton). Heredity 111:248–255. https://doi.org/10.1038/hdy.2013.44
Zas R, Moreira X, Ramos M et al (2015) Intraspecific variation of anatomical and chemical defensive traits in Maritime pine (*Pinus pinaster*) as factors in susceptibility to the pinewood nematode (*Bursaphelenchus xylophilus*). Trees Struct Funct 29:663–673. https://doi.org/10.1007/s00468-014-1143-6

Part III
Pine and Mixed Forest Ecosystems Under Global Climate Change

Gidi Ne'eman and Avi Perevolotsky

Chapter 12
Effects of Global Change on Tree Growth and Vigor of Mediterranean Pines

J. Julio Camarero, Antonio Gazol, and Raúl Sánchez-Salguero

12.1 Introduction

Global change is a multidimensional phenomenon including multiple, interacting drivers such as: (i) changes in land use and land cover, (ii) climate warming and aridification in drought-prone areas, (iii) the modification of major biogeochemical cycles (e.g., C and N cycles), and (iv) the reshuffling of species (including extinctions and invasions), reinforcing existing biotic actions and also leading to new interactions (Vitousek 1994). Some of these drivers have impacted forest ecosystems in the Mediterranean Basin more strongly or evidently, particularly the historical use of these forests, which have been managed or exploited for millennia (Blondel 2006), and the recent impact of climate warming associated with hotter droughts in some areas (Cook et al. 2016).

Historical changes in land-use pressure have been intense and have shifted through time across the Mediterranean Basin for at least the past 3000 years (Stephens et al. 2019). Cultivated and grazed areas expanded during the Roman Empire and became very extensive from the eighteenth century (Klein Goldewijk et al. 2011), as human populations occupied most of the territory, forming urban areas. Urbanization trends peaked in the twentieth century, leaving depopulated areas, mainly in montane and inland (often dry and cold) regions of the Iberian Peninsula, south eastern France, northern Greece, Anatolia and North Africa (UN

J. J. Camarero (✉) · A. Gazol
Instituto Pirenaico de Ecología (IPE-CSIC), Zaragoza, Spain
e-mail: jjcamarero@ipe.csic.es

R. Sánchez-Salguero
Instituto Pirenaico de Ecología (IPE-CSIC), Zaragoza, Spain

Departamento Sistemas Físicos, Químicos y Naturales, Universidad Pablo de Olavide, Sevilla, Spain
e-mail: rsanchez@upo.es

G. Ne'eman, Y. Osem (eds.), *Pines and Their Mixed Forest Ecosystems in the Mediterranean Basin*, Managing Forest Ecosystems 38,
https://doi.org/10.1007/978-3-030-63625-8_12

2019). In 2018, the rural population ranged between 57% and 100% in the northern and western margins of the Mediterranean Basin (Europe) and between 43% and 92% in the southern and eastern margins (Africa and parts of the Middle East), where the urban population is forecasted to rapidly increase from a mean urbanization rate of 41% in 1970 to a projected rate of 66% by 2025 (UN 2019). These geographical changes in the distribution of people between rural and urban locations across the Mediterranean Basin enhance rewilding, including a shift from deforestation to forest expansion and encroachment (Blondel 2006).

In addition to changes in land use and cover, the climate of the Mediterranean Basin became much warmer from 1981 to 2010, compared to the period 1951–1980 when cooling was observed in Iberia, Algeria, the Balkans and South Italy (IPCC 2013). Since the 1980s, climate warming has affected the entire Mediterranean Basin with different intensities across the region, being particularly intense in Bulgaria, Turkey, Greece and the Levant countries. There is evidence for an increase in the frequency of hotter droughts in the Mediterranean Basin, suggesting that the recently observed aridification trend is linked to global warming (Hoerling et al. 2012).

In this chapter we synthesize and analyze some of the effects of two global-change drivers, historical land use and recent global warming, on pine forests in the Mediterranean Basin. We focus on data of tree crown transparency (defoliation) and radial growth, reconstructed using annual rings, as these are both widely used and recognized proxies of changes in tree vigor or vitality at the tree and population levels (Dobbertin 2005; Carnicer et al. 2011). We relate these surrogates of change in forest productivity and vitality to drought severity, and compare them with changes in growth, dieback and mortality data gathered across the Mediterranean Basin and also in continental Spain. In the case of Spain we focus on the comparison of annual defoliation data and decadal changes in stand basal area derived from forest inventory data.

12.2 Study Cases

12.2.1 Defoliation, Growth and Mortality in Mediterranean Pines

We selected the most abundant and widely distributed pine species in the Mediterranean Basin which have been intensively sampled for radial growth, defoliation and canopy dieback data (Fig. 12.1a): *Pinus halepensis* Mill., *P. brutia* Ten., *P. pinea* L., *P. pinaster* Ait., *P. nigra* J.F. Arn., and *P. sylvestris* L. First, we compiled dendrochronological data from 645 locations in which 878 individual growth trends have been analyzed since the 1970s (see Galván et al. 2014). Second, annual crown defoliation was determined for Mediterranean pine species during the 1987–2018 period using the available Level I monitoring plots of the European International

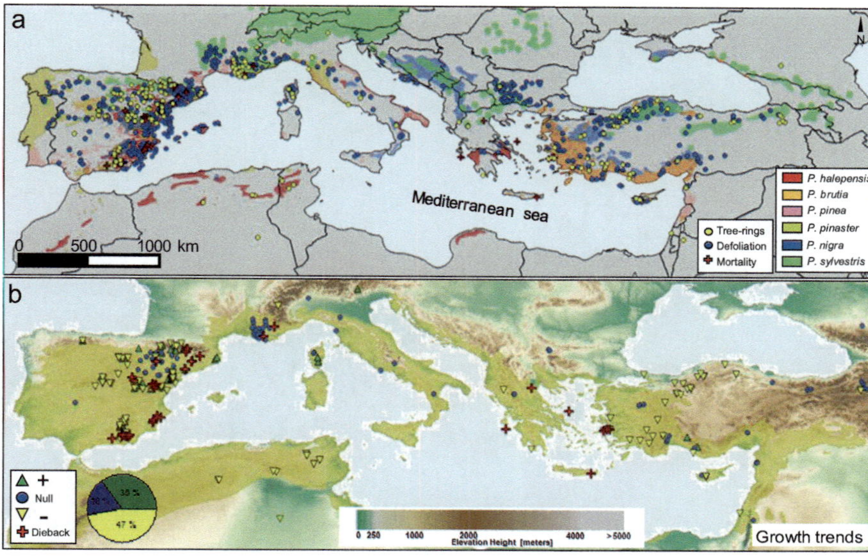

Fig. 12.1 (**a**) Distribution of Mediterranean pine forests (colored areas, source: www.euforgen.org), selected tree-ring width chronologies (yellow circles) and location of the ICP-forest Level I plots used to obtain defoliation data (blue circles) for sites with drought-induced damage. (**b**) Map of post-1970 growth trends showing sites with statistically significant ($p < 0.05$) negative (yellow downward triangles) and positive (green upward triangles) growth trends (slope of linear regressions; see Galván et al. 2014). Neutral trends are shown with blue dots. The trends are summarized in the pie chart. The map also shows elevation. In plot (**b**) the red crosses represent documented cases of drought-related pine stand mortality (Caudullo and Barredo 2019)

Co-operative Programme on Assessment and Monitoring of Air Pollution Effects on Forests (ICP-Forests) network (Fig. 12.1a). These intensive monitoring plots have been monitored since 1985 based on a 16-km grid covering Europe. Individual tree defoliation was annually recorded during summer and autumn by two observers according to international guidelines (Eichhorn et al. 2010). Crown defoliation corresponds to the percentage of foliage missing in comparison with the expected amount for a healthy tree at the site (Dobbertin 2005). For each ICP-Forests plot, we selected those dominant pine species showing drought-induced damage in 536 locations (Fig. 12.2a). Lastly, tree mortality data related to heat and drought stress were gathered from the ICP-Forests network (Neumann et al. 2017) and the European dataset recently assembled by Caudullo and Barredo (2019) (Fig. 12.2a).

Regarding negative growth trends, most cases were located in lower-elevation and more xeric sites in the south western and eastern Mediterranean Basin, whereas positive growth trends corresponded to wet and cold sites situated in high-elevation areas and north western regions (Fig. 12.1b) such as the northern Iberian Peninsula and southern Alps. The warm-dry Atlas sites in north western Africa displayed negative trends. These observations agree with the crown defoliation trends (calculated as the slope of the linear regression between tree defoliation and calendar year) where positive defoliation trends since 1987 tend to be located in continental and

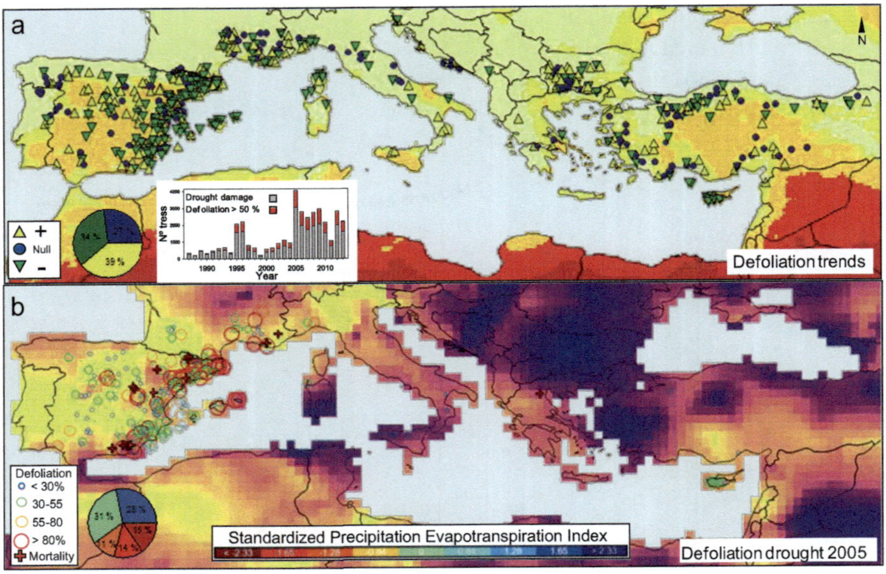

Fig. 12.2 (**a**) Map of the drought-induced damage assessed by crown defoliation trends (slope) from 1987 to 2018. The map shows sites with statistically significant ($p < 0.05$) negative (green downward triangles) and positive (yellow upward triangles) trends. Neutral trends are shown with blue dots. These trends are summarized in the pie chart. The insert shows the temporal evolution of the number of trees with drought-induced damage (grey part of bars) and trees showing defoliation >50% defoliation (red part of bars) for the period 1987–2013. The map also shows aridity with reddish and yellowish shades representing higher and lower aridity, respectively. (**b**) Crown defoliation (%) during the 2005 drought. These crown defoliation groups are summarized in the pie chart. The map shows the 6-month Standardized Precipitation Evapotranspiration Index (SPEI) calculated for June. Shades from yellow to blue indicate driest to wettest regions, respectively. In plot (**b**) the red crosses represent documented cases of drought-related pine stand mortality (Caudullo and Barredo 2019)

arid sites (Fig. 12.2a). In contrast, more humid areas of the Mediterranean Basin often display negative defoliation trends (Carnicer et al. 2011). In the dataset of drought-induced damage, we observed a significant ($p < 0.05$) positive trend in the amount of severely defoliated trees since 1987 (Fig. 12.2a). Defoliation peaked in years during and after severe spring-summer droughts showing low values of the Standardized Precipitation Evapotranspiration Index (SPEI; Vicente-Serrano et al. 2010), such as 1994–1995–1996, 2005–2006, 2009 and 2012 (Fig. 12.2a, insert).

To determine how extreme droughts affected crown defoliation, we selected the 2005 drought, which coincided with sharp reductions in forest productivity and tree growth across the Mediterranean Basin, particularly in Iberia (Sánchez-Salguero et al. 2012; Camarero et al. 2015a; Gazol et al. 2017, 2018). Drought damage peaked in 2005 as manifested by severe defoliation in 29% of the study sites, higher than 80% in many trees (Fig. 12.2a, insert). The highest proportion of affected trees corresponded to xeric sites from the western Mediterranean Basin, mainly in eastern Spain and south-eastern France (Fig. 12.2b), whereas most mortality cases corresponded to pine populations located at the warmest and driest limits of the species distribution area in the Iberian Peninsula (Sánchez-Salguero et al. 2012, 2017b; Camarero et al. 2015a). Drought-induced defoliation above the 55% threshold during 2005 was reported mainly for *P. halepensis* (40% of cases), followed by *P. pinaster* (18%), *P. nigra* (17%), *P. sylvestris* (10%) and *P. pinea* (7%) (Fig. 12.2b).

12.2.2 Tree Defoliation and Mortality in Spain: A Hotspot of Pine Dieback

According to ICP-Forests data, *P. pinaster* was the species with the highest mean percentage of dead trees showing complete defoliation in Spain (3.9% year^{-1}), followed by *P. halepensis* (1.8% year^{-1}). These two species also showed maximum percentages of dead trees in 2000 (*P. pinaster*, 9.5% year^{-1}) and 1994 (*P. halepensis*, 8.0% year^{-1}) (Figs. 12.2a and 12.3a). In these two species, the main mortality factor was drought, which killed ca. 36% and 50% of the highly defoliated trees in *P. pinaster* (16% of trees) and *P. halepensis* (9% of trees), respectively (IDF 2018). For *P. pinaster* only, there was a significant trend of increasing mortality for the period 1987–2018 (Kendall tau $\tau = 0.27$, $p = 0.03$). As for the Mediterranean Basin, the mean percentage of dead trees peaked in years with severe spring-summer droughts (low June SPEI values), such as 1994–1995–1996, 2000, 2005–2006, 2009–2010, 2013 and 2016 (Fig. 12.3).

Considering the Caudullo and Barredo (2019) dataset, the initial year of mortality of Mediterranean pine forests was related to long droughts. Specifically, the association between the year of pine mortality and SPEI across the Mediterranean Basin peaked (in absolute terms) in June when considering 11-month SPEI values ($r = -0.59$, $p = 0.0006$), i.e. representing droughts starting in the previous August. Of the 49 cases reported by Caudullo and Barredo (2019), 36% of them began in the very dry years, 1994 and 2005, and 26% in the 2002 and 2012 droughts. These mortality episodes lasted, on average, 1–2 years and mainly involved *P. sylvestris* (43% of cases) and *P. halepensis* (24% of cases).

The 2012 drought that affected the Iberian Peninsula was characterized by very warm and dry conditions during the 2011–2012 winter (Trigo et al. 2013). These extreme dry conditions in extent and magnitude, which led to a sharp drop in SPEI values in mid-March followed by minimum SPEI values in September when cumulative drought stress peaked (Camarero et al. 2015a), affected most Iberian pine forests (e.g., *P. sylvestris*, *P. halepensis*). According to tree-ring reconstructions, the 2012 drought was among the most severe dry spells affecting the Mediterranean Basin for the last 900 years (Touchan et al. 2008; Cook et al. 2016). In two sites showing intense canopy dieback and high mortality rates located in Aragón, northeastern Spain (see Camarero et al. 2015a), the 12-month SPEI values were negatively associated with the mortality rates of *P. halepensis* (Spearman $r_s = -0.72$, $p = 0.005$) but not with those of *P. sylvestris* ($r_s = 0.14$, $p = 0.580$), which peaked in

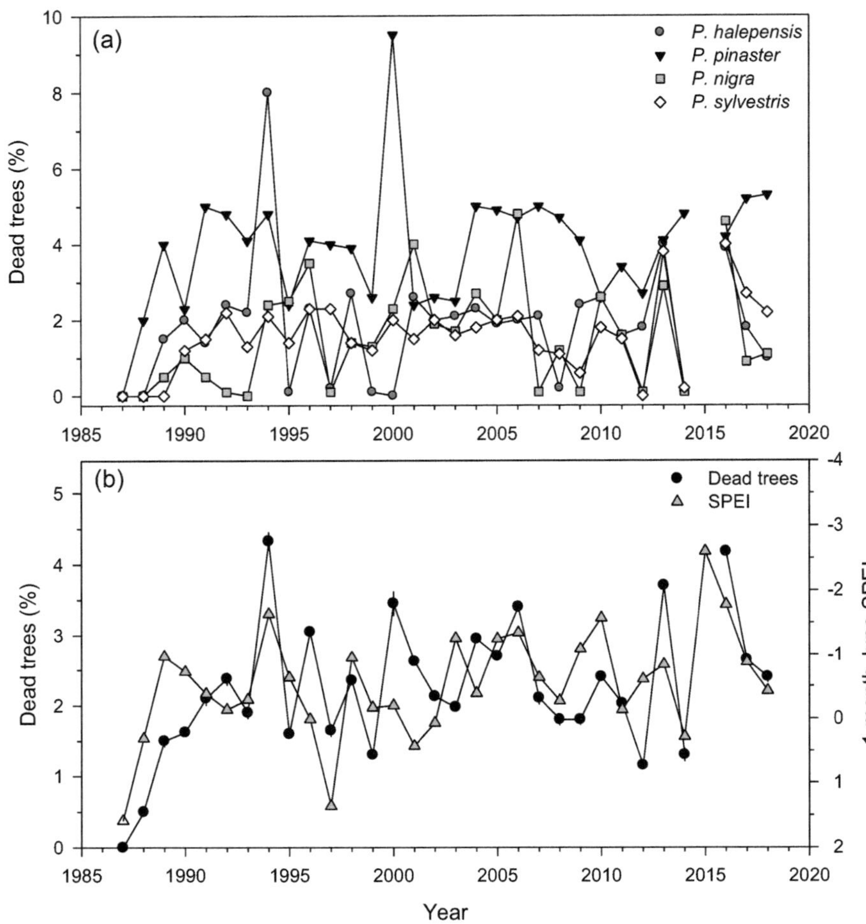

Fig. 12.3 (**a**) Temporal evolution of the percentage of dead trees, i.e. trees showing very high defoliation degree, in ICP-Forests Level I plots of the main pine species across Spain for the period 1987–2018, and (**b**) association between the mean percentage of dead trees and the SPEI drought index (note the reverse scale of the SPEI). In (**b**) error bars are standard errors. No defoliation data were taken in 2015. The SPEI was obtained for the region including most of continental Spain with coordinates 4.25° W, 44.25° and 9.25° E, 49.25° N

2011 and 2012. This difference suggests a prolonged or continuous response of mortality to drought in the *P. halepensis* stand, where dry conditions are common, but a punctual and strong response in the *P. sylvestris* stand, where such hot, dry conditions were not so common in the past.

12.2.3 Composition and Growth Dynamics of Spanish Pine Forests

The Spanish Forest Inventory (hereafter SFI) is carried out every ca. 10 years and provides a harmonized quantification of the composition, structure and diversity of wooded areas in Spain (Alberdi et al. 2017). The SFI is based on the systematic location of sampling plots every 1 km^2 in forested regions (tree cover >5%) in which trees are tagged, measured and classified within 25-m radius circular plots (Villaescusa and Díaz 1998). However, harmonized forest inventories have been carried out systematically only since the 1980s; the second (SFI2) and third (SFI3) inventories are available for most of the territory (Gómez-Aparicio et al. 2011). In the SFI2 (carried out from 1986 to 1996), each tree was identified, georeferenced and permanently tagged. In the SFI3 (carried out from 1997 to 2007) repeated measures on the same individuals were taken, allowing quantification of forest growth between the SFI2 and the SFI3. We excluded SFI plots that had any sign of management or those sampled only in the SFI3.

For each SFI plot, we selected those trees belonging to the three pine species which occupy the greatest area of continental Spain, namely: *P. sylvestris* (ca. 1 million ha), *P. nigra* (ca. 0.7 million ha), and *P. halepensis* (ca. 2 million ha). These three species are mainly distributed across the eastern Iberian Peninsula (Fig. 12.4). *Pinus halepensis* dominates dry and semiarid regions near the Mediterranean coast and in the Ebro Valley and is present in 7092 plots. *Pinus nigra* is frequently found in Mediterranean mountains (present in 4779 plots) where it often coexists with *P. sylvestris* (present in 5139 plots). All together, these three species were present in 15,116 plots, representing approximately 30% of the total SFI plots. The three species coexist in 97 SFI plots, where *P. sylvestris* and *P. nigra* coexist in 1186 plots, *P. sylvestris* co-occurs with *P. halepensis* in 161 plots, and *P. nigra* and *P. halepensis* appear together in 644 plots. Thus, most plots contained only one of these species (*P. halepensis* in 6384 plots; *P. nigra* in 3046 plots, and *P. sylvestris* in 3889 plots). A simple linear model of *P. sylvestris* growth (difference in basal area of living trees between SFI2 and SFI3) showed its dependency on the initial stand basal area of the species in SFI2 but also its negative relationship with the initial basal area in SFI2 of *P. halepensis* and *P. nigra* in the stand (Fig. 12.4). The model explained 67% of the variability in basal area of *P. sylvestris*. For more analyses of the SFI data see Herrero et al. this volume (Chap. 17).

12.3 Discussion

Current pine forests in the Mediterranean Basin are just a shadow of what they were millennia and centuries ago, because anthropogenic deforestation has led to the local disappearance of many pine populations (Morales-Molino et al. 2017). The productivity, cover and vitality of pine populations in the Mediterranean Basin have

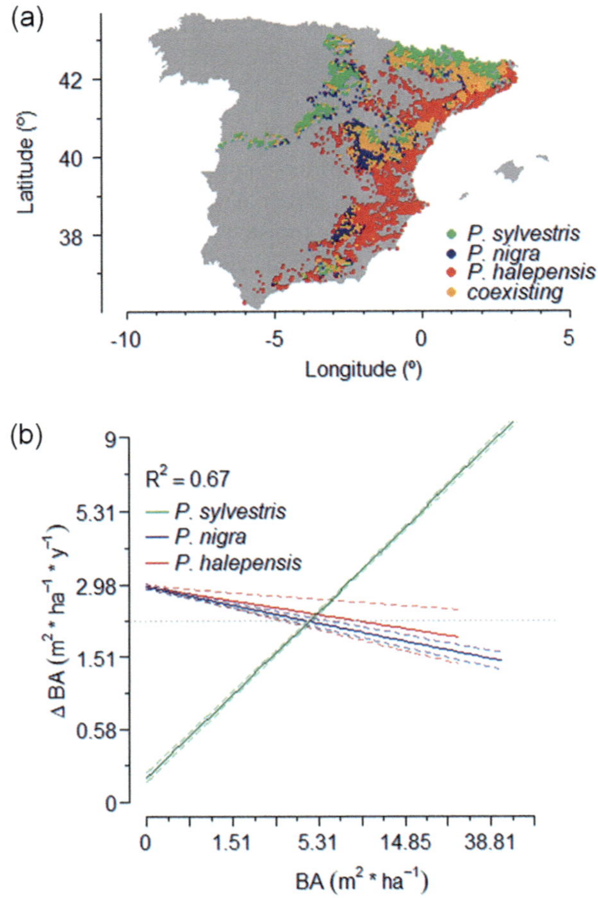

Fig. 12.4 Plots of the Spanish Forest Inventory (SFI) showing (**a**) distribution range of three major pine species (*P. sylvestris*, green dots; *P. nigra*, blue dots; *P. halepensis*, red dots; plots with the three species, orange dots) and (**b**) effects of SFI2 basal area of *P. sylvestris* (green), *P. nigra* (blue) and *P. halepensis* (red) on changes in basal area (ΔBA) of *P. sylvestris* between the second and third SFIs carried out in 1986–1996 and 1997–2007, respectively. This plot shows how the changes in stand basal area (BA) of *P. sylvestris* depend on its previous basal area (green lines) but are also negatively related to the BA of coexisting *P. halepensis* (red lines) and *P. nigra* (blue lines). Continuous and dashed lines show the predicted BA values and 95% confidence intervals

changed rapidly in response to warmer and drier conditions during the last four decades (Carnicer et al. 2011; Gazol et al. 2018). Summer drought characterized by low soil moisture levels and high evapotranspiration rates and spring heat stress trigger defoliation of Mediterranean pine species (Sánchez-Salguero et al. 2017a), which often exhibit a slower recovery than the coexisting, angiosperm shrubs (Peñuelas et al. 2001). Species with an ample Eurosiberian distribution reaching their southern limit of distribution in the Mediterranean Basin and nearby regions, such as *P. sylvestris*, are among the most vulnerable in terms of canopy dieback, needle shedding and mortality rates (Bigler et al. 2006; Vennetier et al. 2007; Gea-Izquierdo et al. 2014; Camarero et al. 2015a). This is expected since *P. sylvestris* is considered to demonstrate low tolerance to drought stress (Martínez-Vilalta and Piñol 2002; Rubio-Cuadrado et al. this volume, Chap. 9). However, pine species that are considered very tolerant to drought such as *P. halepensis*, *P. brutia*, *P. pinea* and *P. pinaster* have also suffered from canopy dieback (dead branches, shoots and needles), increased mortality and reduced reproductive output in response to drought (Girard et al. 2012; Sánchez-Salguero et al. 2012; Camarero et al. 2015a; Klein et al. 2019). We have focused on drought stress, but another major constraint of pine forests is wildfire; the interactions between these two disturbances cannot be ignored (Moreno et al. this volume, Chap. 21) since they have caused historical deforestation across the Mediterranean Basin and could convert many forests into shrublands (Karavani et al. 2018).

The observation that drought affects pine species with different tolerance to drought and contrasting hydraulic traits confirms that all of them are vulnerable to some degree of water shortage, and this vulnerability also depends on regional (e.g., climate seasonality, drought occurrence) and local (e.g., topography, soil type and capacity to absorb or retain water) factors. For instance, the 2003 heat wave mainly affected central European forests and barely influenced pine forests in the Mediterranean Basin, whereas the 1994–1995, 2005 and 2012 droughts triggered defoliation, growth decline and mortality in eastern Iberia (Camarero et al. 2015a). The interacting regional and local factors must play different roles at distinct scales, which complicates forecasting forest resilience after drought. Post-drought resilience depends on climate seasonality, aridity, the characteristic time scale of response to drought, stand density, competition, provenance, biotic stressors (mistletoe, bark beetles, insect defoliators), interactions between tree species, soil features, and so on (Martín-Benito et al. 2011; Dorman et al. 2015a, b; García de la Serrana et al. 2015; Gazol et al. 2018; Sánchez-Salguero et al. 2018; Sangüesa-Barreda et al. 2013, 2018; Rubio-Cuadrado et al. this volume, Chap. 9).

Overall, pines seem to show increased vulnerability to drought compared with coexisting broadleaved species (Anderegg et al. 2016). They also show slower post-drought growth recovery rates. Pine species require, on average, 2 years to recover to pre-drought growth levels, whereas for oak species only 1 year is required (Anderegg et al. 2015). Drought-induced mortality of pine species seems to have increased since the 1990s across the Mediterranean Basin, as our analyses and previous research confirm (Neumann et al. 2017; Klein et al. 2019). It is unclear how this will affect the balance in mixed pine-oak forests that dominate extensive, seasonally dry parts of the Mediterranean Basin (Zavala et al. 2000; Sheffer 2012). A higher mortality of pines in response to long, severe and hotter droughts should favor oaks which are able to sprout, whereas most pine species are obligate seeders. However, moderate dry conditions and a moderate recurrence of fires could also lead to a dominance of pines in harsh sites where oaks may be outcompeted. In the case of ecotones where several pine species coexist, we could expect a steeper decline in the growth of the species more vulnerable to drought, such as *P. sylvestris* when coexisting with *P. nigra*, as shown by tree-ring width data (Marqués et al.

2016). This has also been confirmed by our analyses of the SFI data. However, we have only considered growth data, whereas forest dynamics also depend on recruitment and mortality rates. Further research is needed to forecast the demographic trends of particular mixtures, such as sites where *P. sylvestris* currently coexists with *P. halepensis*. The existence of *P. sylvestris* in those sub-optimal locations contradicts expectations based solely on its higher vulnerability to drought and heat stress, and suggests interactions with past land use. In addition, these ecotones question the relevance of considering differences between pine species and emphasize the need for functional approaches that consider different drought tolerances, responses to drought or water-use strategies (e.g., preferential use of soil water pools situated at different depths). Special attention should be also paid to Mediterranean species, such as *P. pinaster*, growing in low-productivity sites with shallow, rocky soils (e.g., in eastern Spain) where needle loss and growth decline will portend more frequent and severe canopy dieback and mortality episodes. The growth of this species is particularly responsive to changes in precipitation and soil moisture (Camarero et al. 2015b) making this pine very vulnerable to drought.

12.4 Conclusions

Global-change drivers are differentially impacting pine forests across the Mediterranean Basin. During the recent decades we have focused on the detrimental effects of hotter droughts on forest productivity and tree vigor. Severe and relatively short droughts lasting from 1 to 6 months triggered needle loss (defoliation), canopy dieback, growth decline and increased mortality rates. The Eurosiberian *P. sylvestris* was very vulnerable to these drought episodes, but Mediterranean species with higher tolerance to water shortage such as *P. halepensis, P. pinea* and *P. pinaster* also suffered from canopy dieback and elevated mortality in some locations and under particular site conditions. The ways in which the historical use and management of Mediterranean pine forests (historical legacies) have predisposed these ecosystems to the negative effects of recent warmer and drier conditions requires further study. For instance, many pine plantations performed in wetter decades are suffering dieback in the last warmer and drier decades. Past thinning could also have selected fast-growing trees leaving trees with low growth rates, more prone to die due to drought stress. Overall, we do not know how the past use of Mediterranean pine forests for getting timber, wood, resin or cones influences their current resilience.

Acknowledgments We acknowledge support from the projects FunDiver (CGL2015-69186-C2-1-R), FORMAL (RTI2018-096884-B-C31) and LESENS (RTI2018-096884-B-C33) from the Spanish Ministry of Science, Innovation and Universities; and also VULBOS (UPO-1263216, FEDER Funds, Andalucía 2014-2020) and "Diversidad funcional de suelos en ecosistemas forestales cambiantes" projects (Govt. Aragón, FEDER Funds). We also thank J. Diego Galván for his compilation of Mediterranean growth data, and Paloma Ruiz Benito for her help with the analysis and interpretation of the SFI data. We thank the UNECE ICP Forests PCC Collaborative Database, co-financed by the European Commission, for providing tree defoliation data. We thank two anonymous reviewers for improving an earlier version of this chapter.

References

Alberdi I, Vallejo R, Álvarez-González JG et al (2017) The multi-objective Spanish National Forest Inventory. For Syst 26:e04S

Anderegg WRL, Schwalm C, Biondi F et al (2015) Pervasive drought legacies in forest ecosystems and their implications for carbon cycle models. Science 349:528–532

Anderegg WR, Klein T, Bartlett M et al (2016) Meta-analysis reveals that hydraulic traits explain cross-species patterns of drought-induced tree mortality across the globe. PNAS 113:5024–5029

Bigler C, Bräker OU, Bugmann H, Dobbertin M, Rigling A (2006) Drought as an inciting mortality factor in Scots pine stands of the Valais, Switzerland. Ecosystems 9:330–343

Blondel J (2006) The 'design' of Mediterranean landscapes: a millennial story of humans and ecological systems during the historic period. Hum Ecol 34:713–729

Camarero JJ, Gazol A, Sangüesa-Barreda G et al (2015a) To die or not to die: early-warning signals of dieback in response to a severe drought. J Ecol 103:44–57

Camarero JJ, Gazol A, Tardif JC et al (2015b) Attributing forest responses to global-change drivers: limited evidence of a CO_2-fertilization effect in Iberian pine growth. J Biogeogr 42:2220–2233

Carnicer J, Coll M, Ninyerola M et al (2011) Widespread crown condition decline, food web disruption, and amplified tree mortality with increased climate change-type drought. PNAS 108:1474–1478

Caudullo G, Barredo JI (2019) A georeferenced dataset of drought and heat-induced tree mortality in Europe. One Ecosyst 4:e37753

Cook BI, Anchukaitis KJ, Touchan R et al (2016) Spatiotemporal drought variability in the Mediterranean over the last 900 years. J Geophys Res Atmos 121:260–274

Dobbertin M (2005) Tree growth as indicator of tree vitality and of tree reaction to environmental stress: a review. Eur J For Res 124:319–333

Dorman M, Perevolotsky A, Sarris D et al (2015a) The effect of rainfall and competition intensity on forest response to drought: lessons learned from a dry extreme. Oecologia 177:1025–1038

Dorman M, Svoray T, Perevolotsky A et al (2015b) What determines tree mortality in dry environments? A multi-perspective approach. Ecol Appl 25:1054–1071

Eichhorn J, Roskams P, Ferretti M et al (2010) Visual assessment of crown condition and damaging agents. In: Manual on methods and criteria for harmonized sampling, assessment, monitoring and analysis of the effects of air pollution on forests. UNECE ICP Forests Programme, Hamburg

Galván D, Camarero JJ, Ginzler C et al (2014) Spatial diversity of recent trends in Mediterranean tree growth. Environ Res Lett 9:084001

García de la Serrana R, Vilagrosa A, Alloza JA (2015) Pine mortality in southeast Spain after an extreme dry and warm year: interactions among drought stress, carbohydrates and bark beetle attack. Trees Struct Funct 29:1791–1804

Gazol A, Ribas M, Gutiérrez E et al (2017) Aleppo pine forest from across Spain show drought-induced growth decline and partial recovery. Agric For Meteorol 232:186–194

Gazol A, Camarero JJ, Vicente-Serrano SM et al (2018) Forest resilience to drought varies across biomes. Glob Chang Biol 24:2143–2158

Gea-Izquierdo G, Viguera B, Cabrera M et al (2014) Drought induced decline could portend widespread pine mortality at the xeric ecotone in managed Mediterranean pine-oak woodlands. For Ecol Manag 320:70–82

Girard F, Vennetier M, Guibal F et al (2012) *Pinus halepensis* Mill. crown development and fruiting declined with repeated drought in Mediterranean France. Eur J For Res 131:919–931

Gómez-Aparicio L, García-Valdés R, Ruíz-Benito P et al (2011) Disentangling the relative importance of climate, size and competition on tree growth in Iberian forests: implications for forest management under global change. Glob Chang Biol 17:2400–2414
Herrero A, Ruiz-Benito P, Andivia E et al (this volume) Mediterranean pine forest distribution: assessing vulnerability and resilience under climate change. In: Ne'eman G, Osem Y (eds) Pines and their mixed forest ecosystems in the Mediterranean Basin. Springer, Cham
Hoerling M, Eischeid J, Perlwitz J et al (2012) On the increased frequency of Mediterranean drought. J Clim 25:2146–2161
IDF-Inventario de Daños Forestales en España (2018) Red europea de Seguimiento de Daños en los Bosques (Red de Nivel I). Resultados del muestreo de 2018. Ministerio de Agricultura, Pesca y Alimentación, Madrid. Available at https://distritoforestal.es/images/2idf2018_tcm30-502427.pdf. Accessed 12 Dec 2019
IPCC-Intergovernmental Panel on Climate Change (2013) Climate change 2013: the physical science basis. Contribution of working group I to the fifth assessment report of the Intergovernmental Panel on Climate Change. Cambridge University Press, Cambridge
Karavani A, Boer MM, Baudena M et al (2018) Fire-induced deforestation in drought-prone Mediterranean forests: drivers and unknowns from leaves to communities. Ecol Monogr 88:141–169
Klein Goldewijk K, Beusen A, de Vos M et al (2011) The HYDE 3.1 spatially explicit database of human induced land use change over the past 12,000 years. Glob Ecol Biogeogr 20:73–86
Klein T, Cahanovitc R, Sprintsin M et al (2019) A nation-wide analysis of tree mortality under climate change: Forest loss and its causes in Israel 1948–2017. For Ecol Manag 432:840–849
Marqués L, Camarero JJ, Gazol A et al (2016) Drought impacts on tree growth of two pine species along an altitudinal gradient and their use as early-warning signals of potential shifts in tree species distributions. For Ecol Manag 381:157–167
Martín-Benito D, Kint V, Del Río M et al (2011) Growth responses of West-Mediterranean *Pinus nigra* to climate change are modulated by competition and productivity: past trends and future perspectives. For Ecol Manag 262:1030–1040
Martínez-Vilalta J, Piñol J (2002) Drought-induced mortality and hydraulic architecture in pine populations of the NE Iberian Peninsula. For Ecol Manag 161:247–256
Morales-Molino C, Tinner W, García-Antón M et al (2017) The historical demise of *Pinus nigra* forests in the Northern Iberian Plateau (south-western Europe). J Ecol 105:634–646
Moreno J et al (this volume) Fire in Mediterranean pine forests, past present and future. In: Ne'eman G, Osem Y (eds) Pines and their mixed forest ecosystems in the Mediterranean Basin. Springer, Cham
Neumann M, Mues V, Moreno A et al (2017) Climate variability drives recent tree mortality in Europe. Glob Chang Biol 23:4788–4797
Peñuelas J, Lloret F, Montoya R (2001) Severe drought effects on Mediterranean woody flora in Spain. For Sci 47:214–218
Rubio-Cuadrado Á, López R, Rodríguez-Calcerrada J et al (this volume) Stress and tree mortality in Mediterranean pine forests: anthropogenic influences. In: Ne'eman G, Osem Y (eds) Pines and their mixed forest ecosystems in the Mediterranean Basin. Springer, Cham
Sánchez-Salguero R, Navarro-Cerillo RM, Camarero JJ et al (2012) Selective drought-induced decline of pine species in southeastern Spain. Clim Chang 113:767–785
Sánchez-Salguero R, Camarero JJ, Grau JM et al (2017a) Analysing atmospheric processes and climatic drivers of tree defoliation to determine forest vulnerability to climate warming. Forests 8:13
Sánchez-Salguero R, Camarero JJ, Gutiérrez E et al (2017b) Assessing forest vulnerability to climate warming using process-based model of tree growth: bad prospects for rear-edges. Glob Chang Biol 23:2705–2719
Sánchez-Salguero R, Camarero JJ, Rozas V et al (2018) Resist, recover or both? Growth plasticity in response to drought is geographically structured and linked to intraspecific variability in *Pinus pinaster*. J Biogeogr 45:1126–1139

Sangüesa-Barreda G, Linares JC, Camarero JJ (2013) Drought and mistletoe reduce growth and water-use efficiency of Scots pine. For Ecol Manag 296:64–73
Sangüesa-Barreda G, Camarero JJ, Pironon S et al (2018) Delineating limits: confronting predicted climatic suitability to field performance in mistletoe populations. J Ecol 106:2218–2229
Sheffer E (2012) A review of the development of Mediterranean pine-oak ecosystems after land abandonment and afforestation: are they novel ecosystems? Ann For Sci 69:429–443
Stephens L, Fuller D, Boivin N et al (2019) Archaeological assessment reveals Earth's early transformation through land use. Science 365:897–902
Touchan R, Anchukaitis KJ, Meko DM et al (2008) Long term context for recent drought in northwestern Africa. Geophys Res Lett 35:L13705
Trigo RM, Añel JA, Barriopedro D et al (2013) The record winter drought of 2011–2012 in the Iberian Peninsula. Bull Am Meteorol Soc 94:S41–S45
UN-United Nations (2019) World population prospects 2019: highlights. Department of Economic and Social Affairs, Population Division. UN Publication
Vennetier M, Vila B, Liang EY et al (2007) Impact of climate change on pine forest productivity and on the shift of a bioclimatic limit in a Mediterranean area. Cah Options Méditerr 75:189–197
Vicente-Serrano SM, Beguería S, López-Moreno JI (2010) A multiscalar drought index sensitive to global warming: the standardized precipitation evapotranspiration index. J Clim 23:1696–1718
Villaescusa R, Díaz R (1998) Segundo inventario forestal nacional (1986–1996). Ministerio de Medio Ambiente, Madrid
Vitousek PM (1994) Beyond global warming: ecology and global change. Ecology 75:1861–1876
Zavala MA, Espelta JM, Retana J (2000) Constraints and trade-offs in Mediterranean plant communities: the case of holm oak-Aleppo pine forests. Bot Rev 66:119–149

Chapter 13
Mediterranean Pine Forest Distribution: Assessing Vulnerability and Resilience Under Climate Change

Asier Herrero, Paloma Ruiz-Benito, Enrique Andivia, Jaime Madrigal-González, Pedro Villar-Salvador, Sophia Ratcliffe, and Miguel A. Zavala

13.1 Mediterranean Pine Forests Under Global Change

Mediterranean pine forests comprise 25% of forested areas in the Mediterranean Basin, forming culturally and ecologically important landscapes (Barbéro et al. 1998). The Mediterranean landscape has suffered repeated, enduring changes driven by human development (Antrop 1993; Blondel and Aronson 1999). During the nineteenth and twentieth centuries, industrialization followed by globalization transformed European forests. Pine species were favored with plantations established to control soil erosion, meet timber demand and promote the economy (Mather 2001). Subsequently, pine forests thrived in economically marginal areas where farming activities were no longer profitable and land was abandoned. Land abandonment included not only agricultural land, but also cessation of silvicultural

A. Herrero (✉) · P. Villar-Salvador · M. A. Zavala
Forest Ecology and Restoration Group, Departamento de Ciencias de la Vida, Universidad de Alcalá, Alcalá de Henares, Madrid, Spain

P. Ruiz-Benito
Forest Ecology and Restoration Group, Departamento de Ciencias de la Vida, Universidad de Alcalá, Alcalá de Henares, Madrid, Spain

Remote Sensing Research Group, Departamento de Geología, Geografía y Medio Ambiente, Universidad de Alcalá, Alcalá de Henares, Madrid, Spain

E. Andivia
Biodiversity, Ecology and Evolution Department, Universidad Complutense de Madrid, Madrid, Spain

J. Madrigal-González
Climate Change Impacts and Risks in the Anthropocene (C-CIA), Institute for Environmental Sciences (ISE), University of Geneva, Geneva, Switzerland

S. Ratcliffe
National Biodiversity Network Trust, Nottingham, UK

G. Ne'eman, Y. Osem (eds.), *Pines and Their Mixed Forest Ecosystems in the Mediterranean Basin*, Managing Forest Ecosystems 38,
https://doi.org/10.1007/978-3-030-63625-8_13

intervention or grazing in different forest types (e.g. Olano et al. 2008). Consequently, Mediterranean pine forests drastically increased in abundance and distribution (Vadell et al. 2016), probably expanding their altitudinal and latitudinal limits beyond their optima (Ruiz-Benito et al. 2012). Currently, a significant proportion of pine forests are comprised by highly dense and even-aged stands with low structural heterogeneity (Gómez-Aparicio et al. 2009; Ruiz-Benito et al. 2012), which makes them highly vulnerable to biotic (pest, pathogens) and abiotic (wildfire, droughts) disturbances (Schoennagel et al. 2004; Linares et al. 2009, 2010; Ennos 2015). Thus, pine dominance in these novel ecosystems (*sensu* Hobbs et al. 2006) could change due to increased aridity and warming (Sánchez-Salguero et al. 2012; Herrero and Zamora 2014), changes in the disturbance regime (Matías and Jump 2012; Seidl et al. 2014), and colonization and competition by shade-tolerant oak species (Gómez-Aparicio et al. 2011; Ruiz-Benito et al. 2012; Sheffer 2012; Carnicer et al. 2014).

Under ongoing climate change, shifts in species distribution ranges have already been observed and more are expected, due to the displacement of favorable climatic conditions for species performance (Parmesan and Yohe 2003; Chen et al. 2011). Rear edges of distribution ranges with low persistence probability populations are likely to be found at low altitudes and latitudes in the Mediterranean Basin, where high geographic, genetic and ecological marginality is expected (Vilà-Cabrera et al. 2019). Increased drought stress at rear edges could lead to reduced recruitment and tree growth, and increased mortality (Hampe and Petit 2005; Matías and Jump 2012). At leading edges of distribution ranges, increased temperatures could increase fecundity, recruitment and growth, resulting in upward and northward range expansions (Jump et al. 2009). The high exposure to climate change in the Mediterranean Basin (IPCC 2013) is likely to make drought-sensitive mountain pine species highly vulnerable at their distribution limits (e.g. *Pinus sylvestris* L., Herrero et al. 2013; Greenwood et al. 2017). In this context, the expected increases in extreme climatic events, such as droughts or heat-waves (IPCC 2013), could reduce tree vitality and trigger massive mortality events (Allen et al. 2015), accelerating range shifts (Ruiz-Benito et al. 2017a). Disturbances from wind, pests and wildfires are predicted to increase in the future, adding to the negative impacts on species demographic rates (Seidl et al. 2014), particularly in vulnerable, even-aged reforestations (Sánchez-Salguero et al. 2013). Finally, aging of reforestations and young pine forests could imply a shift in dominance towards *Quercus* species due to successional changes (recruitment limitation; Carnicer et al. 2014) and differences in vulnerability to drought (Ferrio et al. 2003; Zweifel et al. 2009; Rigling et al. 2013). Therefore, climate-induced range shifts and disturbance impacts, coupled with broad-leaved species competition, may lead to a drastic future reduction in the range of Mediterranean pine forests.

Extreme droughts are critical climatic disturbances in Mediterranean-type ecosystems. In recent decades Mediterranean forests have been exposed to a higher frequency of extreme drought events than in previous periods (Spinoni et al. 2019), and the predicted increase in aridity is likely to alter forest functioning (e.g. Allen et al. 2015). Extreme droughts can strongly constrain tree growth and recruitment (Herrero and Zamora 2014; Marqués et al. 2016; Andivia et al. 2017, 2020; Gazol et al. 2018) as well as increase mortality (Allen et al. 2010; Anderegg et al. 2015, 2019; Choat et al. 2018; Archambeau et al. 2020). Drought can also interact with other abiotic and biotic factors. For example, drought stress can either enhance or reduce the negative effect of frost in pine species explaining their segregation in space (Fernández-Pérez et al. 2018). Biotic factors such as competition or herbivory can also modulate tree species responses to drought, particularly at their distribution limits (Gómez-Aparicio et al. 2011; Herrero et al. 2012). Altogether, species-specific demographic responses to drought may result in species and functional compositional shifts (Ruiz-Benito et al. 2017a; Gazol et al. 2018). Thus, future changes in Mediterranean pine forest structure and composition may rely on the capacity of tree species to recover from drought-related impacts (Reyer et al. 2015).

Accordingly, forest ecologists have recently adopted the concepts of resistance and resilience to evaluate forest stability to drought events (e.g. Lloret et al. 2011). Resistance is the capacity of an ecosystem, community or individual to maintain pre-disturbance structure and function during a disturbance, while resilience is defined as the capacity to restore pre-disturbance structure and function after a disturbance (Holling 1996). Assessing resistance and resilience to drought events for co-occurring tree species would provide critical information on the mechanisms underlying species distributions, while improving the forecasting of forest dynamics under climate change scenarios. This information is especially relevant in the dry edges of species distribution ranges where inter-specific differences in demographic responses to drought events can lead to range contraction and extirpation processes (Hampe and Petit 2005; Jump and Peñuelas 2005; Martínez-Vilalta and Lloret 2016).

In the present chapter, we assess potential changes in the distribution of Iberian pine forests using demographic data, accounting for key ecological factors at range edges, and discuss different modeling approximations. The chapter focuses on the Iberian Peninsula, which has high topographic, edaphic and climatic heterogeneity (Zamora and Pugnaire 2001) and hosts the most widespread pine species present in the Mediterranean Basin (Barbéro et al. 1998). Mediterranean *P. halepensis* Mill., *P. pinaster* Ait. and *P. pinea* L., and temperate *P. sylvestris*, *P. nigra* J.F. and *P. uncinata* Ramond Ex D.C. (all three restricted to mountain areas) are considered. Specifically, we analyze demographic responses (growth, recruitment and survival) to climate from forest historical archives and national inventories. As key ecological factors operating at range edges we discuss (1) the role of herbivory in modulating tree responses to drought, and (2) tree frost tolerance and its interactions with drought. We also review existing studies about the resilience capacity of Mediterranean pine forests to understand their ability to maintain their structure and function under extreme droughts. Finally, we discuss modeling approximations to assess potential changes in species distribution.

13.2 Demographic Responses of Pine Forests to Climate

13.2.1 *Forest Historical Archives: Accounting for Management Effects*

The vast majority of pine forests in the Mediterranean Basin are managed with the aim of guaranteeing the provision of functions and services for humans, while ensuring conservation of the historical heritage and biodiversity (FAO 2015). In the Iberian Peninsula, managed pine forests represent 25% of the total forested area (14.7 million ha) and provide services to rural societies such as protection against soil erosion, production of wood and resin, and habitat for hunting species (FAO 2015, 2016). In some cases, management plans were first defined at the turn of the eighteenth century and have persisted, in updated formats, to the present day. This tradition has generated a large amount of systematic data on long-term forest dynamics and productivity stored in forest historical archives (Madrigal-González et al. 2017a; Marqués et al. 2018). The high quality of this information is a direct legacy of the necessity of accurately arranging forest products and services as pervasive economic resources in rural human populations since the turn of the nineteenth century. Nowadays, this information allows us to address important research questions in the fields of ecology and forest science.

These forest historical archives contain accurate estimations of decadal wood volume increment computed using accurate proxies of wood volume stocks and management practices (Madrigal-González et al. 2017a). Wood volume stocks are calculated by sampling the diameter at breast height (DBH) and height of all trees periodically in every logging unit. Subsequently, a subset of trees that are representative of different diameter classes are harvested and measured periodically (approx. every decade) to accurately set the allometric relationship between DBH and height and tree wood volume and calculate precise allometric coefficients (see Madrigal-González et al. 2015 for further description). The allometric coefficients are used to estimate decadal wood volume stocks, from which wood volume increments are calculated, comparing stocks from successive decades and accounting for harvested volume due to thinning or wood extraction. This accurate estimation of wood volume includes not only changes in diameter but also in height, which is a more precise estimation of tree growth than radial growth from dendroecological studies. Wood volume increment data from historical archives can also be used as a proxy for forest productivity over long periods (often covering the entire twentieth century) and extensive areas (thousands of hectares; e.g. Madrigal-González et al. 2017a).

In addition, information on forest management (e.g. thinning) and extraction of forest products (e.g. resin) is also available approximately every decade (Soliño et al. 2018). Knowledge of past forest management activities is critical for the assessment of forest vulnerability and climate change impacts, since management modifies forest structure through changes in tree size, age distribution and tree density. Responses to environmental conditions are known to depend on tree size and age (Cavender-Bares and Bazzaz 2000; Herrero and Zamora 2014). Also, management can alter the microclimatic conditions within the forest stand (Camarero et al. 2011). Finally, exploitation of forest resources, such as resin extraction, could diminish tree vigor due to the removal of carbohydrates (Liu and Chen 2004). Although most of the forests in the Northern Hemisphere have been managed for centuries (Kirby and Watkins 1998), most forest studies do not account for past management activities. The forest historical archives in Spain allow us to consider the effects of past and current management activities on forest productivity and dynamics, and to properly assess climate-induced impacts and vulnerabilities.

Over the following paragraphs, we summarize the results of analyses using data series included in historical forest archives (inventories, management plans and resource exploitation) in the central-northern Iberian Peninsula. Specifically, we focus on how these data can help us to better understand the relationships between climatic variability and forest functioning, and how this can help us forecast the potential impacts of climate change on Mediterranean pine forests over the current century.

Pinus pinaster is a widespread species in the central Iberian Peninsula, where it has been managed for timber and resin for centuries. *P. pinaster* is vulnerable to dry conditions, exhibiting tree growth reductions (Sánchez-Salguero et al. 2010), decreased seed production (Ruano et al. 2015) and increased crown defoliation (Prieto-Recio et al. 2015) in response to drought. Madrigal-González and Zavala (2014) have shown that the frequency of dry years played a critical role compared to other climatic variables in limiting wood volume increment during the twentieth century in an extensive *P. pinaster* forest. The isohydric behavior of this species – closing its stomata when confronted with dry soil conditions, limiting transpiration and avoiding hydraulic failure (Ripullone et al. 2007; Matías et al. 2017) – could explain in part the negative impacts of increasing drought frequency. However, responses to drought are modulated by competition and tree age, with young trees being more sensitive to the frequency of dry years than adults, particularly in dense stands. This high vulnerability of young trees could hamper tree recruitment in dense forest stands in the long term.

Although dry conditions could limit tree growth, increasing atmospheric concentration of CO_2 could boost productivity and partially offset the negative effect of drought (Oechel et al. 1995). A combination of modeling and empirical data in the same *P. pinaster* extensive forest mentioned above, suggested that increased atmospheric CO_2 may have positively affected productivity through carbon fertilization during the second half of the twentieth century (Madrigal-González et al. 2015). However, the positive effect of increased CO_2 was only apparent when accounting for forest management activities such as thinning and resin tapping. In this context, resin tapping reduces a part of the fixed carbon during photosynthesis that could have been allocated to other plant carbon pools besides sapwood. Thus, disentangling the balance between the positive and negative influences of increased CO_2 seems necessary to account for the influence of historical forest management and structure on physiological processes. Some authors have demonstrated how soil nitrogen and water stress can hinder the positive effects of rising CO_2 on productivity (Norby et al. 2010). This could result in unprecedented impacts if warming and dry conditions continue to increase as is currently forecasted for southern Europe.

Despite differences in the ecology and functional strategies between *P. sylvestris*, *P. nigra* and *P. pinaster* (Barbéro et al. 1998), the three species show similarities in the way they respond to climatic variability at both the individual tree and the landscape levels. Evidence from a large-scale study in the central Iberian Peninsula supports the existence of coupled responses to the main sources of climatic variability and drought occurrence in five forest landscapes dominated by *P. sylvestris*, *P. nigra* and *P. pinaster* (Madrigal-González et al. 2017a).

Over the most of Western Europe, there are two main climatic patterns affecting local meteorological conditions. The North Atlantic Oscillation (NAO) is representative of the atmospheric sea level air pressure dynamics between the Azores high and the Iceland low (Vinther et al. 2003). This atmospheric pattern has been correlated to rainfall in Western Europe through control of the westerlies coming from the Atlantic Ocean. At low latitudes in Europe, strengthening of the Azores high blocks the westerlies and thus meteorological conditions are stable, and rainfall is scarce. This situation is known as the positive NAO phase (Trigo et al. 2002). In contrast, a weakening of the Azores high generates more unstable weather conditions with abundant rainfall and less continental influence (negative NAO phase).

The other pattern is known as the Atlantic Multidecadal Oscillation (AMO) and is associated with warming-cooling dynamics of sea surface temperatures in the northern Atlantic Ocean (Knight et al. 2006). The multidecadal pattern has been described to comprise two phases. The positive AMO phase is associated with overwarming of the northern Atlantic over a 30–40-year period, whereas the negative AMO phase represents overcooling periods over 30–40 years. These dynamics have been demonstrated to control temperature over the continents on both sides of the Atlantic Ocean (Sutton and Hodson 2005). Interestingly, these two patterns interact to modulate temperature and precipitation and with this, productivity of Iberian pine forests (Madrigal-González et al. 2017a). The combination of positive phases of NAO and AMO has been associated with a decrease in forest productivity due to increased aridity. Understanding the underlying causes of the responses of tree growth and productivity to climatic variability is critical for predicting climate-induced impacts on forests and designing adequate adaptation actions. The oscillatory nature of the NAO and AMO patterns can allow the detection of high drought risk, allowing the application of anticipatory adaptation measures such as thinning.

13.2.2 National Forest Inventories: Demographic Data Along Latitudinal Gradients

National Forest Inventories (NFI) are a valuable source of data for investigating demographic tree responses at large spatial scales (Ruiz-Benito and García-Valdés 2016; Ruiz-Benito et al. 2020). Although the sampling designs depend on the region or country, most inventories distribute plots systematically across the territory and have been used extensively to obtain statistics on past patterns of forest expansion, growth or state (Chirici et al. 2011). Harmonization of multiple NFIs has enabled European-scale studies on the effects of species composition on tree-tree interactions (Kunstler et al. 2016), demographic outcomes, and forest functions (e.g. Ruiz-Benito et al. 2017b; van der Plas et al. 2018).

To analyze tree growth (cm^2 ha^{-1} year^{-1}), tree mortality (cm^2 ha^{-1} year^{-1}) and sapling abundance (No. saplings ha^{-1}) in European pine forests we used NFI data obtained by Ruiz-Benito et al. (2017b) from Spain, Belgium, Germany, Sweden and Finland (Fig. 13.1, Ratcliffe et al. 2016). These variables showed clear geographic patterns across the European latitudinal gradient (Fig. 13.2). Tree growth was calculated as the basal area increment of live adult trees (i.e. diameter at breast height > 10 cm) in two successive surveys plus the basal area of new adult trees in the most recent survey. Tree mortality was quantified as the basal area of adult trees lost between consecutive surveys. Sapling abundance was the number of trees between 0.3 and 1.3 m height in the second survey of the two consecutive surveys.

Mediterranean pine forests (i.e. dominated by *P. halepensis*, *P. pinea* or *P. pinaster*) and temperate pine forests (i.e. dominated by *P. nigra*, *P. sylvestris* or *P. uncinata*) were found to have different demographic responses to climatic conditions.

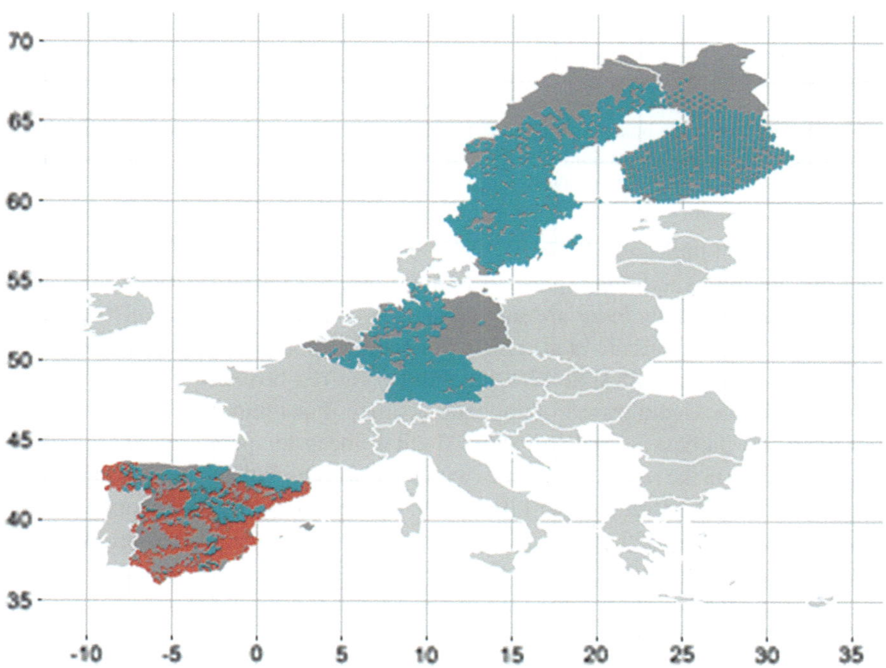

Fig. 13.1 Forest Inventory plots used across Europe, characterized as Mediterranean pine forests (red plots) or temperate pine forests (blue plots). The countries included in our study are shown in the European map in dark grey

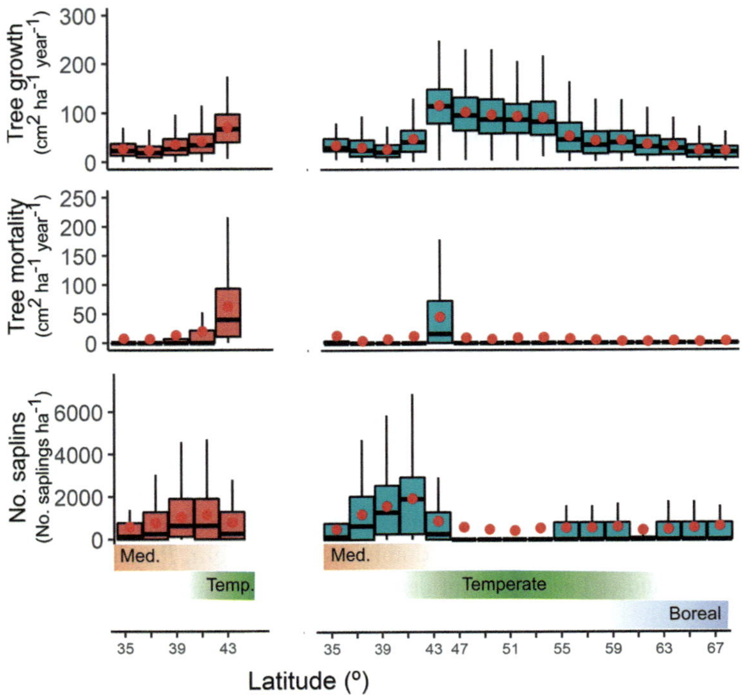

Fig. 13.2 Boxplot and mean values (red dots) of tree growth (cm^2 ha^{-1} year^{-1}), tree mortality (cm^2 ha^{-1} year^{-1}) and sapling abundance (No. saplings ha^{-1}) for Mediterranean (red boxes on the left) and temperate (blue boxes on the right) pine forests along the latitudinal gradient in Europe. Mediterranean pine forests are dominated by *Pinus halepensis*, *P. pinea* or *P. pinaster* and temperate pine forests by *P. nigra*, *P. sylvestris* or *P. uncinata*

P. halepensis, *P. pinea* and *P. pinaster* are pine species commonly distributed in the Mediterranean Basin, *P. nigra* presents a circum-Mediterranean distribution whereas *P. sylvestris* and *P. uncinata* are mountain species more commonly distributed in temperate regions (Barbéro et al. 1998). For Mediterranean pine forests tree growth and mortality were highest at 42–44° latitude (see red boxes in Fig. 13.2), corresponding to the Mediterranean-temperate ecotone. For temperate pine forests, we observed the highest mortality in the Mediterranean-temperate ecotone, but the highest growth in the temperate biome (with the lowest growth in both Mediterranean and boreal latitudinal areas; see blue boxes in Fig. 13.2). Sapling abundance was relatively high at 39–42° latitude with a strong decrease in abundance below 38° for both Mediterranean and temperate pine forests (Fig. 13.2).

The tree high mortality observed in the Mediterranean-temperate ecotone for both Mediterranean and temperate pine species (Fig. 13.2) agrees with previous results reported across the Iberian Peninsula and Europe (e.g. Benito-Garzón et al. 2013; Ruiz-Benito et al. 2013). Using the Spanish National Inventory, Ruiz-Benito et al. (2013) showed a strong interaction between climate and competition on tree mortality, indicating the vulnerability of widespread dense pine reforestations. In addition, smaller and younger trees are generally experiencing increased mortality (Ruiz-Benito et al. 2013). Similarly, it has been found that mortality patterns in *P. sylvestris* are related to the combined effect of drought and competition, with a mortality peaks observed in the northern Spain and the southern French, part of the Mediterranean biome (Galiano et al. 2010; Vilà-Cabrera et al. 2011; Archambeau et al. 2020).

Tree growth was greater in temperate areas and in the temperate-Mediterranean ecotone than in the Mediterranean and boreal areas, consistent with previous results of tree growth studies using NFIs across Europe (Ruiz-Benito et al. 2014, 2017b). Maximum growth in temperate areas could be due, at least in part, to the limited negative effects of dry conditions and extreme temperatures (Ratcliffe et al. 2016). We note that functional diversity is an important driver of tree growth in Mediterranean pine forests (Ratcliffe et al. 2016; Ruiz-Benito et al. 2017b). Globally, it has been observed that increased diversity generally leads to increased tree growth (Liang et al. 2016) and the magnitude of the diversity effect is expected to increase in areas with low productivity, such as Mediterranean regions (Jucker et al. 2016; Ratcliffe et al. 2016). Furthermore, the importance of complementarity (an increase in ecosystem function through niche partitioning; e.g. Morin et al. 2011) and selection (an increase in ecosystem function due to the dominance of particular species; e.g. Roscher et al. 2012) effects can vary between different pine forests due to identity (i.e., species-specific) effects, forest structure and individual tree characteristics. In this context, selection effects were particularly important in Mediterranean pine forests, but had little effect on temperate pines (Ruiz-Benito et al. 2014). Tree size inequality can be key to determining complementarity mechanisms (Yuan et al. 2018; Aponte et al. 2020); Madrigal-González et al. (2016) recorded a lack of complementarity effects in Mediterranean pine forests in small-size trees, which could be associated with trade-offs related to shade and drought stress (Holmgren et al. 1997).

The observed decrease in pine sapling abundance below 38° suggests that the low water availability of this region combined with high temperatures could affect regeneration patterns along the latitudinal gradient. Consistent with this, lower pine recruitment has been recorded at low precipitation levels in Mediterranean and temperate pines (Ruiz-Benito et al. 2012). Furthermore, pine recruitment failure has been commonly reported in the Iberian Peninsula together with an expansion of oak species, which could be due to the process of secondary succession and biotic interactions (e.g. Carnicer et al. 2014; Ruiz-Benito et al. 2017a, b). Notably, Ruiz-Benito et al. (2012) showed that planted pine forests had lower regeneration capacity than natural pine forests in the Iberian Peninsula, suggesting a high vulnerability of reforestations to increased aridity. Increased tree functional diversity has been also linked to increased sapling abundance across Europe via increased complementarity effects and indirectly through modification of responses to climate (Ruiz-Benito et al. 2017b). In fact, several studies have suggested that increased diversity could lead to increased recruitment, particularly under more extreme climatic conditions (e.g. Liang et al. 2007). This has led to suggestions that increasing the functional diversity of forests would help to adapt them to new climatic conditions (Hisano et al. 2017).

13.3 Key Ecological Factors at Range Edges

13.3.1 Herbivory as a Biotic Factor Modulating Drought Responses

Herbivory can modulate climate-induced distributional shifts, ameliorating or exacerbating plant responses to climate. For example, a species at its lowest altitudinal limit may have low establishment due to drought-induced growth reductions, which can be exacerbated by herbivory. The role of herbivores in changing tree distribution ranges generally opposes that of climate warming: whereas warming facilitates the spread to higher latitudes and altitudes, high densities of herbivores can constrain upward and northward migrations (Post and Pedersen 2008; Olofsson et al. 2009; Speed et al. 2010; Kaarlejärvi et al. 2015). A lack of predators, warmer winters and difficulties in performing adequate management actions are increasing populations of wild ungulate herbivores at a global scale (Cote et al. 2004), thus probably also increasing herbivory effects in plant communities (Kaarlejärvi et al. 2015; Herrero et al. 2016).

Although the effects of herbivory exacerbating or constraining climate-induced distributional changes of tree species have been recorded by several studies at high latitudes (Cairns and Moen 2004; Dalen and Hofgaard 2005; Tømmervik et al. 2009; Speed et al. 2010), few studies have been conducted in Mediterranean areas. Ungulate browsing on pine seedlings and saplings in Mediterranean areas is more severe in summer when other food resources are less abundant, leaving less time for tree saplings to recover from damages in comparison with winter or spring browsing (Hester et al. 2004). Repeated browsing reduces sapling height growth, increasing the time needed for saplings to achieve maturity and initiate reproduction (Zamora et al. 2001). This is especially relevant at the edges of species distributions, such as treelines, where climate-driven range expansions can occur and populations comprise mainly young individuals (Matías and Jump 2012).

Herrero et al. (2016) performed a long-term assessment of precipitation and red deer (*Cervus elaphus* L.) herbivory effects on *P. nigra* and *P. sylvestris* sapling height growth in the Southern Iberian Peninsula, revealing strong effects of precipitation on sapling growth at the treeline (Herrero et al. 2016). This could allow upward tree expansion, although recorded increasing ungulate populations and associated browsing damages might limit positive climatic responses of pine saplings in the near future. However, ungulate herbivores preferred *P. sylvestris* over *P. nigra*, leading to growth reductions due to browsing damage for a large number of *P. sylvestris* saplings. Thus, herbivory exacerbates the higher drought sensitivity recorded for *P. sylvestris* compared to *P. nigra* in seasonal dry Mediterranean mountains (Martínez-Vilalta and Piñol 2002; Herrero et al. 2013; Salazar-Tortosa et al. 2018a). Under the scenario of increasing aridity and ungulate populations, upslope migration of *P. sylvestris* could be impeded, while the less browsed and more drought-tolerant *P. nigra* may have an opportunity for altitudinal migration. In the same study area, Matías and Jump (2015) detected strong herbivore damage to saplings at low elevation, where reproduction and growth were lowest, in comparison with medium and high elevations. This browsing damage could enhance the declining performance trend induced by climate, accelerating altitudinal displacement or range contraction (Matías and Jump 2015). Therefore, ungulate herbivory can act as an ecological filter of the pine forest at both upper and lower limits, reinforcing the climatic sensitivity of coexisting *P. sylvestris* and *P. nigra*. However, herbivory acts as a spatially aggregating force through individual-level damage (Herrero et al. 2016). For example, mechanical protection provided by spiny shrubs could reduce herbivory damage to *P. sylvestris* saplings in some areas (Gómez et al. 2001). In this context, the spatially heterogeneous effect exerted by herbivory could result in diverse vegetation structures in ecotones, adding a new dimension to the predictions on climate-driven vegetation shifts.

13.3.2 Frost Tolerance as a Driver of Upper and Northern Limits

Frost is a major driver of plant distribution (Woodward and Williams 1987) as it reduces growth and causes water stress and eventually death (Mayr et al. 2006; McCulloh et al. 2011). Understanding how frost limits pine species distributions can help to forecast potential upward shifts in altitude due to warming. In this sense, the distribution of Mediterranean pine species appears segregated along an altitudinal gradient in a predictable way. Species such as *P. halepensis*, *P. pinaster* and *P. pinea* are located at low altitude with mild to cool winters and hot and dry summers. Temperate pines, such as *P. nigra*, *P. sylvestris* and *P. uncinata*, are located in mountains in cold winter sites (Barbéro et al. 1998). Fernández-Pérez et al. (2018) showed that this altitudinal distribution is in part explained by differences in cold tolerance between these pine species (Fig. 13.3). The temperature at which survival is 50% (LT_{50}) showed a two-fold variation between *P. halepensis* and *P. nigra*, the species with the lowest and highest LT_{50}, respectively. The temperate pines did not show any mortality until −15 °C, while at this temperature survival of *P. halepensis* and *P. pinea* was as low as 20%. Survival was similar for *P. sylvestris* and *P. uncinata*, and their LT_{50} values were slightly higher than those estimated for *P. nigra*. *Pinus pinaster* had lower frost survival than the three temperate pine species, but higher frost survival than *P. halepensis* and *P. pinea*, which showed similar survival. The low frost survival of the low-land Mediterranean pine species is mostly attributed to severe damage to cell membranes in the needles after strong frost in plants with high predawn water potential (Fig. 13.4; Climent et al. 2009; Bachofen et al. 2016; Fernández-Pérez et al. 2018) and cell damage in the roots (Toca et al. 2018).

Soluble sugars increase the stability of cell membranes to frost damage (Uemura and Steponkus 2003) in frost-hardy plants. However, there is no evidence that the differences in frost tolerance among Mediterranean pines can be attributed to differences in the tissue concentration of soluble sugars (Toca et al. 2018;

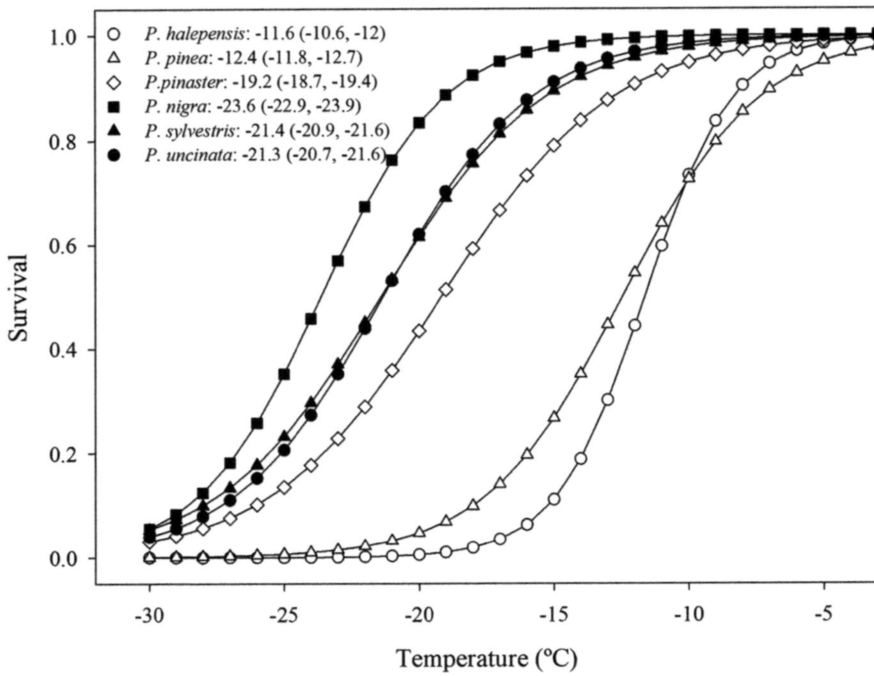

Fig. 13.3 Survival of low-land (open symbols) and temperate (full symbols) pine species distributed in southern Europe in relation to freezing temperatures. Data in the legend represent the mean lethal temperature (°C) at which survival is 50% and the 95% confidence interval (in brackets). Methodological details can be found in Fernández-Pérez et al. (2018). (Figure reproduced with permission of Oxford University Press)

Fernández-Pérez et al. 2018), indicating that other compounds and mechanisms are more important in determining differences in frost tolerance across pine species. Similarly, frost survival differences among seedlings of different pine species have been shown to be unrelated to freezing-induced xylem embolism, probably because tracheids are narrow and more resistant to frost-induced embolisms than angiosperm vessels (Pittermann and Sperry 2003) and the differences in tracheid diameter among seedlings of pine species are small (Fernández-Pérez et al. 2018).

The environmental conditions to which pine species are exposed can change their frost tolerance, with implications for forecasting species distribution changes due to increasing aridity projections for the Mediterranean Basin. Plant species are occasionally exposed to drought during dry winters in Mediterranean locations (Peguero-Pina et al. 2011). Moderate drought stress (seedling predawn water potential up to −1.5 MPa) can affect foliage frost damage (Fig. 13.4). However, this effect does not appear to be related to species ecology (Fernández-Pérez et al. 2018). Thus, drought stress reduces frost damage, particularly in *P. pinea*, and to a lesser extent in *P. sylvestris*. In contrast, drought stress slightly increases frost damage in *P. uncinata* and has no effect on *P. halepensis*, *P. nigra* and *P. pinaster*. The increase

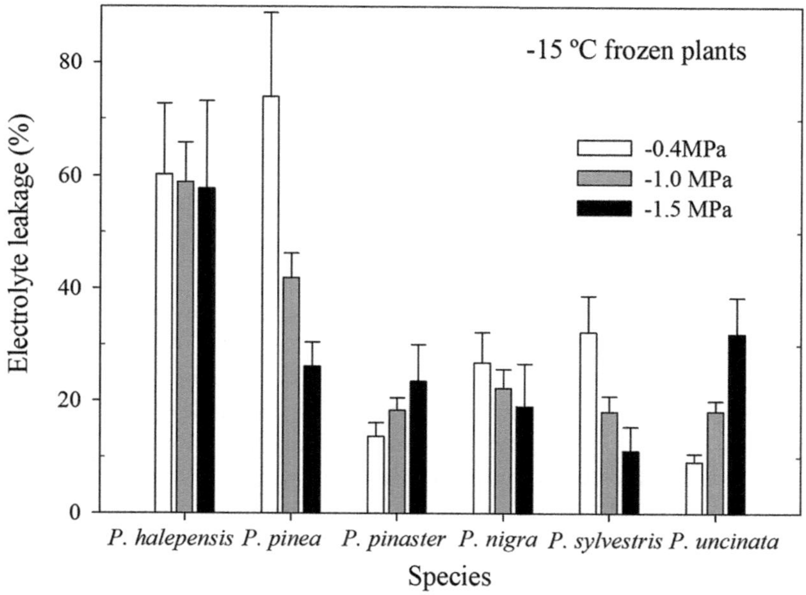

Fig. 13.4 Effect of predawn water potential on foliage electrolyte leakage, a proxy for cell membrane damage, after exposing seedlings of six pine species, native to Southern Europe, to a single −15 °C frost. Data are least square means ± SE. Methodological details can be found in Fernández-Pérez et al. (2018). (Figure reproduced with permission of Oxford University Press)

in frost tolerance of *P. pinea* seedlings when subjected to moderate drought stress may explain why this species can develop in sandy soils in the north of the Iberian Peninsula, where the winters are colder than those experienced by *P. pinea* throughout most of its range. At the same time, the differential response of frost tolerance to winter drought stress in *P. pinea* and *P. halepensis* might affect their capacity to migrate across altitudes under the forecasted climate change scenarios.

Similarly to drought stress, soil nitrogen (N) also affects the frost tolerance of pine species, but unlike drought stress, the N effect depends on species ecology and dormancy physiology and the time of year when N is supplied (Toca et al. 2018). High soil N availability during the growing season increases frost tolerance during the cold season in pine species adapted to cold winter sites such as *P. nigra*. This is likely due to investment of surplus of N into compounds involved in frost tolerance. At the same time frost acclimation in these species begins early in the fall triggered by photoperiod reduction (Charrier et al. 2015) and N availability does not hinder frost acclimation. Conversely, high N decreases frost tolerance in pines species inhabiting mild winter Mediterranean locations such as *P. halepensis* and *P. pinea*. In these pine species, fall and winter temperature controls frost acclimation (Charrier et al. 2015) but high N availability at the end of the summer can arrest shoot growth cessation in the fall reducing thus frost acclimation (Puértolas et al. 2005). Independently of species ecology and dormancy physiology, high N supply in the middle of the autumn, once cold acclimation has started, increases cold tolerance in most cases. These results indicate that modeling pine species distributions under a climate change scenario should consider the response of pine species to winter drought stress and N deposition.

13.4 Resistance and Resilience to Drought: Assessing the Capacity to Maintain Function and Structure

In the Mediterranean Basin, native pine species are spatially segregated along environmental and altitudinal gradients (Ruiz-Benito et al. 2012). As we showed in the previous section, the ability of these pine species to colonize high-altitude locations is related to differences in seedling frost tolerance (Fernández-Pérez et al. 2018). However, at low- and mid-altitudes the main limiting factor for plant life is water, and water stress increase with summer temperatures. Therefore, early survival and persistence of pine species is primarily driven by the capacity of seedlings to grow deep roots and maintain gas exchange under dry conditions (Salazar-Tortosa et al. 2018a, b; Andivia et al. 2019). Consequently, pine species inhabiting low- and mid-altitude locations (e.g. *P. halepensis*, *P. pinea* and *P. pinaster*) are more resistant to drought than temperate pines such as *P. nigra* and *P. sylvestris* (Martín-Benito et al. 2013; Herrero and Zamora 2014; Marqués et al. 2016; Matías et al. 2017; Salazar-Tortosa et al. 2018a, b). However, the ability of a species to persist in dry locations can also depend on its capacity to recover from extreme drought events, as future average conditions are predicted to be close to current extreme events (Battisti and Naylor 2009). In this regard, the evaluation of species' resilience to extreme droughts is key to forecast altitudinal shifts of tree species (Jump and Peñuelas 2005; Peñuelas et al. 2007; Martínez-Vilalta and Lloret 2016). The available evidence in natural populations suggests that species differences in growth resistance and resilience to extreme droughts do not match the distribution of Iberian pine species along water availability gradients. Herrero and Zamora (2014) found similar resistance and resilience to drought in *P. nigra* and *P. sylvestris* adults in Sierra de Baza Mountains in southern Spain, while Marqués et al. (2016) showed a lower resistance to drought in *P. nigra* than in *P. sylvestris* in Gúdar Mountains in eastern Spain. Granda et al. (2018) showed no significant differences in resilience between *P. halepensis*, *P. nigra* and *P. sylvestris* under a Mediterranean continental climate in north-east Spain. Under similar climatic conditions, Andivia et al. (2020) showed lower resistance to drought in *P. nigra* than in *P. sylvestris* and *P. pinaster* in the northern Iberian plateau. Overall, these results suggest that inter-specific differences in resistance and resilience to drought events between Mediterranean pine species are more related to climatic conditions, forest structure and tree age (Linares et al. 2010; Gazol et al. 2018; Granda et al. 2018) than to species' idiosyncratic vulnerability to drought.

The vulnerability of tree species to drought events and their response to climate change scenarios have usually been evaluated in canopy dominant and healthy trees (e.g. Marqués et al. 2016; Sánchez-Salguero et al. 2017; Granda et al. 2018).

However, saplings and small trees are generally more vulnerable to drought than adults because they have shallower root systems (Padilla and Pugnaire 2007; Lloret et al. 2009). In this regard, the performance of juveniles has been overlooked, despite the fact that early survival is a critical determinant of forest regeneration in Mediterranean areas (Andivia et al. 2017). Understanding species-specific vulnerability to drought events in juveniles is crucial to better predict the future structure and composition of forest ecosystems (Madrigal-González et al. 2016). However, to the best of our knowledge only Herrero and Zamora (2014) and Andivia et al. (2020) have evaluated growth resistance and resilience to drought in juveniles of co-occurring Mediterranean pine species. Indeed, inter-specific differences in pine growth resilience and resistance in these studies were revealed in juveniles rather than in adult trees. Research on pine juveniles provides evidence for the relationship between resilience to drought and the drought stress to which species are exposed in their natural range. Herrero and Zamora (2014) reported higher resilience in *P. nigra* than in *P. sylvestris*, whereas Andivia et al. (2020) reported a negative correlation between resilience to drought and the 2.5% annual rainfall percentile in the natural range of the species in the Iberian Peninsula (Fig. 13.5).

Despite recent advances in evaluating forest resilience to drought events, compositional and dominance shifts in forest ecosystems rely not only on growth responses

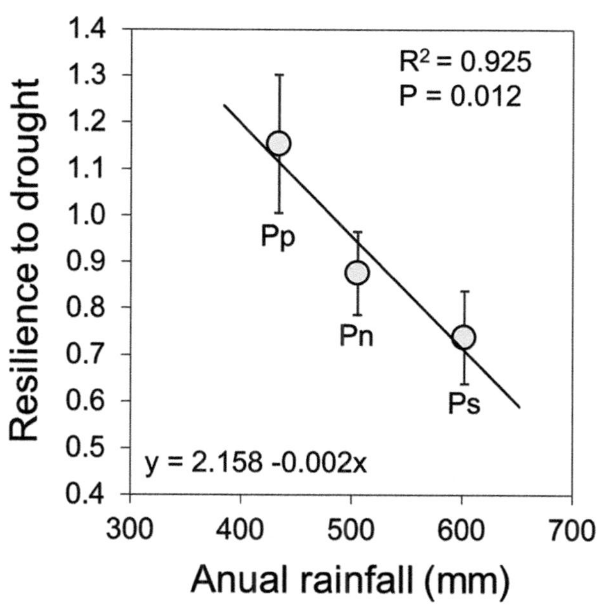

Fig. 13.5 Relationship between pine species' resilience to 2012 drought (30 trees per species sampled in three forest stands in the Iberian Northern Plateau) and the 2.5% annual rainfall percentile in the distribution range of that species in the Iberian Peninsula. Resilience index is calculated as the ratio between tree growth after and before the drought event. Equations of the adjusted linear models are shown together with R^2 and P values. Each point represents mean values ±1 SE. Pp = *P. pinaster*, Pn = *P. nigra* and Ps = *P. sylvestris*

but more importantly on recruitment and mortality (Peñuelas et al. 2007; Allen et al. 2010; Ruiz-Benito et al. 2017a). Evaluating different demographic responses to drought would give a more complete overview of forest resilience by accounting for compensation processes and trade-offs between demographic rates (Madrigal-González et al. 2017b). For example, recruitment limitations can be compensated by high survival rates and longer lifespans (Calama et al. 2017), whereas high mortality rates after disturbances may result in increasing growth and recruitment of surviving individuals due to release from competition (Lloret et al. 2012). However, resilience assessment based on multiple demographic processes has remained challenging, mainly because long-term observations to determine baseline and post-disturbance conditions are required. Using historical forest inventories gathered every 10 years for the period 1912–2002 in an extensive *P. pinaster* forest of approx. 7500 ha in the central Iberian Peninsula, Madrigal-González et al. (2017b) evaluated demographic resilience to an extremely dry decade (1942–1952) based on growth, recruitment and survival. Overall, resilience was high, suggesting high tolerance of dry pine forests to prolonged drought periods. However, resilience in growth was less variable than in recruitment and survival, with some forest stands showing low recruitment and survival resilience. Interestingly, forest stands with the lowest resilience values for recruitment or survival showed high resilience values for the other demographic parameters, suggesting compensation effects and thus optimal recovery of forest functioning after drought. Nevertheless, further investigation is needed, including data from more recent decades in which climate warming has exacerbated dry conditions and the frequency of drought events has increased.

Two studies using data from recent decades in Mediterranean pine forests have reported a reduction in resilience between consecutive drought events (Serra-Maluquer et al. 2018; Andivia et al. 2020). These results suggest that Mediterranean pine forests will become more vulnerable in the context of increasing intensity and frequency of drought events, as expected for this region (Spinoni et al. 2019). Recurrent extreme drought events may trigger legacy effects on trees (Anderegg et al. 2015; Peltier et al. 2016) due to depleted carbohydrate reserves, a loss of leaf area and hydraulic dysfunction associated to xylem cavitation (Galiano et al. 2011; Anderegg and Anderegg 2013). These processes can ultimately jeopardize tree response to subsequent droughts, compromising forest recovery in the long term and posing a challenge for the management of these forest ecosystems. Mediterranean dry pine forests are generally managed following the shelterwood system approach (Calama et al. 2017). Shelterwood thinning enhances seedling establishment by decreasing abiotic stress (i.e. facilitation) under the shelter of adult trees (Bertness and Callaway 1994; Gómez-Aparicio et al. 2008). Several studies have reported facilitative interactions between adult and juvenile pines (González-Alday et al. 2009; Ruano et al. 2009; Rodríguez-García et al. 2011). However, Andivia et al. (2018) showed that the interaction between recurrent drought events and ontogeny shifted the net balance of adult-sapling interactions from facilitation to competition, reducing growth recovery after drought in juveniles growing under the canopy of adult trees (Fig. 13.6). Overall, Mediterranean pine forests show high resilience to

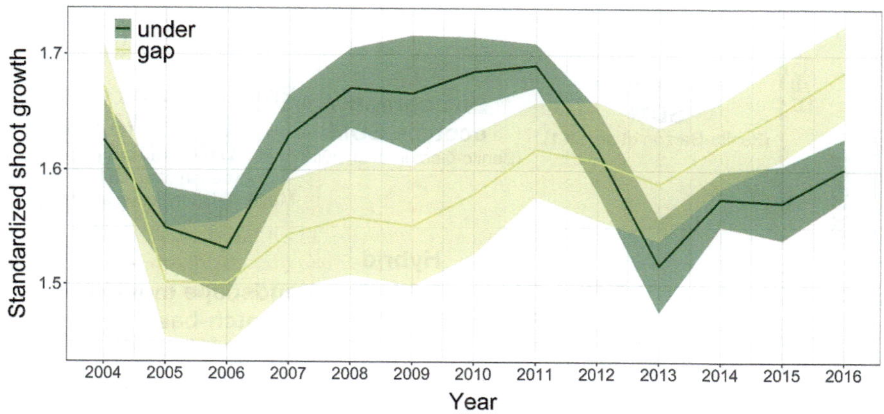

Fig. 13.6 Standardized annual shoot elongation (± 95% confident interval) of *Pinus pinaster* saplings growing under the canopy of adult pines and in open gaps patches ($N = 20$). The study was performed in the Iberian Northern Plateau. 2005 and 2012 were the most intense drought events during the study period with an annual precipitation of 271 and 284 mm, respectively. Standardized growth represents scaled residuals (to 1–2 values) of a linear mixed-effects model which included sapling age (as a second-degree polynomial). Higher values of standardized growth mean higher growth rates after accounting for age effects

extreme droughts, yet recurrent extreme drought events might compromise forest recovery in the long-term, especially for recruitment and survival. Thus, the increased frequency and magnitude of drought events represents novel hazards for the persistence of Mediterranean pine forests, which might imply a paradigm shift regarding the way we manage these valuable forest ecosystems.

13.5 Modelling Pine Ecosystem Dynamics

The importance of understanding forest responses to climate change has stimulated the development of ecological models that examine possible key mechanisms driving pine forest dynamics under changing conditions. These models vary depending on the mechanisms and hypotheses built into the model – and hence their complexity – as well as the spatial scale of application, from stand to region (Fig. 13.7). At the regional level, initial niche models predicted future pine distributions through simple correlational models. These models describe the fate of biodiversity by projecting current space-climate-species distribution associations in time (Benito-Garzón et al. 2011). Thus, they predicted a drastic and rapid reduction in the potential suitable distribution of pine species across Iberian Peninsula, increasing the awareness of the vulnerability of these ecosystems and stimulating the development of models that test specific ecological mechanisms (Keenan et al. 2011; Benito-Garzón et al. 2013; García-Valdés et al. 2013).

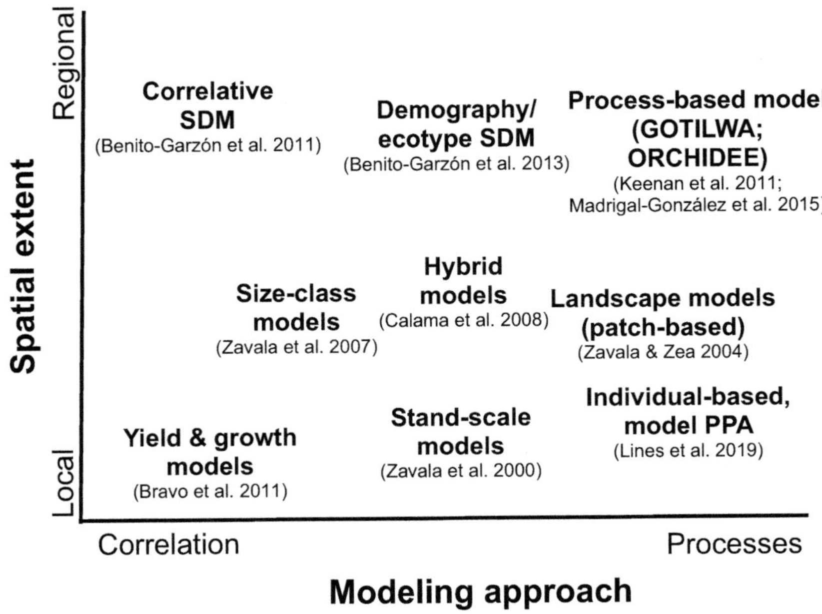

Fig. 13.7 Models of pine forest vulnerability and resilience to climatic change in Mediterranean forests organized across ranges of spatial extent and modeling approaches. Modified from Ruiz-Benito et al. (2020). A model's position represents its relative ranking among all model types

Process-based models, which describe physiological feedbacks through stomatal control, show that pine forests could benefit from CO_2 fertilization effects through an increase in water-use efficiency in a warmer, CO_2-enriched world (Keenan et al. 2011). Yet, CO_2 fertilization effects in pine ecosystems appear to have a limited payoff in relation to concomitant increases in aridity (Madrigal-González et al. 2015). Similarly, niche models calibrated with growth and mortality data, rather than occurrence information, predict pine tree growth enhancement, but also an increase in mortality rates under climate change (Benito-Garzón et al. 2013), suggesting likely weak compensatory effects across the species range. Overall, for the most common pine species in the Iberian Peninsula, both stand densification and aridity and their interactions increase mortality rates (Ruiz-Benito et al. 2013). Mortality responses, however, seem also tightly coupled to local adaptation, and niche models parameterized with mortality data from provenance trials, i.e., including ecotype variation, indicate strong intraspecific genetic potential in the Iberian Peninsula to counteract climatic changes (Benito-Garzón et al. 2011). This suggests that through either gene flow or adaptation measures, such as assisted migration, there is some potential for species resilience if adequate adaptation measures are undertaken.

The spatial dynamics of ecosystem structure are also crucial for the preservation of pine ecosystems due to both environmental heterogeneity and colonizing capability (Zavala and Zea 2004). Current ongoing expansion of pine species′ ranges into the Western Mediterranean is predicted largely due to their pioneer strategy and high colonization capability (García-Valdés et al. 2013). However, climate change may decrease the rate of expansion of temperate pine species and boost the rate of Mediterranean pine species such as *P. halepensis* in newly colonized areas or even substituting other pine species such as *P. pinaster* (García-Valdés et al. 2013). A certain degree of habitat fragmentation is beneficial for pines as their seeds are dispersed by wind, although in highly fragmented habitats species with directed dispersal, such as oaks, can be more successful (Montoya et al. 2008). In general, pines tend to be excluded during most secondary succession scenarios that do not include environmental heterogeneity (Lines et al. 2019; Baudena et al. 2019; Waitz and Sheffer this volume, Chapter 16) and both environmental heterogeneity and disturbances are essential to account for the current, widespread, pine-oak co-dominance (Zavala and Zea 2004). In effect, pines survive during disturbance-free periods in "safe sites" (i.e. their fundamental niche which includes conditions too harsh to be outcompeted by hardwoods, e.g., poor and sandy soils) and can then expand due to alternate favorable climatic cycles (i.e. linked to NAO and AMO oscillations; Madrigal-González et al. 2015) as well as disturbances (Carrión et al. 2001).

Altogether, these models provide useful descriptions for biological filters, but the fate of pine forests is also inextricably linked to the development of human societies and resource management (Gauquelin et al. 2018). Policy decisions will have a great impact on the fate of biodiversity, and pines are no exception. Pines have dominated extensive areas of the Mediterranean Basin under fluctuating climatic conditions and a variety of human societies, and will probably persist in the future despite strong impacts related to climate change.

13.6 Conclusions: Implications for Predicting Potential Changes in Pine Forest Distributions

Evidence from forest demographic responses to climate across prolonged temporal periods (historical archives) or large spatial extents (NFIs), confirms that both Mediterranean and temperate pine species are vulnerable to climate change. Data from NFIs point to the key role of competition in exacerbating the negative effects of drought on demographic rates, which could have critical implications for tree recruitment. However, functional diversity might mitigate, at least in part, climate impacts on growth and recruitment. This is in alignment with managers' efforts to diversify the forests and reduce the tree density of planted pine forests to increase forest resilience to climatic disturbances (Vilà-Cabrera et al. 2018). Furthermore, large-scale spatial patterns across Europe suggest that intense changes are occurring in the Mediterranean-temperate ecotone due to increased pine mortality and changes in tree growth and recruitment patterns. Long-term growth data suggest that incorporating management effects can improve the assessment of climatic, environmental and structural factors influencing tree growth, also pointing to increased vulnerability to drought for small tree sizes. However, we suggest that if warming and more extreme climate episodes continue, the demographic responses of Mediterranean pine forests are likely to be altered.

Key ecological factors, such as herbivory and frost, can influence climate-induced changes in the distribution of pine species affecting pine seedling and sapling performance. Herbivory can modulate plant responses to drought, reinforcing the drought sensitivity of temperate pines such as *P. sylvestris* at both upper and lower range edges. Drought stress could increase or decrease frost damage depending of the pine species considered, favoring or hampering upward and northward migrations. Resilience to drought seems more related to climatic conditions, forest structure and tree age than to species' idiosyncratic vulnerability to drought. In fact, pine saplings seem to be more vulnerable to drought than adult trees, and species-specific differences in growth resilience are more evident in saplings than in adults. Resilience assessments based on multiple demographic rates could be very informative for evaluating pine population persistence in the long term. Finally, the development of ecological models allows us to detect and quantify important processes that could affect pine species distribution, such as CO_2 fertilization, intraspecific genetic variation or interactions between colonization capability and climate.

Acknowledgements We acknowledge support from grant DARE (RTI2018-096884-B-C32; MICIU, Spain) and the project AGL2011-24296 ECOLPIN. The NFI data harmonization was performed thanks to funding from the European Union Seventh Framework Programme (FP7/2007-2013) under grant agreement No. 265171 (project FunDivEUROPE). We thank the MAGRAMA for access to the Spanish NFI, Gerald Kändler and the Johann Heinrich von Thünen-Institut for access to the German NFI, Aleksi Lehtonen and Natural Resources Institute Finland (Luke) for access to the Finish NFI, Jonas Dahlgren and the Swedish University of Agricultural Sciences for making the Swedish NFI data available, and Hugues Lecomte for access to the Walloon NFI. AH was supported by the University of Alcalá Own Research Programme's 2019/20 Postdoctoral Grant and EA by a postdoctoral grant from Complutense University of Madrid. PRB was supported by the Community of Madrid Region under the framework of the multi-year Agreement with the University of Alcalá (Stimulus to Excellence for Permanent University Professors, EPU-INV/2020/010) and the University of Alcalá "Ayudas para la realización de Proyectos para potenciar la Creación y Consolidación de Grupos de Investigación".

References

Allen CD, Macalady AK, Chenchouni H et al (2010) A global overview of drought and heat-induced tree mortality reveals emerging climate change risks for forests. For Ecol Manag 259:660–684

Allen CD, Breshears DD, McDowell NG (2015) On underestimation of global vulnerability to tree mortality and forest die-off from hotter drought in the Anthropocene. Ecosphere 6:1–55

Anderegg WR, Anderegg LD (2013) Hydraulic and carbohydrate changes in experimental drought-induced mortality of saplings in two conifer species. Tree Physiol 33:252–260

Anderegg WR, Hicke JA, Fisher RA et al (2015) Tree mortality from drought, insects, and their interactions in a changing climate. New Phytol 208:674–683

Anderegg WR, Anderegg LD, Kerr KL et al (2019) Widespread drought-induced tree mortality at dry range edges indicates that climate stress exceeds species' compensating mechanisms. Glob Change Biol 25:3793–3802

Andivia E, Villar-Salvador P, Tovar L et al (2017) Multiscale assessment of woody species recruitment in Mediterranean shrublands: facilitation and beyond. J Veg Sci 28:639–648

Andivia E, Madrigal-González J, Villar-Salvador P et al (2018) Do adult trees increase conspecific juvenile resilience to recurrent droughts? Implications for forest regeneration. Ecosphere 9:e02282

Andivia E, Zuccarini P, Grau B et al (2019) Rooting big and deep rapidly: the ecological roots of pine species distribution in southern Europe. Trees 33:293–303

Andivia E, Ruiz-Benito P, Díaz-Martínez P et al (2020) Inter-specific tolerance to recurrent droughts of pine species revealed in saplings rather than adult trees. For Ecol Manag 459:117848

Antrop (1993) The transformation of the Mediterranean landscapes: an experience of 25 years of observations. Landsc Urban Plan 24:3–13

Aponte C, Kasel S, Nitschke C et al (2020) Structural diversity underpins carbon storage in Australian temperate forests. Glob Ecol Biogeogr. https://doi.org/10.1111/geb.13038

Archambeau J, Ruiz-Benito P, Ratcliffe S et al (2020) Similar patterns of background mortality across Europe are mostly driven by drought in European beech and a combination of drought and competition in Scots pine. Agric For Meteorol 280:107772

Bachofen C, Wohlgemuth T, Ghazoul J et al (2016) Cold temperature extremes during spring do not limit the range shift of Mediterranean pines into regions with intermittent frost. Funct Ecol 30:856–865

Barbéro M, Loisel R, Quézel P et al (1998) Pines of the Mediterranean Basin. In: Richardson DM (ed) Ecology and biogeography of *Pinus*. Cambridge University Press, Cambridge, pp 153–170

Battisti DS, Naylor RL (2009) Historical warnings of future food insecurity with unprecedented seasonal heat. Science 323:240–244

Baudena M, Santana VM, Baeza MJ et al (2019) Increased aridity drives post-fire recovery of Mediterranean forests towards open shrublands. New Phytol. https://doi.org/10.1111/nph.16252

Benito-Garzón M, Alía R, Robson TM et al (2011) Intra-specific variability and plasticity influence potential tree species distributions under climate change. Glob Ecol Biogeogr 20:766–778

Benito-Garzón M, Ruiz-Benito P, Zavala MA (2013) Inter-specific differences in tree growth and mortality responses to climate determine potential species distribution limits in Iberian forests. Glob Ecol Biogeogr 22:1141–1151

Bertness MD, Callaway R (1994) Positive interactions in communities. Trends Ecol Evol 9:191–193

Blondel J, Aronson J (1999) Biology and wildlife of the Mediterranean region. Oxford University Press, Oxford

Bravo F, Álvarez-González JG, Rio M et al (2011) Growth and yield models in Spain: historical overview, contemporary examples and perspectives. For Syst 20:315–328

Cairns DM, Moen J (2004) Herbivory influences tree lines. J Ecol 92:1019–1024

Calama R, Gordo FJ, Mutke S et al (2008) An empirical ecological-type model for predicting stone pine (*Pinus pinea* L.) cone production in the northern plateau (Spain). For Ecol Manag 255:660–673

Calama R, Manso R, Lucas-Borja ME et al (2017) Natural regeneration in Iberian pines: a review of dynamic processes and proposals for management. For Syst 26:eR02S

Camarero JJ, Bigler C, Linares JC et al (2011) Synergistic effects of past historical logging and drought on the decline of Pyrenean silver fir forests. For Ecol Manag 262:759–769

Carnicer J, Coll M, Pons X et al (2014) Large-scale recruitment limitation in Mediterranean pines: the role of *Quercus ilex* and forest successional advance as key regional drivers. Glob Ecol Biogeogr 23:371–384

Carrión JS, Munuera M, Dupré M et al (2001) Abrupt vegetation changes in the Segura Mountains of southern Spain throughout the Holocene. J Ecol 89:783–797

Cavender-Bares J, Bazzaz FA (2000) Changes in drought response strategies with ontogeny in *Quercus rubra*: implications for scaling from seedlings to mature trees. Oecologia 124:8–18
Charrier G, Ngao J, Saudreau M et al (2015) Effects of environmental factors and management practices on microclimate, winter physiology, and frost resistance in trees. Front Plant Sci 6:1–18
Chen IC, Hill JK, Ohlemüller R et al (2011) Rapid range shifts of species associated with high levels of climate warming. Science 333:1024–1026
Chirici G, Winter S, McRoberts R (2011) National forest inventories: contributions to forest biodiversity assessments. Managing forest ecosystems. Springer, Dordrecht
Choat B, Brodribb TJ, Brodersen CR et al (2018) Triggers of tree mortality under drought. Nature 558:531–539
Climent J, Costa e Silva F, Chambel MR et al (2009) Freezing injury in primary and secondary needles of Mediterranean pine species of contrasting ecological niches. Ann For Sci 66:407
Cote SD, Rooney TP, Tremblay JP et al (2004) Ecological impacts of deer overabundance. Annu Rev Ecol Evol Syst 35:113–147
Dalen L, Hofgaard A (2005) Differential regional treeline dynamics in the Scandes Mountains. Arct Antarct Alp Res 37:284–296
Ennos RA (2015) Resilience of forests to pathogens: an evolutionary ecology perspective. Forestry 88:41–52
FAO (2016) State of the world's forests (2016) forests and agriculture: land-use challenges and opportunities. FAO, Rome
FAO (Food and Agriculture Organization of the United Nations) (2015) Global forest resources assessment 2015. FAO forestry series 49. FAO, Rome
Fernández-Pérez L, Villar-Salvador P, Martínez-Vilalta J et al (2018) Distribution of pines in the Iberian Peninsula agrees with species differences in foliage frost tolerance, not with vulnerability to freezing-induced xylem embolism. Tree Physiol 38:507–516
Ferrio JP, Florit A, Vega A et al (2003) Delta C-13 and tree-ring width reflect different drought responses in *Quercus ilex* and *Pinus halepensis*. Oecologia 137:512–518
Galiano L, Martinez-Vilalta J, Lloret F (2010) Drought-induced multifactor decline of scots pine in the Pyrenees and potential vegetation change by the expansion of co-occurring oak species. Ecosystems 13:978–991
Galiano L, Martínez-Vilalta J, Lloret F (2011) Carbon reserves and canopy defoliation determine the recovery of scots pine 4 yr after a drought episode. New Phytol 190:750–759
García-Valdés R, Zavala MA, Araújo MB et al (2013) Chasing a moving target: projecting climate-change driven changes in non-equilibrial tree species distributions. J Ecol 101:441–453
Gauquelin T, Michon G, Joffre R et al (2018) Mediterranean forests, land use and climate change: a social-ecological perspective. Reg Environ Chang 18:623–636
Gazol A, Camarero JJ, Vicente-Serrano SM et al (2018) Forest resilience to drought varies across biomes. Glob Change Biol 24:2143–2158
Gómez JM, Hodar JA, Zamora R et al (2001) Ungulate damage on Scots pines in Mediterranean environments: effects of association with shrubs. Can J Bot Can Bot 79:739–746
Gómez-Aparicio L, Zamora R, Castro J et al (2008) Facilitation of tree saplings by nurse plants: microhabitat amelioration or protection against herbivores? J Veg Sci 19:161–172
Gómez-Aparicio L, Zavala MA, Bonet FJ et al (2009) Are pine plantations valid tools for restoring Mediterranean forests? An assessment along abiotic and biotic gradients. Ecol Appl 19:2124–2141
Gómez-Aparicio L, Garcia-Valdes R, Ruiz-Benito P et al (2011) Disentangling the relative importance of climate, size and competition on tree growth in Iberian forests: implications for forest management under global change. Glob Chang Biol 17:2400–2414
González-Alday J, Martínez-Ruiz C, Bravo F et al (2009) Evaluating different harvest intensities over understory plant diversity and pine seedlings, in a *Pinus pinaster* Ait. natural stand of Spain. Plant Ecol 201:211–220

Granda E, Gazol A, Camarero JJ (2018) Functional diversity differently shapes growth resilience to drought for co-existing pine species. J Veg Sci 29:265–275

Greenwood S, Ruiz-Benito P, Martínez-Vilalta J et al (2017) Tree mortality across forest biomes is promoted by drought intensity, lower wood density and higher specific leaf area. Ecol Lett 20:539–553

Hampe A, Petit RJ (2005) Conserving biodiversity under climate change: the rear edge matters. Ecol Lett 8:461–467

Herrero A, Zamora R (2014) Plant responses to extreme climatic events: a field test of resilience capacity at the southern range edge. PLoS One 9:e87842

Herrero A, Zamora R, Castro J et al (2012) Limits of pine forest distribution at the treeline: herbivory matters. Plant Ecol 213:459–469

Herrero A, Castro J, Zamora R et al (2013) Growth and stable isotope signals associated with drought-related mortality in saplings of two coexisting pine species. Oecologia 173:1613–1624

Herrero A, Almaraz P, Zamora R et al (2016) From the individual to the landscape and back: time-varying effects of climate and herbivory on tree sapling growth at distribution limits. J Ecol 104:430–442

Hester AJ, Millard P, Baillie GJ et al (2004) How does timing of browsing affect above- and below-ground growth of *Betula pendula*, *Pinus sylvestris* and *Sorbus aucuparia*? Oikos 105:536–550

Hisano M, Searle EB, Chen HYH (2017) Biodiversity as a solution to mitigate climate change impacts on the functioning of forest ecosystems. Biol Rev 93:439–456

Hobbs RJ, Arico S, Aronson J et al (2006) Novel ecosystems: theoretical and management aspects of the new ecological world order. Glob Ecol Biogeogr 15:1–7

Holling CS (1996) Engineering resilience versus ecological resilience. In: Schulze P (ed) Engineering within ecological constraints. National Academy Press, Washington, DC, pp 31–44

Holmgren M, Scheffer M, Huston MA (1997) The interplay of facilitation and competition in plant communities. Ecology 78:1966–1997

IPCC (Intergovernmental Panel on Climate Change) (2013) Climate change 2013: the physical science basis. Contribution of Working Group I to the fifth assessment report of the Intergovernmental Panel on Climate Change. Cambridge University Press, Cambridge and New York

Jucker T, Avăcăriţei D, Bărnoaiea I et al (2016) Climate modulates the effects of tree diversity on forest productivity. J Ecol 104:388–398

Jump AS, Peñuelas J (2005) Running to stand still: adaptation and the response of plants to rapid climate change. Ecol Lett 8:1010–1020

Jump AS, Mátyás C, Peñuelas J (2009) The altitude-for-latitude disparity in the range retractions of woody species. Trends Ecol Evol 24:694–701

Kaarlejärvi E, Hoset KS, Olofsson J (2015) Mammalian herbivores confer resilience of Arctic shrub-dominated ecosystems to changing climate. Glob Change Biol 21:3379–3388

Keenan T, Maria Serra J, Lloret F et al (2011) Predicting the future of forests in the Mediterranean under climate change, with niche- and process-based models: CO_2 matters! Glob Change Biol 17:565–579

Kirby KJ, Watkins C (1998) The ecological history of European forests. CAB International, Wallingford

Knight JR, Folland CK, Scaife AA (2006) Climate impacts of the Atlantic multidecadal oscillation. Geophys Res Lett 33:L17706

Kunstler G, Falster D, Coomes DA et al (2016) Plant functional traits have globally consistent effects on competition. Nature 529:204–207

Liang J, Buongiorno J, Monserud RA et al (2007) Effects of diversity of tree species and size on forest basal area growth, recruitment, and mortality. For Ecol Manag 243:116–127

Liang BYJ, Crowther TW, Picar N et al (2016) Positive biodiversity-productivity relationship predominant in global forests. Science 354:aaf8957

Linares JC, Camarero JJ, Carreira JA (2009) Interacting effects of changes in climate and forest cover on mortality and growth of the southernmost European fir forests. Glob Ecol Biogeogr 18:485–497

Linares JC, Camarero JJ, Carreira JA (2010) Competition modulates the adaptation capacity of forests to climatic stress: insights from recent growth decline and death in relict stands of the Mediterranean fir *Abies pinsapo*. J Ecol 98:592–603

Lines ER, Zavala MA, Ruiz-Benito P et al (2019) Estimating juvenile demographic processes from count data using approximate Bayesian computation. Ecography 42:1–13

Liu CF, Chen XL (2004) Effect of resin tapping on *Pinus elliottii* forest growth and economic benefit. J Zhejiang For Sci Technol 24:24–26

Lloret F, Peñuelas J, Prieto P et al (2009) Plant community changes induced by experimental climate change: seedling and adult species composition. Perspect Plant Ecol Evol Syst 11:53–63

Lloret F, Keeling EG, Sala A (2011) Components of tree resilience: effects of successive low-growth episodes in old ponderosa pine forests. Oikos 120:1909–1920

Lloret F, Escudero A, Iriondo JM et al (2012) Extreme climatic events and vegetation: the role of stabilizing processes. Glob Change Biol 18:797–805

Madrigal-González J, Zavala MA (2014) Competition and tree age modulated last century pine growth responses to high frequency of dry years in a water limited forest ecosystem. Agric For Meteorol 192–193:18–26

Madrigal-González J, Hantson S, Yue C et al (2015) Long-term wood production in water limited forests: evaluating potential CO_2 fertilization along with historical confounding factors. Ecosystems 18:1043–1055

Madrigal-González J, Ruiz-Benito P, Ratcliffe S et al (2016) Complementarity effects on tree growth are contingent on tree size and climatic conditions across Europe. Sci Rep 6:32233

Madrigal-González J, Ballesteros-Cánovas JA, Herrero A et al (2017a) Forest productivity in southwestern Europe is controlled by coupled North Atlantic and Atlantic multidecadal oscillations. Nat Commun 8:2222

Madrigal-González J, Herrero A, Ruiz-Benito P et al (2017b) Resilience to drought in a dry forest: insights from demographic rates. For Ecol Manag 389:167–175

Marqués L, Camarero JJ, Gazol A et al (2016) Drought impacts on tree growth of two pine species along an altitudinal gradient and their use as early-warning signals of potential shifts in tree species distributions. For Ecol Manag 381:157–167

Marqués L, Madrigal-González J, Zavala MA et al (2018) Last-century forest productivity in a managed dry-edge Scots pine population: the two sides of climate warming. Ecol Appl 28:95–105

Martín-Benito D, Beeckman H, Canellas I (2013) Influence of drought on tree rings and tracheid features of *Pinus nigra* and *Pinus sylvestris* in a Mesic Mediterranean forest. Eur J For Res 132:33–45

Martínez-Vilalta J, Lloret F (2016) Drought-induced vegetation shifts in terrestrial ecosystems: the key role of regeneration dynamics. Glob Planet Change 144:94–108

Martínez-Vilalta J, Piñol J (2002) Drought-induced mortality and hydraulic architecture in pine populations of the NE Iberian Peninsula. For Ecol Manag 161:247–256

Mather AS (2001) Forests of consumption: postproductivism, postmaterialism, and the postindustrial forest. Environ Plan C Gov Policy 19:249–268

Matías L, Jump AS (2012) Interactions between growth, demography and biotic interactions in determining species range limits in a warming world: the case of *Pinus sylvestris*. For Ecol Manag 282:10–22

Matías L, Jump AS (2015) Asymmetric changes of growth and reproductive investment herald altitudinal and latitudinal range shifts of two woody species. Glob Chang Biol 21:882–896

Matías L, Castro J, Villar-Salvador P et al (2017) Differential impact of hotter drought on seedling performance of five ecologically distinct pine species. Plant Ecol 218:201–212

Mayr S, Hacke U, Schmid P et al (2006) Frost drought in conifers at the alpine timberline: xylem dysfunction and adaptations. Ecology 87:3175–3185

McCulloh K, Johnson D, Meinzer F et al (2011) An annual pattern of native embolism in upper branches of four tall conifer species. Am J Bot 98:1007–1015
Montoya D, Zavala MA, Rodríguez MA et al (2008) Animal versus wind dispersal and the robustness of tree species to deforestation. Science 320:1502–1504
Morin X, Fahse L, Scherer-Lorenzen M et al (2011) Tree species richness promotes productivity in temperate forests through strong complementarity between species. Ecol Lett 14:1211–1219
Norby RJ, Warren JM, Iversen CM et al (2010) CO_2 enhancement of forest productivity constrained by limited nitrogen availability. Proc Natl Acad Sci U S A 107:19368–19373
Oechel WC, Hastings SJ, Vourlitis GL et al (1995) Direct effects of CO_2 in chaparral and Mediterranean-type ecosystems. In: Moreno J, Oechel W (eds) Global change and Mediterranean-type ecosystems. Ecological studies, vol 117. Springer, New York
Olano JM, Rozas V, Bartolomé D et al (2008) Effects of changes in traditional management on height and radial growth patterns in a *Juniperus thurifera* L. woodland. For Ecol Manag 255:506–512
Olofsson J, Oksanen L, Callaghan T et al (2009) Herbivores inhibit climate-driven shrub expansion on the tundra. Glob Change Biol 15:2681–2693
Padilla FM, Pugnaire FI (2007) Rooting depth and soil moisture control Mediterranean woody seedling survival during drought. Funct Ecol 21:489–495
Parmesan C, Yohe G (2003) A globally coherent fingerprint of climate change impacts across natural systems. Nature 421:37–42
Peguero-Pina J, Alquézar-Alquézar J, Mayr S et al (2011) Embolism induced by winter drought may be critical for the survival of *Pinus sylvestris* L. near its southern distribution limit. Ann For Sci 68:565–574
Peltier DM, Fell M, Ogle K (2016) Legacy effects of drought in the southwestern United States: a multi-species synthesis. Ecol Monogr 86:312–326
Peñuelas J, Ogaya R, Boada M et al (2007) Migration, invasion and decline: changes in recruitment and forest structure in a warming-linked shift of European beech forest in Catalonia (NE Spain). Ecography 30:829–837
Pittermann J, Sperry J (2003) Tracheid diameter is the key trait determining the extent of freezing-induced embolism in conifers. Tree Physiol 23:907–914
Post E, Pedersen C (2008) Opposing plant community responses to warming with and without herbivores. Proc Natl Acad Sci U S A 105:12353–12358
Prieto-Recio C, Martín-García J, Bravo F et al (2015) Unravelling the associations between climate, soil properties and forest management in *Pinus pinaster* decline in the Iberian Peninsula. For Ecol Manag 356:74–83
Puértolas J, Gil L, Pardos JA (2005) Effects of nitrogen fertilization and temperature on frost hardiness of Aleppo pine (*Pinus halepensis* Mill.) seedlings assessed by chlorophyll fluorescence. Forestry 78:501–511
Ratcliffe S, Liebergesell M, Ruiz-Benito P et al (2016) Modes of functional biodiversity control on tree productivity across the European continent. Glob Ecol Biogeogr 25:251–262
Reyer CP, Brouwers N, Rammig A et al (2015) Forest resilience and tipping points at different spatio-temporal scales: approaches and challenges. J Ecol 103:5–15
Rigling A, Bigler C, Eilmann B et al (2013) Driving factors of a vegetation shift from Scots pine to pubescent oak in dry Alpine forests. Glob Chang Biol 19:229–240
Ripullone F, Guerrieri MR, Nole A et al (2007) Stomatal conductance and leaf water potential responses to hydraulic conductance variation in *Pinus pinaster* seedlings. Trees Struct Funct 21:371–378
Rodríguez-García E, Gratzer G, Bravo F (2011) Climatic variability and other site factor influences on natural regeneration of *Pinus pinaster* Ait. in Mediterranean forests. Ann For Sci 68:811–823
Roscher C, Schumacher J, Gubsch MN et al (2012) Using plant functional traits to explain diversity–productivity relationships. PLoS One 7:e36760

Ruano I, Pando V, Bravo F (2009) How do light and water influence *Pinus pinaster* Ait. germination and early seedling development? For Ecol Manag 258:2647–2653

Ruano I, Manso R, Fortin M et al (2015) Extreme climate conditions limit seed availability to successfully attain natural regeneration of *Pinus pinaster* in sandy areas of Central Spain. Can J For Res 45:1795–1802

Ruiz-Benito P, García-Valdés R (2016) Inventarios forestales para el estudio de patrones y procesos en Ecología. Ecosistemas 25:1–5

Ruiz-Benito P, Gómez-Aparicio L, Zavala MA (2012) Large scale assessment of regeneration and diversity in Mediterranean planted pine forests along ecological gradients. Divers Distrib 18:1092–1106

Ruiz-Benito P, Lines ER, Gómez-Aparicio L et al (2013) Patterns and drivers of tree mortality in Iberian forests: climatic effects are modified by competition. PLoS One 8:e56843

Ruiz-Benito P, Madrigal-González J, Ratcliffe S et al (2014) Stand structure and recent climate change constrain stand basal area change in European forests: a comparison across boreal, temperate and Mediterranean biomes. Ecosystems 17:1439–1454

Ruiz-Benito P, Ratcliffe S, Zavala MA et al (2017a) Climate- and successional-related changes in functional composition of European forests are strongly driven by tree mortality. Glob Change Biol 23:4162–4176

Ruiz-Benito P, Ratcliffe S, Jump A et al (2017b) Functional diversity underlies demographic responses to environmental variation in European forests. Glob Ecol Biogeogr 26:128–141

Ruiz-Benito P, Vacchiano G, Lines ER et al (2020) Available and missing data to model impact of climate change on European forests. Ecol Model 416:108870

Salazar-Tortosa D, Castro J, Villar-Salvador P et al (2018a) The "isohydric trap": a proposed feedback between water shortage, stomatal regulation, and nutrient acquisition drives differential growth and survival of European pines under climatic dryness. Glob Change Biol 24:4069–4083

Salazar-Tortosa D, Castro J, De Casas RR et al (2018b) Gas exchange at whole plant level shows that a less conservative water use is linked to a higher performance in three ecologically distinct pine species. Environ Res Lett 13:045004

Sánchez-Salguero R, Navarro-Cerrillo RM, Camarero JJ et al (2010) Drought-induced growth decline of Aleppo and maritime pine forests in south-eastern Spain. For Syst 19:458–469

Sánchez-Salguero R, Navarro-Cerrillo RM, Camarero JJ et al (2012) Selective drought-induced decline of pine species in southeastern Spain. Clim Chang 113:767–785

Sánchez-Salguero R, Camarero JJ, Dobbertin M et al (2013) Contrasting vulnerability and resilience to drought-induced decline of densely planted vs. natural rear-edge *Pinus nigra* forests. For Ecol Manag 310:956–967

Sánchez-Salguero R, Camarero JJ, Carrer M et al (2017) Climate extremes and predicted warming threaten Mediterranean Holocene firs forests refugia. Proc Natl Acad Sci U S A 114:E10142–E10150

Schoennagel T, Veblen TT, Romme WH (2004) The interaction of fire, fuels, and climate across Rocky Mountain forests. Bioscience 54:661–675

Seidl R, Schelhaas MJ, Rammer W et al (2014) Increasing forest disturbances in Europe and their impact on carbon storage. Nat Clim Chang 4:806–810

Serra-Maluquer X, Mencuccini M, Martínez-Vilalta J (2018) Changes in tree resistance, recovery and resilience across three successive extreme droughts in the northeast Iberian Peninsula. Oecologia 187:343–354

Sheffer E (2012) A review of the development of Mediterranean pine–oak ecosystems after land abandonment and afforestation: are they novel ecosystems? Ann For Sci 69:429–443

Soliño M, Yu T, Alía R et al (2018) Resin-tapped pine forests in Spain: ecological diversity and economic valuation. Sci Total Environ 625:1146–1155

Speed JDM, Austrheim G, Hester AJ et al (2010) Experimental evidence for herbivore limitation of the treeline. Ecology 91:3414–3420

Spinoni J, Barbosa P, De Jager A et al (2019) A new global database of meteorological drought events from 1951 to 2016. J Hydrol Reg Stud 22:100593

Sutton RT, Hodson DL (2005) Atlantic Ocean forcing of North American and European summer climate. Science 309:115–118

Toca A, Oliet JA, Villar-Salvador P et al (2018) Species ecology determines the role of nitrogen nutrition in the frost tolerance of pine seedlings. Tree Physiol 38:96–108

Tømmervik H, Johansen B, Riseth JA et al (2009) Above ground biomass changes in the mountain birch forests and mountain heaths of Finnmarksvidda, northern Norway, in the period 1957–2006. For Ecol Manag 257:244–257

Trigo RM, Osborn TJ, Corte-Real JM (2002) The North Atlantic oscillation influence on Europe: climate impacts and associated physical mechanisms. Clim Res 20:9–17

Uemura M, Steponkus PL (2003) Modification of the intracellular sugar content alters the incidence of freeze-induced membrane lesions of protoplasts isolated from *Arabidopsis thaliana* leaves. Plant Cell Environ 26:1083–1096

Vadell E, de-Miguel S, Pemán J (2016) Large-scale reforestation and afforestation policy in Spain: a historical review of its underlying ecological, socioeconomic and political dynamics. Land Use Policy 55:37–48

van der Plas F, Ratcliffe S, Ruiz-Benito P et al (2018) Continental mapping of forest ecosystem functions reveals widespread synergies. Ecol Lett 21:31–42

Vilà-Cabrera A, Martinez-Vilalta J, Vayreda J et al (2011) Structural and climatic determinants of demographic rates of Scots pine forests across the Iberian Peninsula. Ecol Appl 21:1162–1172

Vilà-Cabrera A, Coll L, Martínez-Vilalta J et al (2018) Forest management for adaptation to climate change in the Mediterranean basin: a synthesis of evidence. For Ecol Manag 407:16–22

Vilà-Cabrera A, Premoli AC, Jump AS (2019) Refining predictions of population decline at species' rear edges. Glob Change Biol 25:1549–1560

Vinther BM, Andersen KK, Hansen AW et al (2003) Improving the Gibraltar/Reykjavik NAO index. Geophys Res Lett 30:2222

Waitz Y, Sheffer E (this volume) Dynamics of mixed pine-oak forests. In: Ne'eman G, Osem Y (eds) Pines and their mixed forest ecosystems in the Mediterranean Basin. Springer, pp 345–362

Woodward F, Williams B (1987) Climate and plant distribution at global and local scales. Vegetatio 69:189–197

Yuan Z, Wang S, Ali A et al (2018) Aboveground carbon storage is driven by functional trait composition and stand structural attributes rather than biodiversity in temperate mixed forests recovering from disturbances. Ann For Sci 75:67

Zamora R, Pugnaire FI (2001) Ecosistemas Mediterráneos: Análisis funcional. Centro Superior de Investigaciones Científicas, Madrid

Zamora R, Gómez JM, Hódar JA et al (2001) Effect of browsing by ungulates on sapling growth of Scots pine in a Mediterranean environment: consequences for forest regeneration. For Ecol Manag 144:33–42

Zavala MA, Zea E (2004) Mechanisms maintaining biodiversity in Mediterranean pine-oak forests: insights from a spatial simulation model. Plant Ecol 171:197–207

Zavala MA, Espelta JM, Retana J (2000) Constraints and trade-offs in Mediterranean plant communities: the case of holm oak-Aleppo pine forests. Bot Rev 66:119–149

Zavala MA, Angulo O, de la Parra RB et al (2007) An analytical model of stand dynamics as a function of tree growth, mortality and recruitment: the shade tolerance-stand structure hypothesis revisited. J Theor Biol 244:440–450

Zweifel R, Rigling A, Dobbertin M (2009) Species-specific stomatal response of trees to drought – a link to vegetation dynamics? J Veg Sci 20:442–454

Chapter 14
Carbon and Energy Balance of Dry Mediterranean Pine Forests: A Case Study

Eyal Rotenberg, Rafat Qubaja, Yakir Preisler, Dan Yakir, and Fyodor Tatarinov

14.1 Semi-arid Forests and Their Contribution to the Climate System

Considerable parts of the Mediterranean region experience semi-arid climatic conditions. Semi-arid lands cover a total of ~18% of the earth's land surface and play a significant role in the direction, magnitude and inter-annual variability of terrestrial carbon storage (Poulter et al. 2014; Ahlström et al. 2015; Ma et al. 2016). While, tropical and boreal forests dominate in the mean global carbon sink accounting for its largest fraction (26% and 10%, respectively, in 1982–2011), semi-arid ecosystems dominate the increasing sink trend (57%, 0.04 Pg C year^{-2}) and in its inter-annual variation (~40%, Ahlström et al. 2015). According to projections, dryland areas will increase considerably under further climate warming (Alpert et al. 2002; Giorgi and Lionello 2008; Sherwood and Fu 2014; Huang et al. 2016).

Uptake of CO_2 from the atmosphere by vegetation is assumed to be an important factor for global warming mitigating. Semi-arid conditions usually support shrublands, dry savannas or grasslands, with a wide range in productivity and water use (Scott et al. 2015; Biederman et al. 2017; Yan et al. 2019), and are rarely considered as offering significant C (carbon) sequestration potential (Lal 2004). However, in recent times considerable efforts have been undertaken to plant forests in the semi-arid zone, e.g., in Israel during the last 100 years and recently in other Middle-Eastern countries (Caliskan and Boydak 2017), in the Sahel region (https://www.greatgreenwall.org/about-great-green-wall) and in China (Cao et al. 2011). If such afforestation was applied to a large area, the contribution of these forests to the global carbon and energy balance could be significant. Qubaja et al. (2020a) showed

E. Rotenberg · R. Qubaja · Y. Preisler · D. Yakir · F. Tatarinov (✉)
Earth and Planetary Sciences, Weizmann Institute of Science, Rehovot, Israel
e-mail: fedor.tatrinov@weizmann.ac.il

G. Ne'eman, Y. Osem (eds.), *Pines and Their Mixed Forest Ecosystems in the Mediterranean Basin*, Managing Forest Ecosystems 38,
https://doi.org/10.1007/978-3-030-63625-8_14

that afforestation of even 10% of the vast semi-arid lands can store ~9 Pg C, with even higher maximum saturation potential, depending on clay content.

The advantage of forests over most other ecosystems types is their ability to store a considerable amount of carbon for decades (Pan et al. 2011), both in trees and in the soil layer below. Therefore, forests mitigate the rise of the CO_2 concentration in the atmosphere. The magnitude of the terrestrial carbon sink over the decade 2007–2016 averaged 3.0 ± 0.8 Pg C year^{-1} (Le Quéré et al. 2018). The increasing accuracy of these estimates of the C sink, its spatial distribution, and the potential to enhance it is critical to supporting future climate change mitigation actions. Due to their long duration, forests can store a large amount of organic matter, which was recently promoted as an important carbon sink for climate mitigation (Minasny et al. 2017). However, its potential has also been challenged, e.g., Amundson and Biardeau (2018) and Schlesinger and Amundson (2019) suggest that under optimal conditions increasing soil carbon could not balance more than 5% of annual emissions of CO_2 from fossil fuel combustion.

14.2 Contrasting Effects of Semi-arid Afforestation on Climate – The Carbon Sink and Energy Balance

The effect of forest ecosystems on climate is not limited to CO_2 sequestration. It is now well understood that land ecosystems influence the climate through physical, chemical and biological processes, by affecting the energy budget and the hydrologic cycle (Bonan 2008) from the local to the global scale. By absorbing CO_2, forests have a cooling effect on the earth's climate system (Houghton et al. 2001). Changes in the landscape, such as from pasture to forest, usually modify the amount of solar radiation absorbed by the ecosystem (Betts 2000), the magnitude of emitted thermal radiation (Rotenberg and Yakir 2010) and the rates of sensible and latent heat exchange with the atmosphere (Baldocchi and Ma 2013), which have a combined effect on the amount of heat transfer to the atmosphere. Forest ecosystems are usually darker than the surrounding landscapes; thus they tend to absorb more solar radiation, creating a heating feedback with the atmosphere. Incoming solar radiation is usually high in many parts of the Mediterranean region, thus feedbacks related to land cover reflectance changes can diminish the cooling feedback on the atmosphere for extended periods of time (Claussen et al. 2013). Similarly, land cover change (LCC) also affects rainfall use by ecosystems; forests tend to evaporate more water than other ecosystem types (Wang and Dickinson 2012) and in dry lands they can drastically reduce the amount of water available for human consumption (Rohatyn et al. 2018). However, forests can enhance cloud cover that also feeds back on the climatic conditions.

14.3 Aleppo Pine – A Dominant Species in Mediterranean Forests

One of the most common forest types in the Mediterranean zone is pine forests. Generally, pine trees (genus *Pinus* (Pinaceae)), include more than 100 species; *Pinus* is the largest genus of existing gymnosperms and the most widely distributed in the Northern Hemisphere (Mirov 1967; Price et al. 1998; Debreczy and Rácz 2011). In the Mediterranean Basin, Aleppo pine (*P. halepensis* Miller) is the most dominant forest species, naturally distributed and widely planted for the last century, showing various drought resistance and avoidance strategies (Ne'eman and Trabaud 2000; Maseyk et al. 2008). Aleppo pine is widespread across most of the Mediterranean from eastern Spain to Israel (Ne'eman and Trabaud 2000). It has high ecological plasticity, surviving a wide range of environmental conditions, from sea level to 2600 m altitude, and from 285 to 1500 mm annual precipitation (Ne'eman and Trabaud 2000). In Israel, Aleppo pine was widely planted during the last century. Currently it grows in most of the non-desert parts of Israel, from the border with Lebanon in the north to the northern edge of the Negev desert in the south. Most Aleppo pine stands in Israel are planted, but in most of those territories natural regeneration can also occur. The physiological mechanisms that allow *P. halepensis* to cope with harsh, hot and dry conditions are yet to be fully understood.

14.4 Yatir Aleppo Pine at the Hot and Dry Timberline

One of the driest forest sites in the world is Yatir Forest, located at the southern edge of the Judean Mountains and the northern edge of the Negev desert in Israel (31°20′N, 35°03′E, 650 ± 150 m above sea level) (Fig. 14.1), between the Mediterranean semi-arid and arid conditions. This forest has been planted from 1964 until the present, mainly by Aleppo pine trees; seeds originate from the entire

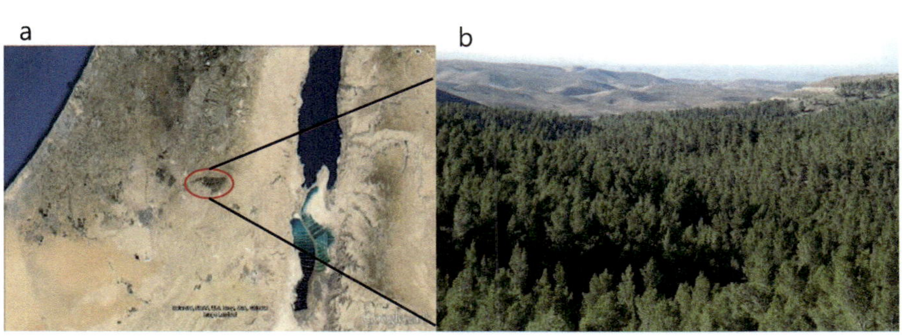

Fig. 14.1 (**a**) Satellite image of the location of Yatir Forest with the Mediterranean Sea to the left (west) and the Dead Sea to the east; (**b**) The surrounding desert as seen from within Yatir Forest

Mediterranean Basin. Currently the forest covers an area of 2800 ha with a density of about 300 trees ha^{-1}. As of 2019, mean tree DBH (diameter at breast height) is ~24 cm and height is ~12 m. Mean annual DBH growth for 2000–2018 is ~2 mm·year^{-1} and mean overstory LAI (leaf area index) is ~1.6. The understory vegetation is sparse, comprising mostly annuals, and active only during the wet season (December to April). The dry season is long, lasting on average from mid-April to late October, with mean annual rainfall (P) of 285 mm, ranging annually between 150 and 450 mm (Fig. 14.2). Since the forest was planted, the shortest and the longest dry periods were 166 and 349 days, respectively (231 days on average). The average daily temperature is 25 °C in July and 10 °C in January. Due to the relatively high temperature and the low humidity values, the annual potential evapotranspiration (ET_O) is ~1600 mm; thus, the aridity index (P/ET_O) is ~0.2, which corresponds to the border between semi-arid and arid conditions. The soil is predominantly light brown rendzina, 25–100 cm deep, with a high level of stoniness. The groundwater depth is >300 m, so the trees must rely only on the seasonal rain feed water. The soil is underlain by chalk and limestone (FAO classification) (Qubaja et al. 2020a).

Although natural regeneration is very sparse, the planted stand successfully survives without irrigation. Yatir Forest has been the object of numerous ecophysiological (Rotenberg and Yakir 2010; Tatarinov et al. 2015), meteorological (Brugger et al. 2018) and physiological (Maseyk et al. 2008; Klein et al. 2013, 2016; Tatarinov et al. 2017) studies. In particular, an eddy covariance (EC) site, unique for the entire semi-arid region of the Middle East and Northern Africa has been functioning in Yatir Forest since 2000. Other EC systems in coniferous (in particular, Aleppo pine) forests in the Mediterranean (e.g., in Spain, Greece or Italy) operate in considerably more humid locations (with annual precipitation above 500 mm, compared to 285 mm at Yatir).

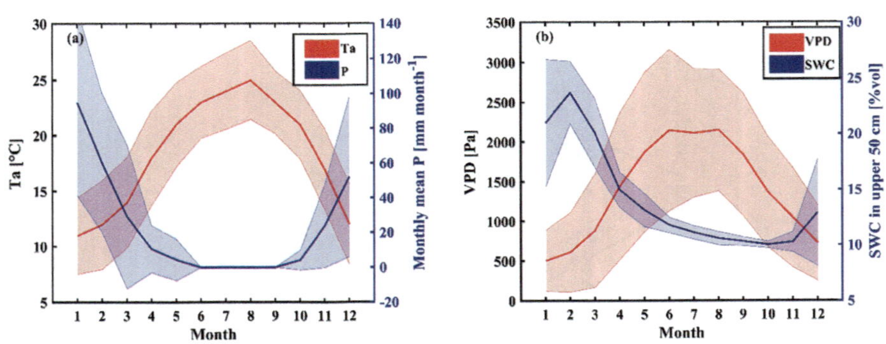

Fig. 14.2 Yatir Forest: (**a**) Long-term (2010–2018) monthly means ± standard deviations of: total precipitation (P) and average air temperature (Ta); (**b**) monthly average VPD and soil moisture in the upper 50 cm (SWC)

14.5 Eddy Covariance Measurements in Yatir Forest

The most direct, and nowadays common, way to measure carbon, water and energy fluxes at the ecosystem scale is the EC approach (Baldocchi 2014). Since 2000, EC and supplementary meteorological measurements have been performed continuously in Yatir Forest (Grünzweig et al. 2007; Rotenberg and Yakir 2011) according to the European methodology (Aubinet et al. 2000) and since then it has been part of the FluxNet (https://fluxnet.fluxdata.org/about/) and EuroFlux (http://gaia.agraria.unitus.it/home) communities. The EC station was established in the middle of Yatir Forest (31°20′42.25″N 35°03′07.01″E, 658 m a.s.l.) in 2000 (Fig. 14.3a) and has operated continuously until today (2020).

The EC approach combines concurrent measurements, at high frequency, of gas concentrations in the atmosphere, air temperature and wind speed components in three dimensions. An algorithm converts the measured values into mass fluxes providing the net ecosystem exchange of carbon. In parallel, radiation fluxes are measured as well, allowing calculation of energy fluxes (see details of the method in Burba 2013). The effectiveness of the EC approach and the urgent need to understand the land surface effect on the Earth's climate system have led to the establishment of instrumental tower stations around the globe to measure those fluxes under the umbrella of the FluxNet community. Currently, there are over 600 participating stations, which provide their data to the FluxNet database; many more such systems use this technique for various applications.

Using the EC method, the station at Yatir measures ecosystem net evapotranspiration (ET), net ecosystem (carbon) exchange (NEE), and the latent (LE) and sensible (H) heat fluxes between the forest and the atmosphere. NEE is the sum of the plant's gross primary production of carbon (GPP) and ecosystem respiration (Re), comprising respiration by plants and the microbial and fungal activity in the soil. According to the agreement by the FluxNet community (https://fluxnet.fluxdata.org/), fluxes are negative if they flow from the atmosphere to ecosystem (sink, e.g., GPP_{atm}) and positive in the opposite direction (for most ecosystems Re_{atm} is positive most times). In the ecological community, the ecosystem net carbon production (NEP), which equals minus NEE, is applied, where NEP is the balance between gross carbon uptake by plants (GPP_{eco}, a positive in the ecology community) and the carbon respired back to the atmosphere by the ecosystem (Re_{eco}, negative in this community). There is no direct way to measure individual NEP components at the ecosystem scale; this deficiency can be overcome by additional measurements (see below) and models (e.g., Reichstein et al. 2005; Lasslop et al. 2010; Keenan et al. 2019), enabling derivation of those fluxes from NEP measured by the EC system and meteorological variables (air and soil temperature and humidity). Sensible heat flux is the vertical heat transfer by air moving between air layers with different temperature. Latent heat flux, calculated from ET as: $LE = \lambda \cdot ET$, where λ is the latent heat of vaporization, is the heat consumed for evaporation (or released with condensation). Footprint area, the area contributing to the measured fluxes, depends on the height of measurement above the canopy and the wind speed; fluxes in Yatir

Fig. 14.3 (**a**) The micrometeorological tower for measuring water, carbon and energy fluxes over Yatir Forest at the Northern edge of the Negev desert in Israel; (**b**) Mobile laboratory with the pneumatic mast (adjustable up to 28 m) for micrometeorological and flux measurements in the background, during measurements at a desert site 4 km outside the Yatir forest

represent the area up to a distance of ~1 km around the measurement tower. Additional measurements include: radiation components above and below the canopy at different wave bands (solar, thermal, photosynthetically active radiation (PAR) and more), meteorological conditions (air temperature and humidity, wind speed and direction etc.), soil conditions (temperature, water content and more) and canopy activity variables (photosynthesis and sap flow rates, tree growth dynamics and more). Sensors work continuously, usually providing average values at 30-min intervals. Based on these measurements, the energy fluxes are calculated, namely net radiation (Rn) as the difference between incoming and outgoing radiation fluxes, sensible heat H, latent heat LE and soil heat (G) fluxes. These fluxes are connected by the energy balance equation: $Rn = H + LE + G$.

14.6 Measurements Along the Precipitation Gradient

In addition to the Yatir EC site, additional EC and supplemental meteorological measurements were performed along the Israeli rainfall gradient, during 2012–2018, using a tailor-made mobile lab system installed on a truck (Fig. 14.3b) (Ramati 2015). This Book Chapter will discuss the results from the rainfall gradient in two extreme sites. The northern site of Birya (33° 00′ N, 35° 30′ E) is characterized by Mediterranean sub-humid conditions. Annual precipitation in Birya is 755 $mm{\cdot}year^{-1}$, and the dry season lasts from June to September. Mean air temperatures in January and July are 8.9 and 24.5 °C, respectively. Forest is dominated by Aleppo pine with small admixture of cypresses. Mean stand age, density and height are 36 years, 600 trees ha^{-1} and 11 m, respectively (Rohatyn et al. 2018). Yatir is the most southern site, located at the transition from semi-arid to arid conditions (285 $mm{\cdot}year^{-1}$). The measurements were performed on a campaign base inside the forests and in nearby open ecosystems in different seasons during the years when all ecosystems were under grazing. Non-forest ecosystems were sparse dwarf shrublands, mostly dominated by *Sarcopoterium spinosum* in a patchy distribution with a wide variety of herbaceous species, mostly annuals, growing in between the shrub patches during the winter to early spring (Rohatyn et al. 2018). Each campaign lasted about 10 days; two to three campaigns per year were performed in each site. The temporal distribution of the campaigns aimed to provide measurements in all seasons. Furthermore, campaign measurements were extrapolated to the whole year based on the relationships between fluxes and meteorological data measured during the campaigns. The fluxes for the entire year were then calculated from independent continuous meteorological data from nearby weather stations using these relationships (Rohatyn et al. 2018).

14.7 Leaf, Branch and Tree Scale and Soil Measurements of Water and CO_2

In addition to the EC measurements, showing ecosystem-level fluxes, detailed smaller-scale measurements are performed at the Yatir site to partition these fluxes into individual components, namely the tree layer and the soil lay. These include continuous monitoring of soil moisture (SWC) at different depths and locations, soil respiration and evaporation rates, tree sap flow with lab-made heat dissipation sensors ('Granier method', Granier and Loustau 1994; Klein et al. 2016), stem diameter variation (point and band dendrometers), stem xylem water content (using calibrated soil moisture FDR sensors) and lab-made branch-scale chamber measurements of CO_2 and H_2O fluxes at a height of 6 m in 14 trees. The chamber contains a few twigs with known needle area (Fig. 14.4). The fluxes are used to calculate branch transpiration (T_{br}) and net assimilation (A_{net}) per leaf area inside the chambers (see, Preisler 2020 in progress, for detailed information) and soil evaporation and respiration fluxes (Qubaja et al. 2020b, c). Occasional ancillary measurements such as branch and twig water potentials have been performed since 2014 during 24-h campaigns.

14.8 Carbon Inventory

In order to obtain independent verification of EC carbon flux measurements a stock-based approach of estimating change in the carbon pools was applied (Qubaja et al. 2020a). The original forest inventory assessment performed in 2001 was repeated

Fig. 14.4 Chambers used for shoot-scale CO_2 and water flux measurements in Yatir Forest

15 years later (2016) on the same five, 30 × 30 m plots in the central part of the forest. Both forest inventories included estimates of carbon storage (CS) in the following four main components: standing biomass, litter, soil, and removal (mortality, thinning and sanitation). Detailed information is reported in Grünzweig et al. (2007) and Qubaja et al. (2020a).

The mean annual net ecosystem production based on CS (NEP_{CS}) over the observation period *t* (15 years) was calculated as follows:

$$NEP_{CS} = \left(\Delta C_{Soil} + \Delta C_{tree} + C_u + C_{tree\_t,m} + C_s\right)/t \tag{14.1}$$

where Δ is the difference in soil or tree CS over the observation period. The equation also accounts for C removed by grazing (*u*), mortality and thinning (*tree_t, m*), and sanitation (*s*). NEP_{CS} was then compared with NEP_{EC} obtained by the integration of EC-based carbon fluxes.

14.9 Annual Carbon and Energy Fluxes

The annual course of ecosystem gross primary production (GPP_{eco}) is characterized by high absolute carbon uptake values for a short period in the wet season (January to May) with a peak in March (monthly mean of ~4 gC m^{-2} d^{-1}) and low values of ~0.5 to 1 gC m^{-2} d^{-1} from June to November (Fig. 14.5a). The course of A_{net} (twig-scale balance between assimilation and respiration), recalculated per unit ground area, showed similar mean values (Fig. 14.5b). During most of the dry period (June to October) A_{net} remained close to 0, meaning that branch-scale magnitudes of daytime photosynthesis and 24-h foliage respiration were similar. Interestingly, the peak of the rain is 2 months earlier than the peak of GPP – in January, whereas the

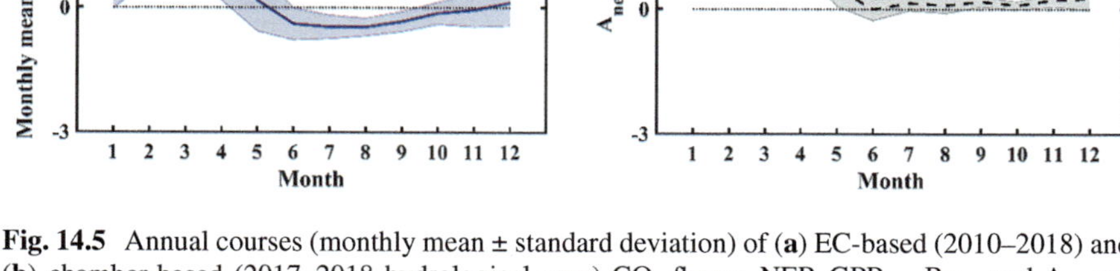

Fig. 14.5 Annual courses (monthly mean ± standard deviation) of (**a**) EC-based (2010–2018) and (**b**) chamber-based (2017–2018 hydrological year) CO_2 fluxes. NEP, GPP_{eco}, Re_{eco} and A_{net} are monthly mean values of net ecosystem production, gross primary production, ecosystem respiration and branch-level net CO_2 flux (average of all chambers), respectively

peak of solar radiation and temperature is 3–4 months later (Fig. 14.2a), in June–July. This timing reflects the optimal period in the trade-off between temperature and solar radiation on one hand and soil moisture on the other hand. Ecosystem respiration (Re_{eco}) generally follows the same dynamics as GPP with a peak at ~2.5 gC m^{-2} d^{-1} on average in March–April and decreases to −0.7 gC m^{-2} d^{-1} in October. The resulting NEP is positive (the ecosystem is a carbon sink) during the wet season from December to May and slightly negative (carbon source) from June to November.

The energy source for ecosystem activity, which also governs its temperature, is the net radiation absorbed by the ecosystem (Rn). The main heat dissipation from the ecosystem is in the form of H representing more than 50% of Rn for all months except February (Fig. 14.6) and above 90% of Rn in autumn. The other heat dissipation component, the LE flux, is very small from May to December because of the lack of water for ET, when it is about 10% of Rn. The LE is higher only during the rainy season, from January to April, and it reaches a maximum of about 50% of Rn from December to February. The Bowen ratio (H/LE) varies from ~1 in February to >9 in August. On a monthly scale, G remains very small throughout the year, being negative in the wet season and positive in the dry season. It varies from ~ −17% of Rn in December to 6% of Rn in July.

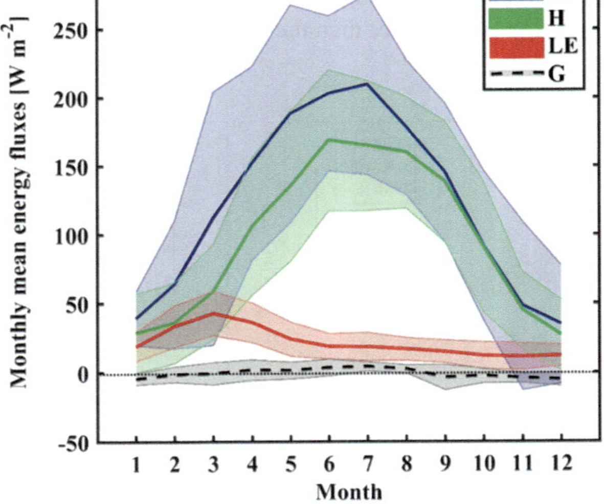

Fig. 14.6 Annual courses (monthly mean of daily means± standard deviation) of energy fluxes (2010–2018) in Yatir Forest. Rn, LE, H and G are mean net radiation, latent heat, sensible heat and soil heat fluxes, respectively

14.10 Diurnal Dynamics of Carbon and Energy Fluxes

On the diurnal timescale, CO_2 fluxes showed two typical seasonal daily cycles (Figs. 14.7a, b, 14.8 and 14.9). As expected, all CO_2 fluxes were higher, on average, during the wet season compared to the dry season by a factor of ~2. However, ecosystem respiration (Re_{eco}) and soil respiration (Rs_{eco}) peaked around midday in both the wet and dry seasons, while NEP and GPP_{eco}, more physiologically controlled by stomata, showed a shift of peak uptake activity from midday (11:00–14:00) in the wet season, to the morning (07:00–10:00) in the dry season, with a strong midday assimilation depression and a smaller, secondary afternoon peak (Fig. 14.7b).

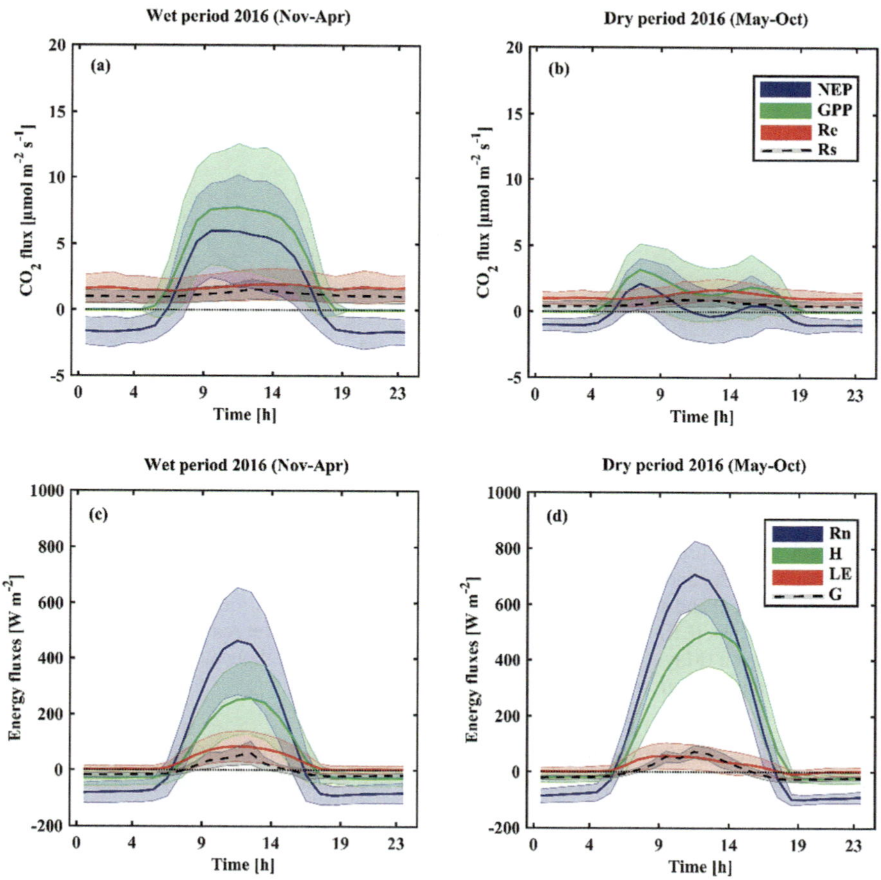

Fig. 14.7 Yatir Forest representative ecosystem diurnal cycles of (**a, b**) net ecosystem carbon production (NEP; canopy scale eddy covariance), gross primary production (GPP_{eco}), ecosystem respiration ($R_{e,eco}$) and soil respiration (R_s,eco) and (**c, d**) energy fluxes – net radiation (Rn), sensible (H) and latent heat (LE) fluxes and soil heat flux (G) for (**a, c**) the wet season (Nov–April) and (**b, d**) the dry season (May–October) in. Based on the EC half-hourly values over the diurnal cycle; shaded areas indicate ±standard deviation

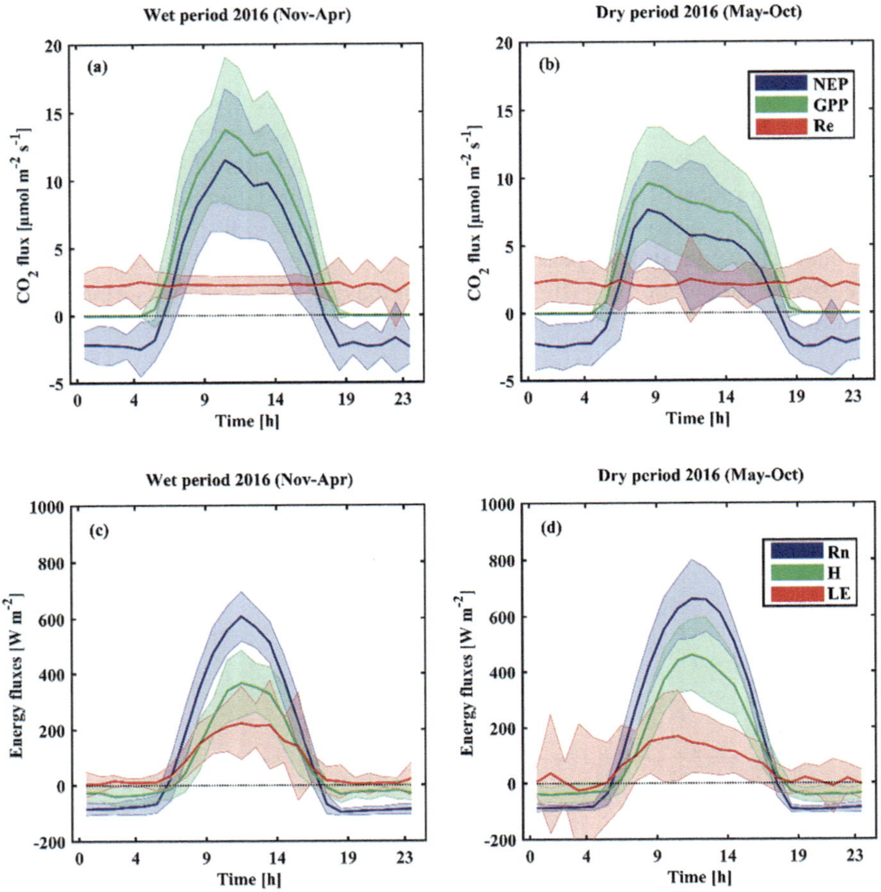

Fig. 14.8 Birya Forest, representative ecosystem diurnal cycles of (**a**, **b**) net ecosystem carbon production (NEP; canopy scale eddy covariance), gross primary production (GPP_{eco}), ecosystem respiration ($Re_{,eco}$) and soil respiration ($R_{s,eco}$) and (**c**, **d**) energy fluxes – net radiation (Rn), sensible (H) and latent heat (LE) fluxes and soil heat flux (G) for (**a**, **c**) the wet season (Nov–April) and (**b**, **d**) the dry season (May–October) in. Based on mobile lab EC half-hourly values over the diurnal cycle; shaded areas indicate ±standard deviation

The diurnal Rn course usually follows the dynamics of incoming solar radiation with a maximum at noon, in both the wet and dry seasons (Fig. 14.7c, d). Nighttime negative Rn values are due to an excess of emitted over received thermal radiation by the ecosystem. Sensible heat flux lags after Rn by 30 min in the wet season and 1 h in the dry season. Latent heat flux (LE) is the most varying among the seasons. In contrast to the other energy fluxes, it is lower in the dry season than in the wet season due to a lack of water for evaporation; compared with the wet season, its peak in the dry season shifts to the morning (9:00–10:00) corresponding to the morning peak of tree physiological activity, demonstrated the biology control on the ecosystem water fluxes. Soil heat flux (G) is negative in the nighttime (i.e., heat emission to the atmosphere by the soil) and positive in the daytime (i.e., heat absorption from the atmosphere). G in the wet season is the lowest of the energy fluxes, but in the dry season it becomes comparable with LE. Heat storage by the canopy layer is not measured in Yatir Forest.

In the northern site of Birya, CO_2 fluxes were much higher than in Yatir and the differences in CO_2 and energy fluxes between the wet and dry seasons were much lower (Fig. 14.8): whereas in Yatir mean GPP_{eco} peak values in the wet and dry seasons were ~7.5 and ~3 μmol·m^{-2} s^{-1}, respectively, in Birya they reached ~14 and ~10 μmol·m^{-2} s^{-1}, respectively. The importance of LE in energy exchange in Birya was much greater than in Yatir. The Bowen ratios in the wet and dry seasons were 2.0 and 4.6, respectively, compared with 4.9 and 14.2 in Yatir (Ramati 2015). A shift of NEP and LE peaks in the dry season from midday to the morning was observed in Birya, as in Yatir.

14.11 Inter-Annual Variability and Dependence of Carbon Fluxes on Precipitation and Land Cover

Yatir Forest GPP, and to a lesser extent, NEP, are highly dependent on the annual rainfall amount, P (Fig. 14.10), were NEP ranging from 33 gC·m^{-2} $year^{-1}$ in the very dry 2010/11 hydrological year to 354 gC·m^{-2} $year^{-1}$ in the wet 2004/2005 year (hydrological year here starts on October 1st usually before the first rain) with

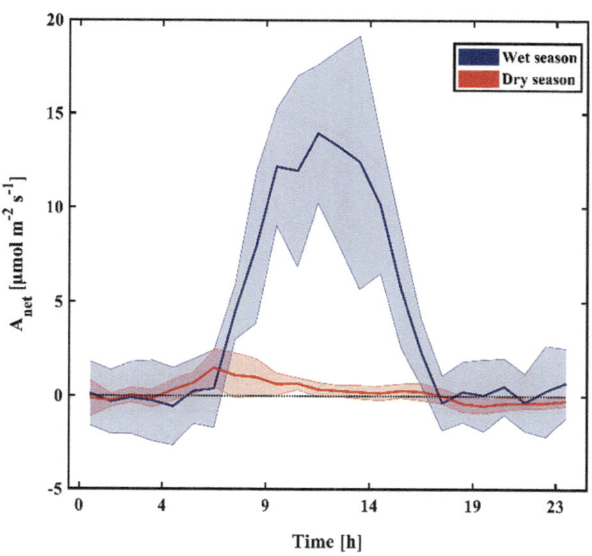

Fig. 14.9 Representative mean diurnal cycles of chamber-based leaf CO_2 fluxes (2017–2018 hydrological year) for the wet (February) and dry (July) seasons in Yatir Forest. Shaded areas indicate ±standard deviation

average for the years 2000–2018 of 154 gC·m^{-2} year^{-1} (Qubaja et al. 2020a). The correlation between annual GPP and P reaches 0.61. The long dry period of 2007–2010 (see Fig. 14.10) also led to increased tree mortality – in total ~8% of trees died. The mortality was patchy – trees mainly died in the plots with deep soil and low stoniness, whereas in the plots with high stoniness trees survived. This is explained by generally lower soil evaporation from the top surface stony soil, water storage in porous stones and slower water movement from depth to ground surface (see details in Preisler et al. 2019). The comparison with the wetter Birya Forest site, as well as the analysis of the FluxNet dataset (https://fluxnet.fluxdata.org/) showed that dependency on P decreases under wetter conditions where others parameters, such as the amount of incoming solar radiation and the air temperature, play a greater role. The sequestration and emission of carbon by ecosystems much depend on the land cover types and management practices. In particular, our measurements along the Israeli rainfall gradient showed that forest NEP was always higher than the NEP of adjacent grassland or shrubland ecosystems. However, the difference between forest and non-forest NEP (ΔNEP) decreased with decreasing rain amount; whereas in Birya ΔNEP was 460 gC·m^{-2} year^{-1} or 97% of forest NEP, in Yatir this difference was 130 gC·m^{-2} year^{-1} or 75% of forest NEP. NEP likely is considerably dependent on the grazing and others management practices, however those topics were not included in this research.

At the end of 2016, the carbon inventories in the forest soil and trees (#13.8 above) showed that the carbon accumulation rate since 2001 was 145 ± 26 g C m^{-2} year^{-1} (Qubaja et al. 2020a). Carbon accumulated mostly in the soil (~71%) while 29% accumulated in the standing biomass, with a long soil carbon turnover time of 59 years.

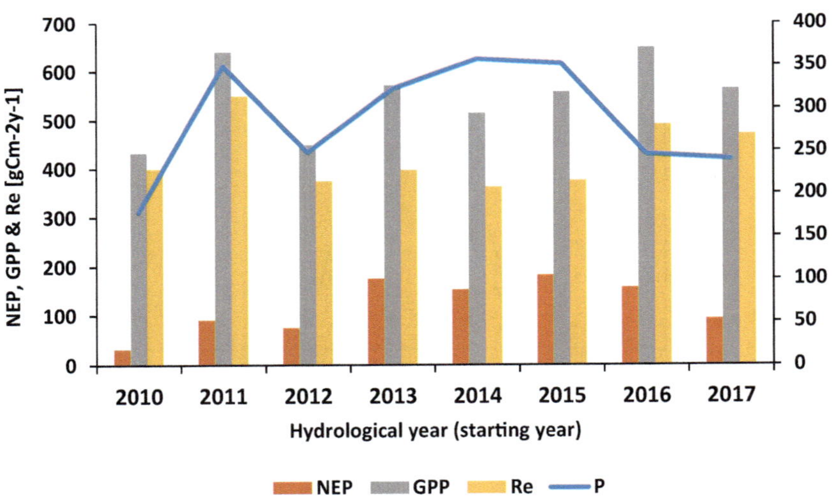

Fig. 14.10 Inter-annual variability of carbon fluxes (NEP; net ecosystem carbon production, GPP_{eco}; gross primary production, R_e ecosystem respiration) and precipitation amount (P) in Yatir Forest

14.12 Flux Variations Along the Rainfall Gradient

The multi-year (2000–2017) annual average net ecosystem production (NEP) at Yatir is 160 gCm^{-2} $year^{-1}$, and the annual average GPP_{eco} and Re_{eco} are 553 and 394 gCm^{-2} $year^{-1}$, respectively (annual values are according to the hydrological year). Surprisingly, despite the harsh growth conditions in Yatir, the net carbon uptake of this ecosystem is close to the average NEP of European pine forests, ~200 gCm^{-2} $year^{-1}$ (Luyssaert et al. 2007). However, NEP in Yatir Forest is only ~40% of the mean NEP of all forests from the Fluxnet global database (https://daac.ornl.gov/get_data/, 2019), equal to 390 gCm^{-2} $year^{-1}$. The NEP similarity to average of European pine forests, despite the very different growing conditions, is due to the differences in the GPP and Re fluxes; those are at least double in wetter sites. The parallel increase of GPP and Re in more humid conditions holds also across the Israeli rainfall gradient; in northern Israel, Birya Forest has 755 mm·$year^{-1}$ P, while other meteorological conditions are comparable to Yatir: GPP is 1.7 higher and Re is 1.4 higher than in Yatir. However, the NEP of 470 gCm^{-2} $year^{-1}$ is 2.7 times greater than in Yatir Forest (Ramati 2015). It is assumed that the high NEP of Birya Forest, beside of climatic factors, partially can be explained by the management effects: this forest was replanted some 20 years before this measurements began, following large scale damage caused by a heavy snow storm in 1992.

Besides total P, its distribution throughout the year differs: rainfall in Birya starts on average about a month earlier and ends a month later than in Yatir. Additionally, annual mean air humidity in Birya is 8% higher than in Yatir. As a result, the annual atmospheric water demand, i.e., VPD (vapor pressure deficit) of Birya is about 70% that of Yatir. Note that other meteorological parameters are very similar: Birya Forest is 1 °C colder than Yatir Forest and its radiation fluxes are ~1% lower than those of Yatir.

14.13 Dryland Forests as Carbon Sinks

More humid Mediterranean pine EC sites (Euroflux database, http://gaia.agraria.unitus.it/home), where rainy season is longer, showed more prolonged vegetation season shifted to late spring and summer (compared to winter and early spring in Yatir, see Rotenberg and Yakir 2010), lower radiation load (with annual mean of ~160–190 W·m^{-2} compared to 240 W·m^{-2} in Yatir), higher NEP magnitude (Fig. 14.11). The very high NEP in Bonis site comparable to that of highly productive temperate forests can be explained by the high site altitude creating optimal growth conditions – high precipitation and moderate temperature, as well as by management. However, H in Yatir is much higher than in more humid sites and LE is much lower. For example, in El Saler EC station (39.346°N, 0.319°W) near Valencia (Spain), with mean annual P of 550 mm, where relatively dry season (with 8–25 mm of rain per month) lasts during June–August, NEP remains positive during

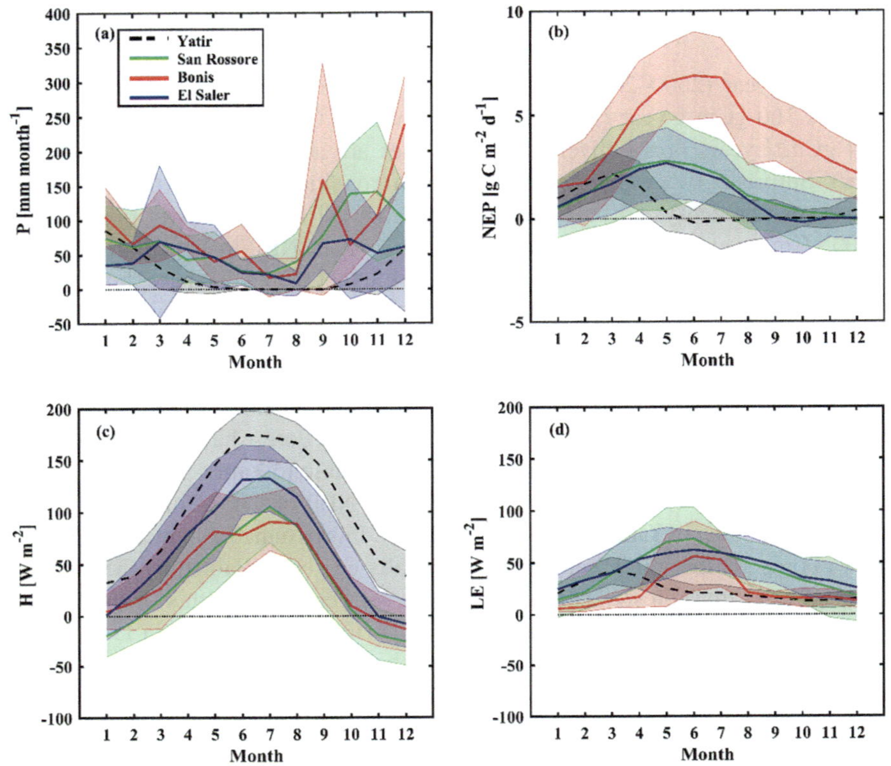

Fig. 14.11 Mean monthly dynamics of precipitation (**a**), NEP (**b**), H (**c**) and LE (**d**) in four Mediterranean pine forest EC sites: El Saler (39.346°N, 0.319°W, Spain), San Rossore (43.732°N, 10.291°E, Italy), Bonis (39.478°N, 16.535°E, Italy) and Yatir (Euroflux database, http://gaia.agraria.unitus.it/home)

the whole year, ranging from 0.3 gCm^{-2} d^{-1} in November to 3.2 gCm^{-2} d^{-1} in May. Mean annual total NEP was 610 gCm^{-2} $year^{-1}$, much higher than in Yatir. In even more humid site of Bonis (39.478°N, 16.535°E, 1130 m a.s.l., southern Italy) with 1170 mm of annual P, NEP is considerably higher, ranging from ~1.5 in January to ~6.5 in July–August. LE flux in Yatir was considerably lower and H was considerably higher than in other sites. In contrast with Yatir, where H > LE during the whole year, in more humid sites in winter LE dominates. The comparison of the EC sites under consideration showed that NEP generally increases with precipitation, but the sensitivity of NEP to P decreases with increasing P (Fig. 14.12a). Water use efficiency (WUE) also increases with P, but under P > 400 mm $year^{-1}$ no more increase observed (Fig. 14.12b).

Forest ecosystems have the highest potential for storing carbon of all biomes (Griscom et al. 2017) and according to the FluxNet database (as for 2019, https://fluxnet.fluxdata.org/) the mean annual NEP of all forest types is four times greater than that of grasslands. The average annual NEP of shrubland ecosystems is

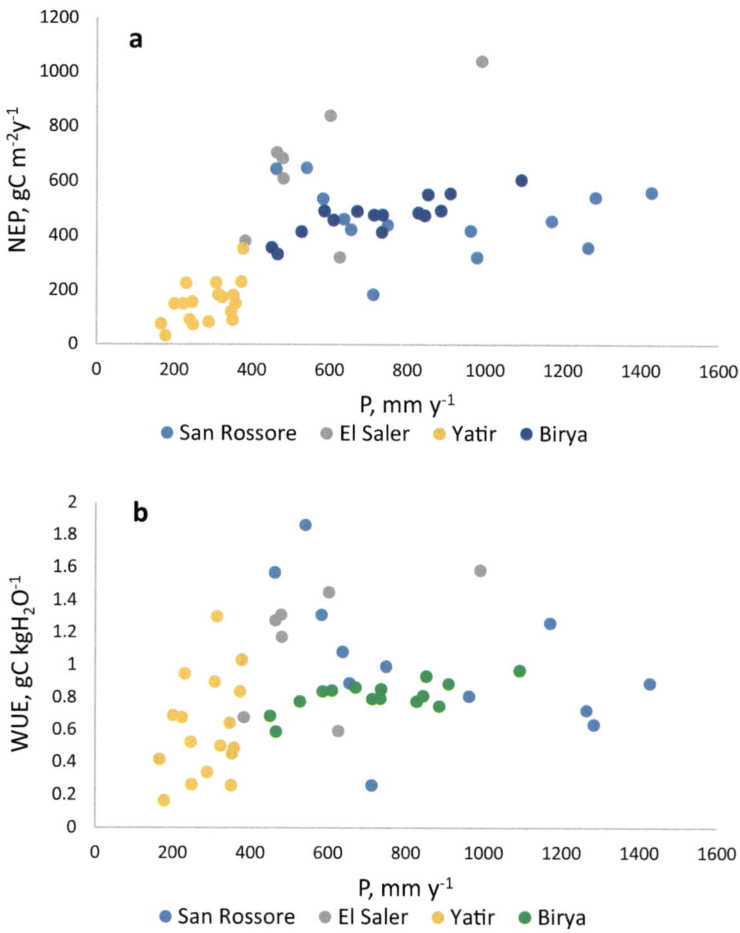

Fig. 14.12 Annual dependence of NEP (**a**) and WUE (**b**) of Mediterranean pine EC sites on annual precipitation

expected to be near zero (carbon neutral), although due to limited number of measurement sites accurate values are still needed. Measurements across the Israeli rainfall gradient with the mobile EC lab system have shown that the mean annual NEP of forests is 39 and 4 times greater than that of the surrounding grassland or shrubland ecosystems in the northern sub-humid (Birya) and semi-arid (Yatir) regions, respectively. As mentioned above, by capturing carbon from the atmosphere, forest ecosystems contribute to a long-term cooling effect on the earth's climatic system. The results of the Yatir carbon inventories support a considerable carbon sink potential in semi-arid soils and forest plantations, and imply that afforestation of even 10% of the semi-arid land area (which covers ~$27*10^6$ km^2 or 18% of the earth's land surface; Lal 2004) under conditions similar to that of the study site, could sequester ~$0.4*10^{15}$ g C year^{-1} (5% of current anthropogenically emitted carbon) over several decades, Qubaja et al. (2020a). Note, that this number is based on Yatir Forest, which is growing in extremely dry conditions, so the real number could be higher. Currently semi-arid lands represent about 20% of global carbon sink (Ahlström et al. 2015). However, the radiation balance of the forest could have the opposite effect to that of carbon sequestration, by direct warming the surface.

14.14 The Paradox of the Yatir Forest Energy Balance and the 'Convertor Effect'

The solar radiation load in Yatir Forest is very high, with an annual average of 240 Wm^{-2}, similar to the radiation load in many parts of the Sahara Desert. The evergreen Yatir Forest is considerably darker than the surrounding desert (see Fig. 14.1); the forest reflected radiation (albedo) is 12% of the incoming solar radiation, compared to ~25% in the surrounding open ecosystem. This translates to the forest trees absorbing ~29 Wm^{-2} more solar radiation on average. Due to the increased radiation absorbance one would expect that the forest foliage temperature would be higher than its surrounding land surface (Lee 2010), however our measurements in Yatir show that the forest surface is on annually average ~5 °C cooler than the surrounding desert soil surface; on mid-summer days a temperature difference of over 30 °C has been recorded.

Thermal emitted radiation is a function of the surface temperature; therefore, compared with the surrounding desert, the thermal radiation emitted by the forest is on average ~25 Wm^{-2} lower (Rotenberg and Yakir 2010, 2011). The combined effect of a higher amount of absorbed solar radiation and reduced emitted thermal radiation means that over the forest the mean annual radiation load is about 46 Wm^{-2} higher than over the desert. As Yatir Forest is growing at the desert edge with annual P of 285 mm, water availability for ET in the summer is highly limited, however the forest 'cools' itself through the highly efficient exchange of H by a process termed the 'convector effect' (Rotenberg and Yakir 2010). While in more wet ecosystems LE is the main cooling mechanism with an H to LE ratio (Bowen ratio, β) of up to 1 most days of the year, in Yatir Forest the average annual β is ~5.4 and in mid-summer it could be well above 10. As already mentioned, by absorbing a larger amount of radiation while releasing higher heat fluxes (H in the case of Yatir), the ecosystem has a warming effect on the lower atmosphere. At Birya, where the ground vegetation cover remains throughout the year and the soil is darker, both the albedo and the thermal radiation differences between the forest and non-forest ecosystems are much lower, only 40% of the differences in Yatir.

14.15 Effect of Afforestation on Cooling/Heating of the Atmosphere

By converting carbon uptake from the atmosphere to a change in the incoming atmospheric thermal radiation, it is possible to compare the two contrasting land cover change (LCC) effects, namely the biogeochemical (carbon sequestration cooling effect) and the biogeophysical (section 14.14, Claussen et al. 2013), on the energy exchange between ecosystems and the atmosphere (Betts 2000; Rotenberg and Yakir 2010) and to assess the overall LLC effect on the climate system. When performing this calculation, the following points should be noted: (1) Although the biogeochemical changes happen locally, its effect is global. This is because the atmosphere is a well-mixed medium for the CO_2 gas; in contrast the biogeophysical effects are probably more local. (2) While for a given land use change the biogeophysical effect, e.g., albedo change, can be assumed as 'fixed' over the years, the biogeochemical change, i.e., CO_2 reduction in the atmosphere, is additive (due to accumulated annual NEP in the ecosystem). (3) The calculation provides the change in the earth's energy balance at the surface. Additional calculations are needed to accurately estimate this effect at the top of the atmosphere and the LCC effect on the earth's climate system (Davin and de Noblet-Ducoudre 2010). In the Yatir region, with the dramatic effect of albedo and thermal changes, the heating effect on the earth's climate means that nearly 80 years of carbon accumulation are required to balance it. The semi-arid Yatir forest is about 55 years old and still accumulate carbon almost linearly with time, but its actual life cycle, and therefore the period of net "cooling" effect, are not known yet. Note however, that the forest is being renewed (mostly by management), and most carbon is accumulated in the soil with long residence time, due to the dry conditions. The estimated soil carbon storage potential at Yatir is of about 14 Kg C m^{-2} (Qubaja et al. 2020a), and because it is estimated to be far from saturation, soil carbon accumulation potential goes well beyand the expected forest life-cycle. The pine woody material is also harvested and used by the local forest owners and residences. At the northern site of Birya, with 3 times higher annual NEP and a considerably lower difference in albedo between forest and non-forest area, the time needed for balancing the two is about one-tenth the time required at Yatir.

Massive LCC affecting the climate in various others ways could create feedbacks that could enhance or relax the effects on the surface. To study the possible impact on climate conditions in a semi-arid region following large scale afforestation with a Yatir-type forest, Yosef et al. (2018) simulated afforestation activity in the Sahel and found that it will enhance cloud cover and rainfall and decrease the temperature burden in that region and beyond. Currently, the authors are not aware of any study on the climatic impacts of large scale afforestation activities across the entire Mediterranean region. We do note that there is evidence that this region's forest cover was much higher than today during the Roman Period (John 2005).

14.16 Conclusions

Yatir Forest is located in a region of severe drought conditions, at the edge of the desert with long term annual P of 285 mm, and it successfully survived and remains productive for more than 50 years. However, long drought events accompanied by bark beetle attacks can lead to significant tree mortality, especially in locations with deep soil, whereas in localities with rocky soil trees survive. It is not yet clear whether Yatir Forest could survive the harsher conditions projected for this region, however this forest provides an indication that Aleppo pine forests currently growing under milder conditions could survive deteriorating conditions, although adjustment to those conditions, e.g., density reduction, will be expected.

Despite the harsh environmental conditions Yatir Forest is a carbon sink with NEP comparable to the NEP of some European pine forests growing in much more favorable conditions because both components of NEP – GPP and Re – are much lower than in regions with wetter conditions. A few adjusting mechanisms by the forest trees were identified which enable the forest existence, including: taking full advantage of a short productive period at the early stage of the spring when soil is considerably wet after the winter rainfall while radiation load and temperature are relatively comfortable, their ability to photosynthesis during almost the whole long dry season in early and late hours of the day while shutting completely this activity during the midday hours, adjustment of the ecosystem leaf area at a low level (compared to humid forests) and other physiological mechanisms, which have not discussed in this paper.

Assessment of the C sequestration potential of afforestation in hot, semi-arid regions could be important to better assess the global potential for C storage in the land biosphere as a mean of mitigation of carbon accumulation in the atmosphere. Our study site provides robust quantitative data from an over-50-year-old, ~30 km^2 semi-arid forest plantation (non-irrigated, non-fertilized) representing the hot, dry edge for pine forest to grow (a timberline edge). The observations of C sequestration are therefore interesting to consider, as a first approximation, in the larger perspective: The results indicate that afforestation of even 10% of the semi-arid lands with the long-term NEE values of Yatir can store ~0.4 Pg C $year^{-1}$. It should be noted that about 70% of the total carbon sequestered (NEP) by Yatir Forest was organic carbon added to the soil layers (Qubaja et al. 2020a). While enhancing C storage in soils has been a contentious topic (Minasny et al. 2017; Amundson and Biardeau 2018; Schlesinger and Amundson 2019), we note that our study provides a distinct perspective in carbon cycle studies, supporting the idea that semi-arid afforestation could be important when considering climate mitigation strategies.

However, we must also consider that the effect of semi-arid afforestation on climate is complex, including both mitigating (carbon fixation) and enhancing (higher albedo) effects, as well as secondary effects, which could be assessed only by modelling, on global warming and local climatic conditions. It should be noted that Yatir conditions are extreme, and the albedo and thermal radiation biogeophysical effects are partially due to over-grazing in the surrounding desert. Under milder conditions, as along the Israeli rainfall gradient, the biogeochemical effect is stronger whereas the physical effect is weaker. All these facts support the idea that semi-arid afforestation could be important when considering climate mitigation strategies. Additionally, any afforestation activities have wide implications on other ecological services and in particular, because of higher evapotranspiration, forests probably decrease water yield, which can be an important issue in dry regions.

Acknowledgments Different parts of this long-term study were supported by KKL-JNF, the Israel Science Foundation (ISF), Israel Ministry of Science and the Ministry of National Education, the Israeli Water Authority (grant 4500962964), Joint German-Israel research program (GLOWA2 Jordan River), German Research Foundation (DFG) as part of the project "Climate feedbacks and benefits of semi-arid forests" (CliFF; grant no. YA 274/1-1 and SCHM 2736/2-1). The long-term operation of the Yatir Forest Research Site is supported by the Cathy Wills and Robert Lewis Program in Environmental Science. The authors thank Efrat Schwartz for assistance with lab work. We thank the entire Yatir team for technical support and the local KKL staff for their cooperation.

References

Ahlström A, Raupach MR, Schurgers G et al (2015) The dominant role of semi-arid ecosystems in the trend and variability of the land CO_2 sink. Science 348(6237):895–899. https://doi.org/10.1126/science.aaa1668

Alpert P, Ben-Gai T, Baharad A et al (2002) The paradoxical increase of Mediterranean extreme daily rainfall in spite of decrease in total values. Geophys Res Lett 29(11):31.1–31.4

Amundson R, Biardeau L (2018) Soil carbon sequestration is an elusive climate mitigation tool. Proc Natl Acad Sci U S A 115:11652–11656

Aubinet M, Grelle A, Ibrom A et al (2000) Estimates of the annual net carbon and water exchange of Europeran forests: the EUROFLUX methodology. Adv Ecol Res 30:113–175

Baldocchi D (2014) Measuring fluxes of trace gases and energy between ecosystems and the atmosphere – the state and future of the eddy covariance method. Glob Chang Biol 20(12):3600–3609. https://doi.org/10.1111/gcb.12649

Baldocchi D, Ma SY (2013) How will land use affect air temperature in the surface boundary layer? Lessons learned from a comparative study on the energy balance of an oak savanna and annual grassland in California, USA. Tellus B 65(1). https://doi.org/10.3402/tellusb.v65i0.19994

Betts RA (2000) Offset of the potential carbon sink from boreal forestation by decreases in surface albedo. Nature 408(6809):187–190

Biederman JA, Scott RL, Bell TW et al (2017) CO_2 exchange and evapotranspiration across dryland ecosystems of southwestern North America. Glob Chang Biol 23(10):4204–4221. https://doi.org/10.1111/gcb.13686

Bonan GB (2008) Forests and climate change: forcings, feedbacks, and the climate benefits of forests. Science 320(5882):1444–1449

Brugger P, Banerjee T, De Roo F et al (2018) Effect of surface heterogeneity on the boundary-layer height: a case study at a semi-arid forest. Bound-Layer Meteorol 169:233–250. https://doi.org/10.1007/s10546-018-0371-5

Burba G (2013) Eddy covariance method for scientific, industrial, agricultural, and regulatory applications: a field book on measuring ecosystem gas exchange and areal emission rates. LI-COR Biosciences, Lincoln, 331 pp

Caliskan S, Boydak M (2017) Afforestation of arid and semiarid ecosystems in Turkey. Turk J Agric For 41(5):317–330

Cao S, Chen L, Shankman D et al (2011) Excessive reliance on afforestation in China's arid and semi-arid regions: lessons in ecological restoration. Earth-Sci Rev 104(4):40–245
Claussen M, Bathiany S, Brovkin V et al (2013) Simulated climate-vegetation interaction in semi-arid regions affected by plant diversity. Nat Geosci 6(11):954–958
Davin EL, de Noblet-Ducoudre N (2010) Climatic impact of global-scale deforestation: radiative versus nonradiative processes. J Clim 23(1):97–112
Debreczy Z, Rácz I (2011) Conifers around the world. Dendropress Ltd, Budapest
Giorgi F, Lionello P (2008) Climate change projections for the Mediterranean region. Glob Planet Chang 63(2–3):90–104. https://doi.org/10.1016/j.gloplacha.2007.09.005
Granier A, Loustau D (1994) Measuring and modeling the transpiration of a maritime pine canopy from sap-flow data. Agric For Meteorol 71:61–81
Griscom BW, Adams J, Ellis PW et al (2017) Natural climate solutions. Proc Natl Acad Sci U S A 114(44):11645–11650
Grünzweig JM, Gelfand I, Fried Y et al (2007) Biogeochemical factors contributing to enhanced carbon storage following afforestation of a semi-arid shrubland. Biogeosciences 4:891–904
Houghton JT, Ding Y, Griggs DJ, Noguer M, van der Linden PJ, Dai X, Maskell K, Johnson CA (eds) (2001) Climate change 2001: the scientific basis. Cambridge University Press, Cambridge/New York, 881pp
Huang JP, Yu HP, Guan XD et al (2016) Accelerated dryland expansion under climate change. Nat Clim Chang 6(2):166–171. https://doi.org/10.1038/nclimate2837
John P (2005) A forest journey: the story of wood and civilization. The Countryman Press, Vermont
Keenan TF, Migliavacca M, Papale D et al (2019) Widespread inhibition of daytime ecosystem respiration. Nat Ecol Evol 3:407–415. https://doi.org/10.1038/s41559-019-0809-2
Klein T, Di Matteo G, Rotenberg E et al (2013) Differential ecophysiological response of a major Mediterranean pine species across a climatic gradient. Tree Physiol 33(1):26–36. https://doi.org/10.1093/treephys/tps116
Klein T, Cohen S, Paudel I et al (2016) Diurnal dynamics of water transport, storage and hydraulic conductivity in pine trees under seasonal drought. Iforest 9:710–719. https://doi.org/10.3832/ifor2046-009
Lal R (2004) Carbon sequestration in dryland ecosystems. Environ Manag 33(4):528–544. https://doi.org/10.1007/s00267-003-9110-9
Lasslop G, Reichstein M, Papale D et al (2010) Separation of net ecosystem exchange into assimilation and respiration using a light response curve approach: critical issues and global evaluation. Glob Chang Biol 16(1):187–208
Le Quéré C, Andrew RM, Friedlingstein P et al (2018) Global carbon budget 2017. Earth Syst Sci Data 10(1):405–448. https://doi.org/10.5194/essd-10-405-2018
Lee X (2010) Forests and climate: a warming paradox. Science 328(5985):1479–1479
Luyssaert S, Inglima I, Jung M et al (2007) CO_2 balance of boreal, temperate, and tropical forests derived from a global database. Glob Chang Biol 13(12):2509–2537
Ma XL, Huete A, Cleverly J et al (2016) Drought rapidly diminishes the large net CO_2 uptake in 2011 over semi-arid Australia. Sci Rep 6:37747. https://doi.org/10.1038/srep37747
Maseyk KS, Lin T, Rotenberg E et al (2008) Physiology-phenology interactions in a productive semi-arid pine forest. New Phytol 178(3):603–616
Minasny B, Malone BP, McBrateny AB et al (2017) Soil carbon 4 per mille. Geoderma 292:59–86
Mirov NT (1967) The genus *Pinus*. The Ronald Press, New York
Ne'eman G, Trabaud L (2000) Ecology, biogeography and management of *Pinus halepensis* and *P. brutia* forest ecosystems in the Mediterranean Basin. Backhuys Pub, Leiden
Pan YD, Birdsey RA, Fang JY et al (2011) A large and persistent carbon sink in the world's forests. Science 333(6045):988–993
Poulter B, Frank D, Ciais P et al (2014) Contribution of semi-arid ecosystems to interannual variability of the global carbon cycle. Nature 509(7502):600–603. https://doi.org/10.1038/nature13376

Preisler Y (2020) Water-use strategies leading to resilience of pine trees to global climatic change [Unpublished PHD thesis]. The Hebrew University Jerusalem

Preisler Y, Tatarinov F, Grünzweig JM et al (2019) Mortality versus survival in drought-affected Aleppo pine forest depends on the extent of rock cover and soil stoniness. Funct Ecol 33(5):901–912

Price RA, Liston A, Strauss SH (1998) Phylogeny and systematics of *Pinus*. In: Richardson DM (ed) Ecology and biogeography of *Pinus*. Cambridge University Press, Cambridge, pp 49–68

Qubaja R, Grunzweig JM, Rotenberg E et al (2020a) Evidence for large carbon sink and long residence time in semiarid forests based on 15-year flux and inventory records. Glob Chang Biol 26(3):1626–1637. https://doi.org/10.1111/gcb.14927

Qubaja R, Amer M, Tatarinov F et al (2020b) Partitioning evapotranspiration and its long-term evolution in a dry pine forest using measurement-based estimates of soil evaporation. Agric For Meteorol 281:107831. https://doi.org/10.1016/j.agrformet.2019.107831

Qubaja R, Tatarinov F, Rotenberg E et al (2020c) Partitioning of canopy and soil CO_2 fluxes in a pine forest at the dry timberline. Biogeosciences 17:699–714. https://doi.org/10.5194/bg-17-699-2020

Ramati E (2015) Tradeoffs between carbon sequestration and radiation budget in influencing climate along the precipitation gradient in Israel. MSc dissertation, The Weizmann Institute of Science, Rehovot, Israel

Reichstein M, Falge E, Baldocci D et al (2005) On the separation of net ecosystem exchange into assimilation and ecosystem respiration: review and improved algorithm. Glob Chang Biol 11–13:1424–1439

Rohatyn S, Rotenberg E, Ramati E et al (2018) Differential impacts of land use and precipitation on "ecosystem water yield". Water Resour Res 54(8):5457–5470

Rotenberg E, Yakir D (2010) Contribution of semi-arid forests to the climate system. Science 327(5964):451–454

Rotenberg E, Yakir D (2011) Distinct patterns of changes in surface energy budget associated with forestation in the semiarid region. Glob Chang Biol 17:1536–1548

Schlesinger WH, Amundson R (2019) Managing for soil carbon sequestration: let's get realistic. Glob Chang Biol 25:386–389

Scott RL, Biederman JA, Hamerlynck EP et al (2015) The carbon balance pivot point of southwestern US semiarid ecosystems: insights from the 21st century drought. J Geophys Res-Biogeosci 120(12):2612–2624. https://doi.org/10.1002/2015jg003181

Sherwood S, Fu Q (2014) A drier future? Science 343(6172):737–739. https://doi.org/10.1126/science.1247620

Tatarinov F, Rotenberg E, Maseyk K, Ogée J, Klein T, Yakir D (2015) Resilience to seasonal heat wave episodes in a Mediterranean pine forest. New Phytol. https://doi.org/10.1111/nph.13791

Tatarinov F, Rotenberg E, Yakir D et al (2017) Forest GPP calculation using sap flow and water use efficiency measurements. Bio-Protocol 7(8):e2221. https://doi.org/10.21769/BioProtoc.2221

Wang KC, Dickinson RE (2012) A review of global terrestrial evapotranspiration: observation, modeling, climatology, and climatic variability. Rev Geophys 50(2). https://doi.org/10.1029/2011RG000373

Yan D, Scott RL, Moore DJ et al (2019) Understanding the relationship between vegetation greenness and productivity across dryland ecosystems through the integration of PhenoCam, satellite, and eddy covariance data. Remote Sens Environ 223:50–62. https://doi.org/10.1016/j.rse.2018.12.029

Yosef G, Walko R, Avisar R et al (2018) Large-scale semi-arid afforestation can enhance precipitation and carbon sequestration potential. Sci Rep 8(996). https://doi.org/10.1038/s41598-018-19265-6

Chapter 15
Soil Properties and Biogeochemical Functioning in Mediterranean Pine Forests

Antonio Gallardo

15.1 Introduction

Mediterranean regions are characterized by a high diversity of soils (Ibáñez et al. 2013). The properties of these soils are shaped by the climatic characteristics, the abrupt topography, dust deposition from deserts, and the sustained and extensive influence of humans (Yaalon 1997). Winter meteorological conditions contribute to the weathering of silicate minerals and the dissolution of limestone and dolomite, while summer conditions are responsible for the formation of reddish minerals such as ferrihydrites and hematites, which are considered typical of Mediterranean soils. Other common properties of Mediterranean soils include the presence of an argillic B horizon due to the leaching of clay particles into deeper horizons (Fedoroff 1997) and the formation of a calcic horizon by dissolution of calcium carbonate minerals and its posterior precipitation at a deeper horizon, a process shared with other semi-arid climates (Nettleton 1991). Another important characteristic of the soils in the Mediterranean Basin is the influence of Saharan dust (Yaalon 1997; Erel and Torrent 2010; Zdruli et al. 2011). For example, Erel and Torrent (2010) estimated that Al-silicates in the fine soil fraction (less than 5μm) contain 33–86% Saharan dust. Finally, soil erosion resulting from the abrupt landscape and intense human intervention have led to the development of shallow soils (e.g. leptosols, regosols and cambisols), where most native pine forest can be found, since deeper soils (e.g. luvisols) are preferentially dedicated to agriculture. Overall, Mediterranean soils

A. Gallardo (✉)
Department of Physical, Chemical and Natural Systems, University Pablo de Olavide, Sevilla, Spain
e-mail: agallardo@upo.es

G. Ne'eman, Y. Osem (eds.), *Pines and Their Mixed Forest Ecosystems in the Mediterranean Basin*, Managing Forest Ecosystems 38,
https://doi.org/10.1007/978-3-030-63625-8_15

have medium to low fertility (Yaalon 1997), although differences can be found between the Mediterranean-type climate regions. Thus, Central Chile is considered nutrient rich (Miller 1981), and the vegetation is basically limited by water availability. California and the Mediterranean Basin are also primarily limited by water and climate, but soils in these regions are less fertile, and nutrient availability may have a clear influence on primary production. In contrast to Chile, the Mediterranean regions of Australia and South Africa are nutrient-poor, thus soil fertility is a critical factor explaining productivity and plant diversity in these regions (Wisheu et al. 2000).

In the absence of major disturbances, three carbon (C) and nutrient input pathways are responsible for soil functioning in Mediterranean pine forests. The first path is litter decomposition (see Grünzweig and Gliksman this volume, Chap. 15) in which C and nutrients are incorporated in soluble and insoluble forms to be further processed by soil biota. The second important C and nutrient source for soil microorganisms comes from root exudates (Shi et al. 2011). Finally, dry and wet C and nutrient deposition from the atmosphere can be also an important input of fuel for the soil functioning engine (Gioda et al. 2008; Reche et al. 2009; De Vicente et al. 2012). As mentioned above, the Mediterranean Basin is frequently affected by dust inputs from the Sahara Desert (Loÿe-Pilot et al. 1986).

In Mediterranean ecosystems, fire (Ne'eman and Arianoutsou this volume), insect outbreaks (Vázquez-González et al. (2021) this volume, Chap. 11) and fungal pests (Oliva this volume, Chap. 10) are major disturbances that disrupt normal soil functioning, including biotic composition and activity and the cycling of essential nutrients for life (Barbero et al. 1990). In addition, the Mediterranean Basin is considered one of the regions that will be more affected by the impacts of climate change on the water cycle (Ozturk et al. 2015). The projected decline in soil water availability in the coming decades may interact with warmer conditions to exacerbate the water deficit (Hernández-Santana et al. 2009; Sardans and Peñuelas 2013; Peñuelas et al. 2018). Warming can also have a direct effect on nutrient cycling during wet winters by increasing soil enzyme activity (Sardans et al. 2006). This positive effect on winters could be counterbalanced by lower summer water availability (Sardans and Peñuelas 2013).

In this chapter I will focus mainly on carbon (C), nitrogen (N) and phosphorus (P), the most essential elements and nutrients for life that also frequently limit primary and secondary production in Mediterranean ecosystems. I will first present the major soil properties typical of Mediterranean regions, including the Mediterranean Basin, and the Mediterranean-type ecosystems in Chile, California, South Africa and Australia. I will then concentrate on major indicators of soil functioning, with a special emphasis on the flux of soil C to the atmosphere (soil respiration rate), soil enzymes activity related to C, N and P cycling, and nutrient availability in pine forests under a Mediterranean climate.

15.2 Soil Properties

15.2.1 pH

Mediterranean pine forests can be found across a wide range of soil pH. However, in Mediterranean regions the dominant pH range is 6–8, in contrast to the more acidic soils of the temperate zones and the more basic soils of the arid zones (Fig. 15.1). This neutral to basic pH range is probably due to the sedimentary origin

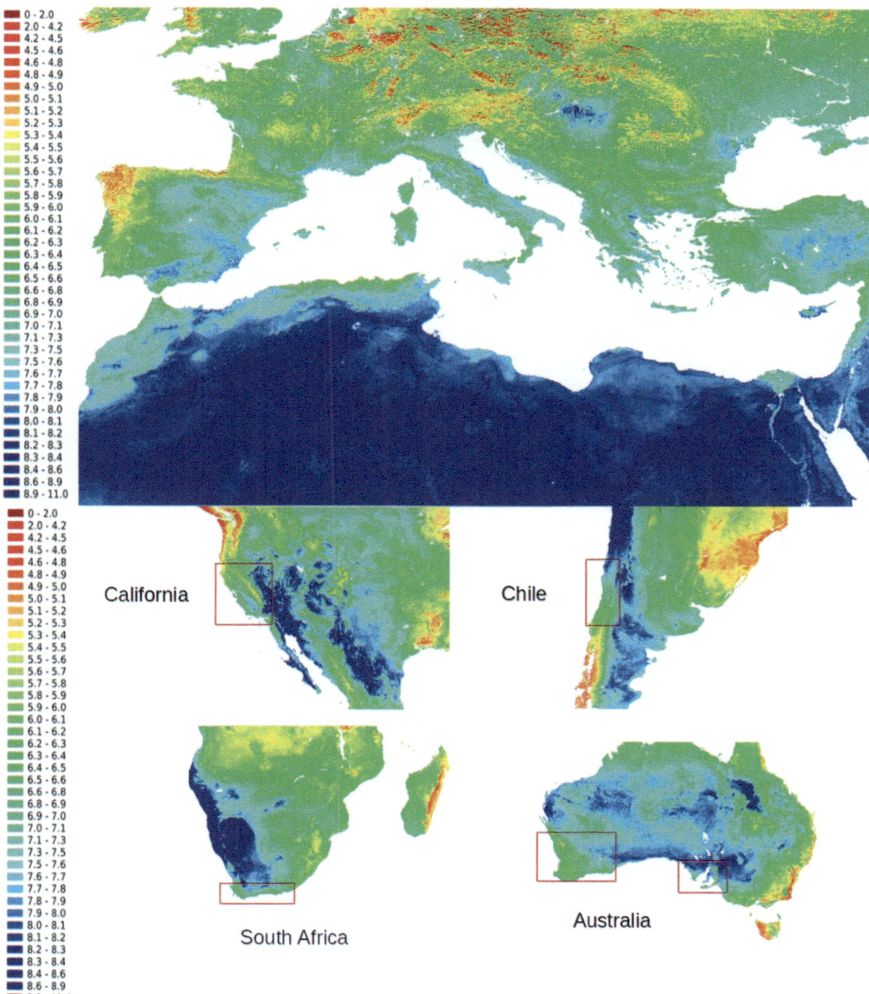

Fig. 15.1 Soil pH (0–5 cm) distribution throughout the Mediterranean Basin (top) and in the Californian, Chilean, South African and Australian Mediterranean regions (bottom). (Maps extracted from ISRIC-World Soil Information, December 2019)

of most Mediterranean soils, particularly limestone and dolomite. Neutral to acid soils are also found where soils develop on igneous and metamorphic rocks. Thus, microbial communities (such as ammonia oxidizing bacteria) or soil processes (such as nitrification rates), which are frequently limited at low pH, should not affect the functioning of most Mediterranean pine forest soils (Nugroho et al. 2007; Lauber et al. 2008; Yao et al. 2011). Although site-specific, in some cases clear soil acidification under pines occurs (Rutigliano et al. 2004; Iovieno et al. 2010). Rapid soil acidification has been observed in pine forests under high atmospheric N deposition (Ochoa et al. 2014).

15.2.2 Cation Exchange Capacity

Cation exchange capacity (CEC) is provided by clay minerals and organic matter, and is also influenced by pH. The combination of these three factors in the Mediterranean regions produces CEC in the range 15–50 mmol(c)/kg (Fig. 15.2). Although these values tended to be lower than those found under temperate climates, cation availability is unlikely to limit primary production in Mediterranean pine forests (but see Sardans and Peñuelas 2015). In some cases, the organic matter from pine stands shows higher CEC than in other forest types under the same conditions and pH (Gruba and Mulder 2015).

15.2.3 Organic Carbon

Soils are the most important terrestrial reservoir of C, with estimates of ca. 1460 Pg of C globally (Scharlemann et al. 2014). Whether soils are acting as a sink or source of atmospheric C is a central issue in global change modeling and mitigation purposes. Soil organic C in Mediterranean ecosystems is usually low compared with temperate ecosystems. For example, the Mediterranean Basin constitutes an intermediate stage between the lower values of North African arid ecosystems and the higher values found at latitudes comprising the temperate and boreal ecosystems of Central and North Europe (Fig. 15.3). Similar gradients of soil C can be found along the Californian and Chilean coasts, where the Mediterranean region is delimited by arid (lower carbon) and temperate (higher carbon) regions. In South Africa and Australia, the Mediterranean regions are delimited mostly by arid regions with lower soil C content (Fig. 15.3). Mediterranean pine forests have contributed to the C increase of soils. For example, Quideau et al. (1998) observed an increase of 2030 g/m^2 of C in the 0–1 m of the soil profile after four decades of soil development under *Pinus coulteri* D.Don in California. Similarly, Martín-Peinado et al. (2016) observed a significant increase in soil organic C (0–5 cm) in pine plantations (*P. sylvestris L., P. pinaster Aiton, P. halepensis Mill., P. radiata D.Don, P. pinea L.*) in relation to unplanted areas, with mean values similar to those found under native forests.

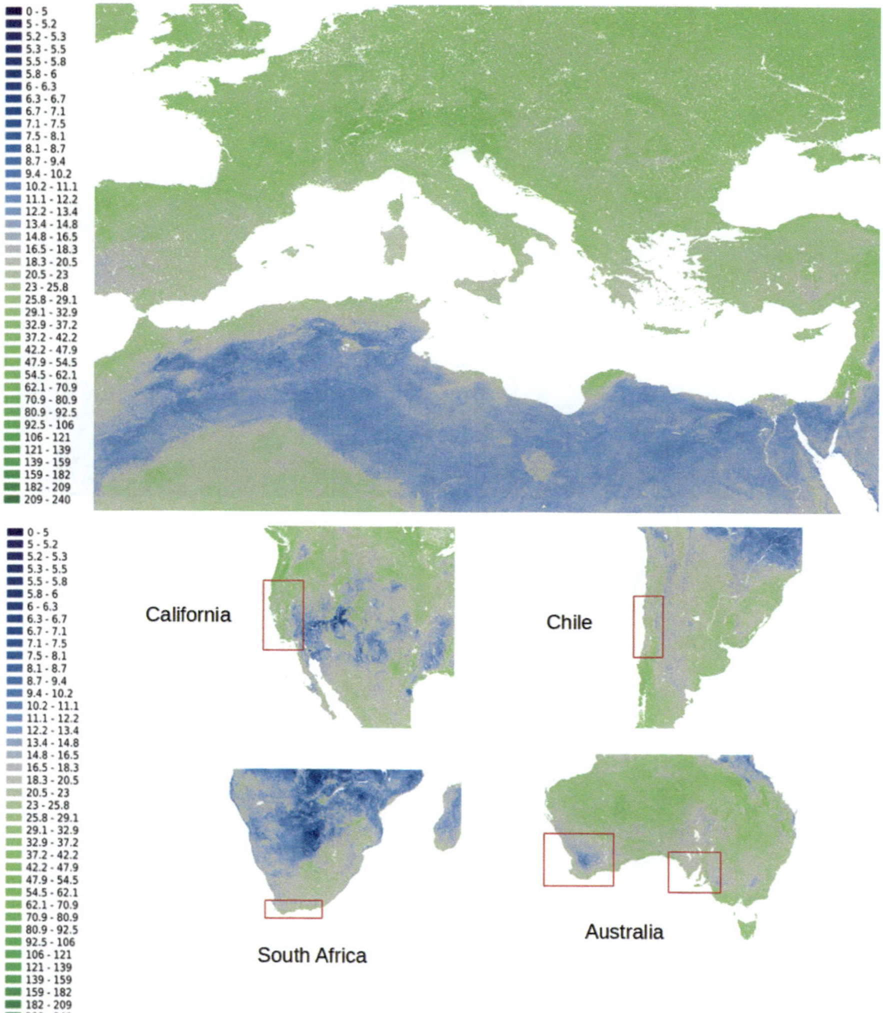

Fig. 15.2 Soil CEC (0–5 cm) distribution throughout the Mediterranean Basin (top) and in the Californian, Chilean, South African and Australian Mediterranean regions (bottom). Units are mmol(c)/kg. (Maps extracted from ISRIC-World Soil Information, December 2019)

15.2.4 Organic Nitrogen

The pattern of organic N is very similar to that of organic C, with intermediate levels in Mediterranean regions compared with the surrounding temperate and drier regions (Fig. 15.4). However Mediterranean pine forests have a characteristically higher C:N ratio that other native species (Cools et al. 2014). Clear differences can be found among the different Mediterranean regions. South Africa and Australia

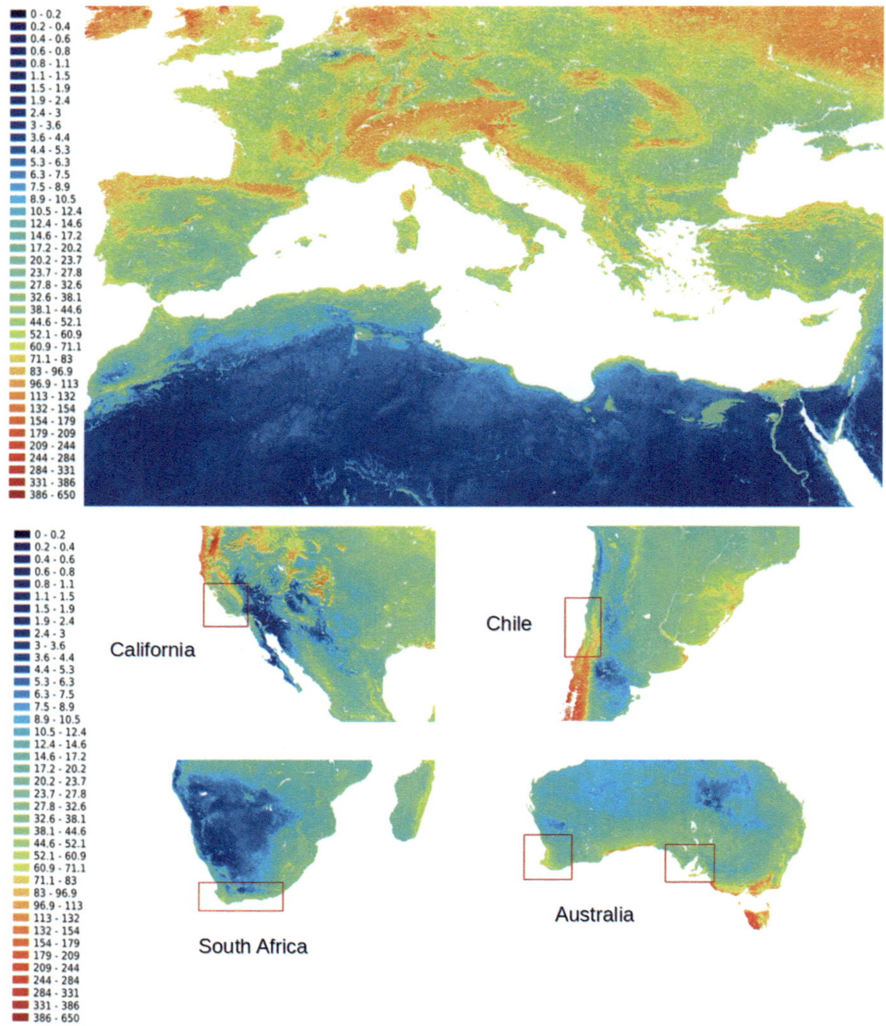

Fig. 15.3 Soil organic C (0–5 cm) distribution around the Mediterranean Basin (top) and in the Californian, Chilean, South African and Australian Mediterranean regions (bottom). Units are dg/kg. (Maps extracted from ISRIC-World Soil Information, December 2019)

exhibit the lowest organic N content, in accordance with the characteristically low nutrient content described for these regions. Contrastingly, the Mediterranean region of Central Chile exhibits the highest N content, with intermediate levels for the Mediterranean Basin and California (Fig. 15.4).

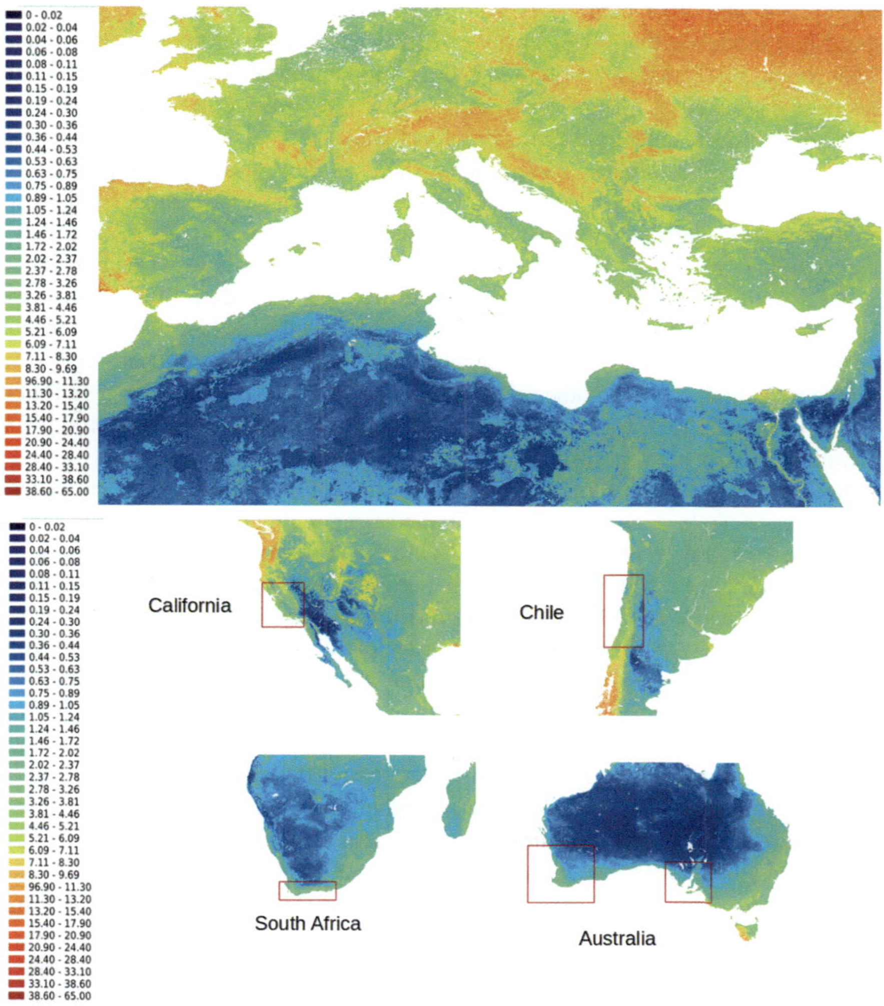

Fig. 15.4 Soil N (0–5 cm) distribution around the Mediterranean Basin (top) and in the Californian, Chilean, South African and Australian Mediterranean regions (bottom). Units are cg/kg. (Maps extracted from ISRIC-World Soil Information, December 2019)

15.3 Carbon Cycling in Mediterranean Pine Forests

Carbon dioxide is produced in soils by roots and soil organisms, although non-biological processes may contribute to soil CO_2 efflux (Rey 2015). The movement of CO_2 from the soil to the atmosphere is controlled by the rate of CO_2 production in the soil, the difference in the CO_2 concentration between the soil and the atmosphere, and other parameters such as soil texture, air temperature, and wind speed that modulate the movement of CO_2 through and out of the soil (Raich and Schlesinger 1992). Mean respiration rate in Mediterranean woodlands and heath has been estimated at 713 g C/m^2/year, a similar rate to those found for temperate coniferous forests, temperate deciduous forests and tropical dry forests (Raich and Schlesinger 1992). Higher mean value (813 g C/m^2/year) was reported for Mediterranean ecosystems (Bond-Lamberty and Thomson 2010), and a similar respiration rate was recently published for temperate evergreen forests (773 g C/m^2/year, Hursh et al. 2017). However, in some high pH soils (above 8) such as those found in some Mediterranean regions, CO_2 efflux may be underestimated, because CO_2 can be dissolved as bicarbonate in water (Angert et al. 2015). The relationship between soil respiration rate and primary production in Mediterranean regions is known to be higher than in other ecosystems with similar rates (see Fig. 1 in Raich and Schlesinger 1992), suggesting that in Mediterranean ecosystems, conditions for microbial activity are relatively more favorable than for plant production.

Soil respiration rates show high temporal variability in Mediterranean ecosystems (Fig. 15.5). In general, soil humidity and temperature drive these rates, but in a complex pattern, where wetting and drying events, seasons and root respiration interact to produce a complex temporal pattern (Carbone et al. 2011; Morillas et al. 2017). In a *P. muricata* D.Don forest in California, Carbone et al. (2011) found that rainfall pulses derived from fog-drip, and precipitation after long dry periods, correlated with the largest soil respiration responses, a phenomenon called the Birch effect (Birch 1958; Jarvis et al. 2007). In the same study, microbial respiration responded within hours to soil respiration drivers, whereas the response of root respiration was delayed for a few days. They concluded that heterotrophic and autotrophic respiration respond differently in timing and quantity to moisture inputs. Sometimes a threshold in soil humidity or rainfall is needed for a response to a pulse of water availability. In a *P. pinaster* forest in Italy, Matteucci et al. (2015) found that soil water availability controlled soil respiration from May to October but this control passed to soil temperature from November to March. High peaks were found only when rain pulses were greater than 10 mm on dry soil. A similar transition where irradiance was also related to respiration was found in a *P. halepensis* plantation by Grünzweig et al. (2009). In the same way, in *P. sylvestris* and *Pinus*-oak stands in Central Spain, showing minimum soil respiration rates in both winter and summer, Inclán et al. (2007) found that those rates were highly influenced by soil temperatures when soil moisture was above a drought threshold of 15%. Below this threshold value, soil moisture was a good predictor of soil respiration in *Pinus* stands.

Although these studies measured soil respiration rates in response to changes in soil water, these rates are also affected by the timing of aboveground and belowground litter inputs, root exudates and changes in the microbial biomass. Also, soil enzymes are directly responsible for the decomposition of soil organic matter, and the resulting metabolites can be respired or assimilated by soil microbes. Soil enzyme activity has been shown to be more intense in the litter layer than in soil in *P. nigra* stands in Central Spain (Lucas-Borja et al. 2019), but considering the soil

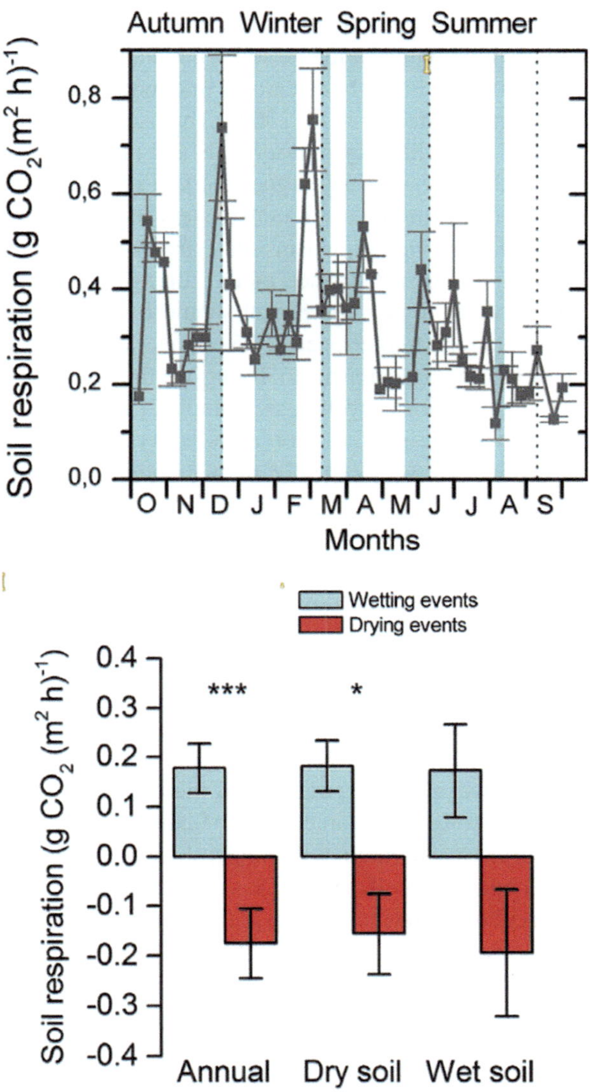

Fig. 15.5 (Top) Weekly soil respiration rates in a *Pinus pinea* stand in southwest Spain. Blue bands indicate wetting events. (Bottom) Changes in soil respiration rate during drying-wetting cycles. Data were analyzed for the entire year and according to initial soil conditions (dry or wet) prior to the drying-wetting cycle. We considered a wetting interval to occur on a previously wet soil when the initial soil moisture in % of the WHC was above 20%. A drying interval was considered to occur within a dry soil when the initial soil moisture in % of the WHC (i.e., immediately before the start of the drying interval) was below 40%. Error bars are ±SE, with N = 8 for the annual average and N = 4 for initially dry or wet soils. Significant differences are indicated as $P < 0.05$ (*), $P < 0.01$ (**), and $P < 0.001$ (***). (From Morillas et al. 2017)

volume, most enzyme activity is found in the bulk soil. The incorporation of these mechanisms provides more information on soil functioning than the mere measurement of soil CO_2 efflux and should be integrated into models of C dynamics during drying–wetting cycles (Lawrence et al. 2009). Thus, Zeglin et al. (2013) found enhanced microbial C use efficiency in drier soils and in soils with prolonged dry periods between wetting events. They argue that soil microbial function may be linked with plant function (litter inputs, root exudates) and that microbial activity in dry soils might be driven by microorganisms surviving or growing efficiently during periods of water deficit. The importance of the timing of precipitation, and the length and intensity of the dry or wet period should be taken into account when evaluating C functioning in Mediterranean ecosystems (Morillas et al. 2017).

The response of soil microorganisms to drying-wetting cycles in Mediterranean forests and its implications for C cycling are complex (Schimel 2018). In general, microbial activity and soil respiration drop as soils dry (Carbone et al. 2011; Wu et al. 2011), but this decline varies greatly among studies. For example, Lu et al. (2017) found little response of soil respiration to large reductions in soil moisture. Sometimes a reduction in soil water content led to the augmentation of oxygen in the soil profile and the reactivation of aerobic microbial activity in saturated soils (Davidson et al. 2012). The main proposed mechanism involved in the reduction of microbial activity under low moisture conditions is a physiological response to stress (Schimel et al. 2007). However, the high diversity of responses to water stress by different groups of microorganisms contradicts the typical pattern of decrease in microbial activity with soil water potential (Manzoni et al. 2014; Schimel 2018). As outlined above, the inclusion of extracellular enzyme activity may shed some light on the relationship between soil respiration and the typical drying-wetting cycles found in Mediterranean forests. A characteristic of extracellular enzymes is the need for smaller water surfaces and volumes than microbial cells, which allows enzyme activity to operate longer with soil drying (Geisseler et al. 2011). However, depolymerization by extracellular enzymes, which is the first step leading to C oxidation, shows variable responses to moisture (Sardans et al. 2008; Alster et al. 2013), perhaps because measurement of enzyme activity in the lab is the result of combining the amount of enzyme in the soil, the individual activity of these molecules, the use of an artificial substrate and optimal conditions of temperature and moisture, which are far from the natural conditions (German et al. 2011; Schimel et al. 2017; Schimel 2018). A review of studies using similar methods to measure extracellular soil enzyme activity in Mediterranean soils showed that the activity of important extracellular enzymes linked to C, N and P cycles, such as urease, protease, β-glucosidase, and phosphatase, was similar to that of other temperate ecosystems, suggesting that Mediterranean ecosystems invest similar amounts of resources to produce soil enzymes as more productive environments (Estiarte et al. 2008; Sardans and Peñuelas 2013).

We are also far from having a deep understanding of the rapid response of microbial activity to re-wetting (the Birch effect, Birch 1958; Jarvis et al. 2007; Göransson et al. 2013). Schimel (2018) reviewed the process involved in this rapid response. The possibilities include the respiration of osmolites excreted by cells to regulate internal osmotic pressure upon re-wetting, the metabolism of microbial corpses that died during the drought or during rewetting, and/or the mobilization of stable C by re-wetting, resulting in a new C source available for microorganisms. Schimel (2018) concluded that if the pulse of C release is from microbial materials, drying–wetting cycles have little impact on long-term C loss from the soil. But if the pulse comes from the mobilization of stable C, then drying-wetting cycles may have a long-term negative impact on soil C storage. Both mechanisms (carbon loss from microbial biomass and from mobilized stable C) may operate simultaneously in a drying-wetting cycle. Thus, microbial biomass C would be the first source of C upon re-wetting, and mobilized C may be used to replace lost biomass C (Schimel 2018).

15.4 Nitrogen Cycling in Mediterranean Pine Forests

Nitrogen cycling is also affected by drying-wetting cycles, and some principles discussed for C cycling are also applicable to N. Morillas et al. (2013) evaluated the effects of the continuous changes (weekly intervals for 1 year) in soil N pools during natural wetting and drying events in a Mediterranean *P. pinea* forest, including microbial biomass [MB-N], dissolved organic N [DON], NH_4^+ and NO_3^- (Fig. 15.6). All soil N pools showed a complex pattern with high intra- and inter-seasonal variability, more noticeable in inorganic than organic N pools.

In the study by Morillas et al. (2013), most soil N pools (MB-N, DON and resin-NH_4) increased during wetting events, but remarkably, soil-NO_3^- and resin-NO_3^- showed the opposite trend, suggesting a low nitrification or high denitrification response to water pulses. NO and N_2O emission pulses upon re-wetting dry soils have been linked with the mobilization of mineral and organic N (Leitner et al. 2017). Morillas et al. (2013) concluded that changes in N pools were not coupled with changes in organic matter or labile C. Increases of N mineralization when dry soils are rewetted has been frequently found (Saetre and Stark 2005; Leitner et al. 2017). A rapid decrease of inorganic N followed by a pulse of ammonification 24 h after re-wetting has also been observed (Placella and Firestone 2013; Göransson et al. 2013; Morillas et al. 2013, 2017). Nitrogen mineralization dominating over immobilization upon re-wetting suggests that N-rich substrates are available for soil microorganisms (Kleber et al. 2007; Liang et al. 2017). Nitrification and denitrification control N losses by leaching and/or emitting N_2O, NO or NO_2 to the atmosphere (Groffman and Tiedge 1988). Therefore, drying-wetting cycles may produce larger N losses compared to C losses (Schimel 2018), which may reveal a mechanism of permanent N limitation in Mediterranean forests.

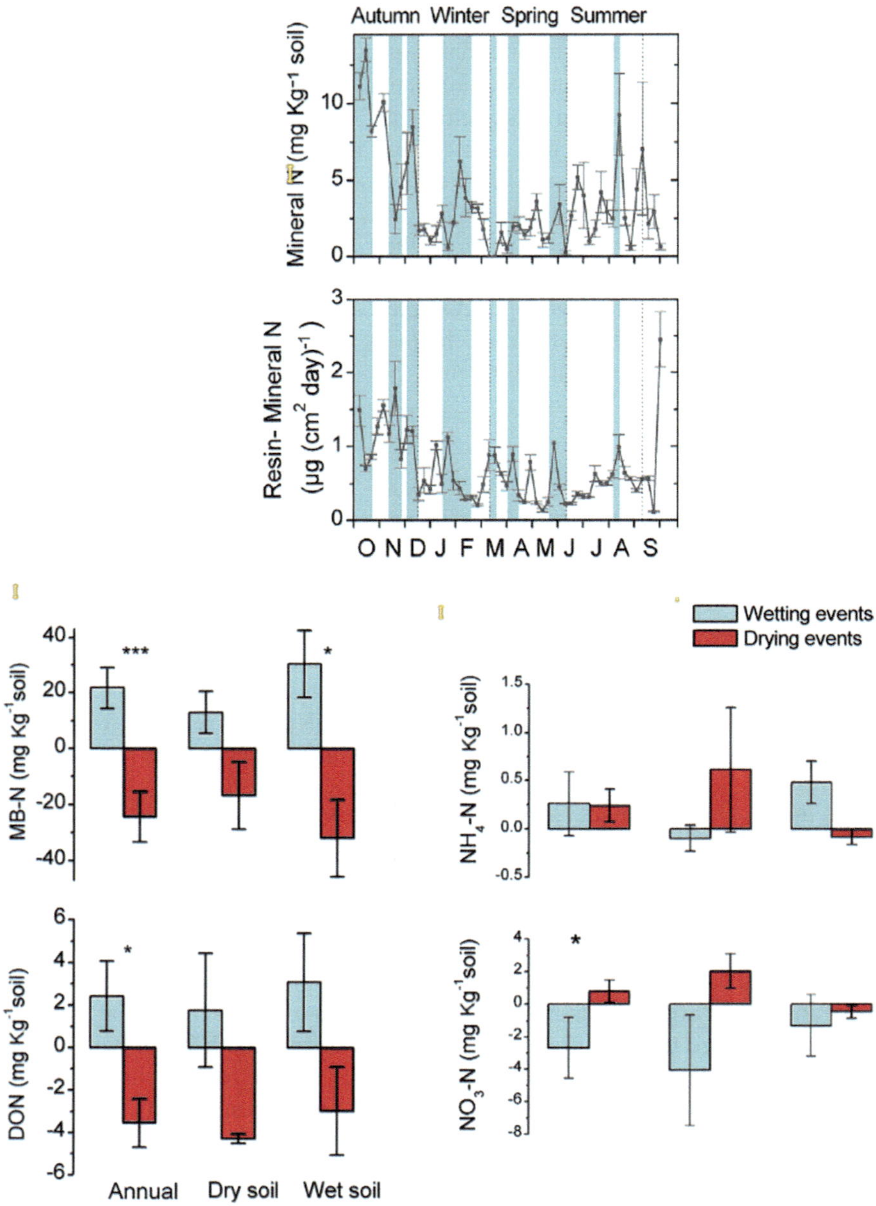

Fig. 15.6 (Top) Weekly soil mineral-N in a *Pinus pinea* stand in southwest Spain. Blue bands indicate wetting events. (Bottom) Changes in soil microbial biomass-N, dissolved organic N (DON), and soil NH_4^+ and NO_3^- concentrations during drying-wetting cycles. Data were analyzed for the entire year and according to initial soil conditions (dry or wet) prior to the drying-wetting cycle. Error bars are ±SE, with N = 8 for the annual average and N = 4 for initially dry or wet soils. Significant differences are indicated as $P < 0.05$ (*), $P < 0.01$ (**), and $P < 0.001$ (***). (From Morillas et al. 2013, 2017)

15.5 The Microbial Community

Microbial biomass and microbial activity are clearly influenced by drying-wetting cycles and seasonality in Mediterranean pine ecosystems (e.g. Figs. 15.5 and 15.6). Microbial communities can be also influenced by weather and climate, either directly or indirectly (e.g. through seasonal changes in litter and exudate inputs, Bardgett et al. 2008). As mentioned above, physiological stress caused by drought or re-wetting may be the main cause of death of some groups of soil microorganisms but not of others, leading to potential changes in the microbial composition of soils (Wu and Brookes 2005). Even if there is no death of microorganisms, some groups can be more active than others. Fungi seem to be more resistant to drought than bacteria (Evans and Wallenstein 2012), thus becoming more abundant and active than bacteria in drier soils (Yuste et al. 2011). Castaño et al. (2018) found that mycorrhizal fungi were less affected by summer drought than free-living fungi in a *P. pinaster* forest, particularly those mycorrhizas of the short-distance exploration type. Species of the extensive mycelia type were more abundant under wetter conditions. Chowdhury et al. (2019) found that soil drying resulted in significant shifts of the soil microbiome, but few changes were found following re-wetting. They found that soil drying not only caused changes in the structure and function of soil microorganisms but also affected metabolic pathways related to the production of osmolites. Schimel (2018) raise the interesting dilemma of whether microbial community dynamics drive the pulses of CO_2 following re-wetting, as suggested by several authors, or conversely, the pulse of CO_2 is initially caused by physical and chemical changes induced by rewetting, which subsequently fuel the response of the microbial community. The different timing of the CO_2 pulse and microbial growth (e.g. Göransson et al. 2013) seem to support the latter opinion. However, the pulse of respiration after wetting seems to be driven first by internal microbial consumption of trehalose and later by the enhanced extracellular C availability (Slessarev and Schimel 2020) giving more prominence to the role of microbial community from the beginning of the rewetting phase.

15.6 Effect of Disturbance on Soil Functioning in Mediterranean Pine Forest

Natural disturbance determines the structure and function of forest ecosystems (Attiwill 1994). Fire is the most extended and intense disturbance affecting Mediterranean ecosystems, with great impacts on soil functioning. Although many Mediterranean ecosystems are considered resilient to fire, pine forests are, with some exceptions, fire sensitive (Pausas et al. 2008). Wildfire in Mediterranean pine forests has a strong effect on soil organic matter and respiration (Hernández et al. 1997; Certini et al. 2011; Uribe et al. 2013; Dadi et al. 2015; López-Serrano et al. 2016; Martínez-García et al. 2017). For example, in *P. halepensis* and *P. pinaster* stands, Hernández et al. (1997) found that 9 months after a fire the contents of organic, extractable and water-soluble C, humic acids, microbial biomass-C, basal soil respiration and dehydrogenase activity were, in general, lower than those of unburnt soils. The impact of wildfire on C cycling is proportional to the burn severity. For example, enzymatic activities and microbial biomass carbon decreased with burn severity in *P. pinaster* and *P. halepensis* in Spain (Fernández-García et al. 2019). Wildfire also has an impact on fungal communities. Thus, Buscardo et al. (2014) in a *P. pinaster* forest found that wildfire significantly reduced the frequency of late stage successional taxa but favored early successional fungal species. Fire in pine forests also affects soil nutrients and fertility. Following a transient increase in soil-available nutrients after fire, nutrient availability decreases due to leaching and erosion (Kutiel and Inbar 1993; Pardini et al. 2003; Durán et al. 2008, 2009, 2010; Ferreira et al. 2016), leading to nutrient limitation and nutrient imbalance (Fig. 15.7). Subsequently, nutrient availability recovers (Yildiz et al. 2010; Marañón-Jiménez and Castro 2013; Marañón-Jiménez et al. 2013). In a *P. canariensis* fire chronosequence in the Canary Islands, Duran et al. (2008, 2009) found important differences between soil N and P recovery after fire. Unlike N, P availability was low in recent burned stands, but showed a rapid recovery, and two decades after the last fire the N:P ratio was close to the unburned areas (Fig. 15.7). The rapid recovery of P is likely related with the high atmospheric dust deposition coming from the nearby Saharan desert (Gallardo et al. 2020). Wildfire also affects the spatial pattern and local variability of nutrient availability (Rodríguez et al. 2009a, b; Durán et al. 2019). Other types of ecological disturbances such as pest, parasite and insect outbreaks are increasing in the Mediterranean region and affecting pine forests; these are sometimes attributed to warming, other consequences of climate change or biological invasions (Barbero et al. 1990; Hódar et al. 2003, 2012; Hódar and Zamora 2004; Luchi et al. 2012; Pernek et al. 2019), with detected impacts on C, nutrient cycles and the microbial communities (Mellado et al. 2016; Mellado and Zamora 2017).

15.7 Concluding Remarks

Soils of the Mediterranean regions showed intermediate characteristics (organic C and N, pH, CEC) between the arid and temperate ecosystems surrounding them. However, apparent differences in soil fertility emerge among the different Mediterranean regions. South Africa and Australia exhibit the lowest soil nutrient content. Contrastingly, the Mediterranean region of Central Chile exhibits the highest soil fertility, with intermediate levels for the Mediterranean Basin and California. Biogeochemical function in these soils seems to be strongly influenced by seasonality and the soil dry-wetting cycles, resulting in an intricate variable temporal pattern. Soil respiration showed similar rates than other most productive temperate forests, which may be related to the Birch effect but suggesting that microbial activity is less limited than plant productivity. Inorganic and organic N is also affected by

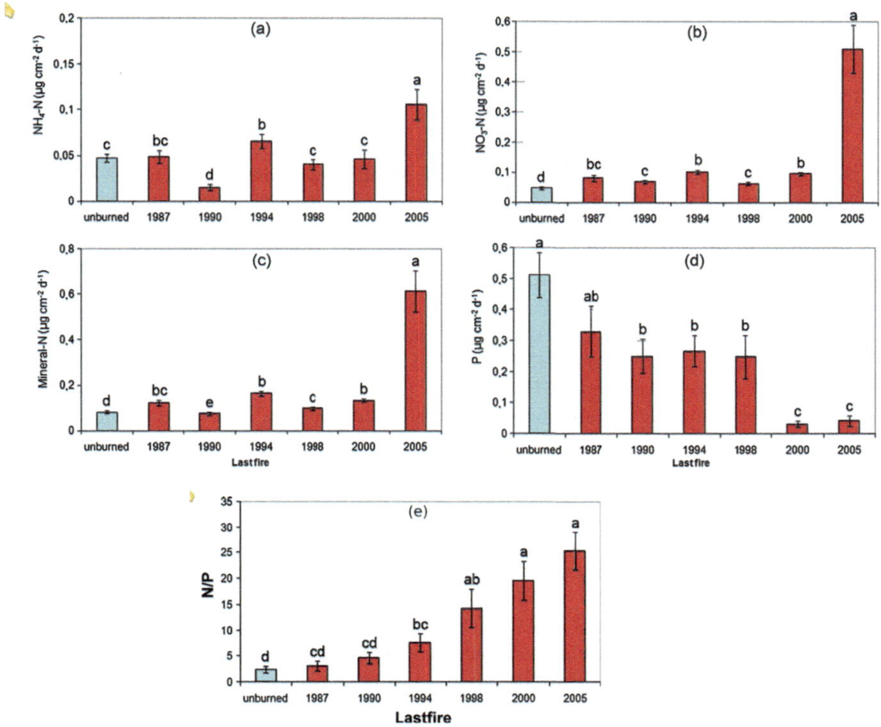

Fig. 15.7 Availability of (**a**) NH_4-N, (**b**) NO_3-N, (**c**) mineral-N, (**d**) PO_4-P and (**e**) available N-to-P ratio in plots burned in different years and in unburned plots. Different letters show significant differences (permutation test, $p < 0.005$). (From Durán et al. 2008)

dry-wetting cycles, and high losses of N uncoupled with labile C losses may be a mechanism keeping Mediterranean ecosystems limited by N availability. Microorganisms are affected unequally by dry-wetting cycles, and at least for soil fungi, changes in successional stages were reported. Forest disturbances, particularly wildfire, strongly influence soil functions in Mediterranean pine ecosystems, resulting in rapid changes in nutrient balance and availability, followed by slower nutrient recovery.

Acknowledgments This paper was supported by FEDER/Ministerio de Ciencia, Innovación y Universidades- Agencia Estatal de Investigación/Proyect (CGL2017-88124-R).

References

Alster CJ, German DP, Lu Y et al (2013) Microbial enzymatic responses to drought and to nitrogen addition in a southern California grassland. Soil Biol Biochem 64:68–79. https://doi.org/10.1016/j.soilbio.2013.03.034

Angert A, Yakir D, Rodeghiero M et al (2015) Using O_2 to study the relationships between soil CO_2 efflux and soil respiration. Biogeosciences 12:2089–2099. https://doi.org/10.5194/bg-12-2089-2015
Attiwill PM (1994) The disturbance of forest ecosystems: the ecological basis for conservative management. For Ecol Manag 63:247–300. https://doi.org/10.1016/0378-1127(94)90114-7
Barbero M, Bonin G, Loisel R et al (1990) Changes and disturbances of forest ecosystems caused by human activities in the western part of the Mediterranean Basin. Vegetatio 87:151–173. https://doi.org/10.1007/BF00042952
Bardgett RD, Freeman C, Ostle NJ (2008) Microbial contributions to climate change through carbon cycle feedbacks. ISME J 2:805–814. https://doi.org/10.1038/ismej.2008.58
Birch HF (1958) The effect of soil drying on humus decomposition and nitrogen availability. Plant Soil 10:9–31. https://doi.org/10.1007/BF01343734
Bond-Lamberty B, Thomson A (2010) A global database of soil respiration data. Biogeosciences 7:1915–1926. https://doi.org/10.5194/bg-7-1915-2010
Buscardo E, Rodríguez-Echeverría S, Freitas H et al (2014) Contrasting soil fungal communities in Mediterranean pine forests subjected to different wildfire frequencies. Fungal Divers 70:85–99. https://doi.org/10.1007/s13225-014-0294-5
Carbone MS, Still CJ, Ambrose AR et al (2011) Seasonal and episodic moisture controls on plant and microbial contributions to soil respiration. Oecologia 167:265–278. https://doi.org/10.1007/s00442-011-1975-3
Castaño C, Lindahl BD, Alday JG et al (2018) Soil microclimate changes affect soil fungal communities in a Mediterranean pine forest. New Phytol 220:1211–1221. https://doi.org/10.1111/nph.15205
Certini G, Nocentini C, Knicker H et al (2011) Wildfire effects on soil organic matter quantity and quality in two fire-prone Mediterranean pine forests. Geoderma 167–168:148–155. https://doi.org/10.1016/j.geoderma.2011.09.005
Chowdhury TR, Lee J-Y, Bottos EM et al (2019) Metaphenomic responses of a native prairie soil microbiome to moisture perturbations. mSystems 4(4):e00061-19. https://doi.org/10.1128/mSystems.00061-19
Cools N, Vesterdal L, De Vos B et al (2014) Tree species is the major factor explaining C: N ratios in European forest soils. For Ecol Manag 311:3–16. https://doi.org/10.1016/j.foreco.2013.06.047
Dadi T, Rubio E, Martínez-García E et al (2015) Post-wildfire effects on carbon and water vapour dynamics in a Spanish black pine forest. Environ Sci Pollut Res 22:4851–4862. https://doi.org/10.1007/s11356-014-3744-4
Davidson EA, Samanta S, Caramori SS et al (2012) The Dual Arrhenius and Michaelis-Menten kinetics model for decomposition of soil organic matter at hourly to seasonal time scales. Glob Chang Biol 18:371–384. https://doi.org/10.1111/j.1365-2486.2011.02546.x
De Vicente I, Ortega-Retuerta E, Morales-Baquero R et al (2012) Contribution of dust inputs to dissolved organic carbon and water transparency in Mediterranean reservoirs. Biogeosciences 9:5049–5060. https://doi.org/10.5194/bg-9-5049-2012
Durán J, Rodríguez A, Fernández-Palacios JM et al (2008) Changes in soil N and P availability in a *Pinus canariensis* fire chronosequence. For Ecol Manag 256:384–387. https://doi.org/10.1016/j.foreco.2008.04.033
Durán J, Rodríguez A, Fernández-Palacios JM et al (2009) Changes in net N mineralization rates and soil N and P pools in a pine forest wildfire chronosequence. Biol Fertil Soils 45:781–788. https://doi.org/10.1007/s00374-009-0389-4
Durán J, Rodríguez A, Fernández-Palacios JM et al (2010) Long-term decrease of organic and inorganic nitrogen concentrations due to pine forest wildfire. Ann For Sci 67:207–207. https://doi.org/10.1051/forest/2009100
Durán J, Rodríguez A, Méndez J et al (2019) Wildfires decrease the local-scale ecosystem spatial variability of Pinus canariensis forests during the first two decades post fire. Int J Wildland Fire 28:288–294. https://doi.org/10.1071/WF18145

Erel Y, Torrent J (2010) Contribution of Saharan dust to Mediterranean soils assessed by sequential extraction and Pb and Sr isotopes. Chem Geol 275:19–25. https://doi.org/10.1016/j.chemgeo.2010.04.007
Estiarte M, Peñuelas J, Sardans J et al (2008) Root-surface phosphatase activity in shrublands across a European gradient: effects of warming. J Environ Biol 29:25–29
Evans SE, Wallenstein MD (2012) Soil microbial community response to drying and rewetting stress: does historical precipitation regime matter? Biogeochemistry 109:101–116. https://doi.org/10.1007/s10533-011-9638-3
Fedoroff N (1997) Clay illuviation in Red Mediterranean soils. Catena 28:171–189. https://doi.org/10.1016/S0341-8162(96)00036-7
Fernández-García V, Miesel J, Baeza MJ et al (2019) Wildfire effects on soil properties in fire-prone pine ecosystems: indicators of burn severity legacy over the medium term after fire. Appl Soil Ecol 135:147–156. https://doi.org/10.1016/j.apsoil.2018.12.002
Ferreira RV, Serpa D, Cerqueira MA et al (2016) Short-time phosphorus losses by overland flow in burnt pine and eucalypt plantations in north-central Portugal: a study at micro-plot scale. Sci Total Environ 551–552:631–639. https://doi.org/10.1016/j.scitotenv.2016.02.036
Gallardo A, Fernández-Palacios JM, Bermúdez A, de Nascimento L, Durán J, García-Velázquez L, Méndez J, Rodríguez A (2020) The pedogenic Walker and Syers model under high atmospheric P deposition rates. Biogeochemistry. http://link.springer.com/article/10.1007/s10533-020-00657-8
Geisseler D, Horwath WR, Scow KM (2011) Soil moisture and plant residue addition interact in their effect on extracellular enzyme activity. Pedobiologia (Jena) 54:71–78. https://doi.org/10.1016/j.pedobi.2010.10.001
German DP, Weintraub MN, Grandy AS et al (2011) Optimization of hydrolytic and oxidative enzyme methods for ecosystem studies. Soil Biol Biochem 43:1387–1397. https://doi.org/10.1016/j.soilbio.2011.03.017
Gioda A, Mayol-Bracero OL, Reyes-Rodriguez GJ et al (2008) Water-soluble organic and nitrogen levels in cloud and rainwater in a background marine environment under influence of different air masses. J Atmos Chem 61:85–99. https://doi.org/10.1007/s10874-009-9125-6
Göransson H, Godbold DL, Jones DL et al (2013) Bacterial growth and respiration responses upon rewetting dry forest soils: impact of drought-legacy. Soil Biol Biochem 57:477–486. https://doi.org/10.1016/j.soilbio.2012.08.031
Groffman PM, Tiedje JM (1988) Denitrification hysteresis during wetting and drying cycles in soil. Soil Sci Soc Am J 52:1626–1629. https://doi.org/10.2136/sssaj1988.03615995005200060022x
Gruba P, Mulder J (2015) Tree species affect cation exchange capacity (CEC) and cation binding properties of organic matter in acid forest soils. Sci Total Environ 511:655–662. https://doi.org/10.1016/j.scitotenv.2015.01.013
Grünzweig J, Gliksman D (this volume) Decomposition in Mediterranean forests. In: Ne'eman G, Osem Y (eds) Pines and their mixed forest ecosystems in the Mediterranean Basin. Springer, Cham
Grünzweig JM, Hemming D, Maseyk K et al (2009) Water limitation to soil CO_2 efflux in a pine forest at the semiarid "timberline". J Geophys Res Biogeosci 114. https://doi.org/10.1029/2008JG000874
Hernández T, García C, Reinhardt I (1997) Short-term effect of wildfire on the chemical, biochemical and microbiological properties of Mediterranean pine forest soils. Biol Fertil Soils 25:109–116. https://doi.org/10.1007/s003740050289
Hernández-Santana V, Martínez-Vilalta J, Martínez-Fernández J et al (2009) Evaluating the effect of drier and warmer conditions on water use by *Quercus pyrenaica*. For Ecol Manag 258:1719–1730. https://doi.org/10.1016/j.foreco.2009.07.038
Hódar JA, Zamora R (2004) Herbivory and climatic warming: a Mediterranean outbreaking caterpillar attacks a relict, boreal pine species. Biodivers Conserv 13:493–500. https://doi.org/10.1023/B:BIOC.0000009495.95589.a7

Hódar JA, Castro J, Zamora R (2003) Pine processionary caterpillar *Thaumetopoea pityocampa* as a new threat for relict Mediterranean Scots pine forests under climatic warming. Biol Conserv 110:123–129. https://doi.org/10.1016/S0006-3207(02)00183-0
Hódar JA, Zamora R, Cayuela L (2012) Climate change and the incidence of a forest pest in Mediterranean ecosystems: can the North Atlantic Oscillation be used as a predictor? Clim Chang 113:699–711. https://doi.org/10.1007/s10584-011-0371-7
Hursh A, Ballantyne A, Cooper L et al (2017) The sensitivity of soil respiration to soil temperature, moisture, and carbon supply at the global scale. Glob Chang Biol 23:2090–2103. https://doi.org/10.1111/gcb.13489
Ibáñez JJ, Zinck JA, Dazzi C (2013) Soil geography and diversity of the European biogeographical regions. Geoderma 192:142–153. https://doi.org/10.1016/j.geoderma.2012.07.024
Inclán R, De La Torre D, Benito M et al (2007) Soil CO_2 efflux in a mixed pine-oak forest in Valsaín (Central Spain). ScientificWorldJournal 7:166–174. https://doi.org/10.1100/tsw.2007.7
Iovieno P, Alfani A, Bååth E (2010) Soil microbial community structure and biomass as affected by *Pinus pinea* plantation in two Mediterranean areas. Appl Soil Ecol 45:56–63. https://doi.org/10.1016/j.apsoil.2010.02.001
Jarvis P, Rey A, Petsikos C et al (2007) Drying and wetting of Mediterranean soils stimulates decomposition and carbon dioxide emission: the "Birch effect". Tree Physiol 27:929–940. https://doi.org/10.1093/treephys/27.7.929
Kleber M, Sollins P, Sutton R (2007) A conceptual model of organo-mineral interactions in soils: self-assembly of organic molecular fragments into zonal structures on mineral surfaces. Biogeochemistry 85:9–24. https://doi.org/10.1007/s10533-007-9103-5
Kutiel P, Inbar M (1993) Fire impacts on soil nutrients and soil erosion in a Mediterranean pine forest plantation. Catena 20:129–139. https://doi.org/10.1016/0341-8162(93)90033-L
Lauber CL, Strickland MS, Bradford MA et al (2008) The influence of soil properties on the structure of bacterial and fungal communities across land-use types. Soil Biol Biochem 40:2407–2415. https://doi.org/10.1016/j.soilbio.2008.05.021
Lawrence CR, Neff JC, Schimel JP (2009) Does adding microbial mechanisms of decomposition improve soil organic matter models? A comparison of four models using data from a pulsed rewetting experiment. Soil Biol Biochem 41:1923–1934. https://doi.org/10.1016/j.soilbio.2009.06.016
Leitner S, Homyak PM, Blankinship JC et al (2017) Linking NO and N_2O emission pulses with the mobilization of mineral and organic N upon rewetting dry soils. Soil Biol Biochem 115:461–466. https://doi.org/10.1016/j.soilbio.2017.09.005
Liang C, Schimel JP, Jastrow JD (2017) The importance of anabolism in microbial control over soil carbon storage. Nat Microbiol 2:17105. https://doi.org/10.1038/nmicrobiol.2017.105
López-Serrano FR, Rubio E, Dadi T et al (2016) Influences of recovery from wildfire and thinning on soil respiration of a Mediterranean mixed forest. Sci Total Environ 573:1217–1231. https://doi.org/10.1016/j.scitotenv.2016.03.242
Loÿe-Pilot MD, Martin JM, Morelli J (1986) Influence of Saharan dust on the rain acidity and atmospheric input to the Mediterranean. Nature 321:427–428. https://doi.org/10.1038/321427a0
Lu H, Liu S, Wang H et al (2017) Experimental throughfall reduction barely affects soil carbon dynamics in a warm-temperate oak forest, central China. Sci Rep 7:15099. https://doi.org/10.1038/s41598-017-15157-3
Lucas-Borja ME, de Santiago J, Yang Y et al (2019) Nutrient, metal contents and microbiological properties of litter and soil along a tree age gradient in Mediterranean forest ecosystems. Sci Total Environ 650:749–758. https://doi.org/10.1016/j.scitotenv.2018.09.079
Luchi N, Mancini V, Feducci M et al (2012) *Leptoglossus occidentalis* and *Diplodia pinea*: a new insect-fungus association in Mediterranean forests. For Pathol 42:246–251. https://doi.org/10.1111/j.1439-0329.2011.00750.x
Manzoni S, Schaeffer SM, Katul G et al (2014) A theoretical analysis of microbial eco-physiological and diffusion limitations to carbon cycling in drying soils. Soil Biol Biochem 73:69–83. https://doi.org/10.1016/j.soilbio.2014.02.008

Marañón-Jiménez S, Castro J (2013) Effect of decomposing post-fire coarse woody debris on soil fertility and nutrient availability in a Mediterranean ecosystem. Biogeochemistry 112:519–535. https://doi.org/10.1007/s10533-012-9744-x

Marañón-Jiménez S, Castro J, Fernández-Ondoño E et al (2013) Charred wood remaining after a wildfire as a reservoir of macro- and micronutrients in a Mediterranean pine forest. Int J Wildland Fire 22:681–695. https://doi.org/10.1071/WF12030

Martínez-García E, López-Serrano FR, Dadi T et al (2017) Medium-term dynamics of soil respiration in a Mediterranean mountain ecosystem: the effects of burn severity, post-fire burnt-wood management, and slope-aspect. Agric For Meteorol 233:195–208. https://doi.org/10.1016/j.agrformet.2016.11.192

Martín-Peinado FJ, Navarro FB, Jiménez MN et al (2016) Long-term effects of pine plantations on soil quality in Southern Spain. Land Degrad Dev 27:1709–1720. https://doi.org/10.1002/ldr.2566

Matteucci M, Gruening C, Goded Ballarin I et al (2015) Components, drivers and temporal dynamics of ecosystem respiration in a Mediterranean pine forest. Soil Biol Biochem 88:224–235. https://doi.org/10.1016/j.soilbio.2015.05.017

Mellado A, Zamora R (2017) Parasites structuring ecological communities: the mistletoe footprint in Mediterranean pine forests. Funct Ecol 31:2167–2176. https://doi.org/10.1111/1365-2435.12907

Mellado A, Morillas L, Gallardo A et al (2016) Temporal dynamic of parasite-mediated linkages between the forest canopy and soil processes and the microbial community. New Phytol 211:1382–1392. https://doi.org/10.1111/nph.13984

Miller PC (1981) Resource use by Chaparral and Matorral. A comparison of vegetation function in two Mediterranean type ecosystems. Springer, New York

Morillas L, Portillo-Estrada M, Gallardo A (2013) Wetting and drying events determine soil N pools in two Mediterranean ecosystems. Appl Soil Ecol 72:160–170. https://doi.org/10.1016/j.apsoil.2013.06.010

Morillas L, Roales J, Portillo-Estrada M et al (2017) Wetting-drying cycles influence on soil respiration in two Mediterranean ecosystems. Eur J Soil Biol 82:10–16. https://doi.org/10.1016/j.ejsobi.2017.07.002

Ne'eman G, Arianoutsou (this volume) Forest fire ecology and management. In: Ne'eman G, Osem Y (eds) Pines and their mixed forest ecosystems in the Mediterranean Basin. Springer, Cham

Nettleton WD (1991) Occurrence, characteristics, and genesis of carbonate, gypsum, and silica accumulations in soils. In: Proceedings of a symposium, Anaheim, CA, November 1988

Nugroho RA, Röling WFM, Laverman AM et al (2007) Low nitrification rates in acid scots pine forest soils are due to pH-related factors. Microb Ecol 53:89–97. https://doi.org/10.1007/s00248-006-9142-9

Ochoa-Hueso R, Arróniz-Crespo M, Bowker MA et al (2014) Biogeochemical indicators of elevated nitrogen deposition in semiarid Mediterranean ecosystems. Environ Monit Assess 186:5831–5842. https://doi.org/10.1007/s10661-014-3822-6

Oliva J (this volume) Fungal diseases of Mediterranean pines. In: Ne'eman G, Osem Y (eds) Pines and their mixed forest ecosystems in the Mediterranean Basin. Springer, Cham

Ozturk T, Ceber ZP, Türkeş M et al (2015) Projections of climate change in the Mediterranean Basin by using downscaled global climate model outputs. Int J Climatol 35:4276–4292. https://doi.org/10.1002/joc.4285

Pardini G, Gispert M, Dunjó G (2003) Runoff erosion and nutrient depletion in five Mediterranean soils of NE Spain under different land use. Sci Total Environ 309:213–224. https://doi.org/10.1016/S0048-9697(03)00007-X

Pausas JG, Llovet J, Rodrigo A et al (2008) Are wildfires a disaster in the Mediterranean Basin? A review. Int J Wildland Fire 17:713–723. https://doi.org/10.1071/WF07151

Peñuelas J, Sardans J, Filella I et al (2018) Assessment of the impacts of climate change on Mediterranean terrestrial ecosystems based on data from field experiments and long-term monitored field gradients in Catalonia. Environ Exp Bot 152:49–59. https://doi.org/10.1016/j.envexpbot.2017.05.012

Pernek M, Lacković N, Lukić I et al (2019) Outbreak of *Orthotomicus erosus* (Coleoptera, Curculionidae) on Aleppo Pine in the Mediterranean Region in Croatia. South-East Eur For 10:19–27. https://doi.org/10.15177/seefor.19-05

Placella SA, Firestone MK (2013) Transcriptional response of nitrifying communities to wetting of dry soil. Appl Environ Microbiol 79:3294–3302. https://doi.org/10.1128/AEM.00404-13

Quideau SA, Graham RC, Chadwick OA et al (1998) Organic carbon sequestration under chaparral and pine after four decades of soil development. Geoderma 83:227–242. https://doi.org/10.1016/S0016-7061(97)00142-0

Raich JW, Schlesinger WH (1992) The global carbon dioxide flux in soil respiration and its relationship to vegetation and climate. Tellus, Ser B 44B:81–99

Reche I, Ortega-Retuerta E, Romera O et al (2009) Effect of Saharan dust inputs on bacterial activity and community composition in Mediterranean lakes and reservoirs. Limnol Oceanogr 54:869–879. https://doi.org/10.4319/lo.2009.54.3.0869

Rey A (2015) Mind the gap: non-biological processes contributing to soil CO_2 efflux. Glob Chang Biol 21:1752–1761. https://doi.org/10.1111/gcb.12821

Rodríguez A, Durán J, Fernández-Palacios JM et al (2009a) Wildfire changes the spatial pattern of soil nutrient availability in *Pinus canariensis* forests. Ann For Sci 66:210–210. https://doi.org/10.1051/forest/2008092

Rodríguez A, Durán J, Fernández-Palacios JM et al (2009b) Short-term wildfire effects on the spatial pattern and scale of labile organic-N and inorganic-N and P pools. For Ecol Manag 257:739–746. https://doi.org/10.1016/j.foreco.2008.10.006

Rutigliano FA, D'Ascoli R, Virzo De Santo A (2004) Soil microbial metabolism and nutrient status in a Mediterranean area as affected by plant cover. Soil Biol Biochem 36:1719–1729. https://doi.org/10.1016/j.soilbio.2004.04.029

Saetre P, Stark JM (2005) Microbial dynamics and carbon and nitrogen cycling following re-wetting of soils beneath two semi-arid plant species. Oecologia 142:247–260. https://doi.org/10.1007/s00442-004-1718-9

Sardans J, Peñuelas J (2013) Plant-soil interactions in Mediterranean forest and shrublands: impacts of climatic change. Plant Soil 365:1–33. https://doi.org/10.1007/s11104-013-1591-6

Sardans J, Peñuelas J (2015) Potassium: a neglected nutrient in global change. Glob Ecol Biogeogr 24:261–275. https://doi.org/10.1111/geb.12259

Sardans J, Rodà F, Peñuelas J (2006) Effects of a nutrient pulse supply on nutrient status of the Mediterranean trees *Quercus ilex subsp. ballota* and *Pinus halepensis* on different soils and under different competitive pressure. Trees 20:619–632

Sardans J, Peñuelas J, Estiarte M (2008) Changes in soil enzymes related to C and N cycle and in soil C and N content under prolonged warming and drought in a Mediterranean shrubland. Appl Soil Ecol 39:223–235. https://doi.org/10.1016/j.apsoil.2007.12.011

Scharlemann JPW, Tanner EVJ, Hiederer R et al (2014) Global soil carbon: understanding and managing the largest terrestrial carbon pool. Carbon Manag 5:81–91. https://doi.org/10.4155/cmt.13.77

Schimel JP (2018) Life in dry soils: effects of drought on soil microbial communities and processes. Annu Rev Ecol Evol Syst 49:409–432. https://doi.org/10.1146/annurev-ecolsys-110617-062614

Schimel J, Balser TC, Wallenstein M (2007) Microbial stress-response physiology and its implications for ecosystem function. Ecology 88:1386–1394. https://doi.org/10.1890/06-0219

Schimel J, Becerra CA, Blankinship J (2017) Estimating decay dynamics for enzyme activities in soils from different ecosystems. Soil Biol Biochem 114:5–11. https://doi.org/10.1016/j.soilbio.2017.06.023

Shi S, Richardson AE, O'Callaghan M et al (2011) Effects of selected root exudate components on soil bacterial communities. FEMS Microbiol Ecol 77:600–610. https://doi.org/10.1111/j.1574-6941.2011.01150.x

Slessarev EW, Schimel JP (2020) Partitioning sources of CO2 emission after soil wetting using high-resolution observations and minimal models. Soil Biol Biochem 143. https://doi.org/10.1016/j.soilbio.2020.107753

Uribe C, Inclán R, Sánchez DM et al (2013) Effect of wildfires on soil respiration in three typical Mediterranean forest ecosystems in Madrid, Spain. Plant Soil 369:403–420. https://doi.org/10.1007/s11104-012-1576-x

Vázquez-González C, Sampedro L, López-Goldar X et al (2021) The genetic and ecological basis of coniferous tree resistance to herbivorous insects. In: Ne'eman G, Osem Y (eds) Pines and their mixed forest ecosystems in the Mediterranean Basin. Springer, Cham

Wisheu IC, Rosenzweig ML, Olsvig-Whittaker L et al (2000) What makes nutrient-poor Mediterranean heathlands so rich in plant diversity? Evol Ecol Res 2:935–955

Wu J, Brookes PC (2005) The proportional mineralisation of microbial biomass and organic matter caused by air-drying and rewetting of a grassland soil. Soil Biol Biochem 37:507–515. https://doi.org/10.1016/j.soilbio.2004.07.043

Wu Z, Dijkstra P, Koch GW et al (2011) Responses of terrestrial ecosystems to temperature and precipitation change: a meta-analysis of experimental manipulation. Glob Chang Biol 17:927–942. https://doi.org/10.1111/j.1365-2486.2010.02302.x

Yaalon DH (1997) Soils in the Mediterranean region: what makes them different? Catena 28:157–169. https://doi.org/10.1016/S0341-8162(96)00035-5

Yao H, Gao Y, Nicol GW et al (2011) Links between ammonia oxidizer community structure, abundance, and nitrification potential in acidic soils. Appl Environ Microbiol 77:4618–4625. https://doi.org/10.1128/AEM.00136-11

Yildiz O, Esen D, Sarginci M et al (2010) Effects of forest fire on soil nutrients in Turkish pine (*Pinus brutia*, Ten) ecosystems. J Environ Biol 31:11–13

Yuste JC, Penuelas J, Estiarte M et al (2011) Drought-resistant fungi control soil organic matter decomposition and its response to temperature. Glob Chang Biol 17:1475–1486

Zdruli P, Kapur S, Çelik I (2011) Soils of the Mediterranean region, their characteristics, management and sustainable use. In: Kapur S, Eswaran H, Blum W (eds) Sustainable land management. Springer, Berlin/Heidelberg, pp 125–142

Zeglin LH, Bottomley PJ, Jumpponen A et al (2013) Altered precipitation regime affects the function and composition of soil microbial communities on multiple time scales. Ecology 94:2334–2345. https://doi.org/10.1890/12-2018.1

Chapter 16
Litter Decomposition in Mediterranean Pine Forests Subjected to Climate Change

José M. Grünzweig and Daniel Gliksman

16.1 Introduction

Decomposition of dead organic matter is an essential process for the sustainment of life in terrestrial, aquatic and marine ecosystems. Berg and Laskowski (2005) stated that "without decay, … the whole land surface of the Earth would be soon covered with a meters-thick layer of undecomposed organic matter. Nutrients would be fixed in a form unavailable to plants, making further production impossible." Therefore, decomposition of organic matter is an indispensable prerequisite for biomass production on Earth. Understanding the decay of organic materials in Mediterranean pine forests is important for assessing carbon (C) storage and for evaluating the dynamics of the release of nutrients back to the soil where they can be reabsorbed by plants and microorganisms. Moreover, knowledge of decomposition processes is equally important for assessing fuel quantities in these fire-prone ecosystems. In this chapter we ask how climate change affects litter decomposition in Mediterranean pine forests by direct climatic influence and by indirect impacts through changes in forest composition, structure and microclimate. We also explore the important insights from litter decomposition in Mediterranean pine forests that are relevant to more mesic forests. We first summarize the drivers and mechanisms of litter decomposition and the projections of climate change in Mediterranean regions, and then discuss the potential effects of climate change on pine forests in these and other regions.

J. M. Grünzweig (✉)
Faculty of Agriculture, Food and Environment, The Hebrew University of Jerusalem, Rehovot, Israel
e-mail: jose.gruenzweig@mail.huji.ac.il

D. Gliksman
Faculty of Environmental Sciences, Institute for Hydrology and Meteorology, Technische Universität Dresden, Tharandt, Germany

G. Ne'eman, Y. Osem (eds.), *Pines and Their Mixed Forest Ecosystems in the Mediterranean Basin*, Managing Forest Ecosystems 38,
https://doi.org/10.1007/978-3-030-63625-8_16

16.2 General Background on Plant Litter and Its Decomposition

Dead organic matter comprises a plethora of materials, such as fresh, undecayed plant litter (plant detritus, including dead roots and woody debris), animal remains, and soil organic matter at different stages of stabilization. Aboveground plant litter in a pine forest comprises mainly needles, and, in lower amounts, branches, inflorescences, cones, bark, bud scales, pine kernels, and a 'miscellaneous' fraction (García-Plé et al. 1995; Segura et al. 2019). Studies on decomposition in Mediterranean pine forests predominantly focus on needle litter (and associated broad-leaved litter) and relatively few studies include woody debris and dead roots.

Litter decomposition, commonly quantified as litter mass loss, is a complex process that is operated by several, at times interacting, degradation mechanisms. In brief, decomposition is influenced by climatic conditions, the chemical composition of the litter, nutrient availability and the decomposer community (Aerts 1997; Berg and McClaugherty 2008). Main relevant climatic variables are moisture, temperature and solar irradiation (Austin and Vivanco 2006; Dirks et al. 2010; Schlesinger and Bernhardt 2013); litter constituents of interest are both structural and non-structural compounds, and include elemental (stoichiometric) relationships (Prescott 2010); relevant nutrients include nitrogen (N), phosphorus (P) and other macronutrients, but also micronutrients such as manganese (Berg et al. 2010; Hättenschwiler et al. 2011); the decomposer community is typically diverse and can include microorganisms, meso-fauna and macro-fauna (Handa et al. 2014; Sagi et al. 2019).

16.3 Litter Decomposition in a Mediterranean Climate

All regions with a Mediterranean climate are affected by drought. Therefore, litter is decomposed by several mechanisms that are primarily known from drylands, in addition to classic microbial degradation driven by rainwater. Litter decay is influenced by a multitude of drivers, differing in their periods of activity, mechanistic influence and target compounds. Most generally, decay can be divided into biotic and abiotic degradation. However, such a distinction does not consider that both forms of decay are tightly coupled. Biotic decay relies on sufficient availability of water and nutrients in the litter or in its environment. Rainfall during the wet season contributes to high rates of microbial degradation, while during the dry season high relative humidity, fog and dew (non-rainfall water sources) enable humidity-enhanced microbial degradation, typically at nighttime and early morning hours (Dirks et al. 2010; Jacobson et al. 2015; Gliksman et al. 2017, 2018a; Wang et al. 2017; Evans et al. 2019).

Abiotic decay during the dry season of the Mediterranean climate is dominated by photochemical degradation (photodegradation). High flux densities of photochemically active radiation (ultraviolet at 280–400 nm and short-wave visible or blue-green at 400–550 nm) induce breakdown of photoreactive compounds, such as lignin (Moorhead and Callaghan 1994; King et al. 2012; Barnes et al. 2015). In Mediterranean regions, photodegradation typically co-occurs with thermal degradation, in which temperatures >30 °C lead to cleavage of chemical bonds in organic materials and the production of reactive compounds (Lee et al. 2012). During the wet season, photodegradation and thermal degradation are possible but the rates are likely to be lower due to the overall reduced intensity of solar radiation, angle of the sun and lower temperatures. On the other hand, leaching can make a substantial contribution, especially during rain events (Ibrahima et al. 1995), but possible also following fog deposition and dew formation.

Photodegradation can alter litter quality and in turn affect microbial degradation in subsequent wet periods ('photopriming', Gallo et al. 2009; Wang et al. 2015; Austin et al. 2016). The relationship between photodegradation and microbial decay is variable; both facilitation and inhibition effects of photodegradation on microbial decay, through alteration of litter chemistry and structure, have been documented (Pancotto et al. 2003; Henry et al. 2008; Gallo et al. 2009; Foereid et al. 2010; Smith et al. 2010; Baker and Allison 2015; Austin et al. 2016; Gliksman et al. 2018a). These interactions can strongly influence the annual mass loss by affecting biotic decay in the wet season. Microbial decay, on the other hand, appears to have mainly a positive effect on photodegradation (but see biofilms below) by exposing components that are easily degraded by solar radiation (Gliksman et al. 2017, 2018a; Lin et al. 2018).

Under conditions of forest degradation and soil erosion, soil-litter mixing can be a critical process that affects photodegradation levels when soil particles and microorganisms form thin biofilms that can block solar radiation from being intercepted by the litter (Barnes et al. 2012). Additionally, soil-litter mixing can enhance microbial degradation by inoculating the litter with microorganisms and temporarily protecting microorganisms from hot and dry conditions (Hewins et al. 2013; Lee et al. 2014).

While climatic conditions play a major role in litter decomposition in Mediterranean regions, litter quality is an additional key player in this process that can dominate decay dynamics. In addition, the chemical composition of highly or moderately sclerophyllous species may have a larger impact on litter decomposition than the structure and morphology of their leaves. Decomposition rates have been shown to decrease with increasing C/N ratio, from *Cistus laurifolius* L. to black pine (*Pinus nigra* J.F. Arnold) and maritime pine (*P. pinaster* Aiton) litter (Moro and Domingo 2000) and to be higher in Aleppo pine (*P. halepensis* Miller) litter, which has less tannins and phenols, than in Palestine oak (*Quercus calliprinos* Webb) litter (Sheffer et al. 2015). Similarly, N and P release from litter has been shown to be rapid in species with high N (e.g. the leguminous *Adenocarpus decorticans* Boiss.) and high P concentrations (e.g. *C. laurifolius*), respectively (Moro and Domingo 2000). In a planted, semi-arid Aleppo pine forest, environmental conditions were more favorable for litter decomposition in the wet season and were equally beneficial in the dry season, compared with the native shrubland (Grünzweig et al. 2007). Nevertheless, because of poorer litter quality, pine needle and root litter in the forest decayed about 25% less than native litter in the shrubland and released only small amounts of N (roots) or even immobilized N (needles) (Gelfand et al. 2012). Finally, concerning the decomposer community, saprophagous arthropods are presumed to have a large impact on decay of organic matter and nutrient cycling in Mediterranean pine forests (Arianoutsou and Radea 2000). However, arthropod effects on litter decomposition are beyond the scope of this review.

Land-use change can influence litter decay through altering vegetation composition, and, thus, litter quality. For example, afforestation of a semi-arid shrubland in Israel resulted in lower decay rates of Aleppo pine needle and root litter resulting from lower litter quality (Grünzweig et al. 2007). Similarly, Ponderosa pine (*P. ponderosa* Douglas ex C. Lawson) afforestation in regions of Patagonia with a Mediterranean-type climate reduced decay rates of needle litter compared with native litter due to exclusion of photodegradation and lower microbial degradation caused by the low litter quality (Araujo and Austin 2015). In secondary successional forests (following agricultural abandonment) in southern France, the decay rate of needle litter was lower in young (10 years old) than in older Aleppo pine stands (30 and >60 years old) (Chomel et al. 2014). The lower decay rates were related to higher concentrations of phenolic compounds in the litter of young stands, compared to older stands, which inhibited colonization of litter by fungi and subsequently also by mesofauna. Needle litter of maritime pine decayed at lower rates in 18–31-year-old planted pine forests in Western Australia than in the native woodland, which might be related to the impoverished mesofauna (less species and lower diversity) in pine stands compared with the native environment (Springett 1976).

Not much is known about the decomposition of woody debris in Mediterranean forests. Wood decay over 5 years in Western Australia was shown to be affected by climate and quality of the debris, with the lowest decay rates at the driest and warmest site (Brown et al. 1996). Wood of maritime pine branches and small logs was low in N, P and potassium (K), and decayed at a lower rate than wood of local *Eucalyptus* species that had higher initial nutrient concentrations. Beside supplying essential ecosystem services, such as providing habitat and sequestering C, woody debris can affect fire behavior. Model simulations have suggested that wood decomposition reduces flame length and fire intensity in an Aleppo pine forest (Palmero-Iniesta et al. 2017).

16.4 Microclimate in Mediterranean Forests

The microclimate in Mediterranean pine forests is highly variable due to the heterogeneity of tree and canopy densities, both strongly dependent on forest management and the regional climate. For biotic litter decay, water availability is the main requirement as temperatures are less of a limiting factor in Mediterranean-type climates, at least in the range of 20–35 °C (Incerti et al. 2011). Interception of rain by the tree canopy can substantially reduce the amount of water reaching the litter and the soil. In an extensive review, Llorens and Domingo (2007) found that rain

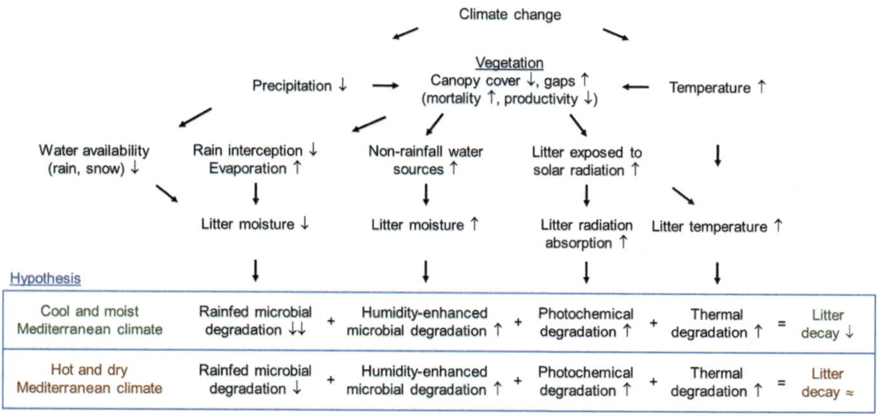

Fig. 16.1 Conceptual model of the direct and indirect influence of climate change on litter microclimate, and a hypothesis (blue boxes) of the consequences for litter decomposition in cool and moist versus hot and dry Mediterranean conditions. ↑ and ↓ denote increase and decrease in properties and processes; ≈ denotes approximately unchanged decay rates. Litter trait (quality) effects and vegetation-climate feedbacks were not considered in this scheme. For an explanation of the hypothesis, see Sect. 16.10

interception varied in the range of 3.4–58%, with an average rate of 26%, for five pine species in 14 Mediterranean sites. In general, interception increased with increasing canopy cover, tree height and tree density (Llorens and Domingo 2007; Granados et al. 2016), thus strongly affecting water availability in the litter layer (Fig. 16.1). Soil evaporation can be higher in open patches due to higher solar irradiance (Raz Yaseef et al. 2010). Moreover, vapor pressure deficit (VPD), which is highly correlated with evapotranspiration, tends to increase with a reduction in canopy cover (Gavinet et al. 2015; Granados et al. 2016; Prévosto et al. 2020). The contrasting influence of tree density on interception and evapotranspiration may prevent any density-induced change in the moisture content of the upper soil layer, which influences litter moisture (Rodríguez-García et al. 2011; Almagro and Martínez-Mena 2012; Granados et al. 2016; Mazza et al. 2019). However, evapotranspiration may also decrease with tree density, depending on intensity of and time since thinning, and on rooting depth and plant available water (Tague and Moritz 2019). Instances of lower soil moisture under higher tree density are more likely to occur in spring and early summer (Johnson-Maynard et al. 2004; Prévosto et al. 2011, 2020; Gavinet et al. 2015, 2016) (but see Lucas-Borja et al. 2016).

Dew deposition is likely to increase with lower canopy cover through enhanced nighttime heat loss as indicated by studies in non-Mediterranean ecosystems (Baier 1966; Jacobs et al. 1990; Xiao et al. 2009; Pan et al. 2010). In this context, although many Mediterranean forests are located in coastal areas with high water vapor levels, proximity to a water source does not necessarily lead to higher dew deposition. Water bodies can alter the energy balance, increasing temperatures and thus reducing litter moisture (Qasemian et al. 2014). Soil depth also has a significant effect on moisture and decomposition. For example, experimental materials [soil + ground

Fig. 16.2 Sparse tree cover in a hot and dry, semi-arid Aleppo pine forest enhancing solar irradiance at the level of the litter layer. Summer mean 24-h maximum temperature (day-time) was higher by 21 °C in forest gaps compared with locations under tree canopies, while mean daily minimum temperature (night-time) was lower by 2 °C in gaps (Gliksman et al. 2018b)

leaf litter of Aleppo pine or holm oak (*Q. ilex* L.)] lost C and N more rapidly in deeper soil layers (0.2 and 0.4 m) than in the top soil (0.05 m) after 2 years of field incubation, likely because of higher moisture contents at depth (Rovira and Vallejo 1997).

Abiotic decay is also likely to be regulated by the tree and canopy densities, which influence the amount of solar radiation reaching the litter layer that causes photochemical and thermal degradation (Figs. 16.1 and 16.2). While the influence on radiation transmittance is rather straight forward and substantial, this is not the case with soil and litter temperatures. Pine tree canopies can reduce the irradiation of the topsoil and the litter by up to 70%, thus reducing daily maximum temperatures considerably (up to 23–30 °C) in relation to unshaded locations (Raz Yaseef et al. 2010; Gliksman et al. 2018b). Other studies have shown that a strong reduction of radiation transmittance by canopies may lead only to a small reduction in the daily temperature (2–4 °C) in relation to unshaded locations (Gavinet et al. 2015; Lucas-Borja et al. 2016; Prévosto et al. 2020). Furthermore, in some cases thinning may have no effect on soil and litter temperature (Delgado et al. 2007; Almagro and Martínez-Mena 2012; Barba et al. 2016a). Thus, the influence of canopy cover and tree density is likely to affect photodegradation, while its influence on thermal degradation is rarer and case specific. Finally, an increase in canopy cover may increase the interception of rain and reduce physical degradation by the impact of rain drops (Whitford 2020).

Shrub canopy and density as understory vegetation may have a unique role in regulating the microclimate of Mediterranean pine forests. Biotic litter decay will likely be influenced in a similar manner by shrubs as it is by trees. Llorens and Domingo (2007) found that shrubs in Mediterranean forests had an average interception rate of 25%, similar to that of pine trees. Thus, the presence of shrubs can further reduce water inputs to litter after rain is intercepted by tree canopies. Shrubs can increase soil moisture (Lucas-Borja et al. 2016), speed up drying of the soil after rain (Rascher et al. 2011), or have no effect (Rodríguez-García et al. 2011). It is possible that the effect of shrubs on soil moisture depends on tree density, as Prévosto et al. (2020) found that shrub presence reduced soil moisture only at lower tree densities (or in open areas). Finally, shrubs may reduce evapotranspiration via a reduction in VPD with greater shrub cover (Prévosto et al. 2020). Regarding abiotic litter decay, shrub presence has been shown to increase leaf area index, reduce gaps in canopies and decrease light transmittance (Rascher et al. 2011; Prévosto et al. 2020), and thus reduce photodegradation. Similarly, shrub cover can reduce maximum daily temperatures by up to 6 °C, lowering the potential for thermal degradation (Rodríguez-García et al. 2011; Lucas-Borja et al. 2016).

16.5 Synopsis of Climate Change Projections for Mediterranean Climate Zones

The five global Mediterranean climate regions are characterized by a mid-latitude temperate climate with a dry and warm (or hot) summer, and a cool and damp winter (Lionello et al. 2012; Polade et al. 2017). The climate in most regions is projected to become drier. Global climate model simulations based on the high-greenhouse-gas-emission forcing scenario of the Intergovernmental Panel on Climate Change (Riahi et al. 2011) project mean reductions of 12–23% in winter precipitation for the Mediterranean Basin and Mediterranean-type climatic zones in South America, South Africa and Australia by the years 2060–2089 (compared to 1960–1989); projections for winter precipitation in California are inconsistent (Polade et al. 2017). The frequency of extremely heavy rain events is expected to increase in the Mediterranean Basin and in California, but not in the other regions.

Winter surface air temperatures are projected to increase in all regions (Polade et al. 2017). In the Mediterranean Basin, warming is expected in all seasons, with a maximum during the summer (Giorgi and Lionello 2008). Different forcing scenarios predict a two- to fivefold increase in the occurrence of hot weather extremes for the Mediterranean Basin by the end of the twenty-first century, while relative air humidity is expected to decrease (Diffenbaugh et al. 2007; Giorgi and Lionello 2008). Simulations based on medium- and high-emission scenarios predict snow accumulation and duration in Mediterranean climate montane regions of Spain, Morocco and California to decline by at least 40% and 15 days, respectively, by the mid-twenty-first century, while the predicted changes in the Andes of northern Chile are considerably smaller (López-Moreno et al. 2017).

Ultimately, water deficits are expected to grow in all Mediterranean regions, even without changes in precipitation. The reasoning is that under global warming land surfaces heat up to a larger degree than ocean surfaces, thus increasing the evaporative demand of the air above the land and decreasing soil moisture (Sherwood and Fu 2014).

16.6 Climate Effects on Litter Decomposition in Mediterranean Pine Forests

Climate influences litter both directly through abiotic drivers (temperature, solar radiation) and its regulation of the activity of decomposers (temperature, moisture), and indirectly through its effects on litter chemistry and the composition of the decomposer community (Fig. 16.1). Therefore, climate change affects litter decomposition and nutrient release from litter via multiple pathways. Modified decomposition dynamics will also affect ecosystem functioning through their impact on plant communities (Facelli and Pickett 1991).

Climatic changes affecting litter moisture are expected to have a large effect on decomposition through the influence of temporal and spatial variation in water availability. For example, lower annual rainfall amount in 1 year (376 mm) than in another year (497 mm) led to 32–47% reductions in annual mass loss and decay constant *k* ($year^{-1}$) of Aleppo pine leaf litter in a hot and moist forest [mean annual temperature (MAT) 17.2 °C, mean annual precipitation (MAP) 602 mm] (García-Plé et al. 1995). Similarly, annual litter mass loss and *k* decreased by 61–65% for maritime pine when comparing a cool and moist with a less moist forest (MAT 14.4 °C, MAP 579 mm vs. MAT 14.9 °C, MAP 539 mm) and by 40–45% for Aleppo pine when comparing a cool and moist with a dry forest (MAT 14.8 °C, MAP 587 mm vs. MAT 15.2 °C, MAP 396 mm) (Kurz-Besson et al. 2006).

In addition to rainfall amounts, a future change in the intra-annual variability in rainfall patterns is also expected to affect litter decay. For example, experimentally manipulated rainfall patterns significantly impacted the moisture content of *P. brutia* Tenore needle litter in a Greek forest, with a regular pattern resulting in higher litter moisture than an irregular pattern (Wilkinson et al. 2002). As a result, fungal biomass was larger than bacterial biomass on the pine litter, and the abundance of both microbial taxa responded more positively to frequent and regular as compared to irregular experimental water additions. Experimentally intercepting all rainfall in the *P. brutia* forest during 2 months of the active season (April–May) decreased soil moisture by 50% during this period, but led to 50% higher needle litter decomposition by the beginning of the wet season (November) (Tsiafouli et al. 2018). The causes for higher mass loss under drier conditions are unclear because litterbags were inserted into the forest floor, thus enhanced abiotic degradation (photochemical and thermal) needs to be ruled out as a cause.

Soil nutrient availability might interact with climate conditions in its effect on litter decomposition. For example, after 3.5 years of decomposition in forests of a similar moisture regime, mass loss from N-poor Aleppo pine needle litter was lower and N immobilization was higher in a forest with lower N availability than in a forest with higher N availability (Garcia-Pausas et al. 2004). In contrast, no difference between these forests was observed for relatively N-rich holm oak leaf litter. Moreover, in forests of presumably similar soil N availability, mass loss and N dynamics were similar, despite a large difference in the moisture regime. In addition to soil nutrients, climate might interact with microorganisms. The response of litter decomposition to a changing climate (temperature, precipitation) in Mediterranean ecosystems (including a mixed pine-oak forest) has been shown to depend on the microbial composition (bacteria, fungi) of the litter (Glassman et al. 2018).

Tree mortality and the predicted prolongation of dry periods due to climate change may enhance litter decay, possibly reducing C stocks in dryland ecosystems. For example, in a study on Aleppo pine forests, maximum litter temperature was almost 20 °C higher in gaps between trees than under the tree canopies of a dry-subhumid and a semi-arid forest Aleppo pine (Gliksman et al. 2018b). Consequently, litter mass loss was greater in forest gaps than under canopies, possibly because of enhanced abiotic degradation in gaps. Additionally, the decay rate slowed down from the first to second wet season, but remained constant during the first and second dry seasons, and the dry seasons contributed 30% to the overall mass loss over 613 days. Thus, decay during the dry season may be more stable than during the wet season and is strongly influenced by canopy density as a regulator of solar radiation interception and dew deposition.

There is very little information on the potential effects of climatic warming on litter decay. Model simulations project that warming by 2 °C would decrease the forest floor needle mass by 11–12% in maritime pine and Aleppo pine forests in eastern Spain and southern France, suggesting an increase in litter decomposition under warmer conditions (Kurz-Besson et al. 2006). However, these simulations do not include the potential impact of climate change on litterfall mass and quality, which in turn may affect needle mass and decomposition.

16.7 Changes in Forest Vegetation as Affected by Climate Change and Management

Climate change can significantly affect the composition and structure of pine forests, which has been described in detail in Chap. 12 of this book (Camarero et al. this volume). The reduction in water availability is expected to cause tree mortality and a decrease in tree and canopy density. Extreme drought events in Mediterranean regions have led to mortality of Aleppo pine trees in Israel, especially in forests with inherently low water availability (semi-arid zone, southern aspect in the Northern Hemisphere, deep soils with low rock volume) (Dorman et al. 2015; Preisler et al. 2019). Similarly, Scots pine (*P. sylvestris* L.) is most vulnerable to drought when located at the dry edge of its distribution in Spain (Gea-Izquierdo et al. 2014), and mortality has been found to be higher in warmer than in cooler sites for the less drought-adapted black pine and Scots pine (Ruiz-Benito et al. 2013). Among various pine species in Spain, mortality rate increases with decreasing tree size and increasing competition among trees (Ruiz-Benito et al. 2013). Furthermore, increased drought in Spain during the period 1987–2007 caused considerable defoliation of various pine species, leading to higher mortality rates and less dense forests (Carnicer et al. 2011).

In addition to defoliation and mortality, forests can also become more open when stands are thinned as a management measure of adaption to climate change-induced drought (Gómez-Aparicio et al. 2011). Thinning is actually often recommended to promote tree survival in the face of climate change (McDowell and Allen 2015). Land management also affects the understory when forests are grazed as a management tool for fire prevention (Ne'eman and Perevolotsky 2000). State transition shifts and cascading ecological responses and feedbacks can further affect forest recovery from tree die-off and management actions (Cobb et al. 2017). Consequently, litter layers become less covered by vegetation if no compensatory growth of more drought-resistant canopy or understory species occurs, or if management prevents such growth responses.

Shifts in forest composition and structure through natural succession or management are additional components of forest responses to climate change. The composition and structure of forest stands can be directly affected by climate change because tree species differ in their response to heat and drought (Ruiz-Benito et al. 2013). In Portugal, the replacement of deciduous, and sometimes even evergreen cork oak (*Q. suber* L.) forests with more drought-adapted trees such as maritime pine has been associated with increasing duration of dry spells during the years 1966–2006 and with management decisions (Acacio et al. 2017).

16.8 Effect of Climate Change-Induced Vegetation Change on Litter Decomposition

Climate change might lead to tree species replacements, which could affect litter decomposition through changes in the microclimate and/or litter quality. Under a warmer and drier climate, native downy oak (*Q. pubescens*) forests in the northern part of the Mediterranean Basin could be replaced by Aleppo pine, resulting in lower decomposition rates and strongly reduce N release from litter (Santonja et al. 2015). A comparison between trophic interactions in Aleppo pine and downy oak litter revealed that fungal decomposer biomass was higher in pine than in oak litter as a consequence of predation by Acari, which reduced fungivorous Collembola abundance (Aupic-Samain et al. 2019). These demographic variations might originate from the structural differences between pine and oak litter, and might play an important role in litter decomposition, if downy oak is replaced by Aleppo pine. In a mixed Scots pine–holm oak forest in Spain, mortality of pine and replacement with more drought-resistant oak would increase leaf litter decay, but decrease root litter decay (Barba et al. 2016b). In the eastern Mediterranean region, the presence of leaf litter from Palestine oak retarded the decay of Aleppo pine leaf litter in mixed assemblages, likely because of the relatively high tannin and phenol content in oak litter (Sheffer et al. 2015). In contrast, the higher the portion of broad-leaved litter in the litter layer, the faster litter of all types decayed, potentially because of the more stable moisture regime imposed by litter from oaks and other native species. On a broader spatial scale, higher relative abundance of deciduous tree species across European forests (Mediterranean to boreal) was shown to increase soil moisture and enhance traits (specific leaf area, magnesium and lignin concentration) of the local litter layer that promote decay of standard cellulose and wood materials (Joly et al. 2017).

Soils in Mediterranean pine forests have a large C storage potential, at least if associated with low-quality litter being added to soil with low organic C stocks (Grünzweig et al. 2007; Qubaja et al. 2020). However, climate change-induced warming combined with vegetation changes might reduce soil C storage through impacts on litter decay, as shown in the following study: Simulating climate change conditions in lab incubations (increased temperatures, vegetation change to ponderosa pine) resulted in a significant enhancement of C losses from white fir [*Abies concolor* (Gordon & Glend.) Lindl. ex Hildebr.] forest soils (Rasmussen et al. 2008).

16.9 Effects of Forest Management on Litter Decomposition

The consequences of forest management (thinning and clear-cutting) for litter decomposition are variable and depend mainly on microclimatic effects and thinning intensity. For example, heavy thinning in a hot and dry Aleppo pine forest (MAT 20.7 °C, MAP 280 mm) to reduce tree density by 65% enhanced decay of needle litter, possibly as a consequence of enhanced abiotic degradation (Gliksman et al. 2018b). Similarly, Aleppo pine litter (needles, twigs, cones) decomposed more rapidly in a hot and dry old-field ecosystem than in a forest (MAT 15.5 °C, MAP 370 mm), possibly mediated by soil-litter mixing enabled by soil erosion (Almagro and Martínez-Mena 2012). In contrast, clear-cutting reduced needle litter decomposition in a relatively cool and moist Aleppo pine forest (MAT 12.5 °C, MAP 457 mm) compared to control plots, which might be explained by mineral soil deposition on litter in clear-cut plots in combination with a major methodological issue (determination of mass loss without ash correction) (Lado-Monserrat et al. 2016). In another study, clear-cutting in a hot and moist Aleppo pine forest (MAT 20 °C, MAP 550 mm) enhanced needle litter decay compared to decay in gaps in very lightly thinned plots, possibly because of enhanced abiotic degradation (Gliksman et al. 2018b).

Forest management also interacted with litter nutrient content and release. For example, thinning by 40% removal of basal area in a cool and moist mixed maritime pine-*Q. pyrenaica* Willd. forest (MAT 12.9 °C, MAP 469 mm) reduced *k* of nutrient-rich oak leaf litter, possibly because of lower soil moisture, but did not affect *k* of relatively nutrient-poor pine needle litter (Bravo-Oviedo et al. 2017). In a cool and moist *P. radiata* D. Don forest (MAT 15.6 °C, MAP 750 mm), clear-cutting reduced *k* of needle litter and the release of N, P and K, possibly because of lower moisture content of the litter layer in the summer and lower soil temperature in the winter, thus decreasing microbial degradation (Cortina and Vallejo 1994). Similarly, thinning by removal of 20% and 30% basal area in a cool and moist Scots pine forest (MAT 12.0 °C, MAP 912 mm) reduced needle litter *k* (calculated over the entire decomposition period) compared to the control, though no thinning effect was found on litter mass loss and nutrient release (N, P, K) at the end of the 5 years of field incubation (Blanco et al. 2011).

16.10 Conclusions and Consequences for Mediterranean Pine Forests and Forests in Other Climatic Zones Subjected to Climate Change

The evidence summarized in this chapter leads us to the following hypothesis regarding climate change effects on litter decomposition in Mediterranean pine forests (mainly based on Sects. 16.6 and 16.9): In forests on the cool and moist side of Mediterranean-type climatic zones, increased solar radiation, temperature and non-rainfall water sources will not be sufficient to compensate for the considerable decrease in litter decay caused by the higher water deficit (Fig. 16.1). On the other hand, such compensation will be more probable in areas within the more xeric range where lack of rainwater already restricts litter decomposition, thus maintaining decay rates at currently observed levels. Nevertheless, even if decay rates remain constant, litter production would most probably decrease with plant biomass, thus reducing overall C storage. There is a critical need for future research to confirm or reject this hypothesis and its consequences for litter decomposition, C storage and nutrient cycling in Mediterranean pine forests.

Models predict a significant expansion of global dryland areas, beginning in the coming decades (Huang et al. 2016). Part of this trend is expected to include a northward and eastward expansion of the Mediterranean-type climate into more mesic Southern European and western North American regions (Alessandri et al. 2014; Huang et al. 2016). The transition of temperate to Mediterranean climate is projected to result in higher rainfall amounts in winter, but less rain in summer (Alessandri et al. 2014). Such a precipitation trend, together with the current and projected warming in these regions (IPCC 2013, 2018), will lead to pronounced dry summer periods.

These climate changes will have consequences for decay rates and for the way that litter decomposes in such 'new' Mediterranean-type forests, some of which are or will be populated in part by pine trees. Enhanced drought severity and duration might limit litter decomposition because of low water availability to decomposers, potentially augmenting the forest floor mass, at least temporarily. On the other hand, these conditions might expose plant litter to 'dryland' mechanisms of decomposition, such as photodegradation, thermal degradation, and humidity-enhanced microbial degradation (enabled by non-rainfall water sources), particularly if drought causes partial canopy defoliation (Carnicer et al. 2011). Therefore, potential drought-induced C storage on the forest floor might be counteracted by 'dryland' mechanisms of decomposition, which might lead to the release of nutrients back to the soil.

References

Acacio V, Dias FS, Catry FX et al (2017) Landscape dynamics in Mediterranean oak forests under global change: understanding the role of anthropogenic and environmental drivers across forest types. Glob Chang Biol 23:1199–1217

Aerts R (1997) Climate, leaf litter chemistry and leaf litter decomposition in terrestrial ecosystems: a triangular relationship. Oikos 79:439–449

Alessandri A, De Felice M, Zeng N et al (2014) Robust assessment of the expansion and retreat of Mediterranean climate in the 21st century. Sci Rep 4:7211

Almagro M, Martínez-Mena M (2012) Exploring short-term leaf-litter decomposition dynamics in a Mediterranean ecosystem: dependence on litter type and site conditions. Plant Soil 358:323–335

Araujo PI, Austin AT (2015) A shady business: pine afforestation alters the primary controls on litter decomposition along a precipitation gradient in Patagonia, Argentina. J Ecol 103:1408–1420

Arianoutsou M, Radea C (2000) Litter production and decomposition in *Pinus halepensis* forests. In: Ne'eman G, Trabaud L (eds) Ecology, biogeography and management of *Pinus halepensis* and *P. brutia* forest ecosystems in the Mediterranean Basin. Backhuys Publishers, Leiden, pp 183–190

Aupic-Samain A, Baldy V, Lecareux C et al (2019) Tree litter identity and predator density control prey and predator demographic parameters in a Mediterranean litter-based multi-trophic system. Pedobiologia 73:1–9

Austin AT, Vivanco L (2006) Plant litter decomposition in a semi-arid ecosystem controlled by photodegradation. Nature 442:555–558

Austin AT, Mendez MS, Ballaré CL (2016) Photodegradation alleviates the lignin bottleneck for carbon turnover in terrestrial ecosystems. Proc Natl Acad Sci U S A 113:4392–4397

Baier W (1966) Studies on dew formation under semi-arid conditions. Agric Meteorol 3:103–112

Baker NR, Allison SD (2015) Ultraviolet photodegradation facilitates microbial litter decomposition in a Mediterranean climate. Ecology 96:1994–2003

Barba J, Curiel Yuste J, Poyatos R et al (2016a) Strong resilience of soil respiration components to drought-induced die-off resulting in forest secondary succession. Oecologia 182:27–41

Barba J, Lloret F, Curiel Yuste J (2016b) Effects of drought-induced forest die-off on litter decomposition. Plant Soil 402:91–101

Barnes PW, Throop HL, Hewins DB et al (2012) Soil coverage reduces photodegradation and promotes the development of soil-microbial films on dryland leaf litter. Ecosystems 15:311–321

Barnes PW, Throop HL, Archer SR et al (2015) Sunlight and soil–litter mixing: drivers of litter decomposition in drylands. In: Lüttge U, Beyschlag W (eds) Progress in botany. Springer Nature, Cham, pp 273–302

Berg B, Laskowski R (2005) Litter decomposition: a guide to carbon and nutrient turnover. Academic, New York

Berg B, McClaugherty C (2008) Plant litter. In: Decomposition, humus formation, carbon sequestration, 2nd edn. Springer, Berlin

Berg B, Davey MP, De Marco A et al (2010) Factors influencing limit values for pine needle litter decomposition: a synthesis for boreal and temperate pine forest systems. Biogeochemistry 100:57–73

Blanco JA, Imbert JB, Castillo FJ (2011) Thinning affects *Pinus sylvestris* needle decomposition rates and chemistry differently depending on site conditions. Biogeochemistry 106:397–414

Bravo-Oviedo A, Ruiz-Peinado R, Onrubia R et al (2017) Thinning alters the early-decomposition rate and nutrient immobilization-release pattern of foliar litter in Mediterranean oak-pine mixed stands. For Ecol Manag 391:309–320

Brown S, Mo J, McPherson JK et al (1996) Decomposition of woody debris in Western Australian forests. Can J For Res 26:954–966

Camarero JJ et al (this volume) Effects of global change on tree growth and vigor of Mediterranean pines. In: Ne'eman G, Osem Y (eds) Pines and their mixed forest ecosystems in the Mediterranean Basin. Springer, Cham

Carnicer J, Coll M, Ninyerola M et al (2011) Widespread crown condition decline, food web disruption, and amplified tree mortality with increased climate change-type drought. Proc Natl Acad Sci U S A 108:1474–1478

Chomel M, Fernandez C, Bousquet-Mélou A et al (2014) Secondary metabolites of *Pinus halepensis* alter decomposer organisms and litter decomposition during afforestation of abandoned agricultural zones. J Ecol 102:411–424

Cobb RC, Ruthrof K, Breshears DD et al (2017) Ecosystem dynamics and management after forest die-off: a global synthesis with conceptual state-and-transition models. Ecosphere 8:e02034

Cortina J, Vallejo VR (1994) Effects of clearfelling on forest floor accumulation and litter decomposition in a radiata pine plantation. For Ecol Manag 70:299–310

Delgado JD, Arroyo NL, Arévalo JR et al (2007) Edge effects of roads on temperature, light, canopy cover, and canopy height in laurel and pine forests (Tenerife, Canary Islands). Landsc Urban Plan 81:328–340

Diffenbaugh NS, Pal JS, Giorgi F et al (2007) Heat stress intensification in the Mediterranean climate change hotspot. Geophys Res Lett 34:L11706

Dirks I, Navon Y, Kanas D et al (2010) Atmospheric water vapor as driver of litter decomposition in Mediterranean shrubland and grassland during rainless seasons. Glob Chang Biol 16:2799–2812

Dorman M, Svoray T, Perevolotsky A et al (2015) What determines tree mortality in dry environments? A multi-perspective approach. Ecol Appl 25:1054–1071

Evans S, Todd-Brown KEO, Jacobson K et al (2019) Non-rainfall moisture: a key driver of microbial respiration from standing litter in arid, semiarid, and mesic grasslands. Ecosystems. https://doi.org/10.1007/s10021-019-00461-y

Facelli JM, Pickett STA (1991) Plant litter: its dynamics and effects on plant community structure. Bot Rev 57:1–32

Foereid B, Bellarby J, Meier-Augenstein W et al (2010) Does light exposure make plant litter more degradable? Plant Soil 333:275–285

Gallo ME, Porras-Alfaro A, Odenbach KJ et al (2009) Photoacceleration of plant litter decomposition in an arid environment. Soil Biol Biochem 41:1433–1441

Garcia-Pausas J, Casals P, Romanyà J (2004) Litter decomposition and faunal activity in Mediterranean forest soils: effects of N content and the moss layer. Soil Biol Biochem 36:989–997

García-Plé C, Vanrell P, Monrey M (1995) Litter fall and decomposition in a *Pinus halepensis* forest on Mallorca. J Veg Sci 6:17–22

Gavinet J, Vilagrosa A, Chirino E et al (2015) Hardwood seedling establishment below Aleppo pine depends on thinning intensity in two Mediterranean sites. Ann For Sci 72:999–1008

Gavinet J, Prévosto B, Fernandez C et al (2016) Introducing resprouters to enhance Mediterranean forest resilience: importance of functional traits to select species according to a gradient of pine density. J Appl Ecol 53:1735–1745

Gea-Izquierdo G, Viguera B, Cabrera M et al (2014) Drought induced decline could portend widespread pine mortality at the xeric ecotone in managed Mediterranean pine-oak woodlands. For Ecol Manag 320:70–82

Gelfand I, Grünzweig JM, Yakir D (2012) Slowing of nitrogen cycling and increasing nitrogen use efficiency following afforestation of semi-arid shrubland. Oecologia 168:563–575

Giorgi F, Lionello P (2008) Climate change projections for the Mediterranean region. Glob Planet Chang 63:90–104

Glassman SI, Weihe C, Li J et al (2018) Decomposition responses to climate depend on microbial community composition. Proc Natl Acad Sci U S A 115(47):11994–11999

Gliksman D, Rey A, Seligmann R et al (2017) Biotic degradation at night, abiotic degradation at day: positive feedbacks on litter decomposition in drylands. Glob Chang Biol 23:1564–1574

Gliksman D, Haenel S, Grünzweig JM (2018a) Biotic and abiotic modifications of leaf litter during dry periods affect litter mass loss and nitrogen loss during wet periods. Funct Ecol 32:831–839

Gliksman D, Haenel S, Osem Y et al (2018b) Litter decomposition in Mediterranean pine forests is enhanced by reduced canopy cover. Plant Soil 422:317–329

Gómez-Aparicio L, García-Valdés R, Ruíz-Benito P et al (2011) Disentangling the relative importance of climate, size and competition on tree growth in Iberian forests: implications for forest management under global change. Glob Chang Biol 17:2400–2414

Granados ME, Vilagrosa A, Chirino E et al (2016) Reforestation with resprouter species to increase diversity and resilience in Mediterranean pine forests. For Ecol Manag 362:231–240

Grünzweig JM, Gelfand I, Fried Y et al (2007) Biogeochemical factors contributing to enhanced carbon storage following afforestation of a semi-arid shrubland. Biogeosciences 4:891–904

Handa IT, Aerts R, Berendse F et al (2014) Consequences of biodiversity loss for litter decomposition across biomes. Nature 509:218–221

Hättenschwiler S, Coq S, Barantal S et al (2011) Leaf traits and decomposition in tropical rainforests: revisiting some commonly held views and towards a new hypothesis. New Phytol 189:950–965

Henry HAL, Brizgys K, Field CB (2008) Litter decomposition in a California annual grassland: interactions between photodegradation and litter layer thickness. Ecosystems 11:545–554

Hewins DB, Archer SR, Okin GS et al (2013) Soil–litter mixing accelerates decomposition in a Chihuahuan Desert grassland. Ecosystems 16:183–195

Huang J, Yu H, Guan X et al (2016) Accelerated dryland expansion under climate change. Nat Clim Chang 6:166–171

Ibrahima A, Joffre R, Gillon D (1995) Changes in litter during the initial leaching phase – an experiment on the leaf-litter of Mediterranean species. Soil Biol Biochem 27:931–939

Incerti G, Bonanomi G, Giannino F, Rutigliano FA, Piermatteo D, Castaldi S, De Marco A, Fierro A, Fioretto A, Maggi O, Papa S, Persiani AM, Feoli E, De Santo AV, Mazzoleni S (2011) Litter decomposition in Mediterranean ecosystems: modelling the controlling role of climatic conditions and litter quality. Appl Soil Ecol 49:148–157

IPCC (2013) Climate change 2013: the physical science basis. Contribution of Working Group I to the fifth assessment report of the Intergovernmental Panel on Climate Change. Cambridge University Press, Cambridge/New York

IPCC (2018) Global warming of 1.5°C. An IPCC Special Report on the impacts of global warming of 1.5°C above pre-industrial levels and related global greenhouse gas emission pathways, in the context of strengthening the global response to the threat of climate change, sustainable development, and efforts to eradicate poverty. Masson-Delmotte V, Zhai P, Pörtner H-O et al (eds). World Meterological Organization, Geneva

Jacobs AFG, van Pul WAJ, van Dijken A (1990) Similarity moisture dew profiles within a corn canopy. J Appl Meteorol 29:1300–1306

Jacobson K, van Diepeningen A, Evans S et al (2015) Non-rainfall moisture activates fungal decomposition of surface litter in the Namib Sand Sea. PLoS One 10:e0126977

Johnson-Maynard JL, Shouse PJ, Graham RC et al (2004) Microclimate and pedogenic implications in a 50-year-old chaparral and pine biosequence. Soil Sci Soc Am J 68:876

Joly FX, Milcu A, Scherer-Lorenzen M et al (2017) Tree species diversity affects decomposition through modified micro-environmental conditions across European forests. New Phytol 214:1281–1293

King JY, Brandt LA, Adair EC (2012) Shedding light on plant litter decomposition: advances, implications and new directions in understanding the role of photodegradation. Biogeochemistry 111:57–81

Kurz-Besson C, Couteaux MM, Berg B et al (2006) A climate response function explaining most of the variation of the forest floor needle mass and the needle decomposition in pine forests across Europe. Plant Soil 285:97–114

Lado-Monserrat L, Lidón A, Bautista I (2016) Litterfall, litter decomposition and associated nutrient fluxes in *Pinus halepensis*: influence of tree removal intensity in a Mediterranean forest. Eur J For Res 135:203–214

Lee H, Rahn T, Throop HL (2012) An accounting of C-based trace gas release during abiotic plant litter degradation. Glob Chang Biol 18:1185–1195

Lee H, Fitzgerald J, Hewins DB et al (2014) Soil moisture and soil-litter mixing effects on surface litter decomposition: a controlled environment assessment. Soil Biol Biochem 72:123–132

Lin Y, Karlen SD, Ralph J et al (2018) Short-term facilitation of microbial litter decomposition by ultraviolet radiation. Sci Total Environ 615:838–848

Lionello P, Abrantes F, Congedi L et al (2012) Introduction: Mediterranean climate – background information. In: Lionello P (ed) The climate of the Mediterranean region. Elsevier, Oxford

Llorens P, Domingo F (2007) Rainfall partitioning by vegetation under Mediterranean conditions. A review of studies in Europe. J Hydrol 335:37–54

López-Moreno JI, Gascoin S, Herrero J et al (2017) Different sensitivities of snowpacks to warming in Mediterranean climate mountain areas. Environ Res Lett 12:074006

Lucas-Borja ME, Candel-Pérez D, García Morote FA et al (2016) *Pinus nigra* Arn. ssp. salzmannii seedling recruitment is affected by stand basal area, shrub cover and climate interactions. Ann For Sci 73:649–656

Mazza G, Agnelli AE, Cantiani P et al (2019) Short-term effects of thinning on soil CO_2, N_2O and CH_4 fluxes in Mediterranean forest ecosystems. Sci Total Environ 651:713–724

McDowell NG, Allen CD (2015) Darcy's law predicts widespread forest mortality under climate warming. Nat Clim Chang 5:669–672

Moorhead DL, Callaghan T (1994) Effects of increasing ultraviolet B radiation on decomposition and soil organic matter dynamics: a synthesis and modelling study. Biol Fertil Soils 18:19–26

Moro MJ, Domingo F (2000) Litter decomposition in four woody species in a Mediterranean climate: weight loss, N and P dynamics. Ann Bot 86:1065–1071

Ne'eman G, Perevolotsky A (2000) The management of burned forests in Israel. In: Ne'eman G, Trabaud L (eds) Ecology, biogeography and management of *Pinus halepensis* and *P. brutia* forest ecosystems in the Mediterranean Basin. Backhuys Publishers, Leiden, pp 321–333

Palmero-Iniesta M, Domènech R, Molina-Terrén D et al (2017) Fire behavior in *Pinus halepensis* thickets: effects of thinning and woody debris decomposition in two rainfall scenarios. For Ecol Manag 404:230–240

Pan Y-X, Wang X-P, Zhang Y-F (2010) Dew formation characteristics in a revegetation-stabilized desert ecosystem in Shapotou area, Northern China. J Hydrol 387:265–272

Pancotto VA, Sala OE, Cabello M et al (2003) Solar UV-B decreases decomposition in herbaceous plant litter in Tierra del Fuego, Argentina: potential role of an altered decomposer community. Glob Chang Biol 9:1465–1474

Polade SD, Gershunov A, Cayan DR et al (2017) Precipitation in a warming world: assessing projected hydro-climate changes in California and other Mediterranean climate regions. Sci Rep 7:10783

Preisler Y, Tatarinov F, Grünzweig JM et al (2019) Mortality versus survival in drought-affected Aleppo pine forest depends on the extent of rock cover and soil stoniness. Funct Ecol 33:901–912

Prescott CE (2010) Litter decomposition: what controls it and how can we alter it to sequester more carbon in forest soils? Biogeochemistry 101:133–149

Prévosto B, Monnier Y, Ripert C et al (2011) Can we use shelterwoods in Mediterranean pine forests to promote oak seedling development? For Ecol Manag 262:1426–1433

Prévosto B, Helluy M, Gavinet J et al (2020) Microclimate in Mediterranean pine forests: what is the influence of the shrub layer? Agric For Meteorol 282–283:107856

Qasemian L, Guiral D, Farnet AM (2014) How do microlocal environmental variations affect microbial activities of a *Pinus halepensis* litter in a Mediterranean coastal area? Sci Total Environ 496:198–205

Qubaja R, Grünzweig JM, Rotenberg E et al (2020) Evidence for large carbon sink and long residence time in semi-arid forests based on 15-year flux and inventory records. Glob Chang Biol. https://doi.org/10.1111/gcb.14927

Rascher KG, Große-Stoltenberg A, Máguas C et al (2011) Understory invasion by *Acacia longifolia* alters the water balance and carbon gain of a Mediterranean pine forest. Ecosystems 14:904–919

Rasmussen C, Southard RJ, Horwath WR (2008) Litter type and soil minerals control temperate forest soil carbon response to climate change. Glob Chang Biol 14:2064–2080

Raz Yaseef N, Rotenberg E, Yakir D (2010) Effects of spatial variations in soil evaporation caused by tree shading on water flux partitioning in a semi-arid pine forest. Agric For Meteorol 150:454–462

Riahi K, Rao S, Krey V et al (2011) RCP 8.5 – a scenario of comparatively high greenhouse gas emissions. Clim Chang 109:33–57

Rodríguez-García E, Bravo F, Spies TA (2011) Effects of overstorey canopy, plant–plant interactions and soil properties on Mediterranean maritime pine seedling dynamics. For Ecol Manag 262:244–251

Rovira P, Vallejo VR (1997) Organic carbon and nitrogen mineralization under Mediterranean climatic conditions: the effects of incubation depth. Soil Biol Biochem 29:1509–1520

Ruiz-Benito P, Lines ER, Gomez-Aparicio L et al (2013) Patterns and drivers of tree mortality in Iberian forests: climatic effects are modified by competition. PLoS One 8:e56843

Sagi N, Grünzweig JM, Hawlena D (2019) Burrowing detritivores regulate nutrient cycling in a desert ecosystem. Proc R Soc B 286:20191647

Santonja M, Baldy V, Fernandez C et al (2015) Potential shift in plant communities with climate change: outcome on litter decomposition and nutrient release in a Mediterranean oak forest. Ecosystems 18:1253–1268

Schlesinger WH, Bernhardt ES (2013) Biogeochemistry. An analysis of global change, 3rd edn. Academic, New York

Segura C, Fernández-Ondoño E, Jiménez MN et al (2019) Carbon and nutrient contents in the miscellaneous fraction of litterfall under different thinning intensities in a semiarid *Pinus halepensis* afforestation. IFOREST 12:375–382

Sheffer E, Canham CD, Kigel J et al (2015) Countervailing effects on pine and oak leaf litter decomposition in human-altered Mediterranean ecosystems. Oecologia 177:1039–1051

Sherwood S, Fu Q (2014) A drier future? Science 343:737–739

Smith WK, Gao W, Steltzer H et al (2010) Moisture availability influences the effect of ultraviolet-B radiation on leaf litter decomposition. Glob Chang Biol 16:484–495
Springett JA (1976) The effect of planting *Pinus pinaster* Ait. on populations of soil microarthropods and on litter decomposition at Gnangara, Western Australia. Austral Ecol 1:83–87
Tague CL, Moritz MA (2019) Plant accessible water storage capacity and tree-scale root interactions determine how forest density reductions alter forest water use and productivity. Front For Glob Chang 2:36
Tsiafouli MA, Monokrousos N, Sgardelis SP (2018) Drought in spring increases microbial carbon loss through respiration in a Mediterranean pine forest. Soil Biol Biochem 119:59–62
Wang J, Liu L, Wang X et al (2015) The interaction between abiotic photodegradation and microbial decomposition under ultraviolet radiation. Glob Chang Biol 21:2095–2104
Wang J, Liu L, Wang X et al (2017) High nighttime humidity and dissolved organic carbon content support rapid decomposition of standing litter in a semi-arid landscape. Funct Ecol 31:1659–1668
Whitford WG (2020) Ecology of desert systems, 2nd edn. Academic, London
Wilkinson SC, Anderson JM, Scardelis SP et al (2002) PLFA profiles of microbial communities in decomposing conifer litters subject to moisture stress. Soil Biol Biochem 34:189–200
Xiao H, Meissner R, Seeger J et al (2009) Effect of vegetation type and growth stage on dewfall, determined with high precision weighing lysimeters at a site in northern Germany. J Hydrol 377:43–49

Part IV
Forest Dynamics, Biodiversity and Biotic Interactions

Efrat Sheffer

Chapter 17
Dynamics of Mixed Pine–Oak Forests

Yoni Waitz and Efrat Sheffer

17.1 Theories of Forest Dynamics

The dynamics of vegetation in general, and forests in particular, have been a focus of ecological and silvicultural research for almost two centuries. Two central theories describe the development of forest vegetation: vegetation succession (primary and secondary) (Clements 1916; Gleason 1927; Thoreau and Emerson 1887) and forest gap dynamics (Cooper 1913; Jones 1945; Watt 1947). Secondary succession refers to the long-term directed temporal change in plant communities following a disturbance event. The theory of secondary succession of abandoned agricultural old fields (*e.g.,* Fisher 1928; Odum 1960) or abandoned fallow and meadow lands (Schmidt and Briibach 1993) was developed during the first half of the twentieth century based on observations of the stages of colonization of abandoned fields. This secondary succession usually begins with land colonization by herbs and then shrubs, through colonization by early-successional, shade-intolerant coniferous trees, eventually replaced by late-successional shade-tolerant trees. In this sense, secondary succession of Mediterranean forests follows the same principles as old-field succession, whereby early-successional pines are replaced by late-successional broad-leaved species such as deciduous oaks (Pausas et al. 2004) or evergreen oaks (Espelta et al. 1995; Urbieta et al. 2011).

The theory of forest gap dynamics (Shugart 1984) describes the maintenance of mature (old-successional) forests as a constant spatial mosaic of patches of forest collapse and regeneration, by the formation of gaps in the forest and stages of regeneration in these gaps. That is, forest gap dynamics can be regarded as complementary to the succession of the forest, or as cyclic succession within the forest. Forest

Y. Waitz · E. Sheffer (✉)
Institute of Plant Science and Genetics in Agriculture, The Robert H. Smith Faculty of Agriculture, Food and Environment, The Hebrew University of Jerusalem, Rehovot, Israel
e-mail: efrat.sheffer@mail.huji.acil

G. Ne'eman, Y. Osem (eds.), *Pines and Their Mixed Forest Ecosystems in the Mediterranean Basin*, Managing Forest Ecosystems 38,
https://doi.org/10.1007/978-3-030-63625-8_17

gap dynamics usually refer to processes at a smaller scale than in secondary succession, i.e., dynamics in local gaps within a forest versus dynamics at the landscape scale (Urban et al. 1991).

One of the main conceptual differences between the two theories is their perception of anthropogenic influences through human activity. Human-induced changes such as agricultural activity and its abandonment or forest clear-cutting are the primary drivers of secondary succession, although secondary succession can also follow natural or semi-natural disturbances such as wildfires. Conversely, natural disturbances such as tree mortality are the primary drivers of forest gap dynamics. The development of Mediterranean forests has been influenced by all such disturbances, both anthropogenic and natural, which have largely determined forest succession and dynamics throughout history.

In this chapter we analyze the dynamics of Mediterranean mixed pine-oak forests. We focus on pine and oak trees as the most dominant and widespread in mediterranean forests, which often appear in different compositions of mixed forests throughout the mediterranean ecosystems of the Mediterranean Basin, California, and Baja-California (the genera *Pinus* and *Quercus* are both naturally distributed in the Northern hemisphere only). *Our goal is to examine whether the framework of the classical theories of secondary forest succession and forest gap dynamics can be used to construe the past, present and future dynamics of Mediterranean mixed forests.* Here, we review the available literature on the different stages of mixed forest dynamics, primarily from the Mediterranean Basin, and explore how they comply with the general framework of forest succession and dynamics.

17.2 Past Dynamics – Formation of Mediterranean Pine and Oak Forests

The history of forest development in Mediterranean ecosystems is tightly linked to centuries of human activity in the region. Different periods can be distinguished along this timeframe, beginning with the paleoclimatic history that contributed to the formation of what is currently known as Mediterranean vegetation, to the potential influences of contemporary global climatic change.

The typical seasonality of the Mediterranean-type climate, with a mild rainy winter and hot, dry summer, first appeared ca. 2.5 million years ago (Ackerly 2009; Jiménez-Moreno et al. 2010; Rundel et al. 2016; Verdú et al. 2003). As a consequence of these past climatic changes, new species of sclerophyllous, drought-adapted plants evolved, which inhabit Mediterranean forests, woodlands (*e.g.*, maquis, matorral, chaparral) and shrublands to this day (Di Castri and Mooney 2012; Valiente-Banuet et al. 2006). During the last two million years Mediterranean vegetation spread and retreated according to climate fluctuations, for example shifting dominance between deciduous and evergreen oaks, between cold-mesic and warm-dry periods, respectively (Thompson 2005).

The Mediterranean attributes of pine and oak species (Barbéro et al. 1998; Ne'eman and Trabaud 2000) have further evolved under the influence of at least 50,000 years of human exploitation, sometimes intensive, in the form of grazing by domesticated livestock, wood-cutting, burning, and land clearing for upland agriculture (Naveh and Dan 1973). Intensive exploitation of the land increased dramatically with the domestication of plants and animals, the intensification of agriculture, sedentarization and the consequent increasing growth of human populations during the Holocene (last 10,000 years). Anthropogenic influences resulted in large-scale shifts in land-use, clearance of much of the natural vegetation and deforestation (Blondel and Aronson 1999; Etienne et al. 1998; Thompson 2005).

Deforestation slowed and human-exploitation began to decrease only by the eighteenth century, following the industrial revolution and consequential demographic and socioeconomic changes, such as urbanization and the shift to fossil fuel. Much of the marginal land that used to be exploited, mainly for agriculture, was abandoned, which created room for the recovery of the natural vegetation. Vegetation either recovered gradually by natural secondary succession (passive sensu Cruz-Alonso et al. 2019), or changed abruptly by active human restoration, mostly forest plantation or re-afforestation. The large-scale abandonment of agriculture and urbanization following the industrial revolution created much "new" land, to be colonized by vegetation.

Active restoration of degraded vegetation has been carried out since the early twentieth century, by active replanting of forests on much of the land, mostly using pioneer pine species (Ne'eman and Trabaud 2000; Sheffer 2012; Vadell et al. 2016). Afforestation efforts aimed to serve several purposes such as soil protection, prevention of soil erosion, timber production and development of rural economy. One of the purposes of pine forest plantation in some of the Mediterranean regions was to facilitate later establishment of oaks (Vadell et al. 2016). This reforestation with pines, and early-successional colonization by pines, resulted in increasing cover of pine stands throughout Mediterranean landscapes, and a spatial contiguity of stands of pines, and of oaks and their early-successional stages.

The secondary succession of vegetation resulted in the development of woodlands or even forests, dominated by oaks and other broad-leaved species that either established initially or eventually replaced the pioneer pines. While most of the passively restored land seems to develop into oak-dominated woodlands, plantations in degraded landscapes were primarily dominated by monospecific pine forests. Therefore, throughout most of the twentieth century, woodlands dominated by oaks (and other broad-leaved species) and forests dominated by pines were regarded as distinct vegetation formations (Sheffer 2012).

Furthermore, these mediterranean pine and oak species have been found throughout the pollen record in many Mediterranean regions (Baruch 1986; Benslama et al. 2010; Carrión et al. 2001; Morales-Molino et al. 2017; Yasuda et al. 2000). However, whether pines and oaks coexisted in the same stands and in what abundance, cannot be deduced from the pollen records.

Only towards the end of the 20th century, after decades of woodland recovery and forest growth, observations of pine colonization outside the planted forests and

Fig. 17.1 Colonization of young *P. halepensis* trees in a woodland of *Q. calliprinos*. Judean Mountains, Israel. (Picture by: Moshe Alon)

oak development in the understory of pine forests, exposed new pathways for the formation of mixed forests of conifers (pines) and broad-leaved (oak) trees. The dynamics of Mediterranean forests in recent decades is therefore strongly influenced by the physiognomy of this landscape mosaic of developing pine stands and oak woodlands. Within these landscapes, mixed pine-oak forests are forming as a result of two reciprocal processes: pine colonization of oak woodlands (Etienne 2001; Lavi et al. 2005; Sheffer 2012; Sheffer et al. 2014; Zavala et al. 2000; see for example Fig. 17.1), and oak colonization of, and/or regeneration in, pine stands (Sheffer et al. 2013; Zavala et al. 2000; see for example Fig. 17.2). The latter is in fact homologous to the classically defined succession. But, the propensity for colonization by the two species, and the likelihood of colonization by each of the species are dynamically changing in space and through time as a result of different species traits, local abundances and stages of development.

17.3 Dynamics of Colonization by Pines and Oaks

Mediterranean pine species inhabit most mediterranean-climate habitats in the range 250–1000 mm annual rainfall, including the Mediterranean Basin and California, and successfully grow (and invade) the mediterranean ecosystems of the Southern hemisphere (Richardson and Rejmánek 2004). These species are typically shallow-rooted but highly adapted to dry conditions, using ecophysiological isohydric regulation (Richardson 2000, see Chap. 7, Klein this volume).

Fig. 17.2 Regeneration of *Q. calliprinos* understory in a gap in planted forest of *P. halepensis*. Judean Mountains, Israel. (Picture by: Yoni Waitz)

Most Mediterranean pines recruit exclusively by seeds. The seeds of pines are dispersed by the wind mainly in the vicinity of adult pine forests, and long-distance dispersal events are scarce but ecologically important (Lavi et al. 2005; Nathan et al. 2000; Osem et al. 2011). Two of the common Mediterranean species – *P. halepensis* Mill. and *P. brutia* Ten. – are fire-adapted and store significant amounts of seeds in serotinous cones, which are dispersed following a fire event or in extreme hot and dry climatic events (Sharqi'a, Mistral, Sharav, Santa Ana winds) (Nathan et al. 2000). Rare, long-distance dispersal events generate a potential but low input of pine propagules throughout mediterranean landscapes and form patches of pines at considerable distances from the seed sources (Nathan and Ne'eman 2004; Sheffer et al. 2014).

The habitat conditions in the sink patches, i.e., where seeds are deposited, have been shown to facilitate or hinder the survival, germination and development of pine seedlings and saplings, therefore influencing the process of woodland colonization by pines. Rocks and small shrubs can facilitate the germination of pine seeds by creating shelters that protect the seeds and seedlings from predation and desiccation. On the other hand, pine seedlings are shade-intolerant and tend to suffer from low light levels and limited water availability when germinating inside tall shrubs or underneath trees (conspecific or broad-leaved), or even in dense herbaceous vegetation (Waitz et al. 2015; Ziffer-Berger et al. 2017). Low or moderate grazing has been shown to reduce the effect of competition by heterospecific plants, while severe grazing and goat browsing tend to damage young pine seedling (Lavi et al. 2005; Osem et al. 2011; Sheffer et al. 2014). We emphasize that the processes that influence pine colonization are natural but the propensity of these processes to occur has been strongly affected by the high abundance of pines in the landscape due to massive afforestation in the past.

Long-living oak trees, sometimes assuming a multi-stemmed shrub structure, inhabit many Mediterranean ecosystems, usually in the more mesic habitats with annual precipitation ranging from 350 mm (in the southern end of the distribution of *Q. calliprinos* Webb in Israel) to over 1000 mm (in the wet-Mediterranean mountains of France and Spain). Oaks in mediterranean ecosystems include evergreen and winter-deciduous phenologies, anatomically and physiologically adapted to the dry summer conditions (Canadell et al. 1996). These oak species typically have a deep-penetrating tap root that allows oak survival in dry conditions, and use anisohydric regulation (Ferrio et al. 2003; Klein 2014; Roman et al. 2015).

Oaks can recruit from seeds (large, short-lived acorns), or regenerate by resprouting from root-stocks. Acorns are typically dispersed by animals, most efficiently by cache-hording jays (*Garrulus glandarius* L.). Jays preferentially disperse acorns into pine stands more than within the oak woodlands, to reduce predation risk (Gómez 2003; Pons and Pausas 2007; Purves et al. 2007). Puerta-Piñero (2010) also presented evidence of pre- and post-dispersal predation of acorns by rodents close to mother plants, and predation by boars in pine forests. However, the spatial pattern of the activity of the jays in the forest was correlated with the recruitment of seedlings (Puerta-Piñero et al. 2012a). A discrepancy exists between inferences of long-distance dispersal of acorns (Sheffer et al. 2013) and evidence of a strong negative effect of distance from oak seed sources on oak regeneration in pine forests (Ruiz-Benito et al. 2012). However, both results relate to the measured pattern of abundance of young oak saplings, which is typically related to both germination from dispersed seeds and regeneration by resprouting.

Oak saplings and resprouts can grow in shrublands and woodlands as part of secondary succession, either with or without an intermediate stage of passive colonization by pines. Alternatively, oaks can establish or resprout in the understory of planted pine forests. Therefore, both secondary succession and colonization in planted forests result in the formation of mixed pine-oak forests. These different forms of mixed forests potentially differ in their composition and initial conditions, and differ from the mixed pine-oak forests that are formed through colonization of oak woodlands by pines.

The dynamics of mixed forest formation are also influenced by differences in habitat preferences between the different developmental stages of the oaks. Germination and seeding survival have been shown to be high underneath shrubs and pine trees, and in forests regenerating after fire under both standing live or dead planted pine trees (Leverkus et al. 2016), and low under adult oaks, mostly due to high seed predation (Gómez and Hódar 2008; Rolo et al. 2013). Moreover, regeneration of *Q. ilex* L. within pine forests has reached even more extreme cold environments than the distribution of adult trees (Carnicer et al. 2014; Urbieta et al. 2011). Furthermore, oak development and establishment from seedlings are limited under dense canopies of either pines or oaks (Cooper et al. 2014; Rodríguez-Calcerrada et al. 2007). These differences are likely to influence the future dynamics of Mediterranean mixed pine-oak forests.

According to Navarro-González et al. (2013) land use legacies are largely more important for oak recruitment below pines, than the density of pines in the canopy or the availability of oak propagules. Legacies of intensive land use in the past, such as cropland or pastures, reduced the chance of oak recruitment compared to less intense land use histories. A similar idea is suggested by Puerta-Piñero et al. (2012b), who found that the abundance of oaks after fire is primarily related to the time passed since the beginning of forest regeneration after cropland abandonment.

Overall, the colonization of pine forests by oaks (*e.g.*, Sheffer et al. 2013) seems to be a more common pathway for the formation of mixed pine-oak forests, compared to colonization of oak woodlands by pines (Sheffer et al. 2014). A review of the two reciprocal pathways of colonization highlighted that pine colonization is more common in vegetation formations with an open canopy, such as grasslands, shrublands or sparse woodlands, but the likelihood of pine colonization decreases in dense woodlands of oaks (Sheffer 2012; Sheffer et al. 2014). The natural transition of pine forests into mixed forests or oak-dominated forests has been typically described in mesic conditions with no further perturbations (Capitanio and Carcaillet 2008; Rouget et al. 2001; Zavala et al. 2000; see for example Fig. 17.3).

Fig. 17.3 Mixed pine-oak forest on Carmel Mountain, Israel. (Picture by: Efrat Sheffer)

17.4 Present and Future Dynamics – The Fate of Mixed Forests in Light of Forest Theories

Mixed pine-oak forests can be considered an intermediate stage of Mediterranean forest succession, which potentially can develop into what has been classically defined as the "climax stage" of Mediterranean oak-dominated forest with no pines. A gradual regeneration of oaks in forest gaps and replacement of declining pine trees by oaks can lead to this transition into oak dominance. Alternatively, the mixed pine-oak forest can remain as a stable stage, resistant to further changes, in which both pine and oak species recruit and therefore remain abundant. An extreme potential trajectory would be a decline of both pine and oak trees, which could result in replacement by shrubs or regeneration of more resilient species, for example in the case of extreme climate warming and drying. In theory, future changes in land-use and climate might also lead to the formation of pine forests as a stable community.

Here, we wish to explore whether in fact the mixed forest is a transitional phase in the succession of Mediterranean vegetation or a resistant and stable stage that can be considered a climax community under current and future conditions.

Next, we evaluate current and future dynamics of mixed pine-oak forests from a combination of the outcomes of models and observations and analyses of recent patterns of recruitment of pines and oaks in Mediterranean forests. We limit our analysis to the mixed forests of the Mediterranean Basin, since this was the focus of all the literature that we found on this topic. We describe the patterns of and expectations for regeneration of pines and oaks in the mixed forests, and the factors that influence these dynamics.

17.5 Succession to Oak Dominance

Much of the recent research supports the theoretical expectations of a gradual progression from mixed pine-oak to oak-dominated forests. That is, in the mixed forests, selection often acts in favor of oak species. This selection could be acting at the adult stage or in the different stages of recruitment and development (seed, seedling, sapling), and depends on the relative success of oaks vs. pines. Vidal-Macua et al. (2017) presented the transition of land cover in the Iberian Peninsula from 1987 to 2012 as a primarily successional process towards oak forests, while the rate of transition to pine forests decreased dramatically during the same period. Reduced pine recruitment and oak expansion is supported by changes in management regimes such as decreases in logging and grazing by domestic livestock (Chauchard et al. 2013; DeSoto et al. 2010; Kouba et al. 2012; Martín-Alcón et al. 2015). For example, according to Gea-Izquierdo et al. (2015), an old stand of *P. sylvestris* L. in central Spain, which is one of the last remnants of an old extensive pine forest in the region, has been colonized by *Q. pyrenaica* Willd., but similar stands remain dominated by pines. Once these pines were no longer used for firewood as the main source of energy, regeneration of oaks became continuous, while pines failed to recruit under the canopy of oaks. Management also influenced the abundance of different oak species, for example the preference of the valuable cork oak, *Q. suber* L., over other local oaks (Urbieta et al. 2008).

Similar patterns of the expansion of the distribution and abundance of oak (primarily *Q. ilex*) have been found in Spain (Gómez-Aparicio et al. 2009; Urbieta et al. 2011; Vayreda et al. 2013) and elsewhere (Vayreda et al. 2016), coinciding with a general negative effect on recruitment of pine species (Carnicer et al. 2014). Analyses of forest inventory data from the last 2–3 decades reveal that the dynamics of regeneration within mixed pine-oak forests select for oak dominance, as oak recruitment increases (*e.g. Q. ilex*) whereas pine regeneration declines (*e.g., P. nigra* J.F.Arnold) (Barreda and Doménech 2013; Martín-Alcón et al. 2015).

From a physiological perspective, although Mediterranean pines are drought resistant, oaks can use deeper water sources in the soil compared to pines (del Castillo et al. 2016; Grossiord et al. 2014; Klein et al. 2013; Rubio-Cuadrado et al. 2018). Thus, pines in mixed stands are more vulnerable to drought, compared to their neighboring oaks and even compared to their own resistance in pine monocultures (Bello et al. 2019). In the mixed forests oaks are less sensitive to water shortage compared to pines in the face of competition (del Castillo et al. 2016; Jucker et al. 2014a). Drought-induced decline and mortality of Mediterranean pines has been reported for *P. pinaster* Ait. (Aldea et al. 2018; Corcuera and Notivol 2015), *P. halepensis* (de la Serrana et al. 2015; Dorman et al. 2015) and *P. sylvestris* (Martínez-Vilalta and Piñol 2002). Water limitation of pines such as *P. pinaster* is associated with low photosynthetic activity and inhibited carbon assimilation, sometimes with additional biotic stress agents (*e.g.*, bark beetles, *Matsucoccus josephi* Bodenhiemer and Harpaz, or mistletoe infection; Gea-Izquierdo et al. 2019), thus enabling the expansion of coexisting oaks such as *Q. ilex*.

Modeling simulations of secondary succession in mixed forests of *Q. ilex* and *P. pinea* L. under expected climate change scenarios has suggested that *Q. ilex* will be favored while the secondary growth of *P. pinea* will be reduced (de-Dios-García et al. 2018). Another model considered the influence of increasing frequency and severity of fire events, and projected that several widespread species of pine (*e.g., P. pinea, P. nigra* and *P. sylvestris*) which lack fire adaptations such as serotinous cones, will eventually decline. Under harsh conditions that combine fire and drought, recruitment is projected to fail even in fire- and drought-adapted species such as *P. halepensis*, particularly in mixed forests, due to competition with the more adapted *Q. ilex* (Gil-Tena et al. 2016). Indeed, in Spain around 60% of the burned area of *P. nigra* changed from pine- to oak-dominated forest (Martín-Alcón and Coll 2016).

The regeneration of oaks is also expected to become limited as the density of the forest increases, and therefore forest gaps are necessary for the maintenance of oak forests such as *Q. ilex*, *Q. faginea* Lam, *Q. cerrioides* Willk. & Costa (Barreda and Doménech 2013; Martín-Alcón et al. 2015), *Q. ithaburensis* Decne. (Cooper et al. 2014), and *Q. pyrenaica* Willd. (Rodríguez-Calcerrada et al. 2007). The gap-phase dynamics of Mediterranean oak forests have barely been studied, because these forests are only now maturing and expanding, and because of the slow rates of oak regeneration dynamics. More research is needed to understand the patterns of gap formation and oak regeneration dynamics in Mediterranean oak forests.

17.6 Enduring Stability of the Mixed Forest

In contrast to the evidence for the relative decline of pines in Mediterranean mixed forests and the processes that lead to oak dominance, several observations indicate a steady state of mixed forests. Mixed pine-oak forests are maintained by the survival and regeneration of both pines and oaks, which is also associated with improved functioning of the entire system compared to monospecific forests. The resilience of ecosystems with high species diversity is well known from different ecosystems (Hooper et al. 2012; Tilman 1999), including Mediterranean forests. Mixed forests often are more productive, and aboveground productivity is more stable (de-Dios-García et al. 2015; Del Río and Sterba 2009; Jucker et al. 2014b) and shows improved microbial activity and nutrient cycling in the soil (Brunel et al. 2017; Guenon et al. 2017) compared to pure pine or oak forests. Niche separation in space and time reduces inter-specific competition, and facilitation has been proposed as a main mechanism to explain the high productivity and stability of mixed forests (Carnicer et al. 2013; Ruiz-Benito et al. 2017).

Another explanation for the resilience of mixed pine-oak forests can be related to evidence for hydraulic redistribution of the water in the soil, especially in dry ecosystems. Water redistribution occurs when water from deep soil layers is lifted by the roots of oaks into shallow soil layers therefore increasing the availability of water to shallow-rooted pines (Prieto et al. 2012; Rodríguez-Robles et al. 2015; Zalloni et al. 2019).

According to Moreno-Fernández et al. (2016) the distribution of mixed forests of *Q. pyrenaica* and *P. sylvestris* increased threefold during the period 1965–2012. The establishment of mixed forests seems minor compared to the creation of pure stands of oaks, which was an order of magnitude larger, but compared to the decline of pure stands of pines it is reasonable to suggest a stabilizing benefit for these mixed forests.

Model projections for four future climatic scenarios for southern Spain showed that the distribution of the two most dominant species in Mediterranean mixed forest, *Q. ilex* and *P. halepensis,* will increase and could spread across the modeled area (*e.g.,* Zavala and de la Parra 2005). However, the same model also predicted that the distribution of eight other species, which are less resilient to aridity (both oak and pine species) is expected to suffer a severe contraction (López-Tirado and Hidalgo 2018).

Active management actions in the forest are also a significant factor influencing the succession of mixed pine-oak forests. For example, absence of thinning in pine plantations can result in a relatively high density of pines that significantly slows colonization by oaks (Gómez-Aparicio et al. 2009; Ruiz-Benito et al. 2012). Grazing in the understory and thinning of pines in mixed forests may be a good choice for limiting inter-specific competition for resources in mixed forests (Gavinet et al. 2015; Osem et al. 2015; Zalloni et al. 2019). Conversely, a decrease in forest management activities such as clear-cutting, thinning and coppicing, is currently reducing the availability of niches for pine regeneration (Martín-Alcón et al. 2015).

Finally, according to Sánchez de Dios et al. (2019) the mapping of mixed forests suffers from underestimation, since they are often described or mapped as monospecific forest types, therefore creating a bias in the estimation of the development of monospecific vs. mixed stands. These authors recently asked whether mixed forests are stable vegetation stages or they will evolve towards monospecific formations, but the question remains to be answered.

17.7 Deterioration of Mixed Forests

A third possible trajectory of Mediterranean mixed pine-oak forests is a decline of the forest to lower vegetation formations, *i.e.*, a regression in the successional process. This kind of regression runs contrary to all the evidence we have summarized so far, and was suggested as a possible result of novel Mediterranean conditions, due to ongoing and intensifying drought scenarios combined with the growing frequency and severity of fire events.

More extreme cases have resulted in the transition from deciduous oak forests to shrublands of more xeric species (Acácio et al. 2017; López-Tirado and Hidalgo 2018). The relative abundance and recruitment of *Q. suber* and other late successional species have been declining, which has been related to a shift towards shrublands and less dense forests (Costa et al. 2010; Costa and Madeira 2011; Ibáñez et al. 2015).

A model of vegetation dynamics in the south-western Mediterranean Basin, which included stochastic fire events and different responses of plants to fire, predicted that a combination of increasing water stress and fire damage may reduce the resilience of the mixed forest and eventually drive post-fire ecosystem dynamics towards open shrublands (Baudena et al. 2020). Another model by Batllori et al. (2019) predicted that the resilience of Mediterranean ecosystems is questionable in the light of the short intervals between disturbances, and that resilience mechanisms may fail and lead to sudden changes in the composition of vegetation towards non-forest. Several studies that evaluated how climate change scenarios will influence the currently dominant oaks, suggested that these oaks might lose their resilience (Torres et al. 2016; Vayreda et al. 2016).

17.8 Synthesis and Conclusions

Evidence from the recent literature on the dynamics of mixed forests demonstrates processes and dynamics homologous to the descriptions of the classical theories of secondary succession and gap dynamics. Oaks, the late-successional component, are becoming dominant in the landscape and are capable of regenerating within mixed forests and in monospecific forests. Pines, on the other hand, in most cases are being gradually excluded by competition, especially in dense forests, or by a relatively lower colonization rate and fitness, compared to oaks.

Our review showed that anthropogenic influences have had a considerable and strong effect on both the long- and short-term dynamics of forest succession and on forest gap dynamics. The distribution of planted pine forests and their high abundance in the landscape is a major anthropogenic change in the landscape that has strongly affected the process of Mediterranean secondary succession. Man-made selection for resistant species for afforestation was designed to enable such forests to withstand harsh climate conditions as well as frequent disturbances such as grazing and fire events. This man-made selection has played a major role in the formation of current Mediterranean mixed pine-oak forests. Management actions in Mediterranean forests (thinning and coppicing to shape the density and structure of forest stands) are influencing the formation of gaps and regeneration of different species within these growing forests.

Forest plantation and management has in fact achieved one of the original objectives for the plantation of pines a hundred years ago – to facilitate the formation of gaps in the planted forests, which are being colonized by oaks (Vadell et al. 2016). Therefore, these planted forests are in fact promoting the shift from monocultures to mixed forests. A significant decrease in forest management actions in recent decades (both stand thinning and grazing in the understory) is still supporting the recruitment of oaks over pines in the mixed forest. In other words, succession seems to be a wide and common process, although many of the cited articles do not use the word "succession", possibly due to scientific trends.

We focused here on the forests of the Mediterranean Basin, although mixed pine-oak mediterranean forests also grow in California (among other conifers and broad-leaved species). The main reason was the lack of treatment of the latter forests in the literature. However, we also assume that this is related to the difference in the anthropogenic influence in California, compared to the history of the Mediterranean Basin, and the relatively higher diversity of formations and tree species in Californian forests compared to the Mediterranean Basin. Some studies have focused on the effects of climatic changes in mediterranean mixed forests in California, which decrease and shift the distribution range of several species (Kueppers et al. 2005) and results in shifts in the dominance of forests from conifers to broad-leaved species (Lenihan et al. 2003).

The predictions for the development of Mediterranean mixed forests in the future also support the deterministic process of succession, but with several knowledge gaps that need to be further explored to better understand future dynamics. One gap is related to limited information on the rate of recruitment of oaks within oak forests, and the processes that determine gap dynamics and regeneration within oak forests. Another important knowledge gap is related to the effects of external factors, such as climatic change (studied to some extent using models) and management, on tree mortality and on the dynamics of the forest, especially with respect to understanding whether it is possible to direct the forest trajectory with active management.

References

Acácio V, Dias FS, Catry FX, Rocha M, Moreira F (2017) Landscape dynamics in Mediterranean oak forests under global change: understanding the role of anthropogenic and environmental drivers across forest types. Glob Chang Biol 23:1199–1217

Ackerly DD (2009) Evolution, origin and age of lineages in the Californian and Mediterranean floras. J Biogeogr 36:1221–1233

Aldea J, Bravo F, Vázquez-Piqué J, Rubio-Cuadrado A, del Río M (2018) Species-specific weather response in the daily stem variation cycles of Mediterranean pine-oak mixed stands. Agric For Meteorol 256:220–230

Barbéro M, Loisel R, Quézel P, Richardson DM, Romane F (1998) Pines of the Mediterranean basin. Ecol Biogeogr Pinus:153–170

Barreda SG, Doménech SR (2013) Short-term dynamics of *Quercus ilex* advance regeneration in a *Pinus nigra* plantation after the creation of small canopy gaps. For Syst 22:179–188

Baruch U (1986) The late Holocene vegetational history of Lake Kinneret (sea of Galilee), Israel. Paléorient 12:37–48

Batllori E, De Cáceres M, Brotons L, Ackerly DD, Moritz MA, Lloret F (2019) Compound fire-drought regimes promote ecosystem transitions in Mediterranean ecosystems. J Ecol 107:1187–1198

Baudena M, Santana VM, Baeza MJ, Bautista S, Eppinga MB, Hemerik L, Garcia Mayor A, Rodriguez F, Valdecantos A, Vallejo VR (2020) Increased aridity drives post-fire recovery of Mediterranean forests towards open shrublands. New Phytol 225:1500–1515

Bello J, Hasselquist NJ, Vallet P, Kahmen A, Perot T, Korboulewsky N (2019) Complementary water uptake depth of *Quercus petraea* and *Pinus sylvestris* in mixed stands during an extreme drought. Plant Soil 437:93–115

Benslama M, Andrieu-Ponel V, Guiter F, Reille M, Migliore J, Djamali M (2010) Pollen analysis from two littoral marshes (Bourdim and Garaat El-Ouez) in the El-Kala wet complex (North-East Algeria). Lateglacial and Holocene history of Algerian vegetation C R Biol 333:744–754

Blondel J, Aronson J (1999) Biology and wildlife of the Mediterranean region. Oxford University Press, USA

Brunel C, Gros R, Ziarelli F, Da Silva AMF (2017) Additive or non-additive effect of mixing oak in pine stands on soil properties depends on the tree species in Mediterranean forests. Sci Total Environ 590:676–685

Canadell J, Jackson RB, Ehleringer JB, Mooney HA, Sala OE, Schulze E-D (1996) Maximum rooting depth of vegetation types at the global scale. Oecologia 108:583–595

Capitanio R, Carcaillet C (2008) Post-fire Mediterranean vegetation dynamics and diversity: a discussion of succession models. For Ecol Manag 255:431–439

Carnicer J, Barbeta A, Sperlich D, Coll M, Peñuelas J (2013) Contrasting trait syndromes in angiosperms and conifers are associated with different responses of tree growth to temperature on a large scale. Front Plant Sci 4:409

Carnicer J, Coll M, Pons X, Ninyerola M, Vayreda J, Peñuelas J (2014) Large-scale recruitment limitation in Mediterranean pines: the role of *Quercus ilex* and forest successional advance as key regional drivers. Glob Ecol Biogeogr 23:371–384

Carrión JS, Riquelme JA, Navarro C, Munuera M (2001) Pollen in hyaena coprolites reflects late glacial landscape in southern Spain. Palaeogeogr Palaeoclimatol Palaeoecol 176:193–205

Chauchard S, Guibal F, Carcaillet C (2013) Land-use legacies: multi-centuries years-old management control of between-stands variability at the landscape scale in Mediterranean mountain forests, France. J For Sci 59:1–7

Clements FE (1916) Plant succession: an analysis of the development of vegetation. Carnegie Institution of Washington

Cooper WS (1913) The climax forest of Isle Royale, Lake Superior, and its development I–III. Bot Gaz 55:189–235

Cooper A, Moshe Y, Zangi E, Osem Y (2014) Are small-scale overstory gaps effective in promoting the development of regenerating oaks (*Quercus ithaburensis*) in the forest understory? New For 45:843–857

Corcuera L, Notivol E (2015) Differences in photosynthetic activity might explain the large-scale shifts in pine recruitment in favour of oaks in continental Mediterranean climates. For An Int J For Res 88:248–256

Costa A, Madeira M (2011) Assessment of Mediterranean evergreen oak (*Quercus suber L.*) woodlands loss. Consequences of climate change effects: a case study at South-Western Portugal, in: proceedings of the global conference on global warming

Costa A, Pereira H, Madeira M (2010) Analysis of spatial patterns of oak decline in cork oak woodlands in Mediterranean conditions. Ann For Sci 67:204

Cruz-Alonso V, Ruiz-Benito P, Villar-Salvador P, Rey-Benayas JM (2019) Long-term recovery of multifunctionality in Mediterranean forests depends on restoration strategy and forest type. J Appl Ecol 56:745–757

de la Serrana RG, Vilagrosa A, Alloza JA (2015) Pine mortality in Southeast Spain after an extreme dry and warm year: interactions among drought stress, carbohydrates and bark beetle attack. Trees 29:1791–1804

de-Dios-García J, Pardos M, Calama R (2015) Interannual variability in competitive effects in mixed and monospecific forests of Mediterranean stone pine. For Ecol Manag 358:230–239

de-Dios-García J, Manso R, Calama R, Fortin M, Pardos M (2018) A new multifactorial approach for studying intra-annual secondary growth dynamics in Mediterranean mixed forests: integrating biotic and abiotic interactions. Can J For Res 48:333–344

del Castillo J, Comas C, Voltas J, Ferrio JP (2016) Dynamics of competition over water in a mixed oak-pine Mediterranean forest: Spatio-temporal and physiological components. For Ecol Manag 382:214–224

Del Río M, Sterba H (2009) Comparing volume growth in pure and mixed stands of *Pinus sylvestris* and *Quercus pyrenaica*. Ann For Sci 66:1–11

DeSoto L, Olano JM, Rozas V, De la Cruz M (2010) Release of *Juniperus thurifera* woodlands from herbivore-mediated arrested succession in Spain. Appl Veg Sci 13:15–25

Di Castri, F., Mooney, H.A., 2012. Mediterranean type ecosystems: origin and structure. Springer

Dorman M, Svoray T, Perevolotsky A, Moshe Y, Sarris D (2015) What determines tree mortality in dry environments? A multi-perspective approach. Ecol Appl 25:1054–1071

Espelta JM, Riba M, Javier R (1995) Patterns of seedling recruitment in West-Mediterranean *Quercus ilex* forest influenced by canopy development. J Veg Sci 6:465–472

Etienne M (2001) Pine trees-invaders or forerunners in Mediterranean-type ecosystems: a controversial point of view. J Mediterr Ecol 2:221–232

Etienne M, Aronson J, Le Floc'h E (1998) Abandoned lands and land use conflicts in Southern France, in: landscape disturbance and biodiversity in Mediterranean-type ecosystems. Springer, pp 127–140

Ferrio JP, Florit A, Vega A, Serrano L, Voltas J (2003) Δ 13 C and tree-ring width reflect different drought responses in *Quercus ilex* and *Pinus halepensis*. Oecologia 137:512–518

Fisher RT (1928) Soil changes and silviculture on the Harvard Forest. Ecology 9:6–11
Gavinet J, Vilagrosa A, Chirino E, Granados ME, Vallejo VR, Prévosto B (2015) Hardwood seedling establishment below Aleppo pine depends on thinning intensity in two Mediterranean sites. Ann For Sci 72:999–1008
Gea-Izquierdo G, Montes F, Gavilán RG, Cañellas I, Rubio A (2015) Is this the end? Dynamics of a relict stand from pervasively deforested ancient Iberian pine forests. Eur J For Res 134:525–536
Gea-Izquierdo G, Férriz M, García-Garrido S, Aguín O, Elvira-Recuenco M, Hernandez-Escribano L, Martin-Benito D, Raposo R (2019) Synergistic abiotic and biotic stressors explain widespread decline of *Pinus pinaster* in a mixed forest. Sci Total Environ 685:963–975
Gil-Tena A, Aquilué N, Duane A, De Cáceres M, Brotons L (2016) Mediterranean fire regime effects on pine-oak forest landscape mosaics under global change in NE Spain. Eur J For Res 135:403–416
Gleason HA (1927) Further views on the succession-concept. Ecology 8:299–326
Gómez JM (2003) Spatial patterns in long-distance dispersal of *Quercus ilex* acorns by jays in a heterogeneous landscape. Ecography (Cop) 26:573–584
Gómez JM, Hódar JA (2008) Wild boars (*Sus scrofa*) affect the recruitment rate and spatial distribution of holm oak (*Quercus ilex*). For Ecol Manag 256:1384–1389
Gómez-Aparicio L, Zavala MA, Bonet FJ, Zamora R (2009) Are pine plantations valid tools for restoring Mediterranean forests? An assessment along abiotic and biotic gradients. Ecol Appl 19:2124–2141
Grossiord C, Gessler A, Granier A, Pollastrini M, Bussotti F, Bonal D (2014) Interspecific competition influences the response of oak transpiration to increasing drought stress in a mixed Mediterranean forest. For Ecol Manag 318:54–61
Guenon R, Day TA, Velazco-Ayuso S, Gros R (2017) Mixing of Aleppo pine and Holm oak litter increases biochemical diversity and alleviates N limitations of microbial activity. Soil Biol Biochem 105:216–226
Hooper DU, Adair EC, Cardinale BJ, Byrnes JEK, Hungate BA, Matulich KL, Gonzalez A, Duffy JE, Gamfeldt L, O'Connor MI (2012) A global synthesis reveals biodiversity loss as a major driver of ecosystem change. Nature 486:105
Ibáñez B, Gómez-Aparicio L, Stoll P, Ávila JM, Pérez-Ramos IM, Marañón T (2015) A neighborhood analysis of the consequences of *Quercus suber* decline for regeneration dynamics in Mediterranean forests. PLoS One 10:e0117827
Jiménez-Moreno G, Fauquette S, Suc J-P (2010) Miocene to Pliocene vegetation reconstruction and climate estimates in the Iberian Peninsula from pollen data. Rev Palaeobot Palynol 162:403–415
Jones EW (1945) The structure and reproduction of the virgin forest of the north temperate zone. New Phytol 44:130–148
Jucker T, Bouriaud O, Avacaritei D, Dănilă I, Duduman G, Valladares F, Coomes DA (2014a) Competition for light and water play contrasting roles in driving diversity–productivity relationships in Iberian forests. J Ecol 102:1202–1213
Jucker T, Bouriaud O, Avacaritei D, Coomes DA (2014b) Stabilizing effects of diversity on aboveground wood production in forest ecosystems: linking patterns and processes. Ecol Lett 17:1560–1569
Klein T (2014) The variability of stomatal sensitivity to leaf water potential across tree species indicates a continuum between isohydric and anisohydric behaviours. Funct Ecol 28:1313–1320
Klein T (this volume) Carbon allocation dynamics in Mediterranean pines under stress. In: Ne'eman G Osem Y (eds) Pines and theirmixed forest ecosystems in the Mediterranean Basin. Springer Nature Berlin
Klein T, Shpringer I, Fikler B, Elbaz G, Cohen S, Yakir D (2013) Relationships between stomatal regulation, water-use, and water-use efficiency of two coexisting key Mediterranean tree species. For Ecol Manag 302:34–42

Kouba Y, Camarero JJ, Alados CL (2012) Roles of land-use and climate change on the establishment and regeneration dynamics of Mediterranean semi-deciduous oak forests. For Ecol Manag 274:143–150
Kueppers LM, Snyder MA, Sloan LC, Zavaleta ES, Fulfrost B (2005) Modeled regional climate change and California endemic oak ranges. Proc Natl Acad Sci 102:16281–16286
Lavi A, Perevolotsky A, Kigel J, Noy-Meir I (2005) Invasion of *Pinus halepensis* from plantations into adjacent natural habitats. Appl Veg Sci 8:85–92
Lenihan JM, Drapek R, Bachelet D, Neilson RP (2003) Climate change effects on vegetation distribution, carbon, and fire in California. Ecol Appl 13:1667–1681
Leverkus AB, Rey Benayas JM, Castro J (2016) Shifting demographic conflicts across recruitment cohorts in a dynamic post-disturbance landscape. Ecology 97:2628–2639
López-Tirado J, Hidalgo PJ (2018) Predicting suitability of forest dynamics to future climatic conditions: the likely dominance of Holm oak [*Quercus ilex* subsp. *ballota* (Desf.) Samp.] and Aleppo pine (*Pinus halepensis* Mill.). Ann For Sci 75:19
Martín-Alcón S, Coll L (2016) Unraveling the relative importance of factors driving post-fire regeneration trajectories in non-serotinous *Pinus nigra* forests. For Ecol Manag 361:13–22
Martín-Alcón S, Coll L, Salekin S (2015) Stand-level drivers of tree-species diversification in Mediterranean pine forests after abandonment of traditional practices. For Ecol Manag 353:107–117
Martínez-Vilalta J, Piñol J (2002) Drought-induced mortality and hydraulic architecture in pine populations of the NE Iberian Peninsula. For Ecol Manag 161:247–256
Morales-Molino C, Colombaroli D, Valbuena-Carabaña M, Tinner W, Salomón RL, Carrión JS, Gil L (2017) Land-use history as a major driver for long-term forest dynamics in the Sierra de Guadarrama National Park (central Spain) during the last millennia: implications for forest conservation and management. Glob Planet Change 152:64–75
Moreno-Fernández D, Hernández L, Sánchez-González M, Cañellas I, Montes F (2016) Space–time modeling of changes in the abundance and distribution of tree species. For Ecol Manag 372:206–216
Nathan R, Ne'eman G (2004) Spatiotemporal dynamics of recruitment in Aleppo pine (*Pinus halepensis* Miller). Plant Ecol 171:123–137
Nathan R, Safriel UN, Noy-Meir I, Schiller G (2000) Spatiotemporal variation in seed dispersal and recruitment near and far from *Pinus halepensis* trees. Ecology 81:2156–2169
Navarro-González I, Pérez-Luque AJ, Bonet FJ, Zamora R (2013) The weight of the past: land-use legacies and recolonization of pine plantations by oak trees. Ecol Appl 23:1267–1276
Naveh Z, Dan J (1973) The human degradation of Mediterranean landscapes in Israel, in: Mediterranean type ecosystems. Springer, pp 373–390
Ne'eman G, Trabaud L (2000) Ecology, biogeography and management of *Pinus halepensis* and *P. brutia* forest ecosystems in the Mediterranean Basin. Backhuys Publishers
Odum EP (1960) Organic production and turnover in old field succession. Ecology 41:34–49
Osem Y, Lavi A, Rosenfeld A (2011) Colonization of *Pinus halepensis* in Mediterranean habitats: consequences of afforestation, grazing and fire. Biol Invasions 13:485–498
Osem Y, Fogel T, Moshe Y, Brant S (2015) Managing cattle grazing and overstorey cover for the conversion of pine monocultures into mixed Mediterranean woodlands. Appl Veg Sci 18:261–271
Pausas JG, Bladé C, Valdecantos A, Seva JP, Fuentes D, Alloza JA, Vilagrosa A, Bautista S, Cortina J, Vallejo R (2004) Pines and oaks in the restoration of Mediterranean landscapes of Spain: new perspectives for an old practice—a review. Plant Ecol 171:209–220
Pons J, Pausas JG (2007) Acorn dispersal estimated by radio-tracking. Oecologia 153:903–911
Prieto I, Armas C, Pugnaire FI (2012) Water release through plant roots: new insights into its consequences at the plant and ecosystem level. New Phytol 193:830–841
Puerta-Piñero C (2010) Intermediate spatial variations on acorn predation shapes Holm oak establishment within a Mediterranean landscape context. Plant Ecol 210:213–224

Puerta-Piñero C, Pino J, Gómez JM (2012a) Direct and indirect landscape effects on *Quercus ilex* regeneration in heterogeneous environments. Oecologia 170:1009–1020
Puerta-Piñero C, Espelta JM, Sánchez-Humanes B, Rodrigo A, Coll L, Brotons L (2012b) History matters: previous land use changes determine post-fire vegetation recovery in forested Mediterranean landscapes. For Ecol Manag 279:121–127
Purves DW, Zavala MA, Ogle K, Prieto F, Benayas JMR (2007) Environmental heterogeneity, bird-mediated directed dispersal, and oak woodland dynamics in Mediterranean Spain. Ecol Monogr 77:77–97
Richardson DM (2000) Mediterranean pines as invaders in the southern hemisphere. Ecol Biogeogr Manag 131–142
Richardson DM, Rejmánek M (2004) Conifers as invasive aliens: a global survey and predictive framework. Divers Distrib 10:321–331
Rodríguez-Calcerrada J, Pardos JA, Gil L, Aranda I (2007) Acclimation to light in seedlings of *Quercus petraea* (Mattuschka) Liebl and *Quercus pyrenaica* Willd planted along a forest-edge gradient. Trees 21:45
Rodríguez-Robles U, Arredondo JT, Huber-Sannwald E, Vargas R (2015) Geoecohydrological mechanisms couple soil and leaf water dynamics and facilitate species coexistence in shallow soils of a tropical semiarid mixed forest. New Phytol 207:59–69
Rolo V, Plieninger T, Moreno G (2013) Facilitation of holm oak recruitment through two contrasted shrubs species in Mediterranean grazed woodlands. J Veg Sci 24:344–355
Roman DT, Novick KA, Brzostek ER, Dragoni D, Rahman F, Phillips RP (2015) The role of isohydric and anisohydric species in determining ecosystem-scale response to severe drought. Oecologia 179:641–654
Rouget M, Richardson DM, Lavorel S, Vayreda J, Gracia C, Milton SJ (2001) Determinants of distribution of six *Pinus* species in Catalonia, Spain. J Veg Sci 12:491–502
Rubio-Cuadrado Á, Camarero JJ, Aspizua R, Sánchez-González M, Gil L, Montes F (2018) Abiotic factors modulate post-drought growth resilience of scots pine plantations and rear-edge Scots pine and oak forests. Dendrochronologia 51:54–65
Ruiz-Benito P, Gómez-Aparicio L, Zavala MA (2012) Large-scale assessment of regeneration and diversity in Mediterranean planted pine forests along ecological gradients. Divers Distrib 18:1092–1106
Ruiz-Benito P, Ratcliffe S, Jump AS, Gómez-Aparicio L, Madrigal-González J, Wirth C, Kändler G, Lehtonen A, Dahlgren J, Kattge J (2017) Functional diversity underlies demographic responses to environmental variation in European forests. Glob Ecol Biogeogr 26:128–141
Rundel PW, Arroyo MTK, Cowling RM, Keeley JE, Lamont BB, Vargas P (2016) Mediterranean biomes: evolution of their vegetation, floras, and climate. Annu Rev Ecol Evol Syst 47:383–407
Sánchez de Dios R, Velázquez JC, Sainz Ollero H (2019) Classification and mapping of Spanish Mediterranean mixed forests. iForest-Biogeosciences For 12:480
Schmidt W, Briibach M (1993) Plant distribution patterns during early succession on an artificial protosoil. J Veg Sci 4:247–254
Sheffer E (2012) A review of the development of Mediterranean pine–oak ecosystems after land abandonment and afforestation: are they novel ecosystems? Ann For Sci 69:429–443
Sheffer E, Canham CD, Kigel J, Perevolotsky A (2013) Landscape-scale density-dependent recruitment of oaks in planted forests: more is not always better. Ecology 94:1718–1728
Sheffer E, Canham CD, Kigel J, Perevolotsky A (2014) An integrative analysis of the dynamics of landscape-and local-scale colonization of Mediterranean woodlands by *Pinus halepensis*. PLoS One 9:e90178
Shugart HH (1984) A theory of forest dynamics: the ecological implications of forest succession models. Springer, New York/Berlin/Heidelberg/Tokyo
Thompson JD (2005) Plant evolution in the Mediterranean. Oxford University Press on Demand
Thoreau HD, Emerson RW (1887) The succession of forest trees: and wild apples. Houghton, Mifflin
Tilman D (1999) The ecological consequences of changes in biodiversity: a search for general principles. Ecology 80:1455–1474

Torres I, Pérez B, Quesada J, Viedma O, Moreno JM (2016) Forest shifts induced by fire and management legacies in a *Pinus pinaster* woodland. For Ecol Manag 361:309–317
Urban DL, Bonan GB, Smith TM, Shugart HH (1991) Spatial applications of gap models. For Ecol Manag 42:95–110
Urbieta IR, Zavala MA, Maranón T (2008) Human and non-human determinants of forest composition in southern Spain: evidence of shifts towards cork oak dominance as a result of management over the past century. J Biogeogr 35:1688–1700
Urbieta IR, García LV, Zavala MA, Marañón T (2011) Mediterranean pine and oak distribution in southern Spain: is there a mismatch between regeneration and adult distribution? J Veg Sci 22:18–31
Vadell E, de-Miguel S, Pemán J (2016) Large-scale reforestation and afforestation policy in Spain: a historical review of its underlying ecological, socioeconomic and political dynamics. Land Use Policy 55:37–48
Valiente-Banuet A, Rumebe AV, Verdú M, Callaway RM (2006) Modern quaternary plant lineages promote diversity through facilitation of ancient tertiary lineages. Proc Natl Acad Sci 103:16812–16817
Vayreda J, Gracia M, Martinez-Vilalta J, Retana J (2013) Patterns and drivers of regeneration of tree species in forests of peninsular Spain. J Biogeogr 40:1252–1265
Vayreda J, Martinez-Vilalta J, Gracia M, Canadell JG, Retana J (2016) Anthropogenic-driven rapid shifts in tree distribution lead to increased dominance of broadleaf species. Glob Chang Biol 22:3984–3995
Verdú M, Dávila P, García-Fayos P, Flores-Hernández N, Valiente-Banuet A (2003) "Convergent" traits of mediterranean woody plants belong to pre-mediterranean lineages. Biol J Linn Soc 78:415–427. https://doi.org/10.1046/j.1095-8312.2003.00160.x
Vidal-Macua JJ, Ninyerola M, Zabala A, Domingo-Marimon C, Pons X (2017) Factors affecting forest dynamics in the Iberian Peninsula from 1987 to 2012. The role of topography and drought. For Ecol Manage 406:290–306
Waitz Y, Cohen Y, Dorman M, Perevolotsky A (2015) From microsite selection to population spatial distribution: *Pinus halepensis* colonization in mediterranean-type ecosystems. Plant Ecol 216:1311–1324
Watt AS (1947) Pattern and process in the plant community. J Ecol 35:1–22
Yasuda Y, Kitagawa H, Nakagawa T (2000) The earliest record of major anthropogenic deforestation in the Ghab Valley, Northwest Syria: a palynological study. Quat Int 73:127–136
Zalloni E, Battipaglia G, Cherubini P, Saurer M, De Micco V (2019) Wood growth in pure and mixed *Quercus ilex* L forests: drought influence depends on site conditions. Front Plant Sci 10:397
Zavala MA, de la Parra RB (2005) A mechanistic model of tree competition and facilitation for Mediterranean forests: scaling from leaf physiology to stand dynamics. Ecol Model 188:76–92
Zavala MA, Espelta JM, Retana J (2000) Constraints and trade-offs in Mediterranean plant communities: the case of holm oak-Aleppo pine forests. Bot Rev 66:119–149
Ziffer-Berger J, Weisberg PJ, Cablk ME, Moshe Y, Osem Y (2017) Shrubs facilitate pine colonization by controlling seed predation in dry Mediterranean dwarf shrublands. J Arid Environ 147:34–39

Chapter 18
Bird Diversity in Mediterranean Pine and Mixed Forests

Mario Díaz, Mario Soliño, and María Martínez-Jáuregui

18.1 Introduction

Although the original vegetation of the Mediterranean region comprised mainly forests, most Mediterranean forest birds tend to select open woodland or shrubland, together with closed forests. Out of the 73 species of Mediterranean forest birds listed by Brotons et al. (2018) for Europe, only 18 (25%) select closed canopy forests, while 38 (52%) select open woodlands, and 17 (23%) are mostly found in different types of shrubland. This generalist pattern of habitat selection seems to be due to the north-south movements of bird distributions following the climatic changes occurring along the Pliocene and Pleistocene periods (Blondel 2018). During the glacial maxima, forests and forest birds that now occupy temperate latitudes in central Europe were mostly found in the southernmost parts of the Mediterranean peninsulas (Iberia, Italy, the Balkans) and islands (Balearics, Corsica, Sardinia, Sicily, Crete, Cyprus). Subsequent climate warming presumably led to a northwards movement of specialist forest birds, leaving behind generalist forest birds at the southern borders of their distribution areas together with shrubland-adapted southern specialists (Blondel 2018; Brotons et al. 2018). This process would have been more significant for birds more adapted to pine forests, as pines are better suited to colder and drier climates, than for birds from more mesic,

M. Díaz (✉)
Department of Biogeography and Global Change, National Museum of Natural Sciences (BGC-MNCN-CSIC), Madrid, Spain
e-mail: Mario.Diaz@ccma.csic.es

M. Soliño
Department of Economic Analysis, Complutense University of Madrid, Madrid, Spain

M. Martínez-Jáuregui
National Institute for Agriculture and Food Research and Technology (INIA), Forest Research Centre (CIFOR), Madrid, Spain

G. Ne'eman, Y. Osem (eds.), *Pines and Their Mixed Forest Ecosystems in the Mediterranean Basin*, Managing Forest Ecosystems 38,
https://doi.org/10.1007/978-3-030-63625-8_18

broadleaved forests (Blondel et al. 2010). In this way, out of the 44 bird species defined by Martínez-Jáuregui et al. (2016a, b) as 'pine-dwelling forest birds' on the basis of their distribution with respect to the pine forest distribution in mainland Spain, only five species scattered over the bird phylogeny can be considered pine-forest specialists. The remaining species are thus forest generalists of either closed forests or open woodland (Fig. 18.1).

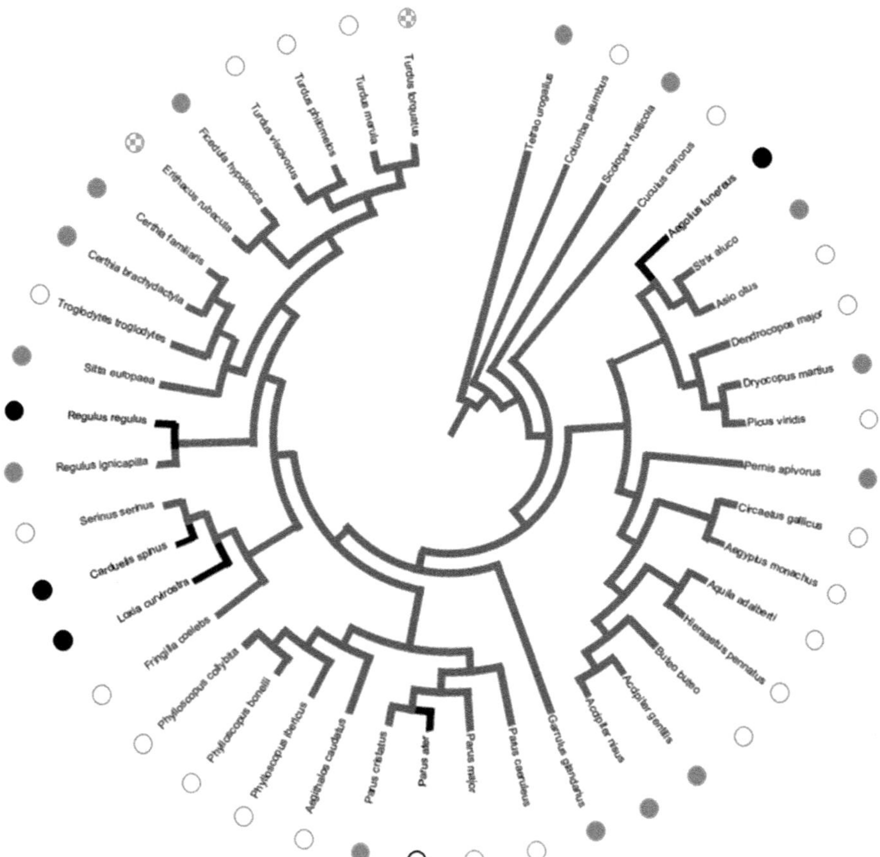

Fig. 18.1 Consensus phylogeny (Maddison and Maddison 2011; Jetz et al. 2012) for the 44 species defined by Martínez-Jáuregui et al. (2016a, b) as 'pine-dwelling forest birds' on the basis of their breeding distribution with respect to pine forest distribution in mainland Spain. Pine-forest specialists are indicated in black. Open circles: open woodland birds; closed circles: closed forest birds; checkered circles: shrubland birds. (After Martí and del Moral 2003; Martínez-Jáuregui et al. 2016a, b)

18.2 Responses to Forest Management

Generalist habits somewhat imply flexibility in the responses of birds to pine forest management, and some resilience to such management by bird communities. Nevertheless, both bird species and bird communities show clear responses to the main management practices of pine forests, such as plantations, clear-cutting, selective tree felling, pruning or shrub removal (reviews by Calladine et al. 2018 and references therein). Responses to management are usually not measured directly, but by modeling the associations between the distributions of species or communities and the landscape and vegetation structure of pine forests resulting from management practices (Mikusinski et al. 2018).

18.2.1 Stand-Level Bird–Habitat Associations

The traditional pine forest traits used to model bird distributions are stand size, vegetation structure (stand density, tree and shrub cover, tree size and height), vegetation composition (tree and shrub species richness, pine species), and whether the stand has been planted or not (Díaz et al. 1998; Calladine et al. 2018; Castaño-Villa et al. 2019). Larger stands maintain more species than smaller stands, especially when considering forest specialists, with minor effects of vegetation structure (Díaz et al. 1998). Plantations tend to maintain less species than natural pine forests, especially when pines are planted outside their natural range (Calladine et al. 2018; Castaño-Villa et al. 2019). Forest-plantation differences arise even after controlling for structural and compositional differences between non-planted forests and plantations (Martínez-Jáuregui et al. 2016a), suggesting that the time needed for the assembly of mature bird communities forms an important part of the difference between plantations and natural forests (Sweeney et al. 2010). Differences among planted and nearby natural forest vary, however, geographically and among different pine species, being smaller southwestwards in mainland Spain and for species used for restoration purposes (e.g. *Pinus halepensis* Miller in mainland Spain) than for commercial species (*P. pinaster* Aiton or *P. sylvestris* L.; Martínez-Jáuregui et al. 2016b; Fig. 18.2). Thus, disturbance rates associated with commercial management of plantations also seem to be important for the maintenance of species-rich bird communities in Mediterranean pine forests.

Vegetation structure and composition consistently explain bird species richness in both natural forests and plantations (Table 1; see also Calladine et al. 2018), although effect sizes for these relationships are usually small to moderate, probably because of the generalist habitat selection of most Mediterranean pine forest birds (Fig. 18.1). Species-rich mixed forests or plantations tend to maintain more species than monospecific stands, with further positive effects of stand development (basal area) and negative effects of high shrub cover (see also Calladine et al. 2018), although a minimum amount of shrub cover is required by several bird species

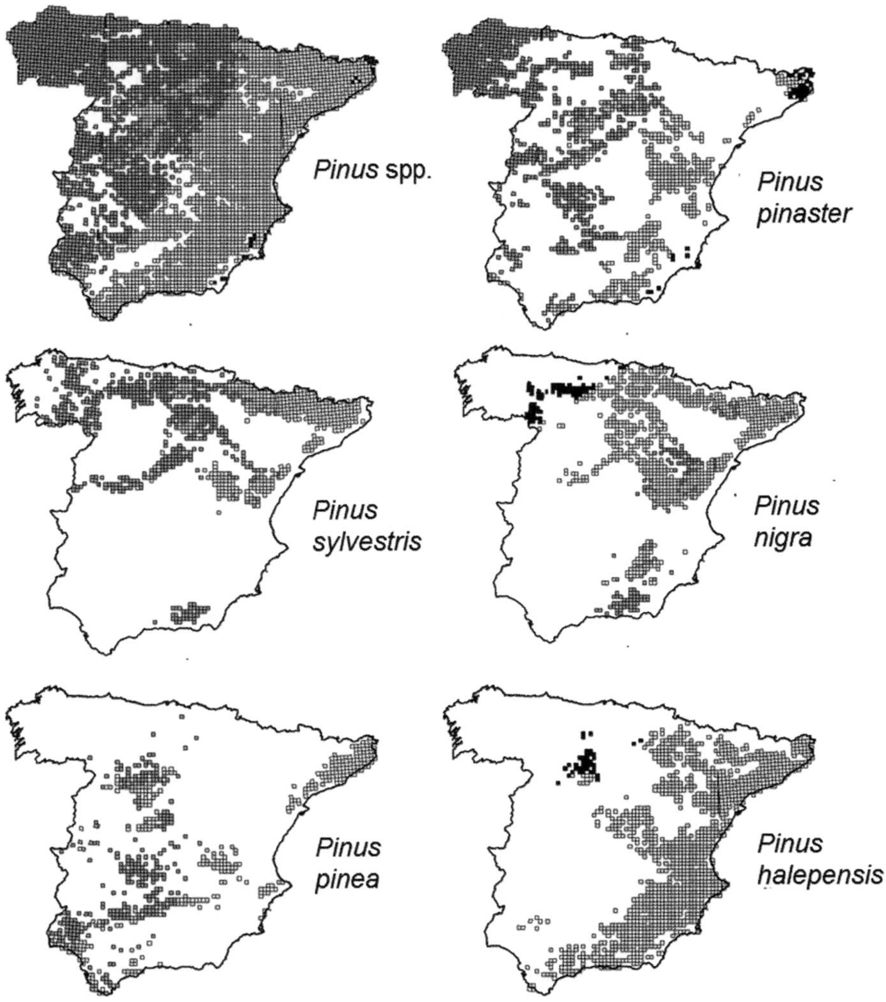

Fig. 18.2 Geographical and pine-species differences in bird species richness among pine plantations and nearby natural pine forests, expressed as the expected changes in the numbers of pine forest bird species per 10 × 10 km UTM (Universal Transverse Mercator) square after transforming current planted pine forests into natural forest of the same species. Light grey: no effect; dark grey: gain of at least one species; black: loss of at least one species. (Details in Martínez-Jáuregui et al. 2016b)

(Dagan and Izhaki 2019). These relationships vary, however, for early successional species more dependent on short and shrubby, open stands (e.g. robins, *Erithacus rubecula* (Linnaeus, 1758), or wrens, *Troglodytes troglodytes* (Linnaeus, 1758)), holding true especially for late-successional forest specialist such as crossbills, *Loxia curvirostra* Linnaeus, 1758, goskawks, *Accipiter gentilis* (Linnaeus, 1758), or European jays, *Garrulus glandarius* (Linnaeus, 1758) (Calladine et al. 2018).

Increasing availability of large-scale databases has allowed testing for non-linear relationships between bird abundance and diversity and habitat traits, as well as for trait x geography interactions (i.e., the geographic variation of bird-habitat relationships; Martínez-Jáuregui et al. 2016b; Table 18.1). Results have demonstrated that many bird-habitat relationships are non-linear, with maximum values of abundance or richness at intermediate rather than at extreme values of habitat traits. For instance, bird species richness peaks at intermediate values of tree species richness and stand development (basal area), with lower values for monospecific and young stands but also for highly diverse and old forests. This pattern might be due to the exclusion of late successional and early successional species in younger and older stands, respectively, that may however thrive in middle-aged forests of intermediate tree diversity (Martínez-Jáuregui et al. 2016a; see also Izhaki 2000, 2012 for the role of other major disturbance of pine forests, fire). This explanation may also hold for the non-linear effect of the ratio between pine plantations and natural pine forests on bird diversity at the scale of 10 × 10 km UTM squares. If plantations tend to be occupied by earlier-successional species than natural forests, then mixtures of forests and plantations would support more species than pure plantations or pure natural forests at landscape scales (Martínez-Jáuregui et al. 2018).

18.2.2 Landscape-Scale Bird–Habitat Associations

Increasing availability of large-scale databases has also allowed us to incorporate the effects on bird communities of landscape-scale variables such as the proportional cover of land uses other than pines around pine forests, as well as climate variables, into more complex bird-habitat models (Martínez-Jáuregui et al. 2016b, 2018; see Table 1 for a recent example). Effect sizes of landscape-scale variables are small to moderate, similar to the effect sizes of stand-level traits (Table 1). The effect of pine forest cover in landscapes, either natural or planted, on landscape-scale pine forest bird diversity is non-linear, as it is the effect of the proportion of planted *vs.* natural pine forests (Table 1). The cover of broadleaved forests (other than *Eucalyptus* spp.), water surfaces and urban areas has consistent positive effects, while the cover of *Eucalyptus* stands and arable land has consistent negative effects. Strong positive effects of broadleaved forests (other than eucalypts) would surely reflect, once more, the generalist habitat selection of birds occupying pine forests (most are just generalist forest birds), whereas the negative effects of eucalypt stands are most likely due to the low quality of intensive plantations of exotic trees for most Mediterranean birds (Calladine et al. 2018; Castaño-Villa et al. 2019; Goded et al. 2019).

The negative effects of arable land on forest bird diversity at landscape scales is the complementary result of the negative effects of woody vegetation on the species-rich communities of open-country birds typical of Mediterranean landscapes (Díaz et al. 1998; Reino et al. 2009; Díaz and Concepción 2016; Concepción and Díaz 2019; Concepción et al. 2020). The uniqueness of these open-land communities

Table 18.1 Model estimates (SE: standard error) of vegetation, landscape, climate, and geographical variables influencing the species richness of pine forest-dwelling birds during the breeding season in the 3950 10 × 10 km UTM squares with some pine forest cover in mainland Spain. All variables were standardized prior to analyses (details in Martínez-Jáuregui et al. 2018). Boldface indicates statistically significant effects ($P < 0.05$). Adjusted R-squared: 67.02%, df = 3906. Effect sizes were computed from t statistics following Becker (1999)

Effect	Estimate	SE	t	P	effect size (r^2*100)
Intercept	2.17E+01	7.04E−01	30.86	<0.001	
Tree density (n° stems per ha)	**−1.39E−03**	**3.93E−04**	**−3.55**	**<0.001**	**0.32**
Mean coefficient of variation of tree diameter (DBH)	−6.60E−02	8.02E−01	−0.08	0.934	0.00
Monospecific character of stands	**−1.25E+00**	**3.20E−01**	**−3.90**	**<0.001**	**0.39**
Shrub diversity	1.02E−01	5.50E−02	1.86	0.063	0.09
Shrub height (m)	4.49E−02	2.69E−02	1.67	0.096	0.07
Shrub canopy cover (%)	**−7.65E−03**	**3.62E−03**	**−2.11**	**0.035**	**0.11**
Tree diversity	**1.17E+02**	**8.00E+00**	**14.65**	**<0.001**	**5.21**
(Tree diversity)^2	**−5.34E+01**	**4.84E+00**	**−11.05**	**<0.001**	**3.03**
Assman dominant height (m)	−4.90E−02	3.38E−02	−1.45	0.147	0.05
Tree basal area (m^2 per ha)	**6.47E+01**	**1.01E+01**	**6.43**	**<0.001**	**1.05**
(Tree basal area (m^2 per ha))^2	**−2.39E+01**	**4.98E+00**	**−4.79**	**<0.001**	**0.58**
***Pinus* forest type: % of the area planted**	**−2.08E+01**	**8.02E+00**	**−2.59**	**0.009**	**0.17**
(*Pinus* forest type: % of the area planted)^2	**−2.20E+01**	**5.83E+00**	**−3.78**	**<0.001**	**0.36**
Pine cover in the 10 × 10 km UTM cell (ha)	**3.83E+01**	**7.99E+00**	**4.79**	**<0.001**	**0.58**
(Pine cover in the 10 × 10 km UTM cell (ha))^2	**−2.86E+01**	**5.00E+00**	**−5.73**	**<0.001**	**0.83**
***Eucalytus* spp. cover in the UTM cell (ha)**	**−6.55E−04**	**1.05E−04**	**−6.22**	**<0.001**	**0.98**
Other broadleaves cover in the UTM cell (ha)	**4.62E−04**	**4.84E−05**	**9.55**	**<0.001**	**2.28**
Shrubland cover in the UTM cell (ha)	8.84E−06	7.28E−05	0.12	0.903	0.00
Pasture land use cover in the UTM cell (ha)	1.60E−04	8.87E−05	1.81	0.071	0.08
Agriculture land use cover in the UTM cell (ha)	**−1.02E−04**	**4.29E−05**	**−2.37**	**0.018**	**0.14**
Urban land use cover in the UTM cell (ha)	**6.60E−04**	**1.67E−04**	**3.95**	**<0.001**	**0.40**
Water cover in the UTM cell (ha)	**6.10E−04**	**2.57E−04**	**2.37**	**0.018**	**0.14**
Other land use cover in the UTM cell (ha)	1.39E−04	1.36E−04	1.02	0.307	0.03
Mean annual precipitation (mm)	**3.00E+01**	**9.53E+00**	**3.15**	**0.002**	**0.25**
(Mean annual precipitation (mm))^2	**−2.73E+01**	**5.89E+00**	**−4.64**	**<0.001**	**0.55**
Mean annual temperature (°C)	**−1.55E+02**	**8.19E+00**	**−18.96**	**<0.001**	**8.43**

(continued)

Table 18.1 (continued)

Effect	Estimate	SE	t	P	effect size (r^2*100)
(Mean annual temperature (°C))^2	**2.04E+01**	**6.01E+00**	**3.40**	**<0.001**	**0.30**
Longitude	−1.64E−01	1.30E−01	−1.26	0.208	0.04
Latitude	**7.82E−01**	**1.30E−01**	**6.02**	**<0.001**	**0.92**
Temperature × pine area planted	1.46E+02	4.18E+02	0.35	0.728	0.00
Temperature^2 × pine area planted	5.11E+02	3.18E+02	1.61	0.108	0.07
Temperature × pine area planted^2	−2.15E+02	3.65E+02	−0.59	0.556	0.01
Temperature^2 × pine area planted^2	**−8.46E+02**	**3.03E+02**	**−2.80**	**0.005**	**0.20**
Precipitation × pine area planted	4.05E+01	5.11E+02	0.08	0.937	0.00
Precipitation^2 × pine area planted	**7.52E+02**	**3.18E+02**	**2.37**	**0.018**	**0.14**
Precipitation x pine area planted^2	7.03E+02	5.17E+02	1.36	0.174	0.05
Precipitation^2 × pine area planted^2	1.98E+01	3.48E+02	0.06	0.954	0.00
Pine area planted × Longitude	**2.46E+01**	**7.08E+00**	**3.48**	**<0.001**	**0.31**
Pine area planted^2 × Longitude	**2.96E+01**	**5.52E+00**	**5.36**	**<0.001**	**0.73**
Pine area planted × Latitude	−1.10E+01	7.86E+00	−1.40	0.160	0.05
Pine area planted^2 × Latitude	−1.12E+01	6.48E+00	−1.73	0.084	0.08

(Blondel and Farré 1988) and its high conservation value at continental, national and regional levels (Brotons et al. 2018; Díaz et al. 2020) must be taken into account when designing landscape-scale conservation measures for both forests and overall bird diversity (Díaz et al. 1998; Santos et al. 2006; Brotons et al. 2018). In fact, the relative conservation value of Mediterranean forest assemblages tends to be much lower than that of open-land communities (Santos et al. 2006; Díaz et al. 2020) due to the historical biogeography of the Mediterranean Basin (Blondel et al. 2010; Blondel 2018; Brotons et al. 2018). Core areas for forest organisms lie mostly at mid-European latitudes, whereas southern, Mediterranean populations are generally marginal and hence their long-term viability is expected to be low (Brotons et al. 2018). Nevertheless, the genetic uniqueness and local adaptation of southern, rear-edge populations are expected to be high, as well as their contribution to within-species overall diversity (Hampe and Petit 2005). Pine stands planted within the distribution area of the main, original Mediterranean forest vegetation (sclerophyllous and deciduous oak woodland) may thus play a role in preserving southern populations of forest birds due to their better recruitment and higher growth rates under dry environmental conditions (Santos et al. 2006). Long-term, landscape-scale management plans based on regional biodiversity targets are the best tool for tackling conservation trade-offs arising from the contrasting habitat requirements of open-country and forest communities under Mediterranean climate and soil conditions (Santos et al. 2006; Valladares et al. 2014; Díaz and Concepción 2016; Brotons et al. 2018).

18.3 Climate Change: Mitigation Through Forest Management

Bird distributions are expected to change rapidly during the twenty-first century due to the fast and strong effects of global change drivers on their habitat requirements (Huntley et al. 2007; Dunn and Møller 2019). Expectations of northwards and upwards changes in distribution areas due to climate warming are in fact being observed (Stephens et al. 2016), although the speed of change tends to be slower than expected from simple models of climate-change effects (Devictor et al. 2012). Other global change drivers, such as land-use changes, may account for the observed deviations, a fact that opens the possibility to mitigate, and hopefully even reverse, the negative effects of climate change on bird distributions by means of land-use and habitat management (Greenwood et al. 2016). Models such as the one parameterized in Table 1 can be used to explore this possibility because they integrate climate, vegetation structure and landscape-scale effects on current bird distributions. Assuming that spatial patterns of bird–habitat associations are good surrogates for temporal associations reflecting global change effects (Gregory et al. 2009), this model can be used to forecast expected effects of management options under the well-established climate change scenarios developed in recent years (Stocker et al. 2013).

Naturalization of pine forest would not suffice to maintain current levels of diversity of pine forest birds (red line in Fig. 18.3) in mainland Spain by 2050 under any climate change scenario (grey and black lines in Fig. 18.3). Here we define naturalization as the replacement of pine forests planted outside the species' ranges by the corresponding pine forest species for each location, either by plantation or by promotion of natural regeneration (Martínez-Jáuregui et al. 2018). Naturalization may increase local diversity only if current climatic conditions do not change (blue lines in Fig. 18.3). Note the non-linear relationship between the proportion of natural *vs*. planted pine forests in 10 × 10 km squares and regional bird diversity, peaking at 55% of naturalized plantations, as explained above). Just to maintain current diversity levels it would be necessary to convert all current eucalypt *Eucalyptus* spp. plantations (around 760,000 ha in Spain almost free of pine forest-dwelling birds; Calladine et al. 2018) into natural pine forest (green line in Fig. 18.3). The effort needed to achieve this goal would vary geographically (Fig. 18.4), as the effects of plantations (compared to nearby natural pine forest) vary geographically and interact with climate variables (Table 1). Much stronger efforts should be made in mesic areas dominated by intensive plantations (e.g. northern and southwestern Spain) than in dry areas occupied mainly by natural forests of restoration plantations (Fig. 18.4; Martínez-Jáuregui et al. 2018).

The large-scale modeling example developed here shows that a change in the exploitation strategy of Mediterranean forest plantations towards more nature-based management would mitigate expected negative effects of climate change on forest bird diversity (Martínez-Jáuregui et al. 2018). Nevertheless, full mitigation would only be achieved under the milder climate change expectations by converting ca. half (55%) of the pine and all eucalypt plantations in each 10 × 10 km UTM square

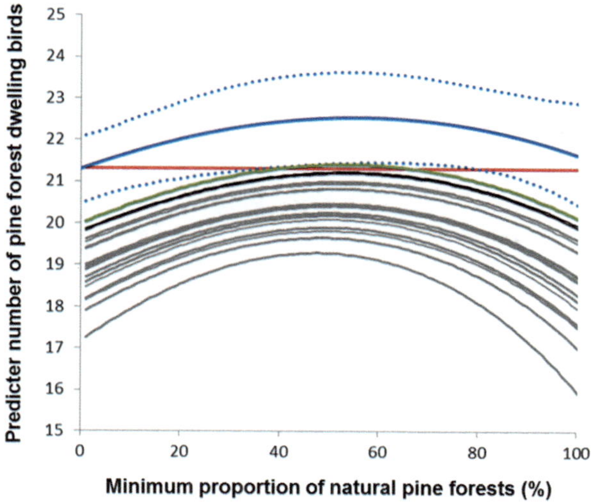

Fig. 18.3 Predicted number of pine forest-dwelling birds present in 10 × 10 km UTM squares according to the minimum proportion of natural pine forest under different management and climatic scenarios. Red line: observed curent mean; grey lines: predictions under climate change models and first step of naturalization management action (pine forest naturalization only); black line: predictions under the MRI-CGCM3 global change model (RCP4.5 and RCP8.5) and first step of naturalization management action; blue line: predictions under current climate and first step of naturalization management action (blue dotted lines indicate the 95% confidence intervals for this prediction); green curve: predictions under the MRI-CGCM3 global change model (RCP 4.5 and RCP8.5) after the two steps of naturalization (pine naturalization and transformation of eucalypt plantations to natural pine forests). (See Martínez-Jáuregui et al. 2018 for details)

into the corresponding natural pine forest, in the case of mainland Spain. This could be accomplished by favoring natural regeneration for several tree rotations to facilitate the assembly of a full set of functional groups (Calladine et al. 2018 and references therein). Strategies based on the naturalization of pine and the transformation of eucalypt forests into natural pine forests could better mitigate climate-driven bird diversity loss in the most productive areas in the north and northwest of Spain, than in the south and southwest of Spain, where pine plantations are mostly aimed at restoration rather than production goals, due to harsh climate and soil conditions. This regional variation in the effectiveness of the strategy would be incorporated into regional plans aimed at integrating realistic conservation and development goals, whose efficiency should be routinely evaluated (Díaz and Concepción 2016).

Modeling and optimization efforts can be improved following two main lines of applied research. First, incorporating the effects of changing disturbance levels associated with global change will surely modify model predictions, as these disturbances depend to a great deal on forest management. The most relevant of such disturbances in the Mediterranean region is fire, either wild or prescribed (Izhaki 2012; Valladares et al. 2014; Doblas-Miranda et al. 2015; Richardson and Nsikani this volume, Chap. 5), so that strategies of fire management will be incorporated into land use and climatic models to improve its usefulness (Regos et al. 2018). Second, specific management prescriptions aimed at improving stand size, tree and

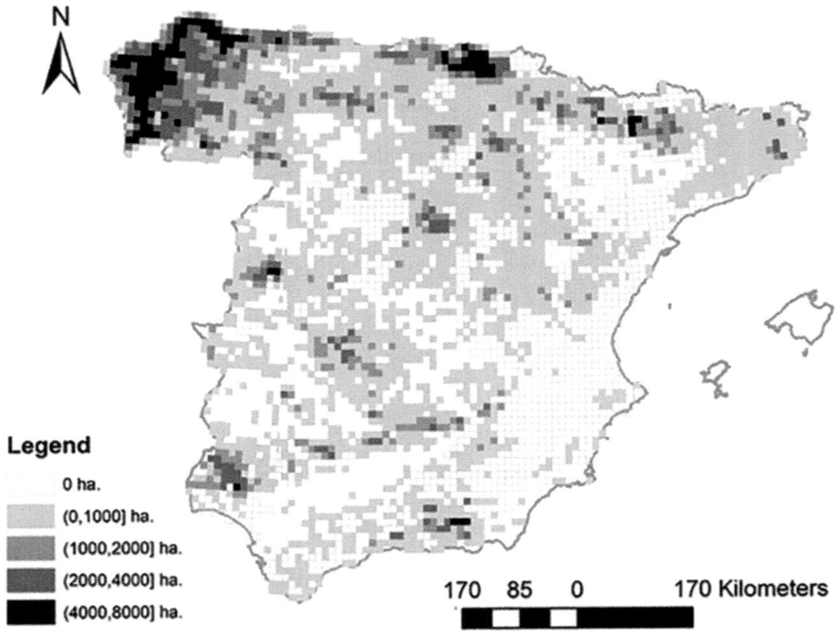

Fig. 18.4 Geographical variation of the forest management effort (in ha) needed to maintain current levels of pine forest-dwelling bird richness in mainland Spain under the MRI-CGCM3 global change model (RCP4.5) and the most effective naturalization strategy for commercial forest plantations (conversion of ca. half of the natural forests in each 10 × 10 km UTM square and conversion of all eucalypt plantations into pine forests). Details in Martínez-Jáuregui et al. (2018)

shrub diversity, and age and vegetation structure may further improve regional diversity (Table 1) and, particularly, the local abundance of selected species, whose responses to global change in terms of demography, dispersal, contemporary evolution, or species interactions will surely vary both in space and in time (Urban et al. 2016). Naturalization strategies should be complemented by considering such species-specific responses, together with the role of mitigation strategies applied to other global change drivers, such as fire, land-use change and pollution. Modeling species-specific responses is particularly relevant when considering the ecosystem services provided by birds, as these services usually depend more on particular species or species groups rather than on overall bird diversity (Gaston et al. 2018).

18.4 Ecosystem Services Provided by Forest Birds

The conservation of biodiversity and its associated ecosystem services in nowadays one of the key priorities for sustainable management (Munang et al. 2011), and biodiversity indicators are commonly used to monitor the objectives and successes of conservation programs (Mace and Baillie 2007). Birds have often been used for this purpose, particularly with respect to regulating and supporting ecosystem services (Whelan et al. 2008). Nevertheless, other social and economic services provided by birds in forests have also been considered, such as provisioning (e.g. hunting, Herruzo et al. 2016) and cultural services (e.g. bird watching, Sekerçioglu 2002), which are less commonly considered in the literature (Feld et al. 2009). Species richness (Martínez-Jáuregui et al. 2018), abundance (Sweeney et al. 2010) or presence and abundance of functional groups (Gil-Tena et al. 2007) are the most usual measures of bird diversity.

Few studies have been conducted, however, on the identification and quantification of the ecosystem services provided by birds in Mediterranean pine forests. As in other systems, most research has focused on physically measuring ecosystem services related to regulating and supporting services, for example, whether insectivorous birds have a role in controlling insect pests of pine forests (Sanz 2001; Martínez-Abraín and Jiménez 2019) or how frugivorous or scatter-hoarding birds contribute to seed dispersal (Morán-López et al. 2015; Pesendorfer et al. 2016). More direct estimates of the value of the ecosystem services linked to wildlife are just starting to be performed in Mediterranean forests (Herruzo et al. 2016; Campos et al. 2019; Martínez-Jáuregui et al. 2019).

Services with associated market values (e.g., game meat) are usually easy to quantify, but most services associated with biodiversity are much harder to valuate (e.g. landscape or existence values; MacMillan and Phillip 2008; Martínez-Jáuregui et al. 2014). Campos et al. (2019) and Martínez-Jáuregui et al. (2019) found significant willingness to pay (a common estimate of the value of non-market services) for preserving biodiversity in pine forests, and that the perceived role of this biodiversity in delivering ecosystem services is a key determinant of the respondents' support for conservation (Martínez-Jáuregui et al. 2019). In particular, they found that the willingness to pay for preserving the number of native birds in Mediterranean pine forests was significantly linked to perceptions of cultural and regulating ecosystem services, but much less to provisioning ecosystem services, perhaps due to low societal concern regarding the role of birds as valuable game or as pest control agents. Further studies using other possible biodiversity indicators based on forest birds are needed, as well as the quantification of possible costs of birds to society, e.g., as pests or as agents of disease dispersal (Viana et al. 2016). Understanding and quantifying both positive and negative ecosystem services will allow us to define the potential benefits and problems associated with forest birds, and thus to increase the effectiveness of plans for conservation management of forest ecosystems.

18.5 Summary and Conclusions

Knowledge of the variable relationships between bird diversity, societal attitudes and the effects of global change drivers on pine forests is essential for designing realistic management strategies aimed at enhancing the multifunctional role of Mediterranean pine forests beyond the market. Recent and ongoing research is providing the parameterized models and ecosystem service valuations required to design these management strategies. It is clear from the results available that no single strategy would universally optimize the outputs of any multifunctionally oriented design. This is because such outputs will vary strongly both in space and in time due to interactions between management effects on diversity, geography, and climate. Furthermore, most diversity–management relationships are non-linear, with local maxima and threshold effects, implying strong uncertainty for the likelihood of management schemes to reach their goals (Díaz and Concepción 2016). Both the so-called 'rewilding', 'back to nature' or 'wilder is better' approaches, aimed at improving biodiversity in managed forests (Navarro and Pereira 2015), or the alternative land-sparing strategy based on no-use reserves surrounded by land under highly intensive management (Gamborg and Larsen 2003; Green et al. 2005), would not work equally better even within a country, as demonstrated for the case of mainland Spain developed throughout this chapter. Fortunately, the databases needed to parameterize prospective models (Martínez-Jáuregui et al. 2016a, b, 2018), and the strategy to analyze these databases to develop, test, and monitor model predictions and the management strategies derived from them are already available (Díaz and Concepción 2016). Finally, incorporating the dimension of forest bird values to humans is critical to address effectively the difficulties associated with their management and conservation.

Acknowledgements We are grateful to Prof. Gidi Ne'eman and Dr. Efrat Sheffer for their kind invitation to write this chapter, and Dr. Ido Izhaki for constructive comments during revision. We also acknowledge the staff of SEO/BirdLife for their work in the management of the bird atlas data, and the Spanish Meteorological Agency (AEMET) and the Climate Research Foundation (FIC) for producing and making available the biodiversity and bioclimatic data used in the papers that are the basis of this review. This paper is a contribution to the projects RTA2013-00048-C03-01 and RTI2018-096348-R-C21, funded by the Spanish Ministry of Economy and Competitiveness; BiodivERsA 3-2015-180, funded by the Spanish National Research Agency, the Bulgarian Science Fund, the German Federal Ministry of Education and Research, and the European Commission within the ERA-Net BiodivERsA Co-Fund scheme; and to the thematic network REMEDINAL3-CM (S2013/MAE-2719).

References

Becker LA (1999) Effect size calculators. University of Colorado, Colorado Springs, UCCS. https://www.uccs.edu/lbecker/

Blondel J (2018) Origins and dynamics of forest birds of the northern hemisphere. In: Mikusinski G, Roberge J-M, Fuller RJ (eds) Ecology and conservation of forest birds. Cambridge University Press, Cambridge, pp 11–63

Blondel J, Farré H (1988) The convergent trajectories of bird communities along ecological successions in European forests. Oecologia 75:83–93

Blondel J, Aronson J, Boudiou JY et al (2010) The Mediterranean basin: biological diversity in space and time. Oxford University Press, Oxford

Brotons L, Herrando S, Sirami C et al (2018) Mediterranean forest bird communities and the role of landscape heterogeneity in space and time. In: Mikusinski G, Roberge J-M, Fuller RJ (eds) Ecology and conservation of forest birds. Cambridge University Press, Cambridge, pp 318–349

Calladine J, Díaz M, Reino L et al (2018) Plantations of non-native tree species. In: Mikusinski G, Roberge J-M, Fuller RJ (eds) Ecology and conservation of forest birds. Cambridge University Press, Cambridge, pp 350–386

Campos P, Caparrós A, Oviedo JL et al (2019) Bridging the gap between national and ecosystem accounting: application in Andalusian forests, Spain. Ecol Econ 157:218–236

Castaño-Villa GJ, Estevez JV, Guevara G et al (2019) Differential effects of forestry plantations on bird diversity: a global assessment. For Ecol Manag 440:202–207

Concepción ED, Díaz M (2019) Varying potential of conservation tools of the common agricultural policy for farmland bird preservation. Sci Total Environ 694:133618

Concepción ED, Aneva I, Jay M et al (2020) Optimizing biodiversity gain of European agriculture through regional targeting and adaptive management of conservation tools. Biol Conserv 241:108384

Dagan U, Izhaki I (2019) Understory vegetation in planted pine forests governs bird community composition and diversity in the eastern Mediterranean region. For Ecosyst 6:29

Devictor V, Van Swaay C, Brereton T et al (2012) Differences in the climatic debts of birds and butterflies at a continental scale. Nat Clim Chang 2:121

Díaz M, Concepción ED (2016) Enhancing the effectiveness of CAP greening as a conservation tool: a plea for regional targeting considering landscape constraints. Curr Land Ecol Rep 1:168–177

Díaz M, Carbonell R, Santos T et al (1998) Breeding bird communities in pine plantations of the Spanish Plateaux: biogeography, landscape and vegetation effects. J Appl Ecol 35:562–574

Díaz M, Concepción ED, Oviedo JL et al (2020) A comprehensive index for threatened biodiversity valuation. Ecol Indic 108:105696

Doblas-Miranda E, Martínez-Vilalta J, Lloret F et al (2015) Reassessing global change research priorities in Mediterranean terrestrial ecosystems: how far have we come and where do we go from here? Glob Ecol Biogeogr 24:25–43

Dunn PO, Møller AP (eds) (2019) Effects of climate change on birds. Oxford University Press, Oxford

Feld CK, Martins da Silva P, Paulo Sousa J et al (2009) Indicators of biodiversity and ecosystem services: a synthesis across ecosystems and spatial scales. Oikos 118:1862–1871

Gamborg C, Larsen JB (2003) 'Back to nature' - a sustainable future for forestry? For Ecol Manag 179:559–571

Gaston KJ, Cox DT, Canavelli SB et al (2018) Population abundance and ecosystem service provision: the case of birds. Bioscience 68:264–272

Gil-Tena A, Saura S, Brotons L (2007) Effects of forest composition and structure on bird species richness in a Mediterranean context: implications for forest ecosystem management. For Ecol Manag 242:470–476

Goded S, Ekroos J, Domínguez J et al (2019) Effects of eucalyptus plantations on avian and herb species richness and composition in North-West Spain. Global Ecol Conserv 19:e00690

Green RE, Cornell SJ, Scharlemann JP et al (2005) Farming and the fate of wild nature. Science 307:550–555

Greenwood O, Mossman HL, Suggitt AJ et al (2016) Using in situ management to conserve biodiversity under climate change. J Appl Ecol 53:885–894

Gregory RD, Willis SG, Jiguet F et al (2009) An indicator of the impact of climatic change on European bird populations. PLoS One 4:e4678

Hampe A, Petit RJ (2005) Conserving biodiversity under climate change: the rear edge matters. Ecol Lett 8:461–467

Herruzo AC, Martínez-Jáuregui M, Carranza J (2016) Commercial income and capital of hunting: an application to forest estates in Andalucía. For Pol Econ 69:53–61

Huntley B, Green RE, Collingham YC et al (2007) A climatic atlas of European breeding birds. Lynx Edicions, Barcelona

Izhaki I (2000) Passerine bird communities in Mediterranean pine forests. In: Ne'eman G, Trabaud L (eds) Ecology, Biogeography and Management of *Pinus halepensis* and *P. brutia* Forest Ecosystems in the Mediterranean Basin. Buckhuys Publishers, Leiden, pp 237–250

Izhaki I (2012) The impact of fire on vertebrates in the Mediterranean Basin: an overview. Is J Evol Ecol 58:221–233

Jetz W, Thomas GH, Joy JB et al (2012) The global diversity of birds in space and time. Nature 491:444–448

Mace GM, Baillie JE (2007) The 2010 biodiversity indicators: challenges for science and policy. Conserv Biol 21:1406–1413

MacMillan DC, Phillip S (2008) Consumptive and non-consumptive values of wild mammals in Britain. Mamm Rev 38:189–204

Maddison WP, Maddison DR (2011) Mesquite: a modular system for evolutionary analysis. Version 2.75. Available from: http://mesquiteproject.org

Martí R, del Moral JC (2003) Atlas de las aves reproductoras de España. Dirección General de Conservación de la Naturaleza-Sociedad Española de Ornitología, Madrid

Martínez-Abraín A, Jiménez J (2019) Dealing with growing forest insect pests: the role of top-down regulation. J Appl Ecol 56:2574–2576

Martínez-Jáuregui M, Pardos M, Balogh P et al (2014) Hunting in European mountain systems: an economic assessment of game gross margins in nine case study areas. Eur J Wildl Res 60:933–936

Martínez-Jáuregui M, Díaz M, Sánchez de Ron D et al (2016a) Plantation or natural recovery? Relative contribution of planted and natural pine forests to the maintenance of regional bird diversity along ecological gradients in Southern Europe. For Ecol Manag 376:183–192

Martínez-Jáuregui M, Soliño M, Díaz M (2016b) Geographical variation in the contribution of planted and natural pine forests to the conservation of bird diversity. Diver Distrib 12:1255–1265

Martínez-Jáuregui M, Serra-Varela MJ, Díaz M et al (2018) Mitigation strategies for biodiversity conservation under climate change scenarios: the role of forest naturalization. PLoS One 13:e0202009

Martínez-Jáuregui M, White PC, Touza J et al (2019) Untangling perceptions around indicators for biodiversity conservation and ecosystem services. Ecosyst Serv 38:100952

Mikusinski G, Roberge J-M, Fuller RJ (eds) (2018) Ecology and conservation of forest birds. Cambridge University Press, Cambridge

Morán-López T, Alonso CL, Díaz M (2015) Landscape effects on jay foraging behavior decrease acorn dispersal services in dehesas. Acta Oecol 69:52–64

Munang RT, Thiaw I, Rivington M (2011) Ecosystem management: tomorrow's approach to enhancing food security under a changing climate. Sustainability 3:937–954

Navarro LM, Pereira HM (eds) (2015) Rewilding abandoned landscapes in Europe. Springer, Cham

Pesendorfer MB, Sillett TS, Koenig WD et al (2016) Scatter-hoarding corvids as seed dispersers for oaks and pines: a review of a widely distributed mutualism and its utility to habitat restoration. Condor 118:215–237

Regos A, Clavero M, D'Amen M et al (2018) Wildfire–vegetation dynamics affect predictions of climate change impact on bird communities. Ecography 41:982–995

Reino L, Beja P, Osborne PE et al (2009) Distance to edges, edge contrast and landscape fragmentation: interactions affecting farmland birds around forest plantations. Biol Conserv 142:824–838

Richardson DM, Nsikani MM (this volume) Mediterranean pines as invasive species in the Southern Hemisphere. In: Ne'eman G, Osem Y (eds) Pines and their mixed forest ecosystems in the Mediterranean Basin. Springer, pp 83–99

Santos T, Tellería JL, Díaz M et al (2006) Evaluating the environmental benefits of CAP reforms: can afforestations restore forest bird communities in Mediterranean Spain? Basic Appl Ecol 7:483–495

Sanz JJ (2001) Experimentally increased insectivorous bird density results in a reduction of caterpillar density and leaf damage to Pyrenean oak. Ecol Res 16:387–394

Sekerçioglu CH (2002) Impacts of birdwatching on human and avian communities. Environ Conserv 29:282–289

Stephens PA, Mason LR, Green RE et al (2016) Consistent response of bird populations to climate change on two continents. Science 352:84–87

Stocker TF, Qin D, Plattner GK et al (2013) Climate change 2013: the physical science basis. Cambridge Universtiy Press, Cambridge

Sweeney OFM, Wilson MW, Irwin S et al (2010) Breeding bird communities of second-rotation plantations at different stages of the forest cycle. Bird Study 57:301–314

Urban MC, Bocedi G, Hendry AP et al (2016) Improving the forecast for biodiversity under climate change. Science 353:aad8466

Valladares F, Benavides R, Rabasa S et al (2014) Global change and Mediterranean forests: current impacts and potential responses. In: Coomes DA, Burslem DFRP, Simonson WD (eds) Forests and global change. Cambridge University Press, Cambridge, pp 47–75

Viana DS, Santamaría L, Figuerola J (2016) Migratory birds as global dispersal vectors. Trends Ecol Evol 31:763–775

Whelan CJ, Wenny DG, Marquis RJ (2008) Ecosystem services provided by birds. Ann N Y Acad Sci 1134:25–60

Chapter 19
Soil Micro- and Macrofauna in Mediterranean Pine and Mixed Forests

Canella Radea and Margarita Arianoutsou

19.1 Introduction

Soil is one of the most significant reservoirs of biological diversity on Earth and hosts the vast majority of microorganisms, plants, invertebrate and vertebrate animals (Barrios 2007; Gardi et al. 2009; Jeffery et al. 2010; Geisen et al. 2019). Soil organisms generally thrive in organic, topsoil and subsoil horizons and their community composition plays a crucial role in regulating key ecosystem functions and providing ecosystem services (Bardgett and Van der Putten 2014; Wagg et al. 2014; Geisen et al. 2019).

No less than one-quarter of described living species of vertebrates and invertebrates are strictly soil inhabitants. Invertebrates are extremely variable and represent as much as 99% of soil animals (Lavelle 1997; Decaëns et al. 2006). They are classified by body width into three groups: microfauna (< 100 μm), mesofauna (100 μm – 2 mm) and macrofauna (> 2 mm) (Swift et al. 1979). Soil invertebrates are providers and mediators of a plethora of ecosystem services (e.g., organic matter decomposition, nutrient cycling, humus formation, bioturbation, water regime control, provision of food for humans and aboveground animals, carbon storage and climate regulation) that they generate from the soil (Brussaard 1997; Swift et al. 1998; Lavelle et al. 2006; Baveye et al. 2016) and are essential to the sustainable function of natural and managed ecosystems (Barrios 2007). Additionally, soil invertebrates are sensitive and reliable biomonitors for evaluating the impacts of various disturbances linked to natural forces and to human activities on soil quality (e.g., Van Straalen 1998; David et al. 1999; Knoepp et al. 2000; Yan et al. 2012; Blasi et al. 2013; Paz-Ferreiro and Fu 2016).

C. Radea (✉) · M. Arianoutsou
Department of Ecology and Systematics, National and Kapodistrian University of Athens, Athens, Greece
e-mail: kradea@biol.uoa.gr

G. Ne'eman, Y. Osem (eds.), *Pines and Their Mixed Forest Ecosystems in the Mediterranean Basin*, Managing Forest Ecosystems 38,
https://doi.org/10.1007/978-3-030-63625-8_19

Various groups (here we use the term "group" as defined by Geisen et al. 2019, i.e., a gross taxonomically-distinct assemblage of soil organisms) of soil invertebrates have been studied extensively in boreal and temperate pine forests of Europe, for example: Nematoda, Acari and Collembola (Huhta et al. 1986; Sohlenius and Boström 2001; Widenfalk et al. 2016; Kamczyc et al. 2019), Oligochaeta (Huhta et al. 1998), Chilopoda and Diplopoda (Lock et al. 2001; Bachvarova 2011). The available information concerning the soil invertebrates of pine and mixed forests (with pines, oaks and shrubs) growing around the Mediterranean Basin is surprisingly limited although mature single pine and mixed forests are the last refuges for various Mediterranean plant and animal species (Naveh 2000; Fattorini 2010).

Pine forests are emblematic of the areas located on the thermo- to supra-Mediterranean and mountainous Mediterranean bioclimatic zones. The main species forming these forests are *P. halepensis* Mill., *Pinus brutia* Ten., *P. pinaster* Ait., *P. pinea* L., *P. nigra* Arn., *P. uncinata* Raymond ex A. DC and *P. heldreichii* H. Christ (Quézel 1980; Barbéro et al. 1998; Garavaglia and Besacier 2014).

The present chapter deals with the spatiotemporal diversity of soil invertebrate communities in Mediterranean pure pine and mixed forests (hereafter MPF and MMF, respectively).

19.2 Technical Restrictions for the Comparative Study of Soil Invertebrates

The problems that have emerged in studying and comparing the structure of soil communities in Mediterranean ecosystems since the 1970s (Di Castri 1973; Di Castri and Vitali-Di Castri 1981; Legakis 1994) remain unresolved:

(i) Lack of detailed knowledge of systematics and ecology at the species level despite the ongoing discovery and description of new species (e.g., Sammler et al. 2006; Baquero et al. 2010; Giuga and Jordana 2013; Duran and Urhan 2015).

(ii) Lack of comparability of the methods used for quantitative sampling and extracting soil fauna. A variety of direct and indirect collection techniques (e.g., collection by hand, collection by shovel, soil corers, pitfall traps etc.) are available to the researchers (Table 19.1). However, the estimated diversity and density of collected invertebrates depends greatly on the method used and soil sample size; this is also true for the extraction techniques (such as Berlese-Tullgren apparatus and its modifications, O' Connor wet extractor, Kempson funnels, flotation techniques etc.) (André et al. 2002; Savin et al. 2007). Consequently, an underestimation of diversity and density of various groups is frequently observed. Additionally, the different expressions of abundance, either as density [individuals per volume (individuals/dm^3, individuals/cm^3) (e.g., Di Castri and Vitali-Di Castri 1981; Bonari et al. 2017, respectively), individuals per area (individuals/m^2) (e.g., Legakis 1986)], relative abundance

Table 19.1 Publications concerning the soil invertebrate communities in Mediterranean pine and mixed forests (plantations and monocultures are included) during the last 35 years

Soil invertebrates	Pine & mixed forests	Site/Country	Methods of collection/ extraction	Results	References in chronological order
Macroarthropods, Gastropoda, Annelida	*P. nigra forest, P. heldreichii forest*	Mt. Olympus/ Greece	By hand	Qualitative	Stamou et al. (1984)
Macroarthropods	*P. halepensis* & *P. brutia* forest reforestation	Attica/Greece	Organic horizons samples/ Berlese-Tullgren apparatus	Quantitative	Legakis (1986)
Micro- and macroarthropods	*P. halepensis* forest	Scopelos Island / Greece	Organic horizons samples, pitfall traps/ Berlese-Tullgren apparatus	Quantitative, semi-quantitative	Radea (1989)
Macroarthropods	*P. halepensis* forest	Sophico/Greece	Organic horizons samples/ Berlese-Tullgren apparatus	Quantitative	Karamaouna (1990a)
Diplopoda: Penicillata: *Polyxenus lagurus* (Linnaeus, 1758)	*P. halepensis* forest	Sophico/Greece	Organic horizons samples/ Berlese-Tullgren apparatus	Quantitative	Karamaouna (1990b)
Micro- and macroarthropods	*P. halepensis* forest	Euboea Island / Greece	Organic horizons samples, pitfall traps/ Berlese-Tullgren apparatus	Quantitative, semi-quantitative	Marmari (1991)
Diplopoda: Iulida: *Symphyoiulus impartitus* (Karsch, 1888)	*P. halepensis* forest	Sophico/Greece	Organic horizons samples/ Berlese-Tullgren apparatus	Quantitative	Karamaouna (1992)
Gastropoda	*P. halepensis* forest	Scopelos Island/ Greece	Organic horizons samples / combination of sieving and flotation	Quantitative	Radea and Mylonas (1992)
Predatory macroarthropods	*P. halepensis* forest	Scopelos Island/ Greece	Organic horizons samples/ Berlese-Tullgren apparatus	Quantitative	Radea (1993)
Micro- and macroarthropods	Mixed *P. halepensis* and oak forest	Mt Carmel/Israel	Soil samples/ Berlese-Tullgren apparatus	Quantitative	Broza and Izhaki (1997)

(continued)

Table 19.1 (continued)

Soil invertebrates	Pine & mixed forests	Site/Country	Methods of collection/ extraction	Results	References in chronological order
Collembola	*P. uncinata* forests	Pyrenees/Spain	Soil samples/ Berlese-Tullgren apparatus	Quantitative	Benito and Sánchez (2000)
Micro- and macroarthropods	*P. halepensis* forest	Mt Carmel/Israel	Soil samples/ Berlese-Tullgren apparatus	Quantitative	Broza (2000)
Coleoptera	*P. halepensis, P. nigra, P. sylvestris* forests	Sierra de Baza/ Spain	Pitfall traps	Hemi-quantitative	Romero-Alcaraz & Ávila (2000)
Coleoptera: Carabidae	*P. pinaster forest*	Spain	Pitfall traps	Semi-quantitative	Fernández Fernández and Salgado Costas (2004)
Coleoptera	*P. brutia, P. brutia afforestation*	Dadia/Greece	Pitfall traps	Semi-quantitative	Argyropoulou et al. (2005)
Chilopoda, Diplopoda	*P. halepensis* forest	Balearic Islands/ Spain	By hand, sifting of leaf litter, pitfall traps	Qualitative, semi-quantitative	Sammler et al. (2006)
Gastropoda, Nematoda, Oligochaeta, Arthropoda	*P. uncinata*	Northern French Alps/France	Humus samples, by hand, pitfall traps/ Berlese-Tullgren apparatus	Quantitative, semi-quantitative	Cassagne et al. (2008)
Acari	*P. brutia* forests	Muğla/Turkey	Litter and soil samples/ modified Berlese-Tullgren apparatus	Quantitative	Urhan et al. (2008)
Macroarthropods	*P. pinaster* monoculture	Portugal	Pitfall traps	Semi-quantitative	Antunes et al. (2009)
Ants	*P. halepensis* forest	Mt Carmel/Israel	Pitfall traps	Semi-quantitative	Izhaki et al. (2009)
Ants	*P. pinea*	Rome/Italy	Pitfall traps	Semi-quantitative	Castracani et al. (2010)
Coleoptera: Tenebrionidae	Mixed *P. pinea*-oaks forest	An Urban Park near Rome/Italy	Pitfall traps	Semi-quantitative	Fattorini (2010)
Gastropoda	*P. brutia* forest *P. nigra* forest	Cyprus	By hand	Qualitative	Vardinoyannis et al. (2012)

Micro- and macroarthropods	*P. nigra* plantation	Instabul/Turkey	Litter and soil samples/ modified Berlese-Tullgren apparatus	Quantitative	Cakir and Makinesi (2013)
Gastropoda	Mixed *P. halepensis* and *Quercus ilex* forest	Collserola Natural Park/ Spain	By hand	Qualitative	Torre et al. (2014)
Micro- and macroarthropods	*P. pinaster* forest	Tuscany/Italy	Soil samples/ Berlese-Tullgren apparatus	Quantitative	Lisa et al. (2015)
Arthropods	*P. halepensis* reforestation	Chbika/Algeria	Pitfall traps	Semi-quantitative	Souttou et al. (2015)
Macroarthropods	*P. brutia*	Muğla/ Turkey	Pitfall traps	Semi-quantitative	Kaynas (2016)
Acari: Oribatida	*P. pinea* forest	Tuscany/Italy	Soil samples/ Berlese-Tullgren apparatus	Quantitative	Bonari et al. (2017)
Platyhelminthes	*P. sylvestris* forest, *P. uncinata*	National Parks/ Spain	By hand	Qualitative	Álvarez-Presas et al. (2018)
Araneae	*P. halepensis* forest, *P. pinea and P. canariensis* plantations	Game reserve of Zéralda/Algeria	Pitfall traps	Semi-quantitative	Touchi et al. (2018)
Protura	*Pinus* spp.	Italy and Corsica	Soil samples/ Berlese-Tullgren apparatus	Quantitative/ no data	Galli et al. (2019)

(e.g., Cassagne et al. 2008) or individuals per soil core (e.g., Broza and Izhaki 1997) make the situation even more complicated.

(iii) Exclusion of some groups of soil invertebrates (such as Acari, Collembola, Nematoda, Mollusca, Oligochaeta, ants and termites) from the sampling efforts (Table 19.1). This fact is attributed to the specific methods required for the effective sampling and extraction of these animals and to their highly contagious distribution.

Another issue hindering the comparative study of Mediterranean soil invertebrates is the ambiguity in what the soil samples contain. In most cases, the depth of samples is reported without any evaluation of its adequacy for effective collection of soil invertebrates as a whole (André et al. 2002). It is also remarkable that very few data are provided about the characteristics of soil horizons and layers included in the soil samples although the diversity, density and biomass of soil invertebrates are affected by them and vice versa (Hasegawa 2001; Izadi and Habashi 2016; Zanella et al. 2018).

The locomotor activity of soil invertebrates on the litter horizon surface has been studied in MPF and MMF, by using pitfall traps that is a semi-quantitative method (Table 19.1). Traps of various sizes containing water or killing-preservative liquids, with or without attracting baits, are buried just below the litter surface (e.g., Fernández Fernández and Salgado Costas 2004; Argyropoulou et al. 2005; Souttou et al. 2015). The pitfall catch represents a combination of activity intensity and population density of soil invertebrates. The sampling design, the duration of collection and the expression of abundance of invertebrates depend on the research purposes. The lack of methodological standardization is obvious once again.

19.3 Species Richness and Community Structure of Soil Invertebrates

Data on diversity are fundamental to our understanding of ecosystems, and dealing with diversity loss requires detailed knowledge of systematics and ecology at the species level (Behan-Pelletier 1993). However, due to the already mentioned high diversity of soil invertebrates and the limited taxonomic expertise, ecological research in MPF and MMF has been based on higher taxonomic levels and on functional groups. At this point, the lack of linkages between the systematic and ecological research should be mentioned. Several of the newly described and the already known species could be soil inhabitants of the Mediterranean pine forests but very limited information is provided about the vegetation type and the humus forms of the type localities and/or the collection sites (e.g., Baquero et al. 2010).

The available data on soil invertebrate species richness in MPF and MMF are presented in Table 19.2. According to Kaynaş (2016), Bonari et al. (2017) and Touchi et al. (2018), the richness of soil invertebrate species (Fig. 19.1) in MPF does not depend on the pine species; however, it is positively related to the degree of

Table 19.2 Species number of various soil invertebrates in Mediterranean pine and mixed forests

Groups	A	B	C	D	E	F	G	H	I	J	K
Isopoda		4		6		6					
Protoura											5
Collembola						23			17–20		
Acari: Oribatida							100				
Araneae			10–25		16			12			
Mollusca	11										
Platyhelminthes										4	

A: Mature *P. halepensis* forest with understory shrubs, Greece (Radea 1989); B: *P. halepensis* forest, Tunisia (Khila et al. 2018); C: *P. halepensis* forest, Algeria (Touchi et al. 2018); D: *P. halepensis* plantation, Israel (Hornung and Warburg 1995); E: *P. pinea,* Algeria (Touchi et al. 2018); F: *P. pinea* mature plantation, Italy (Pitzalis et al. 2005); G: *P. pinea*, Italy (Bonari et al. 2017); H: *P. canariensis*, Algeria (Touchi et al. 2018); I: *P. uncinata*, Spain (Benito and Sánchez 2000); J: *P. sylvestris* and *P. uncinata*, Spain (Álvarez-Presas et al. 2018); K: *Pinus* spp., Italy (Galli et al. 2019)

canopy cover, the understory plant species richness and, consequently, the habitat complexity. In some cases, *P. halepensis* forests have lower species richness than sclerophyllous shrublands and oaks (Marcuzzi 1968; Hornung and Warburg 1995) but no generalization can be made at the present state of knowledge.

Thirty-four supra-specific taxa (of different taxonomic levels or stages of the life cycle) of soil invertebrates have been recorded in MPF and MMF after quantitative samplings and extraction using Berlese–Tullgren apparatus (Table 19.3). The micro- and macro-arthropods constitute the great majority of soil invertebrates, and their densities differ considerably between and within the forests under consideration. Obviously, this fact can be partly explained by large-scale climatological and geological factors (Di Castri and Vitali-Di Castri 1981; Berg et al. 1998), the historical heritage and the anthropogenic pressures (Di Castri and Vitali-Di Castri 1981). On the other hand, various factors significantly affect and differentiate the structure of soil arthropod communities in pine forests from that of other Mediterranean ecosystems such as: (1) Local environmental conditions and micro-spatial factors including the structural complexity of the litter horizon (OL) [i.e., its compressibility and the contribution of various plant species to its formation (Radea 1989)]. (2) Presence of a surface rock layer or fragments (Romanyà et al. 2000). (3) Water and organic matter content at a given time (Di Castri and Vitali-Di Castri 1981; Radea 1989; Karamaouna 1992). (4) Litter character (i.e., the quantity and quality of litter) (Urhan et al. 2008; Aupic-Samain et al. 2019). (5) Chemical characteristics of the humus form (Cakir and Makinesi 2013). (6) Canopy cover (Touchi et al. 2018).

Two microarthropod groups, namely Acari and Collembola, constitute more than 83% of soil arthropods. The dominance of these groups is mentioned in all MPF and MMF in every Mediterranean bioclimatic zone from the thermo- to the mountainous (Radea 1989; Broza 2000; Doblas-Miranda et al. 2007; Cassagne et al. 2008; Lisa et al. 2015 among others). Acari and Collembola also dominate the soil of two

Fig. 19.1 Members of the soil macrofauna in mediterranean pine and mixed forests. **Chilopoda**: Scolopendromorpha: *Scolopendra cingulata* Latreille, 1829 (**a**) and Lithobiomorpha: *Lithobius* sp. (**b**). **Gastropoda**: Stylommatophora: *Lindholmiola lens* (A. Férussac, 1832) (**c**). **Diplopoda**: Julida: *Pachyiulus flavipes* (C. L. Koch, 1847) (**d**) and Callipodida: *Acanthopetalum blanci* (Brölemann, 1932) (**e**). **Malacostraca**: Isopoda: *Porcellionides pruinosus* (Brandt, 1833) (**f**). (Photos by C. Radea)

other types of Mediterranean ecosystems, the evergreen oak forests (Andrés et al. 1999) and phrygana (Sgardelis and Margaris 1993).

The Acari:Collembola ratio (A:C) is an indication of the soil condition: a low ratio means either soil degradation due to human pressure or more xeric conditions (Bachelier 1963; Andrés et al. 1999; Menta et al. 2001). This ratio equals 2 in temperate forests and exceeds 2 in Mediterranean ecosystems (Di Castri and Vitali-Di Castri 1981). In MPF and MMF (Table 19.3), the lowest ratio (0.67) was estimated in a forest with high human pressure (i.e., fires, logging and trampling), near Athens, the capital of Greece (Legakis 1986). The highest value (3.45) was found in an undisturbed mature pine forest on an island in the central Aegean Sea (Radea 1989). In the Mediterranean Basin, the A:C ratio is usually lower in pine and mixed forests than in oak forests where values from 2.66 to 15.30 have been estimated (Di Castri and Vitali-Di Castri 1981; Andrés et al. 1999).

Table 19.3 Density (ind/m^2) of arthropods and mollusks extracted from soil samples using the Berlese-Tullgren apparatus and flotation method respectively, in Mediterranean pine and oak forests

Groups	A	B	C	D	E	F	G	H	I	J	K	L
Acari	4966.4	5326.7				41738	24754				66009	17063–104490
Collembola	1208.3	4530.4				61482	13630				38003	6413–19000
Protura	0.3					3906	340					20–470
Diplura	17.2										340	34–366
Palpigradida						P						
Pauropoda						340	424				1316	67–485
Symphyla	2.2	18.8								946	85	62–831
Araneae	68.7	38.3	39.0	14.4	4.8							
Pseudoscorpionida	133.5	30.1	68.8	14.4	20.8					141	339	
Scorpionida	0.4											
Phalangida	1.2		1.4									
Lithobiomorpha	28.8		18.3	8.0	20.8						169	56–256
Geophilomorpha	8.3	10.8										
Iulida	68.9	10.0	39.0	9.6	4.8					271	764	59–540
Polydesmida	49.7											
Polyxenida	134.1	34.3										
Lysiopetalida	0.7											
Isopoda	17.6		5.7					1.01	0.91		42	
Diplura	17.2										340	34–366
Hymenoptera (except ants)	P											
Ants	P										127	
Hemiptera	22.8	71.9	8.4	16.0	19.2							
Psocoptera	139.0	106.6										308–1017
Isoptera	P											
Thysanoptera	116.1	324.5								592		

(continued)

Table 19.3 (continued)

Groups	A	B	C	D	E	F	G	H	I	J	K	L
Diptera (mature)	19.8	23.5										
Coleoptera (mature)	30.2	12.9	25.6		16.0							
Diptera (larvae)	369.5	31.3	224.0		8.0							
Coleoptera (larvae)	20.6	99.6		16.0								1040–1312
Lepidoptera (larvae)	1.7											
Dictyoptera	1.8		9.7		4.8							
Thysanura	0.3		1.8									
Embioptera	1.8											
Dermaptera	0.3											
Neuroptera (larvae)	0.1											
Mollusca	76.4											

A: Mature *P. halepensis* forest with understory shrubs, Greece (Radea 1989); B: *P. halepensis* forest with understory shrubs, Greece (Marmari 1991); C: *P. halepensis* forest with understory shrubs, Greece (Karamaouna 1990a); D: *P. halepensis*, *P. brutia* afforestation, East slope, Greece (Legakis 1986); E: *P. halepensis*, *P. brutia* afforestation, West slope, Greece (Legakis 1986); F: *P. halepensis*, February, Israel (Broza 2000); *P. halepensis*, July, Israel (Broza 2000); H: *P. halepensis* plantation, Winter, Israel (Hornung and Warburg 1995); I: *P. halepensis* plantation, Spring, Israel (Hornung and Warburg 1995); J: *P. pinaster forest*, Italy (Lisa et al. 2015); K: *P. nigra*, Turkey (Cakir and Makinesi 2013); L: *Quercus ilex* forests, Mediterranean Basin (Andrés et al. 1999); P = present

Diplopoda constitute one of the most important saprophagous macroarthropod groups in temperate forests where their density reaches high values (30–120 ind/m^2) (Bertrand and Lumaret 1991). In MPF and MMF their density ranges widely from 4.8 ind/m^2 (in degraded *P. halepensis* – *P. brutia* forest) to 2166 ind/m^2 (in natural, unburned *P. pinaster* forest) (Table 19.3). In *P. halepensis* forests in Greece, the horizontal distribution of diplopods show a consistent pattern, i.e., their mean annual densities are higher under the shrubs than under the pine trees (Radea 1989; Karamaouna 1992). Nevertheless, *Symphyoiulus impartitus* (Karsch, 1888) (Diplopoda: Julida) shows an unusual preference for the litter produced by *P. halepensis* (Karamaouna 1992) although most Julida avoid pine litter (Arpin et al. 1986). This species spread in Greece and Asia Minor and, having a mean annual density of 8 ind/m^2 and biomass (dry weight) of 100 mg/m^2 (Karamaouna 1992), is one of the dominant macroarthropods in mainland and insular *P. halepensis* forests of Greece. A consistent pattern is also observed in the vertical distribution of diplopods under pine trees since they clearly prefer the OF and OH horizons over the OL horizon (Radea 1989).

In *P. halepensis* forests, Mollusca and predatory macroarthropods, namely Pseudoscorpionida, Araneae, Lithobiomorpha and Geophilomorpha, show similar patterns of horizontal and vertical distribution to those of Diplopoda (Radea 1989, 1993; Karamaouna 1990a; Radea and Mylonas 1992).

The seasonality and variation in annual precipitation and soil water balance recorded in *P. halepensis* forests are pivotal factors for soil organisms and they superimpose the traditional pattern of soil arthropod vertical distribution (Broza 2000).

Apart from the marked surface activity of ants that is 13.7% of the total trapped animals in *P. pinea* reforestations (Castracani et al. 2010), Arachnida, Diplopoda, Chilopoda, Collembola and Coleoptera are the majority of soil invertebrates collected by pitfall traps in MPF and MMF (Radea 1989; Marmari 1991; Cassagne et al. 2008; Antunes et al. 2009; Souttou et al. 2015). The composition of invertebrate fauna trapped depends on vegetation and soil type, plant cover and the microhabitat formed beneath the canopy of different plant species (Romero-Alcaraz and Ávila 2000; Izhaki et al. 2009; Souttou et al. 2015).

19.4 Temporal Variation in Soil Invertebrate Density and Locomotor Activity

For soil invertebrates of Mediterranean ecosystems, Di Castri and Vitali-Di Castri (1981) report one or two peaks of density in winter, one in spring and the other in winter or in autumn. The temporal pattern of density variation is more complicated if we focus not only on one species but on groups of species (Legakis 1994). In *P. halepensis* forests, the most hygrophilous groups, such as Collembola, Isopoda, Julida, Polydesmida, Pseudoscorpionida, Lithobiomorpha, Geophilomorpha and

Diptera larvae, show a density peak during the rainy period of the year (late autumn to early spring) (Radea 1989, 1993; Hornung and Warburg 1995; Broza 2000) and their monthly density is positively correlated with the amount of precipitation (Radea 1989, 1993; Karamaouna 1990a; Marmari 1991). In contrast, the monthly densities of chitinized or hairy xerophilous-mesophilous groups, namely Psocoptera, Thysanoptera and Polyxenida (Di Castri 1973), are not influenced by precipitation (Radea 1989; Marmari 1991).

Although no data exist concerning MPF and MMF specifically, it is well known that in the Mediterranean Basin, species richness and seasonal variations in richness, abundance and biomass of soil invertebrates are more marked in the litter layer (OL horizon) than belowground (Di Castri and Vitali-Di Castri 1981; Legakis 1994; Doblas-Miranda et al. 2007).

In MPF and MMF, the temporal variation in locomotor activity of invertebrates on the litter surface is influenced by temperature and rainfall (Radea 1989; Marmari 1991; Cassagne et al. 2008). In general, the pattern of surface activity of soil invertebrates is similar to the pattern of their density, but both patterns could be better understood only by long-term, seasonal and multi-sample, in-depth, species-level studies (Legakis 1994).

19.5 Conclusions

The significant spatiotemporal complexity of soil invertebrate communities, which has already been documented in broadleaf forests, sclerophyllous and phryganic formations of Mediterranean Basin, has also been documented in pine and mixed forests in these regions. This complexity reflects the spatial patchiness of vegetation, rocks and humus forms and the accentuated temporal variation of climatological parameters. However, the limited number of taxonomists, the lack of comparability of the sampling and extraction methods and the nonexistent linkages between the systematic and ecological research obstruct the deep and integrated investigation of soil invertebrate communities in pine and mixed forests of the Mediterranean Basin.

References

Álvarez-Presas M, Mateos E, Riutort M (2018) Hidden diversity in forest soils: characterization and comparison of terrestrial flatworm's communities in two national parks in Spain. Ecol Evol 8:7386–7400

André HM, Ducarme X, Lebrun P (2002) Soil biodiversity: myth, reality or conning? Oikos 96:3–24

Andrés P, Mateos E, Ascaso C (1999) Soil arthropods. In: Rodà F, Retana J, Gracia CA et al (eds) Ecology of mediterranean evergreen oak forests. Springer, New York, pp 341–354

Antunes SC, Curado N, Castro BB et al (2009) Short-term recovery of soil functional parameters and edaphic macro-arthropod community after a forest fire. J Soil Sedim 9:267–278

Argyropoulou MD, Karris G, Papatheodorou EM et al (2005) Epiedaphic Coleoptera in the Dadia forest reserve (Thrace, Greece): the effect of human activities on community organization patterns. Belg J Zool 135:127–133

Arpin P, David J-F, Guittonneau G-G, et al (1986) Influence du peuplement forestier sur la faune et la microflore du sol et des humus. I. Description des stations et étude de la faune du sol. Rev Ecol Biol Sol 23:89–118

Aupic-Samain A, Baldy V, Lecareux C et al (2019) Tree litter identity and predator density control prey and predator demographic parameters in a Mediterranean litter-based multi-trophic system. Pedobiologia 73:1–9

Bachelier G (1963) La vie Animale Dans les sols I. Déterminisme de la Faune des Sols. ORSTOM: Paris, France

Bachvarova D (2011) Myriapoda (Chilopoda, Diplopoda) of Shumen City and Shumen Plateau (NE Bulgaria): taxonomic structure and zoogeographical analysis. Acta Zool Bulg 63:245–262

Baquero E, Arbea J, Jordana R (2010) New species of Entomobryini from the Mediterranean Palaearctic (Collembola, Entomobryidae) and a new name for *Folsomia potapovi* (Collembola, Isotomidae). Soil Org 82:285–300

Barbéro M, Loisel R, Quezel P et al (1998) Pines of the mediterranean basin. In: Richardson DM (ed) Ecology and biogeography of *Pinus*. Cambridge University Press, Cambridge, pp 153–170

Bardgett RD, Van der Putten WH (2014) Belowground biodiversity and ecosystem functioning. Nature 515:505–511

Barrios E (2007) Soil biota, ecosystem services and land productivity. Ecol Econom 64:269–285

Baveye PC, Baveye J, Gowdy J (2016) Soil "ecosystem" services and natural capital: critical appraisal of research on uncertain ground. Front Environ Sci 4:41–49

Behan-Pelletier VM (1993) Diversity of soil arthropods in Canada: systematic and ecological problems. Mem Ent Soc Can 165:11–50

Benito JCS, Sánchez MJL (2000) Ecology of soil springtails (Collembola, Insecta) from pine woods and *Rhododendron* shrublands in the Central and Eastern Pyrenees (North Spain). Pedobiologia 44:430–441

Berg MP, Kniese JP, Bedaux JJM et al (1998) Dynamics and stratification of functional groups of micro- and mesoarthropods in the organic layer of a Scots pine forest. Biol Fert Soils 26:268–284

Bertrand M, Lumaret JP (1991) The role of Diplopoda litter grazing activity on recycling processes in a Mediterranean climate. Vegetatio 99–100:289–297

Blasi S, Menta C, Balducci L et al (2013) Soil microarthropod communities from Mediterranean forest ecosystems in Central Italy under different disturbances. Environ Monit Assess 185:1637–1655

Bonari G, Migliorini M, Landi M et al (2017) Concordance between plant species, oribatid mites and soil in a Mediterranean stone pine forest. Arthropod-Plant Interact 11:61–69

Broza M (2000) Soil arthropods in east Mediterranean *Pinus halepensis* forests. In: Ne'eman G, Trabaud L (eds) Ecology, biogeography and management of *Pinus halepensis* and *P. brutia* forest ecosystems in the Mediterranean basin. Backhuys, Leiden, pp 203–216

Broza M, Izhaki I (1997) Post-fire arthropod assemblages in Mediterranean forest soils in Israel. Int J Wild Fire 7:317–325

Brussaard L (1997) Biodiversity and ecosystem function in soil. Ambio 26:563–570

Cakir M, Makinesi E (2013) Humus characteristics and seasonal changes of soil arthropod communities in a natural sessile oak (*Quercus petraea* L.) stand and adjacent Austrian pine (*Pinus nigra* Arnold) plantation. Environ Monit Assess 185:8943–8955

Cassagne N, Spiegelberger T, Cécillon L et al (2008) The impact of soil temperature increase on organic matter and faunal properties in a frozen calcareous scree in the French Alps. Geoderma 146:239–247

Castracani C, Grasso DA, Fanfani A et al (2010) The ant fauna of Castelporziano Presidential Reserve (Rome, Italy) as a model for the analysis of ant community structure in relation to environmental variation in Mediterranean ecosystems. J Insect Conserv 14:585–594

David J-F, Devernay S, Loucougaray G et al (1999) Belowground biodiversity in a Mediterranean landscape: relationships between saprophagous macroarthropod communities and vegetation structure. Biodivers Conserv 8:753–767

Decaëns T, Jiménez JJ, Gioia C et al (2006) The values of soil animals for conservation biology. Eur J Soil Biol 42:S23–S38

Di Castri F (1973) Soil animals in latitudinal and topographical gradients of Mediterranean ecosystems. In: Di Castri F, Mooney HA (eds) Mediterranean-type ecosystems: origin and structure. Springer, Berlin, pp 171–190

Di Castri F, Vitali-di Castri V (1981) Soil fauna of Mediterranean-climate regions. In: Di Castri F, Goodall F, Specht RL (eds) Ecosystems of the world. II. Mediterranean shrublands. Elsevier, Amsterdam, pp 445–478

Doblas-Miranda E, Sánchez-Pinero F, González-Megías A (2007) Soil macroinvertebrate fauna of a Mediterranean arid system: composition and temporal changes in the assemblage. Soil Biol Biochem 39:1916–1925

Duran EH, Urhan R (2015) A new species of the genus *Prozercon* Sellnick, 1943 (Acari: Mesostigmata: Zerconidae) from Turkey. Zool Middle East 61:174–178

Fattorini S (2010) Effects on tenebrionid communities of a *Pinus pinea* plantation: a case study in a Mediterranean site. Biodivers Conserv 19:1237–1250

Fernández Fernández MM, Salgado Costas JM (2004) Recolonization of a burnt pine forest (*Pinus pinaster*) by Carabidae (Coleoptera). Eur J Soil Biol 40:47–53

Galli L, Capurroa M, Molyneux T et al (2019) Ecology of Italian Protura. Pedobiologia 73:20–28

Garavaglia V, Besacier C (2014) The state of Mediterranean forests in 2013. Unasylva 65:3–15

Gardi C, Montanarella L, Arrouays D et al (2009) Soil biodiversity monitoring in Europe: ongoing activities and challenges. Eur J Soil Sci 60:807–819

Geisen S, Briones MJI, Ganc H et al (2019) A methodological framework to embrace soil biodiversity. Soil Biol Biochem 136. https://doi.org/10.1016/j.soilbio.2019.107536

Giuga R, Jordana L (2013) Three new species of Collembola from soils of Mediterranean cork-oak forests of Sicily (Italy). Zootaxa 3664:283–300

Hasegawa M (2001) The relationship between the organic matter composition of a forest floor and the structure of a soil arthropod community. Eur J Soil Biol 37:281–284

Hornung E, Warburg MR (1995) Seasonal changes in the distribution and abundance of isopod species in different habitats within the Mediterranean region of northern Israel. Acta Oecol. 16:431–445

Huhta V, Hyvönen R, Kaasalainen P et al (1986) Soil fauna of Finnish coniferous forests. Ann Zool Fennici 23:345–360

Huhta V, Persson T, Setälä H (1998) Functional implications of soil fauna diversity in boreal forests. Appl Soil Ecol 10:277–288

Izadi M, Habashi H (2016) Changes in diversity, biomass and abundance of soil macrofauna, Parrotio-Carpinetum forest at organic and semi-organic horizons. Eurasian J Soil Sci 5:166–171

Izhaki I, Idelovich B, Laster R et al (2009) The impact of macro-vs. micro-environmental factors on the structure of ant communities inhabiting East-Mediterranean Aleppo pine forests. Isr J Entomol 39:129–146

Jeffery S, Gardi C, Jones A et al (eds) (2010) European Atlas of soil biodiversity. European Commission, Publications Office of the European Union, Luxembourg

Kamczyc J, Dyderski MK, Horodecki P et al (2019) Mite communities (Acari, Mesostigmata) in the initially decomposed 'Litter Islands' of 11 tree species in Scots Pine (*Pinus sylvestris* L.). Forest 10. https://doi.org/10.3390/f10050403

Karamaouna M (1990a) On the ecology of the soil macroarthropod community of a Mediterranean pine forest (Sophico, Peloponnese, Greece). Bull Ecol 21:33–42

Karamaouna M (1990b) Aspects of ecology of *Polyxenus lagurus* in Mediterranean conifer formations of Greece (Diplopoda: Penicillata). In: Minelli A (ed) Proceedings of the 7th International Congress of Myriapodology. EJ Brill, Leiden, pp 255–263

Karamaouna M (1992) On the ecology of the iulid millipede *Symphyoiulus impartitus* (Karch) in a Mediterranean pine forest of Greece. Ber nat-med Verein Innsbruck Suppl 10:191–196
Kaynaş BY (2016) Long-term changes in surface-active beetle communities in a post-fire successional gradient in *Pinus brutia* forests. iForest 10:376–382
Khila M, Zaabar W, Bouslama MF et al (2018) Comparison of terrestrial isopod (Crustacea: Oniscidea) assemblages from two preserved areas (Bouhedma and Chambi) in arid regions. Eur Zool J 85:158–168
Knoepp JD, Coleman DC, Crossley DA Jr et al (2000) Biological indices of soil quality: an ecosystem case study of their use. For Ecol Manag 138:357–368
Lavelle P (1997) Faunal activities and soil processes: adaptive strategies that determine ecosystem function. Adv Ecol Res 27:93–132
Lavelle P, Decaens T, Aubert M et al (2006) Soil invertebrates and ecosystem services. Eur J Soil Biol 42:3–15
Legakis A (1986) Comparative studies of the soil arthropods of three ecosystems on Mount Hymettos (Attica, Greece). Biol Gallo-Hellenica 12:371–375
Legakis A (1994) Community structure and species richness in the Mediterranean-type soil fauna. In: Arianoutsou M, Groves RH (eds) Plant-animal interactions in Mediterranean-type ecosystems, Tasks for vegetation science. Springer, Dordrecht, pp 37–45
Lisa C, Paffetti D, Nocentini S et al (2015) Impact of wildfire on the edaphic microarthropod community in a *Pinus pinaster* forest in central Italy. iForest 8:874–883
Lock K, De Bakker D, De Vos B (2001) Centipede communities in the forests of Flanders. Pedobiologia 45:27–35
Marcuzzi G (1968) Osservazioni ecologiche sulla fauna del suolo di alcune regioni forestali italiane. Ann Centro Econ Mont Venezie 7:209–331
Marmari A (1991) Effects of human activities on soil arthropods in a *Pinus halepensis* Mill. ecosystem of N. Euboea (Greece). Doctoral thesis, Faculty of Biology, National and Kapodistrian University of Athens
Menta C, Leoni A, Gardi C et al (2001) Are grasslands important habitats for soil microarthropod conservation? Biodivers Conserv 20:1073–1087
Naveh Z (2000) Mediterranean pine forests facing the challenges of multiple stresses. In: Ne'eman G, Trabaud L (eds) Ecology, biogeography and management of *Pinus halepensis* and *P. brutia* forest ecosystems in the Mediterranean Basin. Backhuys Publisher, Leiden, pp ix–xii
Paz-Ferreiro J, Fu S (2016) Biological indices for soil quality evaluation: perspectives and limitations. Land Degrad Dev 27:14–25
Pitzalis M, Fattorini S, Trucchi E et al (2005) Comparative analysis of species diversity of Isopoda Oniscidea and Collembola communities in burnt and unburnt habitats in central Italy. Ital J Zool 72:127–140
Quézel P (1980) Biogéographie et écologie des conifers sur le pourtour méditerranéen. In: Pesson P (ed) Actualités d'Ecologie Forestière, Bordas, Paris, pp 205–256
Radea C (1989) Study on the litter production, the decomposition rate of organic matter and the arthropod community in ecosystems with *Pinus halepensis* Mill. of insular Greece. Doctoral thesis, Faculty of Biology, National and Kapodistrian University of Athens
Radea C (1993) Environmental factors that influence the temporal variation of predatory macroarthropods in the organic horizon of Mediterranean pine forest. Biol Gallo-Hellenica 20:249–258
Radea C, Mylonas M (1992) Landsnails in the organic horizon of a Mediterranean coniferous forest. Pedobiologia 36:187–192
Romanyà J, Casals P, Cortina J et al (2000) CO_2 efflux from a Mediterranean semi-arid forest soil. II. Effects of soil fauna and surface stoniness. Biogeochemistry 48:283–306
Romero-Alcaraz E, Ávila J (2000) Landscape heterogeneity in relation to variations in epigaeic beetle diversity of a Mediterranean ecosystem. Implications for conservation. Biodivers Conserv 9:985–1005
Sammler S, Voigtländer K, Stoev P et al (2006) New studies on myriapods (Chilopoda, Diplopoda) from Ibiza with a checklist for the Balearic Islands. Norw J Entomol 53:299–309

Savin FA, Pokarzhevskii D, Gongalsky K (2007) Distribution of large soil invertebrates as related to soil parameters. Eurasian Soil Sci 40:64–69
Sgardelis SP, Margaris NS (1993) Effects of fire on soil microarthropods of a phryganic ecosystem. Pedobiologia 37:83–94
Sohlenius B, Boström S (2001) Annual and long-term fluctuations of the nematode fauna in a Swedish Scots pine forest soil. Pedobiologia 45:408–429
Souttou K, Choukri K, Sekour M et al (2015) Ecologie des arthropodes en zone reboisée de Pin d'Alep dans une région présaharienne à Chbika (Djlefa, Algérie). Entomol Faun–Faun Entomol 68:159–172
Stamou GP, Sgardelis S, Margaris NS (1984) Arthropods distribution pattern on a mountain gradient (Mt. Olympus, Greece). Rev Ecol Biol Sol 21:491–505
Swift MJ, Heal OW, Anderson JM (1979) Decomposition in terrestrial ecosystems. Blackwell Scientific, Oxford
Swift MJ, Andren O, Brussaard L et al (1998) Global change, soil biodiversity, and nitrogen cycling in terrestrial ecosystems: three case studies. GCB Bioenergy 4:729–743
Torre I, Bros V, Santos X (2014) Assessing the impact of reforestation on the diversity of Mediterranean terrestrial Gastropoda. Biodivers Conserv 23:2579–2589
Touchi W, Kherbouche-Abrous O, Saadi A et al (2018) Spider communities (Arthropoda, Araneae) in different pine forests of Zéralda Game Reserve (Algiers, Algeria): taxonomy and biodiversity. Rev Ecol (Terre Vie) 73:269–282
Urhan R, Katılmış Y, Kahveci AO (2008) Vertical distribution of soil mites (Acari) in Dalaman (Muğla Prov.-Turkey). Mun Ent Zool 3:333–341
Van Straalen NM (1998) Evaluation of bioindicator systems derived from soil arthropod communities. Appl Soil Ecol 9:429–437
Vardinoyannis K, Demetropoulos S, Mylonas M et al (2012) Terrestrial slugs (Gastropoda, Pulmonata) in the NATURA 2000 areas of Cyprus island. ZooKeys 174:63–77
Wagg C, Bender SF, Widmer F et al (2014) Soil biodiversity and soil community composition determine ecosystem multifunctionality. Proc Natl Acad Sci USA 111:5266–5270
Widenfalk LA, Malmström A, Berg MP et al (2016) Small-scale Collembola community composition in a pine forest soil – overdispersion in functional traits indicates the importance of species interactions. Soil Biol Biochem 103:52–62
Yan S, Singh AN, Fu S et al (2012) A soil fauna index for assessing soil quality. Soil Biol Biochem 47:158–165
Zanella A, Ponge J-F, Briones MJI (2018) Humusica 1, article 8: Terrestrial humus systems and forms – biological activity and soil aggregates, space-time dynamics. App Soil Ecol 122:103–137

Chapter 20
Mycorrhizae in Mediterranean Pine and Mixed Forests

Ana María Rincón, Leticia Pérez-Izquierdo, Sergio de Miguel, and Javier Parladé

20.1 Introduction

Soils are a hidden and dynamic resource of terrestrial ecosystems that host crucial physical, chemical and biological processes (Jackson et al. 2017). Soils are a central component of the global carbon cycle, which has triggered increasing attention to soil ecology as a topical research discipline under the ongoing threat of climate change (Baldrian 2019). In parallel, the importance of belowground, microbial-mediated processes and plant-soil feedbacks in terrestrial ecosystem dynamics is being progressively recognized (Bennett et al. 2017; Pugnaire et al. 2019; Soudzilovskaia et al. 2019; Steidinger et al. 2019). Soil microbial communities represent the trophic base of detrital food webs, drive global carbon and nutrient cycles, improve the structure and water regulation of soils, and establish fundamental interdependent relationships with plants and soil fauna (Baldrian 2017; Bahram et al. 2018). Litter decomposition and soil organic matter mineralization, mainly driven by soil fungi and bacteria, together with biotrophic plant–microbial associations, are key processes and interactions directly involved in the cycling of nutrients

A. M. Rincón (✉)
Institute of Agronomic Sciences – Spanish Research Council (ICA-CSIC), Madrid, Spain
e-mail: ana.rincon@csic.es

L. Pérez-Izquierdo
Swedish University of Agricultural Sciences (SLU), Uppsala, Sweden

S. de Miguel
Department of Crop and Forest Sciences, University de Lleida (UdL), Lleida, Spain

Joint Research Unit CTFC – AGROTECNIO – CERCA, Solsona, Spain

J. Parladé
Institut de Recerca i Tecnologia Agroalimentaries (IRTA), Barcelona, Spain

G. Ne'eman, Y. Osem (eds.), *Pines and Their Mixed Forest Ecosystems in the Mediterranean Basin*, Managing Forest Ecosystems 38,
https://doi.org/10.1007/978-3-030-63625-8_20

and thus, in the productivity of ecosystems (Morris and Blackwood 2007; Smith and Read 2008).

In forest ecosystems, direct and indirect interactions between the aboveground and belowground compartments drive multiple processes and functions (Kardol and Wardle 2010; Bennett et al. 2017). Soil microbiota greatly influence plant communities and *vice versa*, and changes in any of these ecosystem compartments lead to reciprocal shifts in community composition and functioning (Fig. 20.1). Multiple factors such as tree phenology, seasonality or disturbances contribute to the high spatial and temporal heterogeneity in forest ecosystem processes. These processes can occur at timescales from days to many years, affecting above- and belowground compartments and their complex interactions (Baldrian 2017). However, the relative combined effects of above- and belowground biodiversity, or how the environmental conditions and disturbances mediate and influence those relationships in forest ecosystems, are still poorly understood (van der Plas et al. 2016).

Fungi comprise most of the microbial biomass among forest soil microbiota (Baldrian 2017), and display great ecological and taxonomic diversity, e.g., just one gram of forest soil may hold 600–1000 fungal species (Buée et al. 2009; Wu et al.

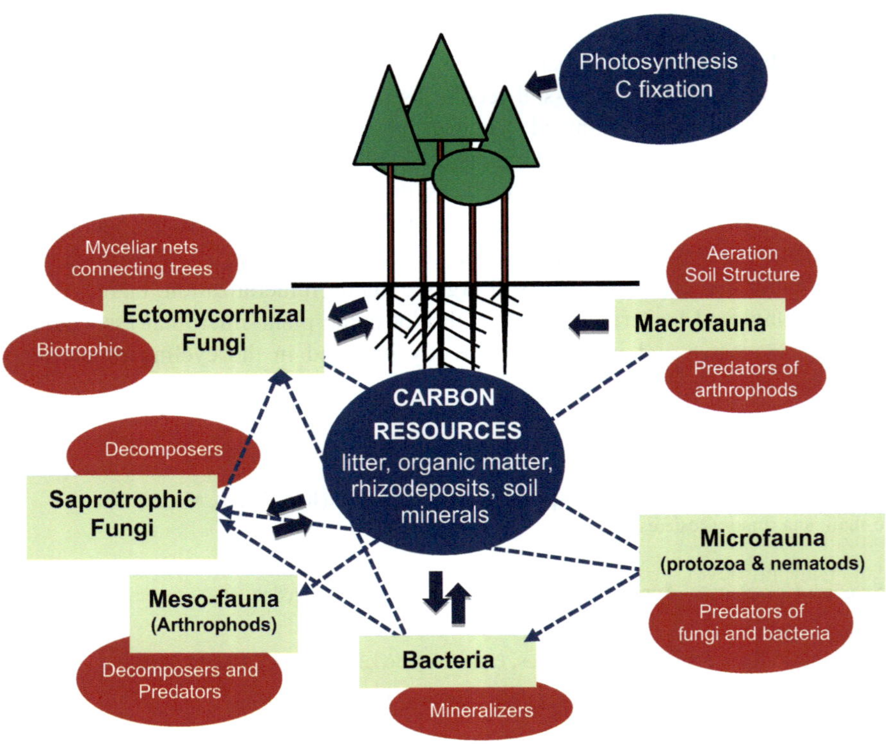

Fig. 20.1 Schematic illustration of aboveground and belowground interactions and processes in forest ecosystems. The great biodiversity of soils and main functional guilds i.e., decomposers, mineralizers, predators and biotrophic mycorrhizal fungi is represented

2019). Continued molecular technological advancement has drastically changed our understanding of this hidden soil biodiversity, although it is still limited compared to what is known about aboveground communities (Bardgett and van der Putten 2014; Wu et al. 2019). At present, the overall diversity of fungi is estimated at 2–4 million species, of which only ~120,000 species have been described to date (Hawksworth and Lücking 2017).

Fungi display key ecological strategies and roles in forest ecosystems as decomposers, mutualists, endophytes and pathogens (Tedersoo and Smith 2013; Zanne et al. 2019). In contrast to pathogens that harm host plants to obtain nutrients, endophytes inhabit host cells, usually providing beneficial effects to plants (Zanne et al. 2019). Saprotrophic fungi are the main decomposers, typically dominating the forest floor surface where they mineralize most of the carbon from dead organic material through the excretion of a potent and diverse array of enzymes (Lindahl et al. 2007; Dighton 2016; Baldrian 2017). Mycorrhizal fungi improve the acquisition of water and nutrients by the host in exchange for photosynthetic carbohydrates, directly affecting the productivity and the response of trees to the surrounding environment (Smith and Read 2008). This mutualistic interaction between the fine roots of plants and mycorrhizal fungi notably spans the aboveground–belowground interface and is particularly critical for plant performance and ecosystem functioning (van der Heijden et al. 2015).

In Mediterranean forest ecosystems, the structure and functioning of both aboveground and belowground communities are highly shaped by the challenging environmental conditions imposed by the Mediterranean climate. Both tree productivity (Benavides et al. 2019) and soil activity (Barba et al. 2018; Flores-Rentería et al. 2018) are strongly constrained by the limited availability and the seasonal and spatial variation of water and nutrients (Sardans and Peñuelas 2013; Collado et al. 2018). Mediterranean forest dynamics are tightly linked to belowground microbial communities, and in particular to mycorrhizal fungal communities, since they directly deliver nutrients and water to trees and mediate interplant connections allowing the efficient redistribution of resources in the ecosystem. This is crucial for increasing the resilience against disturbances commonly found in Mediterranean forest ecosystems, such as fire or drought.

In this chapter, we provide an overview of the key services and functions delivered by mycorrhizal fungi in forest soils, with particular emphasis on Mediterranean pine and mixed forest ecosystems. We place particular focus on mutualistic mycorrhizal fungal communities because all pine species are obligatory mycorrhizal for survival and development and, thus, this symbiosis is critical for the conservation and dynamics of these forests. Furthermore, this close mutualistic association with fungi helps trees to withstand the constrained environmental conditions often found in Mediterranean areas, which is particularly relevant under the current global change context.

20.2 Mycorrhizal Diversity and Functions in Mediterranean Forest Ecosystems

The mycorrhizal symbiosis has evolved independently many times in the plant and fungal kingdoms, and the vast majority of terrestrial plants form this symbiosis (Martin et al. 2016; Spatafora et al. 2017; Tedersoo et al. 2018; Steidinger et al. 2019). Mycorrhizal fungi promote plant nutrition and productivity in return for carbon fixed by photosynthesis, and help plants to cope better with stress (Smith and Read 2008). Potential beneficial effects derived from mycorrhizae usually become more evident under challenging environmental conditions, such as those imposed by the Mediterranean climate (Querejeta et al. 2007; Rincón et al. 2007; Richard et al. 2009; Barea et al. 2011; Ruíz-Gómez et al. 2019).

Up to seven types of mycorrhizae have been described to date, i.e., arbuscular, ectomycorrhizal, ectendomycorrhizal, arbutoid, monotropoid, ericoid and orchid (Smith and Read 2008). Ectomycorrhizae are the dominant type in forests (Steidinger et al. 2019). Ectomycorrhizal fungi colonize the cortical cells of tree root tips forming a highly branched hyphal interface, i.e., the Hartig net, for the bidirectional exchange of nutrients with the plant. The fungus forms a dense hyphal sheath around the root, i.e., the mantle, protecting it from the surrounding environment. The fungal hyphae spread through the surrounding soil forming the external mycelium, which notably increases the root surface area, significantly enlarging the volume of soil that the tree can explore.

Trees can invest up to 15–30% of their primary production to maintaining their associated mycorrhizal fungi in exchange for nutrients (Smith and Read 2008). This can be crucial in Mediterranean forest ecosystems usually constrained by water and nutrient availability (Querejeta et al. 2007; Navarro-Fernández et al. 2016).

Ectomycorrhizal fungi can persist in the soil as free-living vegetative or resistant propagules (e.g., spores, sclerotia), but they are obligate biotrophs dependent on the carbon supplied by the tree to complete their life cycle, i.e., to produce sporocarps and sexual spores (Smith and Read 2008). There are estimates of up to 25,000 different ectomycorrhizal fungal species, although much more fungal diversity remains still cryptic or undescribed (Tedersoo and Smith 2013; Hawksworth and Lücking 2017).

Most ectomycorrhizal research has focused on northern boreal and temperate forests (Uroz et al. 2016; Baldrian 2017), while much less attention has been paid to other forest types such as the Mediterranean. Still, a range of sampling strategies and taxonomic characterisation of fungi e.g., morphological and/or molecular identification of root tips (Pestaña and Santolamazza 2009; Richard et al. 2009; Rincón and Pueyo 2010; Rincón et al. 2014) and/or sporocarps (Oria de Rueda et al. 2010; Pecoraro et al. 2014; Büntgen et al. 2015; Gassibe et al. 2015; Angelini et al. 2016; Collado et al. 2019) can be found in Mediterranean forest surveys, where highly diverse ectomycorrhizal communities have often been reported (Pérez-Izquierdo et al. 2017, 2019; Castaño et al. 2018a; Saitta et al. 2018). Typical ectomycorrhizal fungal communities are hyper-diverse with a few dominant species and many

Fig. 20.2 Ectomycorrizal fungi frequently found in Mediterranean pine forests (**a**) *Rhizopogon roseolus* (Corda) Th. Fr., (**b**) *Lactarius deliciosus* (L. ex Fr.) S.F.Gray, (**c**) *Suillus luteus* (L.: Fries) Gray. Ectomycorrhizae of Mediterranean pine species (**d**) *Pinus pinea* L. and *R. roseolus*, (**e**) *P. pinaster* Ait. and *L. deliciosus*, (**f**) *P. halepensis* Mill. and *Suillus* sp. (Photographs by J. Parladé and A. Rincón)

less-abundant fungi. The ectomycorrhizal symbiosis is relatively species-poor (i.e., at the genus level) (Molina and Trappe 1982), and representative Mediterranean pines such as *Pinus pinaster* Ait. (Pera and Álvarez 1995; Pestaña and Santolamazza 2009; Pérez-Izquierdo et al. 2017, 2020), *P. pinea* L. (Rincón et al. 1999) or *P. halepensis* Mill. (Pérez-Izquierdo et al. 2020) are usually associated with many similar ectomycorrhizal fungal species. Among them, *Rhizopogon* spp., *Suillus* spp., *Tomentella* spp., *Scleroderma* spp., *Russula* spp., or *Lactarius* spp. can be commonly found in these Mediterranean pine forests (Fig. 20.2).

Low-fertility edaphic environments together with water-limiting thermophilic climatic conditions are common ecological features of the habitats where Mediterranean pine species proliferate. These harsh environmental conditions may filter those fungal species with traits that make them capable of better exploiting resources in this specific environment (Treseder and Lennon 2015).

Soil microbial diversity and functioning fluctuate in time (i.e., seasonally, inter-annually) depending on resource availability and temperature and moisture variation (Baldrian 2017; Pérez-Izquierdo et al. 2017; Queralt et al. 2017; Castaño et al. 2018a). These effects can be particularly pronounced in warm and water-limited Mediterranean areas (Scarascia-Mugnozza et al. 2000; Querejeta et al. 2007). In fact, soil moisture is a main factor regulating ectomycorrhizal fungal colonization and fungal community composition (Buée et al. 2005; di Pietro et al. 2007), which can be crucial for Mediterranean forest dynamics, such as natural regeneration and facilitation processes (Richard et al. 2011; Montesinos-Navarro et al. 2019). Because species of ectomycorrhizal fungi can greatly differ in their physiology and ability to promote nutrient uptake and tree productivity, shifts in ectomycorrhizal fungal community composition may have important consequences for host plants and ecosystem functioning in Mediterranean forests (Prieto et al. 2016; Pérez-Izquierdo et al. 2017, 2019).

Besides edaphic-climate conditions, host preferences, inter-specific competition (for both root colonization and nutrients), and host genotype are significant agents structuring the ectomycorrhizal fungal communities (Grogan et al. 2000; Sthultz et al. 2009; Kennedy 2010). Significant differences in the diversity and species composition of ectomycorrhizal fungal communities of *P. pinaster* genotypes differing in their productivity have been reported, e.g., less productive Atlantic genotypes compared with more productive Mediterranean ones (Pérez-Izquierdo et al. 2017, 2019). This study further demonstrated that the variation in ectomycorrhizal fungal diversity associated with *P. pinaster* translated into different abilities of these fungal assemblages to mobilize carbon, nitrogen or phosphorus to pines, highlighting the importance of this symbiosis in nutrient cycling in Mediterranean forests.

The ectomycorrhizal symbiosis plays a crucial role in forest dynamics, especially in Mediterranean ecosystems where plant recruitment processes and outcomes of competitive interactions become particularly critical (Richard et al. 2009; Montesinos-Navarro et al. 2019). A tree root system is usually colonized by many different ectomycorrhizal fungal species and, at the same time, different tree individuals can be linked by shared fungi forming a common mycelial network (Taudier et al. 2015; Pickles and Simard 2017). This has a significant impact on the distribution of resources within the community and can alter the competitive capacity of plants. In Mediterranean ecosystems, seedling establishment can be facilitated by mycorrhizal fungi (Richard et al. 2009; Montesinos-Navarro et al. 2019). Common mycelial networks may also channel signals exchanged by plants, biasing the defence response against herbivores and pathogenic fungi (Johnson and Gilbert 2015).

20.3 Mycorrhizae and Common Stressors in Mediterranean Forest Ecosystems

Mediterranean forest ecosystems are strongly constrained by climate, i.e., high temperature and low precipitation, which usually leads to strong spatial and temporal heterogeneity mainly related to water availability (Valladares et al. 2014). These climatic conditions make forests prone to stressors such as drought or wildfire.

Ectomycorrhizal fungi act as a physical barrier protecting roots from desiccation or pathogen attack and alleviate nutritional and/or water limitations for trees. This symbiosis can provide up to 80% of nutrients to the plants (Kivlin et al. 2013; van der Heijden et al. 2015), and can thus provide a great advantage in the severe Mediterranean environments (Querejeta et al. 2007; Barea et al. 2011; Oliet et al. 2013). Ectomycorrhizal symbiosis has great potential in the restoration of natural ecosystems and disturbed sites (Pickles et al. 2020). Numerous reforestation and afforestation trials with pines inoculated with selected ectomycorrhizal fungi have demonstrated successful results in Mediterranean areas e.g., *P. halepensis* in semi-arid and gypsum soils (Querejeta et al. 1998; El Karkouri et al. 2004; Rincón et al. 2007), or *P. pinea* in formerly arable lands (Parladé et al. 2004).

The Mediterranean region is particularly vulnerable to the effects of climate change, and new climatic scenarios might drive plant-soil-fungal feedbacks in different directions with unknown consequences (Gómez-Aparicio et al. 2017). Climate changes predicted for Mediterranean areas, i.e., increased temperature and annual reduction of rainfall by up of 30%, would probably have significant impacts on ectomycorrhizal fungal communities and their associated vegetation (Ibáñez et al. 2015; Castaño et al. 2018a).

20.3.1 Drought

Drought is among the most important stressors in Mediterranean forests. Long-term evolutionary adaptation of Mediterranean trees and ectomycorrhizal fungi to drought allows them to cope with moderate increases in water limitation without significant losses for ecosystem productivity (Richard et al. 2011; Sardans and Peñuelas 2013). Water limitation affects ectomycorrhizal fungal communities mainly by constraining the primary productivity of trees, decreasing the diffusion of substrates from roots to soil, and by reducing the performance of microbial enzymatic activity and the proliferation of other microbial guilds (Canarini et al. 2016; Fernandez and Kennedy 2016). Different studies have reported highly divergent ectomycorrhizal communities under dry or wet conditions in spatial and temporal Mediterranean gradients with different dominant trees, i.e., *Pinus* spp. (Alday et al. 2017a; Castaño et al. 2018a; Pérez-Izquierdo et al. 2020), *Quercus* spp. (Azul et al. 2010; Scattolin et al. 2014), and mixed stands (Saitta et al. 2018).

Ectomycorrhizal ascomycetes, such as the highly melanised *Cenococcum geophilum* Fr., are commonly found in drought-prone forests (Richard et al. 2005; Querejeta et al. 2009; Gehring et al. 2014; Scattolin et al. 2014; Taniguchi et al. 2018). Fungal traits like melanin or mycelial exploration types and hyphal water repellence have been related to protection of roots from drought (Agerer 2001; Treseder and Lennon 2015). For example, Prieto et al. (2016) reported that ectomycorrhizal fungal species diverging in their hyphal traits differently affected hydraulic redistribution patterns for *P. halepensis*, e.g., *Thelephora terrestris* Ehrh. (short-distance exploration hydrophilic hyphae) enhanced water transfer between pine individuals through common hyphal networks more than *Suillus granulatus* (L.) Roussel (long-distance hydrophobic hyphae). On the other hand, Castaño et al. (2018a) showed that ectomycorrhizal species of the short-distance exploration type increased in relative abundance under dry summer conditions in a *P. pinaster* forest, while long-distance exploration types did so during wet periods.

Acclimation and sensitivity to temperature are also important traits for fungi thriving in Mediterranean systems. For example, *C. geophilum* is able to adapt to temperature increases by altering its hyphal respiration (Malcolm et al. 2008). In fact, it has been proposed that species of the contact and short-distance exploration types, as well as stress-resistant mycorrhizal fungi, e.g., ascomycetes, can be less carbon-demanding from the host and, thus, be favored under drought or other stresses (Gehring et al. 2014; Fernandez et al. 2017; Castaño et al. 2018a; Mueller et al. 2019).

Rainfall exclusion experiments have revealed that fungal communities in Mediterranean forests dynamically adapt to drought without losing diversity, but shift the composition of species in the community, including the variation in relative abundance of dominant fungal taxa (Richard et al. 2011; Yuste et al. 2011; Bastida et al. 2019). Similarly, decreases in ectomycorrhizal formation and fungal diversity during dry periods have been observed in a conifer-oak mixed forest with a Mediterranean climate (Taniguchi et al. 2018). In Mediterranean ecosystems, the hydraulic lift, i.e., the mechanism by which deeper tree roots with access to groundwater redistribute water to the surface under dry conditions seems to be the key for trees to maintain their fungal symbionts in exchange for nutrients (Allen 2007; Querejeta 2017). Different studies suggest that preferential movement of hydraulically redistributed water from adults through ectomycorrhizal hyphal networks can enhance seedling survival during drought (Egerton-Warburton et al. 2007; Querejeta et al. 2007; Warren et al. 2008; Prieto et al. 2016). Drought-driven shifts in ectomycorrhizal community composition could notably affect these interplant water transfers (Prieto et al. 2016; Querejeta 2017).

20.3.2 Fire

In Mediterranean ecosystems, drought is intimately associated with wildfire activity. In fact, the climatic conditions make these ecosystems some of the most fire-prone in the world (Keeley et al. 2011). Mediterranean vegetation, with pines as a good example, has evolved adaptive strategies to fire, fostering the high resilience capacity of these forest ecosystems (Pausas and Verdú 2008). The rise in temperature and decrease in rainfall forecasted by climate change models are expected to increase the risk and intensity of wildfires in Mediterranean areas with unknown consequences for these ecosystems (Lindner et al. 2010).

Post-fire secondary forest succession depends on multiple factors such as the identity of previous tree species, fire intensity, seed bank availability, and the response of soil microbial communities including the ectomycorrhizal fungi (Hart et al. 2005). In fact, typical Mediterranean trees such as pines and oaks are obligate ectomycorrhizal species, i.e., they depend on the fungal inoculum for establishment and survival. The structure and successional dynamics of these forests is also highly dependent on the type of fire, fire intensity and severity, as well as fire frequency. Buscardo et al. (2010, 2012) have suggested that high fire frequencies in the

Mediterranean region could delay the ectomycorrhizal fungal re-colonization process. Indeed, Pérez-Izquierdo et al. (2020) have shown that mature *P. pinaster* and *P. halepensis* forests with low fire recurrence history have higher local and regional ectomycorrhizal fungal diversity and different fungal species composition than forests under a high fire frequency regime.

Fire impacts the ectomycorrhizal communities by direct combustion of fungal biomass, through vegetation damage and by altered soil properties, such as loss of the organic soil layer where most ectomycorrhizal fungi develop and alterations of pH and nutrient availability (Hart et al. 2005; Rincón and Pueyo 2010; Buscardo et al. 2015). Following fire, the persistence of viable ectomycorrhizal inoculum is critical for facilitating natural regeneration and succession processes, and fungal re-colonization mainly occurs from wind-, or animal-borne spores from adjacent unburned forest areas and from resistant fungal propagules surviving in deeper mineral soil layers (Torres and Honrubia, 1997; Taylor and Bruns 1999; Dahlberg 2002; Rincón et al. 2014; Glassman et al. 2016; Livne-Luzon et al. 2017). In turn, post-fire ectomycorrhizal fungi are highly dependent on the availability of roots to establish symbiosis (Grogan et al. 2000). In this sense, pioneer shrubs can represent an important inoculum source for post-fire emerging seedlings, acting as potential bridges facilitating the establishment of new ectomycorrhizal fungal networks (Kipfer et al. 2011; Buscardo et al. 2012; Raudabaugh et al. 2020); examples include *Quercus* spp., *Cistus* spp. and *Pinus* spp. in Mediterranean areas (Martín-Pinto et al. 2006; Mediavilla et al. 2014; Buscardo et al. 2015).

Fire usually reduces the mycorrhizal colonization of trees as well as fungal species richness (Dove and Hart 2017). However, seedlings of fire-adapted Mediterranean pines such as *P. pinaster* and *P. halepensis*, with great post-fire germination rates, have shown high ectomycorrhizal colonization levels after fire (Torres and Honrubia 1997; Rincón and Pueyo 2010). More generally, fire drives drastic disruptions of the ectomycorrhizal fungal species composition, usually leading to a simplified community dominated by a relatively small number of pioneer fungi, usually ascomycetes, in fire-affected Mediterranean forests (Torres and Honrubia 1997; de Román and de Miguel 2005; Martín-Pinto et al. 2006; Rincón and Pueyo 2010; Rincón et al. 2014; Buscardo et al. 2015; Vásquez-Gassibe et al. 2016; Pérez-Izquierdo et al. 2020). Generalist fungi such as species of *Tomentella* or *Wilcoxinia* are commonly found associated with *P. pinaster, P. halepensis* and oak hosts (de Román and de Miguel 2005; Buscardo et al. 2010; Rincón et al. 2014; Pérez-Izquierdo et al. 2020), whereas typical late-stage species e.g., *Russula, Lactarius, Cortinarius*, and *Amanita* are usually lost after fire (Martín-Pinto et al. 2006; Rincón and Pueyo 2010). Fungal traits, such as the production of abundant fruiting bodies and spores that can remain in soil spore banks (Glassman et al. 2015), or long-distance exploration type mycelium, can be significantly advantaged after fire. A good example is *Rhizopogon* spp., which typically dominate the ectomycorrhizal fungal communities of different pine species after fires (Baar et al.

1999; Taylor and Bruns 1999; Izzo et al. 2006; Buscardo et al. 2010; Rincón and Pueyo 2010; Rincón et al. 2014; Franco-Manchón et al. 2019).

20.3.3 Other Stressors

The mycorrhizal symbiosis generally improves the nutritional and physiological status of plants, enhancing their resistance to a range of stresses such as drought, soil degradation (e.g., fire or heavy metals), insect outbreaks and pathogens among others (van der Heijden et al. 2015). Mechanisms such as direct competition and inhibition are likely operating, for example, ectomycorrhizal strains of *R. roseolus*, *S. luteus* or *L. deliciosus* have been demonstrated to be successful as biological agents against *Fusarium* under *in vitro* conditions (Olaizola Suárez et al. 2018a), and when it causes damping off on *P. pinea* seedlings in the nursery (Mateos et al. 2017; Olaizola Suárez et al. 2018b).

Defoliation of *P. pinaster* (Pestaña and Santolamazza 2011) or *P. edulis* (Gehring and Whitham 1991) by insect infestation triggers, in both cases, shifts in their associated ectomycorrhizal fungal communities, probably in relation to reduced photosynthetic capacity of the trees. Additionally, tree decline and mortality can also affect ectomycorrhizal fungal communities by inducing changes in the abiotic conditions e.g., light availability, soil fertility and nutrient availability (Breshears et al. 2018), which could further result in the displacement of ectomycorrhizal fungi by saprotrophic ones (Štursová et al. 2014).

Reports of increased forest decline and tree mortality driven by different global change factors are becoming more frequent (Breshears et al. 2018). This is a complex phenomenon in which multiple abiotic and/or biotic factors probably concur and even interact. Decreases in tree health and cover have great potential to affect ectomycorrhizal communities and plant-soil feedbacks (Ibáñez et al. 2015; Mueller et al. 2019), although this a very poorly explored issue, particularly in Mediterranean ecosystems (Gómez-Aparicio et al. 2017; Pugnaire et al. 2019). In the Mediterranean region the oomycete, *Phytophthora* sp., has been identified as the main biotic cause of oak decline. Differences in the relative abundance of ectomycorrhizal fungal species have been observed between healthy and *Phytophthora*-infected chestnut forests in Italy, with *C. geophilum* prevailing in infected trees (Blom et al. 2009; Scattolin et al. 2012). Lower percentages of ectomycorrhizal root-tips and reduced abundance of certain morphotypes such as *Russula* have been observed in declining *Quercus ilex* L. trees affected by the soil-borne pathogen *Phytophtora cinammomi* (Corcobado et al. 2014). In contrast, ectomycorrhizal fungi have been shown to protect seedlings from *P. cinammomi*, i.e., these seedlings survived and grew better than non-mycorrhizal ones (Azul et al. 2014; Corcobado et al. 2015).

20.4 Forest Management and Structure in Relation to Ecosystem Services Provided by Mycorrhizal Fungi

Forests provide multiple provisioning (timber, mushrooms, food, fuel), cultural (recreation space, landscape), regulatory (carbon sequestration, air and water purification) and supporting (soil formation, primary production, nutrient cycling) ecosystem services (MEA 2005; Bakker et al. 2019). In a simplified scheme, ecosystem services are generated through ecosystem functions, which in turn are supported by biophysical processes and driven by biodiversity that can be considered the foundation of all ecosystem services. The fact that only a few services, such as wood or food, have real market value (Seppelt et al. 2012), may cause degradation of other goods without a market price, as is the case with biodiversity (Gómez-Baggethun et al. 2010). Mediterranean ecosystems are hotspots of biodiversity, whose ecological (and economic) value deserves greater recognition and protection (Lecina-Díaz et al. 2019). Increased awareness of the relationship between biodiversity (above- and belowground) and ecosystem functions and services is essential to their preservation (Rodríguez-Loinaz et al. 2015), and to the efficient management of forest ecosystems.

Soil fungal diversity promotes multiple ecosystem functions, e.g., decomposition, mineralization, and nutrient cycling, ensuring the delivery of key regulatory ecosystem services in forests (Bardgett and van der Putten 2014; Delgado-Baquerizo et al. 2016; Bakker et al. 2019) (Fig. 20.3). Fungal biomass can comprise a substantial portion of the carbon pool in soils (Clemmensen et al. 2013) and it also represents an important food source for macro- and micro-fauna, e.g., collembola (Heděnec et al. 2013). Additionally, edible fungal sporocarps provide highly appreciated provisioning and cultural ecosystem services (Tomao et al. 2017).

Many of the most widely appreciated edible mushrooms (e.g., *Boletus, Lactarius* and *Tuber* species) are ectomycorrhizal. The relationship between fungal sporocarp production and tree growth has been established using dendrochronological techniques (Büntgen et al. 2013, 2015), but the interaction between fungal diversity and tree growth and its consequent carbon fixation is still to be explored. In plots of *P. sylvestris* L., Bonet et al. (2004) observed greater fungal diversity related to higher productivity of edible mushrooms.

In addition to the economic value of fungi as a food resource, mushroom harvesting is a cultural ecosystem service highly valued in the Mediterranean region (Martínez-Peña et al. 2011; Górriz et al. 2014), for which collectors may be willing to pay fees as long as the benefits have an impact on improving ecosystem management (Bonet et al. 2015). The design of a multifunctional lifeline that includes integration into management regimes is one of the strategic research priorities set by the Mediterranean Forest Research Agenda (Palahí and Birot 2009). The results obtained by Palahí et al. (2009), Aldea et al. (2014) and de-Miguel et al. (2014) show that it is possible to design a silviculture regime that integrates different ecosystem services through the use of numerical optimization techniques applied to mycological production and tree growth models. Numerical optimization of models

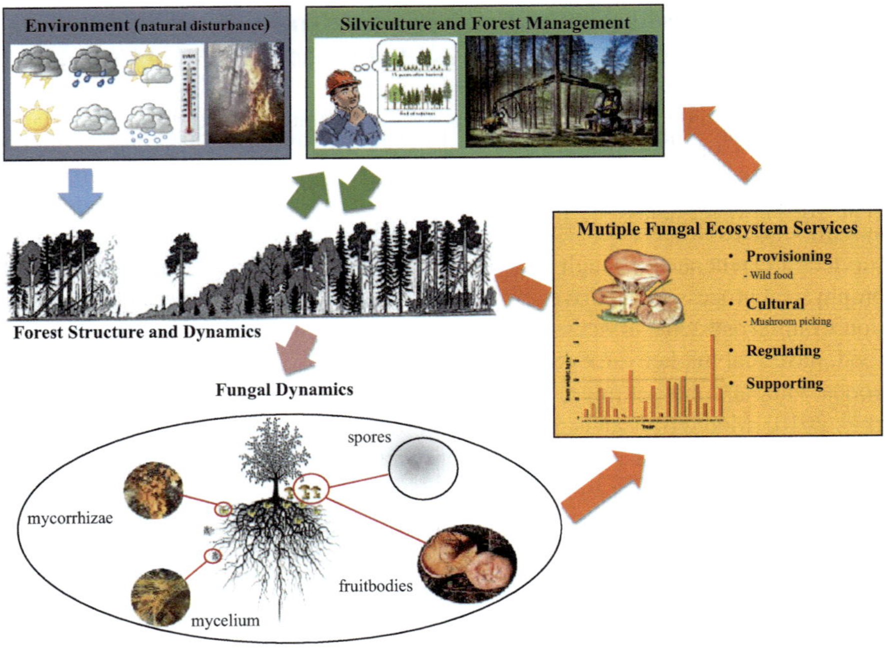

Fig. 20.3 Factors influencing above- and belowground forest dynamics, and the ecosystem services provided by fungi in forest ecosystems

using combined heuristic algorithms will allow development of different planning models suitable for the simulation of forest management alternatives for end users (Palahí et al. 2009).

Previous research has documented a positive relationship between mycorrhizal fungal diversity and stand attributes such as canopy cover, stand basal area and tree species composition (Bonet et al. 2010). Existing scientific knowledge about the effects of forest management practices on ectomycorrhizal fungi suggests that excessively intense silvicultural operations can have a negative impact on the diversity of ectomycorrhizal fungi, at least in the short term (Paillet et al. 2010; Goldmann et al. 2015; Lewandowski et al. 2015). Still, recent findings by Castaño et al. (2018b) from a thinning experiment conducted without causing soil disturbance in Mediterranean pine forests, found that belowground fungal diversity and community composition, including mycorrhizal fungi, were not significantly impaired by a rather wide range of forest thinning intensities. This suggests that belowground mycorrhizal fungal communities may also be able to resist relatively intense forest management regimes in the absence of severe soil disturbance. Therefore, forestry practices such as low-impact logging, enhancing stand structural diversity and complexity and promoting landscape heterogeneity and connectivity at both the stand and landscape levels, can contribute to maintaining mycorrhizal fungal diversity and conserving forest fungi (Tomao et al. 2020).

Forest stand structure can also influence the aboveground productivity (i.e., mushroom yield) of certain mycorrhizal fungal species. Whether this effect is positive or negative will depend on different features such as the ecology of each fungal species and guild, the forest type and the management regime (Tomao et al. 2017). A considerable proportion of the global research efforts devoted to understanding the links between forest stand structure, management and mushroom productivity have been conducted in Mediterranean forest areas. Although climatic and weather conditions (mainly precipitation and temperature), together with soil and site characteristics, represent key drivers of mushroom yield and diversity in water-limited Mediterranean pine forest ecosystems and scrublands (Hernández-Rodríguez et al. 2015; Taye et al. 2016; Alday et al. 2017a, b; Karavani et al. 2018), several studies have also reported that stand basal area, age and canopy cover are key forest stand structural features also related to mushroom productivity of edible mycorrhizal fungi of socioeconomic interest. Thus, stand basal areas ranging from 10–15 to 40–45 m^2 ha^{-1}, depending on the tree species and growing conditions, have been reported as optimal for maximizing edible and commercial mushroom yield of (mostly) ectomycorhizal fungi in pure and mixed pine forest stands (Bonet et al. 2009; Martínez-Peña et al. 2012; de-Miguel et al. 2014; Sánchez-González et al. 2019) (Fig. 20.4). These results are consistent with findings for other biomes and species, such as *Picea abies* (L.) H. Karst in boreal forests (Tahvanainen et al. 2016). Stand age and canopy cover have also been linked to the productivity of mycorrhizal mushrooms of socioeconomic importance in pine ecosystems and scrublands of Mediterranean countries. Thus, maximum yields of *Lactarius* group *deliciosus* and *Boletus* group *edulis* mushrooms seem to occur in relatively young stands and/or in the absence of complete canopy closure in different forest ecosystem types (Salerni and Perini 2004; Fernández-Toirán et al. 2006; Martínez de Aragón et al. 2007; Ortega-Martínez et al. 2011; Martínez-Peña et al. 2012; Ágreda et al. 2014; Hernández-Rodríguez et al. 2015; Taye et al. 2016; Küçüker and Baskent 2018).

The reasons behind the correlations between certain forest stand characteristics and mycorrhizal mushroom yields may be twofold. On one hand, changes in stand structure may contribute to modifying the within-stand microclimatic conditions driving mushroom fruiting, e.g., soil moisture, temperature, radiation reaching the soil, relative humidity. These changes may affect the fruiting patterns not only of mycorrhizal fungi but also of other fungal guilds such as saprotrophs. However, in the particular case of mycorrhizal fungi, different stand structures may also be related to different rates of carbohydrate allocation from the host trees to their fungal symbionts. In this regard, it seems logical that mushroom production of some ectomycorrhizal species is maximized in rather young and open pine stands since these represent stand development stages where the growth of Mediterranean pine trees (and, therefore, carbon uptake by trees) is also maximized (Primicia et al. 2016; Collado et al. 2018, 2019). Similarly, the reported optimal ranges of stand basal areas for mushroom production also tend to match those stand structures where timber production (and tree growth) is also maximized. From the perspective of the supply of multiple ecosystem services from Mediterranean pine forests,

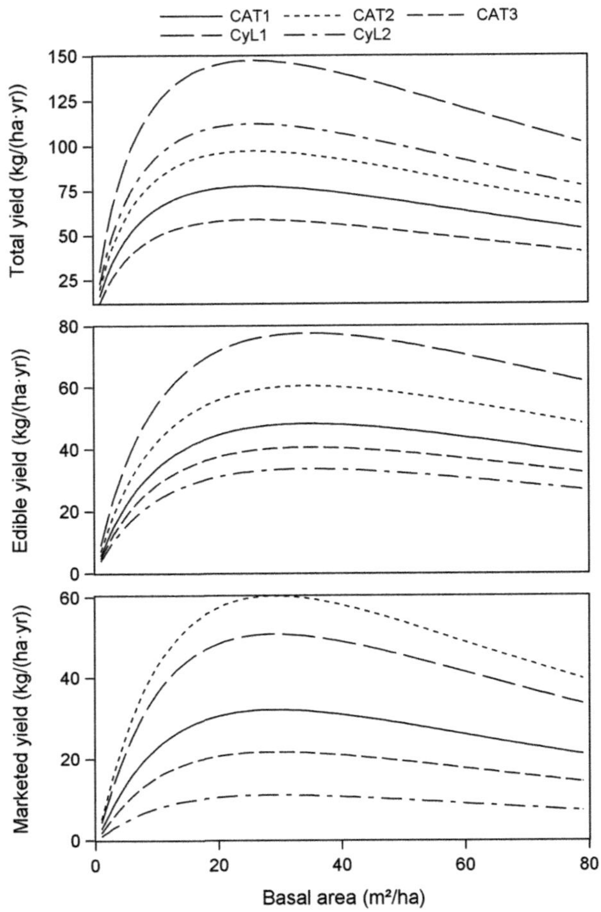

Fig. 20.4 Relationship between stand basal area and mushroom yield in *Pinus pinaster* Ait. and *P. sylvestris* L. forests of northern Spain ("CAT" refers to different areas of Catalonia, northeastern Spain, while "CyL" refers to Castilla y León, central and northwestern Spain). The relationship follows an increasing-decreasing trend suggesting a range of optimal stand basal areas that maximize the overall productivity of mushrooms as well as of the group of edible and marketed species. (From Sánchez-González et al. 2019)

promoting such forest structures through forest management can therefore contribute to maintaining the productivity of mycorrhizal mushrooms of socioeconomic interest in addition to increased tree growth due to reduced tree competition for water and nutrients. In a climate change context, such structures may also contribute to reducing wildfire risk and facilitating the adaptation of forest ecosystems to increasingly arid conditions (de-Miguel et al. 2014).

Non-timber forest products, such as edible mushrooms, have not typically been included in forest management where timber production is the main objective. However, in Mediterranean forests, wild edible mushrooms can reach a significant level of production, which may exceed 4–10 times the value of timber production (Palahí et al. 2009; Aldea et al. 2012; Frutos et al. 2012). Thus, the production of edible fungi in managed pine forests is of paramount importance for the economic development of Mediterranean forest areas where mycological tourism, restoration and picking regulated by permits can generate complementary revenues. Consequently, the current trend in forest management planning is to make non-wood forest products and their related ecosystem services (carbon sequestration, soil protection, water production, biodiversity, ecosystem regulation and recreation use) compatible with timber products (Küçüker and Baskent 2017). For instance, the removal of photosynthetic host trees, which are the main energy sources for sporocarp production, may cause a decrease in ectomycorrhizal fungi in the short term (Amaranthus et al. 1994). In one study, tree cutting at several intensities caused a sharp decrease in *Boletus edulis* Bull. soil mycelial biomass in *P. sylvestris* stands in Spain, and no recovery was observed 3 years after tree cutting (Parladé et al. 2017). Other studies showed that moderate stand thinning conducted without causing soil disturbance produced a temporal increase in sporocarp fruiting of certain species such as *Lactarius* spp. (Bonet et al. 2012; Collado et al. 2018). Parladé et al. (2019) reported a strong effect of tree harvest on the relative abundance of ectomycorrhizal fungi during the first years after clear-cutting of a *P. sylvestris* forest in central Spain. However, the levels of fungal diversity were comparable to the undisturbed forest, thus suggesting potential further recovery of ectomycorrhizal fungi through colonization of the regenerated seedlings. A decline in edible sporocarps was observed in a *P. sylvestris* forest in Spain during 20 years of monitoring, correlating with periods of summer drought, warmer temperatures and declining forest growth (Büntgen et al. 2015). Inventories of truffle yield from Spain, France and Italy, showed that winter truffle harvests significantly depend on previous summer rainfall, which is variable (Büntgen et al. 2015).

When focusing on ectomycorrhizal fungi in the Mediterranean area, the valuable black truffle (Tuber melanosporum Vittad.) merits special mention. This fungus is among the most appreciated edible fungi worldwide, and it is naturally produced mainly in France, Italy and Spain, where it has also been traditionally cultivated (Bonet et al. 2009; Le Tacon et al. 2014; Murat 2015; Zambonelli et al. 2015; Chen et al. 2016). Due to its economic importance, cultivation of the black truffle has expanded in recent decades across all Mediterranean climate regions, including Australia, New Zealand, Chile, North America, and South Africa (Reyna and García-Barreda 2009, 2014). The black truffle has a wide range of potential ectomycorrhizal tree host species (Chen et al. 2016), including pines such as P. sylvestris and P. nigra (García-Montero et al. 2007), although it is most often found naturally associated with oaks (e.g. Quercus ilex L., Q. faginea Lam., Q. coccifera L.) in mixed stands. Truffle cultivation is, thus, being increasingly considered as a promising economic policy tool for rural development in Mediterranean regions. It provides multiple ecosystem services and promotes the involvement of the rural communities in the protection and management of typically Mediterranean forest ecosystems, often characterized by rather low economic profitability based on timber products and highly vulnerable to climate change.

20.5 Conclusions

It is well accepted that biodiversity promotes multiple ecosystem functions ensuring the delivery of benefits essential for humans. Managed Mediterranean forest and agro-forest systems can play an important role in the delivery of global ecosystem services (Burkhard et al. 2016). Enhancing non-timber forest products, such as edible mushrooms, may represent an excellent management strategy for Mediterranean forests. Moreover, the coexistence and functional complementarity of different mycorrhizal plant species, as well as beneficial microorganisms and entomofauna, can be modified to design sustainable and economically viable multiproductive agroforestry systems. It is necessary to develop strategies oriented to sustainable and efficient exploitation models through informed management actions based on knowledge and innovation.

Acknowledgements We thank funding support from the Ministerio de Ciencia y Universidades of Spain, grants "TUBERSYSTEMS" RTI2018093907-B-C21-C22 and "FUMANCHU" RTI2018-099315-A-I00.

References

Agerer R (2001) Exploration types of ectomycorrhizae. A proposal to classify extomycorrhizal mycelial systems according to their patterns of differentiation and putative ecological importance. Mycorrhiza 11:107–114

Ágreda T, Cisneros Ó, Águeda B et al (2014) Age class influence on the yield of edible fungi in a managed Mediterranean forest. Mycorrhiza 24(2):143–152

Alday JG, de Aragón JM, de-Miguel S et al (2017a) Mushroom biomass and diversity are driven by different spatio-temporal scales along Mediterranean elevation gradients. Sci Rep 7:45824

Alday JG, Bonet JA, Oria-de-Rueda JA et al (2017b) Record breaking mushroom yields in Spain. Fungal Ecol 26:144–146

Aldea J, Martínez-Peña F, Díaz-Balteiro L (2012) Integration of fungal production in forest management using a multi-criteria method. Eur J For Res 131:1991–2003

Aldea J, Martínez-Peña F, Romero C et al (2014) Participatory goal programming in forest management: an application integrating several ecosystem services. Forests 5(12):3352–3371

Allen MF (2007) Mycorrhizal fungi: highways for water and nutrients in arid soils. Vadose Zone J 6:291–297

Amaranthus M, Trappe JM, Bednar L et al (1994) Hypogeous fungal production in mature Douglas-fir forest fragments and surrounding plantations and its relation to coarse woody debris and animal mycophagy. Can J For Res 24:2157–2165

Angelini P, Compagno R, Arcangeli A et al (2016) Macrofungal diversity and ecology in two Mediterranean forest ecosystems. Plant Biosyst 150(3):540–549

Azul AM, Soussa JP, Agerer R et al (2010) Land use practices and ectomycorrhizal fungal communities from oak woodlands dominated by *Quercus suber* L. considering drought scenarios. Mycorrhiza 20(2):73–88

Azul AM, Nunes J, Ferreira I et al (2014) Valuing native ectomycorrhizal fungi as a Mediterranean forestry component for sustainable and innovative solutions. Botany 92:161–171

Baar J, Horton TR, Kretzer AM et al (1999) Mycorrhizal colonization of *Pinus muricata* from resistant propagules after a stand-replacing wildfire. New Phytol 143:409–418

Bahram M, Hildebrand F, Forslund SK et al (2018) Structure and function of the global topsoil microbiome. Nature 560:233–237

Bakker MR, Brunner I, Ashwood F et al (2019) Belowground biodiversity relates positively to ecosystem services of European forests. Front For Glob Chang 2:6

Baldrian P (2017) Microbial activity and the dynamics of ecosystem processes in forest soils. Curr Opin Microbiol 37:128–134

Baldrian P (2019) The known and the unknown in soil microbial ecology. FEMS Microbiol Ecol 95:fiz005

Barba J, Lloret F, Poyatos R et al (2018) Multi-temporal influence of vegetation on soil respiration in a drought affected forest. iForest Biogeosci For 11:189–198

Bardgett RD, van der Putten WH (2014) Belowground biodiversity and ecosystem functioning. Nature 515:505–511

Barea JM, Pelenzuela J, Cornejo P et al (2011) Ecological and functional roles of mycorrhizas in semi-arid ecosystems of Southeast Spain. J Arid Environ 75:1292–1301

Bastida F, López-Mondéjar R, Baldrian P et al (2019) When drought meets forest management: effects on the soil microbial community of a Holm oak forest ecosystem. Sci Total Environ 662:276–286

Benavides R, Scherer-Lorenzen M, Valladares F (2019) The functional trait space of tree species is influenced by the species richness of the canopy and the type of forest. Oikos 128:1435–1445

Bennett JA, Maherali H, Reinhart K et al (2017) Plant-soil feedbacks and mycorrhizal type influence temperate forest population dynamics. Science 355(6321):181–184

Blom JM, Vannini A, Vettraino AM et al (2009) Ectomycorrhizal community structure in a healthy and a *Phytophthora*-infected chestnut (*Castanea sativa* Mill.) stand in Central Italy. Mycorrhiza 20:25–38

Bonet JA, Fischer CR, Colinas C (2004) The relationship between orientation and forest age on the production of sporocarps of ectomycorrhizal fungi in *Pinus sylvestris* forests of the Central Pyrenees. For Ecol Manag 203:157–175

Bonet JA, Oliach D, Fischer CR et al (2009) Cultivation methods of the black truffle, the most profitable Mediterranean non-wood forest product; a state of the art review. In: Palahi M, Birot Y, Bravo F et al (eds) Modelling, valuing, and managing Mediterranean forest ecosystems for non-timber goods and services, vol 57. EFI Proceedings, pp 57–71

Bonet JA, Palahí M, Colinas C et al (2010) Modelling the production and species richness of wild mushrooms in pine forests of the Central Pyrenees in northeastern Spain. Can J For Res 40:347–356

Bonet JA, de-Miguel S, Martínez de Aragón J et al (2012) Immediate effect of thinning on the yield of *Lactarius* group *deliciosus* in *Pinus pinaster* forests in North-Eastern Spain. For Ecol Manag 265:211–217

Bonet JA, Górriz E, Evard C et al (2015) El projecte StarTree emfatitza la importancia dels productes forestals no fustaners al Mediterrani. Silvicultura 71:48–50

Breshears DD, Carroll CJW, Redmond MD et al (2018) A dirty dozen ways to die: metrics and modifiers of mortality driven by drought and warming for a tree species. Front For Glob Chang 1:4

Buée M, Vairelles D, Garbaye J (2005) Year-round monitoring of diversity and potential metabolic activity of the ectomycorrhizal community in a beech (*Fagus silvatica*) forest subjected to two thinning regimes. Mycorrhiza 15(4):235–245

Buée M, Reich M, Murat C et al (2009) 454 Pyrosequencing analyses of forest soils reveal an unexpectedly high fungal diversity. New Phytol 184:449–456

Büntgen U, Martínez-Peña F, Aldea J et al (2013) Declining pine growth in Central Spain coincides with increasing diurnal temperature range since the 1970s. Glob Planet Chang 107:177–185

Büntgen U, Egli S, Diego Galván J et al (2015) Drought-induced long-term changes in the phenology, productivity and diversity of Spanish mushrooms. Fungal Ecol 16:6–18

Burkhard B, Hotes S, Wiggering H (2016) Agro(Eco)system services-supply and demand from fields to society. Land 5:1–4

Buscardo E, Rodríguez-Echeverría S, Martín MP et al (2010) Impact of wildfire return interval on the ectomycorrhizal resistant propagules communities of a Mediterranean open forest. Fungal Biol 114:628–636

Buscardo E, Rodríguez-Echeverría S, Barrico L et al (2012) Is the potential for the formation of common mycorrhizal networks influenced by fire frequency? Soil Biol Biochem 46:136–144

Buscardo E, Rodriguez-Echeverria S, Freitas H et al (2015) Contrasting soil fungal communities in Mediterranean pine forests subjected to different wildfire frequencies. Fungal Div 70:85–99

Canarini A, Merchant A, Dijkstra FA (2016) Drought effects on *Helianthus annuus* and *Glycine max* metabolites: from phloem to root exudates. Rhizosphere 2:85–97

Castaño C, Lindahl BD, Alday J et al (2018a) Soil microclimate changes affect soil fungal communities in a Mediterranean pine forest. New Phytol 220(4):1211–1221

Castaño C, Alday JG, Lindahl BD et al (2018b) Lack of thinning effects over inter-annual changes in soil fungal community and diversity in a Mediterranean pine forest. For Ecol Manag 424:420–427

Chen J, Murat C, Oviatt P et al (2016) The black truffles *Tuber melanosporum* and *Tuber indicum*. In: Zambonelli A, Iotti M, Murat C (eds) True truffle (*Tuber* spp.) in the world: soil ecology, systematics and biochemistry. Springer Soil Biology Series, pp 19–32

Clemmensen KE, Bahr A, Ovaskainen O et al (2013) Roots and associated fungi drive long-term carbon sequestration in boreal forest. Science 339:1615

Collado E, Camarero JJ, Martínez de Aragón J et al (2018) Linking fungal dynamics, tree growth and forest management in a Mediterranean pine ecosystem. For Ecol Manag 422:223–232

Collado E, Bonet JA, Camarero JJ et al (2019) Mushroom productivity trends in relation to tree growth and climate across different European forest biomes. Sci Total Environ 689(1):602–615

Corcobado T, Vivas M, Moreno G et al (2014) Ectomycorrhizal symbiosis in declining and non-declining *Quercus ilex* trees infected with or free of *Phytophthora cinnamomi*. For Ecol Manag 324:72–80

Corcobado T, Moreno G, Azul AM (2015) Seasonal variations of ectomycorrhizal communities in declining *Quercus ilex* forests: interactions with topography, tree health status and *Phytophthora cinnamomi* infections. Forestry 88:257–266

Dahlberg A (2002) Effects of fire on ectomycorrhizal fungi in fennoscandian boreal forests. Silva Fennica 36:69–80

de Román M, de Miguel AM (2005) Post-fire, seasonal and annual dynamics of the ectomycorrhizal community in a *Quercus ilex* L. forest over a 3-year period. Mycorrhiza 15:471–482

Delgado-Baquerizo M, Maestre FT, Reich PB et al (2016) Microbial diversity drives multifunctionality in terrestrial ecosystems. Nat Commun 7(10541):1–8

de-Miguel S, Bonet JA, Pukkala T et al (2014) Impact of forest management intensity on landscape-level mushroom productivity: a regional model-based scenario analysis. For Ecol Manag 330:218–227

di Pietro M, Churin JL, Garbaye J (2007) Differential ability of ectomycorrhizas to survive drying. Mycorrhiza 17(6):547–550

Dighton J (2016) Fungi in ecosystem processes, 2nd edn. CRC Press, Boca Raton, 431 pp

Dove NC, Hart SC (2017) Fire reduces fungal species richness and in situ mycorrhizal colonization: a meta-analysis. Fire Ecol 13(2):37–65

Egerton-Warburton LM, Querejeta JI, Allen MF (2007) Common mycorrhizal networks provide a potential pathway for the transfer of hydraulically lifted water between plants. J Exp Bot 58:1473–1483

El Karkouri K, Martin F, Mousain D (2004) Diversity of ectomycorrhizal symbionts in a disturbed *Pinus halepensis* plantation in the Mediterranean region. Ann For Sci 61:705–710

Fernandez C, Kennedy P (2016) Revisiting the Gadgil effect: do interguild fungal interactions control carbon cycling in forest soils ? New Phytol 209:1382–1394

Fernandez CW, Nguyen NH, Stefanski A et al (2017) Ectomycorrhizal fungal response to warming is linked to poor host performance at the boreal-temperate ecotone. Glob Chang Biol 23:1598–1609

Fernández-Toirán LM, Ágreda T, Olano JM (2006) Stand age and sampling year effect on the fungal fruit body community in *Pinus pinaster* forests in Central Spain. Botany 84(8):1249–1258

Flores-Rentería D, Curiel J, Valladares F et al (2018) Soil legacies determine the resistance of an experimental plant-soil system to drought. Catena 166:271–278

Franco-Manchón I, Salo K, Oria-de-Rueda JA et al (2019) Are wildfires a threat to fungi in European *Pinus* forests? A case study of boreal and Mediterranean forests. Forests 10:1–12

Frutos P, Martínez-Peña F, Esteban S (2012) Edible wild mushroom tourism as a source of income and employment in rural areas. The case of Castilla y León. For Syst 21(1):81–98

García-Montero L, Manjón JL, Martín-Fernández S et al (2007) Problems of using pines in *Tuber melanosporum* culture: soils and truffle harvest associated with *Pinus nigra* and *P. sylvestris*. Agrofor Syst 70:243–249

Gassibe PV, Oria de Rueda JA, Martín-Pinto P (2015) *P. pinaster* under extreme ecological conditions provides high fungal production and diversity. For Ecol Manag 337:161–173

Gehring CA, Whitham TG (1991) Herbivore-driven mycorrhizal mutualism in insect-susceptible pinyon pine. Nature 353:556–557

Gehring C, Flores-Rentería D, Sthultz CM et al (2014) Plant genetics and interspecific competitive interactions determine ectomycorrhizal fungal community responses to climate change. Mol Ecol 23:1379–1391

Glassman SI, Peay KG, Talbot JM et al (2015) A continental view of pine-associated ectomycorrhizal fungal spore banks: a quiescent functional guild with a strong biogeographic pattern. New Phytol 205:1619–1631

Glassman SI, Levine CR, Dirocco AM et al (2016) Ectomycorrhizal fungal spore bank recovery after a severe forest fire: some like it hot. ISME J 10:1228–1239

Goldmann K, Schöning I, Buscot F et al (2015) Forest management type influences diversity and community composition of soil fungi across temperate forest ecosystems. Front Microbiol 6:1–11

Gómez-Aparicio L, Domínguez-Begines J, Kardol P et al (2017) Plant-soil feedbacks in declining forests: implications for species coexistence. Ecology 98:1908–1921

Gómez-Baggethun E, de Groot R, Lomas PL et al (2010) The history of ecosystem services in economic theory and practice: from early notions to markets and payment schemes. Ecol Econ 69:1209–1218

Górriz E, Martínez de Aragón J, Bonet JA et al (2014) La societat catalana acepta una regulació per la recollida de bolets i un pagament finalista cap a la gestió del bosc. Silvicultura 70:24–26

Grogan P, Baar J, Bruns TD (2000) Below-ground ectomycorrhizal community structure in a recently burned bishop pine forest. J Ecol 88:1051–1062

Hart SC, DeLuca TH, Newman GS et al (2005) Post-fire vegetative dynamics as drivers of microbial community structure and function in forest soils. For Ecol Manag 220(1):166–184

Hawksworth DL, Lücking R (2017) Fungal diversity revisited: 2.2 to 3.8 million species. Microbiol Spectr 5(4):1–17

Hedĕnec P, Radochová P, Nováková A et al (2013) Grazing preference and utilization of soil fungi by *Folsomia candida* (Isotomidae: Collembola). Eur J Soil Biol 55:66–70

Hernández-Rodríguez M, de-Miguel S, Pukkala T et al (2015) Climate-sensitive models for mushroom yields and diversity in *Cistus ladanifer* scrublands. Agric For Meteorol 213:173–182

Ibáñez B, Gómez-Aparicio L, Ávila JM et al (2015) Impact of tree decline on spatial patterns of seedling-mycorrhiza interactions: implications for regeneration dynamics in Mediterranean forests. For Ecol Manag 353:1–9

Izzo A, Canright M, Bruns TD (2006) The effects of heat treatments on ectomycorrhizal resistant propagules and their ability to colonize bioassay seedlings. Mycol Res 110:196–202

Jackson RB, Lajtha K, Crow SE et al (2017) The ecology of soil carbón: pools, vulnerabilities and biotic and abiotic controls. Ann Rev Ecol Evol Syst 48:419–445

Johnson D, Gilbert L (2015) Interplant signalling through hyphal networks. New Phytol 205(4):1448–1453

Karavani A, de Cáceres M, Martínez de Aragón J et al (2018) Effect of climatic and soil moisture conditions on mushroom productivity and related ecosystem services in Mediterranean pine stands facing climate change. Agric For Meteorol 248:432–440

Kardol P, Wardle DA (2010) How understanding aboveground-belowground linkages can assist restoration ecology. Trends Ecol Evol 25(11):670–679

Keeley JE, Bond WJ, Bradstock RA et al (2011) Fire in Mediterranean ecosystems: ecology, evolution and management. Cambridge University Press, Cambridge

Kennedy P (2010) Ectomycorrhizal fungi and interspecific competition: species interactions, community structure, coexistence mechanisms and future research directions. New Phytol 187:895–910

Kipfer T, Moser B, Egli S, Wohlgemuth T et al (2011) Ectomycorrhiza succession patterns in *Pinus sylvestris* forests after stand-replacing fire in the Central Alps. Oecologia 167:219–228

Kivlin SN, Emery SM, Rudgers JA (2013) Fungal symbionts alter plant responses to global change. Am J Bot 100(7):1445–1447

Küçüker DM, Baskent EZ (2017) Sustaining the joint production of timber and *Lactarius* mushroom: a case study of a forest management planning unit in northwestern Turkey. Sustainability 9:92

Küçüker DM, Baskent EZ (2018) Modeling the productivity of commercial *Lactarius* mushrooms: a case study in the Kizilcasu planning unit, Turkey. Nat Res Model 32(1):e12178

Le Tacon F, Marçais B, Courvoisier M et al (2014) Climatic variations explain annual fluctuations in French Périgord black truffle wholesale markets but do not explain the decrease in black truffle production over the last 48 years. Mycorrhiza:S115–S125

Lecina-Díaz J, Álvarez A, de Cáceres M et al (2019) Are protected areas preserving ecosystem services and biodiversity? Insights from Mediterranean forests and shrublands. Landsc Ecol 34(10):2307–2321

Lewandowski TE, Forrester JA, Mladenoff DJ et al (2015) Soil microbial community response and recovery following group selection harvest: temporal patterns from an experimental harvest in a US northern hardwood forest. For Ecol Manag 340:2–94

Lindahl BD, Ihrmark K, Bober J et al (2007) Spatial separation of litter decomposition and mycorrhizal nitrogen uptake in boreal forest. New Phytol 173:611–620

Lindner M, Maroschek M, Netherer S et al (2010) Climate change impacts, adaptive capacity, and vulnerability of European forest ecosystems. For Ecol Manag 259:698–709

Livne-Luzon S, Avidan Y, Weber G et al (2017) Wild boars as spore dispersal agents of ectomycorrhizal fungi: consequences for community composition at different habitat types. Mycorrhiza 27:165–174

Malcolm GM, López-Gutiérrez JC, Koide RT et al (2008) Acclimation to temperature and temperature sensitivity of metabolism by ectomycorrhizal fungi. Glob Chang Biol 14:1169–1180

Martin F, Kohler MC et al (2016) Unearthing the roots of ectomycorrhizal symbioses. Nat Rev Microbiol 14(12):760–773

Martínez de Aragón J, Bonet JA, Fischer CR et al (2007) Productivity of ectomycorrhizal and selected edible saprotrophic fungi in pine forests of the pre-Pyrenees mountains, Spain: predictive equations for forest management of mycological resources. For Ecol Manag 252(1):239–256

Martínez-Peña F, Oria de Rueda JA, Agreda T (2011) Manual para la gestión del recurso micológico forestal en Castilla y León. Serie Técnica de la Consejería de Medio Ambiente de la Junta de Castilla y León, 448 p

Martínez-Peña F, de-Miguel S, Pukkala T et al (2012) Yield models for ectomycorrhizal mushrooms in *Pinus sylvestris* forests with special focus on *Boletus edulis* and *Lactarius* group *deliciosus*. For Ecol Manag 282:63–69

Martín-Pinto P, Vaquerizo H, Peñalver F et al (2006) Early effects of a wildfire on the diversity and production of fungal communities in Mediterranean vegetation types dominated by *Cistus ladanifer* and *Pinus pinaster* in Spain. For Ecol Manag 225:296–305

Mateos E, Olaizola Suárez J, Pajares JA et al (2017) Influence of *Suillus luteus* on *Fusarium* damping-off in pine seedlings. Afr J Biotechnol 16(6):268–273

MEA (Millennium Ecosystem Assessment) (2005) Ecosystems and human wellbeing: synthesis report. Island Press, Washington, DC

Mediavilla O, Oria-de-Rueda JA, Martín-Pinto P (2014) Changes in sporocarp production and vegetation following wildfire in a Mediterranean Forest Ecosystem dominated by *Pinus nigra* in Northern Spain. For Ecol Manag 331:85–92

Molina R, Trappe JM (1982) Patterns of ectomycorrhizal host specificity and potential among Pacific Northwest conifers and fungi. For Sci 28:423–458

Montesinos-Navarro A, Valiente-Banuet A, Verdú M (2019) Plant facilitation through mycorrhizal symbiosis is stronger between distantly related plant species. New Phytol 224(2):928–935

Morris SJ, Blackwood CB (2007) The ecology of soil organisms. In: Paul EA (ed) Soil microbiology, ecology, and biochemistry. Academic, Burlington, pp 195–229

Mueller RC, Scudder CM, Whitham TG et al (2019) Legacy effects of tree mortality mediated by ectomycorrhizal fungal communities. New Phytol 224:155–265

Murat C (2015) Forty years of inoculating seedlings with truffle fungi: past and future perspectives. Mycorrhiza 25(1):77–81

Navarro-Fernández CM, Pérez-Ramos IM, de la Riva EG et al (2016) Functional responses of Mediterranean plant communities to soil resource heterogeneity: a mycorrhizal trait-based approach. J Veg Sci 27:1243–1253

Olaizola Suárez J, Villada D, Oria de Rueda JA (2018a) Effects of *Lactarius deliciosus* and *Rhizopogon roseolus* ectomycorrhyzal fungi on seeds and seedlings of Scots and stone pines inoculated with *Fusarium oxysporum* and *Fusarium verticillioides*. For Chron 94(2):127–134

Olaizola Suárez J, Pajares JA, Díez JJ (2018b) In vitro antagonism of edible ectomycorrhizal fungi against *Fusarium oxysporum* and *Fusarium verticillioides*. For Chron 94(2):117–125

Oliet J, Puértolas J, Planelles R et al (2013) Nutrient loading of forest tree seedlings to promote stress resistance and field performance: a Mediterranean perspective. New For 11:649–669

Oria de Rueda JA, Hernández-Rodriguez M, Martín-Pinto P et al (2010) Could artificial reforestations provide as much production and diversity of fungal species as natural forest stands in marginal Mediterranean areas? For Ecol Manag 260:171–180

Ortega-Martínez P, Águeda B, Fernández-Toirán LM (2011) Tree age influences on the development of edible ectomycorrhizal fungi sporocarps in *Pinus sylvestris* stands. Mycorrhiza 21(1):65–70

Paillet Y, Bergès L, Hjältén J et al (2010) Biodiversity differences between managed and unmanaged forests: meta-analysis of species richness in Europe. Conserv Biol 24:101–112

Palahí M, Birot I (eds) (2009) A Mediterranean forest research agenda (2010–2020). European Forest Institute, 32 pp

Palahí M, Pukkala T, Bonet JA et al (2009) Effect of the inclusión of mushroom values on the optimal management of even-aged pine stands of Catalonia. For Sci 55(6):503–511

Parladé J, Luque J, Pera J et al (2004) Field performance of *Pinus pinea* and *Pinus halepensis* seedlings inoculated with *Rhizopogon* spp. and outplanted in formerly arable land. Ann For Sci 61:507–514

Parladé J, Martínez-Peña F, Pera J (2017) Effects of forest management and climatic variables on the mycelium dynamics and sporocarp production of the ectomycorrhizal fungus *Boletus edulis*. For Ecol Manag 390:73–79

Parladé J, Queralt M, Pera J et al (2019) Temporal dynamics of soil fungal communities after partial and total clearcutting in a managed *Pinus sylvestris* stand. For Ecol Manag 449:117456

Pausas JG, Verdú M (2008) Fire reduces morphospace occupation in plant communities. Ecology 89:2181–2186

Pecoraro L, Angelini P, Arcangeli A et al (2014) Macrofungi in Mediterranean maquis along a seashore and altitudinal transects. Plant Biosyst 148(2):367–376

Pera J, Álvarez I (1995) Ectomycorrhizal fungi of *Pinus pinaster*. Mycorrhiza 5:193–200

Pérez-Izquierdo L, Zabal-Aguirre M, Flores-Rentería D et al (2017) Functional outcomes of fungal community shifts driven by tree genotype and spatial-temporal factors in Mediterranean pine forests. Environ Microbiol 19(4):1639–1652
Pérez-Izquierdo L, Zabal-Aguirre M, González-Martínez SC (2019) Plant intraspecific variation modulates nutrient cycling through its below-ground rhizospheric microbiome. J Ecol 107(4):1594–1605
Pérez-Izquierdo L, Zabal-Aguirre M, Bueé M, Verdú M, Rincón A (2020) Ectomycorrhizal fungal diversity decreases in Mediterranean pine forests adapted to recurrent fires. Mol Ecol 13:2463–2476
Pestaña M, Santolamazza S (2009) Characterization of juvenile maritime pine (*Pinus pinaster* Ait.) ectomycorrhizal fungal community using morphotyping, direct sequencing and fruitbodies sampling. Mycorrhiza 19:91–98
Pestaña M, Santolamazza S (2011) Defoliation negatively affects plant growth and the ectomycorrhizal community of *Pinus pinaster* in Spain. Oecologia 165(3):723–733
Pickles BJ, Simard S (2017) Mycorrhizal networks and forest resilience to drought. In: Collins Johnson N, Gehring C, Jansa J (eds) Mycorrhizal mediation of soil: fertility, structure and carbon storage. Elsevier, Amsterdam
Pickles BJ, Truong C, Watts-Williams SJ (2020) Mycorrhizas for a sustainable world. New Phytol 225:1065–1069
Prieto I, Roldan A, Huygens D et al (2016) Species-specific roles of ectomycorrhizal fungi in facilitating interplant transfer of hydraulically redistributed water between *Pinus halepensis* saplings and seedlings. Plant Soil 406:15–27
Primicia I, Camarero JJ, Martínez de Aragón J et al (2016) Linkages between climate, seasonal wood formation and mycorrhizal mushroom yields. Agric For Meteorol 228:339–348
Pugnaire FI, Morillo JA, Peñuelas J et al (2019) Climate change effects on plant-soil feedbacks and consequences for biodiversity and functioning of terrestrial ecosystems. Sci Adv 5:eaaz1834
Queralt M, Parladé J, Pera J et al (2017) Seasonal dynamics of extraradical mycelium and mycorrhizas in a black truffle (*Tuber melanosporum*) plantation. Mycorrhiza 27:565–576
Querejeta JI (2017) Soil water retention and availability as influenced by mycorrhizal symbiosis: consequences for individual plants, communities, and ecosystems. Elsevier
Querejeta JI, Roldán A, Albadalejo J et al (1998) The role of mycorrhizae, site preparation, and organic amendment in the afforestation of a semi-arid Mediterranean site with *Pinus halepensis*. For Sci 44:203–211
Querejeta JI, Egerton-Warburton LM, Allen MF (2007) Hydraulic lift may buffer rhizosphere hyphae against the negative effects of severe soil drying in a California oak savana. Soil Biol Biochem 39(2):407–417
Querejeta JI, Egerton-Warburton LM, Allen MF (2009) Topographic position modulates the mycorrhizal response of oak trees to interannual rainfall variability. Ecology 90:649–662
Raudabaugh DB, Matheny PB, Hughes KW et al (2020) Where are they hiding? Testing the body snatchers hypothesis in pyrophilous fungi. Fungal Ecol 43:100870
Reyna S, Garcia-Barreda S (2009) European black truffle: its potential role in agroforestry development in the marginal lands of Mediterranean calcareous mountains. In: Rigueiro-Rodriguez A, McAdam J, Mosquera-Losada MR (eds) Agroforestry in Europe: current status and future prospects, Advances in agroforestry, vol 6. Springer, Dordrecht, pp 295–317
Reyna S, García-Barreda S (2014) Black truffle cultivation: a global reality. For Syst 23:317–328
Richard F, Millot S, Gardes M et al (2005) Diversity and specificity of ectomycorrhizal fungi retrieved from an old-growth Mediterranean forest dominated by *Quercus ilex*. New Phytol 166:1011–1023
Richard F, Selosse MA, Gardes M (2009) Facilitated establishment of *Quercus ilex* in shrub dominated communities within a Mediterranean ecosystem: do mycorrhizal partners matter? FEMS Microbiol Ecol 68:14–24

Richard F, Roy M, Shahin O et al (2011) Ectomycorrhizal communities in a Mediterranean forest ecosystem dominated by *Quercus ilex*: seasonal dynamics and response to drought in the surface organic horizon. Ann For Sci 68:57–68

Rincón A, Pueyo JJ (2010) Effect of fire severity and site slope on diversity and structure of the ectomycorrhizal fungal community associated with post-fire regenerated *Pinus pinaster* Ait. seedlings. For Ecol Manag 260(3):361–369

Rincón A, Álvarez IF, Pera J (1999) Ectomycorrhizal fungi of *Pinus pinea* L. in northeaster Spain. Mycorrhiza 8:271–276

Rincón A, de Felipe MR, Fernández-Pascual M (2007) Inoculation of *Pinus halepensis* Mill with selected ectomycorrhizal fungi improves seedling establichemnet two years after planting in a degraded gypsumsoil. Mycorrhiza 18:23–32

Rincón A, Santamaría-Pérez B, Ocaña L et al (2014) Structure and phylogenetic diversity of post-fire ectomycorrhizal communities of maritime pine. Mycorrhiza 24:131–141

Rodríguez-Loinaz G, Alday JG, Onaindia M (2015) Multiple ecosystem services landscape index: a tool for multifunctional landscapes conservation. J Environ Manag 147:152–163

Ruíz-Gómez FJ, Perez de Luque A, Navarro-Cerrillo RM (2019) The involvement of *Phytophthora* root rot and drought stress in Holm Oak Decline: from ecophysiology to microbiome influence. Curr For Rep 5:251–266

Saitta A, Anslan S, Bahran M et al (2018) Tree species identity and diversity drive fungal richness and community composition along an elevational gradient in a Mediterranean ecosystem. Mycorrhiza 28:39–47

Salerni E, Perini C (2004) Experimental study for increasing productivity of *Boletus edulis* s.l. in Italy. For Ecol Manag 201(2):161–170

Sánchez-González M, de-Miguel S, Martin-Pinto P et al (2019) Yield models for predicting aboveground ectomycorrhizal fungal productivity in *Pinus sylvestris* and *Pinus pinaster* stands of northern Spain. For Ecosyst 6:52

Sardans J, Peñuelas J (2013) Plant-soil interactions in Mediterranean forest and shrublands: impacts of climate change. Plant Soil 365:1–33

Scarascia-Mugnozza G, Oswald H, Piussi P et al (2000) Forests of the Mediterranean region: gaps in knowledge and research needs. For Ecol Manag 132:97–109

Scattolin L, Dal Maso E, Mutto Accordi S (2012) Detecting asymptomatic ink-diseased chestnut trees by the composition of the ectomycorrhizal community. For Pathol 42:501–509

Scattolin L, Lancellotti E, Franceschini A et al (2014) The ectomycorrhizal community in Mediterranean old-growth *Quercus ilex* forests along an altitudinal gradient. Plant Biosyst 148:74–82

Seppelt R, Fath B, Burkhard B et al (2012) Form follows function? Proposing a blueprint for ecosystem service assessments based on reviews and case studies. Ecol Indic 21:145–154

Smith SE, Read DJ (2008) Mycorrhizal symbiosis, 3rd edn. Academic, London

Soudzilovskaia NA, van Bodegom PM, Terrer C et al (2019) Global mycorrhizal plant distribution linked to terrestrial carbon stocks. Nat Commun 10:5077

Spatafora JW, Aime MC, Grigoriev IV et al (2017) The fungal tree of life: from molecular systematics to genome-scale phylogenies. Microbiol Spectrum 5(5):FUNK-0053-2016

Steidinger BS, Crowther TW, Liang J et al (2019) Climatic controls of decomposition drive the global biogeography of forest-tree symbioses. Nature 569:404–408

Sthultz CM, Whitham TG, Kennedy K et al (2009) Genetically based susceptibility to herbivory influences the ectomycorrhizal fungal communities of a foundation tree species. New Phytol 184:657–667

Štursová M, Šnajdr J, Cajthaml T (2014) When the forest dies: the response of forest soil fungi to a bark beetle-induced tree dieback. ISME J 8:1920–1931

Tahvanainen V, Miina J, Kurttila M et al (2016) Modelling the yields of marketed mushrooms in *Picea abies* stands in Eastern Finland. For Ecol Manag 362:79–88

Taniguchi T, Kitajima K, Douhan GW et al (2018) A pulse of summer precipitation after the dry season triggers changes in ectomycorrhizal formation, diversity, and community composition in a Mediterranean forest in California, USA. Mycorrhiza 28:665–677

Taudier A, Munoz F, Lesne A et al (2015) Beyond ectomycorrhizal bipartite networks: projected networks demonstrate contrasted patterns between early and late successional plants in Corsica. Front Plant Sci 6:881

Taye ZM, Martínez-Peña F, Bonet JA et al (2016) Meteorological conditions and site characteristics driving edible mushroom production in *Pinus pinaster* forests of Central Spain. Fungal Ecol 23:30–41

Taylor DL, Bruns TD (1999) Community structure of ectomycorrhizal fungi in a *Pinus muricata* forest: minimal overlap between the mature forest and resistant propagule communities. Mol Ecol 8:1837–1850

Tedersoo L, Smith ME (2013) Lineages of ectomycorrhizal fungi revisited: foraging strategies and novel lineages revealed by sequences from belowground. Fungal Biol Rev 27(3–4):83–99

Tedersoo L, Sánchez-Ramírez S, Kõlag U et al (2018) High-level classification of the Fungi and a tool for evolutionary ecological analyses. Fungal Div 90:135–159

Tomao A, Bonet JA, Martínez de Aragón J et al (2017) Is silviculture able to enhance wild forest mushroom resources? Current knowledge and future perspectives. For Ecol Manag 402:102–114

Tomao A, Bonet JA, Castaño C et al (2020) How does forest management affect fungal diversity and community composition? Current knowledge and future perspectives for the conservation of forest fungi. For Ecol Manag 457:117678

Torres P, Honrubia M (1997) Changes and effects of a natural fire on ectomycorrhizal inoculum potential of soil in a *Pinus halepensis* forest. For Ecol Manag 96:189–196

Treseder KK, Lennon JT (2015) Fungal traits that drive ecosystem dynamics on land. Microbiol Mol Biol Rev 79(2):243–262

Uroz S, Buée M, Deveau A et al (2016) Ecology of the forest microbiome: highlights of temperate and boreal ecosystems. Soil Biol Biochem 103:471–488

Valladares F, Benavides R, García-Rabasa S et al (2014) Global change and Mediterranean forests: current impacts and potential responses. In: Coomes DA, Burslem D, William D (eds) Forests and global change. Cambridge University Press, Cambridge, pp 47–75

van der Heijden MG, Martin FM, Selosse AM et al (2015) Mycorrhizal ecology and evolution: the past, the present, and the future. New Phytol 205(4):1406–1423

van der Plas F, Manning P, Allan E et al (2016) Jack-off-all-trades effects drive biodiversity ecosystem multifunctionality relationships in European forests. Nat Commun 7(11109):1–11

Vásquez-Gassibe P, Oria-de-Rueda JA, Santos-del-Blanco L et al (2016) The effects of fire severity on ectomycorrhizal colonization and morphometric features in *Pinus pinaster* Ait. seedlings. For Syst 25:1–12

Warren JM, Brooks JR, Meinzer FC et al (2008) Hydraulic redistribution of water from *Pinus ponderosa* trees to seedlings: evidence for an ectomycorrhizal pathway. New Phytol 178:382–394

Wu B, Hussain M, Zhang W et al (2019) Current insights into fungal species diversity and perspective on naming the environmental DNA sequences of fungi. Mycology 10:27–140

Yuste JC, Peñuelas J, Estiarte M et al (2011) Drought-resistant fungi control soil organic matter decomposition and its response to temperature. Glob Chang Biol 17:1475–1486

Zambonelli A, Iotti M, Hall I (2015) Current status of truffle cultivation: recent results and future perspectives. Micol Ital 44:31–40

Zanne AE, Abarenkov K, Afkhami ME et al (2019) Fungal functional ecology: bringing a trait-based approach to plant-associated fungi. Biol Rev 95(2):409–433

Part V
Forest Fire Ecology and Management

Gidi Ne'eman

Chapter 21
Fire in Mediterranean Pine Forests: Past, Present and Future

José M. Moreno, César Morales-Molino, Iván Torres, and Margarita Arianoutsou

21.1 Introduction

Pine forests and woodlands are a major component of extant vegetation in the Mediterranean region. Pinewoods have played that role for millennia, notably as the climate warmed after the last glaciation. Their extent has varied in response to both climate and human occupation (Carrión et al. 2010; Rubiales et al. 2010). In fact, some Mediterranean areas acted as refugia for pine species during glacial times, after which they colonized other parts of Europe (Salvador et al. 2000; Gómez et al. 2005; Cheddadi et al. 2006; Rodríguez-Sánchez et al. 2010). Throughout history, pine woodlands have provided, and continue to provide, various ecosystem services such as timber, firewood, edible nuts, resin, recreation and cultural and spiritual values, among others (see Orenstein this volume, Chap. 30; Rincón et al. this volume, Chap. 20). Pines have been a major source of wood for ship construction and other warfare materials since ancient times (Veal 2017), and woodlands have often been managed by the marine army (Martínez-González 2015). Pines have expanded because they have been used for afforestation projects since the seventeenth century, owing to their capacity to withstand adverse climate conditions and grow on poor soils, including degraded ones (Quézel et al. 1990). Despite their prominence in Mediterranean landscapes, the ecological role of *Pinus* species has been

J. M. Moreno (✉) · I. Torres
Departamento de Ciencias Ambientales, Universidad de Castilla-La Mancha, Toledo, Spain
e-mail: josem.moreno@uclm.es

C. Morales-Molino
Institute of Plant Sciences and Oeschger Centre for Climate Change Research, University of Bern, Bern, Switzerland

M. Arianoutsou
Department of Ecology and Systematics, National and Kapodistrian University of Athens, Athens, Greece

G. Ne'eman, Y. Osem (eds.), *Pines and Their Mixed Forest Ecosystems in the Mediterranean Basin*, Managing Forest Ecosystems 38,
https://doi.org/10.1007/978-3-030-63625-8_21

undervalued by some schools of thought in vegetation science (Carrión and Fernández 2009). Moreover, despite the adaptations they exhibit to cope with fire (see Ne'eman and Arianoutsou this volume, Chap. 22), the role of fire in shaping these adaptations was not acknowledged until the latter part of the twentieth century. Fires are nowadays ubiquitous throughout the Mediterranean region and many burn pine forests and woodlands.

In this chapter we review the current extent of pines and how the area they cover has changed in recent decades. This is important for understanding current fire risks. Subsequently, we evaluate how fires affected pine woodlands in the past (since the onset of the Holocene), using pollen, charcoal and other fossil remains in sedimentary deposits, and during historical times, using dendroecological records. Next, we evaluate the recent fire regime, focusing on the extent to which pine forests and other wooded areas have been affected by fires. Finally, we assess how climate change will affect future pine distribution as well as fire weather danger and fire activity.

21.2 Mediterranean Pines Today

The countries surrounding the Mediterranean Sea, as well as Portugal, Bulgaria, and the Atlantic islands of Portugal and Spain, cover a total of 8.7 Mkm^2. Forests and other woodland areas (OWLAs) cover 13.4% of the land (FAO (Food and Agriculture Organization) 2015). This figure includes territories of the larger countries that have sizable portions of their land in non-Mediterranean-type climate areas (e.g., Algeria, Libya in the south, France in the north). Countries in which most of the territory is dominated by Mediterranean-type vegetation have much higher proportions of forests and OWLAs (e.g., Spain [54%], Italy [37%], and Greece [50%]). This is a consequence of the Mediterranean Sea being surrounded by mountains, creating a very rugged landscape, much of which is unsuitable for agriculture. Forests represent approximately 73% of the total wooded areas, while OWLAs represent the remaining 27%. This means that, while shrublands and open woodlands are important, forests dominate. Moreover, during recent decades, forested areas have increased, while OWLAs have decreased (Fig. 21.1). This trend has continued since the middle of the twentieth century (Le Houerou 1987), meaning that, in general, the amount of fuel has increased in the countries surrounding the Mediterranean Sea. This is particularly true in the southern EU (European Union) countries (Portugal, Spain, southern France, Italy and Greece, from here on referred to as EUMED countries).

Compared to the rest of the world, Mediterranean countries have a much higher proportion of planted forests and OWLAs than natural ones. In 1990, natural areas in Mediterranean countries comprised 85.4% of forested areas; this figure slightly decreased during the following decades, reaching 84.0% in 2015. Planted forests and OWLAs comprised 14.6% in 1990 and increased to 16% in 2015. The percentage of natural versus planted forests and OWLAs varied largely among countries. In

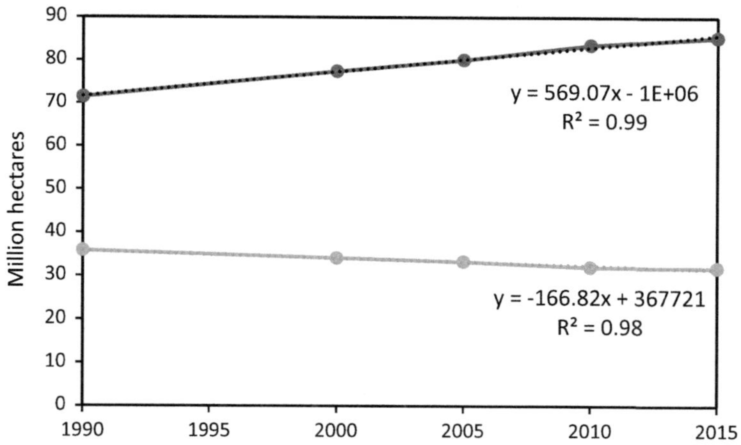

Fig. 21.1 Trends in forests (black) and other woodland areas (OWLAs) (grey) in Mediterranean countries during 1990–2015 (Mha), based on FAO country reports (FAO 2015). The dotted lines show the adjusted linear trends

northern African countries, this figure varied from 13.4% to 24.3% and 63.4% for Morocco, Algeria and Tunisia, respectively. In the southern European countries, these figures were 24.5% for Portugal, 15.2% for Spain, 11.0% for France, 7.0% for Italy and 3.6% for Greece. In Asia, Turkey had 22.2%, Syria 44.2%, Lebanon 7.7% (this country, however, did not always report the proportion of each type of forest) and Israel 55.2%.

Similar figures for the rest of the world show that in 1990 natural areas comprised 96.1% of total forested areas, steadily decreasing during the following decades to 93.2% in 2015. In other words, deforestation has been the dominant trend. In contrast, the planted areas increased from 3.9% in 1990 to 6.8% in 2015, indicating that natural forests are being substituted by plantations. These data show that the Mediterranean region not only has a larger proportion of planted forests and OWLAs, but, contrary to the rest of the world, its proportion of natural wooded areas is increasing (data from FAO 2015, see summary in Keenan et al. 2015).

Out of the total area covered by forests, the proportion occupied by coniferous forests varies among countries: in more mesic countries, broad-leaved forests are dominant (e.g., Italy), while in drier countries, nearly half of the forested areas may be coniferous (e.g., Turkey) (Table 21.1). Most coniferous forests comprise species of *Pinus*, with Turkey leading the area covered by these species (Table 21.2). The species that grow in each country vary from west to east. For example, *P. pinaster* Ait. is abundant in the west, *P. halepensis* Mill. is widespread throughout the region, and *P. brutia* Ten. is in the east; the latter two form the majority of pines across the region.

Extant *Pinus* cover is the result of historical deforestation and, at least since the seventeenth and eighteenth centuries (Barbero et al. 1998; Caplat et al. 2006), of reforestation. Indeed, pines have been the preferred species for reforestation

Table 21.1 Percentage area covered by forests of different types in 2015 in a representation of Mediterranean countries

Forest type	Portugal	Spain	Italy	Turkey
Coniferous	29.7	37.6	13.4	47.6
Broad leaved	70.3	62.4	67.8	32.9
Mixed	n.a.	n.a.	9.6	19.5
Other	n.a.	n.a.	9.2	n.a.

Source: FAO (2015)

Table 21.2 Area covered (kha; % relative to the total area covered by forests) by *Pinus* species in a representation of Mediterranean countries based on data from 2005 (Italy), 2015 (Portugal and Turkey) and 2016 (Spain)

	Portugal		Spain		Italy		Turkey	
Species	kha	%	kha	%	kha	%	kha	%
P. halepensis			2065	11.3	105	1.2		
P. brutia							5610	25.1
P. pinea	194	6.0	400	2.2	46	0.5	162	0.7
P. pinaster	713	22.1	1067	5.8	63	0.7		
P. nigra			709	3.9	235	2.7	4245	19.0
P. sylvetris			1030	5.6	152	1.7	1519	6.8
P. heldreichii					2	0.1		
P. radiata			264	1.4				
Other			191	1.0	382	4.4		
Mixed			1142	6.3				
Total	907	28.1	6867	37.6	984	11.3	11,536	51.6

Sources: INFC (2007), ICNF (2015), OGM (2015) and MAPA (2019)

projects due to their rapid growth, endurance to poor soils and the summer drought of the Mediterranean climate, and easy cultivation in nurseries. The example of Spain is illustrative. Between 1877 and 1939, the area planted by the forestry service reached 128,930 ha, although there is disagreement among authors about this figure. The species used are unknown but, at least before 1913, part of the administrative units used autochthonous pine species. Immediately after the civil war, in 1940, a program of reforestation was put in place to create jobs and add economic value to the poorest areas of the country. In later periods, managing erosion and soil conservation was added to the commercial interest for planting pines. More recently, due to the Common Agricultural Policy of the EU, environmental purposes have prevailed in the objectives of reforestation projects, and new species were used. Between 1940 and 1984, 3,678,522 ha were planted; the species planted are known for approximately 90% of the forested area. These were mainly pines (72% of the area planted), with *P. pinaster*, *P. sylvestris* L. and *P. halepensis* being the most commonly planted species (Fig. 21.2). During 1985–2013, an additional 1,468,980 ha were planted, for which the species planted are known for 86% of the area. Of this area, 49% was planted with pines, the rest was planted with species such as *Quercus* and other broad-leaved species (Vadell Guiral et al. 2019).

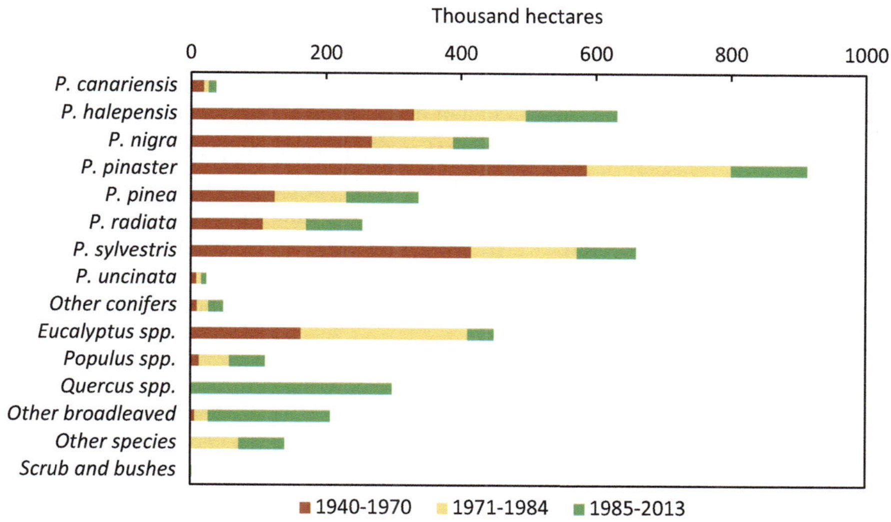

Fig. 21.2 Reforestation in Spain (ha) during 1940–1970, 1971–1984 and 1985–2013 (kha). Note that these are minimum values, since the species planted was not always known. (Data from Vadell Guiral et al. 2019)

The species used for reforestation during the twentieth century in other Mediterranean countries are varied, but pines dominated. Unfortunately, statistics are not readily available for all countries. Values ranging from 45% to 94% have been reported from personal and written sources (Pausas et al. 2004) for reforestations using pines in Algeria, Morocco, Tunisia, Greece and Turkey.

As indicated above, the objective of the plantations has varied over time. In northern Africa, Spain and elsewhere, *P. halepensis* has often been used for restoration purposes in the fight against desertification, given that this species is the most tolerant to drought (Ciancio 1986; Maestre and Cortina 2004). In other cases, the economic value of timber or nuts was behind plantations. For example, *P. pinea* L. nuts are of high commercial value, particularly nowadays, and this has led to extensive planting of this species. For example, in Portugal, *P. pinea* has increased in area by 46% since 1995, to reach 6% of the total forested area in 2015 (Valdiviesso et al. 2017); in Turkey, the area of *P. pinea* increased from 69,294 ha in 2004 to 197,550 ha in 2015 (Küçüker and Baskent 2017).

Planting is not the only mechanism contributing to enhancing pine-covered areas in the region. Pines are expanding to nearby areas due to land abandonment (Gerard et al. 2010) and discontinued grazing, and in some cases this is being aided by regional warming. For example, in the Apennines, *P. nigra* J.F.Arn. is expanding to areas that it occupied in ancient times (Piermattei et al. 2012). In northern Spain, *P. sylvestris* and *Q. ilex* L. are expanding to areas that were thought to sustain *Juniperus thurifera* L. open forest in their climax state (DeSoto et al. 2010). *Pinus sylvestris* and *P. uncinata* Mill. ex Mirbel are also colonizing areas that were formerly cultivated (Poyatos et al. 2003; Améztegui et al. 2010); forest plantations are acting as seed sources from which pines (e.g., *P. halepensis*) are expanding (Osem et al. 2011). Densification and expansion of pines, including above the tree line, have also been reported for *P. uncinata*, *P. sylvestris*, *P. heldreichii* Christ and *P. peuce* Griseb. at sites in Spain, Italy and the Balkans (Vitali et al. 2019).

In other words, the pioneer character of many pines is helping them to expand into areas from which intensive agriculture or grazing halted in the second half of the twentieth century. But recent regional warming and water stress are causing decreases in growth and increases in plant mortality in the least favorable areas (e.g., *P. sylvestris* at low elevations in central Spain; Gea-Izquierdo et al. 2014), or at the limit of the distribution range (e.g., *P. nigra* in Sierra de Cazorla, Jaén, Spain; Navarro-Cerrillo et al. 2018).

21.3 Pines and Fire in the Mediterranean Region During the Holocene

21.3.1 The Role of Pines in Mediterranean Post-glacial Landscapes

According to the available fossil pollen evidence, pines were a major component of the woodlands that spread throughout the northern Mediterranean region during the Lateglacial, i.e. ca. 14,600–11,700 years ago (e.g., Vescovi et al. 2010; Connor et al. 2012; Gassner et al. 2020). The onset of warmer and moister conditions in the early Holocene (ca. 11,700 years ago) triggered a widespread and rapid replacement of pinewoods by forests and woodlands dominated by broad-leaved trees and, in some cases, also mesophilous conifers, such as *Abies alba* Mill. (e.g., Rubiales et al. 2010; Vescovi et al. 2010; Camuera et al. 2019). This replacement occurred particularly in sites featuring relatively oceanic conditions or high moisture availability. In contrast, pinewoods persisted as the (co-)dominant vegetation for several millennia during the early and mid-Holocene over extensive areas of inland Iberia, northern Greece, western Anatolia and inland Morocco, characterized by drier climates, usually with pronounced seasonality (e.g. Lamb et al. 1991; Carrión and van Geel 1999; Eastwood et al. 1999; Carrión et al. 2001; Lawson et al. 2005; Morales-Molino et al. 2013; Morales-Molino and García-Antón 2014; Gassner et al. 2020). Indeed, these pinewoods and mixed forests were quite stable in structure and composition until intensified human activities, usually farming (in some cases alongside metallurgy and mining), drove their decline during recent millennia (e.g. Eastwood et al. 1999; Carrión et al. 2003; Rubiales et al. 2012; Morales-Molino et al. 2017a; Gassner et al. 2020). In certain mountainous regions of the eastern Iberian Peninsula, Corsica and the Taurus Mountains (Turkey), Mediterranean mountain pinewoods (mostly comprising *P. sylvestris* and *P. nigra*) have persisted even until today under particularly favorable conditions (i.e. continental and relatively dry climate, steep slopes, shallow soils) and a low-to-moderate human impact (e.g., Eastwood et al.

1999; Reille et al. 1999; Stevenson 2000; Carrión 2002; Leys et al. 2014; Morales-Molino et al. 2017b). These 'surviving' pinewoods correspond to the bulk of the current natural distribution range of pines in the Mediterranean region (Quézel and Médail 2003; Costa et al. 2005).

In general, pollen analysis cannot provide precise information about the exact pine species present at a given site. Nevertheless, detailed morphological analyses of modern and fossil *Pinus* pollen grains have shown that it is possible to attain more precise identifications, usually to group level, i.e., Mediterranean (*P. pinaster, P. halepensis, P. pinea*) vs. mountain pines (*P. sylvestris*, *P. nigra*) and, in some cases, even to species level (see Carrión et al. 2000; Desprat et al. 2015). Consequently, several pine pollen types are often found within a single pollen diagram, usually *P. pinaster* type, *P. halepensis* type or *P. sylvestris/nigra* type (e.g. Carrión 2002; López-Merino et al. 2009; Tinner et al. 2009; López-Sáez et al. 2010). Likewise, the interpretation of *Pinus* pollen abundances is not straightforward, given the high production and outstanding dispersal ability of pine pollen. However, the study of modern pollen-vegetation relationships permits more refined and robust interpretation of *Pinus* in pollen assemblages (e.g. Fall 2012; López-Sáez et al. 2013; Morales-Molino et al. 2020). However, despite the aforementioned significant progress in pine pollen taxonomy and taphonomy, macrofossil analysis is still the most reliable tool for unequivocally and unambiguously assessing the local presence of pines, as well as for precisely determining the species involved (e.g. Hernández et al. 2011; Morales-Molino et al. 2013; Leys et al. 2014; Rubiales and Génova 2015; Servera-Vives et al. 2018). For instance, findings of wood and charcoal in natural and archaeological contexts informed about the presence of *P. sylvestris* or *P. nigra* until historical times in parts of Sierra de Gredos (central Spain; Rubiales and Génova 2015) and the northern Iberian Plateau (Hernández et al. 2011), where it is currently absent. Similarly, charcoal identification has shown how the purely Mediterranean *P. pinaster* replaced mountain pines (most likely *P. sylvestris*) during the early Holocene (ca. 9000–8500 years ago) at the foothills of Sierra de Francia (central Spain), when the climate became warmer and drier (Morales-Molino et al. 2013).

To complete the picture about the past dynamics of pines in the Mediterranean region, we must mention those territories where pines have played a rather minor role during Holocene vegetation history. In this context, the rather low abundances of *Pinus* pollen found in most of the available palaeoecological records from many Mediterranean islands, such as Crete (Bottema and Sarpaki 2003), Sardinia (Beffa et al. 2016), Sicily (e.g. Sadori and Narcisi 2001; Tinner et al. 2009, 2016) and Minorca (Burjachs et al. 2017, but see Servera-Vives et al. 2018) are remarkable. Similarly, pines also seem to have been a minor component of the Holocene vegetation of mainland Italy at most elevations (e.g. Magri 1999; Allen et al. 2002; Colombaroli et al. 2008; Vescovi et al. 2010; de Beaulieu et al. 2017) and in the Levant (e.g. Cheddadi and Khater 2016; Schiebel and Litt 2018).

21.3.2 Long-Term Fire Regimes and Fire Ecology: The Sedimentary Record

The analysis of sedimentary charcoal archived in lakes and mires is probably the most widely used tool for reconstructing fire history at multi-decadal to millennial timescales. A great advantage of this proxy for fire activity is that palaeoecological sites are quite numerous and offer relatively good spatial and temporal coverage. The main disadvantage lies in the relatively coarse temporal resolution: one sample typically comprises several decades, and it is not rare that samples are spaced several centuries apart. However, contiguous sampling of high-resolution (ca. 10 years cm^{-1}) sequences may overcome these limitations, enabling quite detailed fire regime reconstructions.

Firstly, we will focus on the few available detailed fire regime reconstructions based on high-resolution charcoal records from sites in the Mediterranean region where pines have been relevant during the Holocene. Corsican *P. nigra* subsp. *laricio* mountain forests have persisted throughout the Holocene under a fire regime characterized by frequent surface fires, with a mean fire return interval (mFRI) of ca. 100 years and up to 12.5 fires/millennium (Leys et al. 2014). Similarly, in the Northern Iberian Plateau, Morales-Molino et al. (2017a) found that *P. nigra* forests dominated in this region under a mixed fire regime with frequent surface fires and occasional crown fires (mFRI = 200 years). However, too frequent crown fires linked to the intensification of farming during the Middle Ages triggered the demise and final regional extirpation of these pine forests (Fig. 21.3). Continuing with mountain pine forests, *P. sylvestris* stands in Sierra de Gredos (central Spain) have withstood an mFRI of 290 years during the last ca. 2000 years, but have usually been replaced with pasturelands under enhanced fire activity, probably crown fires (López-Merino et al. 2009; López-Sáez et al. 2018). In the lowlands of this mountain range, *P. pinaster* woods were resilient to fires occurring with an mFRI of 350 years, although fire episodes caused transient retreats of pines and encroachment of *Cistus* shrubs (López-Sáez et al. 2018).

Less detailed charcoal records do not provide specific details about fire regimes, but still give relevant insights into fire activity and long-term fire ecology of pines. For instance, they show a similar scenario for mountain pinewoods (*P. sylvestris*, *P. nigra*) in eastern and southeastern Iberia, as they often retreated and were in some cases partially replaced by oak woodlands and pasturelands under enhanced fire activity (probably crown fires) (Carrión and van Geel 1999; Carrión et al. 2007). One of the most paradigmatic sites from this region, Sierra de Gádor, revealed that fire frequency and intensity interacted to drive the dynamics between *Pinus cf. nigra* and evergreen *Quercus* species (Carrión et al. 2003). In recent times, more frequent and severe fires have triggered the demise of the local mountain pinewoods (Carrión et al. 2003), in close resemblance to the Northern Iberian Plateau (see previous paragraph). The palaeoecological records of Sierra de Guadarrama in central Spain have also shown that *P. sylvestris* forests persisted under low fire activity but slightly retreated when fire became more severe or extensive (Morales-Molino et al. 2017b).

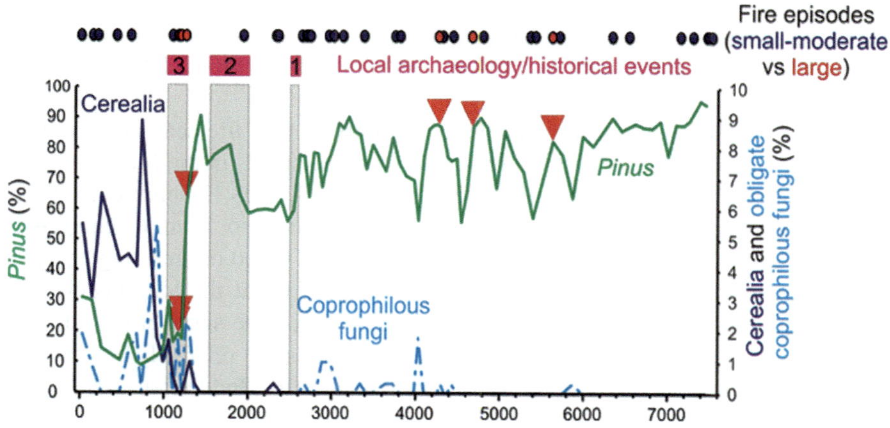

Fig. 21.3 Fire-induced decline of *Pinus nigra* forests in the Northern Iberian Plateau (Tubilla del Lago site, Burgos, Spain) during the Middle Ages related to intensified farming. Fire episodes: local to extra-local fire episodes as reconstructed from peak detection analysis of the high-resolution macroscopic charcoal series during the last 7000 years. "Small-moderate" peaks are interpreted as surface fires, while "large" peaks are related to crown fires. "Large" peaks are represented as red triangles in the main plot. Local archaeology/historical events: 1. Iron Age settlements; 2. Roman settlements; 3. Christian/Muslim medieval wars. *Pinus*, Cerealia (pollen from cereals) and obligate coprophilous fungi percentages are proxies for the abundance of pines in the extra-local to regional vegetation, local agriculture and local pastoralism, respectively. (Modified from Morales-Molino et al. 2017a)

In northern Greece, *P. sylvestris* stands around Lake Zazari have declined during the last 3000 years because of increased fire activity related to humans, demonstrating that mid- to late Holocene declines of mountain pinewoods were not just an Iberian 'specialty' (Gassner et al. 2020). At lower elevation in southeastern Iberia, increased fire activity favored the spread of *P. halepensis* (Carrión et al. 2001). In another purely Mediterranean setting, *P. pinaster* stands expanded on the island of Pantelleria (Italy) under higher fire activity during the Little Ice Age (1300–1900 CE (Common Era); Calò et al. 2013).

21.3.3 Long-Term Fire Regimes and Fire Ecology: Dendroecological and Historical Evidence

Both tree-ring analysis and historical documents have a clear advantage with respect to sedimentary charcoal analysis: their data are at annual or even seasonal resolution and this allows elucidating very detailed and accurate fire history and fire regime reconstructions. In contrast, the timespan they cover is usually restricted to several centuries at the most, at least in the Mediterranean region (but see Swetnam 1993 for a >1500-years fire history reconstruction based on dendrochronology of fire scars in *Sequoiadendron* groves). Furthermore, the few available fire history reconstructions from documentary sources in general do not provide enough details about the composition of the burnt vegetation to reliably assess the nature of the fire regimes affecting pinewoods (e.g. Montiel-Molina 2013; Montiel-Molina et al. 2019; Sequeira et al. 2019).

To the best of our knowledge, there are only a few fire-scar and growth-suppression fire regime reconstructions related to pine stands in the Mediterranean region, mainly from *P. nigra* tree-ring records (Fulé et al. 2008; Touchan et al. 2012; Christopoulou et al. 2013; Camarero et al. 2019) but also from *P. halepensis* (Fournier et al. 2013). The rarity of fire reconstructions based on tree-ring analysis in the Mediterranean region may be related to the fact that the long-lasting and heavy exploitation of pine forests would have prevented the survival of old trees until the present. Moreover, stand-replacing crown fires are relatively common in *P. halepensis and P. pinaster* stands (e.g. Torres et al. 2016; Budde et al. 2017), thus limiting the usefulness of a method that is particularly suited to identifying surface fire. Nevertheless, the high interest in the information provided by these reconstructions compensates for their paucity. For instance, in Sierra de Gredos (central Spain), Camarero et al. (2019) inferred a major shift in the fire regime towards higher-severity fires by the end of the nineteenth century CE based on growth suppressions found in local *P. nigra* stands. An increase in charcoal accumulation in a mire nearby together with historical documents supported the robustness of this reconstruction (Camarero et al. 2019). In this case, a major reorganization of land use in the region seems to have been responsible for this notable fire regime change (Camarero et al. 2019). The rest of the tree ring-based fire regime reconstructions from *P. nigra* stands rely on fire scars and show a relatively high frequency of surface fires affecting *P. nigra* forests during the dry season throughout the northern Mediterranean. For instance, in Sierra Turmell (northeastern Spain), Fulé et al. (2008) reconstructed the occurrence of 11 surface fires in a 172-year fire-scar record, with FRI ranging from 2 to 57 years. A similar picture arises from the 162-years fire-scar record of the Mount Taygetos (southern Greece), where surface fires occurred every 4.9 years on average during summer and early autumn (Christopoulou et al. 2013). Further north, a ca. 700-years fire-scar record from the Pindus Mountains (northwestern Greece) indicated the occurrence of 16 fire events with FRI between 5 and 119 years (Touchan et al. 2012). Finally, the only fire scar-based fire regime reconstruction available from *P. halepensis* showed that 11 surface fires separated by 2 to 57 years occurred in southern France during the past 172 years (Fournier et al. 2013).

21.4 Pines and Fire in Recent Times

21.4.1 Current Fire Regime

Fires occur throughout the Mediterranean region. During the last decade (2011–2018), there were on average nearly 40,000 fires per year in the EUMED countries, burning on average nearly 350 kha. In contrast, the countries of northern Africa (Morocco, Algeria), the eastern Mediterranean (Cyprus), Asia (Lebanon, and Turkey), the Balkans (Slovenia, Croatia, and North Macedonia) and Bulgaria reported 3656 fires in 2018. In the last decade, the area burned in these countries annually amounted to 67 kha (San-Miguel-Ayanz et al. 2018). This area has changed over the last half century. Prior to the 1960s, the area burned was small in most countries. After that, notably in EUMED countries, the area burned began increasing, multiplying by several fold (Le Houerou 1987; Moreno et al. 1998), and reaching maximum values two decades later, except in Portugal. Thereafter, the area burned has been decreasing to values that are two- to threefold lower than in the 1960s, with the exception of Portugal (Urbieta et al. 2015, 2019). Indeed, while the trend of decreasing area burned is clear in Spain, Italy and France, with Greece lagging behind them, Portugal has demonstrated an increasing trend in recent decades, leveling off in the last decade (Fig. 21.4).

The trend for the northern African and Eastern Mediterranean countries does not reflect such a clear picture, except in the case of Turkey. Yet, the maximum values were reached a decade earlier than in EUMED. It appears that the low absolute figures can be overridden by a severe season (e.g., Algeria, 2012, 99,000 ha; Fig. 21.4). Noteworthy is the magnitude of fires relative to the total area of forests and OWLAs. By far, Portugal is the country with the highest incidence, with approximately 4.5 ha burned/100 ha of land, or 4.5% of the forests and OWLAs. Italy and Greece follow, although at great distance, with 0.7% and 0.6%, respectively. Although Spain is just behind Portugal in absolute numbers, on relative terms it burns much less. The figure for France cannot be interpreted along the same lines, since it includes the entire European territory of the country, despite the fact that fires occur mainly in the southern Mediterranean part and in Corsica; thus, the figure is much lower than if it were corrected for the areas where most fires actually occur.

The absolute figures of some of the larger countries of northern Africa or Turkey contrast with those of EUMED and are not in line with what their climate would predict. It has been shown that under-reporting has been common in some northern African and eastern Mediterranean countries, the actual area burned amounting to nearly twice the area reported (e.g., Tunisia; Belhadj-Khedher et al. 2018), or 1000 to 5000 times greater (e.g., Bulgaria and Turkey; Turquety et al. 2014). Often, underreporting results from how a country defines a forest fire, since some countries may only report fires that affect a high forest, and no other woodland areas. Actually, a statistical effect is included in the EUMED figures since in earlier decades reporting was commonly done only for forests. Nevertheless, since large fires tend to

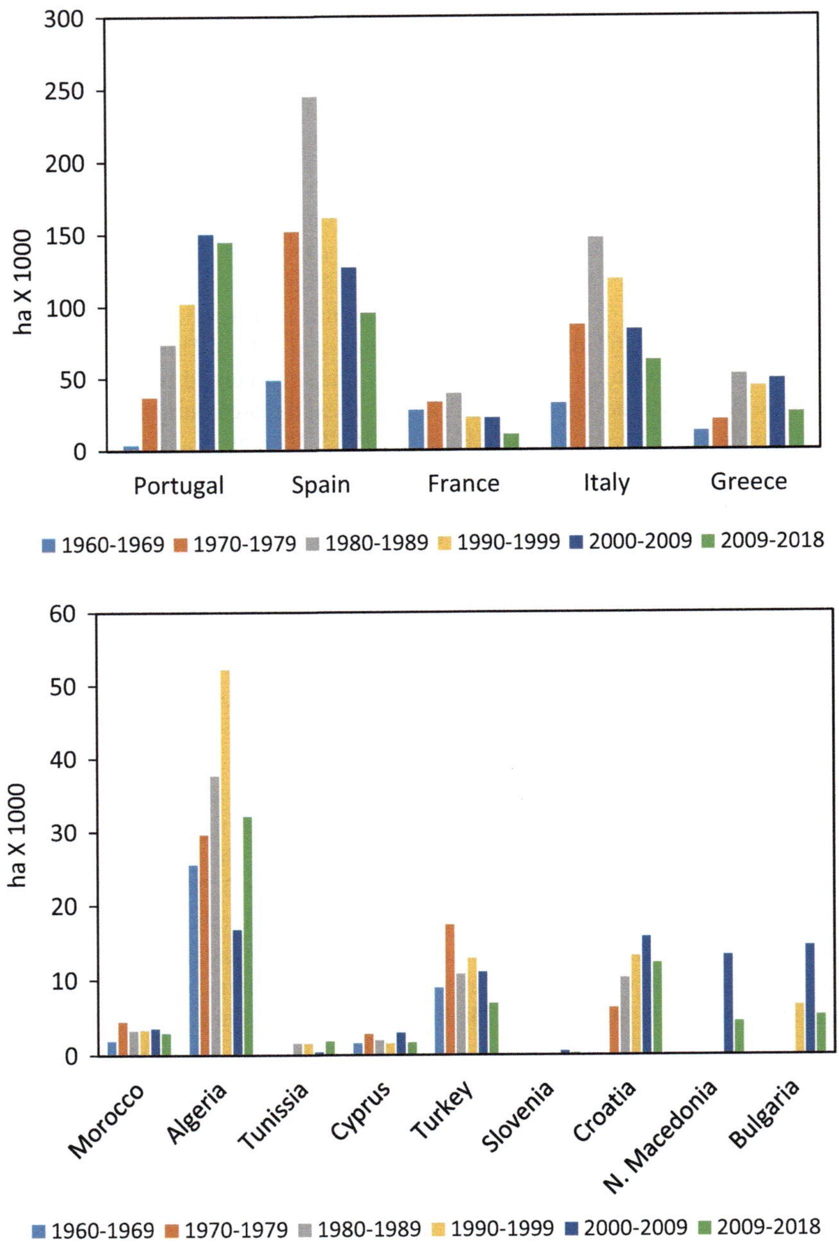

Fig. 21.4 Area burned per decade during the last six decades in Portugal, Spain, Italy and Greece (top) and Morocco, Algeria, Tunisia, Cyprus, Turkey, Slovenia, Croatia, N. Macedonia and Bulgaria (bottom). (Sources: Alexandrian and Esnault 2001; FAO 2015; San-Miguel-Ayanz et al. 2018)

mainly affect forests, globally these figures indicate the order of magnitude of the area burned in the various countries (Le Houerou 1987).

The trend for the area burned in EUMED countries is general, and occurs at different scales (entire EUMED, regions within a country, and provincial level (only Spain was studied) (Urbieta et al. 2015), and affects the number of fires (>1 ha), the area burned by these fires and the number of large fires (i.e., fires >100 ha are also decreasing). This general trend contrasts with the fact that the climate is warming, even at a pace greater than the rest of the world (Seneviratne et al. 2016; Lionello and Scarascia 2018) and rainfall is generally decreasing in the region (Norrant and Douguédroit 2006; Philandras et al. 2011; Tang et al. 2018). Fire danger indices correlate well with the area burned in the region and globally (Bedia et al. 2014), explaining sizeable portions of either the number of fires, the number of large fires or the area burned, albeit with decreasing power at lower scales (Urbieta et al. 2015). Fire danger indices have been increasing in southern Europe (Venäläinen et al. 2014; Jolly et al. 2015; Dupire et al. 2017; Urbieta et al. 2019). In addition, as shown above, forests and OWLAs have increased in virtually all countries, a trend that has continued since the 1960s (Le Houerou 1987), thus increasing the fire hazard (Pausas and Fernández-Muñoz 2012). Changes in policy (Ruffault and Mouillot 2015; Curt and Frejaville 2018), and increasing firefighting resources, particularly aerial means (Urbieta et al. 2019), seem to be responsible for the decreasing trend in EUMED countries, counteracting other fire-driving factors. Notwithstanding, most extreme fire events continue to occur in the region, coinciding with the most severe fire weather (e.g., extreme values of the Canadian FWI) (San-Miguel-Ayanz et al. 2013), driven by particular synoptic weather patterns (Trigo et al. 2006; Amraoui et al. 2013). Additionally, while changes in landscape fire hazard may have affected fires, more so in southern Europe than in other Mediterranean areas, particularly the northern African and eastern Mediterranean countries, large fires can also occur during extreme weather events, irrespective of the absence of significant landscape changes (Viedma et al. 2017).

21.4.2 *Burning of Pine Forests*

21.4.2.1 Resource Selection Approaches

A common approach to assess how pine forests are burning entails evaluating the selectivity of fires for these land-use/land-cover (LULC) types in comparison with their availability in the landscape. To do this, fire perimeters for a specified period are crossed with LULC maps, from which calculations are made about how much each LULC type was burned. Fire perimeters are often based on satellite images (e.g., Landsat, MODIS), whereas LULC maps can be based on actual cartography or satellite images. A number of studies have done this at supranational, national or local scales.

Fire selectivity for coniferous forests has been studied across all Mediterranean-type areas (MTA) of the world, including the Mediterranean region proper, showing that in all areas fires burned certain LULC types more than would be expected by chance based on their abundance in the landscape; that is, fires were selective. Coniferous forests burned in greater proportion to their abundance only in the Mediterranean region, while shrublands and other LULC types burned more often than coniferous forests in the other MTAs (Syphard et al. 2009). This study was based on MODIS images from one single year, so extrapolating the results to longer periods could be misleading, given the variability of fires even at this large scale.

An assessment at the level of the entire European continent, but focusing on Portugal, Spain, France and Italy, using a longer period (2000–2013), and two LULC maps (CORINE Land Cover 2000 and 2006) showed that the percentage area of burned coniferous forests varied among countries (e.g., 13.4% Portugal, 11.1% Spain, 10.4% France and 3.2% Italy). Coniferous forests ranked third (10.9%), after sclerophyllous woodlands (42.5%) and agricultural areas (17.4%). This led to a decrease in the area covered by coniferous forests between 2000 and 2006 in all four countries. Overall, the reduction was minor in Spain, France and Italy (less than 5% cover), but not so in Portugal, where it exceeded 20%. We note that the pattern of selectivity was similar and constant over the years in all four countries, except during extreme years with very large fires, when forests became the dominant LULC type burned. This pattern suggests that small fires tended to burn more shrubby vegetation while larger ones burned more forests (Pereira et al. 2014).

This study confirmed earlier findings of a study that was also carried out at the European scale, including the countries in the Balkans, Bulgaria, Cyprus and Turkey, although over a shorter period (2000–2008) (Oliveira et al. 2014). These authors, however, found that there were differences among countries, particularly between EUMED countries and the others. While Bulgaria and the Balkan countries formed a distinct unit in which agricultural and grassland fires dominated, in Cyprus and Turkey, fires in coniferous forest were dominant. In the case of these two countries, *P. nigra* forests were dominant, hence their higher selection ratios.

Studies at the national level, using a similar approach, showed comparable results. For example, in Portugal it was found that shrublands were the LULC type most selected for, followed by forests, of which pines were first. As mentioned above, larger fires were not selective whereas smaller ones were. In other words, large fires, which tend to occur under more severe weather, are more controlled by factors other than fuels, whereas fuel plays a greater role in smaller fires. Notwithstanding, in the case of large fires, *P. pinaster* was selected for more strongly than the other forests types, which concurs with observations of these stands having high fuel loads that can promote strong fire behavior (Nunes et al. 2005). Similar results were obtained in a second study (Moreira et al. 2009; Barros and Pereira 2014). Shrublands were again the most selected LULC type. Forests were selected for at an intermediate level, with conifer stands (pure or mixed) being more susceptible to burning than eucalypt and other broad-leaved forests. In contrast, in Italy, burnt Mediterranean pines represented 9.7% of the areas burned, and were also the most strongly selected for, followed by high maquis and transitional woodland-shrubland (Mancini et al. 2017). Nevertheless, if all shrublands were combined, then this LULC type would have also been the most strongly selected for.

As different countries may exhibit different levels of selectivity by fire, different parts of a country may differ in the LULC type affected by fire, depending on the dominant LULC types or other physical or human features of the landscape (Nunes et al. 2005). Moreover, the same area may present different selectivity depending on the weather conditions under which it is burned (Fernandes et al. 2012; Koutsias et al. 2012).

When fires burn sizeable areas, their leverage needs to be accounted for in this type of analysis (Duane et al. 2019). Accounting for it requires that the selectivity of LULC types by fires should be crossed with LULC maps updated annually (Moreno et al. 2011). These authors studied a series of 16 years using Landsat images to map fire scars and LULC in Sierra de Gredos (Spain). They found that, contrary to the studies mentioned earlier, *P. pinaster* woodlands showed consistent and significant positive fire selectivity, whereas deciduous woodlands showed consistent and significant negative selectivity. Shrublands were generally not selected. Although pines were generally selected, the degree of selectivity varied over the years. Additionally, no differences in selectivity were found between large (>100 ha) and smaller fires. Moreover, fires positively selected areas at small or intermediate distances to towns and intermediate distances to roads. Topographic features were less strongly selected for, which is consistent with the fact that these were considered fixed, and not aligned with the fire propagation front (Viedma et al. 2015b). In contrast, Carmo et al. (2011) found that selectivity for areas with different slope varied among several areas in northern Portugal. In this case, shrubs followed by forests (mainly *P. pinaster* and *Eucalyptus*) were the most strongly selected for.

The fact that distance to towns and roads can be a factor in determining what gets burned is particularly relevant in the context of the wildland-urban interface (WUI) (Lampin-Maillet et al. 2011). As mentioned earlier, pines or other highly flammable vegetation near towns and villages could have a higher probability of burning than at distant locations. Indeed, Calviño-Cancela et al. (2016) found that forest plantations, notably those of *P. pinaster*, shrublands and open woodlands showed the highest ignition risk at the WUI. They argued that the high flammability of such LULC types resulted in a disproportionate increase in the risk of ignition when the density of human population increased. This finding has important management implications because, pines and eucalypts have often been planted in highly productive areas, such as Galicia (Spain) and northern Portugal, where villages are interspersed throughout the territory, hence leading to a very high increment in fire hazard and actual risk.

21.4.2.2 Other Assessments

Forest inventories are a valuable source for estimating the probability of fire for various forests, including conifer species. Based on inventories and fire perimeters for a 12-year period the probability of fire was calculated for *Pinus* species in Catalonia (Spain; González et al. 2006): *P. nigra* (0.23 fire probability in 12 years), *P. halepensis* (0.12), *P. pinaster* (0.06), *P. pinea* (0.05), *P. sylvestris* (0.01) and *P. nigra* (<0.01). Additionally, they modeled fire probability for the various forest stands. In their model, the species dominating the forest plot (e.g., pines or other broad-leaved species) was one of the variables, together with other stand structural or topographic characteristics. While the period is too short to account for the large variability in the area burned from year to year, this approach provides a valuable tool for initially assessing fire hazard and, furthermore, guides management towards concentrating on more hazardous plots. The time-dependency of burn probability was analyzed in Portugal, applying survival analysis to mapped fires, showing that the median fire-free interval varies across regions (18–47 years) and that the probability of burning increased almost linearly with time. That is, there was not a clear fuel-age dependency. Large fires, which occurred under more severe fire weather, selected for older fuels (Fernandes et al. 2012).

A comparison of forestry plots with repeated inventories can also be valuable for showing the rate of change due to harvest and, when combined with fire maps, assessing the impact of fire. The probabilities of harvest in Castilla y León and Catalonia (Spain) are among the lowest in the EU (1.4% and 1.2%, respectively). Mortality is higher in Catalonia than in Castilla-León (0.4% and 0.2%). In Catalonia, if we exclude the plots within the perimeters of the severe fires of 1994 and 1998, the overall harvest probability decreases from 1.2% to 0.3%, and the mortality probability from 0.4% to 0.2% (Schelhaas et al. 2018). That is, fire plays an important role in the provision of timber services by pines in these regions, and probably elsewhere.

21.5 Pine Fires Towards the End of the Twenty-First Century in the Context of Climate and Other Global Changes

21.5.1 Projections of Species Distribution Change

The projections of the distribution of Mediterranean pine species under climate change support the overall claim that the most drought- and fire-sensitive mountain pines (*P. sylvestris* and *P. nigra*) will suffer a strong reduction in their distribution range, while the more Mediterranean species will be less affected or may even increase their range, particularly *P. halepensis*.

For widespread Euro-Siberian species, such as like *P. sylvestris*, which have their southernmost populations in the Mediterranean Basin, the general expectation is that, in spite of a range expansion at the pan-European scale as a consequence of climate change, at a more regional Mediterranean scale this species is projected to lose half of its current distribution (Bombi et al. 2017; Noce et al. 2017). The growth of the species around the Mediterranean Basin is projected to decrease under the RCP8.5 emission scenario, while under the milder RCP2.6 scenario only some Italian populations would experience growth limitations (Dorado-Liñán et al. 2019). Other researchers (Gómez-Aparicio et al. 2011; Sánchez-Salguero et al. 2017) have also projected a reduction of growth in Spain. Similarly, recruitment is expected to be reduced in the southernmost populations (Matías and Jump 2014). More local studies project a strong reduction of the suitable area for *P. sylvestris*, with the subsequent loss of populations in Spain (Benito Garzón et al. 2008a,b; García-López and Allué 2010; Keenan et al. 2011; Ruiz-Labourdette et al. 2012; Lloret et al. 2013; Ruiz-Labourdette et al. 2013; García-Valdés et al. 2015) and in Italy (Attorre et al. 2011; Di Traglia et al. 2011). Alongside this reduction in suitable area, most models project an upward shift in elevation, which can already be observed in the Pyrenean mountain range in Spain (Hernández et al. 2014, 2017) where *P. sylvestris* is expanding upwards and increasingly coexisting with other species. However, not all projections agree with such negative trends, and when models account for CO_2 fertilization, local adaptation and plasticity they find that this species might be able to maintain most of its current distribution range in Spain (Benito Garzón et al. 2011; Keenan et al. 2011; Valladares et al. 2014). Similarly, a spatially explicit patch-occupancy model showed that *P. sylvestris* would maintain most of its distribution range (García-Valdés et al. 2013).

Pinus nigra is also a mountain Mediterranean species whose growth is sensitive to drought and thus vulnerable to climate change (Gómez-Aparicio et al. 2011; Camarero et al. 2013). Most models project a strong decrease in suitable areas for this species or an upward shift in elevation in Italy (Marchi et al. 2016), Greece (Fyllas et al. 2017) and Spain (Benito Garzón et al. 2008a; Lloret et al. 2013; García-Valdés et al. 2015; Navarro-Cerrillo et al. 2018). However, as with *P. sylvestris*, not all models concur with such strong reductions, and demographic or patch-occupancy models project that *P. nigra* might maintain most of the distribution range (García-Valdés et al. 2013), or at least habitat suitability in northern Spain (Benito Garzón et al. 2011). *Pinus sylvestris* and *P. nigra* are fire-sensitive species without serotiny, so in addition to the constraints imposed by more unsuitable climatic conditions, they are likely to be negatively affected by an increase in fire frequency or intensity caused by climate change. An assessment of the effects of different fire regimes on landscape dynamics in north-eastern Spain found that fire-sensitive pine species, such as *P. sylvestris* and *P. nigra* would lose cover under the worst scenarios of climate change and large area burned (Gil-Tena et al. 2016).

The more Mediterranean species, like *P. pinaster* and *P. pinea,* will likely be less affected by climate change. In the case of *P. pinaster*, some studies project a slight decrease of habitat suitability at the scale of the Mediterranean region (Noce et al. 2017) and in Spain (Bede-Fazekas et al. 2014; García-Valdés et al. 2015), but most models project stability or even a slight increase in Italy (Attorre et al. 2011) and Spain (Benito Garzón et al. 2008a; Ruiz-Labourdette et al. 2012; Lloret et al. 2013). *Pinus pinea* is also a Mediterranean species for which some studies mostly anticipate reductions of suitable habitat (Benito Garzón et al. 2008a, 2011) and others expect a light increase (García-Valdés et al. 2013, 2015; Lloret et al. 2013; Bede-Fazekas et al. 2014). In the case of García-Valdés et al. (2015), the increase in area was found when climate change with or without habitat creation was considered, but if habitat destruction was also considered the model projected a reduction in species occupancy. Similarly, a loss of forest cover is anticipated for *P. pinea* under the fire regime scenarios that involved higher amounts of area burned (Gil-Tena et al. 2016).

Finally, the most Mediterranean, drought- and fire-adapted *P. halepensis* and *P. brutia* are expected to be favored by the new conditions. *Pinus halepensis* will very likely increase its distribution range, expanding towards new areas, but also losing some of its current locations (Benito Garzón et al. 2008a; Attorre et al. 2011; Ruiz-Labourdette et al. 2012; García-Valdés et al. 2013; Lloret et al. 2013; Bede-Fazekas et al. 2014; García-Callejas et al. 2016; Gil-Tena et al. 2016). Only one study (Keenan et al. 2011) found that ensemble niche models projected a light reduction, but this trend was reversed when they used a growth model that accounted for the possible effects of CO_2 fertilization. As a fire-adapted species, *P. halepensis* would increase its occupied area in north-eastern Spain even under harsher fire regime scenarios and without firefighting (Gil-Tena et al. 2016). Fewer studies are available for *P. brutia*, but they point to maintenance of the current distribution, with potential increases in suitable area and a replacement of broad-leaved forests by evergreen conifers, which include *P. brutia* (Bede-Fazekas et al. 2014; Donmez et al. 2016), although it has been argued that conifer forests in Mediterranean southern Turkey might be highly vulnerable to climate change (Türkeş et al. 2017).

21.5.2 Fires in the Future

Fire is a complex phenomenon that is affected by different processes at various spatial and temporal scales. Climate and its underlying meteorology are critical for processes that vary from seconds and minutes (fine dead fuel moisture content, lightning ignition), to days and weeks (live fuel moisture, fire propagation) and years and decades (vegetation, landscapes) (Moritz et al. 2005). Humans also intervene in fire in many ways: directly (ignition sources, fire extinction) or indirectly (forest management, planning and hence land use and land cover) (Brotons et al. 2013). Fire regime, i.e., fire frequency, size, season, intensity and extent (Archibald et al. 2013), is determined by the interactions of many drivers, each of which has many uncertainties, to accurately model fire occurrence and spread at any given place. This is more challenging with respect to the future, in which climate will change, but also the land use and land cover of the territory, the socioeconomic conditions of the areas affected by fires and technology, among others. Therefore, modeling the future fire regime is a complicated exercise, plagued with uncertainties (Harris et al. 2016). Moreover, feedbacks to fire caused by changes in ecosystem processes due to climate change, atmospheric change and biological change (species invasions or extinctions) need to be accounted for. Next, we will focus on reviewing the various approaches that have been used in the Mediterranean region to anticipate how climate change could affect fires in the future.

21.5.3 Conceptual Models

Conceptual models help us understand which factors control fires. Knowing them, one can venture into the future, assuming changes in one or more of the controlling factors, to qualitatively estimate how forest fires will be affected. Among these models, the so-called fire triangle (vegetation [fuels], oxygen, heat) (Byram 1957) is commonly used to explain the principles of combustion. This model has been extended to incorporate several triangles to clarify how drivers vary across temporal and spatial scales (from seconds to decades, and from leaf level to entire regions), and the feedbacks and controls between different processes and scales, to more accurately reflect forest fire controls (Moritz et al. 2005). The model can explain how fires can occur and spread in certain places and not in others, or during certain times of the year, or how changes in climate can directly or indirectly affect fires.

Other models are based on comparisons of fires across space in order to understand their constraints across resource gradients. There are places in which vegetation abounds all year round, but weather conditions for it to burn may not be available during the year (e.g., tropical evergreen moist forests), or at certain times (e.g., boreal forest). At other locations, conditions may be amenable to burning during the entire year; however, vegetation may not be abundant enough to support fire (e.g., arid areas and deserts). Only when moisture is available, vegetation may grow sufficiently to sustain fire (Meyn et al. 2007). The relationship between direct (conditions to burn, moisture) and indirect (biomass accumulation and structure) effects of climate and fire enables us to explain patterns of fire across the globe (Krawchuk and Moritz 2011). Pausas and Ribeiro (2013) further developed this idea to formulate the intermediate fire-productivity hypothesis, whereby fires are controlled by productivity, thus they are most frequent at some intermediate levels of productivity, decreasing towards lower and higher levels.

Related to the above models is the so-called four-switch model. Biomass production (B), its availability to burn (A), fire weather (S, for spread) and ignition (I) are a set of conditional processes driving fire. For a fire to occur, all four "switches" must be simultaneously activated (Bradstock 2010). Note that biomass production is not enough to create a fire, but whether it is in a state (dry enough, that is, available) to allow the spread of fire. This model integrates the concepts of the previous models into one. The application of these ideas has led to the development of the pyrome concept (Archibald et al. 2013), that is, areas that share a similar fire regime.

Conceptual frameworks allow us to anticipate that if some fire drivers change (e.g., temperature, wind, fuels, climate, etc.), the fire regime will change. While they provide only a qualitative estimate of how changes in the future could affect fires, they set the basis for a more in-depth analysis by way of understanding how the various fire drivers operate at the relevant scales. For example, based on the intermediate fire-productivity hypothesis, increasing severe weather at places limited by fuel will not increase fires, whereas where fires are limited by weather they will likely do so, provided that a certain weather threshold is reached (Bedia et al. 2015). Thus, climate change may lead to reduced fire activity in more arid areas (southeastern Spain, northern Africa), but increased fires at the northern limits of the Mediterranean, in more humid areas, and in the upper mountains.

Furthermore, the strength of the fire-climate relationship will decrease as seasonal weather severity rises, leading to lower predictability of fire in relation to weather/climate (Urbieta et al. 2015). That is, as regional climate warms and precipitation decreases in the southern Mediterranean, the capacity for anticipating a fire season based on weather will be more limited than in the more temperate and mesic areas in the North. Nevertheless, see Vázquez and Moreno (1995) for the complexities that may arise in areas dominated by fires ignited by people. Lagged relationships between rainfall and fire activity are more likely to occur where water limits fuel availability (Pausas 2004). Consequently, droughts followed by rainfall spells might have a delayed fire response. Again, this is more likely to occur in semiarid areas than in more mesic ones.

Additionally, different drivers will have varying consequences for fires. For example, climate, CO_2 or land-use change may differentially affect each of the switches (B, A, S, I), and thus fire regime (Bradstock 2010). Consequently, the factors important in a given place need to be considered to anticipate the response of fire as they change. Nevertheless, conceptual models can help us to understand the factors that control fire occurrence in a given place. They are, however, limited to estimating other variables of the fire regime (Harris et al. 2016).

21.5.4 Correlative Models

The relationship between fire activity (number of fires, area burned, seasonality, etc.) and weather/climate in a given area has been tested at various levels (regional or subregional, countries, provinces/departments, etc.). Weather/climate variables are used as predictors either by themselves (de la Cueva et al. 2012), as fire weather indices (Pereira et al. 2013), or as other meteorological/climatic indices, of which the most common are drought indices (e.g., standardized precipitation-evaporation index [SPEI], Palmer Drought Severity Index [PDSI], or the DC code of the Canadian Fire Weather Index [CFWI]) (Gudmundsson et al. 2014; Urbieta et al. 2015; Turco et al. 2017). The CFWI is routinely used by several countries in the Mediterranean region, including the EU; it comprises combinations of several daily meteorological variables. In addition, other biophysical or socioeconomic variables have also been used by themselves or in combination with some of the above to explain fire activity (Viedma et al. 2015a). These relationships help to explain past patterns of fires, and are based on the assumption of stationarity during the period studied. Once a relationship is established for the past, the relevant independent variables are modeled using specified changes that are supposed to represent the future, or, most commonly, one or various emission scenarios to infer or quantitatively project future fire activity for specified periods during the twenty-first century. The literature in this field in Europe has experienced notable growth in recent years, much of which has been produced in Mediterranean countries (Dupuy et al. 2020). Next, we review a representation of models that use various approaches, concentrating on regional and national levels. Global models will be normally omitted unless they have a specific output for the Mediterranean region.

21.5.4.1 Future Fires and Weather

Several studies have related fire activity (number of fires, area burned, number of large fires, etc.) to fire weather indices, normally based on the CFWI, either using one single code or several of the ones composing this system. Later, based on various emission scenarios, changes in codes are calculated for various periods of the year, including the whole year, and one or more time slices, normally until the end of the twenty-first century. Future fire activity is only inferred, not specifically calculated. The first of this type of study for the whole Mediterranean region was conducted by Moriondo et al. (2006). They found that fire danger will increase markedly in the region, with some differences between and within countries. Fire danger increased with increasing emissions (A2 vs. B2 scenarios) (Nakicenovic et al. 2000). Climate change implied a higher number of seasons with high danger values, increased fire season length, and increased number of extreme events. Based on this, they argued that area burned, fire intensity and severity will increase in the region by the end of the current century.

Accounting for the variability in projections among models is an important step forward in this type of analysis, because variability among models can be large, as is differentiating between more proximal and distant times. Using six RCM-GCM ensembles for the A1B scenario (Nakicenovic et al. 2000), for four 30-year periods, starting in 1971–2000 and ending in 2071–2100, and the entire Mediterranean region (Bedia et al. 2014), a consistent increase in fire danger and seasonal severity was found over time, with magnitude varying among countries. Increases up to midcentury were small and became larger with time (Fig. 21.5). While the uncertainty for continuous variables (e.g., FWI) was low, the use of frequency over thresholds (e.g., number of days above certain FWI values, which is done to characterize extreme days) suffered from larger uncertainties, and the use of such indices in impact studies was not recommended. Modeling (Bedia et al. 2015) for the whole world based on the RCP4.5 and RCP8.5 scenarios produced similar results. We note the increase in fire seasonal severity towards the north and east (Fig. 21.5). Similar studies based on CFWI have been carried out for the EU + Turkey, confirming that

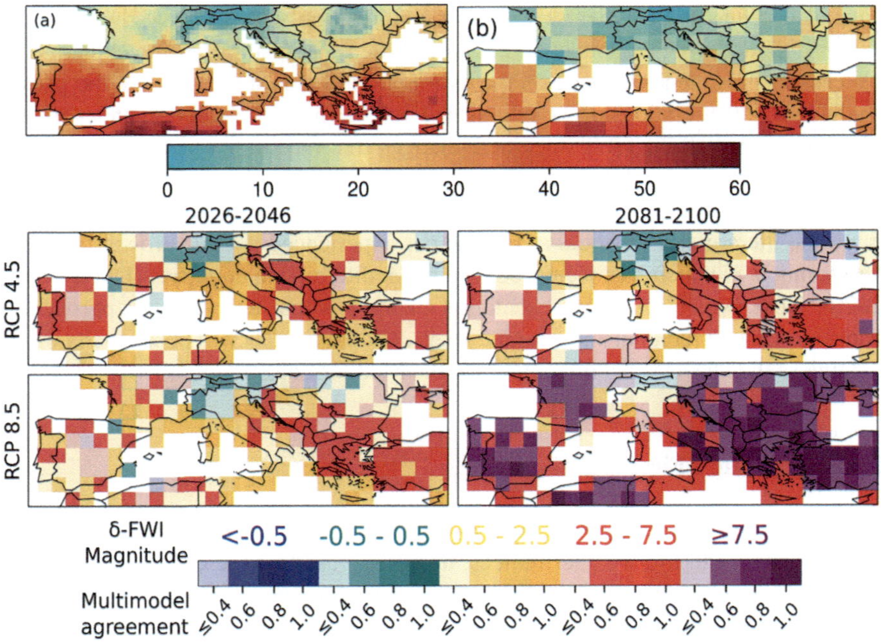

Fig. 21.5 Upper panel: (**a**) Seasonal fire severity (mean FWI) for the Mediterranean region and the reference period (1981–2000) based on a grid cell resolution of 0.5°; (**b**) historical simulation of the multimodel ensemble for the same period. Lower panel: Projected FWI anomalies (δ-FWI) for the multimodel ensemble for the fire season (JJAS) for two periods, (proximal [left], 2026–2046 and distant, 2081–2100 [right]), and two forcing scenarios, moderate (RCP 4.5, top) and high (RCP 8.5, bottom). Color tones denote the magnitude and sign of the anomaly, while transparency indicates the fraction of models agreeing with a given anomaly. (From Bedia et al. 2015)

fire seasonal severity will increase in the south, expanding to the north of the region and beyond (de Rigo et al. 2017). Similar studies based on CFWI have been carried out at the EU level, including Turkey, confirming that fire seasonal severity will increase in the south, expanding to the north of the region and beyond (de Rigo et al. 2017). Similar modeling at the national level, e.g., Italy (Faggian 2018), or Greece (Karali et al. 2014), reached similar conclusions.

21.5.4.2 Modeling Fire Frequency and Area Burned

Some authors have calculated future fire regimes (fire frequency, area burned) based on the past relationships of these variables with weather. Fire danger and area burned were modeled using the A2 and B2 scenarios and three RCMs for 2071–2100 for the EUMED countries (Amatulli et al. 2013). On average, seasonal fire danger severity increased by 25% and 38% for the B2 and A2 scenario, respectively, for all five countries. There were important differences among countries, with Portugal having the largest increase by the end of the century (51%, A2). Changes in area burned amounted to 66% (B2) and 140% (A2) of the reference period for all countries. Differences in the projections of area burned among countries were important, with Spain being projected to undergo the largest increase (839%, A2).

Future fire occurrence was modeled based on fire scars from remote sensing and a projection of five meteorological variables related to precipitation and temperature from six GCMs and the A2 scenario, and for three periods (2010–2039, 2040–2069, 2070–2099) (Batllori et al. 2013). The analysis was carried out for each of the five MTAs. They found that changes in fire probability might not equally affect the whole Mediterranean region. They defined two climate syndromes: warmer-drier, and warmer-wetter. Fire trends over the century were dominated by increased probability of fire for both syndromes. However, reduced fire probability gradually increased, notably under the warmer-drier syndrome. Increased/decreased fire probabilities were concentrated at high/low latitudes for the warmer-drier syndrome, whereas the opposite was found for the warmer-wetter one. This study demonstrates that changes in climate will affect fires in a complex way. Thus, where today it is warm and dry, but will become warmer and drier, fires may not increase due to fuel limitations.

Other studies have focused on the national level, and extrapolations were made for area burned based on past weather/climate fire relationships either using the CFWI, e.g., Pereira et al. (2013), and Carvalho et al. (2010) for Portugal, or weather variables, e.g., de la Cueva et al. (2012) for Spain and Sousa et al. (2015) for the Iberian Peninsula. The area burned was projected to increase substantially in all these modeling exercises. By the end of the century, the area burned could be two to three times larger than the baseline periods used, always under the assumption that everything else will remain unaltered, e.g., Sousa et al. (2015). While differences among models were large, the spatial resolution of the models did not significantly alter these projections (Carvalho et al. 2010).

A problem with models that are based on past weather/climate-fire relationships is that these may not be stationary, as supported by the fact that climate and land use may facilitate fires, but policies and firefighting may do the opposite (Urbieta et al. 2019). To account for non-stationarity, Turco et al. (2018) built a model in which the climate-fire relationship was modified to account for this. They used one drought index (SPEI) to feed future projections, based on two ensembles of nine regional climate simulations (four RCMs and five GCMs), and calculated changes for specific global temperature increases (1.5 °C, 2 °C and 3 °C), rather than for specified periods. Their results support the fact that accounting for non-stationarity reduced the projections of area burned, particularly for large temperature increments. Increases were larger with a greater degree of global warming, with increments from 40–54% (1.5 °C), to 62–87% (2 °C), and to 96–187% (3 °C). Still, projections were purely correlative, and did not account for the complex climate-vegetation interactions that may emerge over time (Brotons and Duane 2019), or for the role of changes in policies and their impacts on ignitions, in an area where most of them are caused by people (Ganteaume et al. 2013).

21.5.5 *Process-Based Models*

21.5.5.1 Landscape Fire Modeling

Landscape fire behavior modeling is a tool that is often used to help managers assess critical fire locations in the landscape to help manage fires (Sakellariou et al. 2020). Thus, they have the capacity to anticipate how changes in the various drivers operating at the landscape scale could affect fire. Normally, outputs calculate fire probabilities (i.e., fire frequency), flame length, and fire size. The potential effects of climate change on fires were studied in Italy and Corsica (France), using the A1B scenario. Results showed that burn probabilities increased, as well as fire size, whereas flame length varied little (Lozano et al. 2017). Similar studies have been carried out in Greece, where Mitsopoulos et al. (2015) used FLAMMAP and specific scenarios of increasingly severe weather, without basing their predictions on particular emission scenarios, and a detailed reconstruction of fuel types. They found that fire risk increased markedly in dense shrublands and *P. halepensis* forests under the most extreme burning scenario. A study in the region of Messinia (Greece) based on one RCM and the A1B scenario and the MMT algorithm found that larger fires resisting initial control will increase in the future, as will fire probabilities and fire intensities for large portions of their study area (Kalabokidis et al. 2015).

These studies support the claim that changes in climate will bring more severe fire seasons. Given similar fuels and landscapes as in the past, fires will be more frequent, more severe and larger. While illustrative of what could happen in the future, the realization of how fuels will change is still missing, as are the roles of fire prevention and mitigation, among other processes. Moreover, the interaction between climate and vegetation is not included, neither are the direct leverage effects of fires themselves, or more indirect effects due to reduced soil fertility resulting from increased fire frequency or drought (Hinojosa et al. 2019).

21.5.5.2 Ecosystem and Fire Modeling

The limitations of correlative or fire behavior modeling that do not account for more realistic representations of the feedbacks between climate, vegetation and other ecosystem processes, including the response to fire, have been addressed in part in models that specifically attempt to account for some of these processes. To this end, Dynamic Global Vegetation Models (DGVMs) attempt to simulate vegetation dynamics in response to climate change by modeling processes such as photosynthesis, soil moisture, plant growth, soil decomposition, regeneration after fire, and plant competition, among other processes, in a more realistic approximation. An important issue for Mediterranean ecosystems is the interaction between CO_2 and water use efficiency. Furthermore, water limitations are important, since increasing drought (Vicente-Serrano et al. 2014) can increase dead fuels. These models now incorporate fire modules. To achieve these goals, these models use a simplified representation of plants and the landscape. Nevertheless, they can provide insights into how various limiting factors could affect fire.

As indicated, water is a limiting factor in the region, and climate change is projected to reduce rainfall, with fewer rain events of greater intensity. How water stress may affect plant stress, and through it fire, was modeled by Mouillot et al. (2002) in a Corsican maquis dominated by several shrubs and *P. pinaster*. They used a semi-empirical model (SIERRA) that accounts for some basic plant assimilation processes, as well as regeneration after fire, in combination with climates derived from one GCM model. They found that increasing water stress with climate change led to increased fire frequency. A decrease in the time interval between two successive fires in the forest was enough to transform the landscape into one dominated by shrublands.

Wu et al. (2015) conducted a study to explore the effects of CO_2 on plant growth and the interaction with fire. They used two models (LPJ-GUESS-SIMFIRE and LPJmL-SPITFIRE) and eight Earth System Models with two representative concentration pathways (RCP2.6 and RCP8.5) (van Vuuren et al. 2011). Simulations included changes in climate, CO_2 and population density. Results showed that the simulated area burned in the Mediterranean region increased, much more than emissions did (RCP2.6 to RCP8.5). In both models, CO_2 was a significant factor, albeit with differences among models; while in LPJ-GUESS climate had a larger signal, in LPJmL climate change reduced fuel loads, making them a limiting factor for fires. Thus, in spite of the increasing fire danger, climate change imposed a negative effect on fires through limited plant growth, and reduced fire spread and area burned.

This analysis demonstrates the importance of including feedbacks and interactions among the various drivers that affect fire. Weighting each is complicated, hence the value of using different assumptions and models. Notwithstanding, there are limitations that arise from the type of vegetation. As we have shown earlier, *Pinus* species are a major component of the Mediterranean vegetation and were not specifically included in these models.

The negative feedbacks between climate change and fires were explored by Loepfe et al. (2012) in three areas in north-eastern Spain, dominated by shrublands and pine woodlands of *P. halepensis*, *P. nigra* and *P. sylvestris*. They used a fire-regime model (FIRE LADY) in which biomass and moisture drive fire spread; climate was based on one GCM and the A2, B2 scenarios. Tree growth and competition were simulated. They found that changes in fire regime differ depending on the respective areas; in drier areas, where fires had been frequent and large, a gradual increase would be expected, and the landscape would be transformed into one dominated by shrubland, while in humid areas, warming and drying will multiply fires, as well as their magnitude. Management will not be able to offset this change.

The various models described here provide examples of how climate change could affect fires. They all show that while climate change will increase fire weather severity, changes in fire activity will not be straightforward. Whether or not fire frequency increases throughout the region because of more severe weather is dubious, unlike the intensity and extent of fires once they break out. In places where water is more limiting, the level of change in fire activity will differ from places where moisture is more available. Limited moisture can reduce fires due to lower vegetation growth, whereas where vegetation growth is more favorable, more severe weather can increase fire activity more than we have ever known. This opens the way to new fire areas in Europe beyond the Mediterranean region (Wu et al. 2015). Changes in vegetation are likely to occur if increased frequency does not allow *Pinus* to establish, with shrubs or other evergreen replacing some pines (Rodrigo et al. 2004; Torres et al. 2016). The stress and demographic attrition due to heat and lack of moisture (Ruiz-Benito et al. 2015) caused by climate change can threaten the persistence of pine populations in places where they will be retreating.

21.6 Conclusions

Mediterranean countries have a much higher proportion of forests and other woodland areas than the rest of the world and it is increasing. Pines cover up to nearly half of forest and woodland areas, either natural or afforested (e.g., in Spain, nearly 4 Mha of pines were planted during the twentieth century). *Pinus pinaster* (west), *P. brutia* (east) and *P. halepensis* (throughout) are the dominant species in the lowlands and foothills, while *P. sylvestris* and *P. nigra* dominate the upper mountains.

The palaeoecological record shows that pines have played a major role in the vegetation of the Mediterranean region during the Holocene, particularly in mountainous areas, under continental climates, and on ‘challenging’ soils. Fires have been a rather common disturbance affecting pinewoods under natural conditions, with mountain pines resisting or recovering from quite frequent surface and even rare crown fires. However, human-induced changes in fire regimes involving increasing incidence of crown fires to clear forests for farming have caused widespread demise of pine forests across the Mediterranean region. The palaeoecological record therefore attests to the high sensitivity of pinewoods to fires related to human activities.

Forest fires nowadays occur throughout the region. The fire frequency and area burned are greater in the EUMED countries than in northern Africa or in the eastern Mediterranean countries. In general, fire activity increased from the mid-twentieth century until the late 1980s, decreasing thereafter. This trend is clear for the EUMED countries, despite increases in forest and other woodland areas or regional warming and drying, probably due to fire policies and increased firefighting capacity. However, very large, catastrophic fires still occur. Fires primarily affect shrublands and other open vegetation, followed by forests. Within the latter, pine forests are usually the ones that burn the most. Nevertheless, during extreme weather or at particular locations, notably at the WUI, pine forests can be the most affected by fire.

Climate change will very likely reduce the suitable habitat of the most drought- and fire-sensitive mountain pines (*P. sylvestris*, *P. nigra*), but local adaptation and plasticity may enable them to maintain larger parts of their decreasing ranges. On the other hand, more drought-tolerant and fire-adapted species will be less affected, or will even be able to increase their range, particularly *P. halepensis*. The changes in climate projected for the twenty-first century support the claim that fire weather will increase much more than the level of greenhouse emissions and thus of global warming. Increments in fire weather will rise over time, particularly beyond the middle of the century. This will cause a fire season with more severe, longer, and more extreme fire days. How this will affect fire frequency and the area burned is more complicated to anticipate, given the many uncertainties that exist. Empirical models that assume burning rates similar to those of the past project a several-fold increment in the area burned. However, mechanistic and conceptual models suggest that increases will vary depending on the factor controlling the fires. In areas with lower temperatures but high water availability, more to the north and up in the mountains, and with abundant fuels, where fires are limited by temperature, fires will very likely increase; however, in warmer and drier areas, more towards the south, fuel limitations will very likely reduce the size of the area burned as time goes by. Observations and modeling support the claim that increases in fire frequency will turn pine forests into shrublands. Higher fire intensities will threaten mountain pines due to their reduced regeneration capacity.

Acknowledgments We (JMM, IT) acknowledge funding from the Government of Spain (project CGL2016-78357-R).

References

Alexandrian D, Esnault F (2001) Politiques nationales ayant une incidence sur les incendies de forêt dans le Bassin Méditerranéen. In: Etude FAO Forêts 138 – Réunion de la FAO sur les Politiques Nationales Ayant une Incidence sur les Incendies de Forêt, FAO, Rome. http://www.fao.org/3/x2095f08.htm#TopOfPage

Allen JRM, Watts WA, McGee E et al (2002) Holocene environmental variability – the record from Lag Grande di Monticchio, Italy. Quat Int 88:69–80

Amatulli G, Camia A, San-Miguel-Ayanz J (2013) Estimating future burned areas under changing climate in the EU-Mediterranean countries. Sci Total Environ 450–451:209–222

Améztegui A, Brotons L, Coll L (2010) Land-use changes as major drivers of mountain pine (Pinus uncinata Ram.) expansion in the Pyrenees. Glob Ecol Biogeogr 19:632–641

Amraoui M, Liberato M, Calado T et al (2013) Fire activity over Mediterranean Europe based on information from Meteosat-8. For Ecol Manag 294:62–75

Archibald S, Lehmann CER, Gómez-Dans JL et al (2013) Defining pyromes and global syndromes of fire regimes. Proc Natl Acad Sci U S A 110:6442–6447

Attorre F, Alfò M, De Sanctis M et al (2011) Evaluating the effects of climate change on tree species abundance and distribution in the Italian peninsula. Appl Veg Sci 14:242–255

Barbero M, Loisel R, Quézel P et al (1998) Pines of the Mediterranean Basin. In: Richardson DM (ed) Ecology and biogeography of *Pinus*. Cambridge University Press, Cambridge, pp 153–170

Barros AMG, Pereira JMC (2014) Wildfire selectivity for land cover type: does size matter? PLoS One 9:e84760

Batllori E, Parisien M-A, Krawchuk MA et al (2013) Climate change-induced shifts in fire for Mediterranean ecosystems. Glob Ecol Biogeogr 22:1118–1129

Bede-Fazekas Á, Horváth L, Kocsis M (2014) Impact of climate change on the potential distribution of Mediterranean pines. Időjárás/Quart J Hung Met Serv 118:41–52

Bedia J, Herrera S, Camia A et al (2014) Forest fire danger projections in the Mediterranean using ENSEMBLES regional climate change scenarios. Clim Chang 122:185–199

Bedia J, Herrera S, Gutiérrez JM et al (2015) Global patterns in the sensitivity of burned area to fire-weather: implications for climate change. Agric For Meteorol 214:369–379

Beffa G, Pedrotta T, Colombaroli D et al (2016) Vegetation and fire history of coastal North-Eastern Sardinia (Italy) under changing Holocene climates and land use. Veg Hist Archaeobot 25:271–289

Belhadj-Khedher C, Koutsias N, Karamitsou A et al (2018) A revised historical fire regime analysis in Tunisia (1985–2010) from a critical analysis of the National Fire Database and remote sensing. Forests 9:59

Benito Garzón M, Sánchez de Dios R, Sainz Ollero H (2008a) Effects of climate change on the distribution of Iberian tree species. Appl Veg Sci 11:169–178

Benito Garzón M, Sánchez de Dios R, Sainz Ollero H (2008b) The evolution of the *Pinus sylvestris* L. area in the Iberian Peninsula from the last glacial maximum to 2100 under climate change. The Holocene 18:705–714

Benito Garzón M, Alía R, Robson TM et al (2011) Intra-specific variability and plasticity influence potential tree species distributions under climate change. Glob Ecol Biogeogr 20:766–778

Bombi P, D'Andrea E, Rezaie N et al (2017) Which climate change path are we following? Bad news from Scots pine. PLoS One 12:e0189468

Bottema S, Sarpaki A (2003) Environmental change in Crete: a 9000-year record of Holocene vegetation history and effect of the Santorini eruption. The Holocene 13:733–749

Bradstock RA (2010) A biogeographic model of fire regimes in Australia: current and future implications. Glob Ecol Biogeogr 19:145–158

Brotons L, Duane A (2019) Correspondence: uncertainty in climate–vegetation feedbacks on fire regimes challenges reliable long-term projections of burnt area from correlative models. Fire 2(1):8

Brotons L, Aquilué N, De Cáceres M et al (2013) How fire history, fire suppression practices and climate change affect wildfire regimes in Mediterranean landscapes. PLoS One 8(5):e62392

Budde KB, González-Martínez SC, Navascués M et al (2017) Increased fire frequency promotes stronger spatial genetic structure and natural selection at regional and local scales in *Pinus halepensis* Mill. Ann Bot 119:1061–1072

Burjachs F, Pérez-Obiol R, Picornell-Gelabert L et al (2017) Overview of environmental changes and human colonization in the Balearic Islands (Western Mediterranean) and their impacts on vegetation composition during the Holocene. J Archaeol Sci Rep 12:845–859

Byram GM (1957) Some principles of combustion and their significance in forest fire behaviour. Fire Control Notes 18:47–57

Calò C, Henne PD, Eugster P et al (2013) 1200 years of decadal-scale variability of Mediterranean vegetation and climate at Pantelleria Island, Italy. The Holocene 23:1477–1486

Calviño-Cancela M, Chas-Amil ML, García-Martínez ED et al (2016) Wildfire risk associated with different vegetation types within and outside wildland-urban interfaces. For Ecol Manag 372:1–9

Camarero JJ, Manzanedo RD, Sanchez-Salguero R et al (2013) Growth response to climate and drought change along an aridity gradient in the southernmost Pinus nigra relict forests. Ann For Sci 70:769–780

Camarero JJ, Sangüesa-Barreda G, Pérez-Díaz S et al (2019) Abrupt regime shifts in post-fire resilience of Mediterranean mountain pinewoods are fuelled by land use. Int J Wildland Fire 28:329–341

Camuera J, Jiménez-Moreno G, Ramos-Román MJ et al (2019) Vegetation and climate changes during the last two glacial-interglacial cycles in the western Mediterranean: a new long pollen record from Padul (southern Iberian Peninsula). Quat Sci Rev 205:86–105

Caplat P, Lepart J, Marty P (2006) Landscape patterns and agriculture: modelling the long-term effects of human practices on *Pinus sylvestris* spatial dynamics (Causse Mejean, France). Landsc Ecol 21:657–670

Carmo M, Moreira F, Casimiro P et al (2011) Land use and topography influences on wildfire occurrence in northern Portugal. Landsc Urban Plann 100:169–176

Carrión JS (2002) Patterns and processes of Late Quaternary environmental change in a montane region of southwestern Europe. Quat Sci Rev 21:2047–2066

Carrión JS, Fernández S (2009) The survival of the 'natural potential vegetation' concept (or the power of tradition). J Biogeogr 36:2202–2203

Carrión JS, van Geel B (1999) Fine-resolution Upper Weichselian and Holocene palynological record from Navarrés (Valencia, Spain) and a discussion about factors of Mediterranean forest succession. Rev Palaeobot Palynol 106:209–236

Carrión JS, Navarro C, Navarro J et al (2000) The distribution of cluster pine (*Pinus pinaster*) in Spain as derived from palaeoecological data: relationships with phytosociological classification. The Holocene 10:243–252

Carrión JS, Andrade A, Bennett KD et al (2001) Crossing forest thresholds: inertia and collapse in a Holocene sequence from South-Central Spain. The Holocene 11:635–653

Carrión JS, Sánchez-Gómez P, Mota JF et al (2003) Holocene vegetation dynamics, fire and grazing in the Sierra de Gádor, Southern Spain. The Holocene 13:839–849

Carrión JS, Fuentes N, González-Sampériz P et al (2007) Holocene environmental change in a montane region of southern Europe with a long history of human settlement. Quat Sci Rev 26:1455–1475

Carrión JS, Fernández S, González-Sampériz P et al (2010) Expected trends and surprises in the Lateglacial and Holocene vegetation history of the Iberian Peninsula and Balearic Islands. Rev Palaeobot Palynol 162:458–475

Carvalho A, Flannigan MD, Logan KA et al (2010) The impact of spatial resolution on area burned and fire occurrence projections in Portugal under climate change. Clim Chang 98:177

Cheddadi R, Khater C (2016) Climate change since the last glacial period in Lebanon and the persistence of Mediterranean species. Quat Sci Rev 150:146–157

Cheddadi R, Vendramin GG, Litt T et al (2006) Imprints of glacial refugia in the modern genetic diversity of *Pinus sylvestris*. Glob Ecol Biogeogr 15:271–282

Christopoulou A, Fulé PZ, Andriopoulos P et al (2013) Dendrochronology-based fire history of *Pinus nigra* forests in Mount Taygetos, Southern Greece. For Ecol Manag 293:132–139

Ciancio O (1986) Sylviculture du pin d'Alep. Le pin d'Alep et le pin brutia dans la sylviculture méditerranéenne, vol 1986-I. CIHEAM, Paris, pp 47–54

Colombaroli D, Vannière B, Chapron E et al (2008) Fire-vegetation interactions during the Mesolithic-Neolithic transition at Lago dell'Accesa, Tuscany, Italy. The Holocene 18:679–692

Connor SE, Araújo J, van der Knaap WO et al (2012) A long-term perspective on biomass burning in the Serra da Estrela, Portugal. Quat Sci Rev 55:114–124

Costa M, Morla C, Sainz H (eds) (2005) Los Bosques Ibéricos. Una interpretación geobotánica, 2nd edn. Editorial Planeta, Barcelona

Curt T, Frejaville T (2018) Wildfire policy in Mediterranean France: how far is it efficient and sustainable? Risk Anal 38:472–488

de Beaulieu J-L, Brugiapaglia E, Joannin S et al (2017) Lateglacial-Holocene abrupt vegetation changes at Lago Trifoglietti in Calabria, Southern Italy: the setting of ecosystems in a refugial zone. Quat Sci Rev 158:44–57

de la Cueva AV, Quintana JR, Cañellas I (2012) Fire activity projections in the SRES A2 and B2 climatic scenarios in peninsular Spain. Int J Wildland Fire 21:653–665

de Rigo D, Libertà G, Houston Durrant T et al (2017) Forest fire danger extremes in Europe under climate change: variability and uncertainty. EUR 28926. Publication Office of the EU, Luxembourg

DeSoto L, Olano JM, Rozas V et al (2010) Release of Juniperus thurifera woodlands from herbivore-mediated arrested succession in Spain. Appl Veg Sci 13:15–25

Desprat S, Díaz Fernández PM, Coulon T et al (2015) *Pinus nigra* (European black pine) as the dominant species of the last glacial pinewoods in south-western to Central Iberia: a morphological study of modern and fossil pollen. J Biogeogr 42:1998–2009

Di Traglia M, Attorre F, Francesconi F et al (2011) Is cellular automata algorithm able to predict the future dynamical shifts of tree species in Italy under climate change scenarios? A methodological approach. Ecol Model 222:925–934
Donmez C, Berberoglu S, Cilek A et al (2016) Spatiotemporal modeling of net primary productivity of Eastern Mediterranean biomes under different regional climate change scenarios. Int J Environ Res 10:342–356
Dorado-Liñán I, Piovesan G, Martínez-Sancho E et al (2019) Geographical adaptation prevails over species-specific determinism in trees' vulnerability to climate change at Mediterranean rear-edge forests. Glob Chang Biol 25:1296–1314
Duane A, Kelly L, Giljohann K et al (2019) Disentangling the influence of past fires on subsequent fires in Mediterranean landscapes. Ecosystems 22:1338–1351
Dupire S, Curt T, Bigot S (2017) Spatio-temporal trends in fire weather in the French Alps. Sci Total Environ 595:801–817
Dupuy J-L, Fargeon H, Martin-StPaul N et al (2020) Climate change impact on future wildfire danger and activity in southern Europe: a review. Ann For Sci 77:35
Eastwood WJ, Roberts N, Lamb HF et al (1999) Holocene environmental change in Southwest Turkey: a palaeoecological record of lake and catchment-related changes. Quat Sci Rev 18:671–695
Faggian P (2018) Estimating fire danger over Italy in the next decades. Euro Med J Environ Int 3:15
Fall PL (2012) Modern vegetation, pollen and climate relationships on the Mediterranean island of Cyprus. Rev Palaeobot Palynol 185:79–92
FAO (Food and Agriculture Organization) (2015) Global forest resources assessment 2015 – desk reference, Rome
Fernandes PM, Loureiro C, Magalhães M et al (2012) Fuel age, weather and burn probability in Portugal. Int J Wildland Fire 21:380–384
Fournier TP, Battipaglia G, Brossier B et al (2013) Fire-scars and polymodal age-structure provide evidence of fire-events in an Aleppo pine population in southern France. Dendrochronologia 31:159–164
Fulé PZ, Ribas M, Gutiérrez E et al (2008) Forest structure and fire history in an old *Pinus nigra* forest, Eastern Spain. For Ecol Manag 255:1234–1242
Fyllas NM, Christopoulou A, Galanidis A et al (2017) Predicting species dominance shifts across elevation gradients in mountain forests in Greece under a warmer and drier climate. Reg Environ Chang 17:1165–1177
Ganteaume A, Camia A, Jappiot M et al (2013) A review of the main driving factors of forest fire ignition over Europe. Environ Manag 51:651–662
García-Callejas D, Molowny-Horas R, Retana J (2016) Projecting the distribution and abundance of Mediterranean tree species under climate change: a demographic approach. J Plant Ecol 10:731–743
García-López JM, Allué C (2010) Effects of climate change on the distribution of *Pinus sylvestris* L. stands in Spain. A phytoclimatic approach to defining management alternatives. For Syst 19:329
García-Valdés R, Zavala MA, Araújo MB (2013) Chasing a moving target: projecting climate change-induced shifts in non-equilibrial tree species distributions. J Ecol 101:441–453
García-Valdés R, Svenning J-C, Zavala MA et al (2015) Evaluating the combined effects of climate and land-use change on tree species distributions. J Appl Ecol 52:902–912
Gassner S, Gobet E, Schwörer C et al (2020) 20,000 years of interactions between climate, vegetation and land use in Northern Greece. Veg Hist Archaeobot 29:75–90
Gea-Izquierdo G, Viguera B, Cabrera M et al (2014) Drought induced decline could portend widespread pine mortality at the xeric ecotone in managed Mediterranean pine-oak woodlands. For Ecol Manag 320:70–82
Gerard F, Petit S, Smith G et al (2010) Land cover change in Europe between 1950 and 2000 determined employing aerial photography. Prog Phys Geo Earth Environ 34:183–205

Gil-Tena A, Aquilué N, Duane A et al (2016) Mediterranean fire regime effects on pine-oak forest landscape mosaics under global change in NE Spain. Eur J For Res 135:403–416

Gómez A, Vendramin GG, González-Martínez SC et al (2005) Genetic diversity and differentiation of two Mediterranean pines (*Pinus halepensis* Mill. and *Pinus pinaster* Ait.) along a latitudinal cline using chloroplast microsatellite markers. Divers Distrib 11:257–263

Gómez-Aparicio L, García-Valdés R, Ruiz-Benito P et al (2011) Disentangling the relative importance of climate, size and competition on tree growth in Iberian forests: implications for forest management under global change. Glob Chang Biol 17:2400–2414

González JR, Palahí M, Trasobares A et al (2006) A fire probability model for forest stands in Catalonia (North-East Spain). Ann For Sci 63:169–176

Gudmundsson L, Rego FC, Rocha M et al (2014) Predicting above normal wildfire activity in southern Europe as a function of meteorological drought. Environ Res Lett 9:084008

Harris RMB, Remenyi TA, Williamson GJ et al (2016) Climate–vegetation–fire interactions and feedbacks: trivial detail or major barrier to projecting the future of the Earth system? WIREs Clim Chang 7:910–931

Hernández L, Rubiales JM, Morales-Molino C et al (2011) Reconstructing forest history from archaeological data: a case study in the Duero basin assessing the origin of controversial forests and the loss of tree populations of great biogeographical interest. For Ecol Manag 261:1178–1187

Hernández L, Cañellas I, Alberdi I et al (2014) Assessing changes in species distribution from sequential large-scale forest inventories. Ann For Sci 71:161–171

Hernández L, Sánchez de Dios R, Montes F et al (2017) Exploring range shifts of contrasting tree species across a bioclimatic transition zone. Eur J For Res 136:481–492

Hinojosa MB, Laudicina VA, Parra A et al (2019) Drought and its legacy modulate the post-fire recovery of soil functionality and microbial community structure in a Mediterranean shrubland. Glob Chang Biol 25:1409–1427

ICNF (Instituto da Conservacao da Naturaleza e das Florestas) (2015) 6° Inventário Florestal Nacional – 2015 Relatório Final. Instituto da Conservacao da Naturaleza e das Florestas, p 18

INFC (Inventario Nazionale delle Foreste e dei Serbatoi Forestali di Carbonio) (2007) Le stime di superficie 2005 – Prima parte. Inventario Nazionale delle Foreste e dei Serbatoi Forestali di Carbonio. Corpo Forestale dello Stato – Ispettorato Generale. CRA – ISAFA, Trento, p 409

Jolly WM, Cochrane MA, Freeborn PH et al (2015) Climate-induced variations in global wildfire danger from 1979 to 2013. Nat Commun 6:7537

Kalabokidis K, Palaiologou P, Gerasopoulos E et al (2015) Effect of climate change projections on forest fire behavior and values-at-risk in southwestern Greece. Forests 6:2214–2240

Karali A, Hatzaki M, Giannakopoulos C et al (2014) Sensitivity and evaluation of current fire risk and future projections due to climate change: the case study of Greece. Nat Hazards Earth Syst Sci 14:143–153

Keenan T, Maria Serra J, Lloret F et al (2011) Predicting the future of forests in the Mediterranean under climate change, with niche- and process-based models: CO2 matters! Glob Chang Biol 17:565–579

Keenan RJ, Reams GA, Achard F et al (2015) Dynamics of global forest area: results from the FAO Global Forest Resources Assessment 2015. For Ecol Manag 352:9–20

Koutsias N, Arianoutsou M, Kallimanis AS et al (2012) Where did the fires burn in Peloponnisos, Greece the summer of 2007? Evidence for a synergy of fuel and weather. Agric For Meteorol 156:41–53

Krawchuk MA, Moritz MA (2011) Constraints on global fire activity vary across a resource gradient. Ecology 92:121–132

Küçüker DM, Baskent EZ (2017) State of stone pine (*Pinus pinea*) forests in Turkey and their economic importance for rural development. In: Mutke S, Carrasquinho I, Correia AC (eds) Mediterranean pine nuts from forests and plantations, vol 122. CIHEAM, Zaragoza, pp 111–117

Lamb HF, Damblon F, Maxted RW (1991) Human impact on the vegetation of the Middle Atlas, Morocco, during the last 5000 years. J Biogeogr 18:519–532

Lampin-Maillet C, Long-Fournel M, Ganteaume A et al (2011) Land cover analysis in wildland–urban interfaces according to wildfire risk: a case study in the South of France. For Ecol Manag 261:2200–2213

Lawson IT, Al-Omari S, Tzedakis PC et al (2005) Lateglacial and Holocene vegetation history at Nisi Fen and the Boras mountains, northern Greece. The Holocene 15:873–887

Le Houerou HN (1987) Vegetation wild fires in the Mediterranean basin: evolution and trends. Ecol Mediterr 13:13–24

Leys B, Finsinger W, Carcaillet C (2014) Historical range of fire frequency is not the Achilles' heel of the Corsican black pine ecosystem. J Ecol 102:381–395

Lionello P, Scarascia L (2018) The relation between climate change in the Mediterranean region and global warming. Reg Environ Chang 18:1481–1493

Lloret F, Martinez-Vilalta J, Serra-Diaz JM et al (2013) Relationship between projected changes in future climatic suitability and demographic and functional traits of forest tree species in Spain. Clim Chang 120:449–462

Loepfe L, Martinez-Vilalta J, Piñol J (2012) Management alternatives to offset climate change effects on Mediterranean fire regimes in NE Spain. Clim Chang 115:693–707

López-Merino L, López-Sáez JA, Alba-Sánchez F et al (2009) 2000 years of pastoralism and fire shaping high-altitude vegetation of Sierra de Gredos in Central Spain. Rev Palaeobot Palynol 158:42–51

López-Sáez JA, López-Merino L, Alba-Sánchez F et al (2010) Late Holocene ecological history of *Pinus pinaster* forests in the Sierra de Gredos of Central Spain. Plant Ecol 206:195–209

López-Sáez JA, Sánchez-Mata D, Alba-Sánchez F et al (2013) Discrimination of Scots pine forests in the Iberian Central System (*Pinus sylvestris* var. *iberica*) by means of pollen analysis. Phytosociological considerations. Lazaroa 34:191–208

López-Sáez JA, Vargas G, Ruiz-Fernández J et al (2018) Paleofire dynamics in Central Spain during the late Holocene: the role of climatic and anthropogenic forcing. Land Degrad Dev 29:2045–2059

Lozano OM et al (2017) Assessing climate change impacts on wildfire exposure in Mediterranean areas. Risk Anal 37:1898–1916

Maestre FT, Cortina J (2004) Are Pinus halepensis plantations useful as a restoration tool in semiarid Mediterranean areas? For Ecol Manag 198:303–317

Magri D (1999) Late Quaternary vegetation history at Lagaccione near Lago di Bolsena (Central Italy). Rev Palaeobot Palynol 106:171–208

Mancini LD, Barbati A, Corona P (2017) Geospatial analysis of woodland fire occurrence and recurrence in Italy. Ann Silvic Res 41:41–47

MAPA (Ministerio de Agricultura, Pesca y Alimentación) (2019) Anuario de Estadística Forestal 2016. Ministerio de Agricultura, Pesca y Alimentación, Gobierno de España, Madrid, p 96

Marchi M, Nocentini S, Ducci F (2016) Future scenarios and conservation strategies for a rear-edge marginal population of Pinus nigra Arnold in Italian central Apennines. For Syst 25:e072

Martínez-González AJ (2015) Masas forestales para las Armadas: las áreas jurisdiccionales de montes y plantíos (siglos XVI-XVIII). Navegamérica 14:36

Matías L, Jump AS (2014) Impacts of predicted climate change on recruitment at the geographical limits of Scots pine. J Exp Bot 65:299–310

Meyn A, White PS, Buhk C et al (2007) Environmental drivers of large, infrequent wildfires: the emerging conceptual model. Prog Phys Geo Earth Environ 31:287–312

Mitsopoulos I, Mallinis G, Arianoutsou M (2015) Wildfire risk assessment in a typical Mediterranean wildland–urban interface of Greece. Environ Manag 55:900–915

Montiel-Molina C (ed) (2013) Presencia histórica del fuego en el territorio. Ministerio de Agricultura, Alimentación y Medio Ambiente, Madrid

Montiel-Molina C, Vilar L, Romão-Sequeira C et al (2019) Have historical land use/land cover changes triggered a fire regime shift in Central Spain? Fire 2:44

Morales-Molino C, García-Antón M (2014) Vegetation and fire history since the last glacial maximum in an inland area of the western Mediterranean Basin (Northern Iberian Plateau, NW Spain). Quat Res 81:63–77

Morales-Molino C, García-Antón M, Postigo-Mijarra JM et al (2013) Holocene vegetation, fire and climate interactions on the westernmost fringe of the Mediterranean Basin. Quat Sci Rev 59:5–17

Morales-Molino C, Tinner W, García-Antón M et al (2017a) The historical demise of *Pinus nigra* forests in the Northern Iberian Plateau (South-Western Europe). J Ecol 105:634–646

Morales-Molino C, Colombaroli D, Valbuena-Carabaña M et al (2017b) Land-use history as a major driver for long-term forest dynamics in the Sierra de Guadarrama National Park (Central Spain) during the last millennia: implications for forest conservation and management. Glob Planet Chang 152:64–75

Morales-Molino C, Devaux L, Georget M et al (2020) Modern pollen representation of the vegetation of the Tagus Basin (Central Iberian Peninsula). Rev Palaeobot Palynol 276:104193

Moreira F, Vaz P, Catry F et al (2009) Regional variations in wildfire susceptibility of land-cover types in Portugal: implications for landscape management to minimize fire hazard. Int J Wildland Fire 18:563–574

Moreno JM, Vázquez A, Vélez R (1998) Recent fire history of forest fires in Spain. In: Moreno JM (ed) Large forest fires. Backhuys, Leiden, pp 159–185

Moreno JM, Viedma O, Zavala G et al (2011) Landscape variables influencing forest fires in Central Spain. Int J Wildland Fire 20:678–689

Moriondo M, Good P, Durao R et al (2006) Potential impact of climate change on fire risk in the Mediterranean area. Clim Res 31:85–95

Moritz MA, Morais ME, Summerell LA et al (2005) Wildfires, complexity, and highly optimized tolerance. Proc Natl Acad Sci U S A 102:17912–17917

Mouillot F, Rambal S, Joffre R (2002) Simulating climate change impacts on fire frequency and vegetation dynamics in a Mediterranean-type ecosystem. Glob Chang Biol 8:423–437

Nakicenovic N, Alcamo J, Grubler A et al (2000) Special report on emissions scenarios (SRES), a special report of Working Group III of the Intergovernmental Panel on Climate Change. Cambridge University Press, Cambridge

Navarro-Cerrillo R, Duque-Lazo J, Manzanedo R et al (2018) Climate change may threaten the southernmost *Pinus nigra* subsp. *salzmannii* (Dunal) Franco populations: an ensemble niche-based approach. iForests Biog For 11:396–405

Ne'eman G, Arianoutsou M (this volume) Mediterranean pines – adaptations to fire. In: Ne'eman G, Osem Y (eds) Pines and their mixed forest ecosystems in the Mediterranean Basin. Springer, pp 457–480

Noce S, Collalti A, Santini M (2017) Likelihood of changes in forest species suitability, distribution, and diversity under future climate: the case of Southern Europe. Ecol Evol 7:9358–9375

Norrant C, Douguédroit A (2006) Monthly and daily precipitation trends in the Mediterranean (1950–2000). Theor Appl Climatol 83:89–106

Nunes MCS, Vasconcelos MJ, Pereira JMC et al (2005) Land cover type and fire in Portugal: do fires burn land cover selectively? Landsc Ecol 20:661–673

OGM (General Directorate of Forestry) (2015) The Turkish forests, vol 2020. Republic of Turkey, General Directorate of Forests

Oliveira S, Moreira F, Boca R et al (2014) Assessment of fire selectivity in relation to land cover and topography: a comparison between Southern European countries. Int J Wildland Fire 23:620–630

Orenstein D (this volume) The cultural ecosystem services of Mediterranean pine forests. In: Ne'eman G, Osem Y (eds) Pines and their mixed forest ecosystems in the Mediterranean Basin. Springer, pp 621–655

Osem Y, Lavi A, Rosenfeld A (2011) Colonization of Pinus halepensis in Mediterranean habitats: consequences of afforestation, grazing and fire. Biol Invasions 13:485–498

Pausas JG (2004) Changes in fire and climate in the Eastern Iberian Peninsula (Mediterranean Basin). Clim Chang 63:337–350
Pausas JG, Fernández-Muñoz S (2012) Fire regime changes in the Western Mediterranean Basin: from fuel-limited to drought-driven fire regime. Clim Chang 110:215–226
Pausas JG, Ribeiro E (2013) The global fire–productivity relationship. Glob Ecol Biogeogr 22:728–736
Pausas JG, Bladé C, Valdecantos A et al (2004) Pines and oaks in the restoration of Mediterranean landscapes of Spain: new perspectives for an old practice – a review. Plant Ecol 171:209–220
Pereira MG, Calado TJ, DaCamara CC et al (2013) Effects of regional climate change on rural fires in Portugal. Clim Res 57:187–200
Pereira MG, Aranha J, Amraoui M (2014) Land cover fire proneness in Europe. For Syst 23:598–610
Philandras C, Nastos P, Kapsomenakis J et al (2011) Long term precipitation trends and variability within the Mediterranean region. Nat Hazards Earth Syst Sci 11:3235
Piermattei A, Renzaglia F, Urbinati C (2012) Recent expansion of *Pinus nigra* Arn. above the timberline in the central Apennines, Italy. Ann For Sci 69:509–517
Poyatos R, Latron J, Llorens P (2003) Land use and land cover change after agricultural abandonment. Mt Res Dev 23:362–368
Quézel P, Médail F (2003) Écologie et biogéographie des forêts du basin méditerranéen. Elsevier, Paris
Quézel P, Barbero M, Loisel R (1990) Les reboisements en region mediterraneenne. Incidences biologiques et economiques. Foret Mediterraneenne 12:14
Reille M, Gamisans J, Andrieu-Ponel V et al (1999) The Holocene at Lac de Creno, Corsica, France: a key site for the whole island. New Phytol 141:291–307
Rincón A, Pérez-Izquierdo L, de Miguel S et al (this volume) Mycorrhizae in Mediterranean pine and mixed forests. In: Ne'eman G, Osem Y (eds) Pines and their mixed forest ecosystems in the Mediterranean Basin. Springer, pp 395–418
Rodrigo A, Retana J, Picó FX (2004) Direct regeneration is not the only response of Mediterranean forests to large fires. Ecology 85:716–729
Rodríguez-Sánchez F, Hampe A, Jordano P et al (2010) Past tree range dynamics in the Iberian Peninsula inferred through phylogeography and palaeodistribution modelling: a review. Rev Palaeobot Palynol 162:507–521
Rubiales JM, Génova M (2015) Late Holocene pinewoods persistence in the Gredos Mountains (Central Spain) inferred from extensive megafossil evidence. Quat Res 84:12–20
Rubiales JM, García-Amorena I, Hernández L et al (2010) Late Quaternary dynamics of pinewoods in the Iberian Mountains. Rev Palaeobot Palynol 162:476–491
Rubiales JM, Morales-Molino C, Álvarez G et al (2012) Negative responses of highland pines to anthropogenic activities in inland Spain: a palaeoecological perspective. Veg Hist Archaeobot 21:397–412
Ruffault J, Mouillot F (2015) How a new fire-suppression policy can abruptly reshape the fire-weather relationship. Ecosphere 6:199
Ruiz-Benito P, Gómez-Aparicio L, Lines ER et al (2015) Efectos del clima y la estructura del roda sobre procesos de mortalidad en los bosques ibéricos. In: Herrero A, Zavala MA (eds) Los bosques y la biodiversidad frente al cambio climático: Impactos, Vulnerabilidad y Adaptación en España. Ministerior de Agricultura, Alimentación y Medio Ambiente, Madrid, pp 225–233
Ruiz-Labourdette D, Nogués-Bravo D, Ollero HS et al (2012) Forest composition in Mediterranean mountains is projected to shift along the entire elevational gradient under climate change. J Biogeogr 39:162–176
Ruiz-Labourdette D, Schmitz MF, Pineda FD (2013) Changes in tree species composition in Mediterranean mountains under climate change: indicators for conservation planning. Ecol Indic 24:310–323
Sadori L, Narcisi B (2001) The Postglacial record of environmental history from Lago di Pergusa, Sicily. The Holocene 11:655–670

Sakellariou S, Parisien M-A, Flannigan M et al (2020) Spatial planning of fire-agency stations as a function of wildfire likelihood in Thasos, Greece. Sci Total Environ 729:139004

Salvador L, Alía R, Agúndez D et al (2000) Genetic variation and migration pathways of maritime pine (*Pinus pinaster* Ait) in the Iberian peninsula. Theor Appl Genet 100:89–95

Sánchez-Salguero R, Camarero JJ, Gutiérrez E et al (2017) Assessing forest vulnerability to climate warming using a process-based model of tree growth: bad prospects for rear-edges. Glob Chang Biol 23:2705–2719

San-Miguel-Ayanz J, Moreno JM, Camia A (2013) Analysis of large fires in European Mediterranean landscapes: lessons learned and perspectives. For Ecol Manag 294:11–22

San-Miguel-Ayanz J, Durrant T, Boca R et al (2018) Forest fires in Europe, Middle East and North Africa 2018. Technical report by the Joint Research Centre (JRC). Joint Research, Centre (JRC), European Union

Schelhaas M-J, Fridman J, Hengeveld G et al (2018) Actual European forest management by region, tree species and owner based on 714,000 re-measured trees in national forest inventories. PLoS One 13(11):e0207151

Schiebel V, Litt T (2018) Holocene vegetation history of the southern Levant based on a pollen record from Lake Kinneret (Sea of Galilee), Israel. Veg Hist Archaeobot 27:577–590

Seneviratne SI, Donat MG, Pitman AJ et al (2016) Allowable CO2 emissions based on regional and impact-related climate targets. Nature 529:477–483

Sequeira CR, Montiel-Molina C, Rego FC (2019) Historical fire records at the two ends of Iberian Central Mountain System: Estrela massif and Ayllón massif. Investigaciones Geográficas 72:31–52

Servera-Vives G, Riera S, Picornell-Gelabert L et al (2018) The onset of islandscapes in the Balearic Islands: a study-case of Addaia (northern Minorca, Spain). Palaeogeogr Palaeoclimatol Palaeoecol 498:9–23

Sousa PM, Trigo RM, Pereira MG et al (2015) Different approaches to model future burnt area in the Iberian Peninsula. Agric For Meteorol 202:11–25

Stevenson AC (2000) The Holocene forest history of the Montes Universales, Teruel, Spain. The Holocene 10:603–610

Swetnam TW (1993) Fire history and climate change in giant sequoia groves. Science 262:885–889

Syphard AD, Radeloff VC, Hawbaker TJ et al (2009) Conservation threats due to human-caused increases in fire frequency in Mediterranean-climate ecosystems. Conserv Biol 23:758–769

Tang T, Shindell D, Samset B et al (2018) Dynamical response of Mediterranean precipitation to greenhouse gases and aerosols. Atmos Chem Phys 18:8439–8452

Tinner W, van Leeuwen JFN, Colombaroli D et al (2009) Holocene environmental and climatic changes at Gorgo Basso, a coastal lake in southern Sicily, Italy. Quat Sci Rev 28:1498–1510

Tinner W, Vescovi E, van Leeuwen JFN et al (2016) Holocene vegetation and fire history of the mountains of Northern Sicily (Italy). Veg Hist Archaeobot 25:499–519

Torres I, Pérez B, Quesada J et al (2016) Forest shifts induced by fire and management legacies in a *Pinus pinaster* woodland. For Ecol Manag 361:309–317

Touchan R, Baisan C, Mitsopoulos I et al (2012) Fire history in European black pine (*Pinus nigra* Arn.) forests of the Valia Kalda, Pindus Mountains, Greece. Tree-Ring Res 68:45–50

Trigo RM, Pereira JMC, Pereira MG et al (2006) Atmospheric conditions associated with the exceptional fire season of 2003 in Portugal. Int J Climatol 26:1741–1757

Turco M, von Hardenberg J, AghaKouchak A et al (2017) On the key role of droughts in the dynamics of summer fires in Mediterranean Europe. Sci Rep 7:1–10

Turco M, Rosa-Cánovas JJ, Bedia J et al (2018) Exacerbated fires in Mediterranean Europe due to anthropogenic warming projected with non-stationary climate-fire models. Nat Commun 9:1–9

Türkeş M, Musaoğlu N, Özcan O (2017) Assessing the vulnerability of a forest ecosystem to climate change and variability in the western Mediterranean sub-region of Turkey: future evaluation. J. For Res 29:1177–1186

Turquety S, Menut L, Bessagnet B et al (2014) APIFLAME v1.0: high-resolution fire emission model and application to the Euro-Mediterranean region. Geosci Model Dev 7:587–612

Urbieta IR, Zavala G, Bedia J et al (2015) Fire activity as a function of fire–weather seasonal severity and antecedent climate across spatial scales in southern Europe and Pacific Western USA. Environ Res Lett 10:114013

Urbieta IR, Franquesa M, Viedma O et al (2019) Fire activity and burned forest lands decreased during the last three decades in Spain. Ann For Sci 76:90

Vadell Guiral E, De Miguel MS, Pemán García J (2019) La repoblación forestal en España: las especies utilizadas desde 1877 a partir de las cartografías forestales. Historia Agraria 77:28

Valdiviesso T, Pimpao M, Trindade CS et al (2017) Reproductive phenology of *Pinus pinea*. In: Mutke S, Carrasquinho I, Correia AC (eds) Mediterranean pine nuts from forests and plantations, vol 122. CIHEAM, Zaragoza, pp 63–68

Valladares F, Matesanz S, Guilhaumon F et al (2014) The effects of phenotypic plasticity and local adaptation on forecasts of species range shifts under climate change. Ecol Lett 17:13511364

van Vuuren DP, Edmonds J, Kainuma M et al (2011) The representative concentration pathways: an overview. Clim Chang 109:5

Vázquez A, Moreno JM (1995) Patterns of fire occurrence across a climatic gradient and its relationship to meteorological variables in Spain. In: Moreno JM, Oechel WC (eds) Global change and Mediterranean-type ecosystems, vol 117. Springer, New York, pp 408–434

Veal R (2017) The politics and economics of ancient forests: timber and fuel as levers of Greco-Roman control. In: Derron P (ed) Economie et inégalité: Ressources, échanges et pouvoir dans l'Antiquité classique, vol 63. Hart Foundation, pp 317–367

Venäläinen A, Korhonen N, Hyvärinen O et al (2014) Temporal variations and change in forest fire danger in Europe for 1960–2012. Nat Hazards Earth Syst Sci 14:1477

Vescovi E, Ammann B, Ravazzi C et al (2010) A new Late-glacial and Holocene record of vegetation and fire history from Lago del Greppo, northern Apennines, Italy. Veg Hist Archaeobot 19:219–233

Vicente-Serrano SM, Lopez-Moreno J-I, Beguería S et al (2014) Evidence of increasing drought severity caused by temperature rise in southern Europe. Environ Res Lett 9:044001

Viedma O, Moity N, Moreno JM (2015a) Changes in landscape fire-hazard during the second half of the 20th century: agriculture abandonment and the changing role of driving factors. Agric Ecosyst Environ 207:126–140

Viedma O, Quesada J, Torres I et al (2015b) Fire severity in a large fire in a Pinus pinaster forest is highly predictable from burning conditions, stand structure, and topography. Ecosystems 18:237–250

Viedma O, Moreno JM, Güngöroglu C et al (2017) Recent land-use and land-cover changes and its driving factors in a fire-prone area of southwestern Turkey. J Environ Manag 197:719–731

Vitali A, Garbarino M, Camarero JJ et al (2019) Pine recolonization dynamics in Mediterranean human-disturbed treeline ecotones. For Ecol Manag 435:28–37

Wu M, Knorr W, Thonicke K et al (2015) Sensitivity of burned area in Europe to climate change, atmospheric CO_2 levels, and demography: a comparison of two fire-vegetation models. Biogeosciences 120:2256–2272

Chapter 22
Mediterranean Pines – Adaptations to Fire

Gidi Ne'eman and Margarita Arianoutsou

22.1 Fire in the Mediterranean Basin

Fire is an ancient, recurrent, natural disturbance worldwide, and in the Mediterranean Biome in particular (Pausas and Keeley 2009). Mediterranean-type climates are ideal for ignition and spreading of fires due to their long, hot and dry summer, short and mild winter, and seasonal hot and dry desert gusts. Fires have been an integral component of the Mediterranean Basin environment at least since the Miocene, probably even more so after Mediterranean seasonality intensified during the Pliocene (Rundel et al. 2016).

The occurrence of fire is strongly induced by the prevailing weather conditions (Moritz et al. 2005), while human activities are also related to fire incidents and their spread over the landscape, especially in the Mediterranean Basin (Moreira et al. 2011; Penuelas et al. 2017). The earliest anthropogenic fire in the Mediterranean Basin was documented to have occurred in Israel, 79,000 years ago (Goren-Inbar et al. 2004).

About 65,000 fires occur every year in Europe, burning 0.5 million ha of forested areas, 85% of which are in the Mediterranean region (San-Miguel-Ayanz et al. 2013). The number of anthropogenic fires in Eastern Spain, for example, has varied throughout history due to environmental and socioeconomic factors. Since the early 1970s, annual fire frequency has doubled and the burned area has increased almost ten-fold; the main driver of this shift was the increase in fuel amount following farm abandonment (Pausas and Fernández-Muñoz 2012). Recently, fire events in Mediterranean Europe have become a more serious problem, burning over larger

G. Ne'eman (✉)
Department of Biology and the Environment, University of Haifa-Oranim, Tivon, Israel

M. Arianoutsou
Department of Ecology and Systematics, National and Kapodistrian University of Athens, Athens, Greece

G. Ne'eman, Y. Osem (eds.), *Pines and Their Mixed Forest Ecosystems in the Mediterranean Basin*, Managing Forest Ecosystems 38,
https://doi.org/10.1007/978-3-030-63625-8_22

areas (San-Miguel-Ayanz et al. 2013) as a consequence of climate shifts (Moriondo et al. 2006; Lindner et al. 2010) or land use changes (Moreira et al. 2011). For a detailed review of fires in the Mediterranean Basin see Moreno et al. (this volume), Chap. 21.

22.2 The Early Evolutionary History of Fire-Adaptive Traits

Fire adaptation is a fire-adaptive trait that was shaped by natural selection under conditions prevailing in a fire-prone environment. Fire exaptation, also termed pre-adaptation (Bock 1959), is a trait that was previously selected by factors other than fire (Bradshaw et al. 2011; Keeley et al. 2011). Since both adaptation types currently contribute to plant fitness, it is difficult to distinguish between them. Therefore, to avoid confusing the readers, we will not distinguish between them in this chapter.

Fire adaptive traits should not be examined with respect to a single fire, but rather in relation to the fire regime (Keeley et al. 2012). This is because any single trait, such as serotiny (delayed post-fire seed dispersal) in obligate seeding species, can be adaptive under a low frequency fire regime with long fire free periods, but non-adaptive under a frequent fire regime due to 'immaturity risk'. Immaturity risk is when plants are burned while still in their juvenile stage, and before producing a sufficient quantity of seeds required for their post-fire regeneration (Lamont 1991; Lamont et al. 1991; Keeley et al. 1999).

Fire has affected plants since the origin of terrestrial plants in the early Silurian (440 mya (million years ago) (Pausas and Keeley 2009). Fire adaptive traits appeared first among conifers 350 mya (He et al. 2016). During the Cretaceous (65–145 mya), conifer forests were widespread, and the atmospheric oxygen concentration was higher than today, thus fires probably facilitated the expansion of early angiosperms (Pausas and Keeley 2009; Bond and Scott 2010; Bond and Midgley 2012).

Considerable advances in tracking the evolutionary history of fire adaptive traits have been facilitated by the use of trait-assignment techniques (Lamont et al. 2019a). The new evidence strongly supports the hypothesis that fire has been a pre-emptive evolutionary trigger for fire adaptive traits among seed plants (He et al. 2016; Lamont and He 2017; Lamont et al. 2019b).

Human activities and climate change are increasing fire risk in the Mediterranean Basin (Moriondo et al. 2006; Pausas and Keeley 2009; Pausas and Fernández-Muñoz 2012; San-Miguel-Ayanz et al. 2013; Chap. 21), and in other Mediterranean-type ecosystems (Rundel et al. 2016). For example, a correlation between short-term changes in the frequency of anthropogenic fires in Chile and inter-population variation in seed traits indicated that anthropogenic fires can drive the evolution of seed traits (Gómez-González et al. 2011). Another study reported that a single fire significantly increased the average percentage of serotiny in *P. halepensis* from 39.3% in unburned sites to 84.5% in burned sites (Goubitz et al. 2004). Both studies indicate that anthropogenic fire regimes are currently selecting and affecting fire adaptive traits.

22.3 Fire Adaptations

The Mediterranean climate is ideal for ignition and rapid spread of fires, thus fire adaptive traits are common and have been studied extensively in plants of Mediterranean-type environments (Bradshaw et al. 2011; Keeley et al. 2011; Bond and Van Wilgen 2012; Moreno and Oechel 2012; Groom and Lamont 2015; Rundel et al. 2016). Fire adaptive traits are numerous and affect various stages of the plant life cycle, as detailed below.

22.3.1 Germination and Establishment

Germination-regulating mechanisms have been selected to direct germination to the time and habitat that optimize plant growth, maturation and seed production (Baskin and Baskin 2014). Various environmental factors serve as germination-regulating cues, many of which act to break seed dormancy (Baskin and Baskin 2014).

22.3.1.1 Temperature

Pinus halepensis and *P. brutia* share the characteristic of partial cone serotiny (Thanos 2000). Seeds of *P. halepensis* and *P. brutia* germinate optimally at 20 °C in the dark. Experiments under daily alternating conditions of light and temperature resembling those in nature indicate that field germination is feasible throughout the rainy season, mainly in open, sunny sites (Thanos and Skordilis 1987). Stratification considerably promotes germination of *P. brutia* seeds, mainly from a northern provenance in Greece. Consequently, seed germination in cold and moist, northern climates is timed to spring, while in mild and dry, southern climates it is timed to autumn and early winter (Skordilis and Thanos 1995). In Turkey, germination of *P. brutia* seeds occurs throughout the rainy season if temperatures are suitable, with two frequency peaks: the major one in the spring and the minor one in the autumn (Boydak 2004; Boydak et al. 2006).

Generally, temperature alters seed germination of Mediterranean pines mainly with respect to their distribution range and altitude. Seeds of northern and high altitude species that grow in cold and moist climates require winter stratification and are thus timed to spring, while southern, lowland species that grow in warmer, dry areas germinate in autumn and early winter to avoid summer drought (Skordilis and Thanos 1997).

22.3.1.2 Heat Shock

Heat shock, brief exposure to high temperatures, is a typical effect of fire that scarifies impermeable hard seed coats, permits water imbibition and breaks physical seed dormancy (Baskin and Baskin 2014). Indeed, hard-seeded legumes are typical of fire-prone Mediterranean regions (Arianoutsou and Thanos 1996; Lamont et al. 2019b). Hard-seeded species are currently not restricted to currently fire-prone environments (Thanos et al. 1992), but dated phylogeny research indicates that this trait was already selected by fire early in angiosperm evolutionary history (Lamont et al. 2019b).

Heat shock typically enhances germination of hard seeds stored in soil seed banks (Izhaki et al. 2000), but not of seeds of serotinous species (Hanley and Lamont 2000) that are insulated from the heat of the fire and dispersed after fire. The highly nutritious seeds of pines are highly predated and almost absent from the soil seed bank (Ne'eman and Izhaki 1999; Nathan and Ne'eman 2000; Keeley et al. 2012). Post-fire germination, mainly of serotinous pines, is from canopy-stored seeds, which are protected from the fire's heat shock by the insulation of the cones, and are dispersed only after fire (Moya et al. 2008).

Martínez-Sánchez et al. (1995) found that heating for 1–5 min at 90–200 °C did not affect the average germination of *P. halepensis* but reduced that of *P. pinaster*. The temperature increase that can be reached in soils during a fire does not enhance germination of either species.

Escudero et al. (1999) tested the effect of heat shock (50–150 °C for 1–15 min) on the germination of *P. halepensis, P. pinaster, P. pinea, P. nigra, P. sylvestris, P. uncinata* and *P. canariensis*. Seeds of all species germinated best with no heat shock, and decreased with an increase in temperature or duration. In contrast, the germination of *P. pinea* and *P. pinaster* was similar in control and treatments and decreased only after the most severe treatment (130 °C for 3 min), probably because the seeds are protected by a thick seed coat (Escudero et al. 1999; Herrero et al. 2007). This may explain partial post-fire germination even in cases of weak or even absence of serotiny.

Habrouk et al. (1999) demonstrated that the percentage of germination of *P. halepensis*, *P. nigra* and *P. sylvestris* seeds decreased with an increase in temperature and time of heat shock. *P. halepensis* showed higher germination rates than *P. nigra*, and both were higher than *P. sylvestris*. That may explain the successful post-fire seedling recruitment of *P. halepensis*, and the failure of *P. nigra* and *P. sylvestris* after fire (Habrouk et al. 1999).

Boydak and Caliskan (2016) studied the effects of heat shock (70–110 °C for 30 s up to 30 min) on germination of *P. brutia* seeds. Heat shock (up to 130 °C for 3 min) increased germination by 5% relative to the control (65%). They suggested that under low or moderate intensity fires, *P. brutia* seeds can survive in the partially burned scorched cones, or if they fall on the ground, and germinate. Seeds that are protected in closed cones also survive and germinate once they are dispersed after fire from the cones (Boydak and Caliskan 2016). The thick seed coat contribute to the tolerance of seeds to high tenperatures 30 min at 100 °C and 2 min at 150 °C (Boydak et al. 2006).

22.3.1.3 Smoke

Charred wood and smoke are probably the most typical fire-related environmental cues (Keeley et al. 2012). The effect of smoke on germination was first discovered by Brown (1993) and its stimulation of germination is known from many post-fire recruiting species in South Africa (Brown 1993), Western Australia (Dixon et al. 1995), California (Keeley and Fotheringham 1998) and the Mediterranean Basin (Moreira et al. 2010; Moreira and Pausas 2018).

The most well known active compound responsible for inducing smoke-responsive germination later identified as a butenolide ring and named 'karrikinolide' (Dixon et al. 2009). Karrikinolides are now included in the 'trigolactones', a new group of plant hormones (Zwanenburg et al. 2016), and their molecular mode of action in stimulating germination of soil-stored seeds is known (Lamont et al. 2019b). Recently, it was discovered that they have a general positive effect on germination, growth, photosynthesis and the nutritional value of carrots (Akeel et al. 2019). However, much of the literature suggests that smoke-stimulated germination is a far more complex trait and supports the idea that multiple compounds in smoke can stimulate germination (Keeley and Pausas 2018).

Smoke mainly affects the germination of soil-stored seeds and much less, if at all, canopy-stored seeds in serotinous plants, because they are not directly exposed to the fire's smoke (Brown 1993). Therefore, the germination of pine seeds is not expected to be affected by smoke. Indeed, smoke treatments of 5, 10, 15 and 20 min and control, applied to seeds of *P. sylvestris, P. nigra, P. uncinata* and *P. pinaster* had no significant effects on seed germination in any of the four species tested (Reyes and Casal 2006).

22.3.1.4 Ash

Soil seed banks are exposed to the combined effect of ash and heat (Izhaki et al. 2000), while canopy-stored seed banks are insulated from the fire's heat and are exposed only to ash during their post-fire germination.

Ash is composed mainly of the mineral residuals of plant combustion, thus it provides improved mineral nutrition to regenerating plants in the post-fire environment. In addition to its effect on mineral nutrition, ash induces low water potential equivalent to aridity even during the first rains. Thus, adaptation to aridity may also help plants grow in post-fire ash beds. Accordingly, the germination of *P. halepensis*, that grows in semi-arid Mediterranean areas, and of *P. brutia*, is inhibited only at very low osmotic potentials (Thanos and Skordilis 1987). Henig-Sever et al. (1996) reported that a water potential of –0.1 MPa, typical of the ash layer, did not affect the germination of *P. halepensis*.

As a result of oxidation during fire, ash contains many cations that induce a high pH (~9) on contact with water. A positive effect of ash on germination has been found for *Rhus coriaria* (Ne'eman et al. 1999a) and some Mexican species (Zuloaga-Aguilar et al. 2011). In contrast, no effect of ash in combination with temperature up to 70 °C was found on germination of *P. nigra* ssp. *salzmannii* or *P. sylvestris* var. *iberica* seeds (Escudero et al. 1997), or on germination of *P. pinaster* with short heat-shock temperatures up to 130 °C for 1 min (Herrero et al. 2007). Negative effects of ash on seed germination have also been reported: two studies reported that seed germination of *P. halepensis* at pH 9 was reduced by half (Ne'eman et al. 1993; Henig-Sever et al. 1996), and a negative effect of ash on seed germination has also demonstrated for *P. pinaster*, *P. sylvestris*, *P. nigra* and *P. radiat*a (Reyes and Casal 2004).

The differential effect of ash on germination that have been reported, might reflect real differences in the habitats of the tested species, but might also result from experimental faults, such as not using buffered solutions, or monitoring pH changes, during seed incubation.

After fire, ash accumulates mainly under large burned trees, decreasing the germination of many species. However, for seeds that manage to germinate, ash provides an excellent, low-competition, high-nutrition regeneration site. This is advantageous mainly for obligate seeding trees whose adults die during canopy fires, as demonstrated for *P. halepensis* (Ne'eman and Izhaki 1998; Ne'eman 2000) and *Cupressus sargentii* (Ne'eman et al. 1999b).

22.3.1.5 Serotiny

The germination percentage of seeds from serotinous cones decreases with their age, but 9-year-old cones of *P. brutia* have been reported to still contain viable seeds (Thanos 2000). Seeds of *P. halepensis* from serotinous cones have been found to be relatively more tolerant to ash and simulated fire heat than seeds from non-serotinous cones, demonstrating better adaptation to the post-fire environment (Goubitz et al. 2002). These differences seem to be adaptive because seeds from serotinous cones germinate in post-fire environment, while seeds from non-serotinous cones in non-fire environment. In Spain, differences in the germination of seeds from serotinous and non-serotinous cones were related to the fact that seeds from non-serotinous cones are larger and heavier, and have higher Si and Ca concentrations in their seed coat (Salvatore et al. 2010).

22.3.2 *Bark Thickness*

Tree bark consists of air-containing dead cork cells that insulate the cambium from the heat of the fire and increase survival potential after fire. Bark thicker than 30 mm protects from intense surface fires with temperatures over 800 °C for 10 min (Van

Mantgem and Schwartz 2003). Trees with thick bark survive frequent low-intensity ground fires that are typical of Californian giant *Sequoia* forests (Keeley and Stephenson 2000), savannas (Bond 2008), and forests of some pine species (Pausas 2015a). Such trees are capable of epicormic post-fire resprouting (Pausas and Keeley 2017), and their thick bark has been selected by fire (Pausas 2015a). Indeed, early pines from the Lower Cretaceous (135 mya) already possessed thick bark (He et al. 2012). Comparing bark thickness and bud tolerance to heat, Fernandes et al. (2008) ranked pine resistance to heat in the following order: P. *pinaster* > *P. pinea* > *P. nigra* > *P. halepensis* > *P. brutia* > *P. sylvestris* > *P. uncinata*. However, bark thickness varies among populations, as reported for *P. pinaster* by Climent et al. (2004).

22.3.3 Self-pruning (Branch Shedding)

Plants commonly shed their dead branches; in the event of fire this creates a fuel gap that limits the ability of surface fires to climb up to the trees' canopies. In contrast, retention of dead biomass increases fire intensity and turns ground fire into canopy fire that may result in plant death (Schwilk 2003). Branch retention in fire-prone systems is typical of post-fire obligate seeders with soil- or canopy-stored seed banks of serotinous cones or fruits (Schwilk and Ackerly 2001). *Pinus nigra* is a good example of a self-pruning pine that survives surface fires, being also equipped with thick bark. Fire scar dendrochronology studies show that historically these trees were exposed to repeated fires, providing information on past fire frequencies as well as demonstrating that these fires were of sufficiently low intensity that trees survived repeated fires (Fulé et al. 2008; Christopoulou et al. 2013). Mature trees of other Mediterranean pines, mainly those that can survive surface fires, such as *P. pinaster* and *P. pinea*, or those that form dense forests, such as *P. sylvestris*, also lack lateral branches on the lower part of their trunk. In contrast, the serotinous *P. halepensis* and *P. brutia* are not self-pruning trees.

22.3.4 Flammability

Flammability is the potential of plant fuel to ignite and burn. Many parameters are used to characterize flammability, such as fuel type and its flammability, ignitability, fuel load, heat release, fire spread rate, burning time, and more (Pausas et al. 2017).

In general, pines are considered very flammable, because of their high content of flammable resin and their needle-like leaves that permit free oxygen flow during fire. However, pines are not equally flammable and differences have been found in the flammability of the needles of eight American species (Fonda 2001).

Experiments on time to ignition at 700 °C revealed that this parameter was correlated mainly with fuel moisture, and differences were found among 24 dominant

Mediterranean forest shrubs and trees that were divided into four groups. The group of flammable species included *Cupressus sempervirens*, *P. brutia* and *P. halepensis* (Dimitrakopoulos and Papaioannou 2001).

Recently, Pausas et al. (2017) proposed that flammability should not be viewed as a single quantitative trait, but rather one that includes three dimensions: ignitability, heat release and fire spread rate, which are not necessarily correlated. These parameters are controlled by different plant traits and have differing ecological impacts during fire. Pausas et al. (2017) propose three plant flammability strategies in fire-prone ecosystems: (1) **Non-flammable** plants that have very low ignitability and barely burn in nature. This strategy includes succulent plants, but also pine trees with thick basal bark and self-pruning of lower branches, which protects them in the event of understory ground fires and decreases the risk of crown fires. (2) **Fast-flammable** plants that ignite easily, burn quickly, produce rapidly spreading fires with low heat release; they are burned but survive fires. This strategy is typical of Australian grass trees, palms and the juvenile 'grass-stage' of a few pine species (Pausas et al. 2017). (3) **Hot-flammable** plants that ignite easily and generate extremely hot and long-lasting crown fires. These plants do not survive fires but regenerate from soil- or canopy-stored seed banks. This category of plants, also termed 'post-fire obligate seeders', which use the favorable conditions (sun, minerals and low competition) in the newly opened post-fire space, includes several pine species, as well as small trees and shrubs of Mediterranean-type climates (Pausas et al. 2017).

Pinus halepensis and ***P. brutia*** belong to the hot flammable group mainly because the trees retain old dead branches and empty cones and any surface fire quickly becomes a hot crown fire that consumes dead and live fine fuel.

Pinus pinaster is a species that belongs to the hot flammable group in North Africa because of its thin bark and serotinous cones, but to the non-flammable group with surface fires in the Atlantic coast, where it has thick bark and survives surface fires even when its canopy is partially scorched (Keeley et al. 2012).

Pinus pinea resembles *P. pinaster* and belongs to the non-flammable group that survives surface fires, as it has thick bark and endures partial encroachment of its canopy.

Pinus heldreichii, P. mugo and P. uncinata all grow at high elevation, where fire is rare in their original habitats and they are not well adapted to fire. They are non-serotinous, have thin bark, do not survive even rapid surface fires, and do not regenerate after fires, but may invade burned areas from unburned trees.

Pinus nigra is also a mountainous species, which could be considered more adapted to surface fires, as it has thick bark which sustains fire, as well as being self-pruning, thus decreasing the risk of crown fires.

Pinus sylvestris is a mountainous species in the North Mediterranean where fires are rare. It grows mainly in the boreal forests of Europe and Asia (Willis et al. 1998) where fires are more frequent. Adámek et al. (2016) studied semi-natural forests dominated by *P. sylvestris* in Central Europe. They revealed fast recovery of the vegetation cover, but the return to pre-fire species composition took

140 years. They concluded that periodic wildfires once every 200 years seem to be a factor in maintaining *P. sylvestris* (Adámek et al. 2016).

Pinus canariensis does not fit well into any flammability group. It has thick bark and survives fast surface fires. In many cases it is also subjected to hot canopy fires, after which it presents epicormic resprouting (Pausas and Keeley 2017). It has also serotinous cones that are typical to post-fire obligate seeders that are killed by fire, but no massive post-fire seedling recruitment (Climent et al. 2004).

22.3.5 Resprouting

Many plants overcome aboveground damages by vegetative regrowth, termed 'resprouting'. Most post-fire plants resprout from the root crown, while some others do so from the trunk and the main branches (epicormic resprouting), both of which are protected by thick heat-insulating bark (Pausas and Keeley 2017). Post-fire epicormic resprouting is a fire adaptive trait allowing for rapid recovery of forest canopies in highly productive sites with high fire frequency of moderate to high fire intensity (Pausas and Keeley 2017). The grass stage in juvenile pines with a bunchgrass form is a unique pine life-history trait that delays stem elongation (Keeley and Zedler 1998). During this stage the needles accumulate over the apex, insulating it against fire heat (Lamont et al. 2019b).

Resprouting is optimal in fertile environments where vigorous resprouts rapidly occupy post-fire gaps. In less fertile sites post-fire resprouting is slower, leaving open gaps that allow post-fire seedling recruitment (Keeley and Zedler 1998). Such conditions favor selection for delayed post-fire recruitment and the evolution of obligate seeders (Pausas and Keeley 2014). Intermediate conditions might have resulted in the evolution of facultative regeneration strategies, and in the coexistence of plants differing in their post-fire regeneration strategies.

Although rare in the Pinaceae, resprouting also occurs among pines in sites with low productivity and high fire frequency (Keeley and Zedler 1998). The oldest lineages with fire-related resprouting date to at least 61 mya (Lamont et al. 2019b). However, it was recently found that resprouting is not ancestral to non-sprouting, and transitions between these two life forms were common throughout plant evolutionary history (Lamont et al. 2019b).

Out of 34 pine species, seven are resprouters and two are reported as weakly so (Keeley and Zedler 1998). However, only a few pine species are able to resprout as adults, namely, *P. canariensis, P. echinata, P. leiophylla, P. merkusii* and *P. rigida* (Climent et al. 2004).

Pinus canariensis trees resprout epicormically after fire (except juvenile seedlings) from heat-insulated buds located under its thick bark. The average bark thickness in adult trees ranges from 22 to 49 mm and is unrelated to tree age or diameter. Thick bark provenances of *P. canariensis* coincide with moderately to highly productive areas where forest fires have been frequent and intense during recent decades (Climent et al. 2004; Pausas and Keeley 2017).

22.3.6 Fire-Stimulated Flowering

Fire-stimulated flowering uses the direct sunlight, improved mineral nutrition (via ash) and smoke in burned forest gaps to increase reproduction in the post-fire environment (Keeley 1993; Lamont et al. 2019b). Fire-stimulated flowering has been found mainly in geophytes of 50 families on all continents (Lamont et al. 2019b). To the best of our knowledge, there are no data on increased pine cone production in trees that survive fires.

22.3.7 Cone Opening and Serotiny

Serotiny is the delayed release of mature seeds from the plant canopy-stored seed bank; it is also termed 'bradyspory' (Lamont 1991; Lamont et al. 1991). Serotiny occurs in fire-prone environments and is correlated with flammability (Schwilk and Ackerly 2001). Serotiny is typical of fire-prone, nutrient-poor and seasonally dry sclerophyllous vegetation in Australia, South Africa, North America and the Mediterranean Basin. All of the various seed-storing cones or fruits release their seeds in reaction to fire heat (pyriscence); some release their seeds following aging and extreme drought (xeriscence), or the death of the tree. The massive seed release after fire provides optimal timing for ensuring germination and establishment in the first post-fire winter, when moisture, light and nutrients are optimal and competition with resprouting adults is minimal (Lamont et al. 1991; Keeley et al. 2011). Serotiny is common in fire-prone environments and in the absence of fire there is only negligible recruitment; this clearly points to the fitness benefits of serotiny in such ecosystems, particularly for fire-killed, obligate seeding species (Lamont et al. 1991; Keeley et al. 2011; He et al. 2012).

The proportion of serotinous cones varies among pine species, among conspecific populations and among individuals. There is evidence of increasing serotiny with the frequency of crown fires (Pausas 2015a), and even between once-burned stands and non-burned stands of the same population (Goubitz et al. 2002). This micro-evolutionary evidence is in concert with the macro-evolutionary data that fire has preceded the selection for serotiny in pines since 350 mya (He et al. 2016).

Pinus halepensis and ***P. brutia*** are partially serotinous as expected from post-fire obligate seeder and pioneer tree species in disturbed Mediterranean areas. *Pinus halepensis* is probably the most studied pine concerning serotiny. Serotiny is a heritable trait in *P. halepensis*, and selection acts on it, giving rise to contrasting serotiny levels among populations depending on the fire regime, and supporting the role of fire in generating genetic divergence for adaptive traits (Hernández-Serrano et al. 2014). In an analysis of 14 studies, including seven pine species, the strength of the relationship between serotiny and fire regime varied among species (Hernández-Serrano et al. 2014). However, the fact that a tree may bear both types of cones, and that the degree of serotiny of a tree varies with its age (Goubitz et al. 2004) probably reflects effects of phenotypic plasticity and epigenetics that are yet to be explored.

Serotinous cones open mainly as a result of fire heat and drought. However, some of them may open as a result of severe drought or increasing age (Nathan and Ne'eman 2000; Espelta et al. 2011). Loss of connection between the cones and the branch through abscission of the peduncles or the absence of water supply induces faster cone opening in *P. halepensis* in weakly serotinous trees. Thus cone serotiny in *P. halepensis* involves the costs of water allocation to the cones, while older cones lose more water and open at lower temperatures (Martín-Sanz et al. 2017).

In a study by Hernández-Serrano et al. (2014), the average (±SD) percentage of serotiny of 1227 individual *P. halepensis* trees growing in a common garden was 60.37% ± 35.60, and that of the 29 provenances, covering its entire natural range in the Iberian Peninsula was 59.48% ± 18.75. In Israel, the mean level of serotiny among trees up to 8 m high was found to be significantly higher in post-fire regenerated stands (84.5% ± 7.02) than in unburned stands (39.3% ± 9.31). The percentage of serotiny was significantly negatively correlated to tree size; small, young trees had relatively more serotinous cones than large, adult trees (Goubitz et al. 2004). In Greece *P. halepensis* was found to have 95% serotiny in young trees and 48% in adult trees (Thanos and Daskalakou 2000). The early age of sexual reproduction and the fact that young trees are more serotinous than adults are important for the early accumulation of a large canopy-stored seed bank.

Pinus brutia produces less cones and has a lower percentage of serotinous cones than *P. halepensis* (Chambel et al. 2013). In Greece it has 86% serotiny in young trees, but there are no data for adult trees (Thanos and Daskalakou 2000).

Pinus pinaster in Spain has a lower proportion of serotinous cones than *P. halepensis*. Hernández-Serrano et al. (2013) found that the maximum age of the serotinous cones was higher in *P. pinaster* than in *P. halepensis*, and for both species, it was much higher in populations with high fire frequency.

Pinus cnariensis has about 40% of serotinous trees with high variability both within and between stands. The coexistence of serotinous and non-serotinous individuals in the same population may be related to the extremely variable morphology of cone apophyses (Climent et al. 2004).

Pinus nigra and ***P. sylvestris*** non-serotinous cones opend similarly to heat treatments that was considerably higher than cones of *P. halepensis* (Habrouk et al. 1999)

22.4 Fire-Adaption Syndromes

A shift from light surface fires to intense crown fires in the early conifer forests occurred during the Cretaceous (about 89–96 mya), probably due to an increase in atmospheric O_2 concentration (He et al. 2012, 2016). This shift in fire regime has been proposed as the main selective force for the evolution of alternative fire syndromes in pines: (1) Thick bark with branch shedding as an adaption to high frequency of low-intensity surface fires (fire-avoiders); (2) Thin bark with serotiny and branch retention as an adaptation to low frequency of high-intensity crown fires (obligate seeders), which has an evolutionary legacy of about 350 my (He et al. 2016); (3) Thick bark, branch retention with epicormic resprouting (resprouters, fire-tolerators). Later, some species developed adaptations to non-fire-prone habitats, especially if fire adaptive traits reduced fitness in those habitats, and so would be killed if they were burnt (Lamont and He 2017). Badik et al. 2018 have recently suggested that adaptations to fire evolved repeatedly, possibly due to localized changes in climate and environment, rather than resulting from large dispersal events of pre-adapted individuals, and that fire-adaptive syndromes seem to have evolved more frequently in New World taxa and probably are related to the uplift of major North American mountain ranges.

Correlations have been found among fire adaptive traits (Keeley and Zedler 1998; Schwilk and Ackerly 2001) and formed the basis for the definition of three global fire-adaptive syndromes among pines by Pausas (2015b): (1) **Fire-tolerators** (resprouters) are species with thick bark and protected buds; they survive frequent, mostly understory, fires and resprout from their stem base, root crowns or lignotubers after fire. (2) **Fire-embracers** (obligate seeders) are species that retain their lower dead branches and empty cones and are killed by crown fires. Their populations recover quickly after fire by massive germination of seeds from a canopy-stored serotinous seed bank (or from a soil seed bank in other taxa). Early reproduction is advantageous for serotinous species for avoiding 'immaturity risk', namely, being burned before the production of a sufficient large seed bank for eventual post-fire recruitment (Lamont 1991; Lamont et al. 1991; Keeley et al. 1999). (3) **Fire-avoiders** live in environments that rarely burn and thus lack fire adaptive traits (Pausas 2015b). Interestingly, there is evidence that the fiery environments of the Cretaceous already selected for the 'fire tolerator' and 'fire embracer' syndromes in pines (He et al. 2012).

The contrasting traits between 'fire tolerators' and 'fire embracers' are explained by the cost of producing either thick protective bark or serotinous cones, each providing an advantage under different fire regimes (Pausas 2015b). However, several plant genera (mainly in Australia) include serotinous taxa that are either obligate seeders or resprouters, while non-serotinous genera with a dormant soil seed bank include both obligate seeders and resprouters (Ne'eman et al. 2009). Testing the germination responses to heat shock, ash and smoke of 27 Australian species (resprouting vs non-resprouting obligate seeders from soil vs canopy seed banks), Ne'eman et al. (2009) found no correlation between the effects of the fire-related cues and post-fire regeneration strategy. A possible explanation is that fire frequency or intensity may change through time in any given site due to human effects (e.g. abandonment of agricultural lands) or other biotic effects under natural conditions (e.g. seed predation, extreme herbivory of new resprouts). Such changes may cause coexistence of species belonging to contrasting fire syndromes (e.g. Clarke et al. 2016) and even to proliferation of contrasting traits within a single species, resulting in 'intermediate fire syndromes'.

Partial serotiny, in which trees simultaneously bear serotinous and non-serotinous cones, in proportions that vary with fire frequency, such as in *P. halepensis* and

P. brutia (and several American pines), may result in a 'dual life strategy' of fire, with both fire embracing and pioneer species in open unburned disturbed sites (Ne'eman et al. 2004; Boydak et al. 2006).

Pinus canariensis has thick bark typical of fire avoiders but it produces serotinous cones like fire embracers. The question arising is, how can serotiny increase fitness in *P. canariensis* if adult trees survive fire? The unpredictable fire regime deriving from volcanism, the contrasting intensity and frequency of fires caused by different vegetation structures, and human activity seem to have selected for this combination of traits and the differences among ecological regions in *P. canariensis* (Climent et al. 2004). Intraspecific variability among *P. canariensis* populations regarding bark thickness and level of serotiny in relation to fire regime has also been documented (Climent et al. 2004).

Using multivariate cluster analysis, Climent et al. (2004) analyzed and classified native Spanish pines by life history and fire adaptive traits. Four main ecological groups were identified: (1) Serotiny and young age of first cone production in *P. halepensis* and *P. pinaster* reflect their reproductive strategy in relation to fire as both enhance post-fire recruitment. (2) Older age of first cone production and absence of serotinous cones in *P. nigra*, *P. sylvestris* and *P. uncinata* reflects evolution under non-frequent crown fire regime. (3) Post-fire epicormic resprouting of *P. canariensis*. (4) Large seed size in *P. pinea*.

22.5 Fire Adaptions of Mediterranean Pines

The Mediterranean pines can be clearly divided into two ecological groups: coastal and island pines versus mountainous pines (Klaus 1989). As a consequence of cooler and wetter climatic conditions and lower density of towns, villages, tourist resorts, and abundant agricultural lands, high mountain pines are exposed to lower fire risk than those growing in lowlands (Keeley et al. 2012). For details about the fire risks in the various Mediterranean pines see Moreno et al. (this volume, Chap. 21). To minimize redundancy with the previous chapter, here we briefly present the major fire adaptive traits of Mediterranean pines (Table 22.1).

22.5.1 Pinus halepensis *Mill. (Aleppo pine)*

Pinus halepensis is a west Mediterranean species with scattered populations in the Middle East growing at low altitudes (Barbero et al. 1998, Quézel 2000), exposed to extremely high fire risk (Moreno et al. this volume).

Pinus halepensis is a typical post-fire obligate seeder (fire embracer), but it is also most successful in occupying open disturbed sites in Mediterranean-type climate regions, thus exhibiting a dual life strategy (Ne'eman et al. 2004). Trees bear dead branches and numerous empty cones that increase their flammability and serve

Table 22.1 Major Mediterranean pines, their habitats, risk of fire in their natural habitats (in dense plantations there is a high risk of canopy fire for all species) and their post-fire regeneration mode. NA- North Africa, AC- Atlantic coast

Pine species	Habitat	Fire risk in nature	Fire adaptation syndrome	Post-fire regeneration
P. halepensis ***P. brutia***	Low altitudes with Mediterranean climate	Extremely high risk of canopy fire	Thin-barked, killed by fire highly serotinous[1]	Obligate seeders, massive post-fire seedling recruitment[1]
P. pinaster	Low altitudes with Mediterranean climate	Mostly surface fire; canopy fire in dense stands and plantations	NA: thin-barked, highly serotinous AC: thick-barked, weakly serotinous, survives surface fires	NA: unknown AC: recruitment is relatively good even after canopy fires
P. pinea	Low altitudes with Mediterranean climate	Mostly surface fire; canopy fire in dense stands and plantations	Thick-barked, non-serotinous, self-pruning, survives surface fires	Recruits seedlings even after canopy fires
P. nigra	High mountains in the north, cold winter, cool summer with some rain	Rare fires in natural populations	Thick-barked, non-serotinous, survives low severity, frequent surface fires, but not crown fires	Seed dispersal from unburned trees; low seedling recruitment
P. sylvestris	High mountains, cold winter, cool summer with some rain	Rare fires in natural populations	Thin-barked, non-serotinous, does not survive crown fires	Seed dispersal from unburned trees; low seedling recruitment
P. canariensis	Mid altitudes with Mediterranean climate	Frequent lighting and volcanic fires	Thick-barked, serotinous	Adults have strong epicormic resprouting; no massive seedling recruitment
P. heldreichii ***P. mugu*** ***P. unicata***	High mountains, cold winter, cool summer with some rain	Very rare fires	Thin-barked, non-serotinous, does not survive crown fires	Seed dispersal from unburned trees; low seedling recruitment

[1]*Pinus halepensis* produces more cones, has a higher percentage of serotiny and recruits denser seedlings after canopy fires

as a ladder for ground fires to become canopy fires. This species' thin bark does not protect the cambium from the fire's heat, resulting in death of almost all trees, even following only partial canopy scorching. As a result most fires in *P. halepensis* forests are stand-replacing fires (Keeley et al. 2012).

Pinus halepensis has no soil seed bank and it regenerates after fire only by seed germination from its canopy-stored seed bank (Izhaki et al. 2000). *Pinus halepensis* is partially serotinous with higher percentage of serotiny in young trees and in post-fire populations than in large trees and in unburned expanding populations (Goubitz et al. 2004). The production of cones, most of which are serotinous, begins already at a young age (3–10 year) depending on site quality and tree size; this rapid accumulation of a sufficient canopy-stored seed bank is advantaveous when post fire (10–15 year) regeneration ocuures (Keeley et al. 2012).

Seed germination occurs in the first post-fire rainy season (Thanos 2000), and results in extremely high densities, up to 10^5 ha^{-1} (Keeley et al. 2012). The most favorable regeneration niche is under large burned pine trees because of the low competition and high mineral nutrition of the ash bed (Ne'eman 2000). However, seedlings found under individual burned trees were not necessarily their sole offspring (Gershberg et al. 2016).

22.5.2 **Pinus brutia** ***Ten. (Turkish or Calabrian pine)***

Pinus brutia is a north-east Mediterranean species closely related to *P. halepensis* (Barbero et al. 1998; Quézel 2000); it is the most common pine in the Middle East, ranging from Turkey to Lebanon at altitudes from sea level to more than 2000 m asl, where mean annual temperatures are 12–20 °C and mean annual precipitation is 400–2000 mm (Boydak 2004; Boydak et al. 2006).

Concerning its life history and fire adaptive traits, *P. brutia* highly resembles the closely related *P. halepensis* and is also serotinous, non-self-pruning and obligate seeder. However, it has thicker bark (7.8 cm) when harvested (Boydak 2004; Boydak et al. 2006), which may protect it in the event of ground fires, and has up to 20% survival in the event of canopy fires (Fernandes et al. 2008). *Pinus brutia* produces fewer cones per tree and has a lower percentage of serotinous cones than *P. halepensis* (Chambel et al. 2013), resulting in less vigorous post-fire seedling recruitment and expansion into open disturbed sites. Stands must be older than 13 y to gain 1–13 post-fire seedlings m^{-2} (Kavgaci et al. 2016). However, even a density of 0.3–1 seedlings m^{-2} can be accepted as successful regeneration in certain parent materials and poor soil conditions (Boydak 2004). Afte fire 43 seeds m^{-2} were found on the ground, 2 and 6 years later seedlng density m^{-2} was 0.3 and 0.15 respectively (Boydak et al. 2006).

Seeds from northern latitudes, and probably also from high altitudes, may require winter (cold) stratification and germinate in spring, while those from warmer sites germinate in spring and require no stratification (Skordilis and Thanos 1995). Brief heat shock may enhance seed germination (Boydak and Caliskan 2016).

22.5.3 Pinus pinaster *Aiton (Maritime pine)*

Pinus pinaster is a low- to medium-altitude species, growing in the Western Mediterranean and North Africa, with many sub-species that are difficult to distinguish (Barbero et al. 1998). *Pinus pinaster* populations from North Africa are highly serotinous with thin bark, similar to fire-embracing, obligate seeders adapted to stand-replacing crown fires. In contrast, Atlantic coast populations are non- or weakly-serotinous with thicker bark typical of fire tolerators adapted to low-intensity ground fires. Intermediate types have also also recorded (Climent et al. 2004). Under such a fire regime, trees have been recorded to survive fire once every 15 y during the last 180 years (Vega 2000).

22.5.4 Pinus pinea *L. (Stone pine)*

Pinus pinea is a low- and medium-altitude pine species growing on sandstone substrates in the northern part of the Mediterranean Basin, in Lebanon, but not in Africa. It has been widely planted for many decades, thus it is difficult to define its natural range (Barbero et al. 1998). *Pinus pinea* is non-serotinous, has thick bark, and has no lateral low branches due to self-pruning, typical of fire-tolerant pines that survive ground fires (Keeley et al. 2012). In addition, it can survive fires that scorch up to 80% of its canopy (Rigolot 2004). It has some post-fire regeneration probably due to seed dispersal from non-serotinous cones and its thick seed coat protecting seeds from heat (Escudero et al. 1999). The short seed dispersal distance (<20 m) and the low seedling survival in post-fire conditions would explain the low natural regeneration of *P. pinea* in burned areas (Pausas et al. 2008).

22.5.5 Pinus nigra *Arnold (Black pine)*

Pinus nigra is a high mountain pine species with many local subspecies raging from Morocco to Turkey, with maximum distribution in the southern Balkans and Anatolia (Barbero et al. 1998). *Pinus nigra* is a long-lived tree with thick bark and non-serotinous cones containing heat-sensitive seeds (Escudero et al. 1999; Habrouk et al. 1999). It can survive low intensity surface fires but not crown fires in high density forests resulting from forest management (Pausas et al. 2008). Excellent dendrochronological studies have proved the occurrence of repeated surface fires both in West (Fulé et al. 2008) and East Mediterranean (Christopoulou et al. 2013) black pine populations. Post-fire natural regeneration occurs from unburned plants and patches remaining intact within burned areas (Ordonez et al. 2006; Christopoulou et al. 2014).

22.5.6 **Pinus sylvestris** *L. (Scots pine)*

Pinus sylvestris is a northern pine species with local populations in Spain and southern France that are well-adapted to the high Mediterranean mountain environment (Barbero et al. 1998). *Pinus sylvestris* has thin bark and non-serotinous cones containing heat-sensitive seeds (Escudero et al. 1999; Habrouk et al. 1999), and it cannot survive crown fires (Pausas et al. 2008).

22.5.7 **Pinus canariensis** *C. Sm. (Canary Island pine)*

Pinus canariensis is a unique pine species restricted to the volcanic Canary Islands, and is not found naturaly growing in the Mediterranean Basin (Barbero et al. 1998). However, it grows in mid-elevation zones with Mediterranean-type climate and is exposed to frequent fires due to lightning and volcanic activity, and exhibits unique fire adaptive traits (Keeley et al. 2012).

Pinus canariensis is a long-lived, self-pruning tree with thick bark; it bears serotinous cones but also exhibits a strong epicormic resprouting ability from protected buds on its main trunk and branches (Climent et al. 2004). The average percentage of serotinous trees, throughout its natural range of distribution, varies from 3% to 35%; the average bark thickness in adult trees ranges from 22 to 49 mm and is unrelated to tree age or diameter. The serotinous cones of *P. canariensis* are 'xeriscent' rather than 'pyriscent', since seed dispersal occurs after a time lapse even without fire. Seedling recruitment in the absence of fire is low due to inhibition by the thick cover of neeldles (Climent et al. 2004), but no massive germination was found in stands that were burned in hot canopy fires (Keeley personal communication). Thick bark provenances of *P. canariensis* coincide with productive areas with frequent and intense fires. Thin-barked trees grow in dry areas with sparse pine stands and a scarce understory, with ground fires that burn only the thick litter layer. The resistance to fires of aged individuals is enabled by their extremely thick bark (>15 cm), tree height > 60 m, self-pruning of lower branches, and deep root system. Moreover, their resinous heartwood prevents rotting and allows these monumental trees to live for up to 600 years (Climent et al. 2004).

22.5.8 **Pinus heldreichii** *H. Christ*, **P. mugo** *Turra* and **P. uncinata** *Ramond ex DC.*

These pine species grow in restricted high-altitude areas where fires are relatively rare. *Pinus heldreichii* grows in southern Italy, *P. mugo* grows in the northwestern Balkans and eastern Greece, and *P. uncinata* grows in northern Spain, southern France, the western Alps, the Pyrenees and Sierra de Javalambre (Barbero et al.

1998). All of these pine species have thin bark and non-serotinous cones. *Pinus uncinata* has heat-sensitive seeds (Escudero et al. 1999; Habrouk et al. 1999) that cannot survive crown fires (Pausas et al. 2008).

22.6 Conclusions

The leaf morphology and high resin content of pines makes them more flammable than many other tree species. The increasing human population in Mediterranean lowlands and its spread into natural and forested areas also increases fire risk for nearby forests. Global climate change is predicted to further increase the frequency and severity of forest fires. Thus the effects of fires must be considered when planting new forests, when managing natural and planted pine forests and when restoring pine forests. Knowing and understanding quantitative aspects of fire adaptive traits and post-fire demography are crucial for predicting the responses of pine forests during different fire types, intensities and frequencies (Table 22.1).

Based on morphological traits and lethal heating experiments, Fernandes et al. (2008) proposed a model for rating the fire resistance of European pine species. The classification is based on the probability of cambium kill by ground fire and crown kill by canopy fire (Fig. 22.1). *Pinus canariensis* is the most resistant and *P. uncinata* the most sensitive to both dangers, while other species are intermediates.

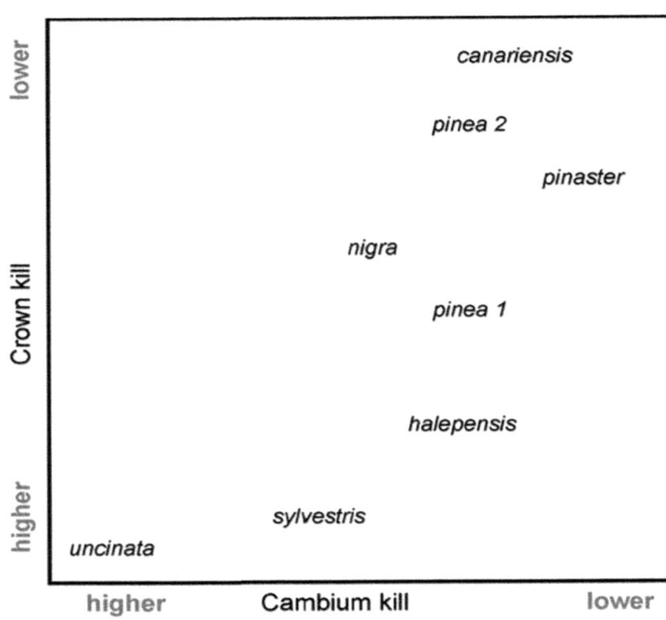

Fig. 22.1 Fire resistance to crown and cambium death of European pine species. *Pinea 1* does not accommodate the effect of canopy architecture, *Pinea 2* accommodates the effect of canopy architecture (By permission of Fernandes et al. 2008)

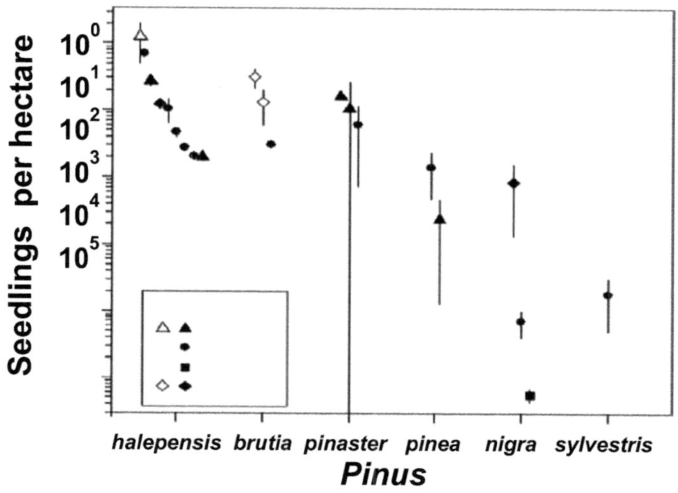

Fig. 22.2 Seedling density (seedlings ha^{-1}; log scale) after crown fire for different pine species and different post-fire ages, in different localities around the Mediterranean Basin. Symbols indicate the mean values, solid vertical lines are standard deviation, s.d. (in solid symbols), or range values (in open symbols), post-fire years: triangle represtns 1 year, circle from 2 to 3 years, square 4–8 years and diamond more than 8 years (by permission from: Pausas et al. 2008)

Once a forest is burned, predicting the potential of natural seedling recruitment is essential for planning successful restoration. Pausas et al. (2008) compiled data on natural seedling densities of naturally regenerating Mediterranean pine forests. *Pinus halepensis* demonstrates the highest post-fire seedling density and *P. uncinata* the lowest, while other species are intermediates (Fig. 22.2).

We note that genetic variation within and among populations, and local environmental factors, cause high variability in both tree mortality and regeneration capacity of pine forests. Therefore, foresters and natural resource managers should consider local environmental conditions when planning their actions. Finally, acting in concert with the natural processes should be the preferred default because it increases the chances of success and reduces costs, while acting in contrast to the natural processes increases costs and reduces the chances of success.

References

Adámek M, Hadincová V, Wild J (2016) Long-term effect of wildfires on temperate Pinus sylvestris forests: vegetation dynamics and ecosystem resilience. For Ecol Manag 380:285–295

Akeel A, Khan MMA, Jaleel H et al (2019) Smoke-saturated water and karrikinolide modulate germination, growth, photosynthesis and nutritional values of carrot (*Daucus carota* L.). J Plant Growth Regul 38:1387–1401

Arianoutsou M, Thanos CA (1996) Legumes in the fire-prone Mediterranean regions: an example from Greece. Int J Wildland Fire 6:77–82

Badik KJ, Jahner JP, Wilson JS (2018) A biogeographic perspective on the evolution of fire syndromes in pine trees (Pinus: Pinaceae). R Soc Open Sci 5:172412

Barbéro M, Loisel R, Quézel P et al (1998) Pines of the Mediterranean basin. In: Richardson DM (ed) Ecology and biogeography of *Pinus*. Cambridge University Press, Cambridge, pp 153–170

Baskin CC, Baskin JM (2014) Seeds: ecology biogeography and evolution of dormancy and germination, 2nd edn. St Louis Academic Press

Bock WJ (1959) Preadaptation and multiple evolutionary pathways. Evolution 13:194–211

Bond WJ (2008) What limits trees in C_4 grasslands and savannas? Ann Rev Ecol Evol 39:641–659

Bond W, Midgley JJ (2012) Fire and the angiosperm revolutions. Int J Plant Sci 173:569–583

Bond WJ, Scott AC (2010) Fire and the spread of flowering plants in the cretaceous. New Phytol 188:1137–1150

Bond WJ, Van Wilgen BW (2012) Fire and plants. Chapman & Hall London

Boydak M (2004) Silvicultural characteristics and natural regeneration of *Pinus brutia* Ten. A review. Plant Ecol 171:153–163

Boydak M, Caliskan S (2016) Effects of heat shock on seed germination of Turkish red pine (*Pinus brutia*). Bosque 37:327–333

Boydak M, Dirik H, Çalikikoğlu M (2006) Biology and silviculture of Turkish red pine (*Pinus brutia* Ten.). Ogem-Vak, Ankara

Bradshaw S, Dixon KW, Hopper SD et al (2011) Little evidence for fire-adapted plant traits in Mediterranean climate regions. Trends Plant Sci 16:69–76

Brown NAC (1993) Promotion of germination of fynbos seeds by plant-derived smoke. New Phytol 123:575–583

Chambel MR, Climent J, Pichot C et al (2013) Mediterranean pines (*Pinus halepensis* Mill. and *P. brutia* Ten.). In: Pâques LE (ed) Forest tree breeding in Europe: current state-of-the-art and perspectives, managing forest ecosystems. Springer, Dordrecht, pp 229–265

Christopoulou A, Fulé PZ, Andriopoulos P et al (2013) Dendrochronology-based fire history of *Pinus nigra* forests in Mount Taygetos, Southern Greece. For Ecol Manag 293:132–139

Christopoulou A, Fyllas NF, Andriopoulos P et al (2014) Post-fire regeneration patterns of *Pinus nigra* in a recently burned area in Mount Taygetos, Southern Greece: the role of unburned forest patches. For Ecol Manag 327:148–156

Clarke PJ, Lawes MJ, Midgley J et al (2016) Fire regime, soil fertility and growth form interact to shape fire and growth traits in two co-occurring Banksia species. Evol Ecol 30:35–45

Climent J, Tapias R, Pardos JA et al (2004) Fire adaptations in the Canary Islands pine (*Pinus canariensis*). Plant Ecol 171:185–196

Dimitrakopoulos AP, Papaioannou KK (2001) Flammability assessment of Mediterranean forest fuels. Fire Technol 37:143–152

Dixon KW, Roche S, Pate JS (1995) The promotive effect of smoke derived from burnt native vegetation on seed germination of Western Australian plants. Oecologia 101:185–192

Dixon KW, Merritt DJ, Flematti GR et al (2009) Karrikinolide–a phytoreactive compound derived from smoke with applications in horticulture, ecological restoration and agriculture. Acta Hortic 813:155–170

Escudero A, Barrero S, Pita JM (1997) Effects of high temperatures and ash on seed germination of two Iberian pines (*Pinus nigra* ssp *salzmannii*, *P sylvestris* var *iberica*). Ann Sci For 54:553–562

Escudero A, Sanz MV, Pita JM et al (1999) Probability of germination after heat treatment of native Spanish pines. Ann For Sci 56:511–520

Espelta JM, Arnan X, Rodrigo A (2011) Non-fire induced seed release ina weakly serotinous pine: climatic factors, maintenance costs or both? Oikos 120:1752–1760

Fernandes PM, Vega JA, Jimenez E et al (2008) Fire resistance of European pines. For Ecol Manag 256:246–255

Fonda RW (2001) Burning characteristics of needles from eight pine species. For Sci 47:390–396

Fulé PZ, Ribas M, Gutierrez E et al (2008) Forest structure and fire history in an old Pinus nigra forest eastern Spain. For Ecol Manag 255:1234–1242

Gershberg A, Ne'eman G, Ben-Shlomo R (2016) Genetic structure of a naturally regenerating post-fire seedling population: *Pinus halepensis* as a case study. Front Plant Sci 7:549

Gómez-González S, Torres-Díaz C, Bustos-Schindler C et al (2011) Anthropogenic fire drives the evolution of seed traits. PNAS 108:18743–18747

Goren-Inbar N, Alperson N, Kislev ME et al (2004) Evidence of hominin control of fire at Gesher Benot Yaaqov, Israel. Science 304:725–727

Goubitz S, Werger MJA, Ne'eman G (2002) Germination response to fire-related factors of seeds from non-serotinous and serotinous cones. Plant Ecol 169:195–204

Goubitz S, Nathan R, Roitemberg R et al (2004) Canopy seed bank structure in relation to: fire, tree size and density. Plant Ecol 173:191–201

Groom PK, Lamont B (2015) Plant life of southwestern Australia: adaptations for survival. Walter de Gruyter GmbH and Co KG

Habrouk A, Retana J, Espelta JM (1999) Role of heat tolerance and cone protection of seeds in the response of three pine species to wildfires. Plant Ecol 145:91–99

Hanley ME, Lamont BB (2000) Heat pre-treatment and the germination of soil-and canopy-stored seeds of south-western Australian species. Acta Oecol 21:315–321

He T, Pausas JG, Belcher CM et al (2012) Fire-adapted traits of *Pinus* arose in the fiery cretaceous. New Phytol 194:751–759

He T, Belcher CM, Lamont BB et al (2016) A 350-million-year legacy of fire adaptation among conifers. J Ecol 104:352–363

Henig-Sever N, Eshel A, Ne'eman G (1996) pH and osmotic potential of pine ash as post-fire germination inhibitors. Pysiol Plant 96:71–76

Hernández-Serrano A, Verdú M, González-Martínez SC et al (2013) Fire structures pine serotiny at different scales. Am J Bot 100:2349–2356

Hernández-Serrano A, Verdú M, Santos-del-Blanco L et al (2014) Heritability and quantitative genetic divergence of serotiny, a fire-persistence plant trait. Ann Bot 114:571–577

Herrero C, San Martin R, Bravo F (2007) Effect of heat and ash treatments on germination of *Pinus pinaster* and *Cistus laurifolius*. J Arid Environ 70:540–548

Izhaki I, Henig-Sever N, Ne'eman G (2000) Soil seed banks in Mediterranean Aleppo pine forests: the effect of heat, cover and ash on seedling emergence. J Ecol 88:667–675

Kavgaci A, Örtel E, Torres I et al (2016) Early postfire vegetation recovery of *Pinus brutia* forests: effects of fire severity, prefire stand age, and aspect. Turk J Agric For 40:723–736

Keeley JE (1993) Smoke-induced flowering in the fire-lily *Cyrtanthus ventricosus*. S Afr J Bot 59:638

Keeley JE, Fotheringham CJ (1998) Smoke-induced seed germination in California chaparral. Ecology 79:2320–2336

Keeley JE, Pausas JG (2018) Evolution of 'smoke' induced seed germination in pyroendemic plants. S Afr J Bot 115:251–255

Keeley JE, Stephenson NL (2000) Restoring natural fire regimes to the Sierra Nevada in an era of global change. In: Cole DN, SF MC, Borrie WT et al (eds) Wilderness science in a time of change conference volume 5: wilderness ecosystems, threats, and management. 1999 Proceedings RMRS-P-15-VOL-5. US Department of Agriculture, vol 15. Forest Service, Rocky Mountain Research Station, pp 255–265

Keeley JE, Zedler PH (1998) Evolution of life histories in Pinus. In: Richardson DM (ed) Ecology and biogeography of *Pinus*. Cambridge University Press, Cambridge, pp 219–250

Keeley JE, Ne'eman G, Fotheringham CJ (1999) Immaturity risk in a fire-dependent pine. J Mediterr Ecol 1:41–48

Keeley JE, Pausas JG, Rundel PW et al (2011) Fire as an evolutionary pressure shaping plant traits. Trends Plant Sci 16:406–411

Keeley JE, Bond WJ, Bradstock RA et al (2012) Fire in Mediterranean ecosystems: ecology, evolution and management. Cambridge University Press Cambridge, Cambridge

Klaus W (1989) Mediterranean pines and their history. Plant Syst Evol 162:133–163

Lamont BB (1991) Canopy seed storage and release: what's in a name? Oikos 62:266–268

Lamont BB, He T (2017) Fire-proneness as a prerequisite for the evolution of fire-adapted traits. Trends Plant Sci 22:278–288
Lamont BB, Le Maitre DC, Cowling RM et al (1991) Seed canopy storage in woody plants. Bot Rev 57:277–317
Lamont BB, He T, Yan Z (2019a) Fire as a pre-emptive evolutionary trigger among seed plants. Perspect Plant Ecol 36:13–23
Lamont BB, He T, Yan Z (2019b) Evolutionary history of fire-stimulated resprouting, flowering, seed release and germination. Biol Rev 94:903–928
Lindner M, Maroschek M, Netherer S et al (2010) Climate change impacts, adaptive capacity, and vulnerability of European forest ecosystems. For Ecol Manag 259:698–709
Martínez-Sánchez JJ, Marín A, Herranz JM et al (1995) Effects of high temperatures on germination of *Pinus halepensis* Mill. and *P. pinaster* Aiton subsp. pinaster seeds in Southeast Spain. Vegetatio 116:69–72
Martín-Sanz RC, Callejas-Díaz M, Tonnabel J et al (2017) Maintenance costs of serotiny in a variably serotinous pine: the role of water supply. PLoS One 12(7):e0181648
Moreira B, Pausas JG (2018) Shedding light through the smoke on the germination of Mediterranean Basin flora. S Afr J Bot 115:244–250
Moreira B, Tormo J, Estrelles E et al (2010) Disentangling the role of heat and smoke as germination cues in Mediterranean Basin flora. Ann Bot 105:627–635
Moreira F, Viedma O, Arianoutsou M et al (2011) Landscape–wildfire interactions in southern Europe: implications for landscape management. J Environ Manag 92:2389–2402
Moreno JM, Oechel WC (2012) The role of fire in Mediterranean-type ecosystems, vol 107. Springer Science and Business Media, New York
Moreno JM, Morales-Molino C, Torres I, Arianoutsou M (this volume) Fire in Mediterranean Pine Forests, Past, Present and Future. In: Ne'eman G Osem Y (eds) Pines and their mixed forest ecosystems in the Mediterranean Basin. Springer, Berlin
Moriondo M, Good P, Durao R et al (2006) Potential impact of climate change on fire risk in the Mediterranean area. Clim Res 31:85–95
Moritz MA, Morais ME, Summerell LA et al (2005) Wildfires, complexity, and highly optimized tolerance. PNAS 102:17912–17917
Moya D, Saracino A, Salvatore R et al (2008) Anatomic basis and insulation of serotinous cones in *Pinus halepensis* Mill. Trees 22:511–519
Nathan R, Ne'eman G (2000) Serotiny, seed dispersal and seed predation in *Pinus halepensis*. In: Ne'Eman G, Trabaud L (eds) Ecology, biogeography and management of *Pinus halepensis* and *P. brutia* forest ecosystems in the Mediterranean Basin. Backhuys Publishers, Leiden, pp 105–119
Ne'eman G (2000) The effect of burned pine trees on post fire regeneration. In: Ne'eman G, Trabaud L (eds) Ecology, biogeography and management of *Pinus halepensis* and *P. brutia* forest ecosystems in the Mediterranean Basin. Backhuys Publishers, Leiden, pp 292–303
Ne'eman G, Izhaki I (1998) Stability of pre-and post-fire spatial structure of pine trees in Aleppo pine forest. Ecography 21:535–542
Ne'eman G, Izhaki I (1999) The effect of stand age and microhabitat on soil seed banks in Mediterranean Aleppo pine forests after fire. Plant Ecol 144:115–125
Ne'eman G, Meir I, Ne'eman R (1993) The influence of pine ash on the germination and early growth of *Pinus halepensis* Mill. and *Cistus salviifolius* L. Water Sci Technol 27:525–532
Ne'eman G, Henig-Sever N, Eshel A (1999a) Regulation of the germination of *Rhus coriaria*, a post-fire pioneer, by heat, ash, pH, water potential and ethylene. Physiol Plant 106:47–52
Ne'eman G, Fotheringham CJ, Keeley JE (1999b) Patch to landscape patterns in post fire recruitment of a serotinous conifer. Plant Ecol 145:235–242
Ne'eman G, Goubitz S, Nathan R (2004) Reproductive traits of *Pinus halepensis* in the light of fire–a critical review. Plant Ecol 171:69–79
Ne'eman G, Ne'eman R, Keith DA et al (2009) Does post-fire plant regeneration mode affect the germination response to fire-related cues? Oecologia 159:483–492

Ordonez JL, Molowny-Horas R, Retana J (2006) A model of the recruitment of *Pinus nigra* from unburned edges after large wildfires. Ecol Model 197:405–417

Pausas JG (2015a) Bark thickness and fire regime. Funct Ecol 29:315–327

Pausas JG (2015b) Evolutionary fire ecology: lessons learned from pines. Trends Plant Sci 20:318–324

Pausas JG, Fernández-Muñoz S (2012) Fire regime changes in the Western Mediterranean Basin: from fuel-limited to drought-driven fire regime. Clim Chang 110:215–226

Pausas JG, Keeley JE (2009) A burning story: the role of fire in the history of life. Bioscience 59:593–601

Pausas JG, Keeley JE (2014) Evolutionary ecology of resprouting and seeding in fire-prone ecosystems. New Phytol 204:55–65

Pausas JG, Keeley JE (2017) Epicormic resprouting in fire-prone ecosystems. Trends Plant Sci 22:1008–1015

Pausas JG, Llovet J, Rodrigo A et al (2008) Are wildfires a disaster in the Mediterranean basin?–A review. Inter J wildland Fire 17:713–723

Pausas JG, Keeley JE, Schwilk DW (2017) Flammability as an ecological and evolutionary driver. J Ecol 105:289–297

Penuelas J, Sardans J, Fillela I et al (2017) Impact of global change on Mediterranean forests and their services. Forests 8:463. https://doi.org/10.3390/f8120463

Quézel P (2000) Taxonomy and biogeography of Mediterranean pines. In: Ne'eman G, Trabaud L (eds) Ecology, biogeography and management of *Pinus halepensis* and *P. brutia* forest ecosystems in the Mediterranean Basin. Backhuys Publishers, Leiden, pp 1–12

Reyes O, Casal M (2004) Effects of forest fire ash on germination and early growth of four *Pinus* species. Plant Ecol 175:81–89

Reyes O, Casal M (2006) Can smoke affect the germination of *Pinus sylvestris, P. nigra, P. uncinata and P. pinaster*? In: Viegas DX (ed) International conference on forest fire research. ADAI/CEIF University of Coimbra, Coimbra

Rigolot E (2004) Predicting post fire mortality of *Pinus halepensis* Mill. and *Pinus pinea* L. Plant Ecol 171:139–151

Rundel PW, Arroyo MTK, Cowling RM et al (2016) Mediterranean biomes: evolution of their vegetation, floras, and climate. Ann Rev Ecol Evol Syst 47:383–407

Salvatore R, Moya D, Pulido L et al (2010) Morphological and anatomical differences in Aleppo pine seeds from serotinous and non-serotinous cones. New For 39:329–341

San-Miguel-Ayanz J, Moreno JM, Camia A (2013) Analysis of large fires in European Mediterranean landscapes: lessons learned and perspectives. For Ecol Manag 294:11–22

Schwilk DW (2003) Flammability is a niche construction trait: canopy architecture affects fire intensity. Am Nat 162:725–733

Schwilk DW, Ackerly DD (2001) Flammability and serotiny as strategies: correlated evolution in pines. Oikos 94:326–336

Skordilis A, Thanos CA (1995) Seed stratification and germination strategy in the Mediterranean pines Pinus brutia and P. halepensis. Seed Sci Res 5:151–160

Skordilis A, Thanos CA (1997) Comparative ecophysiology of seed germination strategies in the seven pine species naturally growing in Greece. In: Ellis RH, Black M, Murdoch AJ et al (eds) Basic and applied aspects of seed biology. Current plant science and biotechnology in agriculture, vol 30. Springer, Dordrecht, pp 623–632

Thanos CA (2000) Ecophysiology of seed germination *Pinus halepensis* and *P. brutia*. In: Ne'eman G, Trabaud L (eds) Ecology, biogeography and management of *Pinus halepensis* and *P. brutia* forest ecosystems in the Mediterranean Basin. Backhuys Publishers, Leiden, pp 37–50

Thanos CA, Daskalakou EN (2000) Reproduction in *Pinus halepensis* and *P.brutia*. In: Ne'eman G, Trabaud L (eds) Ecology, biogeography and management of *Pinus halepensis* and *P. brutia* forest ecosystems in the Mediterranean Basin. Backhuys Publishers, Leiden, pp 79–90

Thanos CA, Skordilis A (1987) The effects of light, temperature and osmotic stress on the germination of *Pinus halepensis* and *P. brutia seeds*. Seed Sci Technol 15:163–174

Thanos CA, Georghiou K, Kadis C et al (1992) Cistaceae: a plant family with hard seeds. Isr J Plant Sci 41:251–263

Van Mantgem P, Schwartz M (2003) Bark heat resistance of small trees in Californian mixed conifer forests: testing some model assumptions. For Ecol Manag 178:341–352

Vega JA (2000) Resistencia vegetativa ante el fuego a través de la historia de los incendios. In: Vélez R (ed) La Defensa Contra Incendios Forestales, Fundamentos y Experiencias. McGraw-Hill, Madrid, pp 466–485

Willis KJ, Bennett KD, Birks HJB (1998) The late quaternary dynamics of pines in Europe. In: Richardson DM (ed) Ecology and biogeography of *Pinus*. Cambridge University Press, Cambridge, pp 107–121

Zuloaga-Aguilar S, Briones O, Orozco-Segovia A (2011) Seed germination of montane forest species in response to ash, smoke and heat shock in Mexico. Acta Oecol 37:256–262

Zwanenburg B, Pospíšil T, Zeljković SĆ (2016) Strigolactones: new plant hormones in action. Planta 243:1311–1326

Chapter 23
Post-fire Soil Erosion – The Mediterranean Perception

Lea Wittenberg

23.1 Introduction: Fires as a Geomorphological Agent

Fire is a natural and important part of the disturbance regime that has shaped the Mediterranean landscape for centuries; its outcomes depend upon the complex interplay of weather, topography, geology, vegetation cover and fire characteristics. The effects of fire on geosystems include changes to soil properties, hydrologic processes and weathering rates, potentially resulting in disruption and damage to critical infrastructures and human safety (Shakesby and Doerr 2006). Given the enhanced nature of fire-induced geomorphic processes and the generally low background yields of erosion from non-disturbed forest soils, there is a growing recognition that fires, specifically around the Mediterranean Basin, have played a central role in landscape evolution, soil formation and sediment production (Shakesby 2011). For example, Swanson (1981), in his seminal paper, estimated that more than 70% of the total annual erosion yields in the Mediterranean climate region are caused by wildfires. Sankey et al. (2017), using an ensemble of climate, fire and erosion models, showed that in the western USA, post-fire sedimentation rates are projected to increase for nearly nine tenths of watersheds by >10% and for more than one third of watersheds by >100%, by the 2040s, due to increased fire activity.

Although the links between wildfire, flooding, erosion and sedimentation have been observed worldwide, post-wildfire studies, particularly regarding fire-induced erosion responses, only began in the early 1980s in the Mediterranean Basin, following increased fire activity (Shakesby 2011). The purpose of this chapter is to review the current knowledge and identify research gaps in the issues associated with fire-induced erosion processes in the Mediterranean Basin. A discussion of the

L. Wittenberg (✉)
Department of Geography and Environmental Studies, University of Haifa, Haifa, Israel
e-mail: leaw@geo.haifa.ac.il

G. Ne'eman, Y. Osem (eds.), *Pines and Their Mixed Forest Ecosystems in the Mediterranean Basin*, Managing Forest Ecosystems 38,
https://doi.org/10.1007/978-3-030-63625-8_23

effects of fire on erosion with particular attention to the major driving forces and future trajectories will be included.

23.2 Soil Erosion: Natural vs. Fire-Affected Rates

> Fire changes forest ecosystems and interacts with geomorphic processes, climate, and landform in a variety of ways to alter the landscape and temporarily increase the potential for erosion (McNabb and Swanson, 1990).

Soil erosion is a complex process involving many factors including soil properties, slope gradient, vegetation cover, and rainfall characteristics (Römkens et al. 2002). Naturally, it is a two-phase process – initial detachment of soil particles, followed by the downslope transport and deposition of the material by erosive agents such as wind, water and gravity (Lal 2001). Soil erosion is a major environmental problem worldwide and is the major cause of land degradation and reduced soil functioning, because the most fertile topsoil, where organic substances and vital nutrients are stored, is removed.

Mediterranean soils are particularly prone to erosion (García-Ruiz et al. 2015), due to: (a) sharp relief (45% of the Mediterranean Basin has slopes steeper than 8%), (b) intense rainfall events (> 100 mm h^{-1}) mainly in autumn and winter, (c) existence of poor, shallow and skeletal soils, and (d) sparse natural vegetation which is often subjected to severe seasonal droughts (Raclot et al. 2016). In addition to these natural drivers, the long history of intensive agriculture, burning, overgrazing and deforestation has accelerated regional erosion rates. Nonetheless, natural soil losses from forested or densely vegetated Mediterranean landscapes are often negligible due to the protective nature of the vegetation cover (Wittenberg and Inbar 2009; Wittenberg et al. 2014). Four main erosion-mitigation mechanisms are facilitated by the vegetation: (1) rain drop interception through the plant canopy, which reduces splash erosion and soil detachment, (2) funneling the intercepted rain-drops along the stem toward the base of the plant where the soil is more permeable, (3) increasing surface roughness which reduces runoff flow velocity, (4) increasing soil aggregate stability through the binding of the soil organic matter (Mataix-Solera et al. 2011). The sum effect of these processes is the formation of a source-sink mosaic, in which bare soil patches function as sources, providing available sediments for transport, whilst vegetation-covered patches act as sinks by trapping the sediments and enriching the underlying soil with nutrients and organic compounds (Puigdefábregas 2005). These processes and patterns are altered by fires which increase the risk of erosion of the burned soils.

Vegetation cover patterns and vegetation type have significant effects on runoff and erosion rates. This concept was demonstrated in a plot-scale (1 m^2) study where the impacts of pine trees (*Pinus halepensis* Mill.) versus oaks (*Quercus ilex* L.) were compared in a 30-year old plant cover area recovering from a forest fire. Results indicated higher runoff coefficients and sediment concentrations from the pine (8.31%, 4.9 g l^{-1}, respectively) than the oak (1.4%, 2.6 g l^{-1}) plots; moreover, the annual erosion yields were ten times higher under pine trees (2.6 t ha^{-1} $year^{-1}$) than under Holm oak plots (0.26 t ha^{-1} $year^{-1}$) (Cerdà et al. 2017).

Using a fallout radionuclide ^{137}Cs technique to assess soil erosion magnitude, the net erosion rate in a mixed Mediterranean forest yielded values of about 4.2 t ha^{-1} $year^{-1}$ (Meliho et al. 2019). A meta-analyses on data from 81 experimental sites in 19 European countries (including the Mediterranean), covering 2741 plot-years in total, provided plot erosion rates for various land uses (Cerdan et al. 2010). The results emphasized that annual soil losses from agricultural uses (4.4–15.1 T ha^{-1} $year^{-1}$) are 1–2 orders of magnitude higher than the relatively low soil losses from grasses (0.3 T ha^{-1} $year^{-1}$), shrublands (0.5 T ha^{-1} $year^{-1}$) and forests (0.14 T ha^{-1} $year^{-1}$). Another set of studies conducted in various sub-climates and vegetation types along the Mediterranean demonstrated particularly low rates of natural erosion (0.0001–0.001 T h^{-1} $year^{-1}$) (Inbar et al. 1998; Kutiel and Inbar 1993; Pausas et al. 2009; Wittenberg et al. 2014; Zituni et al. 2019).

23.3 Fire Effects on Soils and Erosion

Fires can lead to substantial hydro-geomorphological changes, directly by weathering bedrock surfaces and changing soil structure and properties, and indirectly during the post-fire recovery period, via enhanced soil erosion and depositional processes. The explicit effect is site-specific and depends on multiple factors including the temporal occurrence of the fire (e.g. the interval between fires and the duration of each event) and its spatial pattern and behavior (Moody et al. 2013). Fire reduces vegetation cover and increases net precipitation at the surface (Williams et al. 2019).

Elevated rates of fire-induced weathering are often reported in the fire-prone landscape of the Mediterranean Basin. One of the recently suggested mechanisms is the accelerated weathering of the carbonate rock, mainly during severe fires, by spalling and exfoliation, which creates abundant peels, flakes and spalls that cover 80–100% of the exposed burned rocks (Fig. 23.1). The extent of the physical weathering depends on rock composition; the most severe response (determined by the depth of exfoliation) was found in chalk formations covered by calcrete (nari crusts) where an average depth of 7.7–9.6 cm and a maximum depth of 20 cm were measured. In the years after a fire, the disintegrated rock fractures trap and store fine particles of dust, ash and organic matter, presumably maintaining higher than average weathering rates (Shtober-Zisu et al. 2018).

During the post-fire recovery period two processes prevail: (1) elevated soil erosion and sediment redistribution, and (2) input of new minerals and charred organic components from the fire residues (ash).

Studies from numerous sites across the Mediterranean Basin have reported large increases in peak flow, erosion rates, downstream flooding and sedimentation following moderate and high severity wildfires (e.g. Wittenberg and Inbar 2009;

Fig. 23.1 Burned rock spalling, Arqan Formation, 2010 fire Mt. Carmel. (Photo: Nurit Shtober-Zisu 10/12/2010)

Shakesby 2011; Ohana-Levi et al. 2018). The elevated fire-induced response is driven by four ecosystem state changes: (1) reduction in plant canopy; (2) changes in soil properties; (3) loss of surface cover; and (4) consumption of soil organic matter (Wagenbrenner et al. 2015). These state changes might accelerate other interacting processes such as higher soil erodibility, infiltration and erosion. The loss of vegetation cover affects the overall water balance; this is generally regarded as having a smaller effect on hillslope erosion rates than the changes to soil properties (Shakesby and Doerr 2006). Figure 23.2 shows the state changes and processes occurring after wildfires (Wittenberg et al. 2019, after Wagenbrenner et al. 2015).

Fires determine the rates, magnitudes and timing of overland flow and soil losses through direct and indirect effects on each of the following parameters (Fig. 23.2): soil organic matter (SOM), soil water repellency (SWR), soil roughness, ash type and amount, ground and canopy cover (Stoof et al. 2014; Santín and Doerr 2016). The direct heat transfer from the combustion of canopy and SOM compounds creates multilevel changes to the chemical, physical and biological properties of the soil. These changes are associated not only with the burning temperature (fire severity) but also with heat duration and oxygen availability (Santín and Doerr 2016). At temperatures below 200 °C, fire mostly affects biological activity, although some physical characteristics such as SWR might be affected. At higher temperatures (>300 °C), physical, chemical and mineralogical properties are modified through the burning of the SOM and the production of pyrogenic carbon. For example, soil bulk density, clay and silt content increase, whereas SOM and essential nutrients are lost (Stoof et al. 2010; Gómez-Rey et al. 2013), soil structure breaks down

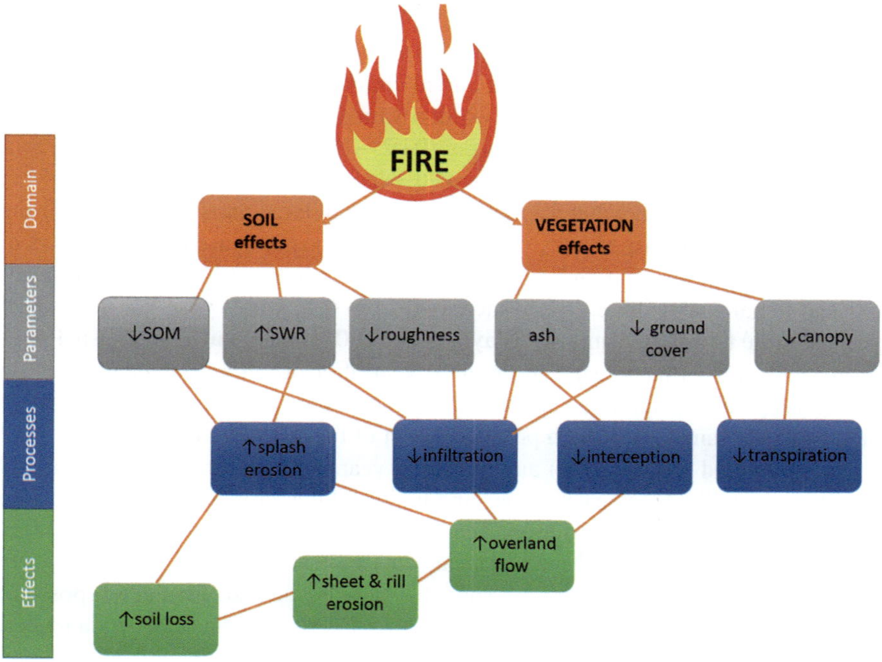

Fig. 23.2 Domains of fire effects on soil erosion: parameters, processes and net effects

(Mataix-Solera et al. 2011), soil moisture is reduced and SWR often increases (Bodí et al. 2011). Roughness of the soil surface can also vary in response to fire by the formation of weathered rock spalls, ash deposits and stony armoring. Early work by Lavee et al. (1995) suggested that low- to moderate-temperature fires produce a mosaic-like surface containing both rough patches with almost no runoff generation and relatively smooth patches which can enable higher runoff and erosion rates.

It is widely agreed that ash plays a key role in the hydraulic response of soil to fire (Bodí et al. 2014); however, the specific effects remain inconclusive as studies showing reduced and increased overland flow rates have been published. The physical and chemical properties of ash residues are highly variable, as are the properties of the subsurface soil. This combination makes it difficult to assess the immediate effect of ash on infiltration and erodibility. This variability in response is also related to ash depth and type (which in turn depends on pre-fire vegetation and fire severity), soil type and rainfall characteristics (Brook and Wittenberg 2016). In addition to its effect on hydro-geomorphological processes, the ash layer serves as a key source of fine sediments and provides nutrients to the soil and solutes in runoff (Bodí et al. 2011).

Owing to the complexity of the interacting forces and the erratic nature of the fire itself, studies addressing fire-induced erosion have yielded a wide spectrum of results, sometimes with opposing trends (Shakesby 2011). For example, Moody and Martin (2009) reviewed sediment data from 135 sites over the US and indicated that annual post-fires yields span five orders of magnitude, with the lower yields equal to the non-burned yields. Smith et al. (2011) analyzed first-year post-fire suspended sediment at 17 locations representing varied climates including the Mediterranean, and found that suspended sediment exports varied from 0.017 to 50 t ha^{-1} year^{-1} across a large range of catchment sizes (0.021–1655 km^2), representing an estimated increase of 1–1459 times unburned exports. In Mt. Carmel, Nahal Oren (drainage area = 18 km^2) that was severely burned in large parts of its basin during the 2010 fire, mean suspended sediment concentrations increased by 2–3 orders of magnitude during the first and second years after the fire in relation to pre-fire values and the annual sediment yield increased from 0.17 t ha^{-1} year^{-1} to 0.22 t ha^{-1} year^{-1} for the first year. Similarly, Mayor et al. (2007) documented an unusual sediment yield of 3.7 t ha^{-1} year^{-1} in the third year after fire in a 2.1-ha catchment in southern Spain. The authors explained this high rate by the occurrence of a single high-intensity rainstorm, which produced most of the annual eroded sediments.

Shakesby and Doerr (2006) analyzed first-year, post-fire erosion data from 25 Mediterranean hillslope-scale studies and indicated that erosion rates span a range of 0.1 to 414 t ha^{-1} year^{-1}. Later on, Shakesby (2011) compiled first-year, post-fire, field-plot data from tens of Mediterranean sites and indicated that soil losses were determined by fire severity and the spatial scale investigated. In general, post-fire erosion rates were found to increase with fire severity. Mean and median rates, respectively, were 0.39 and 0.10 for low, 3.28 and 2.10 for moderate, and 10.8 and 3.70 t ha^{-1} year^{-1} for high severity fires. Only in five cases did erosion rates exceed 10 t ha^{-1} in any single year of measurement and in half of the field plots studied, annual post-wildfire rates did not exceed 1 t ha^{-1}. Addressing the spatial scale, first year soil erosion at point (45–56 t ha^{-1}) and plot-scales (< 10 t ha^{-1}) were similar to, or even lower than those reported for other fire-affected lands or those with other uses (e.g. cultivated) and natural poorly-vegetated (e.g. badlands or rangeland) in the Mediterranean Basin. The limited data regarding soil loss at larger-scales (hillslope and catchment) were inconstant.

The key question that arises is why there are such large differences in estimated post-fire erosion rates measured within a defined bio-climatic region.

23.4 Factors Controlling Post-fire Erosion Rates

Sheridan et al. (2019) provided four possible explanations for the observed variability in post-fire erosion rates:

1. ***Climate***: precipitation amounts, timing and form (e.g. rainfall vs. snow).
2. ***Landscape properties***: site-specific traits such as soil type, geology, topography, aspect, land use, fire severity and time since last fire.
3. ***Stochasticity:*** the random nature of precipitation and other weather properties in the first year after the fire.

4. ***Erosion processes***: the combined effects of climate, landscape properties and stochasticity result in different erosional processes, timing and magnitudes.

23.4.1 Climate

Four meteorological variables determine the propagation and severity of wildfires, and the post-fire responses: wind speed, ambient temperature, relative humidity and the number of dry days since the last rainfall (Wittenberg and Kutiel 2016). Post-fire erosion rates are strongly related to climatic drivers during the "window of risk" following the fire, and site-specific precipitation (amount, intensity, form and timing) is the key factor. However, the stochastic nature of the climatic variables mean that even in the same climate and landscape, considerable inter- and intra-annual variations are expected. With frequent and intense rain storms, the Mediterranean Basin is particularly vulnerable to soil erosion and land degradation, mainly during the first weeks after a fire (Peña-Angulo et al. 2019). Intense rainfall episodes are common during the autumn following the warm, dry summers that are responsible for severe wildfires (Versini et al. 2013; Pereira et al. 2016). Several studies have indicated that soil loss rates are linear during low-intensity rainfall events, while it is non-linear during high-intensity storms. Therefore, it is likely that intense, short duration storms lead to greater soil losses (Bracken and Kirkby 2005; Mohamadi and Kavian 2015). Due to global climate change, extreme autumn rainfall events are expected to increase, potentially triggering flash floods, with a detrimental impact on fire-affected areas (Francos et al. 2016). Long-term climate patterns also drive regeneration dynamics and recovery rates, and increase the time period that the burned soil is susceptible to erosion.

23.4.2 Landscape Properties

Varying slope properties produce diverse runoff and soil loss yields, due to differences in aspect, steepness, shape, lithology and vegetation (Mayor et al. 2007; Poesen 2018). Pre-fire vegetation composition, fire severity, rainstorm characteristics and revegetation rates contribute to the formation of more active sediment-producing slopes.

In the burned slopes of the Mediterranean Basin, fire-induced runoff and erosion rates are significantly higher on south-facing slopes than on north-facing ones (Wittenberg et al. 2014). There are several reasons why aspect is important in the Mediterranean-type climate. First, higher evapotranspiration rates on south-facing slopes lead to less dense vegetation cover and therefore shallow and stony soil with less SOM (Pausas et al. 2009). In addition, the drier conditions on the south-facing slopes increase susceptibility to more frequent fires, elevate post-fire erosion and slow revegetation rates.

Vegetation regeneration may commence immediately after fire, depending on the climate, species composition and fire characteristics (Tsafrir et al. 2019). In their study, Tsafrir et al. (2019) found that fire season plays a fundamental role in the renewal dynamics of perennial plants, due to phenological traits rather than fire severity effects. Altered fire season effects were only detected among obligate seeders, and a more profound decrease in abundance was noted after spring fires than following autumn burns. Such changes in community composition might affect behavior following repeated fires.

Usually, vegetation cover is relatively low (20–45%) in the first winter after a fire, and increases to 50–70% during the second rainy season (Mayor et al. 2007, Wittenberg et al. 2014). Runoff and soil loss decrease dramatically after a threshold of vegetation cover is achieved. In the Mediterranean, 30–40% vegetation cover is sufficient to significantly decrease soil loss and runoff yields (Gimeno-Garcia et al. 2007; Wittenberg et al. 2014).

23.4.3 Fire Severity

Moderate and severe fires increase the potential for accelerated erosion, primarily through their effects on vegetation and soil properties. As fire severity increases, more vegetation and SOM are consumed, and more chemical and physical properties of the soil are changed. Low-severity wildfires generally burn slowly through the understory and have little direct effect on overstory vegetation. Under severe fire conditions, entire crowns and portions of the stem may be burned. If the crowns do not completely burn, dead foliage can fall to the ground and protect the exposed soil against erosion (Swanson 1981). Erosion rates from the complete Mediterranean data set analyzed by Shakesby (2011), indicated a clear link between erosion rates and fire severity, regardless of plot size, precipitation amount or soil type and geology.

A meta-analysis conducted by Vieira et al. (2015) proved a significant effect of fire on runoff and erosional processes. The magnitude of the post-fire runoff and erosion response was determined by four key factors: (1) soil burn severity; (2) time since last fire; (3) rainfall intensity; and (4) soil cover. However, the relationship between fire severity and elevated post-fire rates was only for inter-rill erosion (the process of detachment and transportation of soil particles by the shallow overland flow), and not for runoff. This enhanced impact on erosion rates but not on runoff rates can be explained by heating-induced changes in soil properties that determine soil erodibility, such as aggregate stability and SOM content (Fig. 23.3) (Mataix-Solera et al. 2011). The consumption of ground-covering vegetation and/or litter layer may play a role in reducing flow resistance and thereby increasing the erosive power of the runoff (Shakesby and Doerr 2006).

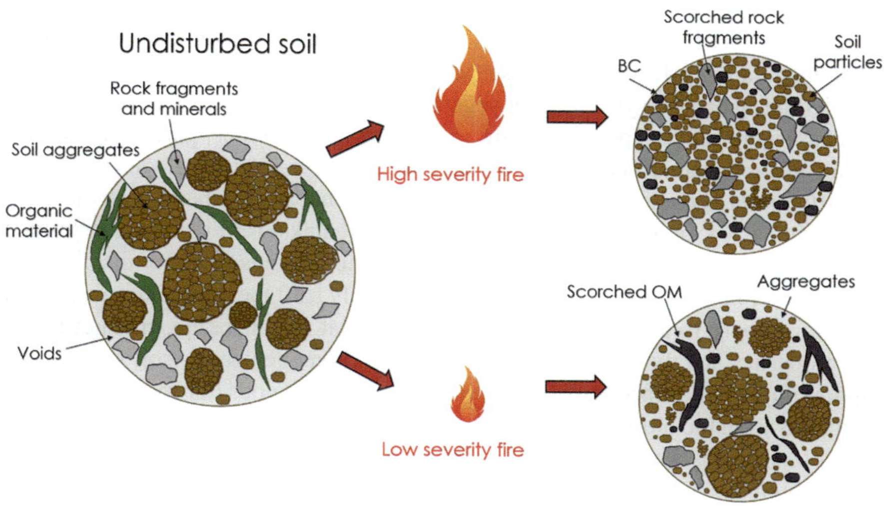

Fig. 23.3 Effects of burn severity on soil organic matter and aggregate stability. During low-severity fires, organic matter (OM) is scorched; while during high-severity fires, organic matter is consumed and black carbon (BC) is deposited and soil aggregates disintegrate, leading to an overall increase in soil erodibility and elevated erosion rates. (Drawing: Nurit Shtober-Zisu)

23.4.4 Spatio-Temporal Scales and Sediment Connectivity

Erosion and sediment transport processes also depend on the spatio-temporal scale and the geomorphic connectivity of the landscape (Wohl et al. 2019). Swanson (1981) described the temporal response of fire-disturbed systems as convex along the time scale. Accordingly, burned ecosystems exhibit a peak in resource loss immediately following a fire, with a rapid return to pre-disturbance levels, particularly in Mediterranean-climate ecosystems. Studies show that the return to background levels following repeated fires or extreme dry conditions make take several years, e.g. Wittenberg and Inbar (2009) suggest that following recurrent fires the response of the systems is delayed with respect to resource losses, essentially shifting the response from a convex to a concave type.

Wilcox et al. (2003) investigated unit-area resource loss (runoff and sediment) at nested spatial scales, ranging from micro-plots to hillslopes. They identified two types of system response to resource loss: (1) conserving (functional) systems, characterized by a rapid decrease in resource loss per unit area as the spatial scale of investigation increases, similar to a convex type response, (2) non-conserving, leaking (dysfunctional) systems, characterized by a concave type of response, where per unit-area resource loss from the system remains high despite the increase in spatial scale.

Conceptually, different landscapes can be placed on a theoretical continuum, at one end the fully functional landscapes capture all water and sediment, whilst at the other, the totally dysfunctional landscapes do not trap any of the eroded material. In the Mediterranean ecosystem, fully vegetated quasi-homogenous landscapes serve as conserving landscapes, and fire might generate either patchy or quasi-homogeneous systems denuded of vegetation cover.

Differences between conserving and leaky systems are attributed to hydrological, sediment, or landscape connectivity. Sediment sinks, for example, are formed by vegetation-covered patches and/or topographic features such as macropores. Alteration of these patterns by fire might temporarily shift the system from a convex type of response to a concave one. Post-fire removal of vegetation, changes to the soil properties and the formation of ash blankets, homogenize the soil surface and support improved connectivity which promotes the continuous transport of soils and sediments from the plots via the hillslopes to the river systems (Fig. 23.4). However, only a few studies have recorded elevated sediment yields at basin (> 1 km^2) scales (Wagenbrenner and Robichaud 2014), indicating that the effects of the fire are mainly related to soil destruction and redistribution of the eroded sediment at the hillslope scale. Since spatial and temporal scales are positively linked, it is assumed that larger scale effects can only be observed and assessed with long-term monitoring schemes.

Fig. 23.4 Increasing numbers of repeated fires impede vegetation regeneration, modify vegetation cover and type, affect soil rehabilitation processes, increase connectivity and potentially change the system from conserving to leaky. In leaky (dysfunctional) systems basin erosion might exceed hillslope rates

23.5 Discussion: Post-fire Erosion in the Mediterranean

Despite the long history of human and natural disturbances in the Mediterranean Basin, fire-induced erosion rates are lower than in other bio-climatic regions of the world (Cerdan et al. 2010, Shakesby 2011). These low rates are attributed to the shallow depth and high stone content of the Mediterranean soils. Surface stoniness impedes the extent or continuity of water-repellent patches, facilitating lower run-off–infiltration ratios compared to less stony patches (Kutiel et al. 1995; Urbanek and Shakesby 2009). The emerging armoring and increased roughness of the soil surface also limits sediment availability and reduces long-term erosional processes (Fig. 23.5).

Nonetheless, following repeated fires, the cumulative damage to soil properties, alteration to vegetation structure and reduction in SOM (Wittenberg and Inbar 2009; Malkisnon et al. 2011), not only increase peak erosion rates, but also the extent of the erosive response and the baseline erosion rate.

Most studies addressing post-fire soil erosion are concerned with specific amounts of eroded material and explicit factors dictating the magnitude and timing of these fire-induced responses. The specific amounts are in fact highly variable, but generally show an increase of several orders of magnitude compared to unburned soils. Since natural erosion rates from Mediterranean forests and maquis are frequently negligible, the comparison between burned and unburned soils is of limited importance as it might imply that fires have limited (if any) effects on the soil

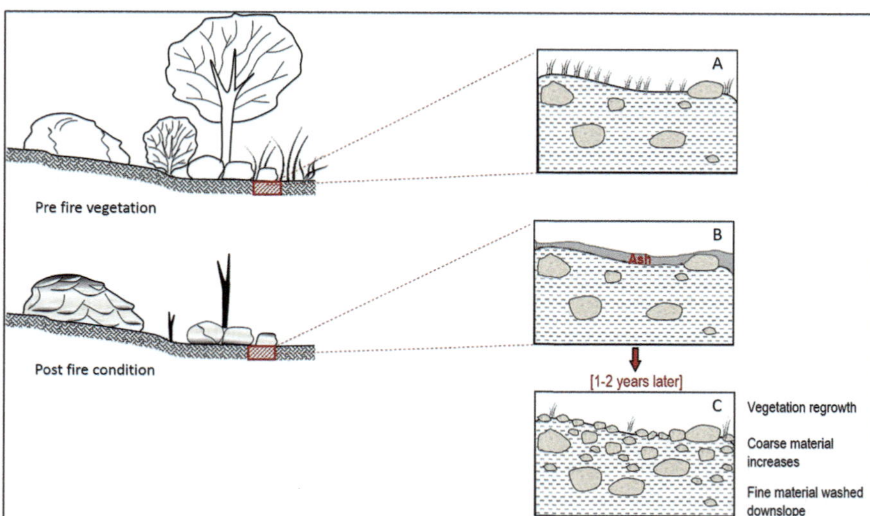

Fig. 23.5 Post-fire armoring of a burned hillslope. (**a**) the unburned soil is composed of coarse and fine particles. (**b**) following a fire ash blankets the soil; (**c**) fine sediments are progressively washed away, and with the addition of the weathered and eroded rock spalls, the soil becomes thinner and stonier. (Drawing: Nurit Shtober-Zisu)

system. The significance of post-fire erosion rates can only be assessed when the rate of soil renewal (which, in Mediterranean systems is relatively slow, at only 0.1 t ha^{-1} per year), is taken into consideration.

Results from local studies indicate that up to 4 years post-fire, erosion rates exceed those required to sustain soil formation and maintain soil productivity in burned plots. In addition to erosion, the preferential loss of organic matter, nutrients and fine sediments, coupled with the exposure of rock fragments and tree root residues, impede rehabilitation processes and weaken the resilience of the soil ecosystem (Wittenberg 2012).

Current climatic and anthropogenic conditions across the Mediterranean Basin favor widespread fires. Despite increasing knowledge about fire dynamics and the efficiency of prevention measures, fire activity is not expected to decline (de Rigo et al. 2017). Climate change around the Mediterranean will increase the weather-driven danger of forest fires, and parts of southern Europe that are already dry, such as Spain, Portugal and Greece, will become drier and more susceptible to burning. Under the hottest climatic predictions of 3 °C warming, the annual burned areas would double, whereas the moderate projections of 1.5 °C warming suggest that 40% more areas will burn annually (Turco et al. 2018).

23.6 Conclusions and Research Gaps

The overall effects of fires on the Mediterranean ecosystem are complex, ranging from the removal of aboveground vegetation and litter to long-term impacts on soil properties and erosion dynamics. The direct heating of the soil, coupled with the addition of charcoal and ash, produces short- and long-term changes to soil properties, often resulting in elevated geomorphological responses and increased soil erosion rates.

The effects of fire on soil properties are spatio-temporally dependent, and include a wide range of complicating factors such as fire history, severity, and behavior; antecedent soil moisture; climatic conditions before, during, and after the fire; soil type and characteristics; vegetation cover and site topography. Owing to the intricacy, diversity and interactions among these factors, forest fires have diverse effects. The mosaic nature of the Mediterranean landscape may partly explain the seemingly inconsistent fire-induced erosion rates at different sites and spatial scales. During the first year after a fire in the Mediterranean Basin, soil loss values vary from 45–55 t ha^{-1} (point) to 1–10 t ha^{-1} (plot), exceeding the natural renewal rates of soils. Removal of nutrients and organic material from the stony degraded soils is presumably as important as the total soil loss.

Data from larger scale studies are inconsistent due to greater spatial heterogeneity and difficulties in assessing the findings. Commonly, the thin and rocky soils covering the hillslopes (before the fire) create a restricted supply system and limit the geomorphic response. Additionally, the relatively rapid regeneration of the vegetation moderates the response of the soil system. In cases of slow regeneration, peak erosion rates might occur several years after the fire event. The eroded sediments are not necessarily transported to the channel system; nonetheless, the destruction of the well-structured soil horizons, and the redistribution and deposition of sediments and nutrients within hillslope sinks, further dictate the recovery of the vegetation, which in turn, affects the rehabilitation of the soil and the consequent erosion. A threshold of 30% vegetation cover is sufficient to reduce these erosional processes.

Since the number of fire events in the Mediterranean region is increasing, it is essential to address the gaps in current knowledge and propose an agenda for future research. However, despite the growing literature, more research effort is suggested to: (a) wind erosion and aeolian transport processes are among the less studied in the post-wildfire environment, though previous measurements indicate that wildfire can convert a relatively wind erosion-resistant landscape into a highly erodible source of particulate emissions; (b) the effect of a fire and recurrent fires on soil fertility and longevity through further quantification of soil nutrient (soil carbon cycle specifically) and organic matter depletion; (c) fires at the wildland urban interface have been growing in frequency and intensity due to population growth and other social factors in driving land-use/cover change. Anthropic or urban soils, are soils extensively influenced by human activities and may exert a strong effect on vegetation and soil organisms. Long-term changes in anthropic soil properties remain almost completely unstudied in the context of Mediterranean WUI fires; (d) Paleo fire–landscape interactions are scarcely understood, mainly because of the limited ability to identify and analyze paleo fire residue (e.g. charcoal). Advanced temporal resolution of sedimentary sequences is required to study palaeofire-erosion–sedimentation links and better understand the regional past fire regime and its effects on soil and vegetation dynamics, particularly in the light of changing climate; (e) soil microbiology is crucial for soil system functioning; the soil microbiome governs biogeochemical cycling of macronutrients, micronutrients and other elements vital for vegetation growth. Furthermore, microbial communities have been found to play important geomorphic roles, mainly in weathering and protecting soil surfaces from erosion. However, the effects of fire on soil microbiota and its relation to geomorphic processes has gained relatively little attention in the Mediterranean.

Comprehensive understanding of these complex interactions, on various scales, is needed to improve our predictions of the response of the land surface to fire.

References

Bodí MB, Mataix-Solera J, Doerr SH et al (2011) The wettability of ash from burned vegetation and its relationship to Mediterranean plant species type, burn severity and total organic carbon content. Geoderma 160:599–607

Bodí MB, Martin DA, Balfour VN et al (2014) Wildland fire ash: production, composition and eco-hydro-geomorphic effects. Earth Sci Rev 130:103–127

Bracken LJ, Kirkby MJ (2005) Differences in hillslope runoff and sediment transport rates within two semi-arid catchments in Southeast Spain. Geomorphology 68:183–200

Brook A, Wittenberg L (2016) Ash-soil interface: mineralogical composition and physical structure. Sci Total Environ 572:1403–1413

Cerdà A, Borja MEL, Úbeda X et al (2017) *Pinus halepensis* M. versus *Quercus ilex subsp. rotundifolia* L. runoff and soil erosion at pedon scale under natural rainfall in eastern Spain three decades after a forest fire. For Ecol Manag 400:447–456

Cerdan O, Govers G, Le Bissonnais Y et al (2010) Rates and spatial variations of soil erosion in Europe: a study based on erosion plot data. Geomorphology 122:167–177

de Rigo D, Libertà G, Houston Durrant T et al (2017) Forest fire danger extremes in Europe under climate change: variability and uncertainty, EUR 28926 EN. Publications Office of the European Union, Luxembourg, 71 p

Francos M, Pereira P, Alcañiz M et al (2016) Impact of an intense rainfall event on soil properties following a wildfire in a Mediterranean environment (North-East Spain). Sci Total Environ 572:1353–1362

García-Ruiz JM, Begueria S, Nadal-Romero E et al (2015) A meta-analysis of soil erosion rates across the world. Geomorphology 239:160–173

Gimeno-García E, Andreu V, Rubio JL (2007) Influence of vegetation recovery on water erosion at short and medium-term after experimental fires in a Mediterranean shrubland. Catena 69:150–160

Gómez-Rey MX, Couto-Vázquez A, García-Marco S et al (2013) Impact of fire and post-fire management techniques on soil chemical properties. Geoderma 195:155–164

Inbar M, Tamir M, Wittenberg L (1998) Runoff and erosion processes after a forest fire in Mount Carmel, a Mediterranean area. Geomorphology 24:17–33

Kutiel P, Inbar M (1993) Fire impacts on soil nutrients and soil erosion in a Mediterranean pine forest plantation. Catena 20:129–139

Kutiel P, Lavee H, Segev M et al (1995) The effect of fire-induced surface heterogeneity on rainfall-runoff-erosion relationships in an eastern Mediterranean ecosystem, Israel. Catena 25:77–87

Lal R (2001) Soil degradation by erosion. Land Degrad Dev 12:519–539

Lavee H, Kutiel P, Segev M et al (1995) Effect of surface roughness on runoff and erosion in a Mediterranean ecosystem: the role of fire. Geomorphology 11:227–234

Malkisnon D, Wittenberg L, Beeri O, Barzilai R (2011) Effects of repeated fires on the structure, composition, and dynamics of Mediterranean maquis: short-and long-term perspectives. Ecosystems 14:478–488

Mataix-Solera J, Cerdà A, Arcenegui V et al (2011) Fire effects on soil aggregation: a review. Earth Sci Rev 109:44–60

Mayor AG, Bautista S, Llovet J et al (2007) Post-fire hydrological and erosional responses of a Mediterranean landscape: seven years of catchment-scale dynamics. Catena 71:68–75

McNabb DH, Swanson FJ (1990) Effects of fire on soil erosion. In: Walstad JW, Sandberg DV, Radosevich SR (eds) Natural and prescribed fire in Pacific Northwest forests. Oregon State University Press, Oregon, pp 159–176

Meliho M, Nouira A, Benmansour M et al (2019) Assessment of soil erosion rates in a Mediterranean cultivated and uncultivated soils using fallout 137Cs. J Environ Radioact 208:106021

Mohamadi MA, Kavian A (2015) Effects of rainfall patterns on runoff and soil erosion in field plots. Int Soil Water Conserv Res 3:273–281

Moody JA, Martin DA (2009) Synthesis of sediment yields after wildland fire in different rainfall regimes in the western United States. Int J Wildl Fire 18:96–115

Moody JA, Shakesby RA, Robichaud PR et al (2013) Current research issues related to post-wildfire runoff and erosion processes. Earth-Sci Rev 122:10–37

Ohana-Levi N, Givati A, Paz-Kagan T et al (2018) Forest composition effect on wildfire pattern and run-off regime in a Mediterranean watershed. Ecohydrology 11:e1936

Pausas JG, Llovet J, Rodrigo A et al (2009) Are wildfires a disaster in the Mediterranean Basin? – a review. Int J Wildl Fire 17(6):713–723

Peña-Angulo D, Nadal-Romero E, González-Hidalgo JC et al (2019) Spatial variability of the relationships of runoff and sediment yield with weather types throughout the Mediterranean Basin. J Hydrol 571:390–405

Pereira P, Oliva M, Misiune I (2016) Spatial interpolation of precipitation of precipitation indexes in Sierra Nevada (Spain): comparing the performance of some interpolation methods. Theor Appl Climatol 126:683–698

Poesen J (2018) Soil erosion in the Anthropocene: research needs. Earth Surf Proc Land 43:64–84

Puigdefábregas J (2005) The role of vegetation patterns in structuring runoff and sediment fluxes in drylands. Earth Surf Process Landf 30:133–147

Raclot D, Le Bissonnais Y, Annabi M et al (2016) Challenges for mitigating Mediterranean soil erosion under global change. In: Sabrié M-L, Gibert-Brunet E, Mourier T (eds) The Mediterranean region under climate change: a scientific update. IRD Editions, Marseille, pp 311–318

Römkens MJ, Helming K, Prasad SN (2002) Soil erosion under different rainfall intensities, surface roughness, and soil water regimes. Catena 46:103–123

Sankey JB, Kreitler J, Hawbaker TJ, McVay JL, Miller ME, Mueller ER, Vaillant NM, Lowe SE, Sankey TT (2017) Climate, wildfire, and erosion ensemble foretells more sediment in western USA watersheds. Geophys Res Lett 44(17):8884–8892

Santín C, Doerr SH (2016) Fire effects on soils: the human dimension. Philos Trans R Soc B 371(1696):20150171

Shakesby RA (2011) Post-wildfire soil erosion in the Mediterranean: review and future research directions. Earth-Sci Rev 105:71–100

Shakesby RA, Doerr SH (2006) Wildfire as a hydrological and geomorphological agent. Earth-Sci Rev 74:269–307

Sheridan G, Lane P, Nyman P (2019) Erosion. In: Pereira P, Cerdà A, Úbeda X et al (eds) . CSIRO Publishing, Fire effects on soil properties, pp 89–115

Shtober-Zisu N, Brook A, Kopel D et al (2018) Fire induced rock spalls as long-term traps for ash. Catena 162:88–99

Smith HG, Sheridan GJ, Lane PN et al (2011) Wildfire effects on water quality in forest catchments: a review with implications for water supply. J Hydrol 396:170–192

Stoof CR, Wesseling JG, Ritsema CJ (2010) Effects of fire and ash on soil water retention. Geoderma 159:276–285

Stoof CR, Slingerland EC, Mol W et al (2014) Preferential flow as a potential mechanism for fire-induced increase in streamflow. Water Resour Res 50:1840–1845

Swanson FJ (1981) Fire and geomorphic processes. In: Mooney, HA; Bonnicksen, TM; Christensen, NL et al (technical coordinators), Proceedings of the Conference Fire Regimes and Ecosystem Properties, December 11–15, 1978, Honolulu, HI. General Technical Report WO-GTR-26. Washington, DC: USDA Forest Service, pp 401–444

Tsafrir A, Osem Y, Shemesh H et al (2019) Fire season modifies the perennial plant community composition through a differential effect on obligate seeders in eastern Mediterranean woodlands. Appl Veg Sci 22:115–126

Turco M, Rosa-Cánovas JJ, Bedia J et al (2018) Exacerbated fires in Mediterranean Europe due to anthropogenic warming projected with non-stationary climate-fire models. Nat Commun 9:3821

Urbanek E, Shakesby RA (2009) Impact of stone content on water movement in water-repellent sand. Eur J Soil Sci 60:412–419

Versini PA, Velasco M, Cabello A et al (2013) Hydrological impact of forest fires and climate change in Mediterranean basin. Nat Hazards 66:609–628

Vieira DCS, Fernández C, Vega JA, Keizer JJ (2015) Does soil burn severity affect the post-fire runoff and interrill erosion response? A review based on meta-analysis of field rainfall simulation data. J Hydrol 523:452–464

Wagenbrenner JW, Robichaud PR (2014) Post-fire bedload sediment delivery across spatial scales in the interior western United States. Earth Surf Proc Land 39:865–876

Wagenbrenner JW, MacDonald LH, Coats RN et al (2015) Effects of post-fire salvage logging and a skid trail treatment on ground cover, soils, and sediment production in the interior western United States. For Ecol Manag 335:176–193
Wilcox BP, Breshears DD, Allen CD (2003) Ecohydrology of a resource-conserving semiarid woodland: effects of scale and disturbance. Ecol Monogr 73:223–239
Williams CH, Silins U, Spencer SA et al (2019) Net precipitation in burned and unburned subalpine forest stands after wildfire in the northern Rocky Mountains. Int J Wildland Fire 28:750–760
Wittenberg L (2012) Post-fire soil ecology: properties and erosion dynamics. Isr J Ecol Evol 58:151–164
Wittenberg L, Inbar M (2009) The role of fire disturbance on runoff and erosion processes–a long-term approach, Mt. Carmel case study, Israel. Geogr Res 47:46–56
Wittenberg L, Kutiel H (2016) Dryness in a Mediterranean-type climate–implications for wildfire burnt area: a case study from Mount Carmel, Israel. Int J Wildl Fire 25:579–591
Wittenberg L, Malkinson D, Barzilai R (2014) The differential response of surface runoff and sediment loss to wildfire events. Catena 121:241–247
Wittenberg L, van der Wal H, Keesstra S et al (2019) Post-fire management treatment effects on soil properties and burned area restoration in a wildland-urban interface, Haifa Fire case study. Sci Total Environ 135190. https://doi.org/10.1016/j.scitotenv.2019.135190
Wohl E, Brierley G, Cadol D et al (2019) Connectivity as an emergent property of geomorphic systems. Earth Surf Proc Land 44:4–26
Zituni R, Wittenberg L, Malkinson D (2019) The effects of post-fire forest management on soil erosion rates 3 and 4 years after a wildfire, demonstrated on the 2010 Mount Carmel fire. Int J Wildl Fire 28:377–385

Chapter 24
Grazing and Understory Fuel Load Reduction: An Exploration Based on Rothermel's Surface-Fire Model and GPS Tracking of Herds

Eugene D. Ungar, Mor Ashkenazi, Reuven Horn, Maya Millet, Maya Zahavi, Yossi Moshe, and Yagil Osem

24.1 Introduction

In the Mediterranean Basin the occurrence of large and severe forest fires is becoming more frequent, primarily because of the accumulation of live fuels (Garcia-Llamas et al. 2019). In the eastern Mediterranean region this trend is mainly the result of establishment and development of pine forest plantations during the twentieth century, i.e., afforestation (Osem et al. 2008), and spontaneous regeneration of native woodland vegetation in the understory of these forests (Osem et al. 2012). Livestock grazing is commonly proposed as a possibly efficient means to reduce vegetative fuel accumulation and to control fire hazard in forests (Mancilla-Leyton

E. D. Ungar (✉) · M. Ashkenazi
Department of Natural Resources, Institute of Plant Sciences, Agricultural Research Organization – The Volcani Center, Rishon LeZion, Israel
e-mail: eugene@volcani.agri.gov.il

R. Horn
Rangeland Service, Ministry of Agriculture and Rural Development, Rishon LeZion, Israel

M. Millet
The Robert H Smith Faculty of Agriculture, Food and Environment, The Hebrew University of Jerusalem, Rehovot, Israel

M. Zahavi
Soil Survey, Land Evaluation and Remote Sensing, Ministry of Agriculture and Rural Development, Rishon LeZion, Israel

Y. Moshe
Department of Natural Resources, Institute of Plant Sciences, Volcani Center, Israel
e-mail: yosi@agri.gov.il

Y. Osem
Department of Natural Resources – Institute of Plant Sciences, Agricultural Research Organization – Volcani Center, Rishon LeZion, Israel

G. Ne'eman, Y. Osem (eds.), *Pines and Their Mixed Forest Ecosystems in the Mediterranean Basin*, Managing Forest Ecosystems 38,
https://doi.org/10.1007/978-3-030-63625-8_24

and Vicente 2012). However, the effectiveness of this putative tool is often questioned in light of both declining availability of grazing herds and the apparent inability of livestock grazing to achieve significant reduction of the vegetative fuel load in areas where the woody vegetation is already well developed.

Addressing this issue, with a view to developing a quantitative and spatially explicit regional grazing plan, presents a number of scientific challenges: (1) to quantify the current state of the forest or woodland vegetation on a regional scale; (2) to model the associated fire hazard in a static but spatially explicit way; (3) to use the model to calculate the changes to the vegetation that would be needed to reduce hazards to acceptable levels; and (4) to estimate whether grazing can be expected to bring about those changes.

Regarding the first challenge – definition of the current vegetation state depends on the fuel classification system, of which there are many (Arroyo et al. 2008). The choice of classification system depends on the type of landscape and the applied methodology. In the work described here – adapted to Mediterranean pine ecosystems – the vegetation is defined by the overstory plus four primary types, or "fuel families" (Morvan and Dupuy 2004; Cohen et al. 2002) of the understory: herbaceous, live shrub, litter, and downed woody material (DWM). For each of these families specific field methods are used to estimate biomass per unit area and other features of the vegetation required by the fire model being used (Note that "fire model" should not be confused with "fuel model" – a stylized set of fuel-bed characteristics). In the model described below, the values calculated for the various fuel families need to be merged into a single set of input parameters that forms the starting point of the model, and the way in which that merger is performed is crucial.

24.2 Fire Modeling

A basic distinction needs to be made between models of surface/understory fires and those of crown fires. In the context of grazing, in which animal impact is largely limited to the understory, the focus is on surface fire behavior and the associated risk of development into a crown fire. Surface-fire behavior is characterized primarily by variables such as its spreading rate (termed widely "rate of spread"), flame length and flame intensity, whereas the associated risk of crown fire is characterized by the open wind speed, i.e., wind speed in the absence of trees, that would be required to initiate a crown fire and sustain it. The latter two factors are designated as the torching index and the crowning index, respectively.

The classic model that computes rate of spread is that of Rothermel (1972) which, to this day forms the core of the widely used BehavePlus fire modeling package (Andrews 2014). As the word "model" may conjure up various connotations, it is worth clarifying that the Rothermel model is a static model, as opposed to a dynamic, simulation model such as the FARSITE model (Finney 2004); it is also spatially dimensionless. Essentially, it computes the instantaneous rate of spread of a surface fire that is initiated in an expanse of forest or woodland, whose characteristics are encapsulated in eleven input parameters (Table 24.1) that need to be inserted by the model user. Two of these input parameters – topographic slope and wind speed – are the only descriptors

Table 24.1 The input parameters of the Rothermel model. Input parameters are listed in the order they are addressed in this chapter. "–" in the units column denotes dimensionless fractions. The acronyms are used in Fig. 24.1 and Table 24.2

	Name	Units	Acronym	Symbol in Wilson (1980)
1	Slope (vertical rise/horizontal run)	–	TP	Tan Φ
2	Wind speed at midflame height	m min^{-1}	WMH	U
3	Fuel heat content	kJ kg^{-1}	FHC	h
4	Fuel effective mineral content	–	FEM	S_e
5	Fuel total mineral content	–	FTM	S_T
6	Fuel moisture of extinction	–	FME	M_X
7	Surface area: Volume ratio	cm^{-1}	S	σ
8	Fuel particle density	kg m^{-3}	FPD	ρ_P
9	Fuel moisture content	–	FMCT	M_f
10	Ovendry fuel loading	kg m^{-2}	OFL	w_o
11	Fuel depth	m	FD	δ

of the abiotic environment. The remaining nine parameters describe "the fuel". This is a noteworthy term that informs us at first glance that the understory vegetation, in its entirety, is treated as a homogeneous and uniform substance that fills the space from the ground up to a certain height, termed the fuel depth. Thus, a key question is how to incorporate the complexity of the field into this abstract description, and thereby to reduce all our observations to a single set of input parameters. In any event, the model could be applied independently to each "pixel" within a certain landscape, assuming the necessary parameter inputs can be provided at that spatial resolution, to construct a regional map of fire behaviour and hazard (Elia et al. 2015).

24.3 The Fire Model

We based the core of the model on the set of 21 equations defined in Wilson's (1980) paper, which itself was a reformulation in SI units of the widely used forest fire-spread equations of Rothermel (1972). Although modifications to the model have been proposed in the light of accumulating empirical data (e.g. Wilson 1990), the broad qualitative behavior of interest to us here is not expected to change. Our implementation was in Microsoft Excel, by means of a standard spreadsheet with macro support for batch runs and sensitivity analysis. The equations generate three characteristics of a surface fire: rate of spread, Byram's fireline intensity, and flame length (Table 24.2, rows 34–54). A relational diagram of the entire model is shown in Fig. 24.1, with this core section shown in blue. Note that all computational pathways of this section of the model trace back to the eleven input parameters that specify biotic and abiotic conditions.

The section of the model dealing with the eleven input parameters is shown in green in Fig. 24.1. Of these, the simplest to measure or derive from existing data sources relate to the abiotic environment: the topographic or terrain slope and wind speed. But the wind speed required by the model is, in fact, the wind speed in the wooded area, for which data are unlikely to be available, as opposed to the open

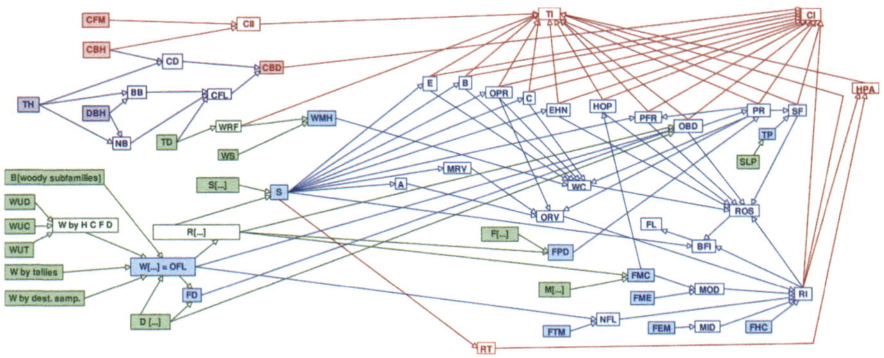

Fig. 24.1 Relational diagram of the simple fire behavior and risk model defined in Table 24.2. Each variable in the model is represented by a boxed acronym, to which are connected all variables of which it is a function. Variables with no antecedents are field measurements or published constants; all computational pathways trace back to them. The model has four layers, which are differentiated by color, as follows: (1) Blue – variables and connections contained in the equations, for computing, primarily, rate of spread (ROS) of a surface fire. The 11 input parameters for this set of equations have blue fill. (2) Green – derivation of 7 of the above input parameters. The input parameters for this layer have green fill. (3) Red – the additional variables and connections for computing torching index (TI) and crowning index (CI). The 3 additional input parameters for this set of equations have red fill. (4) Purple – the additional variables and connections for deriving crown bulk density (CBD). The input parameters for this layer have purple fill. Unshaded boxes are always-computed variables. Shaded boxes are input parameters of the layer to which they belong, which can themselves be computed in a different layer. Square brackets indicate an array of variables according to fuel family (see Tables 24.3, 24.4 and 24.5)

wind speed, for which data – even if only long-term normative values – generally are available. As a first approximation, the former can be derived from the latter by employing a reduction factor based on overstory tree density in a typical *Pinus halepensis* Mill. forest (Calev 2013). A further four of the eleven parameters were treated as physical constants and were based on published data for similar vegetation: the heat content, effective mineral content, total mineral content, and moisture of extinction of "the fuel" (Table 24.2 rows 23–26). The remaining five input parameters differ greatly among fuel families, and need to be specified accordingly.

24.4 Fuel Families

In the complete list of fuel families adopted here, the primary fuel families of live shrub and DWM are subdivided. Live shrub is commonly subdivided into five fuel families: foliage, and four branch thickness ranges (<7 mm, ≥7 and <11 mm, ≥11 and <25 mm, and ≥25 mm) (Ottmar et al. 2007). Downed woody material is subdivided on the basis of thickness into coarse (diameter ≥ 76 mm) and fine (<76 mm) categories, of which the latter is further subdivided into small (<7 mm), medium (<24 mm) and large (<76 mm) subcategories (Woodall and Monleon 2008). Together with litter and herbaceous material, this yields a total of eleven fuel families.

Table 24.2 Specification of the fire behavior and risk model

Table row number	Acronym	Short definition	Equation or value	Units	Source	Symbol in source
Derivation of Wilson (1980) input parameters						
1	TD	Tree density of overstory	Value	per 1000 m^2		
2	WS	Wind speed in open area	Value	km h^{-1}		
3	WRF	Tree density wind reduction factor	= (0.3 + 14 × LN(TD))/100	df		
4	SLP	Local terrain slope	Value	degrees		
5	WUT	Woody understory thickness (height)	Value	m		
6	WUC	Woody understory cover	Value	dl		
7	WUD	Woody understory density coefficient	Value	kg m^{-3}		
8	WUF	Woody understory fill	Value			
9	UWB	Biomass of woody understory	= WUT × WUC × WUD × WUF	kg m^{-2}		
10	B[LF,1,7,11,25]	Allocation of understory woody biomass to 5 fuel families	Array of 5 values (numbers are branch thicknesses)	df		
11	W[LF,1,7,11,25]	Understory woody biomass of 5 fuel families	B[LF,1,7,11,25] × UWB	kg m^{-2}		
12	W[herbaceous]	Herbaceous dry biomass	Value	kg m^{-2}		
13	W[litter]	Litter dry biomass	Value	kg m^{-2}		
14	W[DWMfineSmall]	Biomass of DWMfineSmall	Based on given tallies and set of equation parameters	kg m^{-2}		

(continued)

Table 24.2 (continued)

Table row number	Acronym	Short definition	Equation or value	Units	Source	Symbol in source
15	W[DWMfineMedium]	Biomass of DWMfineMedium	Based on given tallies and set of equation parameters	kg m^{-2}		
16	W[DWMfineLarge]	Biomass of DWMfineLarge	Based on given tallies and set of equation parameters	kg m^{-2}		
17	W[DWMfineTotal]	Total DWMfine biomass		kg m^{-2}		
18	S[…]	Surface-area-to-volume ratio of all fuel families	See Table 24.3	cm^{-1}		
19	PD[…]	Fuel particle density of all fuel families	See Table 24.4	kg m^{-3}		
20	FMC[…]	Fuel moisture content of all fuel families	See Table 24.5	df		
21	IFD[…]	Implicit fuel depths of all fuel families	Ratio of W[…] to given bulk density for each fuel family			
22	R[…]	Relative masses of all fuel families	W[…]/OFL	df		
Wilson (1980) input parameters						
23	FHC	Fuel heat content	18,608	kJ kg^{-1}		h
24	FEM	Fuel effective mineral content	0.01	df		S_e
25	FTM	Fuel total mineral content	0.055	df		S_T
26	FME	Fuel moisture of extinction	0.85875	df		M_x
27	WMH	Wind speed at midflame height	= (1 − WRF) × WS × 1000/60	m min^{-1}		U
28	TP	Slope (vertical rise/horizontal run)	TAN(SLP × PI()/180)	df		Tan ϕ
29	OFL	Ovendry fuel loading	= sum of W[…] (total understory biomass)	kg m^{-2}		w_o

30	S	Surface area: Volume ratio	Sum of (R[…] × S[…]) across all fuel families	cm^{-1}		σ
31	FPD	Fuel particle density	= sum of (R[…] × PD[…]) across all fuel families	$kg\ m^{-3}$		ρ_p
32	FMCT	Fuel moisture content	= sum of (R[…] × FMC[…]) across all fuel families	df		M_f
33	FD	(implicit) fuel depth	= OFL/OBD	m		δ
Wilson (1980) equations						
34	EHN	Effective heating number	=EXP(−4.528/S)	dl	W 14	ε
35	A	undefined in Wilson 1980	=8.9033 × (S^-0.7913)	dl	W 39	A
36	B	undefined in Wilson 1980	= 0.15988 × (S^0.54)	ng	W 49	B
37	C	undefined in Wilson 1980	= 7.47 × EXP(−0.8711 × (S^0.55))	ng	W 48	C
38	E	undefined in Wilson 1980	= 0.715 × EXP(−0.01094 × S)	Ng	W 50	E
39	OPR	Optimum packing ratio	= 0.20395 × (S^-0.8189)	dl	W 37	β_{op}
40	MRV	Maximum reaction velocity	= (0.0591 + 2.926 × (S^-1.5))^-1	min^{-1}	W 36	Γ'_{max}
41	OBD	Ovendry bulk density	= biomass-weighted average bulk density across all fuel families, or OFL/max IFD[…]	$kg\ m^{-3}$	W 40	ρ_b
42	PR	Packing ratio	= OBD/FPD	dl	W 31	β
43	ORV	Optimum reaction velocity	= MRV × ((PR/OPR) × EXP((1 − (PR/OPR))))^A	min^{-1}	W 38	Γ'
44	MOD	Moisture damping coefficient	= 1 − (2.59 × FMCT/FME) + (5.11 × (FMCT/FME)^2) − (3.52 × (FMCT/FME)^3)	df	W 29	η_M
45	MID	Mineral damping coefficient	= 0.174 × FEM^-0.19	df	W 30	η_s
46	NFL	Net fuel loading	= OFL × (1 − FTM)	$kg\ m^{-2}$	W 24	w_n
47	RI	Reaction intensity	= ORV × NFL × FHC × MOD × MID	$KJ\ min^{-1}\ m^{-2}$	W 27	I_R

(continued)

Table 24.2 (continued)

Table row number	Acronym	Short definition	Equation or value	Units	Source	Symbol in source
48	PFR	Propagating flux ratio	= (192 + 7.9095 × S)^-1 × EXP((0.792 + (3.7597 × (S^0.5))) × (PR + 0.1))	df	W 42	ξ
49	WC	Wind coefficient	= C × ((3.281 × WMH)^B) × ((PR/OPR)^-E)	dl	W errata	Φ_w
50	SF	Slope factor	= (5.275 × (PR^-0.3)) × (TP^2)	dl	W 51	Φ_s
51	HOP	Heat of preignition	= 581 + (2594 × FMCT)	KJ kg^{-1}	W 12	Q_{ig}
52	ROS	Rate of spread	= (RI × PFR × (1 + WC + SF))/(OBD × EHN × HOP)	m min^{-1}	W 52	R
53	BFI	Byram's fireline intensity	= (1/60) × RI × ROS × (12.6/S)	kW m^{-1}	W errata	I_B
54	FL	Flame length	= 0.0775 × BFI^0.46	m	W errata	L_f
Derivation of Scott and Reinhardt (2001) input parameters						
55	TH	Overstory tree height	Value	m		
56	DBH	Overstory tree diameter at breast height	Value	cm		
57	NB	Pine overstory needle biomass	= EXP(−0.572 + 11.47 × DBH/(DBH + 18.745) − 1.179 × LN(TH))	kg $tree^{-1}$	D Table 3	
58	BB	Pine overstory branch biomass	=EXP(0.113 + 12.79 × DBH/(DBH + 23.007) − 1.166 × LN(TH))	kg $tree^{-1}$	D Table 3	
59	CFL	Available crown fuel load	=(NB + BB) × TD/1000	kg m^{-2}		
Scott and Reinhardt (2001) input parameters (not already defined)						
60	CFM	Crown foliar moisture content	100	%		
61	CBH	Crown base height	Value	m		
62	CD	Overstory tree crown depth	= TH – CBH	m		
63	CBD	Crown bulk density	= CFL/CD	kg m^{-3}		

Scott and Reinhardt (2001) equations						
64	CII	Crown fire initiation intensity	= ((CBH × (460 + (25.9 × CFM)))/100)^(3/2)	kW m^{-1}	S 11	$I'_{initiation}$
65	RT	Residence time	= 12.595/S	min	S 3	t_R
66	HPA	Heat release per unit area	= RI × RT	KJ m^{-2}	S 2	*HPA*
67	TI	Torching index	=(1/(54.683 × (1 – WRF))) × ((((((60 × CII × OBD × EHN × HOP))/((HPA × PFR × RI))) – SF – 1)/(C × ((PR/OPR) – E)))^(1/B))	km h^{-1}	S 18	TI, $O'_{initiation}$
68	CI	Crowning index	=(1/54.683) × (((((3/CBD) × OBD × EHN × HOP)/(3.34 × PFR × RI)) – SF – 1)/(C × ((PR/OPR)^-E))^(1/B))	km h^{-1}	S 19	U'_{active}

Sources W, S and D are Wilson (1980), Scott and Reinhardt (2001) and De-Miguel et al. (2014), respectively; numbers are equation number in source. For simplicity, the woody understory is treated in the table as a single "average" species. WRF is the factor by which wind speed is reduced, hence our use of (1-WRF) in WMH and TI

dl dimensionless, *df* dimensionless fraction, *ng* not given

Table 24.3 Surface-area-to-volume ratios (S, cm^{-1}) assumed in the model, according to fuel family. Each S value is an average of the sourced values

Fuel family	S	Source
Herbaceous	151.3	Brown 1970; Table 24.3, plant mean, rows 1, 2, 3
Litter	52.37	Brown and Simmerman 1986; p. 44 (49.2); Dimitrakopoulos and Panov 2001; Table 24.4, row 4 (55.54)
DWM-fine-small	14.82	Assumed similar to Shrub-branch-0
DWM-fine-medium	4.378	Assumed similar to Shrub-branch-7
DWM-fine-large	1.247	Assumed similar to Shrub-branch-25
DWM-coarse	1.247	Assumed similar to Shrub-branch-25
Shrub leaves	44.512	Dimitrakopoulos and Panov 2001; Table 24.4, rows 2, 4–8
Shrub-branch-0	14.82	As above, rows 10, 12, 17–19
Shrub-branch-7	4.378	As above, rows 21, 23, 26–28
Shrub-branch-11	4.378	As above, rows 21, 23, 26–28
Shrub-branch-25	1.247	As above; rows 32–34

Table 24.4 Particle density (PD, kg m^{-3}) assumed in the model, according to fuel family. Each PD value is an average of the sourced values

Fuel family	PD	Source
Herbaceous	200	guesstimate; influenced by Shipley and Vu (2002)
Litter	310	Dimitrakopoulos and Panov 2001; Table 24.5; row 5; assumed well-represented by *P. brutia* leaves
DWM-fine-small	610	Assumed similar to Shrub-branch-0
DWM-fine-medium	678	Assumed similar to Shrub-branch-7
DWM-fine-large	760	Assumed similar to Shrub-branch-25
DWM-coarse	760	Assumed similar to Shrub-branch-25
Shrub leaves	445	Dimitrakopoulos and Panov (2001) Table 24.5 rows 1–8 and Papió and Trabaud (1990) Table 24.2, leaves, rows 1–5.
Shrub-branch-0	610	Dimitrakopoulos and Panov 2001; Table 24.5, rows 10–14
Shrub-branch-7	678	As above, rows 15, 17–20
Shrub-branch-11	678	As above, rows 15, 17–20
Shrub-branch-25	760	As above, rows 22–24

Table 24.5 Fuel moisture content (FMC; dimensionless wet/dry matter ratio) assumed in the model, according to fuel family. Based on samples gathered at HaKedoshim Forest. Values for DWM-coarse were copied from DWM-fine-large, and values for Shrub-branch-25 were copied from Shrub-branch-11

Fuel family	Winter	Spring	Autumn
Herbaceous	5.900	3.696	0.308
Litter	0.520	0.086	0.044
DWM-fine-small	0.390	0.152	0.074
DWM-fine-medium	0.704	0.162	0.100
DWM-fine-large	0.481	0.190	0.100
DWM-coarse	0.481	0.190	0.100
Shrub leaves	1.506	1.059	0.747
Shrub-branch-0	0.693	0.619	0.505
Shrub-branch-7	0.682	0.614	0.504
Shrub-branch-11	0.583	0.564	0.460
Shrub-branch-25	0.583	0.564	0.460

24.5 The Fuel-Family-Dependent Input Parameters

The five input parameters that remain to be determined differ among the fuel families. We based two of them – surface-area-to-volume ratio (S) and particle density (PD) – on a broad literature review and comparison with the most closely similar vegetation. Fuel moisture content (FMC) was based on data collected locally (see below). Ovendry fuel loading, i.e., biomass of each fuel family per unit area, is the primary axis of variation that we are interested in exploring. As a guide to the reasonable range of relevance of each of the components of ovendry fuel loading, we drew on unpublished datasets collected in two mature pine forests – HaKedoshim (*Pinus halepensis*) and Beitar (*Pinus brutia* Ten.) forests in the Judean Hills of Israel (by authors M.M. and M.A., respectively). The way fuel depth is handled in our formulation of the model is strongly linked to the way biomass per unit area is modeled and will be elaborated last.

In general, the field methods and ranges of values that we report here are relevant to mature east Mediterranean pine forests that are characterized by a high overstory layer of pines (mostly planted) and an understory of herbaceous and woody vegetation which developed spontaneously over the years. Alongside its contribution to biodiversity and complexity of the forest ecosystem (Osem et al. 2012), this understory also increases fire hazards.

24.5.1 Surface-Area-to-Volume Ratio

We assembled a database of 112 values of S, drawn from seven publications (Brown 1970; Burgan and Rothermel 1984; Brown and Simmerman 1986; Papió and Trabaud 1990; Fernandes and Rego 1998; Dimitrakopoulos and Panov 2001; Liu et al. 2015). The entire set of S values ranged from 1 to 290 cm^{-1} (plus one extreme value of 632 cm^{-1}), with 90% of values below 136 cm^{-1}. We drew S values for shrub leaves and for the four size categories of shrub branch directly from values given by Dimitrakopoulos and Panov (2001). We found no S estimates for DWM of any class, but the estimates of Papió and Trabaud (1990) for Mediterranean shrubs indicated little difference between live and dead stems and branches. As a first approximation, we therefore mapped the four values for shrub branches onto the corresponding categories of DWM, according to their size categories. The values assumed in the model for all eleven fuel families, and their sources, are summarized in Table 24.3. Figure 24.2 shows the six different S values used to represent the 11 fuel families relative to the distribution of values drawn from the literature sources.

24.5.2 Particle Density

We assembled a database of 226 values of vegetation PD, drawn from seven publications; the values differed widely in their relative contributions and relevance to Mediterranean vegetation (Papió and Trabaud 1990; Fernandes and Rego 1998; Dimitrakopoulos and Panov 2001; Porté et al. 2002; Shipley and Vu 2002; Liu et al. 2015; Conti et al. 2019). We assumed the "dry matter concentration" of Shipley and Vu (2002) to be equivalent to our PD. The entire collection spanned PD values ranging from 60 to 1070 kg m^{-3}.

Clear matches could only be found for some of the eleven fuel families defined here (see Table 24.4). The PD values for the four size categories of shrub branch were based directly on values given by Dimitrakopoulos and Panov (2001). We found no PD estimates for DWM of any class, but, as found for S, the estimates of Papió and Trabaud (1990) for Mediterranean shrubs indicated little difference between live and dead stems and branches. As a first approximation, we again mapped the four values for shrub branches onto the corresponding categories of DWM. For litter we adopted the value for *P. brutia* leaves given by Dimitrakopoulos and Panov (2001). For shrub leaves we took the average of the values given by Dimitrakopoulos and Panov (2001) and Papió and Trabaud (1990). For herbaceous vegetation we used the study by Shipley and Vu (2002) as a guideline, although it addressed only leaf material, which would be expected to have a lower value than the whole plant; our guesstimate of 200 kg m^{-3} exceeded 84% of the values given by Shipley and Vu (2002). Figure 24.3 shows how the six different PD values we used to represent the 11 fuel families cut across the distributions of values drawn from the literature sources. Liu et al. (2015) presented a single normative value of 512.5 kg m^{-3} for PD which lies near the center of the values we adopted.

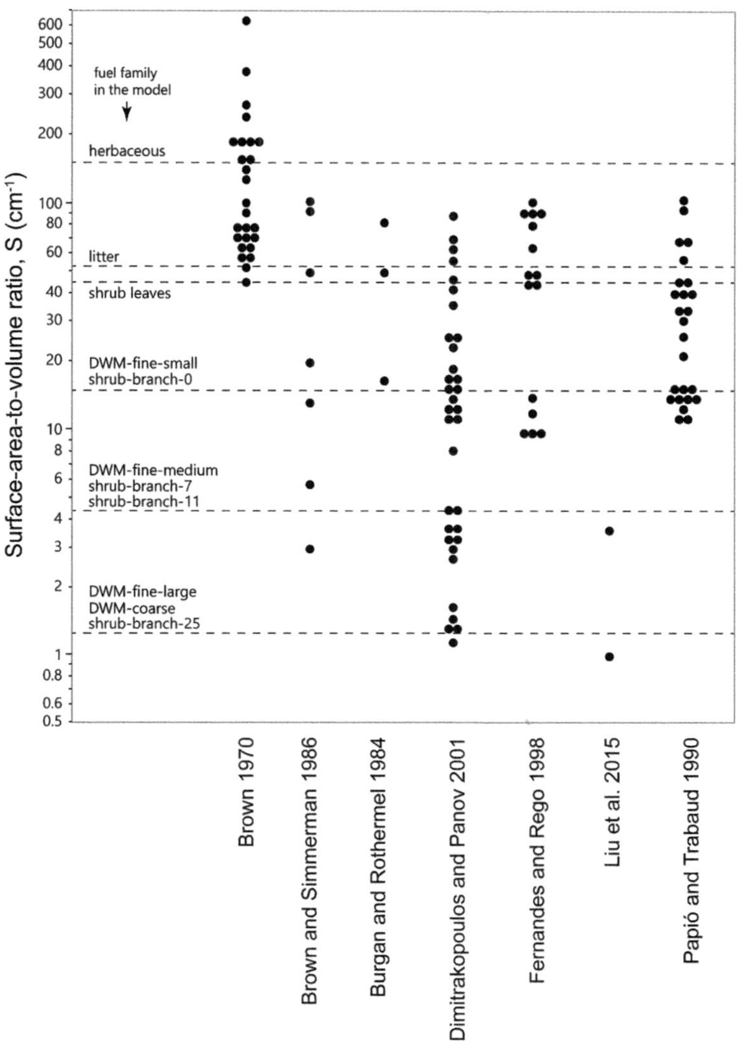

Fig. 24.2 Relative placement of surface-area-to-volume ratios (S) assumed in the model for primary fuel families (some of which served more than one fuel family) and in relation to values found in a literature survey

24.5.3 Fuel Moisture Content

The values for FMC (water/dry matter ratio) that we assumed in the model were based on data from HaKedoshim Forest, sampled in three seasons (n = 1443 samples). The fuel families DWM-coarse and Shrub-branch-25 were all but absent; values for these were copied from those of the closest fuel family. DWM-fine material was collected in winter only; the initial weighing was taken as the basis for winter

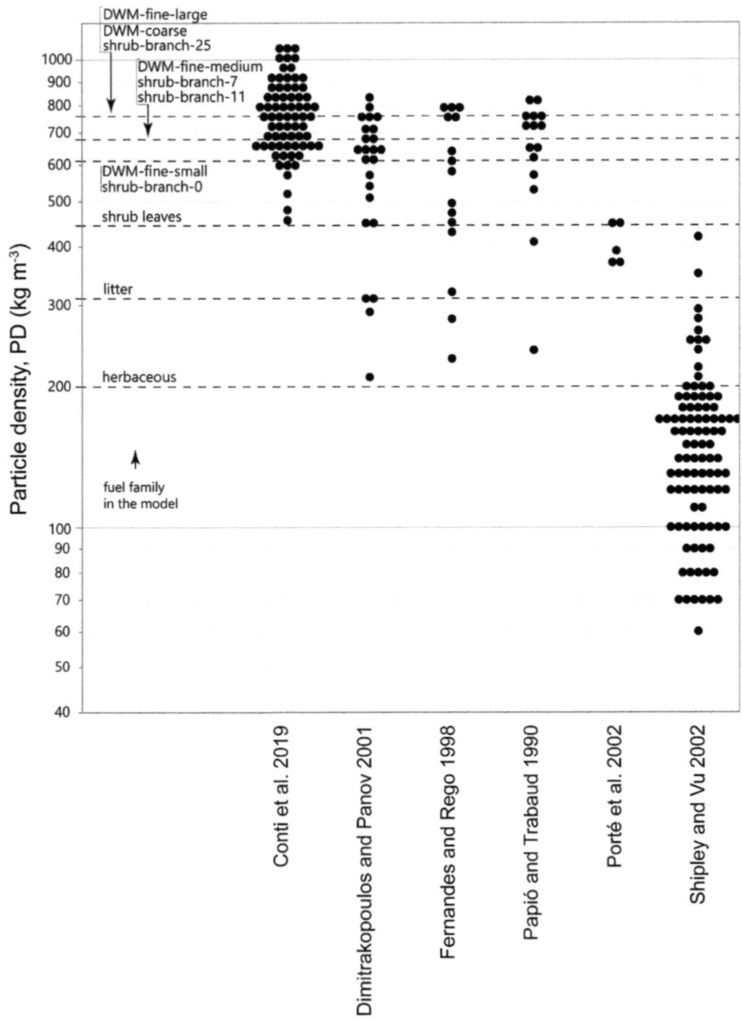

Fig. 24.3 Relative placement of particle density (PD) values assumed in the model for primary fuel families (some of which served more than one fuel family) and in relation to values found in a literature survey

values. Samples were air-dried for 20 days, with intermediate and terminal weighings, and then oven dried. These weighings formed the basis for spring and autumn values (see Table 24.5).

Comparisons with values from the literature were based on three sources: Liu et al. (2015), Papió and Trabaud (1990), and Shipley and Vu (2002) (see Fig. 24.4). The band of values for moisture content of fine fuels given by Papió and Trabaud (1990) is equivalent to 0.7 to 1.6 in our units, and encompassed our values for shrub leaves in all three seasons, and also those for DWM-fine-medium in winter. Liu

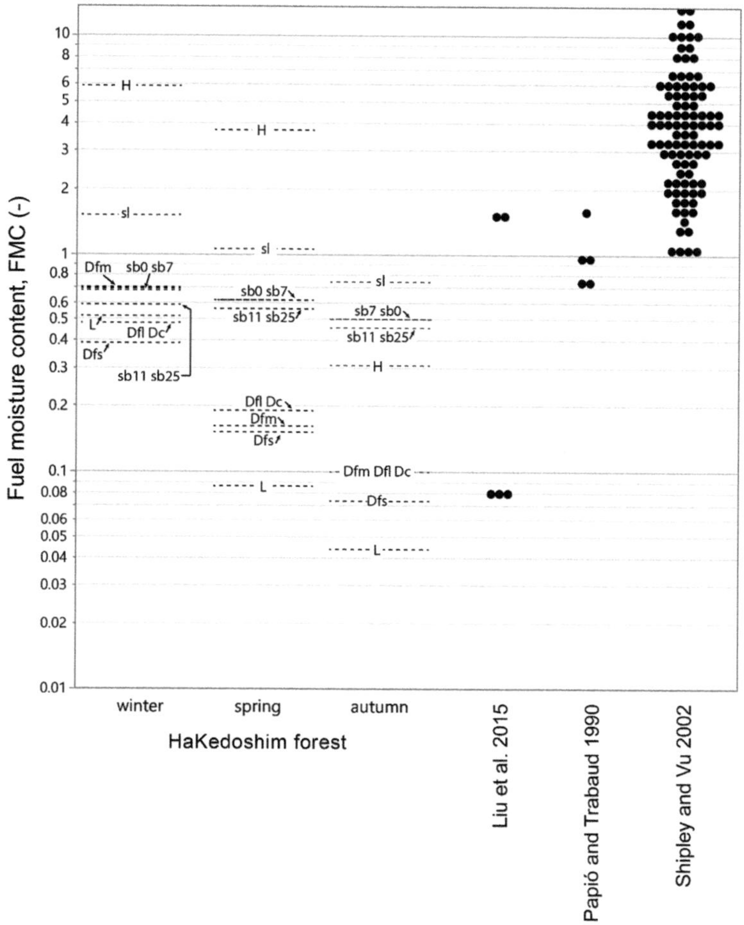

Fig. 24.4 Relative placement of fuel moisture content (FMC; wet/dry matter ratio) assumed in the model for primary fuel families (dashed lines; some of which served more than one fuel family), according to sampling season, and values found in a literature survey (dots). (Fuel family abbreviations: H = herbaceous; sl = shrub leaves; L = litter; Dfs = DWM-fine-small; Dfm = DWM-fine-medium; Dfl = DWM-fine-large; Dc = DWM-coarse; sb0 = shrub-branch-0; sb7 = shrub-branch-7; sb11 = shrub-branch-11; sb25 = shrub-branch-25)

et al. (2015) assumed a value of 0.08 for 1-, 10-, and 100-h categories, which is quite similar to our values for litter (spring) and our three subcategories of DWM-fine (autumn); their value for live herbaceous is within our seasonal range. The largest set of values came from Shipley and Vu (2002); these were all clearly at the upper end of the range but nevertheless encompassed our estimates for herbaceous vegetation in winter and spring, and also for shrub leaves in these seasons. Overall, our values appear broadly consistent with those found in the three sources we examined in our brief literature survey for FMC.

24.5.4 *Ovendry Fuel Loading*

The total understory ovendry fuel loading is derived as the sum of its fuel family components, each estimated with the most appropriate technique (Table 24.2 row 29). The most direct technique is destructive sampling, although this is most commonly used only for the herbaceous and litter components, and for calibration of indirect techniques. For DWM, the planar intersect method of Brown (1971) is most commonly used: the number of elements encountered along a line transect is tallied according to diameter category, and equations are applied to the counts to convert them to estimates of biomass per unit area. The biomass of the understory woody vegetation is most commonly estimated by measuring the dimensions of the vegetation, and applying allometric relationships to convert them to biomass, rather than by destructive sampling (Conti et al. 2019). In our spreadsheet application it is possible to treat all understory woody vegetation as a single category (assumed in Table 24.2 for simplicity), or to specify an unlimited number of woody species, each with its own allometric parameters (a feature excluded from Fig. 24.1 for simplicity).

24.5.4.1 Herbaceous Biomass

We drew on estimates of herbaceous biomass that we conducted at Beitar Forest (n = 555) and HaKedoshim Forest (n = 719). The means and distributions are shown in Fig. 24.5, which illustrates well the high spatial heterogeneity at the scale of the sampling quadrat, and also the strong effects of season and tree density. Based on these results we set an upper limit of 0.6 kg m^{-2} to the range examined.

24.5.4.2 Litter Biomass

Estimates of litter biomass were drawn from data that we collected in Beitar Forest (n = 150) and HaKedoshim Forest (n = 447). The means and distributions are shown in Fig. 24.6, which shows higher levels in the former than in the latter. Based on these results we set an upper limit of 2.0 kg m^{-2} to the range examined.

24.5.4.3 Downed Woody Material Biomass

Biomass estimates for DWM-fine were based on data that we collected in Beitar Forest (n = 32) and HaKedoshim Forest (n = 30) using line intersect sampling. Two primary sources for the equations used to convert tallies to biomass were Brown (1974) and Chojnacky et al. (2004), which are broadly the same although the equations use different symbols. The constant 11.64 in Brown's (1974) equation [Br] parallels the unit conversion factor, *f* in the equation for DWM-fine biomass of

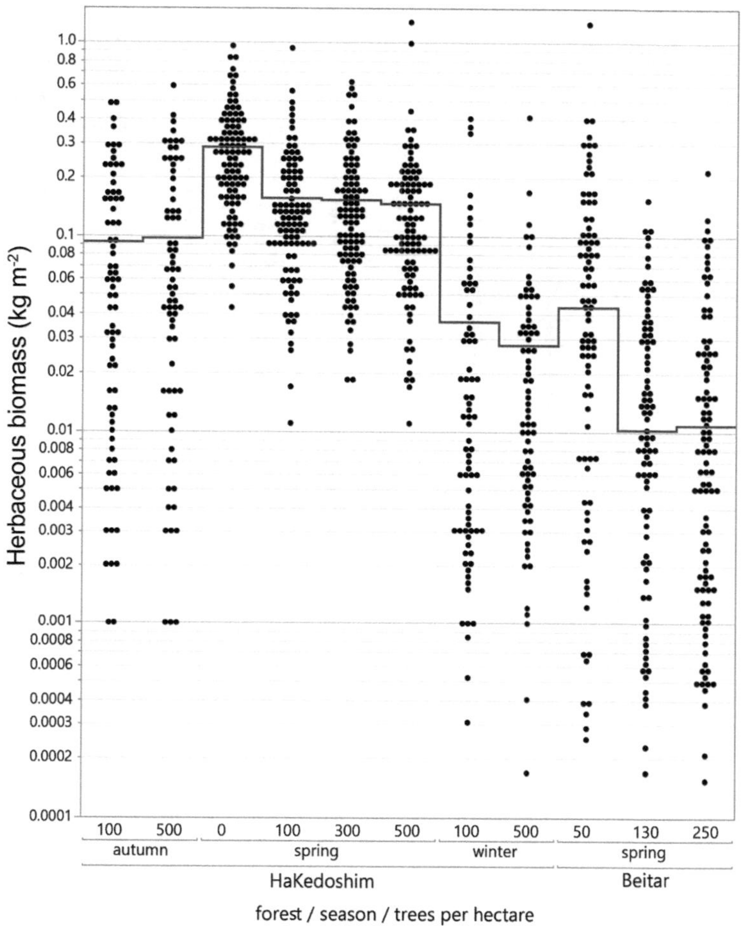

Fig. 24.5 Means and distributions of herbaceous biomass (kg m^{-2}) based on samples collected at Beitar Forest (n = 555) and HaKedoshim Forest (n = 719) in quadrats of area 0.04–0.1 m^2

Chojnacky et al. (2004) [Ch], after correction for typographic errors in the latter citation; similarly, n, d^2, s, c, a and Nl used by Br parallel, respectively, T_i, $dclas_i^2$, ρ, c, a and L_i used by Ch. Some parameters/constants of the equation differ among the three size classes of DWM-fine, and these corresponded between the two sources. One difference between the equations is a decay deduction factor (0.9) introduced by Chojnacky et al. (2004); the dependent variable in both is the total biomass per unit area of all classes of fine DWM. Figure 24.7a shows the individual biomass estimates according to site and tree density after applying the equations of Chojnacky et al. (2004) to the tallies recorded in the field. For comparison, Fig. 24.7b shows the biomass statistics for various classes of DWM (with the three FWM classes combined) given in Table 2 of Chojnacky et al. (2004), which was based on regions in the eastern United States. The values from Israel are lower than the comparable

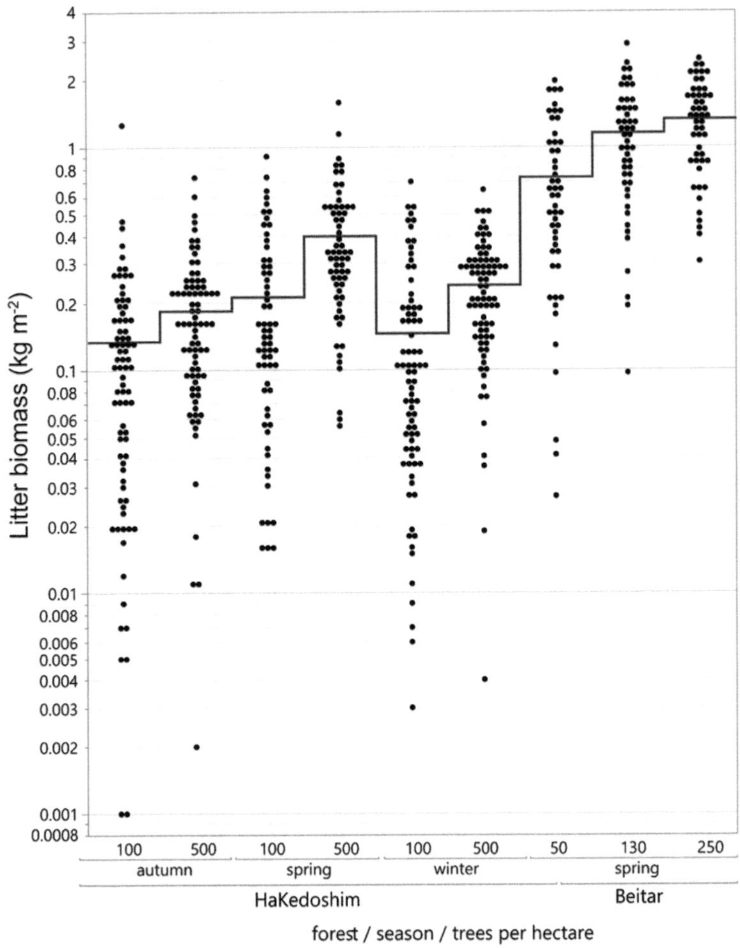

Fig. 24.6 Means and distributions of litter biomass (kg m^{-2}) based on samples collected at Beitar forest (n = 150) and HaKedoshim forest (n = 447). Quadrat sizes used were 0.0625 and 0.1 m^2

values in the US, but are certainly plausible. Downed woody material at the Israel sites was almost entirely of the fine class, and for simplicity we excluded both DWM of the coarse class and cones from our simulation. Based on these results we set an upper limit of 3.0 kg m^{-2} to the range examined.

24.5.4.4 Shrub Biomass

We drew on a dataset that we collected at the Beitar site in Israel for the estimation of understory shrub biomass. We defined biomass per unit area as the product of height (H; m), proportional plant cover (C; dimensionless), canopy fill (F; dimensionless) and bulk density coefficient (D; kg m^{-3}).

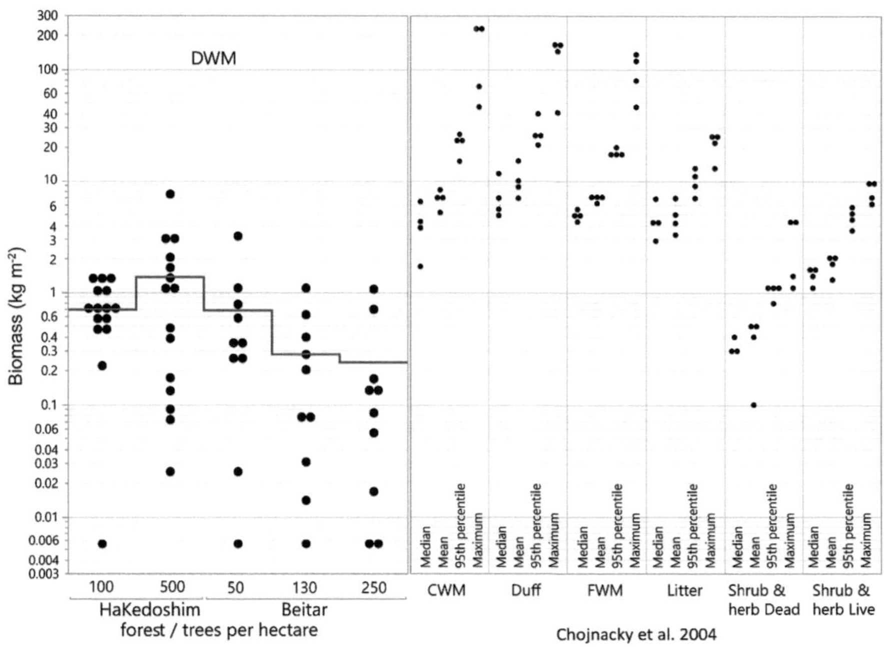

Fig. 24.7 Means and distributions of downed woody material (DWM) biomass (kg m^{-2}) measured by the line intersect method in two studies in Israel (left) and the biomass metrics for various classes of DWM in regions in the eastern United States (right; based on Table 2 in Chojnacky et al. 2004 with the three FWM classes summed)

$$V = H \cdot C \cdot F \cdot D$$

Height, cover, and fill were based on transect data and the bulk-density coefficient was based on destructive-sampling data, as was the allocation of biomass among fuel families. The range and means obtained for shrub total biomass, and for its five constituent fuel families, are shown in Fig. 24.8. Based on these results we set an upper limit of 0.5 kg m^{-2} to the range examined.

24.5.5 Fuel Depth

The remaining input parameter of the Rothermel (1972) model to be discussed is fuel depth. In computing rate of spread, the sole use of fuel depth is in deriving ovendry bulk density, which is defined as the ratio of biomass per unit area to fuel depth. For simplicity, consider first a single fuel family, for which the eq. V = H . C . F . D must always hold. The terms H and C form a pair that relates directly to fuel depth, which is defined as their product; C parallels the proportion of hits in a presence/absence transect, and H is the mean height at the points of presence. In other

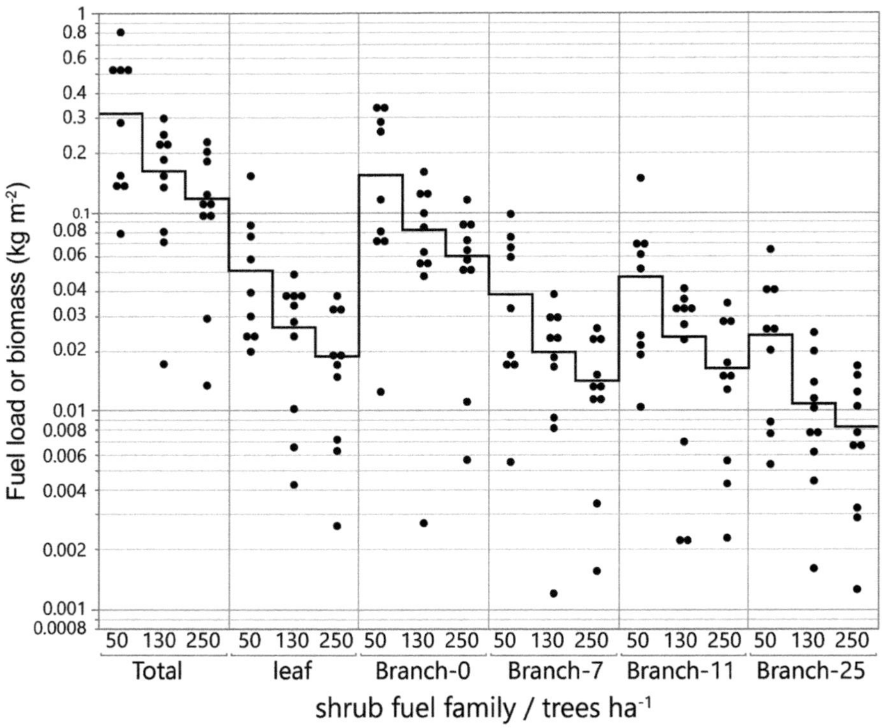

Fig. 24.8 Means and distributions of understory shrub biomass (kg m^{-2}), estimated by indirect methods, at the Beitar site in Israel

words, at constant total volume and bulk density, fuel depth is the height the vegetation would attain if spread out over the entire area; fuel depth equals H when C = 1. That makes intuitive sense for fuel families such as herbaceous, litter, and DWM, which have a diffuse presence and are estimated by random sampling that includes zero values, for which C is implicitly equal to 1. Likewise, the terms F and D form a pair that relates directly to bulk density; D is in units of bulk density and F is a dimensionless modifier that relates to "fill". If one is satisfied with a single mean value for D, then F is set to 1.

Since ovendry bulk density is the ratio of biomass per unit area (= ovendry fuel loading) to fuel depth, we can express it as: (H . C . F . D)/(H . C) = F . D. Thus, for any fuel family whose biomass per unit area is based on the product (H . C . F . D), the parameters needed to compute its ovendry bulk density are already determined. The alternative is for biomass per unit area to be determined directly rather than by computation; in such a case one of the pairs (H . C) or (F . D) must also be provided, and the other is implicit and must be adjusted to maintain the basic equality. Our implementation of the model requires the provision of (F . D), which results in a uniform conceptual approach across all fuel families. The implicit part of the equation is then (H . C), which is a much easier unit than bulk density to track for unrealistic values. Thus, instead of a fuel depth for each fuel family, we need to provide the biomass bulk density (F . D). How values from different fuel families need to be combined in the model is discussed below.

24.5.6 *Biomass Bulk Density*

For shrub vegetation, estimates of D are required as an integral step in using the product (H . C . F . D) to compute biomass per unit area; however, this requires destructive sampling, which is performed sparingly. Common shrub species were destructively sampled to determine D at the Beitar site; the results ranged from 1.5 to 2.2 kg m^{-3}, with an average of 1.9 kg m^{-3}, which is the value we adopted. We allocated this shrub-level value among the constituent five shrub fuel families according to proportions that we also determined during the destructive sampling.

For herbaceous vegetation, we drew on a dataset for which H and C were measured in 100 0.1-m^2 quadrats (selected to span a wide range of biomass) that were then destructively sampled (Osem et al. 2019). By using (H . C) as the denominator, we obtained a mean bulk density of 2.89 kg m^{-3}.

The thickness of the litter layer was measured in the quadrats that were sampled for litter biomass at the Beitar site; we used these data to examine the relationships among biomass per unit area, thickness, and biomass density (computed as the ratio between biomass per unit area and thickness). Since the methodology for estimating biomass was based on random sampling, the notion of plot-level cover was not relevant, and C was implicitly taken to be equal to one. No estimates were made of within-quadrat "fill" (which parallels the notion of vertical fill applied to woody vegetation) therefore this too was implicitly equal to one. There was no indication that D varied with biomass per unit area, and there was a strong, though not linear, relationship between biomass per unit area and thickness. As a first approximation, we adopted the mean D of 46.9 kg m^{-3}.

The depth of the DWM-fine layer was measured in one of the measurement seasons at the HaKedoshim site, under the protocol for determination of FMC of this fuel family. We adopted the average value of 5 cm. The mean biomass per unit area was 4.6 kg m^{-2}, therefore our first approximation of DWM-fine bulk density was 92 kg m^{-3}. When DWM biomass is derived from tallies of three subcategories, it is necessary to distribute this quantity among them, which we did, proportionately to their relative biomasses, i.e. 0.2, 0.3, and 0.5, respectively for small, medium, and large families of DWM-fine fuel.

24.5.7 *Reduction to a Single Value*

Now that we have specified many of the input parameters for Rothermel's (1972) model for each fuel family, we must somehow merge them into a single value to be used in the equations of the model; treating each separately does not appear to be how Rothermel's model was intended to be implemented. Determination of the best way to perform this merger is not trivial: for instance, how does one combine the surface area-to-volume ratios of such contrasting items as herbaceous vegetation and shrubs? Likewise, we defined the fuel depth of the woody understory as the product of its height and its relative cover. But, for instance, is the fuel depth of an understory that contains only a uniform cover of shrubs the same as that of one that also contains a thick layer of litter?

We dealt with the input parameters, S, fuel PD, and FMC, by taking the value for each fuel family and weighting it according to the relative contribution of that fuel family to total biomass per unit area (Table 24.2 rows 30–32). This same logic was applied to ovendry bulk density by computing the single value for insertion into the Rothermel equations from the individual biomass densities (F . D) of the fuel families, weighted by their relative contributions to biomass per unit area. The implicit overall fuel depth could then be reverse-calculated as the ratio of ovendry fuel loading to ovendry bulk density (Table 24.2 row 33).

However, an alternative approach is to view fuel depth as something defined by the uppermost layer among the various vegetation categories, albeit after normalizing their heights by relative cover (Table 24.2 row 41). In this case, we revert to the simple definition of understory bulk density as the ratio between ovendry fuel load and fuel depth, with fuel depth determined by the vegetation category with the maximum implicit fuel depth, i.e., biomass per unit area divided by (F . D).

This problem of fuel-depth definition is not new. Hough and Albini (1978) referred to it as "deceptively simple", and as recently as 2018 an outstanding in-depth report (Andrews 2018) was published that addressed the seemingly problematic Rothermel surface-fire spread model whose calculations "are highly sensitive to the fuel bed depth" (Heinsch and Andrews 2010). To illustrate this, we present (in Sect. 24.7.2 below) the key results obtained by using both definitions.

24.6 Crown Fire Risks

The next section of the model adds five equations (Table 24.2 rows 64–68; Fig. 24.1 in red) elaborated by Scott and Reinhardt (2001) on the basis of Van Wagner's crown-fire transition criteria (Van Wagner 1977, 1989, 1993), to link Rothermel's separate surface-fire and crown-fire spread models (Rothermel 1972, 1991). The equations generate two indices of crown-fire risk, both of which are wind speed thresholds: the torching index (TI) and the crowning index (CI). The torching index is the minimum open wind speed (6.1 m above the ground) required for the fire to climb from the understory to the overstory, thereby initiating a crown fire. As outlined by Cruz and Alexander (2010), this can be computed by using the surface-fire model of Rothermel (1972) in conjunction with the crown-fire initiation criteria of Van Wagner (1977). The crowning index is the minimum open wind speed required for an active crown fire, i.e., one that persists and spreads.

Although TI and CI are both functions of many of the intermediate variables of the core surface-fire model, three additional input parameters are required that describe the overstory vegetation: the foliar moisture content, base height, and bulk density of the crown vegetation (needles and branches). The moisture content of the crown foliage is set to 100% (Agee et al. 2002). Crown base height is measured directly and crown bulk density is derived from the available crown fuel load and crown depth (Table 24.2 rows 63; Fig. 24.1, in purple). The crown biomass or fuel load is computed from DBH, and tree height by using the equations developed by De-Miguel et al. (2014).

24.7 Model Results

24.7.1 Single Fuel Family

Our first exploration of the model avoids the thorny issue of how to define fuel depth, by taking a theoretical approach and treating the vegetation as a single fuel family. This leads to mutual equivalence of the two definitions of fuel depth, and also bypasses the issue of how to derive a single input parameter (e.g. PD) from a set of values appropriate to the respective fuel families; our approach of weighting according to biomass per unit area may be incorrect. The sensitivity analysis examined the responses of rate of spread, TI, and CI to variation of plus-or-minus 10% in each of the 11 input parameters. For simplicity we examined a single baseline scenario even though the relative importance of input parameters can vary with the "fuel model" being examined (Liu et al. 2015). The baseline values for the fuel-family-dependent input parameters were based on biomass per unit area of 0.5 kg m^{-2} for each of herbaceous, litter, DWM-fine and shrub, the latter shared in proportions of 0.50, 0.20, 0.15, and 0.10 to leaves and to the four size categories of branch fuel families, in increasing order. Ovendry bulk density was 5.97 kg m^{-3}, giving an implicit fuel depth of 0.33 m.

All responses were monotonic and linear or close to linear (Table 24.6). For rate of spread, the response was positive for slope, wind speed, fuel heat content, fuel moisture of extinction, S, PD and ovendry fuel loading. It was negative for effective mineral content, fuel total mineral content, FMC, and ovendry bulk density. For CI, the direction of response was opposite to those for all the above input parameters; this was also true for TI, except for its response to changes in S. The three vegetation-related input parameters to which rate of spread was most sensitive were ovendry bulk density, S and ovendry fuel loading. For TI and CI, fuel heat content was more influential than S.

Of course, a 10% variation may be small relative to the changes that occur naturally, but even a PD change from 300 to 600 kg m^{-3} elicited only small responses in the three target variables. Likewise, changes in topographic slope over the range of 5–30 degrees had small impacts. The strongest response of rate of spread was to

Table 24.6 Sensitivity analysis of rate of spread, torching index (TI) and crowning (CI) index to a change of plus-or-minus 10% relative to the standard values. The fuel-family-dependent input parameters of Rothermel's model were based on a biomass of 0.5 kg m^{-2} for each of herbaceous, litter, DWM-fine and shrub

	Name	Units	Sensitivity analysis				
			Baseline value	Sign of slope for ROS, TI, CI	Range in rate of spread	Range in torching index	Range in crowning index
1	Slope (vertical rise/horizontal run)	–	5	+ – –	0.01	0.02	0.01
2	Wind speed at midflame height	m min^{-1}	15	+	3.00	–	–
3	Fuel heat content	kJ kg^{-1}	18,608	+ – –	2.12	2.84	1.10
4	Fuel effective mineral content	–	0.01	– + +	0.43	0.56	0.22
5	Fuel total mineral content	–	0.055	– + +	0.12	0.16	0.06
6	Fuel moisture of extinction	–	0.85875	+ – –	0.51	0.67	0.26
7	Surface area: Volume ratio (S)	cm^{-1}	59.2	+ + –	2.29	0.53	0.97
8	Fuel particle density (PD)	kg m^{-3}	436.2	+ – –	0.55	0.79	0.27
9	Fuel moisture content (FMC)	–	0.2656	– + +	1.74	1.47	0.88
10	Ovendry fuel loading	kg m^{-2}	2	+ – –	2.23	2.96	1.15
11	Fuel depth (implicit value)	m	0.33				
12	Ovendry bulk density	kg m^{-3}	5.9712	– + +	2.60	2.03	1.28

wind speed – over a range of 57 m min^{-1} in response to a wind speed increase from 5 to 50 km hr^{-1}. This was followed by ovendry fuel loading over the range of 0.2–4.0 kg m^{-2}, which also induced the largest responses in TI and CI. Rate of spread was influenced somewhat less strongly by ovendry bulk density over a range of 3.6–8.2 kg m^{-3} and S over a range of 30–90 cm^{-1}. Note that rate of spread responds in opposite directions to changes in ovendry fuel loading and ovendry bulk density. Therefore, in vegetation comprising various fuel families of diverse bulk densities, an increase in fuel loading could result in an increase or a decrease in rate of spread if the relative amounts of the fuel families are not maintained. To illustrate this, we now progress from involvement of a single fuel family to that of multiple fuel families, in the determination of input parameters.

24.7.2 *Multiple Fuel Families*

As an exploratory exercise we chose to generate a population of landscape scenarios that represent permutations of three biomass levels – zero, medium, and high – of each of herbaceous, litter, DWM-fine and shrub fuel families. The absolute values were based on the natural range found in Mediterranean conifer forests in Israel. This generates 80 permutations (excluding the all-zero situation), the results of which can be organized hierarchically and illustrated graphically. These results will of course differ between the two alternative definitions of fuel depth, as shown in Fig. 24.9. When ovendry bulk density is defined as the weighted mean of the values for each fuel family the addition of relatively dense DWM and litter to shrub and herbaceous can have the net effect of reducing rate of spread. Furthermore, as increasing amounts of other categories are added to the relatively high herbaceous component in the left-most combinations in Fig. 24.9a, the impact of various levels of herbaceous vegetation is moderated, especially by DWM. Reduction of shrub biomass also has the greatest impact when DWM levels are low. A quite different pattern is obtained when ovendry bulk density is based on ovendry fuel load and the fuel depth (H . C) of the tallest fuel family (Fig. 24.9b); the opposing effects on rate of spread of ovendry fuel loading and ovendry bulk density are more evenly balanced, and the rate of spread declines moderately from left to right in Fig. 24.9b. The sharp periodic reductions in rate of spread are caused by combinations of low herbaceous biomass and low shrub biomass, which lead to drastic changes in fuel depth, with accompanying increases in bulk density, and result in dramatically reduced rate of spread.

As might be expected, the pattern obtained for TI using the weighting method of calculation (Fig. 24.10a) broadly mirrors that for rate of spread, although the region of greatest impact of herbaceous biomass has moved from the left to the center of the figure. Nevertheless, the broad overall trend is that increasing presence of the denser DWM and litter components reduces risks even though biomass is being added. The opposite trend is obtained when fuel depth is based on the maximum among the fuel families present (Fig. 24.10b); as we progress from left to right, there is a broad trend of increasing hazard, but in all subsets of runs, the dramatic effect of lowering herbaceous and shrub biomasses simultaneously is evident. Crowning index responds very similarly to TI and these results are not shown.

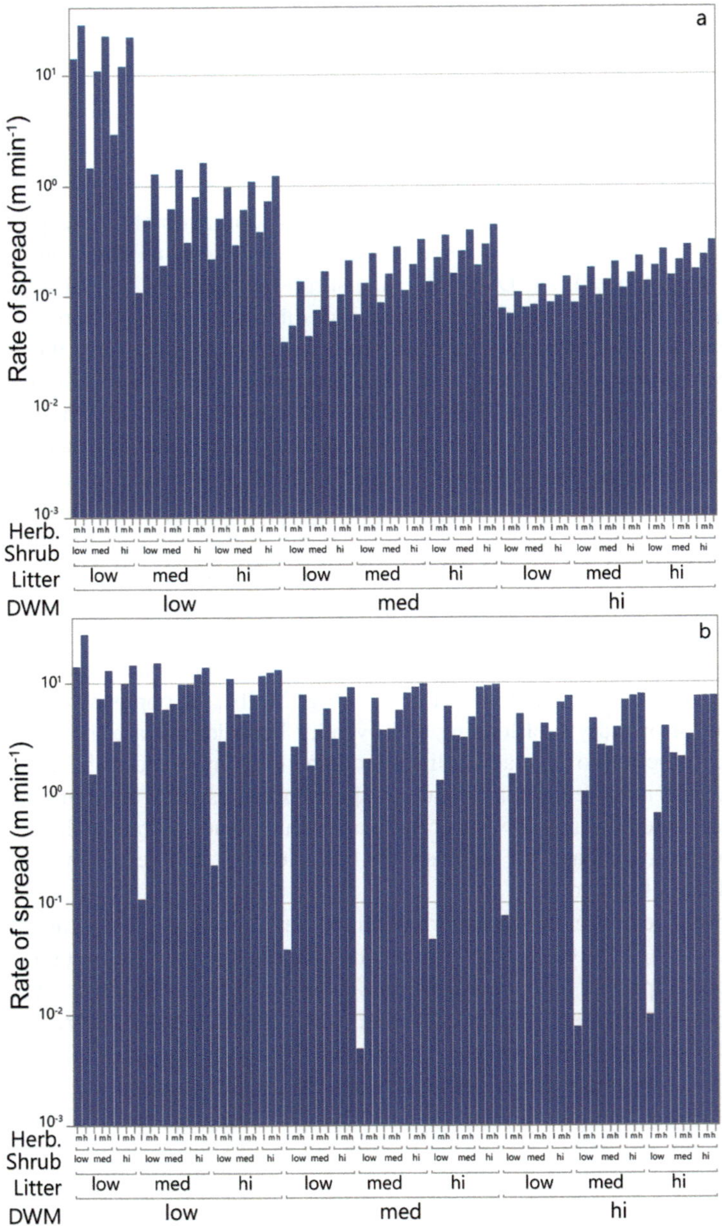

Fig. 24.9 Rate of spread computed by the Rothermel model for permutations of biomass per unit area of DWM-fine, litter, shrub and herbaceous fuel families, using two rules for deriving fuel depth: back-calculated from the biomass-weighted ovendry bulk density of each fuel family (**a**), and the maximum of the implied fuel depths of each fuel family based on its biomass and bulk density (**b**). The fuel families are nested, with herbaceous changing between every column and DWM changing in three blocks

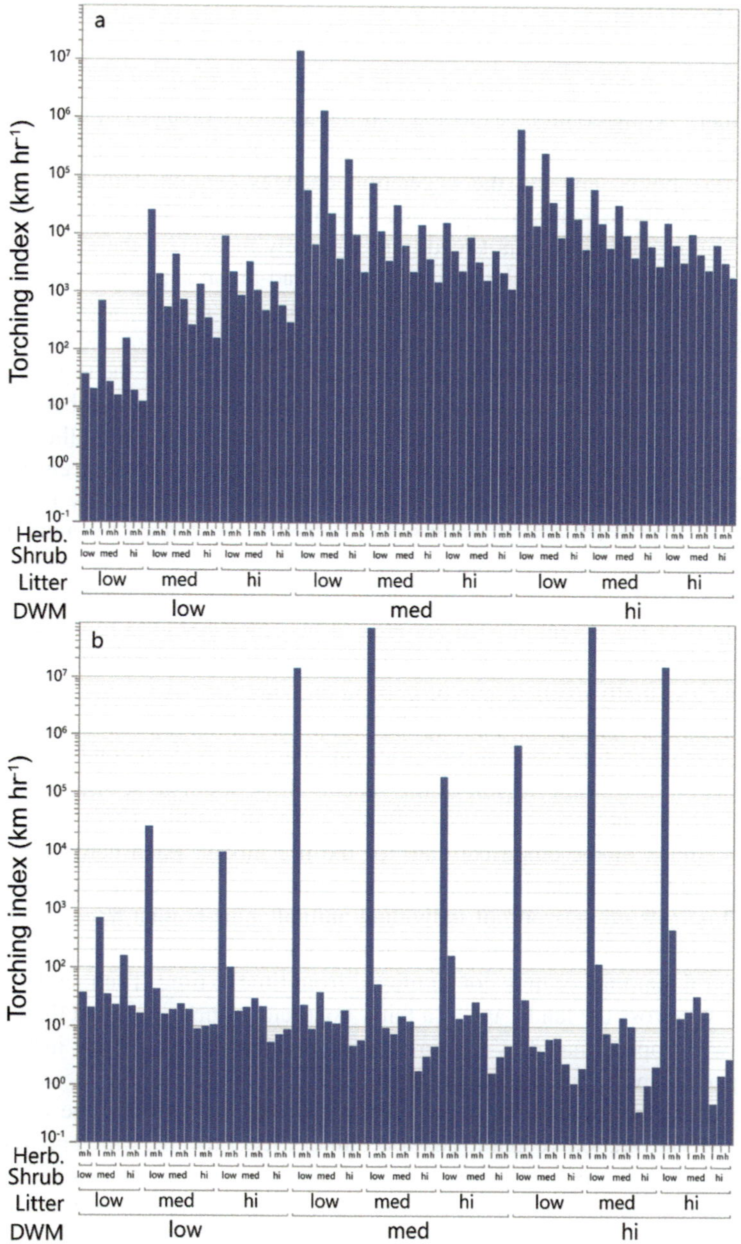

Fig. 24.10 Torching index computed by the Rothermel model for permutations of biomass per unit area of DWM-fine, litter, shrub and herbaceous fuel families, using two rules for deriving fuel depth: (**a**) back-calculated from the biomass-weighted ovendry bulk density of each fuel family, and (**b**) the maximum of the implied fuel depths of each fuel family based on its biomass and bulk density. The fuel families are nested, with herbaceous changing between every column and DWM changing in three blocks

24.8 Grazing

The fourth challenge we presented in the Introduction (Sect. 24.1) was calculation of whether grazing could be expected to bring about the changes in vegetation necessary to reduce fire hazards to acceptable levels; in the absence of arbitrary acceptability thresholds, how do the vegetation changes examined in the fire model compare with those achieved in practice by grazing herds? The consumption rate of a free-ranging herbivore is not readily or routinely measured, therefore normative values are used, which should suffice for the broad-brush approach taken here. But that provides only one component of the information we require; we must be able to convert intake rate expressed per animal to an intake rate expressed per unit area, which would be compatible with the notion of fuel load, once the time dimension was accounted for. That requires us to know animal density – the number of animals per unit area. In fenced areas this is most simply expressed by using the size of the paddock as the denominator. Admittedly, if the animals in fact concentrated their presence in one half of the paddock, so that no area at all actually received the implied average grazing pressure, it would be necessary to track the animals' location to obtain that spatial resolution. In unfenced areas, or highly heterogeneous settings in which it is clear from the outset that grazing pressure will be distributed unevenly over the potential roaming area, it becomes essential to use tracking to obtain a meaningful picture. With the resulting data, it then becomes possible to construct a spatially explicit map of grazing pressure.

We drew on unpublished tracking data that were collected from seven goat herds (three permanent, four nomadic) over an entire annual cycle by one of the authors (R.H.) within a 100-km^2 region of the Judean Hills, south-west of Jerusalem. The region includes the above-mentioned Beitar and HaKedoshim Forests, from which we drew much of the empirical data for the fire model. Each herd of goats was accompanied by a shepherd, who enforced some degree of spatial cohesion on the herd. This interacting group of individual-animals-plus-human traces out a meandering pathway across the landscape, usually ending the day where it started, in a night corral that may or may not be moved from time to time. The fact that the animals move across the landscape as a fairly coherent group means that their location can be well approximated by the position of any one animal, so that just one GPS collar per group can yield reasonable estimates.

Regarding the methodology: the GPS collar was built around the i-gotU GPS logger (Mobile Action Technology, New Taipei City, Taiwan), which proved robust under field conditions and generated good quality data; the system used a third-party external battery (Fig. 24.11). The device was configured to track satellites continuously. The collars were replaced every few weeks throughout a year, to provide – or at least aim for – continuous coverage. Data were filtered to retain only those GPS locations that represented positions that were logged while the herd traced out its daily foraging route across the landscape between leaving the corral and returning to it; the resulting dataset, comprising 2134 daily foraging routes, was sampled at 1-min intervals.

Fig. 24.11 Goat herd grazing along a fuel-break corridor in the Judean Hills. The brown-and-white goat in the center is wearing an in-house assembled GPS collar. The picture was taken along the seasonally active Refaim River in Aminadav Forest

Of course, one year of monitoring is no guarantee that similar grazing pressures are exerted across the region from year to year, and large variations exist, as can be seen in Fig. 24.12, which shows 4 years of curated GPS locations for the HaKedoshim Forest. However data from the full complement of seven goat herds was only available in the first year, which was analyzed here. The next step in that analysis was to convert GPS locations to meaningful units as expressed in the terminology of range management: animal-days per unit area. The concept of "animal-days" encompasses both the number of animals and the duration of their presence. For free-ranging animals, for example, 30 animals for 20 days and 15 animals for 40 days both represent 600 animal-days. For herds corralled at night, the equivalent of 1 day's herbage consumption is concentrated during the foraging hours, so even though the duration of a typical daily foraging route varied among the herds and also across the seasons, it was always regarded as equivalent to "one day" of presence of a free-ranging animal. Location data collected for many individuals in the herd simultaneously showed that the area occupied by a herd at any point in time was well approximated by a circle, and its area was estimated for each herd. Therefore, as a first approximation, we can draw such a circle around a GPS point and reason that the patch of rangeland within the circle accumulated animal presence (animal-days) equal to the (known) number of animals in the herd multiplied by the amount of time represented by one GPS point (1 min). If we divide that value by the area of the circle, we express this grazing pressure in standardized units of animal-days per unit area, with 1 day being based on the duration of a foraging route. By means of suitable GIS tools this reasoning can be applied to the entire

Fig. 24.12 Point-map of herd presence recorded for one goat herd in the HaKedoshim Forest area over 4 consecutive years, based on approximately 520,000 GPS points, each representing 1 min of herd presence. Years are separated by color

collection of GPS points and the result summed across all of them to generate a spatially explicit map of animal-days per unit area. This is shown in Fig. 24.13 at the regional level after 1 year of monitoring seven herds.

The accumulated presence of the herds in the area was highly heterogeneous in space, and accumulated grazing pressure in the grid cells varied from 0.2 to 4930 animal-days ha^{-1} for the pooled dataset. The average grazing pressure in the areas visited at least once (the "area of sustenance") was 80 animal-days ha^{-1}. Distance from the corral was strongly and negatively related to grazing pressure, and most of the grazing pressure imposed by all herds occurred within 1 km of the corral.

Behavioral observations revealed that most of the daily foraging along the route was dedicated to eating (67% of observation time on average). The goats ate primarily woody vegetation, although during the green season there was a greater proportion of time devoted to eating herbaceous vegetation than during the dry season (20 and 7%, respectively). Consistent with these findings, acoustic monitoring showed that 75% of the daily foraging along the route was devoted to active grazing, i.e., eating, which includes biting and chewing.

Grazing pressure, expressed as animal-days per unit area, was converted to units of biomass removed per unit area by assuming a normative value for the daily herbage intake of one animal, i.e., 1.5 kg of dry matter day^{-1} head^{-1}, including all types of vegetation consumed (interestingly, 1.5 kg is roughly the total fuel load per square meter in many of the scenarios examined here; it includes the inedible fractions). The average consumption per unit area for the pooled herd tracking data was

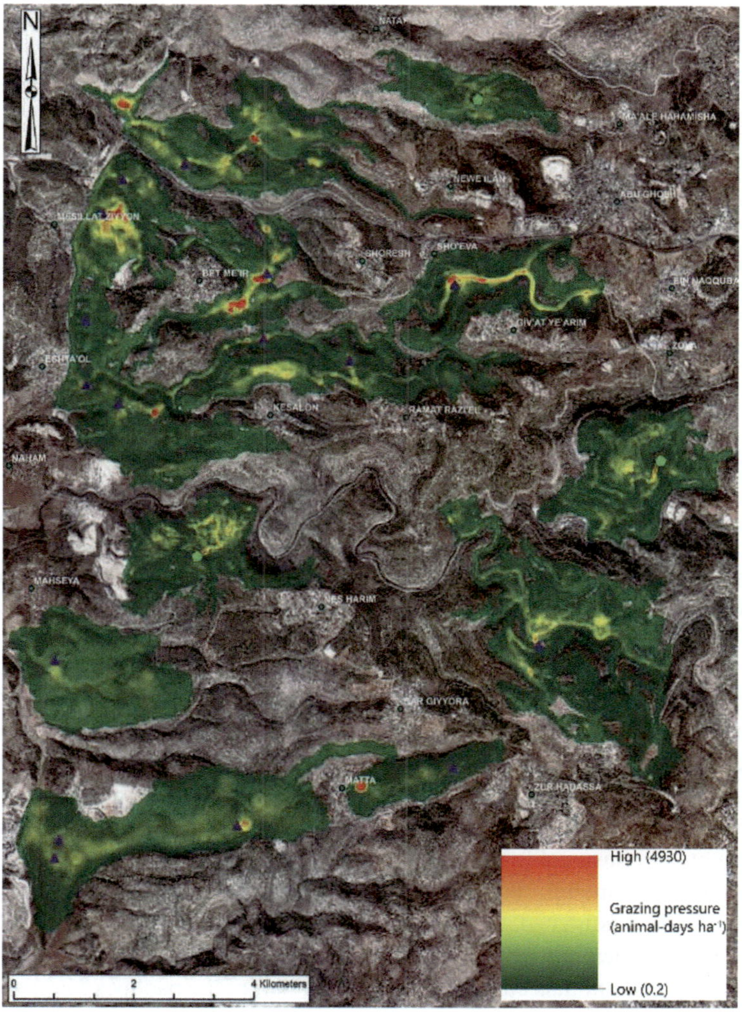

Fig. 24.13 Heat-map of grazing pressure imposed by seven goat herds in the Judean Hills over 1 year, based on approximately 800,000 GPS points, each representing 1 min of herd presence. Grazing pressure is expressed as animal-days per unit area and values are comparable across the entire region despite differences in herd size

118 kg dry matter ha^{-1}, or 0.0118 kg m^{-2} – two orders of magnitude *less* than the values we are dealing with in the understory, as seen in the data presented earlier (coverage of the GPS collars was not perfect, but any correction would increase this value only modestly).

However, the relevant comparison is not with total fuel load, much of which is the standing stock of inedible vegetation characterized by slow dynamics, i.e., DWM, litter, and the largest classes of live shrub branch. Live shrub leaf is certainly relevant, as is herbaceous production. Regarding the season of highest fire risk, in

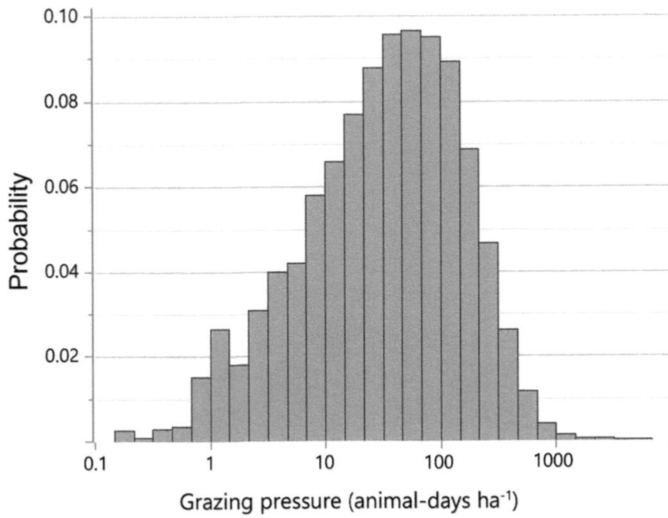

Fig. 24.14 Frequency distribution of grazing pressure, in units of animal-days ha^{-1}. Based on pooled regional data from seven herds over 1 year encompassing an area of 539,576 cells of 100 m^2 with non-zero animal presence

late summer, the herbaceous vegetation is inert and feedbacks of consumption onto growth can be ignored. We assigned a relatively high proportion of shrub biomass to leaf, equivalent to 0.125 and 0.250 kg m^{-2} at the medium and high biomass levels, respectively. If we assume an open shrub architecture that allows unimpeded access to all foliage, and likewise, assume all the herbaceous vegetation to be consumable, the amounts of consumable vegetation are 0.3 and 0.6 kg m^{-2}, respectively. These are more modest amounts, but they are still much larger than the average consumption per unit area estimated above.

Might spatial heterogeneity of animal presence resolve the issue, much like the above example of animals concentrated into one half of the paddock? Perhaps when we consider the entire area of sustenance, the "important" impacts are being diluted by areas utilized only minimally. However, when we examine the frequency distribution of grazing pressure for the entire database (Fig. 24.14; based on 539,576 cells of 100 m^2 with non-zero grazing pressure), only a tiny fraction of the area was subjected to grazing pressures that exceeded, say, 1000 animal-days ha^{-1} – a pressure expected to remove 0.15 kg m^{-2}, which could have a significant impact.

As far as shrub biomass is concerned, a different approach may hold the key to bridging the gap. Pasalodos-Tato et al. (2015) computed an overall average for the main Mediterranean shrub formations in the region of Andalusia in southern Spain. They reported a regional mean value for biomass accumulation of 16.73 Mg ha^{-1} (= 1.7 kg m^{-2}) and an annual growth rate of 1.14 Mg ha^{-1} yr^{-1} (0.114 kg m^{-2} yr^{-1}). In relative terms, that is 0.07 yr^{-1}, which, if applied to our medium shrub biomass of 0.25 kg m^{-2}, yields 0.0175 kg m^{-2} yr^{-1}. Taking leaf to account for 50% of this yields

0.00875 kg m^{-2} yr^{-1}, which is 74% of the value we estimated above for animal off-take, and is also compatible with our estimates of consumption allocated to shrub vegetation.

Such a level of annual biomass removal may be sufficient to arrest further expansion of the shrub component. Thus, "depletion" is hardly the appropriate term to describe the computed amounts removed by grazing, but grazing may maintain a status quo. If mechanical intervention to reduce woody vegetation cover precedes the introduction of grazing – as practiced to various degrees in the GPS study region – then grazing should be able to retard recovery and increase the number of years before further intervention is required.

However, this leaves us little spare consumptive capacity to make a serious dent in the herbaceous biomass, which is mostly seasonal growth. In fact, the low allocation (7%) of eating time to herbaceous vegetation in the dry season also suggests that other processes might be more important. These might include activities that cause physical breakdown, as well as exposure to weathering and insect activity. For example, we used our simple model to explore the effect of trampling on herbaceous bulk density; if the herbaceous vegetation is squashed to half its height there is a doubling of bulk density. Inserting this change into the model had a strong impact: it shifted the overall distribution of results so that the median rate of spread was reduced from 8.4 to 3.7 m min^{-1}; and median TI and CI increased from 5.8 and 9.2 to 22.1 and 21.2 km hr^{-1}, respectively.

24.9 Discussion

Setting up Rothermel's model so as to link the model equations with our forest survey data raised the fundamental question of how to consolidate data from many fuel families into one set of input parameters that treated the vegetation as a uniform substance. As Hough and Albini (1978) wrote: "The mathematical model of fire behavior developed by Rothermel (1972) predicts fire spread and intensity well enough if the fuel is relatively homogeneous". More recently, Schwilk (2015) argued that the response of flame spread rate to an increase in fuel density or packing ratio might be the opposite in "canopy fuels" of the understory versus the "litter fuels". There are trade-offs that can change the direction of response: small litter particles produces litter that is more tightly packed and less flammable, but at the same time smaller particles have greater surface-area-to-mass ratios which can promote fire. The approach of weighting according to relative biomass per unit area led to very shallow fuel depths whenever the DWM and litter layers were included, because their biomass can be high relative to those of shrub and herbaceous vegetation. This is not necessarily wrong, but the term "fuel depth" takes on a quite different meaning from how most people would interpret it. At any given biomass per unit area a reduction in fuel depth increases bulk density, and we saw from the single-fuel-family runs of the model that this is expected to reduce rate of spread and increase TI and CI (Table 24.6). The alternative approach of designating fuel depth according to the top of the vegetation changed the response of these three variables dramatically, and is probably more in line with approaches taken by others. Based on theoretical considerations and empirical data (Ungar and Noy-Meir 1988), selective grazing can reduce heterogeneity of vegetation distribution *within* the herbaceous category (and presumably within the shrub category too), but it would not be expected to impact the integration of fuel depths across fuel categories.

Our use of the product H . C to "level out" each fuel family in defining its height seems in line with published statements: for example, the BehavePlus user's guide (Fig. 6 in Heinsch and Andrews (2010)) shows clearly that a shrub height of 1 m with a cover of 0.5 would be regarded as a fuel depth of 0.5 m. But Heinsch and Andrews' (2010) figure also suggests that overlapping and non-overlapping cover of different fuel families would result in differing average fuel depths. Our more idealized approach treats each fuel family independently; it reverse-calculates H . C for each, and uses the highest value obtained. Note that Heinsch and Andrews (2010) introduced a correction factor of 2/3 in calculating fuel depth for the palmetto-gallberry fuel model and Hough and Albini (1978) referred to the need to "adjust several variables, such as fuel depth and moisture content of extinction, so that the output of Rothermel's model was representative of measured fire behavior". Likewise, they wrote: "By a series of systematic trials, it was determined that a value of 2/3 of the visual understory fuel-bed depth gave the most consistent agreement between observed and predicted rates of spread values and flame lengths." Thus, some degree of model tuning has been deemed necessary by model developers.

But even on a much simpler level it could be argued that height should not be weighted according to cover when we consider the ability of understory vegetation to ignite the overstory canopy. Islands of tall vegetation must surely pose a greater risk than the equivalent volume leveled across the entire area. One approach that could be explored would be to treat the current formulation of the Rothermel model as a basic heuristic tool for qualitative rather than quantitative estimation, and to explore its development by applying different definitions of the vertical dimension to different parts of the model.

Under either of our two extreme approaches to addressing fuel depth, the ability of grazing to bring about changes in rate of spread and risk of occurrence of crown fire can be greatly moderated by the presence of DWM and litter, which are denser than herbaceous and shrub fuel families, and can exceed both in biomass per unit area. If DWM and litter are present at low biomass, a given quantity of herbaceous and shrub vegetation will maintain enhanced rates of spread, which will be strongly reduced by their consumption. However, that presumes changes in biomass level in the order of at least 0.1 kg m^{-2}, which dwarfs our GPS-based estimate of offtake – 0.0118 kg m^{-2}. However, if this estimate of offtake is indeed similar to the annual increment in shrub biomass implied by a plausible annual relative growth rate of 7%, then our animal presence data suggest that the two processes could be in balance, and grazing could be effective in suppressing accumulation of standing biomass. Current grazing practice cannot be expected to bring about a step change in the overall level of shrub biomass by direct consumption in any but the most frequently visited areas. A similar process seems to have been at play in the study of Osem et al. (2017) based on the response of the woody understory to release from cattle grazing. The fact that goat grazing has a clearly visible impact on herbaceous biomass is not very consistent with our estimate of offtake, and we can only speculate that other routes of disappearance (including transfer to litter), perhaps enhanced by trampling (Abdel-Magid et al. 1969), account for a large part of the apparent loss. Measurements would be needed to determine the effect of trampling on bulk density and to re-examine that pathway with the model.

All the above is within the framework of the Rothermel model, which deals with surface-fire spread and intensity. The importance of non-consumptive effects of animal presence on fire *ignition* processes would need to be examined by means of a different model, such as the IGNITE module of the BehavePlus package (Andrews et al. 2005), which addresses the probabilities of ignition by a firebrand (Schroeder 1969) and by lightning (Latham and Schlieter 1989), of which the former is more relevant to our region.

Finally, we ask how large the margins of error might be around our estimate of animal consumptive offtake per unit area, assuming we know the number of animals in the herd. Herein lies a major advantage of GPS tracking; we know from examination of the foraging routes on orthophoto maps that GPS positional error is negligible for our purposes. To the extent that equipment malfunctions are minimal and temporal coverage is complete, the cumulative annual point cloud obtained after eliminating time in corrals is expected to be spatially accurate; certainly in terms of its overall envelope area or number of cells intersected at least once in a GIS grid, both of which can be used to determine area of sustenance. At the level of the individual GPS location the error is well within the range of tens of meters. Thus, we have a reliable map of the time-varying location of one animal in the herd, with some random noise added at the scale of approximately 10 m. The remainder of the herd could be almost anywhere in relation to that point, whereas we assume that point to be at its center, which may be its average relative position, although the animal is rarely there. The assumption of a circular herd footprint is a good approximation on average but ignores large deviations that can develop along a foraging route. This possibility was determined by deploying many collars in the same herd for one-day campaigns of intense monitoring. Thus, there is certainly some error in the precise shape and positioning of the area associated with one GPS location, but it is on the scale of tens of meters. The consumption attributed to the area covered by the herd depends on the assumption that intake occurs within the area associated with every GPS location. Behavioral observations and acoustic monitoring showed that most, but not all of the time along the foraging route is spent in active foraging. This results in some degree of spatial dilution of grazing pressure. But total grazing pressure summed over all circles is not influenced by their spatial position, and hence our estimate of average offtake per unit area is not affected. We also assumed that the herd/shepherd unit behaves rationally when foraging; if a location accumulates more animal-day units, it contains sufficient vegetation to support consumption at the normative rate per head. Likewise, we assumed that the instantaneous intake rate of an individual was constant throughout the day, and that its integral over an entire daily foraging bout was equal to our normative value of 1.5 kg head^{-1}. This parameter could vary greatly in response to animal nutritional demand, the quality of the vegetative supply, and feed supplementation in the corral, but not by more than a factor of two, and not consistently over an annual cycle. Thus, there are many assumptions and many opportunities for error but, even pooled, they seem small in the broader context of the overall balance between the processes of accretion and depletion.

A similar but not identical approach was applied to GPS tracking data gathered from a flock of goats (accompanied by a shepherd) foraging open hilly rangeland that comprised a mosaic of herbaceous vegetation and low shrubs (Arnon et al. 2011). The estimate of average offtake obtained in that study was less than 0.008 kg m^{-2}, which is a magnitude comparable with our present value. Arnon et al. (2011) also found that the proportion of the area visited that received grazing pressures at an order of magnitude greater than the average was very small. In a separate study at the same site, and using an independently developed computational approach to handle the intersect between circles and a grid in GIS, Zahavi et al. (2012) reported mean grazing pressures not exceeding 0.05 animal-hours m^{-2}, which corresponds to an offtake of approximately 0.015 kg m^{-2} – surprisingly similar to our value.

GPS tracking places very clear bounds on average offtake per unit area, which are hard to dispute. This forces us to confront the open questions raised by this exploratory analysis in other places. Given the relative ease and low cost involved in operating such a scheme on a regional level, a strong case can be made for routine adoption of GPS tracking of herds in woodlands and forests. It should become part of the basic information infrastructure needed to understand both the potential and the limitations of grazing as a tool in fire risk management.

Acknowledgements We express our gratitude to the Jewish National Fund for financial support of the work reported here. Special thanks are due to Hagit Baram who developed the computer code for preprocessing the GPS data files, and to Eitan Goldshtein for sharing his GIS expertize.

References

Abdel-Magid AH, Trlica MJ, Hart RH (1969) Soil and vegetation responses to simulated trampling. J Range Manag 40:303–306

Agee JK, Wright CS, Williamson N et al (2002) Foliar moisture content of Pacific Northwest vegetation and its relation to wildland fire behavior. For Ecol Manag 167:57–66

Andrews PL (2014) Current status and future needs of the BehavePlus Fire Modeling System. Int J Wildland Fire 23:21–33

Andrews PL (2018) The Rothermel surface fire spread model and associated developments: a comprehensive explanation. Gen Tech Rep. RMRS-GTR-371. USDA Forest Service, Rocky Mountain Research Station, Fort Collins, CO:121 pp

Andrews PL, Bevins CD, Seli RC (2005) BehavePlus fire modeling system, version 3.0: user's guide. Gen Tech Rep. RMRS-GTR-106WWW Revised. USDA Forest Service, Rocky Mountain Research Station, Ogden, UT:134 pp

Arnon A, Svoray T, Ungar ED (2011) The spatial dimension of pastoral herding: a case study from the northern Negev. Isr J Ecol Evol 57:129–149
Arroyo LA, Pascual C, Manzanera JA (2008) Fire models and methods to map fuel types: the role of remote sensing. For Ecol Manag 256:1239–1252
Brown JK (1970) Ratios of surface area to volume for common fine fuels. For Sci 16:101–105
Brown JK (1971) A planar intersect method for sampling fuel volume and surface area. For Sci 17:96–102
Brown JK (1974) Handbook for inventorying downed woody material. Gen Tech Rep. INT-16. USDA Forest Service, Intermountain Forest and Range Experiment Station, Ogden, UT, 24 pp
Brown JK, Simmerman DG (1986) Appraising fuels and flammability in Western Aspen: a prescribed fire guide. USDA Forest Service, Intermountain Research Station, General Technical Report INT-205
Burgan R, Rothermel R (1984) BEHAVE: fire behavior prediction and fuel modeling system – FUEL subsystem. USDA Forest Service, Intermountain Forest and Range Experiment Station, 126 pp
Calev A (2013) Thinning treatment in mature Aleppo pine forests: influences on tree vitality. M.Sc. thesis, Hebrew University of Jerusalem, 60 pp
Chojnacky DC, Mickler RA, Heath LS et al (2004) Estimates of down woody materials in Eastern US forests. Environ Manag 33(S1):S44–S55
Cohen M, Rigolot E, Etienne M (2002) Modelling fuel distribution with cellular-automata for fuel-break assessment. In: Forest fire research and wildland fire safety. Proceedings of IV International Conference on Forest Fire Research 2002, Wildland Fire Safety Summit, Luso, Coimbra, Portugal, 18–23 November 2002. Millpress Science Publishers, Rotterdam
Conti G, Gorné LD, Zeballos SR et al (2019) Developing allometric models to predict the individual aboveground biomass of shrubs worldwide. Glob Ecol Biogeogr 28:961–975
Cruz MG, Alexander ME (2010) Crown fires. In: VI Short Course on Fire Behaviour, 13–14 November 2010, Coimbra, Portugal. Association for the Development of Industrial Aerodynamics, Forest Fire Research Centre, Coimbra, pp 30–46
De-Miguel S, Pukkala T, Assaf N et al (2014) Intra-specific differences in allometric equations for aboveground biomass of eastern Mediterranean *Pinus brutia*. Ann For Sci 71:101–112
Dimitrakopoulos AP, Panov PI (2001) Pyric properties of some dominant Mediterranean vegetation species. Int J Wildland Fire 10:23–27
Elia M, Lafortezza R, Lovreglio R, Sanesi G (2015) Developing custom fire behavior fuel models for Mediterranean wildland–urban interfaces in Southern Italy. Environ Manag 56:754–764
Fernandes P, Rego F (1998) A new method to estimate fuel surface area-to-volume ratio using water immersion. Int J Wildland Fire 8:121–128
Finney MA (2004) FARSITE: fire area simulator–model development and evaluation. Research Paper RMRS-RP-4 Revised. USDA Forest Service, Rocky Mountain Research Station, Ogden, UT, 47 pp
Garcia-Llamas P, Suarez-Seoane S, Taboada A et al (2019) Environmental drivers of fire severity in extreme fire events that affect Mediterranean pine forest ecosystems. For Ecol Manag 433:24–32
Heinsch FA, Andrews PL (2010) BehavePlus fire modeling system, version 5.0: design and features. Gen Tech Rep. RMRS-GTR-249. USDA Forest Service, Rocky Mountain Research Station, Fort Collins, CO, 111 pp
Hough WA, Albini FA (1978) Predicting fire behavior in palmetto-gallberry fuel complexes. Research 4er SE-RP-174. USDA Forest Service, Southeastern Forest Experiment Station. Asheville, NC, 48 pp
Latham DJ, Schlieter JA (1989) Ignition probabilities of wildland fuels based on simulated lightning discharges. USDA Forest Service Research Paper INT-411, 16 pp
Liu Y, Hussaini MY, Ökten G (2015) Global sensitivity analysis for the Rothermel model based on high-dimensional model representation. Can J For Res 45:1474–1479

Mancilla-Leyton JM, Vicente M (2012) Biological fire prevention method: evaluating the effects of goat grazing on fire-prone Mediterranean scrub. For Syst 21:199–204
Morvan D, Dupuy JL (2004) Modeling the propagation of a wildfire through a Mediterranean shrub using a multiphase formulation. Combust Flame 138:199–210
Osem Y, Ginsberg P, Tauber I et al (2008) Sustainable management of Mediterranean planted coniferous forests: an Israeli definition. J For 106:38–46
Osem Y, Zangy E, Bney-Moshe E et al (2012) Understory woody vegetation structure in manmade Mediterranean pine forests: variation in community structure along a rainfall gradient. Eur J For Res 131:693–704
Osem Y, Fogel T, Moshe Y et al (2017) Understory structure and function following release from cattle grazing and overstory thinning in Mediterranean conifer plantations. Ann For Sci 74:22
Osem Y, Zangy E, Ashkenazi M et al (2019) Long Term Ecological Research (LTER) site for studying the management of planted forests in Israel: the effects of thinning and understory treatments on forest vitality and health, natural regeneration and biodiversity in a planted *Pinus halepensis* forest. Annual Scientific Report submitted to the Forestry Division, KKL Eshtaol
Ottmar RD, Sandberg DV, Riccardi CL et al (2007) An overview of the fuel characteristic classification system – quantifying, classifying, and creating fuel beds for resource planning. Can J For Res 37:2383–2393
Papió C, Trabaud L (1990) Structural characteristics of fuel components of five Mediterranean shrubs. For Ecol Manag 35:249–259
Pasalodos-Tato M, Ruiz-Peinado R, del Río M et al (2015) Shrub biomass accumulation and growth rate models to quantify carbon stocks and fluxes for the Mediterranean region. Eur J For Res 134:537–553
Porté A, Trichet P, Bert DG et al (2002) Allometric relationships for branch and tree woody biomass of maritime pine (*Pinus pinaster* Aït.). For Ecol Manag 158:71–83
Rothermel RC (1972) A mathematical model for predicting fire spread in wildland fuels. Research Paper INT-115 of the USDA Forest Service, Intermountain Forest and Range Experiment Station, Ogden, UT, 40 pp
Rothermel RC (1991) Predicting behavior and size of crown fires in the northern Rocky Mountains. USDA Forest Service Research Paper INT-438
Schroeder MJ (1969) Ignition probability. USDA Forest Service Office Report 2106-1. USDA Forest Service, Rocky Mountain Forest and Range Experiment Station, Fort Collins, CO, 27 pp
Schwilk DW (2015) Dimensions of plant flammability. New Phytol 206:486–488
Scott JH, Reinhardt ED (2001) Assessing crown fire potential by linking models of surface and crown fire behavior. Research Paper RMRS-RP-29. USDA Forest Service, Rocky Mountain Research Station, Fort Collins, CO, 59 pp
Shipley B, Vu T-T (2002) Dry matter content as a measure of dry matter concentration in plants and their parts. New Phytol 153:359–364
Ungar ED, Noy-Meir I (1988) Herbage intake in relation to availability and sward structure: grazing processes and optimal foraging. J Appl Ecol 25:1045–1062
Van Wagner CE (1977) Conditions for the start and spread of crown fire. Can J For Res 7:23–34
Van Wagner CE (1989) Prediction of crown fire in conifer stands. In: MacIver DC, Auld H, Whitewood R (eds) 10th conference on fire and forest meteorology, April 17–21, 1989, Ottawa, Ontario, pp 207–212
Van Wagner CE (1993) Prediction of crown fire behavior in 2 stands of jack pine. Can J For Res 23:442–449
Wilson R (1980) Reformulation of forest fire spread equations in SI Units. Research Note INT-292 of the USDA Forest Service, Intermountain Forest and Range Experiment Station, Ogden, UT, 5 pp
Wilson RA (1990) Reexamination of Rothermel's fire spread equations in no-wind and no-slope conditions, USDA Forest Service Research Paper INT-434. Intermountain Research Station, Ogden

Woodall CW, Monleon VJ (2008) Sampling protocol, estimation, and analysis procedures for the down woody materials indicator of the FIA program. Gen. Tech. Rep. NRS-22. USDA Forest Service, Northern Research Station Newtown Square, PA, 68 pp
Zahavi M, Ungar ED, Svoray T (2012) Uniformity of spatial utilization by small ruminant flocks in a family farm. Ecol Environ 3:78–83. (in Hebrew)

Chapter 25
Post-fire Restoration of Mediterranean Pine Forests

Jorge Castro

25.1 Introduction

Ecological restoration may be defined as a process of intentional modification of a habitat to establish a defined ecosystem that is considered natural and/or historical. In doing so, management based on ecological restoration should seek to create or promote the natural elements that allow the functioning of the ecosystem. This is in clear contrast to, for example, conventional reforestation: ecological restoration may seek to regenerate a forest but, for that undertaking, it might not be necessary to plant a single tree if we can promote the functioning of a particular process (e.g. seed dispersal) that leads to the recovery of the forest. Ecological restoration should focus on imitating and recovering the structure and function of the ecosystem. Additionally, this will mitigate the impact on the managed ecosystems (thereby clearly helping to accelerate its regeneration) and reduce the cost. In this chapter, I will focus on the post-fire restoration of Mediterranean pine forests, from the approach that views ecological restoration as the vehicle to achieve this aim. Also, as this review is devoted to forest restoration, I propose as a primary goal the restoration of ecosystem function and structure in order to provide ecosystem services beyond timber exploitation. This does not mean that timber production cannot be a goal, but one subordinated to other ecosystem services derived from soil protection, water regulation, biodiversity, increased resilience or resistance to future disturbances, and recreational, aesthetic or spiritual values. Along the same lines, the potential economic revenue from harvesting burnt wood should be balanced against the potential repercussions of its exploitation. Overall, post-fire pine-forest restoration should consider the larger cycle of costs and benefits, not simply

J. Castro (✉)
Department of Ecology, University of Granada, Granada, Spain
e-mail: jorge@ugr.es

G. Ne'eman, Y. Osem (eds.), *Pines and Their Mixed Forest Ecosystems in the Mediterranean Basin*, Managing Forest Ecosystems 38,
https://doi.org/10.1007/978-3-030-63625-8_25

economics – that is, ecological and societal implications within the context of ecosystem services (Leverkus and Castro 2017; Thorn et al. 2018; Leverkus et al. 2020).

In this chapter I will consider both natural as well as planted pine forests. During the last century, managers planted pines over a large area in Mediterranean-type ecosystems for different purposes (Allen 2001; Pausas et al. 2004), mainly soil protection and timber production. In many cases, those reforestations ultimately proved unprofitable for the logging industry and were not managed over the medium or long term, creating monospecific and coetaneous stands with a high risk of fire. Although any pine forest in Mediterranean ecosystems may undergo fires and may need restoration, the differentiation between natural and planted stands may have implications for management. Whereas post-fire pine regeneration may be necessary in an area where a particular species is naturally distributed, it might be preferable to replace the pines with other species in areas where those pines were planted outside their natural distribution range. In many cases, pine plantations were imposed on areas where the potential vegetation should be dominated by oaks (Blondel et al. 2010). In addition, pines and oaks are two major natural components of many Mediterranean forests, and can coexist or replace each other depending on successional dynamics (Zavala et al. 2000; Allen 2001; Blondel et al. 2010). Thus, I will consider not only pines but also oaks as key elements for post-fire pine-forest restoration.

Finally, I should highlight that my aim is not to offer a guide of technologies that could be used for restoration. Rather, I will discuss the processes that can help us to work with nature. If, for instance, we know that the reduction of water stress will aid restoration success, I will focus on the factors that may help to reduce water stress by using the natural elements of the post-fire ecosystem. Of course, if we could irrigate the plants, add hydrogels, or use whatever other appropriate technology, we might (perhaps) further ensure establishment success. But this lies outside the scope of this review. Actually, although it is unquestionable that technologies may facilitate forest restoration, I feel reluctant to base our success on them unless they work in the same direction as natural processes. In fact, I believe that too often managers have based their approach on technology without considering ecosystem functioning, resulting in too many failed attempts to recover perturbed forests.

25.2 The Scene: Why We Do What We Do

The management measures taken to correct complex environmental problems almost always have a long history, and it may be difficult to distinguish between what is done because it is considered the most appropriate and what part of that activity is based on outdated knowledge, even based on beliefs or observations that lack scientific evidence (e.g. Beyers 2004; Wuerthner 2006). This is particularly true for post-fire forest management for several reasons. First, managers are impelled by society to repair the damage as fast as possible, which may force the adoption of measures based more on public expectations than on ecological arguments. Second, management efficiency is not usually judged, as the results are so long term that accountability for decisions is unusual, making feedback even more rare. Third, many hazards can ruin the best of plans over the years (e.g. droughts, frosts, fires, pests, and diseases), making it complex and difficult to judge the appropriateness of past management practices. Given that innovation has a risk of failure (although a lack of innovation also has a risk), managers often hesitate to step ahead of what has been done traditionally. All this encourages the adoption of work routines that are hardly questioned and seldom scrutinized for their efficiency.

The construction of log erosion barriers is a clear example. There is no single term for these structures (e.g. contour-felled log erosion barriers, brushwood dikes, etc.; Marques and Mora 1998; Robichaud et al. 2000), but overall they consist of logs and/or branches (single or piled) anchored to the ground by a few vertical stakes (Fig. 25.1). The purpose of these barriers is primarily to reduce erosion, and local forest services consider them to be one of the most urgently needed measures after fires in Mediterranean pine forests, to the point that their implementation is even regulated by specific directives (e.g. Myronidis and Arabatzis 2009). Nonetheless, their construction implies additional disturbance, including the intense trampling and dragging of logs at the sites near the barrier, which in turn might aggravate erosion (Fig. 25.1a). Thus, despite their broad implementation, we should ask what scientific evidence, properly tested, verifies their usefulness.

In 2005, Raftoyannies and Spanos published one of the first studies that evaluated the effect of log and branch erosion barriers on regeneration in a burnt pine forest in the Mediterranean Basin. Their conclusion was that the barriers were ineffective for post-fire recovery of the vegetation (see also Fig. 25.1b). Marques and Mora (1998) also concluded that similar structures failed to reduce erosion in a burnt area of south-eastern Spain. Robichaud et al. (2000, 2008, 2010) have presented comprehensive reviews evaluating the effectiveness of post-fire rehabilitation treatments, including log erosion barriers. They consistently conclude that no clear consensus confirms the usefulness of these structures, for which the effect varies from positive to negative, and depends on a complex set of site characteristics such as climate, soil type and texture, slope, sporadic torrential storms, installation quality, etc. Moreover, most of the reports offer qualitative assessments, based on managers' experiences and impressions, but not quantitative data (Robichaud et al. 2000).

Studies published over the last decade continue to report contrasting results, with a clear negative effect of log and branch barriers in some cases, positive in others, and inconclusive in still others (reflecting a shifting effect depending on year, site, conditions, etc.) in a roughly similar proportion (e.g. Fox 2011; Aristeidis and Vasiliki 2015; Badía et al. 2015; Fernández and Vega 2016a; Fernández et al. 2019). Also, I note that some of the studies assuming positive effects were conducted without true replication at the scale of management operations or without a randomized allocation of treatments (e.g. Lee et al. 2014; Gómez-Sánchez et al. 2019), a common problem in post-fire management evaluation given that researchers often have to work with what has been left after management by the forest services or timber companies, without a previous, planned experimental design (Leverkus et al.

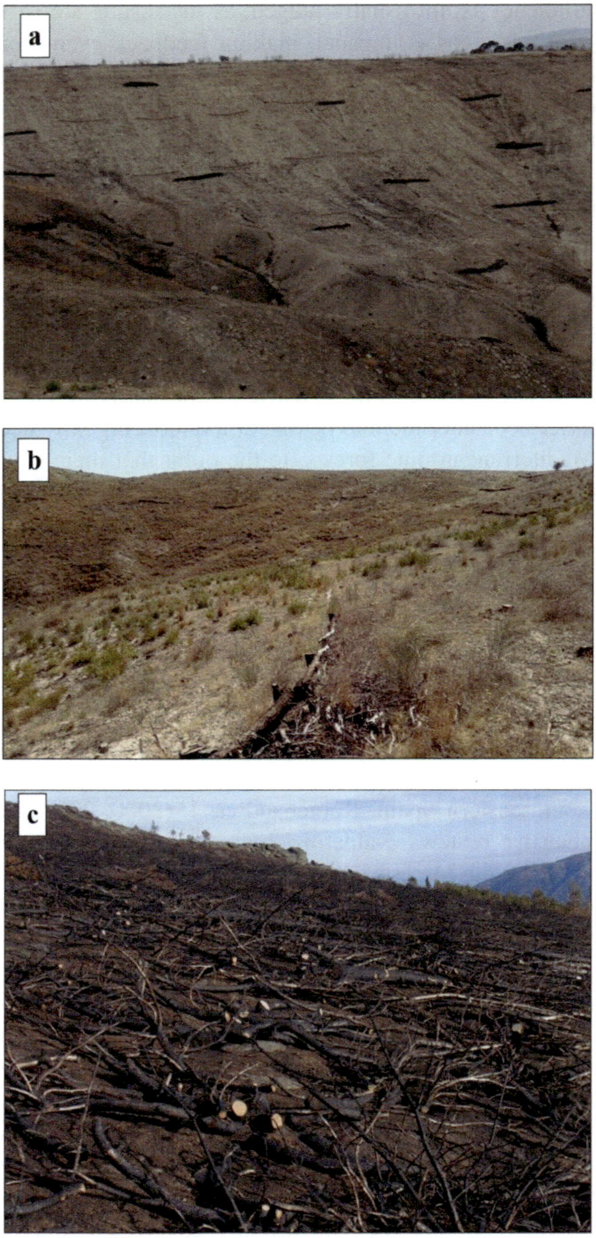

Fig. 25.1 Intense management of the burnt wood is common after forest fires, in an effort to restore the ecosystem. However, the effectiveness of these actions has not been sufficiently tested in many cases, and may even result in a negative impact. (**a**) The construction of log erosion barriers may increase erosion instead of reducing it; erosion rills may be detected under most of these barriers (Quesada-Huesa fire, Spain, year 2015). (**b**) Log erosion barriers are often constructed at a large spatial scale even though their usefulness is not clear; here we may see how, in addition, there are no apparent differences in plant cover produced by the barriers 2 years after their construction (Quesada-Huesa fire, Spain, year 2015). (**c**) Leaving logs and/or branches spread on the ground creates a dense mesh that reduces water splash and runoff, which might be used instead of log erosion barriers (Lanjarón fire, Spain, year 2005). (Photos (**a**) and (**b**) courtesy of Agustín Bermejo (Montevivo S.L.))

2018a). Moreover, the control treatment is frequently an area where the barriers have not been constructed, but where the burnt wood has been similarly removed by salvage logging (i.e. removal of trunks; see definition in Sect. 25.4). In reality, this is not a control; rather, the effect of erosion barriers constructed – in order to reduce erosion – in areas where the burnt trees have been cut and removed, should be compared with areas where trees are left *in situ*, constituting a true non-intervention area. A Master's thesis conducted recently in the Department of Soil Sciences at the University of Granada (Spain) compared erosion in non-managed burnt pine stands versus stands where log erosion barriers were constructed after salvage logging, while controlling for slope and other site characteristics (Martín-Lorenzo 2017): soil erosion was 7.5 times higher in the stands with the barriers. Additional soil disturbance by both logging operations and barrier construction was likely the reason for the increased erosion. In summary, log erosion barriers offer no consistent benefit, their construction implies an additional disturbance that may further hamper the recovery of the ecosystem, and they are expensive and technically demanding. Despite all these drawbacks, they are still widely implemented in burnt Mediterranean pine forests.

However, the most surprising aspect of all this is not whether the log and branch barriers are useful or not. Let us assume for a moment that they are and that their construction will help reduce erosion and regenerate vegetation. In this case, it might be easier and cheaper to leave the remains of burnt vegetation scattered on the ground. It would suffice to cut down the trees (not even necessarily to cut all of them), maybe lopping the main branches to increase contact with the ground (Fig. 25.1c; see also Sect. 25.8). Or perhaps leave the branches scattered on the ground if logs are removed for commercial purposes. Both coarse and fine woody debris, when spread over the ground, create an intricate mesh that buffers rainfall splash and the kinetic energy of the water, retaining sediment. Thus, the question arises as to why managers should spend money and disturb the site even more if something equivalent could be accomplished with far less impact and cost. Many managers would argue that what appears in Fig. 25.1c is inadmissible because it hinders the movement of personnel and machinery, or because of pest or fire risk. However, such arguments are controversial. This is precisely the starting point for the post-fire restoration of forests (either pines or other forests), what to do with the burnt wood, a biological legacy with enormous relevance in the post-disturbance succession, that represents the largest portion of the post-fire biomass.

25.3 Burnt Wood as a Biological Legacy

The post-fire pine-forest scenario is dominated by burnt wood, usually at high log density and with most burnt trees still upright, characteristics that usually apply both to natural pine forests as well as to pine plantations. This burnt wood will collapse and decompose over time (Molinas-González 2017a, b; Fig. 25.2a), during which it hosts a diverse array of organisms and fosters multiple processes, including

Fig. 25.2 Burnt wood is a biological legacy with an important role in post-fire ecosystem functioning and recovery. (**a**) Through decomposition it releases its nutrients, increasing soil fertility in the medium to long term. It is also a resource for many organisms, such as the xylophagous insects

forest regeneration. This scenario has far-reaching implications for management. Thus, what to do with this wood is a critical first step in managing post-fire regeneration and restoration.

From an ecological standpoint, the burnt wood is a biological legacy of enormous relevance. To begin with, a tree undergoes continuous growth, accumulating biomass and, with it, nutrients. A tree is ultimately a sink of nutrients drawn from the soil and exported to the trunks, branches, and leaves. During a fire, a large portion of these nutrients are volatilized and released to the atmosphere or transferred to sediment (Whelan 1995), but a large portion remains locked in the wood. In this sense, Marañón-Jiménez et al. (2013a) documented that the macro- and micronutrient contents in trunks of burnt pine proved similar to concentrations in live specimens. Furthermore, the nutrient pool remaining in burnt wood was 2–9 times higher than in the first 10 cm of the soil for elements such as Na, Mn, Fe, Zn or Cu (Marañón-Jiménez and Castro 2013). These nutrients, being protected from sudden losses from runoff or wind after the fire, are gradually released into the soil during decomposition, enhancing soil fertility for the recovering vegetation. For example, Marañón-Jiménez et al. (2013b) found that the nutrient pool in 3-year-old *Pinus pinaster* Ait. seedlings that had regenerated after a fire was higher for plants where all burnt wood was left *in situ* compared with areas where trees were removed. The reproductive status of the pines was also better after 6 years, and the analysis of stable isotopes pointed to greater water-use efficiency in pines growing in the presence of burnt wood due to better nutritional status. The loss of soil fertility as a consequence of burnt-wood removal may last for decades after post-fire management (Bowd et al. 2019).

The burnt wood also acts as a nurse object. Mediterranean ecosystems are characterized precisely by the abundance of facilitative interactions among plants (Gómez-Aparicio et al. 2004; Gómez-Aparicio 2009; Siles et al. 2010a), which involve the mitigation of abiotic stress for seedlings by growing under the shelter of other plant species, especially shrubs. A fundamental aspect to consider is that facilitation in Mediterranean-type ecosystems arises primarily from the reduction of high radiation intensity by the canopy of the nurse plant, which improves water status in the beneficiary plants by, for instance, reducing temperature, wind speed, and increasing soil moisture and air water vapor content. However, at the soil level a negative outcome of the interaction may occur due to competition for resources between the nurse plant and the facilitated seedling (Maestre et al. 2003; Gómez-Aparicio et al. 2005; Rey-Benayas et al. 2005). In this sense, the burnt wood can act as a nurse object of extraordinary quality, as it reduces water stress without underground competition (Castro et al. 2011; Ginzburg and Steinberger 2012; Marcolin

Fig. 25.2 (continued) whose larvae made the holes that can be seen in the picture (log 7 years after the fire). (**b**) Burnt wood acts as an excellent nurse object in Mediterranean-type ecosystems, as it reduces radiation intensity and water stress, and provides nutrients. Seedling recruitment is often higher if burnt wood is left *in situ*. (**c**) Burnt wood may also act as a physical structure that protects seedlings and saplings against herbivores; a 3 years old *Pinus pinaster* sapling growing protected by a barrier of burnt branches. (Photos from the Lanjarón fire (Spain)) et al. 2019: Fig. 25.2b). Moreover, the wood provides nutrients during decomposition, as mentioned above, and can retain moisture, organic matter and fine sediments by acting as a barrier against runoff. Burnt trees also reduce wind speed (Serrano-Ortiz et al. 2011), and can act as a physical barrier against herbivores (Castro 2013; Fig. 25.2c). All this may promote post-fire regeneration in the presence of burnt wood in contrast to areas where the wood is removed (Castro et al. 2011; Marzano et al. 2013; Taboada et al. 2018; Urretavizcaya and Defossé 2019).

The presence of burnt wood can also favor key mutualistic interactions for the natural regeneration of the forest after a fire, either for autosuccession (stand replacing fires) or for succession towards other species composition, such as oaks or mixed forests. Many frugivorous birds use standing burnt trees, felled trees or piles of branches as perches (e.g. Castro et al. 2010; Rost et al. 2010). This increases the entry of tree and shrub seeds from intermediate stages of succession – many of them producing fleshy fruits – with respect to areas where burnt wood has been removed (e.g. Rost et al. 2010; Cavallero et al. 2013). A noteworthy case of mutualistic interaction for post-fire regeneration is the dispersal of acorns by the European jay (*Garrulus glandarius* L.). This bird is the main disperser of acorns in Mediterranean forests, acting as a scatterer-hoarder that caches seeds in a large number of locations across the landscape for later consumption. A fraction of the acorns may remain un-recovered, providing the opportunity for seed germination and tree recruitment (Pesendorfer et al. 2016). The Eurasian jay dwells in relatively dense forests and avoids open landscapes, and in fact pine forests, including dense plantations, are among the habitats most used by jays to hide acorns (Mosandl and Kleinert 1998; Gómez 2003). Notably, the Eurasian jay uses the burnt pine forest to cache acorns (Castro et al. 2012), perhaps because burnt trees still provide protection from predators or provide visual landmarks for the jay to memorize the position of hidden acorns. As a result, areas with standing burnt trees act as an attractant for the dispersal of acorns, with a subsequent increase in seedling recruitment compared to sites where logs were extracted. This interaction is lost if trees are removed, and therefore post-fire burnt wood elimination reduces the natural regeneration of oak species (Leverkus et al. 2016).

So far, the above-mentioned mechanisms by which the burnt wood may favor natural regeneration are based on a physical component – because it reduces abiotic stress (facilitation), provides nutrients through decomposition, or creates a landscape that attracts seed dispersers. However, the role of burnt wood as a biological legacy goes beyond its physical features, as the wood itself is a resource for many organisms that simultaneously form a complex food chain: decomposers such as fungi and bacteria, plus the wood itself, are the resource for extensive and complex communities of arthropods, which simultaneously provide resources for reptiles, birds, mammals, and the trophic interactions they maintain (Brown et al. 2003; Cobb et al. 2010; Molinas-González et al. 2019). As such a resource, the burnt wood increase biodiversity, and as a physical structure that increases habitat complexity, it further helps to increase niches and biodiversity (e.g. Castro et al. 2010; Leverkus et al. 2014). For example, Leverkus et al. (2014) found that burnt wood retention augmented the diversity of woody plant species, and recent global studies support the ability of burnt wood to sustain biodiversity and ecosystem services (Thorn et al. 2018; Leverkus et al. 2020). In short, burnt wood, in Mediterranean environments at least, acts as a biological legacy that reduces environmental stress and promotes biodiversity as well as natural regeneration.

25.4 First Management Actions: Post-fire Salvage Logging

Despite the potential benefits that burnt wood may provide for regeneration (and for ecosystem functioning in general), a common and widespread post-fire management practice – particularly in the case of conifer forest fires in Mediterranean ecosystems – is salvage logging, i.e. the removal of burnt trunks, often including elimination of the remaining woody debris by chopping, mastication or burning (Ne'eman et al. 1997; Castro et al. 2010). The reasons given for removing burnt wood vary in different regions of the world. A major justification is to salvage the natural capital still available in the forest (Akay et al. 2007; Castro et al. 2010). In the case of non-commercial pine forests, the need for salvage logging is still defended for other reasons, arguing that it: (i) favors future work in the area by facilitating the transit of personnel and machinery, (ii) avoids the risk for people and goods derived from possible future accidents caused by falling trees, (iii) reduces fire risk and the fuel load for future fires, and (iv) reduces the risk of pests that might affect the unburnt or partially burnt portions of the forest. In addition, a more emotional than ecological stance (although no less important for that reason), argues for burnt-wood removal to alleviate the visual and emotional shock to the public, especially the local population who, in addition, frequently demand actions for forest restoration (Castro et al. 2010; Zituni et al. 2019).

These reasons have, however, been questioned by studies performed in different parts of the world, especially over the last 20 years, which have shown that the arguments used for post-fire salvage logging depend on the context and are not always justified. For example, a causal relationship between the presence of burnt wood and the increased incidence of fire has not been demonstrated (McIver and Starr 2000; Thompson et al. 2007). Reasons other than fuel load such as wind regime, access, or population density may be more important for fire risk (Kulakowskii and Veblen 2007; Thompson et al. 2007), and even greater fire risk has been suggested after wood removal (Donato et al. 2006). On the other hand, the fire risk of burnt wood remaining on site may be managed without the need to remove all the wood, as for instance creating areas with different salvage intensity, breaking the connectivity between coarse woody debris at stand level, etc. In the end, fire risk elimination in Mediterranean-type ecosystems is impossible, and it will persist at different times and under different forms, even if burnt wood is removed (e.g. Thompson et al. 2007). Moreover, plant regeneration after the fire will create new fuels, with or without salvage logging, at the same time that burnt wood keeps decomposing, reaching a point at which the fire risk of salvaged and non-salvaged areas will converge (McIver and Ottmar 2018). Thus, the benefits of post-fire salvage logging in relation to fire risk reduction should be balanced against its negative impact on other ecosystem processes and services (Mauri and Pons 2019). The risk related to wood-boring insects depends more on the presence of partially damaged trees than burnt ones (Toivanen et al. 2009; Santolamazza-Carbone et al. 2011; Mauri and Pons 2019), which would not justify total removal as a general solution. The emotional reasons related to the public can be addressed with communication and environmental education, explaining to the populace the ecological and forestry rationale for not removing the burnt wood after the fire. Finally, accidents can be avoided by taking safety measures such as cutting down burnt trees in the most visited or trafficked areas (e.g. burnt trees bordering paths and roads) or carrying out restoration when there is less risk of snag fall. Overall, salvage logging represents an additional disturbance that follows the fire, thereby increasing the impact on the ecosystem (Hernández-Hernández et al. 2017; Leverkus et al. 2018b). In summary, researchers have recently found strong evidence for the negative impact of post-fire salvage logging (Thorn et al. 2018; Bowd et al. 2019; Leverkus et al. 2020) and, in contrast, compelling evidence for the ecological benefits of burnt wood (Sect. 25.3). Thus, management actions including the retention of burnt wood, either total or partial, should be considered in post-fire forest restoration.

25.5 Chopping, Mulching, and Other Ways to Eliminate Woody Debris

Another common post-fire management approach involving burnt wood is the mastication of non-merchantable debris creating small size wood pieces ("woodchips"), often spreading the product over the soil surface as a mulch in order to reduce erosion (Fig. 25.3a). A mulch cover may in fact be an effective way to reduce post-fire soil erosion (Robichaud et al. 2010; Prats et al. 2019). However, post-fire mulching is often presented as a necessary management method to reduce erosion in places where intense management has previously employed heavy machinery, usually salvage logging (e.g. Prats et al. 2019). Ground-based logging equipment and salvage logging operations compact the soil and may delay the recovery of vegetation (Leverkus et al. 2014; Wagenbrenner et al. 2016; García-Orenes et al. 2017), reducing infiltration and facilitating runoff and erosion with respect to unlogged sites (Wagenbrenner et al. 2016; Lewis et al. 2019). The implementation of mulching in these circumstances is usually positive, but this is a remedy to counteract the negative impact of previous actions.

Mulching can be conducted with *in situ* material (from the burnt forest) or with foreign material (from outside the burnt forest). In the first case, mulching inevitably requires the use of heavy ground-based machinery, with its associated impact. In the second case, mulching can be done from the air (e.g. straw spread with helicopters), a measure that can be applied prior to any further management action and without soil compaction. Straw application has received strong criticism for its

Fig. 25.3 (**a**) Mastication with heavy machinery (like the rotary chain mulcher used to chop the branches in this picture) destroys part of the seedling bank and resprouts, reducing the regeneration potential. (**b**) However, mastication and mulching are often done routinely as the final point to eliminate the woody debris, even if there is no erosion risk (as in this terraced area), and even if the woody debris does not comprise a significant fuel load: the branches in the picture, of small size and diameter, will reduce water splash and erosion, and will be virtually decomposed in 3–4 years. (**c**) Woody debris is even eliminated by burning, thereby ignoring their potential use as biological legacies to prompt regeneration. (**d**) Wood pile burning has a long-lasting effect on regeneration: gaps where piles were burnt persist without regeneration for years. (Photos (**a**) – (**c**): Lanjarón fire, 2005; photo (**d**): a large fire in Seville and Huelva provinces, Spain (usually referred to as "Aznalcóllar" fire), 2004)

potential impact on the regeneration of woody plants or the introduction of invasive species (Beyers 2004; Shive et al. 2017; Bontrager et al. 2019), but positive effects have also been widely reported provided that the mulch cover and depth do not exceed certain limits (e.g. Dodson and Peterson 2010; Bontrager et al. 2019; Jonas et al. 2019; Lucas-Borja et al. 2019). Note that mulching with *in situ* materials prior to salvage logging has also been reported to have a positive impact in controlled experiments (Fernández and Vega 2016b). However, the problem here is how to implement broad-scale mulching with local material (mastication of burnt wood) without ground disturbance from transit and heavy machinery.

All in all, although different types of mulching may reduce soil erosion, it is not yet clear whether this practice offers advantages over other less aggressive ways to handle woody debris. Most studies focus on the effect of mulching on either salvage logged areas or control (without logging) areas. However, intermediate management seeking to keep a certain proportion of burnt wood spread on the ground might act in a similar way to slashing and mulching (e.g. Fig. 25.3b). This situation may be reached from objectives as different as conservationist and commercial ones. For example, in the case of harvesting merchantable logs, the crowns and branches may be cut off and spread on the ground without mastication. In the case of unproductive forests, burnt trees or a portion of them might be felled and their main branches lopped off to create a network of woody debris on the ground (e.g., as shown in Fig. 25.1c). As mentioned above (Sect. 25.3), many studies have reported the positive effect of these biological legacies for forest regeneration due to microclimatic amelioration, nutrient supplies, or protection against herbivores. We still lack a comprehensive comparison of the effect of mulching vs. preservation of post-fire biological legacies for forest regeneration. Nonetheless, the second option has the advantage of almost guaranteeing reduced cost (ultimately mulching is often implemented as the last step for the mastication of woody debris). In the end, the key factor for protecting the soil is clearly plant cover (Ne'eman et al. 1997), and post-fire plant cover regeneration has enormous potential in Mediterranean-type ecosystems through seeder and resprouter species (Médail 2008; Keeley et al. 2012; Leverkus et al. 2014). Post-fire management intended to promote forest restoration should therefore focus on promoting the regeneration of plant cover. In this regard, mulching activities that hamper regeneration (e.g. mastication with a rotary chain mulcher; Fig. 25.3a) would be counterproductive. Furthermore, at least in the case of the Mediterranean Basin, soil erosion after fire may not be critical for the environment (e.g. Ne'eman et al. 1997), especially when compared to usual erosion levels in agricultural lands, and in addition tend to stabilize (reaching values similar to pre-fire conditions) within a period of 3–10 years (Shakesby 2011). Thus, it is questionable whether any effort to stop erosion that might affect the ecosystem in a different way is worthwhile. Strong intervention to prevent soil erosion should be limited to very specific, sensitive cases, where the application itself has less impact, for instance at the edges of roads.

A question remains concerning why wood mastication is used. Arguments favoring mulch cover to reduce soil erosion may be reasonable, as discussed above, but the mastication of the woody debris (and consequent mulching) is often also applied in situations of low erosion risk (Fig. 25.3a, b). As indicated in Sect. 25.2 for log erosion barriers, the *raison d'être* for burnt-wood mastication is frequently a mixture of arguments coming from a complex set of situations that do not necessarily apply to post-fire restoration. I have heard managers argue that wood chopping reduces the risk of spreading pests. This might make sense in the case of live woody debris, such as ravages after windstorms, but not in the case of burnt branches that are usually completely charred, hence not material in which wood-boring insects thrive. Another important reason provided for mastication is the reduction of surface fuel loads (Brown et al. 2003), especially seeking to reduce the amount of fuel that could ignite a crown fire in the future. However, I note that: (i) the efficiency of mastication for this purpose depends on complex factors (including decay rate, snow load, amount of woody debris), (ii) the management of the understory regeneration may be more relevant compared to burnt vegetation remains in the context of crown fire prevention, and (iii) the volume of the remaining woody debris may be reduced without the need to reach management as intense as mastication. In addition, mastication plus the spreading of the chips (i.e. mulching) may increase the risk of ignition, as indicated above (Donato et al. 2006; Leverkus et al. 2020), producing the opposite effect to what was planned. The negative perception of coarse and fine woody debris is so deeply rooted in standard management policies that the debris may even be eliminated by burning in case chipping is not possible (Fig. 25.3c). Apart from making no sense, this approach has a strong negative impact on the soils (Rhoades and Fornwalt 2015; Fig. 25.3d).

25.6 Planting or Sowing?

When reforestation is needed to restore a burnt area, a key decision is whether to transplant seedlings previously grown in a nursery or to sow seeds directly ("direct sowing" or "direct seeding": the process of sowing tree seed by hand or machine directly onto the final growing position for the woodland). Planting of nursery-grown seedlings (either bare-rooted seedlings or, more commonly, cultivated in containers) is currently the most widespread method used in restoration and reforestation. Plant-production methods in nurseries have developed considerably in recent decades, and today a plethora of procedures are available to enhance post-planting success, including aspects related to the substrate, plant quality (nutritional status, hardening, morphology), mycorrhization, and the shape, size, and structure of containers (e.g. Kildisheva et al. 2017). However, the success of reforestation in Mediterranean-type ecosystems is often low, with high mortality rates and poor seedling or sapling growth (del Campo et al. 2007; Ruthrof et al. 2013; Kildisheva et al. 2017). Unfortunately, few published data from reforestations verify this impression, since programmed long-term surveys of planting success are rare (Kildisheva et al. 2017). Most data come from academic studies, which may overestimate success since (i) the plants used are often part of experiments that seek to improve the reforestation; (ii) experimental plantations are likely to be conducted with more care and attention than large-scale reforestation; and (iii) data from such studies are often available for only the first or second year after reforestation. In fact, results from studies conducted in southern Spain in the last 20 years already suggest low effectiveness of reforestations. For example, Castro et al. (2004a) reported survival of 10% for *P. sylvestris* L. and 36% for *P. nigra* J.F. Arn. after 4 years in a reforestation attempt made by the local Forest Service (data for seedlings planted according to the protocol used by the Forest Service). Approximately 60% of those pines finally died 4 years later during an episode of severe summer drought (Herrero et al. 2013), virtually eliminating the reforestation success. In another study, Castro et al. (2006a) found survival values of only 5% for *Quercus pyrenaica* Willd. seedlings 6 years after planting (seedlings planted following the protocols of the Forest Service). Leverkus et al. (2015a) reported survival rates of around 55% for *Q. ilex* L. and 45% for *Q. pyrenaica* after 2 years. However, 7 years after planting, the survival rate did not exceed 10% (Castro J., unpublished data). It is worth noting that in all these studies the reforestation was performed by workers of the Forest Service and the studies monitored several thousands of seedlings, the work being therefore representative of a realistic management situation.

A major reason for the low success of seedling reforestations could be poor root-system development as a result of growing in a container, currently the most common cultivation procedure. This has been studied especially for species that develop a tap root – a long, central dominant root growing directly downward from which the rest of the root system sprouts laterally – as in oaks. In the case of nursery-grown oak seedlings, the tap root may be damaged or anomalous-shaped when grown in containers, or pruned in the case of bare-root transplanted seedlings (Fig. 25.4a). This leads to a shallower root system with less access to soil moisture, a result of particular relevance in water-stressed habitats such as Mediterranean-type environments (Pemán et al. 2006; Löf et al. 2019). In fact, it has been suggested that the negative consequences of the anomalous development of the root system derived from planting instead of direct sowing may persist until the adult age of the tree (Zadworny et al. 2014).

The problem of abnormal development of the tap root in oaks can be overcome by direct sowing of the acorn, provided that the seed is protected from predation (see below). In this case, the tap root develops *in situ*, penetrating the ground (even more than 1 m in the first growing season) and reaching access to more permanent and abundant sources of water and nutrients (Löf et al. 2019). Direct sowing in fact has other advantages vs. planting, such as a much lower impact on soil and vegetation during working operations, more flexibility in terms of timing, the possibility of restoration in areas where seedling planting is too costly or difficult, less transference of diseases from nurseries to the field, and lower economic costs (see Löf et al. 2019 for a recent review). We should consider that the success of oak reforestation by direct sowing is not low, with seedling emergence values often exceeding 50%, and seedling survival often exceeding 50% after several growing seasons (e.g. Navarro et al. 2006; Mendoza et al. 2009; Prévosto et al. 2011; Matías et al. 2012). Nonetheless, a key point to make direct sowing a feasible restoration method is to protect the acorns from predators (Castro et al. 2015). Today several techniques have proved efficient for this purpose (summarized in Löf et al. 2019; Fig. 25.4b), and thus there are no substantial reasons not to undertake oak restoration by direct sowing. Also, individual acorn protectors (mechanical devices that could protect acorns from rodents but not from ungulates), in combination with the use of burnt logs and branches that act as physical barriers against ungulates, have been demonstrated to guarantee low predation rates of the sown acorns (e.g. Leverkus et al. 2015b). Thus, once again, burnt woody debris may provide a service that would disappear if masticated or removed.

In the case of pines, it is even more widely assumed that planting is more appropriate than direct sowing. In large areas of Mediterranean-type ecosystems (the Mediterranean Basin being a paramount example), vast pine afforestations and reforestations growing today were planted in the twentieth century with bare-rooted transplanted seedlings. This provides an argument in favor of this method. Also,

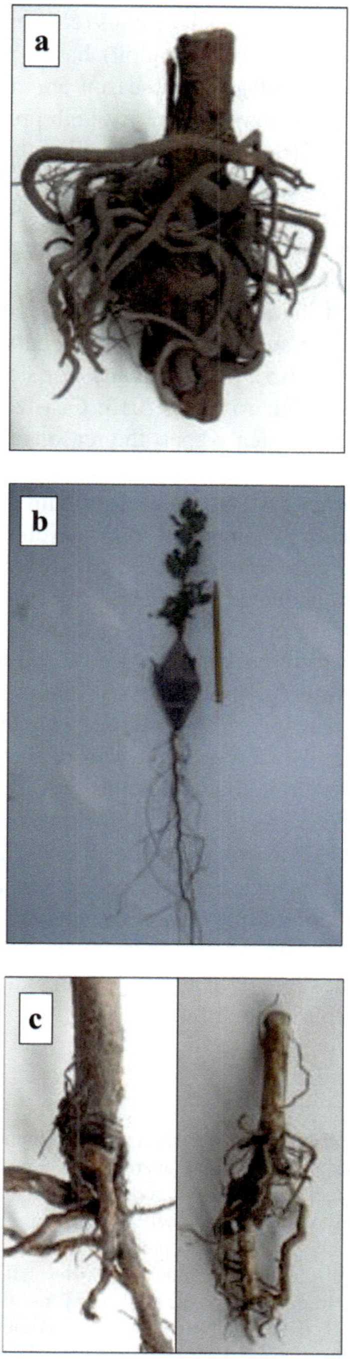

Fig. 25.4 (**a**) Root system of a Holm oak (*Quercus ilex*) 5 years after being transplanted in the field. The plant was cultivated in a nursery in containers with an antispiral system. Note that the

pine seedlings are tiny compared to oak seedlings (creating the impression of empty terrain, even several years after the reforestation), have a much higher mortality rate (values above 90% after the first growing season are common, e.g. Castro et al. 2004b; Tíscar et al. 2017), and may suffer very high predation rates (Castro et al. 1999; Carrillo-Gavilán et al. 2010). For decades, all of these reasons have undoubtedly impelled practitioners to choose planting over sowing. However, even if the transplanted seedlings survive and become trees, they face problems that might be linked to the reforestation method. Some of the current mature reforestations in Mediterranean-type ecosystems are sharply declining in vigor, and even show episodes of forest dieback (Sánchez-Salguero et al. 2010, 2013). Many interacting factors may underlie these processes, such as gradually harsher climatic stress (hotter droughts, *sensu* Allen et al. 2015), pests, parasites, or steadily more intense neighbor competition (Sánchez-Salguero et al. 2013; Gea-Izquierdo et al. 2019). This complexity makes it difficult to tease apart the reasons for massive tree mortality. However, as mentioned above for oaks, some evidence supports the contention that poor development of the root system might also be involved, since container pine seedlings are often stunted by abnormal root development after transplanting (Fig. 25.4c). A particular case was documented in one of the above-mentioned reforestation experiments conducted on two pine species in 1997 in Sierra Nevada National Park, Spain (Castro et al. 2004a). As the conservation of local genetic diversity and singularity was of concern, the seeds used were harvested from local trees (in 1995). The seeds were then sown in conventional containers (300 ml volume) and grown for two growing seasons before being transplanted, again following conventional protocols. However, the growth of the planted pines was not as vigorous as the naturally recruited saplings that were present in the area. Later, massive sapling mortality occurred in 2005 following an acute drought (10-year-old saplings; Herrero et al. 2013), provoking the death of ca. 60% of the surviving pines by that time. However, not a single mortality event was detected among naturally recruited saplings of similar age (personal observation). It might be that naturally recruited saplings were already filtered through demography, and perhaps they occupied the best microsites for survival and growth, but an effect of the reforestation method might also explain the difference.

Fig. 25.4 (continued) main root aborts and the subsequent development is deficient. This is a common situation in species of the genus *Quercus* that are grown in nurseries and then transplanted to the field. (**b**) A device to protect acorns against rodent predation for direct sowing in the field, which also allows the tap-root to grow ("seed shelter"; Löf et al. 2019). Restoration of oaks and other late-successional species that coexist with pine species is nowadays feasible by direct sowing. (**c**) Root system of a 5-year-old maritime pine (*P. pinaster*, left) and a 3-year-old Aleppo pine (*P. halepensis*, right) that were transplanted to the field after 2 and 1 (respectively) years of cultivation in the nursery using 300-ml containers. Note how the main root has aborted, and the entire root system is seriously twisted. The plants were grown in flat terrain, so there is no reason to expect this convoluted development. It is possible that abnormal root development originating from seedling growth in containers might affect future development of the tree, as happens in oaks. (Photo (**a**) courtesy of Francisco B. Navarro Reyes

The success and benefits of direct sowing instead of planting have been suggested for pines in recent studies (*P. albicaulis* Engelm., whose seed mass is ca. 155 mg; Pansing and Tomback 2019). Large-seeded species are almost certainly better candidates than smaller-seeded ones, given that larger seed size provides more resources and therefore increases the chances for initial seedling survival (Castro et al. 2006b). Pine species from arid environments show higher seed mass than pines from more mesic environments (Salazar-Tortosa et al. 2020), supporting a role of seed size for restoration of pine forests in Mediterranean-type ecosystems. In this sense, several pine species from the Mediterranean Basin and Mediterranean climate-type areas of California have large seeds, such as *P. pinea* L. (ca. 440 mg), *P. monophylla* Torr. & Frém. (ca. 400 mg), *P. coultery* D. Don. (ca. 350 mg), *P. sabiniana* Douglas ex D.Don. (ca. 800 mg), and *P. torreyana* Parry ex Carr.(ca. 930 mg; see supplementary material in Salazar-Tortosa et al. (2020) for an updated list of seed mass in *Pinus*). For such cases, direct sowing might have advantages over planting seedlings. Moreover, promising results were also obtained for direct sowing of *P. halepensis* Mill. (seed mass ca. 25 mg) in a harsh, post-fire environment when sowing points were covered with woodchips (García-Morote et al. 2017), a result that was attributed to an increase of soil moisture while not reducing light intensity for seedling growth. Research on direct sowing for pine-forest restoration should be therefore considered a priority given the ecological and economic advantages that this can provide.

In this context, it should be noted that direct sowing is a different concept from broadcast seeding, where seeds are spread in large quantities in the ground without precise control of the site where a seed finally ends up. Although broadcast seeding may be an effective method in more temperate and boreal pine forests, it has been reported to be extremely inefficient in Mediterranean-type ecosystems (Retana et al. 2012). Both seed predators and the usual high seedling mortality during the first summer in these environments have been indicated as the main reasons. As in oaks, a restoration plan for direct sowing of pines should consider protection of the seeds against predators and sowing in specific microhabitats such as nurse plants, or using burnt wood as a nurse structure. Similarly, care should be taken regarding seed origin either for planting or sowing, choosing local provenances unless a major reason against this arises in relation to genetic depletion or assisted migration. Note that broadcast seeding has also been used in different contexts within burnt areas, i.e. as an emergency response to stabilize soils and reduce erosion, generally using herbaceous species (thus with a similar objective as mulching, see Sect. 25.5). Nonetheless, different reviews have shown that the results of broadcast seeding are largely unpredictable and, overall, inefficient (Beyers 2004; Peppin et al. 2010).

Finally, it is worth considering that whatever the method used, the focus needs to be shifted from large-scale reforestations that concentrate the effort and budget at a single moment (the time of the reforestation) but pay little attention to the fate of the seedlings in subsequent years, to more individual-based practices that care for plants over a period long enough to ensure establishment (Castro et al. 2021). Seeds or seedlings should be placed in the most promising areas for their survival and growth, both at the micro-scale (e.g. choosing favorable nurse shrubs or nurse

Fig. 25.5 Silvicultural practices for post-fire reforestation should shift the focus from intense interventionism that is concentrated almost exclusively in the first moments after the disturbance to practices that optimize plant development and ecosystem functioning in the medium to long term. (**a**) Site selection for each particular species should be done with accuracy; here, maritime pine (*Pinus pinaster*) reforestation was done without considering fine-scale shifts in environmental conditions. This area has a mix of limestone (foreground) and dolomite (background) at small spatial scales. Reforestation was successful on limestone, but ca. 95% of the seedlings died in the dolomite (burnt areas in Sierra de Cázulas Mountains, Spain). Issues related to genetic provenance of the seeds could be also important in this case. (**b**) A 22-year-old reforestation (*P. nigra* and *P. sylvestris*) in a burnt area where protection against herbivory was not planned; this stand has not rendered a single mature tree, and probably never will (Loma de los Panaderos, Sierra Nevada, Spain; reforestation done in 1995). (**c**) Reforestation by local volunteers (a group of friends, not even organized as a NGO) in a burnt area. They planted maritime pine seedlings at low density but paid attention to tree survival (large tree to the left and those in the distance); seedlings were individually protected against herbivory with wire mesh, and irrigated occasionally during the first summer. These trees are now producing seeds and colonizing the rest of the area (two smaller trees behind the foreground tree), acting as nuclei that boost the expansion of the forest. This reforestation was much cheaper than conventional reforestations (La Peza fire, Spain, 1994). (**d**) Natural regeneration of maritime pine in a burnt area (Sierra de Cázulas, Spain). Here, post-fire salvage logging and other considered urgent actions were implemented soon after the fire, but there is no more funding to manage the stand. Now, the fire risk associated with the dense shrub cover is very high, and a new fire will destroy the regeneration potential of the pines (still too young to have created a sufficient aerial seed bank). Cutting of shrubs around the pine saplings, pruning of the lower branches of the pines and, in general, management directed to preserve and promote the growth of those saplings might be more efficient and cheap in the long term than large scale reforestation

objects) and the macro-scale (e.g. distributing seeds or seedlings at the best locations at the landscape level; Fig. 25.5a). The seed might also need to be protected from seed predators using devices designed for this purpose (Löf et al. 2019). Control of herbaceous competition might become necessary within a small radius around the seedling during the first 2–3 years (Castro and Leverkus 2019). Protection against herbivores over the medium term (at least until the sapling stage) is also crucial, as this factor alone may ruin reforestation efforts even if plants survive and grow properly (Vallejo et al. 2012; Fig. 25.5b). Finally, for the restoration of the entire area, it might suffice to achieve a relatively low density of mature trees (and not necessarily evenly distributed) that boost natural regeneration through seed production, establishment of mutualistic interactions with seed dispersers, etc. (Fig. 25.5c). The effort should be directed to concentrating these plants in the right places throughout the landscape and taking the necessary care during the first years after reforestation to boost seedling performance and subsequently promote natural regeneration through proper ecosystem functioning (Fig. 25.5d). Several approaches have been proposed for this in last decades (Schönenberger 2001; North et al. 2019) but unfortunately they have rarely been put into practice.

25.7 Precision Restoration: The Need to Shift Management Paradigms

As outlined in the previous sections, current post-fire management in burnt-pine forests emphasizes intense interventionism, using a few treatments as soon as possible, and based largely on the use of hard technology (heavy machinery) with a strong impact on ecosystem structure and functioning. Although the main objective is forest regeneration, the main targets of the management are primarily geophysical processes (soil erosion), preparation for future management (hence the effort to clear the landscape), and potential future risks (that nonetheless could be handled in different ways). This often harms natural regeneration and thus implies the loss of potential for restoration. Even when it finally renders a satisfactory result, the hindrance of natural regeneration and ecosystem functioning might well translate into higher costs of the overall restoration plan (e.g. Leverkus et al. 2012; Leverkus and Castro 2017). The first problem of this approach is the unrealistic simplification of the perception of the post-fire ecosystem. It is virtually impossible for a single management (or few alternatives) to be the best solution for most of the area. Post-fire salvage logging is an example, often being conducted throughout the entire burnt area. The second problem is that the efforts and the budget are concentrated in the time immediately after the fire, with hardly any attention (or funding) to the development of such actions. For example, in serotinous pine forests, natural regeneration is usually extensive (or at least sufficient) irrespective of the management of burnt wood (e.g. Castro et al. 2011; de las Heras et al. 2012). However, a high density of pine seedlings can pose a problem in the short term, with a high risk of fire, competition, or pests. Thus, instead of forceful, urgent management in the first months after the fire, it would be more useful to save time, effort, and funding to manage the dense stand a few years after the fire, once mortality has stabilized and saplings still have a size that facilitates stand thinning (e.g. Ne'eman et al. 1997).

In contrast to the conventional approach, I strongly recommend shifting the paradigm towards the biotic component of the post-fire ecosystem and its ecological processes. Furthermore, management should be based on what could be defined –in parallel to what is being developed for farming– as "precision restoration" (see also St-Denis et al. 2018; Castro et al. 2021). This should consist of reorienting the objectives as well as the spatial and temporal use of budgets in order to boost the effectiveness of the restoration from the microscale to the macroscale, by focusing efforts on promoting the functioning of the ecosystem. At the microscale, we may consider, for example, the use of coarse woody debris as nurse structures to protect seedlings and saplings against environmental stress and herbivores, the use of reforestation methods and technologies that reduce the impact in already disturbed forests, such as direct sowing, seed protectors, etc. At the macroscale, we could define the areas of the landscape where nuclei of trees (or contours bands, e.g. following water courses) should be established according to their ecological requirements as a means of enhancing restoration success (instead of planting trees throughout the entire area). We could also define the areas at the landscape level where mutualistic interactions could promote natural regeneration. For example, the retention of burnt trees might be advisable up to several hundred meters from mature oak trees where jays are present (Sect. 25.3). We could also take management options to handle fire hazards by, for example, implementing a set of patches with different fuel loads, fire risk, and accessibility. There are currently technologies (e.g. LiDAR, multispectral and radar images captured from satellite and airborne sensors, devices to protect individual seeds or seedlings) and ecological-based knowledge (e.g. accurate species distribution or abundance models, plant-plant and plant-animal interaction outcomes) that make these fine approaches feasible, increasing the efficiency of the restoration (e.g. Siles et al. 2010b; Navarro-Cerrillo et al. 2014; Castro et al. 2015; Löf et al. 2019). This is not necessarily more costly; on the contrary, it could be cheaper in the long run, as it takes advantage of the natural capital of the post-fire habitat and concentrates part of the activities in specific locations in the landscape. In addition, this approach will reduce the negative impact that more aggressive management has in the post-fire ecosystem at the landscape level (e.g. Thorn et al. 2018; Bowd et al. 2019; Georgiev et al. 2020; Leverkus et al. 2020).

25.8 Summary and General Guidelines

The previous sections discuss factors that may affect post-fire forest regeneration, and how post-fire management may promote or alter the pace of natural regeneration. Overall, post-fire regeneration of coniferous forests is faced with strong interventionism using routine methods whose application may make little sense in some cases. It should be stated with absolute clarity that this interventionism cannot be considered a mistake committed by managers. On the contrary, managers face extremely complex situations combining the management of nature with social demands and, furthermore, local, regional or national regulations and policies that may constrain their margin for action. In fact, it is common that after a fire an action plan is drawn up that almost takes the rank of law, and that must be scrupulously followed for the restoration of the affected area. How to manage a burnt forest should be defined from a debate among different actors, including (among others) local populations, stakeholders, managers, and researchers, and it should have flexibility. In addition, there is no single recipe, and thus the specific measures may change from one case to another depending on the objectives established. Ultimately, ecological restoration should ensure the functioning of an ecosystem, but we may need to establish multiple plant communities for a particular site.

The following guidelines are recommended for the restoration of post-fire pine forests. Overall, management options should be enriched with more alternatives. There are already excellent up-to-date texts offering specific guidance for some aspects of post-fire pine-forest management, e.g., for burnt wood (Brown et al. 2003; Mauri and Pons 2019), and repeated calls for changes in management approaches have already been made (e.g. Ne'eman et al. 1997; Hutto 2006; Mauri and Pons 2019). Here, I will focus on general aspects, in the belief that, on one hand, it is unrealistic to offer specific, closed working routines that apply to all cases and, on the other hand, that the most important step is to raise awareness that an ecological restoration perspective might foster regeneration and provide advantages vs. more classical management approaches in terms of reduced impact and promotion of the recovery of ecosystem properties (e.g. resilience, multiple ecosystem services, etc.) that will benefit the whole ecosystem. Most of these recommendations may apply to pine forests of other regions, as well as other forest types, and many have been even suggested previously (e.g. in the references above). Their main goal is the restoration of ecosystem functioning, not necessarily the restoration of the pine forest or its productivity. Nonetheless, this guidance may help foster regeneration and thus may also be useful for other goals such as timber production.

1. Avoid the application of the same management across the entire landscape. A single management approach (or a limited number of alternatives) cannot possibly be the best solution for most of the area. Managements should be based on specific site characteristics (regeneration capacity, topography, risk of accidents, etc.) but, when possible, they should be combined to create a patchy landscape. This will increase regeneration niches at landscape scales, fit requirements for different management approaches (including intense interventionism if needed), and will provide conditions to mitigate future hazards. No intervention at all is also an option. In short, diversify the management.
2. Overall, reduce disturbance after the disturbance. Intense post-fire management creates a scenario of compound disturbances that may further hinder forest regeneration and, in general, ecosystem functioning. Be aware that certain disturbances may favor regeneration in particular cases. For instance, in burnt

serotinous pine forests, regeneration may be more intense in areas where workers have felled trees than in a control area (no action after the fire), likely because trampling and tree fall help to break the layer of ash and the release of seeds from the serotinous cones. However, more intense management such as salvage logging, mastication of woody debris, etc. will generally hamper regeneration.

3. Leave (totally or partially) burnt woody debris on the site, without mastication. Branches and other coarse woody debris spread on the ground will act as a barrier against soil erosion, and will also provide seedlings other benefits such as shade or protection against herbivores. In productive forests, branches may remain on site after log removal. In the event that branches are also merchantable for energy purposes (e.g. pellet production) some of them could still be left in consideration of the economic benefits derived from ecosystem services (e.g. promotion of regeneration). Overall, take advantage of burnt wood as nurse objects that reduce environmental stress for seedlings and resprouts, protect against herbivores, reduce erosion, and promote mutualistic interactions that boost regeneration.
4. If clearing the terrain for future work is a concern for burnt-wood removal or elimination, consider that the degree of impediment from coarse woody debris depends on factors such as tree density, tree size, decay rate, topography, etc. For example, in a non-productive pine forest in southern Spain with medium-sized trees, mechanized reforestation was possible 4 years after the fire (Leverkus et al. 2012). Thus, once again, diversify the management. Besides, patchy management will create a landscape where different plans can be implemented throughout the landscape.
5. If the presence of burnt wood is considered a risk for future fire, consider the following: (i) the ignition will require the fine material provided by the regenerating vegetation, which requires a minimum time period, and the wood will not be the source of the ignition, (ii) chipping and mulching may exacerbate the risk of ignition, as it will provide fine material, (iii) by the time an intense fire may be a risk, the burnt wood will have decomposed to a certain degree and its fuel characteristics will have changed. The role of burnt wood as a potential fuel that might intensify fire severity will be variable and should be evaluated according to site-specific characteristics, (iv) if the wood remaining on site may increase fire risk and/or intensity, there are many management alternatives to salvage logging and mastication across the whole area, for example, breaking the continuity of burnt wood at the landscape level by creating patches with different management intensity, strips, etc.
6. Avoid the construction of log erosion barriers. Their usefulness is questionable and, in any case, their purported function can be provided with different alternatives, such as felling trees and lopping the branches and leaving them scattered on the ground, spreading the crown trees in case logs are extracted, cutting the branches of the lower part of standing trees and leaving them spread *in situ*, etc. This may be possible even in cases where logs are removed for commercial purposes, so long as the branches are not masticated.

7. Reduce the use of heavy machinery, as it compacts the soil and encourages erosion. In certain cases, the transit of heavy vehicles may be inevitable, such as log-forwarders for salvage logging. Nevertheless, further machinery transit can be avoided. In this sense, elimination of the woody debris that remains after salvage logging operations by mastication or related techniques (chopping, chipping, etc., which needs heavy machinery) should be avoided not only because of its impact on seedlings, resprouts, or fire risk, but also because of soil compaction.
8. For oaks and other large-seeded species (perhaps including pines with large seeds, although further research is needed on this point), use direct sowing instead of transplanting seedlings for reforestation. Seeds may need individual protection against rodents. Protection against ungulates may be achieved with habitat structures; sowing below branches and other coarse woody debris is an excellent option for this purpose.
9. Overall, if the objective is forest restoration, try to think from the perspective of a seed or seedling. Current post-fire restoration is based on massive and immediate intervention that focuses on the physical aspect of the landscape, but hardly considers the biological components of the regeneration. Both are important, but need to be more balanced.

Acknowledgements This study is the result of years of research and classroom work, as well as discussions with colleagues, students, and friends, too numerous to be listed. Nonetheless, I would like to express my gratitude to Alex Leverkus, Pepe Quirós (the "tree planter"), J. Ramón Guzmán, Craig Allen, Pedro Antonio Tíscar, and Pipo (FB Navarro) for their generous support. An anonymous reviewer provided comments that improved the manuscript. Special thanks to David Nesbitt, who apart of correcting my English has had the ability to read my mind for years and find the way to express critical ideas in what for me is a foreign tongue. Special thanks also to my daughter, Julia, for our field trips and her patient while in listening to my enthusiastic (but I guess sometimes boring) explanations about how forests function, including photosynthesis and the wonder of its pathways. This work is dedicated to her.

References

Akay AE, Erdas O, Kanat M et al (2007) Post-fire salvage logging for fire-killed Brutian pine (*Pinus brutia*) trees. J Appl Sci 7:402–406

Allen HD (2001) Mediterranean ecogeography. Prentice Hall, Harlow

Allen CD, Breshears DD, McDowell NG (2015) On underestimation of global vulnerability to tree mortality and forest die-off from hotter drought in the Anthropocene. Ecosphere 6:129. https://doi.org/10.1890/ES15-00203.1

Aristeidis K, Vasiliki K (2015) Evaluation of the post-fire erosion and flood control works in the area of Cassandra (Chalkidiki, North Greece). J For Res 26:209–217

Badía D, Sánchez C, Aznar JM et al (2015) Post-fire hillslope log debris dams for runoff and erosion mitigation in the semiarid Ebro Basin. Geoderma 237–238:298–307

Beyers JL (2004) Postfire seeding for erosion control: effectiveness and impacts on native plant communities. Conserv Biol 18:947–956

Blondel J, Aronson J, Bodiou J-Y et al (2010) The Mediterranean region, 2nd edn. Oxford University Press, Oxford

Bontrager JD, Morgan P, Hudak AT et al (2019) Long-term vegetation response following post-fire straw mulching. Fire Ecol 15:22
Bowd EJ, Banks SC, Strong CL et al (2019) Long-term impacts of wildfire and logging on forest soils. Nat Geosci 12:113–118
Brown JK, Reinhardt ED, Kramer KA (2003) Coarse woody debris: managing benefits and fire hazards in the recovering forest. USDA General Technical Report RMRS-GTR-105
Carrillo-Gavilán MA, Lalagüe H, Vilà M (2010) Comparing seed removal of 16 pine species differing in invasiveness. Biol Invasions 12:2233–2242
Castro J (2013) Postfire burnt-wood management affects plant damage by ungulate herbivores. Int J For Res 2013:965461
Castro J, Leverkus AB (2019) Effect of herbaceous layer interference on the post-fire regeneration of a serotinous pine (*Pinus pinaster* Aiton) across two seedling ages. Forests 10:74. https://doi.org/10.3390/f10010074
Castro J, Gómez JM, García D, Zamora R, Hódar JA (1999) Seed predation and dispersal in relict Scots pine forests in southern Spain. Plant Ecol 145:115–123
Castro J, Zamora R, Hódar JA et al (2004a) Benefits of using shrubs as nurse plants for reforestation in Mediterranean mountains: a 4-year study. Restor Ecol 12:352–358
Castro J, Zamora R, Hódar JA et al (2004b) Seedling establishment of a boreal tree species (*Pinus sylvestris*) at its southernmost distribution limit: consequences of being in a marginal Mediterranean habitat. J Ecol 92:266–277
Castro J, Zamora R, Hódar JA (2006a) Restoring *Quercus pyrenaica* forest using pioneer shrubs as nurse plants. Appl Veg Sci 9:137–142
Castro J, Hódar JA, Gómez JM (2006b) Seed size. In: Basra A (ed) Handbook of seed science and technology. Haworth's Food Products Press, New York, pp 397–427
Castro J, Moreno-Rueda G, Hódar JA (2010) Experimental test of postfire management in pine forests: impact of salvage logging versus partial cutting and nonintervention on bird-species assemblages. Conserv Biol 24:810–819
Castro J, Allen CD, Molina-Morales M et al (2011) Salvage logging versus the use of burnt wood as a nurse object to promote post-fire tree seedling establishment. Restor Ecol 19:537–544
Castro J, Puerta-Piñero C, Leverkus AB et al (2012) Post-fire salvage logging alters a key plant-animal interaction for forest regeneration. Ecosphere 3:90
Castro J, Leverkus AB, Fuster F (2015) A new device to foster oak forest restoration via seed sowing. New For 46:919–929
Castro J, Morales-Rueda F, Navarro FB, Löf M, Vacchiano G, Alcaraz-Segura D (2021) Precision restoration: a necessary approach to foster forest recovery in the 21st century. Restoration Ecology (in press). https://doi.org/10.1111/rec.13421
Cavallero L, Raffaele E, Aizen MA (2013) Birds as mediators of passive restoration during early post-fire recovery. Biol Conserv 158:342–350
Cobb TP, Hannam KD, Kishchuk BE et al (2010) Wood-feeding beetles and soil nutrient cycling in burned forests: implications of post-fire salvage logging. Agric For Entomol 12:9–18
de Las Heras J, Moya D, Vega JA et al (2012) Post-fire management of serotinous pine forests. In: Moreira F, Arianoutsou M, Corona P et al (eds) Post-fire management and restoration of southern European forests. Springer, London, pp 121–150
del Campo AD, Navarro-Cerrillo RM, Hermoso J et al (2007) Relationship between site stock quality in *Pinus halepensis* Mill. reforestation on semiarid landscapes in eastern Spain. Ann For Sci 64:719–731
Dodson EK, Peterson DW (2010) Mulching effects on vegetation recovery following high severity wildfire in non-central Washington State, USA. For Ecol Manag 260:1816–1823
Donato DC, Fontaine JB, Campbell JL et al (2006) Post-fire logging hinders regeneration and increases fire risk. Science 311:352
Fernández C, Vega JA (2016a) Are erosion barriers and straw mulching effective for controlling soil erosion after a high severity wildfire in NW Spain? Ecol Eng 87:132–138

Fernández C, Vega JA (2016b) Effects of mulching and post-fire salvage logging on soil erosion and vegetative regrowth in NW Spain. For Ecol Manag 375:46–54

Fernández C, Fontúrbel T, Vega JA (2019) Effects of pre-fire site preparation and post-fire erosion barriers on soil erosion after a wildfire in NW Spain. Catena 172:691–698

Fox DM (2011) Evaluation of the efficiency of some sediment trapping methods after a Mediterranean forest fire. J Environ Manag 92:258–265

García-Morote F, Martínez-García E, Andrés-Abellán M et al (2017) Direct seeding of *Pinus halepensis* Mill. for recovery of burned semi-arid forests: implications for post-fire management for improving natural regeneration. Forests 8:353

García-Orenes F, Arcenegui V, Chrenková K et al (2017) Effects of salvage logging on soil properties and vegetation recovery in a fire-affected Mediterranean forest: a two year monitoring research. Sci Total Environ 586:1057–1065

Gea-Izquierdo G, Férriz M, García-Garrido S et al (2019) Synergistic abiotic and biotic stressors explain widespread decline of *Pinus pinaster* in a mixed forest. Sci Total Environ 685:963–975

Georgiev K, Chao A, Castro J et al (2020) Salvage logging changes the taxonomic, phylogenetic and functional successional trajectories of forest bird communities. J Appl Ecol 00:1–10. https://doi.org/10.1111/1365-2664.13599

Ginzburg O, Steinberger Y (2012) Salvage logging versus natural regeneration post-fire practices in a forest: soil chemical and microbial aspects. Open J Ecol 2:29–37

Gómez JM (2003) Spatial patterns in long-distance dispersal of *Quercus ilex* acorns by jays in a heterogeneous landscape. Ecography 26:573–584

Gómez-Aparicio L (2009) The role of plant interactions in the restoration of degraded ecosystems: a meta-analysis across life-forms and ecosystems. J Ecol 97:1202–1214

Gómez-Aparicio L, Zamora R, Gómez JM et al (2004) Applying plant facilitation to forest restoration: a meta-analysis of the use of shrubs as nurse plants. Ecol Appl 14:1128–1138

Gómez-Aparicio L, Gómez JM, Zamora R et al (2005) Canopy vs. soil effects of shrubs facilitating tree seedlings in Mediterranean montane ecosystems. J Veg Sci 16:191–198

Gómez-Sánchez E, Lucas-Borja ME, Plaza-Álvarez PA et al (2019) Effects of post-fire hillslope stabilisation techniques on chemical and microbiological soil properties in Mediterranean forest ecosystems. J Environ Manag 246:229–238

Hernández-Hernández R, Castro J, Del Arco-Aguilar M et al (2017) Post-fire salvage logging imposes a new disturbance that retards succession: the case of bryophyte communities in a Macaronesian laurel forest. Forests 8:252

Herrero A, Castro J, Zamora R et al (2013) Growth and stable isotope signals associated with drought-related mortality in saplings of two coexisting pine species. Oecologia 173:1613–1162

Hutto RL (2006) Toward meaningful snag-management guidelines for post-fire salvage logging in North American conifer forests. Conserv Biol 20:984–993

Jonas JL, Berryman E, Wolk B et al (2019) Post-fire wood mulch for reducing erosion potential increases tree seedlings with few impacts on understory plants and soil nitrogen. For Ecol Manag 453:117567

Keeley JE, Bond WJ, Bradstock RA et al (2012) Fire in Mediterranean ecosystems: ecology, evolution and management. Cambridge University Press, Cambridge

Kildisheva OA, Aghai MM, Bouazza K et al (2017) Improving restoration success through research-driven initiatives: case studies targeting *Pinus pinea* reforestation stock development in Lebanon. Plant Ecol 218:39–53

Kulakowski D, Veblen TT (2007) Effect of prior disturbance on the extent and severity of wildfire in Colorado subalpine forests. Ecology 88:759–769

Lee CW, Seo JI, Youn HJ et al (2014) Effectiveness of rehabilitation treatments on a slowly revegetating hillslope in a recently burned coastal forest, Republic of Korea. Landsc Ecol Eng 10:249–260

Leverkus AB, Castro J (2017) An ecosystem services approach to the ecological effects of salvage logging: valuation of seed dispersal. Ecol Appl 24:1057–1063

Leverkus AB, Puerta-Piñero C, Guzmán JR et al (2012) Post-fire salvage logging increases restoration costs in a Mediterranean mountain ecosystem. New For 43:601–613
Leverkus AB, Lorite J, Navarro FB et al (2014) Post-fire salvage logging alters species composition and reduces cover, richness, and diversity in Mediterranean plant communities. J Environ Manag 133:323–331
Leverkus AB, Castro J, Delgado-Capel MJ et al (2015a) Restoring for the present or restoring for the future: enhanced performance of two sympatric oaks (*Quercus ilex* and *Quercus pyrenaica*) above the current forest limit. Restor Ecol 23:936–946
Leverkus AB, Rojo M, Castro J (2015b) Habitat complexity and individual acorn protectors enhance the post-fire restoration of oak forests via seed sowing. Ecol Eng 83:276–280
Leverkus AB, Rey-Benayas JM, Castro J (2016) Shifting demographic conflicts across recruitment cohorts in a dynamic post-disturbance landscape. Ecology 97:2628–2639
Leverkus AB, Rey-Benayas JM, Castro J et al (2018a) Salvage logging effects on regulating and supporting ecosystem services – a systematic map. Can J For Res 48:983–1000
Leverkus AB, Lindenmayer DB, Thorn S et al (2018b) Salvage logging in the world's forests: interactions between natural disturbance and logging need recognition. Glob Ecol Biogeogr 27:1140–1154
Leverkus AB, Gustafsson L, Lindenmayer DB et al (2020) Salvage logging effects on regulating ecosystem services and fuel loads. Front Ecol Environ 18:391-400
Lewis J, Rhodes JJ, Bradley C (2019) Turbidity responses from timber harvesting, wildfire, and post-fire logging in the Battle Creek watershed, northern California. Environ Manag 63:416–432
Löf M, Castro J, Engman M et al (2019) Tamm review: direct seeding to restore oak (*Quercus* spp.) forests and woodlands. For Ecol Manag 448:474–489
Lucas-Borja ME, González-Romero J, Plaza-Álvarez PA et al (2019) The impact of straw mulching and salvage logging on post-fire runoff and soil erosion generation under Mediterranean climate conditions. Sci Total Environ 654:441–451
Maestre FT, Bautista S, Cortina J (2003) Positive, negative, and net effects in grass-shrub interactions in Mediterranean semiarid grasslands. Ecology 84:3186–3197
Marañón-Jiménez S, Castro J (2013) Effect of decomposing burnt wood on soil fertility and nutrient availability in a Mediterranean ecosystem. Biogeochemistry 112:519–535
Marañón-Jiménez S, Castro J, Fernández-Ondoño E et al (2013a) Charred wood remaining after a wildfire as a reservoir of macro- and micronutrients in a Mediterranean pine forest. Int J Wildland Fire 22:681–695
Marañón-Jiménez S, Castro J, Querejeta JI et al (2013b) Post-fire wood management alters water stress, growth, and performance of pine regeneration in a Mediterranean ecosystem. For Ecol Manag 308:231–239
Marcolin E, Marzano R, Vitali A et al (2019) Post-fire management impact on natural forest regeneration through altered microsite conditions. Forests 10:1014. https://doi.org/10.3390/f10111014
Marques MA, Mora E (1998) Effects on erosion of two post-fire management practices: clear-cutting versus non-intervention. Soil Tillage Res 45:433–439
Martín-Lorenzo D (2017) Evaluación de la evolución del incendio en la sierra de Lújar mediante el uso de indicadores de la calidad del suelo. Master thesis, University of Granada
Marzano R, Garbarino M, Marcolin E et al (2013) Deadwood anisotropic facilitation on seedling establishment after a stand-replacing wildfire in Aosta Valley (NW Italy). Ecol Eng 51:117–122
Matías L, Zamora R, Castro J (2012) Sporadic rainy events are more critical than increasing of drought intensity for woody species recruitment in a Mediterranean community. Oecologia 169:833–844
Mauri E, Pons P (2019) Handbook of good practices in post-wildfire management, 2nd edn. Universitat de Girona, Girona
McIver JD, Ottmar R (2018) Fuel mass and stand structure 13 years after logging of a severely burned ponderosa pine forest in northeastern Oregon, U.S.A. For Ecol Manag 424:505–518

McIver JD, Starr L (2000) Environmental effects of post-fire logging: literature review and annotated bibliography. USDA Technical Report PNW-GTR-486

Médail F (2008) Mediterranean. In: Jørgensen SE, Fath BD (eds) Encyclopedia of ecology. Elsevier, Oxford, pp 2296–2308

Mendoza I, Zamora R, Castro J (2009) A seeding experiment for testing tree-community recruitment under variable environments: implications for forest regeneration and conservation in Mediterranean habitats. Biol Conserv 149:1491–1499

Molinas-González CR, Leverkus AB, Marañón-Jiménez S et al (2017a) Fall rate of burnt pines across an altitudinal gradient in a Mediterranean mountain. Eur J For Res 136:401–409

Molinas-González CR, Castro J, Leverkus AB (2017b) Deadwood decay in a burnt Mediterranean pine reforestation. Forests 8:158. https://doi.org/10.3390/f8050158

Molinas-González CR, Castro J, González-Megías A et al (2019) Effects of post-fire deadwood management on soil macroarthropod communities. Forests 10:1046. https://doi.org/10.3390/f10111046

Mosandl R, Kleinert A (1998) Development of oaks (*Quercus petraea* (Matt.) Liebl.) emerged from bird-dispersed seeds under old-growth pine (*Pinus sylvestris* L.) stands. For Ecol Manag 106:35–44

Myronidis D, Arabatzis G (2009) Evaluation of Greek post-fire erosion mitigation policy trough spatial analysis. Pol J Environ Stud 18:865–872

Navarro FB, Jiménez MN, Ripoll MA et al (2006) Direct sowing of holm oak acorns: effects of acorn size and soil treatment. Ann For Sci 63:961–967

Navarro-Cerrillo RM, del Campo AD, Ceacero CJ et al (2014) On the importance of topography, site quality, stock quality and planting date in a semiarid plantation: feasibility of using low-density LiDAR. Ecol Eng 67:25–38

Ne'eman G, Perevolotsky A, Schiller G (1997) The management implications of the Mt. Carmel research project. Int J Wildland Fire 7:343–350

North MP, Stevens JT, Greene DF et al (2019) Tamm review: reforestation for resilience in dry western U.S. forests. For Ecol Manag 432:209–224

Pansing ER, Tomback DF (2019) Survival of whitebark pine seedlings grown from direct seeding: implications for regeneration and restoration under climate change. Forests 10:677. https://doi.org/10.3390/f10080677

Pausas JG, Bladé C, Valdecantos A et al (2004) Pines and oaks in the restoration of Mediterranean landscapes of Spain: new perspectives for an old practice –a review. Plant Ecol 171:209–220

Pemán J, Voltas J, Gil-Pelegrin E (2006) Morphological and functional variability in the root system of *Quercus ilex* L. subjected to confinement: consequences for afforestation. Ann For Sci 63:425–430

Peppin D, Fulé PZ, Sieg CH et al (2010) Post-wildfire seeding in forest of the western United States: an evidence-based review. For Ecol Manag 260:573–586

Pesendorfer MB, Sillett TS, Koeing WD et al (2016) Scatter-hoarding corvids as seed dispersers for oaks and pines: a review of a widely distributed mutualism and its utility to habitat restoration. Condor 118:215–237

Prats SA, Malvar MC, Coelho COA et al (2019) Hydrologic and erosion responses to compaction and added surface cover in post-fire logged areas: isolating splash, interrill and rill erosion. J Hydrol 575:408–419

Prévosto B, Monnier Y, Ripert C et al (2011) Diversification of *Pinus halepensis* forests by sowing *Quercus ilex* and *Quercus pubescens* acorns: testing the effects of different vegetation and soil treatments. Eur J For Res 130:67–76

Raftoyannis Y, Spanos I (2005) Evaluation of log and branch barriers as post-fire rehabilitation treatments in a Mediterranean pine forest in Greece. Int J Wildland Fire 14:183–188

Retana J, Arnan X, Arianoutsou M et al (2012) Post-fire management of non-serotinous pine forests. In: Moreira F, Arianoutsou M, Corona P et al (eds) Post-fire management and restoration of southern European forests. Springer, London, pp 151–170

Rey-Benayas JM, Navarro J, Espigares T et al (2005) Effects of artificial shading and weed mowing in reforestation of Mediterranean abandoned cropland with contrasting *Quercus* species. For Ecol Manag 212:302–314
Rhoades CC, Fornwalt PJ (2015) Pile burning creates a fifty-year legacy of openings in regenerating lodgepole pine forests in Colorado. For Ecol Manag 336:203–209
Robichaud PR, Beyers JL, Neary DG (2000) Evaluating the effectiveness of postfire rehabilitation treatments. USDA General Technical Report RMRS-GTR-63
Robichaud PR, Wagenbrenner JW, Brown RE et al (2008) Evaluating the effectiveness of contour-felled log erosion barriers as a post-fire runoff and erosion mitigation treatment in the western United States. Int J Wildland Fire 17:255–273
Robichaud PR, Ashmun LE, Sims BD (2010) Post-fire treatment effectiveness for hillslope stabilization. USDA General Technical Report RMRS-GTR-240
Rost J, Clavero M, Bas JM et al (2010) Building wood debris piles benefits avian seed dispersers in burned and logged Mediterranean pine forests. For Ecol Manag 260:79–86
Ruthrof KX, Renton M, Dixon K (2013) Overcoming restoration thresholds and increasing revegetation success for a range of canopy species in a degraded urban Mediterranean-type woodland ecosystems. Aust J Bot 61:139–147
Salazar-Tortosa D, Castro J, Saladin B et al (2020) Arid environments select for larger seeds in pines (*Pinus* spp.). Evol Ecol 34:11–26. https://doi.org/10.1007/s10682-019-10016-1
Sánchez-Salguero R, Navarro-Cerrillo RM, Camarero JJ et al (2010) Drought-induced growth decline of Aleppo and maritime pine forests in south-eastern Spain. For Syst 19:458–469
Sánchez-Salguero R, Camarero JJ, Dobbertin M et al (2013) Contrasting vulnerability and resilience to drought-induced decline of densely planted vs. natural rear-edge *Pinus nigra* forests. For Ecol Manag 310:956–967
Santolamazza-Carbone S, Pestaña M, Vega JA (2011) Post-fire attractiveness of maritime pines (*Pinus pinaster* Ait.) to xylophagous insects. J Pest Sci 84:343–353
Schönenberger W (2001) Cluster afforestation for creating diverse mountain forest structures – a review. For Ecol Manag 145:121–128
Serrano-Ortiz P, Marañón-Jiménez S, Reverter BR et al (2011) Post-fire salvage logging reduces carbon sequestration in Mediterranean coniferous forest. For Ecol Manag 262:2287–2296
Shakesby RA (2011) Post-wildfire soil erosion in the Mediterranean: review and future research directions. Earth-Sci Rev 105:71–100
Shive KL, Estes BL, White AM et al (2017) Rice straw mulch for post-fire erosion control: assessing non-target effects on vegetation communities. Int J Wildland Fire 26:538–549
Siles G, Rey PJ, Alcántara JM (2010a) Post-fire restoration of Mediterranean forests: testing assembly rules mediated by facilitation. Basic Appl Ecol 11:422–431
Siles G, Alcántara JM, Rey PJ et al (2010b) Defining a target map of native species assemblages for restoration. Restor Ecol 18:439–448
St-Denis A, Kneeshaw D, Messier C (2018) Effect of predation, competition, and facilitation on tree survival and growth in abandoned fields: towards precision restoration. Forests 9:692
Taboada A, Fernández-García V, Marcos E et al (2018) Interactions between large high-severity fires and salvage logging on a short return interval reduce the regrowth of fire-prone serotinous forests. For Ecol Manag 414:54–63
Thompson JR, Spies TA, Ganio LM (2007) Reburn severity in managed and unmanaged vegetation in a large wildfire. PNAS 104:10743–10748
Thorn S, Bässler C, Brandl R et al (2018) Impacts of salvage logging on biodiversity –a meta-analysis. J Appl Ecol 55:279–289
Tíscar PA, Candel-Pérez D, Estrany J et al (2017) Regeneration of three pine species in a Mediterranean forest: a study to test predictions from species distribution models under changing climates. Sci Total Environ 584–585:78–87
Toivanen T, Liikanen V, Kotiaho JS (2009) Effects of forest restoration treatments on the abundance of bark beetles in Norway spruce forests of southern Finland. For Ecol Manag 257:117–125

Urretavizcaya MF, Defossé GE (2019) Restoration of burned and post-fire logged *Austrocedrus chilensis* stands in Patagonia: effects of competition and environmental conditions on seedling survival and growth. Int J Wildland Fire 28:365–376
Vallejo VR, Arianoutsou M, Moreira F (2012) Fire ecology and post-fire restoration approaches in southern European forest types. In: Moreira F, Arianoutsou M, Corona P et al (eds) Post-fire management and restoration of southern European forests. Springer, London, pp 93–119
Wagenbrenner JW, Robichaud PR, Brown RE (2016) Rill erosion in burned and salvage logged western montane forests: effects of logging equipment type, traffic level, and slash treatment. J Hydrol 541:889–901
Whelan RJ (1995) The ecology of fire. Cambridge University Press, Cambridge
Wuerthner G (ed) (2006) Wild fire: a century of failed forest policy. Island Press, Sausalito
Zadworny M, Jagodziński AM, Łakomy P et al (2014) The silent shareholder in deterioration of oak growth: common planting practices affect the long-term response of oaks to periodic drought. For Ecol Manag 318:133–141
Zavala MA, Espelta JM, Retana J (2000) Constraints and trade-offs in Mediterranean plant communities: the case of Holm oak-Aleppo pine forests. Bot Rev 66:119–149
Zituni R, Wittenberg L, Malkinson D (2019) The effect of post-fire forest management on soil erosion rates 3 and 4 years after a wildfire, demonstrated on the 2010 Mount Carmel fire. Int J Wildland Fire 28:377–385

Chapter 26
Post-fire Regeneration in Yellow Pine and Mixed Conifer Forests of California: Challenges, Tools and Applications for Fire-Adapted Forests

John N. Williams, Jonathan W. Long, Marc D. Meyer, Becky Estes, Michelle Coppoletta, and Hugh D. Safford

26.1 Introduction

The North American Mediterranean-climate zone (California, SW Oregon, NW-most Baja California, Mexico; hereafter "California") supports a high diversity of forest types and fire regimes (Barbour et al. 2007). Since the mid-nineteenth century when American settlement of California began, forests have been managed for timber production and grazing, human settlements have expanded, fire suppression policies have been imposed, exotic plant species have been introduced, and the global climate has been greatly altered. As in other parts of western North America (Agee 1993; Allen et al. 2002), these influences have resulted in notable changes to forest structure and composition and the fire regimes that influence them (Keeley and Safford 2016; Van Wagtendonk et al. 2018). In California, yellow pine and mixed conifer (YPMC) forests comprise ca. 40% of the forested area (Barbour et al. 2007). In these forests, historical mean fire return intervals of ca. 10–20 years and a preponderance of low severity fire shaped a vegetation type dominated largely by fire-resistant pines (e.g., *Pinus ponderosa* Douglas ex C.Lawson, *P. jeffreyi* Balf., *P. lambertiana* Douglas) and oaks (*Quercus* spp.) (Stephens and Collins 2004; Van de Water and Safford 2011). Frequent, low-severity fires limited seedling recruitment and fuel build-up to favor low-density, large-diameter stands resistant to high-severity fire (Safford and Stevens 2017). In cases where fire was either absent or created small gaps of fire-killed trees, infilling of stands with dense conifer

J. N. Williams (✉) · J. W. Long · M. D. Meyer · B. Estes · M. Coppoletta · H. D. Safford
University of California, Davis, CA, USA

USDA Forest Service, Pacific Southwest Research Station, Albany, CA, USA

USDA Forest Service, Pacific Southwest Region, Vallejo, CA, USA
e-mail: jnwill@ucdavis.edu

G. Ne'eman, Y. Osem (eds.), *Pines and Their Mixed Forest Ecosystems in the Mediterranean Basin*, Managing Forest Ecosystems 38,
https://doi.org/10.1007/978-3-030-63625-8_26

regeneration occurred, creating a heterogeneous patchwork of tree densities and species arrays across the landscape (Fig. 26.1).

With the exclusion of fire from many of these forests for a century or more (Safford and Van de Water 2014, Fig. 1) and in combination with other anthropogenic effects (e.g., climate change, expansion of the wildland–urban interface, timber harvest), fire patterns have changed, as have the heterogeneous vegetation patterns that once characterized these forests (Minnich et al. 1995; Keeley and Safford 2016;

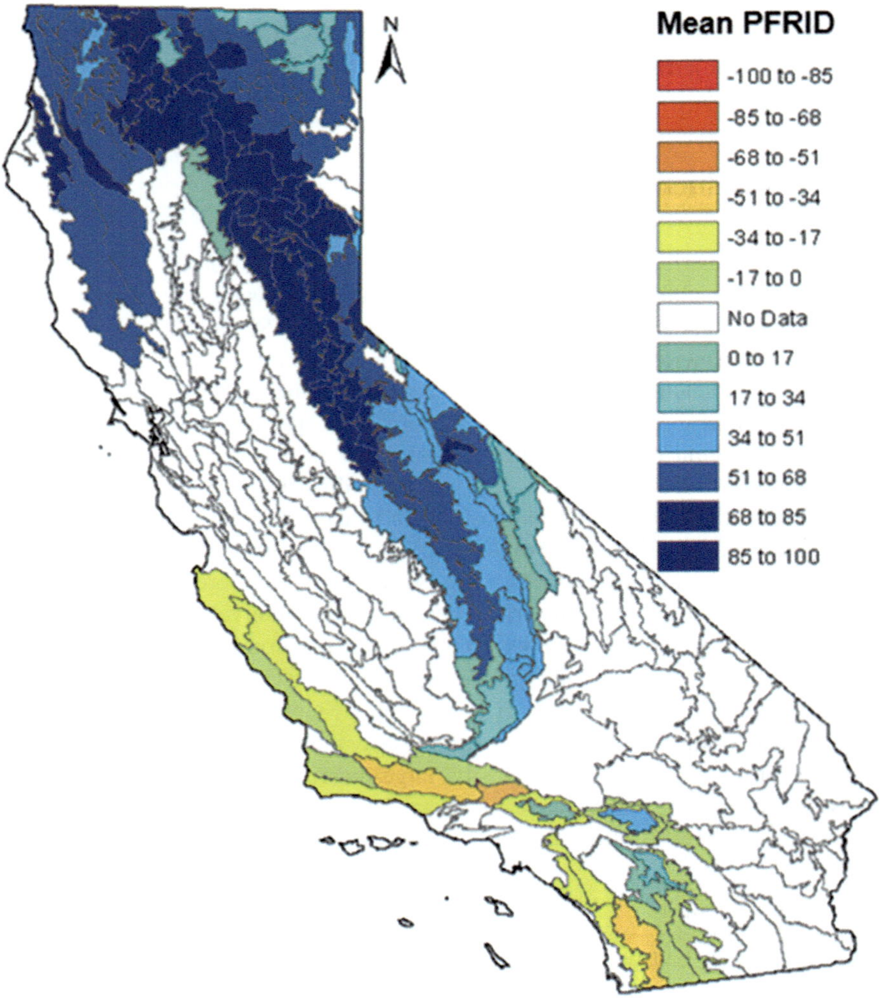

Fig. 26.1 Mean percent fire return interval departure (PFRID), generalized to the ecological subregion (Miles and Goudey 1997) from Forest Service and Park Service lands found within each subregion. Warm colors (yellow to red) are experiencing more fire than under pre-Euroamerican conditions, cool colors (blues and greens) are experiencing less fire. (Figure reprinted with permission from Safford and Van de Water (2014))

Safford and Stevens 2017; North et al. 2019). The lack of regular, low-intensity fires means that fine, coarse and ladder fuels accumulate and become more continuous. This in turn results in conditions that increase the incidence of larger, stand-replacing fires (Stephens et al. 2015; Collins et al. 2019), which subsequently create less hospitable conditions for conifer regeneration (Fig. 26.2).

A growing body of research focused on YPMC forests in California has demonstrated how less frequent but higher severity fires in recent decades are causing shifts in forest structure, composition, and function (Mallek et al. 2013; Steel et al. 2015; Steel et al. 2018). These changes include declines in the more fire-adapted but shade-intolerant yellow and sugar pines, and increases in less fire-adapted but more shade-tolerant taxa such as white fir (*Abies concolor* (Gordon) Lindley ex Hidlebrand), incense cedar (*Calocedrus decurrens* (Torr.) Florin), and Douglas-fir (*Pseudotsuga menziesii* (Mirbel) Franco). Oaks, aspen (*Populus tremuloides* Michx.) and other hardwoods generally respond to stand-replacing fires by resprouting, but frequent intense fire may affect their population dynamics by reducing the availability of mature trees (Hammett et al. 2017; Long et al. 2017). The YPMC forest type – which encompasses dry mixed conifer, moist mixed conifer and yellow pine forest types, and spans a broad elevational band in California (from 300 to 2400 m, depending on latitude, aspect and topographic position; Barbour

Fig. 26.2 Fire severity and conifer regeneration. (**a**) Part of the large, high-severity burn area that resulted from the 2014 King Fire on the El Dorado National Forest in the Sierra Nevada mountains of California. (**b**) Regeneration is lacking in interior parts with high-severity, far from seed trees or dispersal agents. (**c**) In contrast, natural regeneration is present on the periphery of the wildfire near seed trees and in small high-severity patches where only a few overstory trees were killed and openings were created

et al. 2007, Safford and Stevens 2017) – is important for timber production and recreation, as well as for a range of ecosystem services including water provision, carbon sequestration, wildlife habitat, and biodiversity (Kitzmiller and Lunak 2012; Missoula_Fire_Sciences_Laboratory 2018). As such, there is relative consensus among federal, state, and private interests on the need to preserve its integrity through the application of management techniques such as forest thinning, prescribed fire, other fuel reduction interventions, and targeted replanting. These actions attempt to restore historic stand structures (and in some cases fire regimes) and/or create forest conditions that are more resilient to stressors and disturbances that are being exacerbated by climate warming, such as severe fire, drought, and insect and pathogen outbreaks.

26.2 Need for Better Data and Better Tools

Central to understanding long-term forest dynamics and restoring more resilient forest conditions in fire-prone regions is being able to forecast natural forest regeneration after fire. Similarly, promoting forest resilience and sustainability of ecosystem services under global change also requires an understanding of how regeneration varies among species, with climate and stand structure, and across landscapes and through time (e.g., Meigs et al. 2009; Carlson et al. 2012; Dobrowski et al. 2015). One of the main tools available to forest managers for forecasting forest dynamics is the Forest Vegetation Simulator (FVS; Dixon 2002). FVS is a stand-level simulator that models forest dynamics over time, incorporating processes such as mortality, growth, succession and disturbance. Over more than four decades, FVS has expanded to include nearly two-dozen geographic variants and model extensions that simulate dynamics and the effects of factors such as insect outbreaks, disease, and fire. A well-known limitation to FVS, however, is the regeneration stage of forest succession. Currently, only two geographic regions of the United States have FVS submodels that can predict natural regeneration. In all other regions – including California – natural regeneration parameters must be entered by the user. Other forest growth and succession models have similar limitations.

In the conifer forests of California, regeneration is highly heterogeneous both spatially and temporally and is sensitive to site-specific weather, climate and biophysical variables (Zald et al. 2008; Davis et al. 2019; Werner et al. 2019; Young et al. 2019). Post-fire tree regeneration surveys in California have found high variation in natural seedling densities and composition, and regeneration rates that are often insufficient – with median densities often at or around zero seedlings/ha – to meet management objectives (Crotteau et al. 2013; Welch et al. 2016; Tepley et al. 2017; Shive et al. 2018). There is also a need to better understand the medium- and long-term vulnerability of conifer regeneration, as abrupt state shifts may become more frequent under climate change. For example, there is a growing body of evidence suggesting that decreases in post-fire recruitment, particularly for low-elevation tree species, are at least partially explained by long-term climate changes

(Serra-Diaz et al. 2018; Davis et al. 2019; Young et al. 2019). In some of these cases, changes may push conditions past physiological or ecological suitability thresholds, resulting in species' population declines and raising the potential for abrupt ecosystem transitions.

The need for reforestation (the active process of planting seedlings and managing for the survival of specific stands of trees) arises when natural regeneration is insufficient to meet management objectives. In California, a poor understanding of natural regeneration patterns and drivers, a lack of robust models of natural regeneration, and dated seedling density targets that are unsynchronized with changing climate and fire regimes have been impediments to successful reforestation planning and implementation (Welch et al. 2016; North et al. 2019). In addition, budgets cuts and changing priorities have led to reduced investments in reforestation on federal lands, resulting in increasing dependence on natural regeneration. As such, there is a clear need for a better understanding of natural regeneration patterns in YPMC and related forests under both current and future environmental conditions.

26.3 Predicting Conifer Regeneration

Accurate forecasting of vegetation development during regeneration – either natural or assisted – is a critical part of ensuring the sustainability of conifer forests in fire-prone landscapes. In California, a key concern following wildfire is whether a state shift might occur from forest to a persistent shrubland or grassland, or whether fire might actually help to restore vegetation communities that have historically declined, including shrub patches, hardwood-dominated groves, and meadows (White and Long 2019; Meyer et al. 2021). Such shifts can impact flows of ecosystem services, including biodiversity, carbon sequestration, and timber, and provision of food sources or wildlife habitat (Long et al. 2018).

Successful forecasting of future forest conditions depends on having tools based on sufficient regeneration data that can project vegetation development over relevant time frames, including short-term (e.g., 1–5 years after fire), medium-term (e.g., 5–20 years), and longer-term periods (>20 years). Much progress has been made in modeling longer-term growth and yield of forest stands (e.g., FVS, JABOWA) and in forest succession models, whether spatially-explicit or not (e.g., LANDIS, LANDSUM, SIMPPLLE, FM, RMLANDS, VDDT/TELSA, ST-SIM). However, a major limitation to the realism and utility of these models has been the lack of focus on realistic regeneration parameters and spatiotemporal variability in those parameters.

The National Forest Management Act and Forest Service regulations (e.g., Forest Service Handbook (FSH) 2409.13–21.42) require that productive forests be restocked in accordance with land management plans within five years after a major stand altering event, such as a tree harvest or stand-replacing fire. Five years is also a forestry "rule of thumb" threshold, beyond which burned areas require major extra investment in site preparation to plant, and where planting is therefore rarely undertaken. Ecologically, this time period is also critical for seedling establishment and growth, especially on drier sites where fast-growing shrubs frequently outcompete conifer seedlings for resources such as light and moisture (Bohlman et al. 2016; Lutz et al. 2017). For these reasons, predictive tools have focused on regeneration of young trees within this five-year window.

Recently, several tools have been created to predict seedling densities ca. five years after fire (Box 26.1). One developed by Welch et al. (2016), uses data collected on natural regeneration following 14 wildfires that occurred in California YPMC forests to generate a set of statistical models of conifer regeneration density in dry and moist forest types. Welch et al. (2016) sought to develop a post-fire regeneration planning tool that predicts seedling composition and density during the key five-year reestablishment window. They found highly variable regeneration responses where median recruitment numbers were much more informative than mean values. Additionally, the authors found zero recruitment in 43% of their plots, sub-par seedling densities (by Forest Service standards) after most of the fires, a negative correlation between seedling presence and shrub cover, and a predominance of seedlings of shade-tolerant, fire-sensitive species (i.e., non-yellow pine) – findings corroborated elsewhere (e.g., Stevens-Rumann and Morgan 2019). With such high variability and the abundance of zero-recruitment points, Welch et al. (2016) adapted their field assessment tool into one that can be used to field check locations identified by spatial prediction and determine whether seedling densities five years after a fire are likely to meet FSH seedling stocking thresholds or not. They were unable, however, to accurately predict seedling composition or densities across the full range of responses. We note that the authors showed that natural seedling recruitment in areas with basal area mortality <75% was usually sufficient to meet stocking guidelines, suggesting that even under current practices the area of land considered for reforestation could be reduced substantially for certain target species or in cases where species identity is less important (e.g., conifers generally, rather than *P. ponderosa* or *P. jeffreyi*).

Shive et al. (2018) built on the data collected by Welch et al. (2016), and made use of an extensive Forest Service post-fire inventory dataset from California to develop a spatially explicit predictive tool: the Post-fire Spatial Conifer Regeneration Prediction Tool (POSCRPT). This tool uses basal area maps and burn severity data to spatially model seed production of source trees that are likely to contribute germplasm for conifer regeneration. In large, high-severity fires, the distance from a viable seed source is a key limiting factor for regeneration (Welch et al. 2016; Stevens et al. 2017; Haffey et al. 2018; North et al. 2019; Stevens-Rumann and Morgan 2019). The POSCRPT model addresses this issue and highlights distance to seed source as a driver of regeneration (Box 26.2). From a large number of potentially influencing factors, Shive et al. (2018) distilled their prescriptive model into one based on three variables: burn severity, distance to seed source ("seed availability proxy" or SAP), and annual precipitation. Although they modeled species-specific seed production as did Welch et al. (2016), Shive et al. (2018) ultimately lumped seedling species together in order to produce a tool with sufficient predictive power. As such, the principle application of POSCRPT is to identify locations where the probability of conifer seedling presence five years after fire is low enough to warrant management actions.

Box 26.1 – Post-fire Regeneration Decision Support Tools

While there are numerous ways to consider when and how to manage forest regeneration following fire, we present three approaches that guide the user through a different set of considerations that will be potentially useful depending on the user's scope and focus.

(i) Welch et al. (2016) Predictive Conifer Regeneration Tool for the North American Mediterranean-climate Zone (NAMCZ) is designed for users working in the YPMC forests in the first years after fire who want to predict the outcome of regeneration (natural or assisted) following the burn. Users only need basic forestry measurement tools (clinometer, diameter tape or other basal area gauge, range-finder, compass) and familiarity with simple sampling protocols (e.g., common stand exam) to estimate live basal area of a stand. With this data and some basic information about the site (e.g., slope, distance to live seed trees), users use charts provided by the authors to determine within 70–80% accuracy if the site will meet predetermined stocking thresholds five years after the burn. The tool also allows users to predict the number of regenerating conifers at year 5, albeit with less accuracy.

(ii) Shive et al. (2018) Post-fire Spatial Conifer Regeneration Prediction Tool (POSCRPT), also designed for use in YPMC forests, predicts conifer seedling regeneration by modeling annual seed production of regeneration stock from existing basal area and burn severity maps. Unlike the other tools described, POSCRPT is a spatially-explicit habitat suitability model that incorporates site-specific climatic and topographic variables as well as species-specific mean dispersal distances. These attributes make this tool more sophisticated and enable it to produce more site-specific results than the other tools considered here, but they also make it more computationally demanding, difficult to learn and thus less user-friendly to apply.

(iii) Stevens-Rumann and Morgan (2019) Planting Decision Support Tool takes a decision-tree approach to site selection for planting target-species trees with high probability of success. While it is not spatially-explicit in the sense of producing site-specific maps, this tool is nevertheless most appropriate for situations in which managers have specific sites in mind, for which they can answer a successive series of questions such as: what are the natural climate and forest type, followed by distance to living seed tree, followed by aspect and elevation, slope and competing vegetation, etc. Based on the answers at each level, the tool recommends whether or not to plant, or asks for additional information or suggests considering planting species with better odds of success (e.g., more drought-tolerant).

Box 26.2 – Conifer Regeneration Potential in the 2019 Walker Fire (Plumas National Forest) – An Example of POSCRPT Predictive Output

The Sierra Cascade Province Ecology Program (Region 5) of the USDA Forest Service applied POSCRPT, a spatially explicit model developed by Shive et al. (2018), to produce a five-year post-fire predictive map of potential conifer regeneration following the 2019 Walker Fire on the Plumas National Forest. POSCRPT models the probability of natural regeneration as a function of burn severity and other variables, including time since fire, topographic (slope, aspect), climatic (evapotranspiration, annual precipitation, snowpack, climatic water deficit), and seed availability variables. There are five predicted probability classes mapped across the burn area (see map below) that relate to the probability of observing at least one regenerating conifer five years after fire at the 60-m^2 (field plot) scale.

Key Findings

- 50% of the Walker Fire landscape burned at high severity (75–100% basal area mortality).
- 8% of the burned area was within the highest prediction class for seedling density (80–100%)
- In areas that supported yellow pine or mixed conifer forest prior to the fire, 51% were within the two lowest prediction classes for seedling density, where the probability of finding at least one regenerating yellow pine/mixed conifer seedling at the 60-m^2 scale five years post-fire was only 0–20% and 20–40%, respectively.
- While the lowest and next-to-lowest prediction classes had very broad observed seedling density intervals (Table A), the median values for both of these categories was 0 seedlings/ha, meaning over half the burn area would have no short-term conifer regeneration (Figs. A, B, and C).

Table A Key to mapped predicted probability classes based on observed seedling densities in 60 m^2 sample plots for 2019 Walker Fire, Plumas National Forest, California, USA

Predicted Probability Class	Observe Densities (seedlings/ha)			
	Min	Max	Median	Mean (SE)
0 - <20%	0	14,666	0	144(86)
20 - <40%	0	15,333	0	317(55)
40 - <60%	0	17,166	166	672(101)
60 - <80%	0	380,166	333	3,665(985)
80 - 100%	0	210,666	1,333	6,301(755)

POSCRPT offers a distinct improvement over current Forest Service practice in the Pacific Southwest Region, which bases reforestation considerations on basal area mortality maps produced through the MTBS (Monitoring Trends in Burn Severity) program (Finco et al. 2012). In this approach, areas mapped with >50% basal area

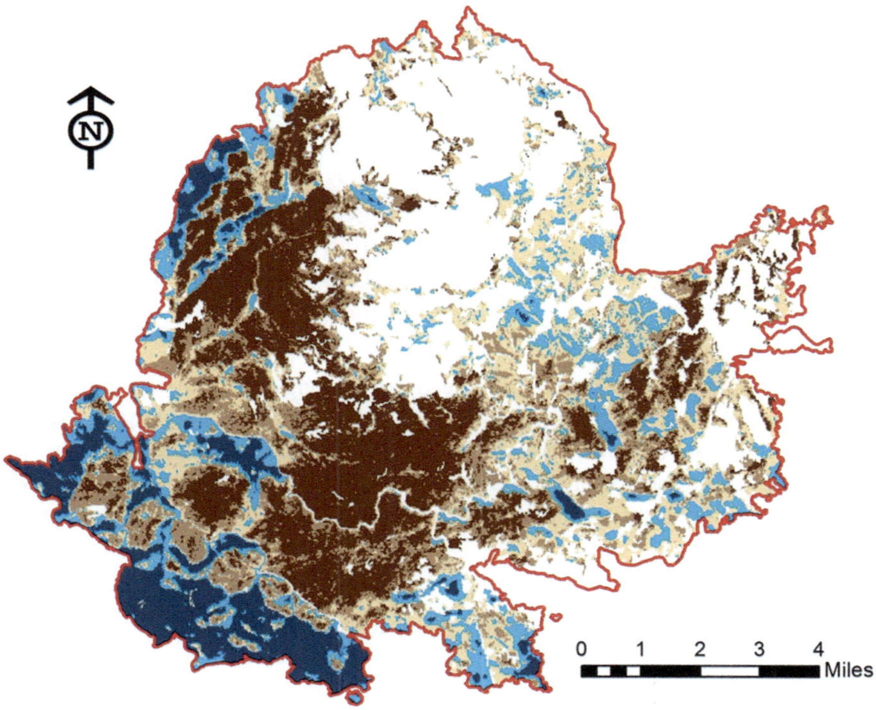

Fig. A Predicted 5-year post-fire conifer regeneration for burn area of the 2019 Walker Fire in the Plumas National Forest, California, USA, based on observed seedling densities in sample plots (Table A). White areas are non-conifer vegetation

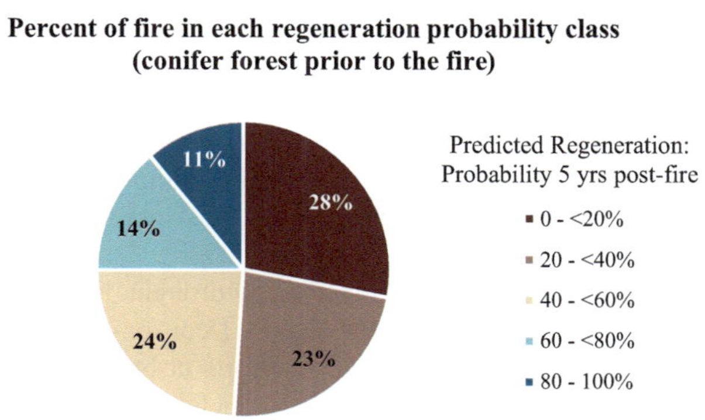

Fig. B Percent of total burn area of the 2019 Walker Fire in the Plumas National Forest (California, USA) that burned in each of the five post-fire predicted regeneration probability classes according to POSCRPT output

mortality are declared "deforested" and reforestation need is assessed using this baseline. However, basal area mortality maps are not predictions of post-fire seedling recruitment and do not include considerations of average climate, burn

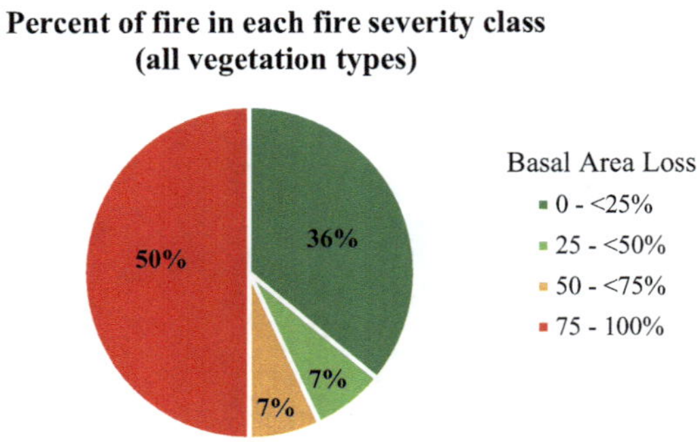

Fig. C Distribution of total burn area of the 2019 Walker Fire in the Plumas National Forest (California, USA) as a percent of each of the four major fire severity classes (defined by percent mortality in basal area)

patch size, or distance to nearest living seed tree (among other things) – all of which are major drivers of tree regeneration. By contrast, POSCRPT incorporates these factors and directly models conifer regeneration. Most recently POSCRPT has been modified to allow consideration of interannual variation in seed availability (driven by masting) and post-fire climate (J. Stewart, unpublished).

26.4 Thinking Longer Term: Post-fire Forest Restoration Strategies

Because managing forested landscapes is an inherently long-term proposition, it is not sufficient to focus only on regeneration densities during a five-year, post-fire time frame. Although five years is within the range of the historic fire return intervals for YPMC forests (Van de Water and Safford 2011), existing fire suppression policies and risks of high severity fire in areas of heavy fuel buildup make it unlikely that wildfire alone will restore the broader suite of historically fire-adapted forest conditions. As such, prescribed fire and/or mechanical treatments that simulate the effects of fire (e.g., mastication, pile burning) may improve regeneration success, channeling regeneration toward desired species composition and stem densities, and increasing resilience in the face of severe fire or stressors such as drought (Zald et al. 2008; Westlind and Kerns 2017; Young et al. 2020).

In Sierra Nevada forest landscapes impacted by drought, bark-beetle outbreaks, and stand-replacing wildfires, the Reforestation Decision-Support Tool (Meyer et al. in Press) can help inform reforestation efforts at broad spatial scales. This spatial prioritization tool allows the user to view data layers related to tree mortality, forest type, fire severity, mechanical treatment constraints, wildlife habitat, and other relevant datasets (e.g., wildland urban interface). Users can select their geographic area of interest, relevant data layers, and relative importance of individual data layers to identify areas of low, moderate, and high reforestation priority. Although this tool does not predict post-disturbance conifer seedling densities, it does highlight areas with low conifer regeneration potential, particularly within areas of high biomass loss such as the interior of high-severity fire patches (based on published estimates of distance to intact forest patches). Users can also view summaries of conifer and hardwood regeneration densities and other stand data within yellow pine and dry mixed conifer forests, based on plot data collected following California's 2012–2016 drought.

Data on the historical or natural range of variation can serve to develop a reference for desirable forest ecosystem conditions (structure, composition, and function) characterized by high ecological integrity and resilience (Landres et al. 1999; Wiens et al. 2012). These dynamics include tree regeneration, forest successional classes, fuel bed development, and other ecosystem characteristics that can inform natural resource management in California's national forests. Several examples of natural range of variation (NRV) assessments currently exist for forested regions in California, including the Sierra Nevada, Southern Cascades, Klamath, and North Coast regions (e.g., Safford and Stevens 2017; Meyer and North 2019; Bohlman et al. In press) which can be supplemented with fire effects and other information (e.g., Fire Effects Information System "https://www.feis-crs.org/feis/").

Meyer et al. (in press) recently developed a broad framework to guide post-fire ecological restoration efforts for the national forests in California. The framework is based on the following ecological restoration principles designed to enhance the recovery of ecosystem integrity across forest and shrubland landscapes:

- Reestablish key ecological processes (e.g., natural fire regimes) to provide for long-term ecosystem integrity and function.
- Plan at the landscape scale with locally implemented restoration projects contributing to landscape restoration goals.
- Support regional native biodiversity and habitat connectivity.
- Seek pragmatic and balanced approaches that sustain diverse ecosystem services.
- Prioritize restoration actions to provide focused and effective management responses.
- Recognize and address agents of change including climate change.

The framework uses a five-step process: (1) form an interdisciplinary team that identifies priority resources, desired conditions, and restoration goals; (2) gather and analyze relevant data to evaluate current and potential future landscape conditions; (3) use a post-fire flow chart to identify restoration opportunities; (4) develop a list of potential management actions that support restoration opportunities; and (5) build a suite of potential restoration actions that support landscape restoration goals by prioritizing actions (called a "restoration portfolio").

The post-fire flow chart evaluates where fire effects improved or degraded ecological conditions, based largely on the natural range of variation and where management actions are feasible. For each distinctive area on the landscape, the restoration "portfolio" identifies one of three outcomes that informs restoration planning and monitoring: (1) maintain or promote desired conditions; (2) take management actions to restore desired conditions; or (3) reevaluate desired conditions where they are likely unattainable due to current and future conditions. A critical step in this framework is the spatial assessment of post-fire landscape condition that is informed by analytical tools and models, including those predicting short- and long-term conifer regeneration or vegetation succession in California's forest ecosystems (e.g., POSCRPT). As such, the framework places analytical models and decision-support tools in the context of post-fire restoration planning efforts in California's national forests.

Although it is widely recognized that YPMC forests that have experienced extended fire suppression need frequent low- to moderate-intensity fire to recover, most evaluations of the effects of fire on YPMC and other forest types have focused on the effects of single burns. Research shows that repeated fire can have a range of potential effects on forest composition, structure and ecosystem functional properties distinct from those caused by a single event (Coppoletta et al. 2016; Stevens-Rumann et al. 2016; Bottero et al. 2017; Westlind and Kerns 2017; Pellegrini et al. 2018). Unfortunately, there are few such studies, and their geographic coverage and the range of forest types considered is patchy. This dissonance between the importance of repeated fire and the lack of related research extends to the available regeneration tools. There is thus an unmet need to incorporate the effects of burn frequency into regeneration models, both for estimating the degree to which current regeneration is approximating historical or desired patterns and for assessing the impact of fire where its frequency exceeds historical averages (Underwood et al. 2018). Similarly, repeated high-severity fires in YPMC forests may induce state shifts away from conifers (Coppoletta et al. 2016; Lyderson et al. 2019), and may limit the recovery of mature hardwoods and associated ecosystem services (Hammett et al. 2017).

26.5 Conclusions

The predictive models and tools described in this chapter have greatly increased our ability to understand post-fire conifer regeneration and successional patterns in California's YPMC forests, especially in the early years following fire. Available tools draw on post-fire conifer regeneration models based on stand productivity, proximity to seed source, and site conditions, and can help predict short-term seedling survival and stand density. Some of these models may also help predict the potential for state shifts to non-forest vegetation types, which among other impacts could result in substantial loss of ecosystem services.

Post-fire conifer regeneration tools can also be incorporated within larger frameworks for post-fire restoration planning, assessing departure from desired landscape conditions, weighing the feasibility of restoration actions, and evaluating the effects of interactive stressors such as drought and insect outbreaks. Other tools developed for post-fire or post-drought forest landscapes inform priority areas for reforestation efforts where natural conifer regeneration is unlikely to meet desired conditions.

There are quite a few obvious directions for future developments in research and technology. These include:

- Post-fire regeneration patterns are not well parameterized in current post-fire regeneration models for species that often occur but are not typically dominant in YPMC forests (e.g. *Abies magnifica* A.Murray, *Pinus contorta* Douglas, *P. monticola* Douglas ex D.Don, *P. monophylla* Torr. & Frém., and *Sequoiadendron giganteum* (Lindl.) J.Buchh.). In addition, models by Welch et al. (2016) and Shive et al. (2018) would benefit from including more plot data from drier forest types (generally dominated by *P. jeffreyi*), the southern Sierra Nevada, and southern California.

- To date, the focus on California YPMC forests has been on the obligate seeding conifers that dominate most of the landscape. A few locally common serotinous conifers also occur, most notably *P. attenuata* Lemmon, as well as a few *Hesperocyparis* species. Hardwood species (especially *Quercus kelloggii* Newberry, *Q. chrysolepis* Liebm., *Q. wislizenii* A.DC., and *Q. garryana* Douglas ex Hook.) are also important members of the community. More needs to be learned about how these species recruit after fire and how their regeneration dynamics interact with the obligate seeding conifers, especially under changing environmental conditions.
- Tracking of regeneration and seedling survival and growth dynamics beyond the five-year, post-fire window should be prioritized.
- A robust regeneration submodule could be developed for the Californian variants of FVS. Currently there are sufficient data to develop such a tool for YPMC forests; other forest types will require more data collection and scientific attention.
- Long-term, post-fire dynamics in riparian areas and historical meadows with and without interventions are not well documented or incorporated into restoration models or strategies.
- Studies of interacting stressors on tree regeneration patterns are relatively rare; drought, repeated fire, and non-fire tree mortality drivers are especially important (see, e.g., Werner et al. 2019, Young et al. 2019)
- There is little available research on conifer regeneration patterns in forest landscapes with partially or completely restored fire regimes, including spatial patterns in regeneration (but see: Stephens and Fry 2005, Stephens and Gill 2005, Dunbar-Irwin and Safford 2016).

Although the research summarized in this chapter is focused on California, similar research and management thinking has been developing in the Mediterranean Basin, which shares strong floristic, physiognomic, and ecological ties with California (e.g., Keeley et al. 2012; Safford and Vallejo 2019). However, most forest management around the Mediterranean occurs in forest stands dominated by serotinous conifers (e.g., *P. halepensis* Mill., *P. brutia* Ten., *P. pinaster* Ait.) and resprouting hardwoods. Serotinous species are rare in California YPMC forests; thus, there has been much less scientific focus on forests or stands dominated by obligate seeding trees. Conservation and management of relatively rare, non-sprouting conifer species, such as *P. nigra* J.F.Arn. and various *Abies* and *Cedrus* species (etc.), would benefit from inventory and monitoring of post-fire regeneration and the development of post-fire regeneration models based on climatic, biophysical, and fire metrics. For these species, rigorous post-fire inventory work has been carried out in some of the Mediterranean nations (see, e.g., Ordóñez et al. 2004, Politi et al. 2009, Christopoulou et al. 2014), but few regeneration models have been published (e.g., Ordóñez et al. 2006). Interest in post-fire regeneration, dynamics, and restoration of Mediterranean Basin forests has increased since the publication of (Moreira et al. 2011), which summarized current knowledge about post-fire forest management and restoration in southern Europe. Strategic thinking about post-fire restoration remains rare in the Mediterranean Basin, however, with one exception being Spain, where a GIS-based, post-fire restoration support tool has been developed (Duguy et al. 2012), as well as a national rubric for post-fire restoration decision making (Vallejo and Alloza 2012; Gobierno_de_España 2013).

References

Agee JK (1993) Fire ecology of Pacific Northwest forests. Island Press, Washington, DC

Allen CD, Savage M, Falk DA, Suckling KF, Swetnam TW, Schulke T, Stacey PB, Morgan P, Hoffman M, Klingel JT (2002) Ecological restoration of southwestern ponderosa pine ecosystems: a broad perspective. Ecol Appl 12

Barbour M, Keeler-Wolf T, Schoenherr AA (2007) Terrestrial vegetation of California. University of California Press, Berkeley

Bohlman G, Skinner C, Safford HD (In press) Natural range of variation (NRV) for yellow pine and mixed conifer forests in northwestern California and southwestern Oregon. in P. S. R. Station, editor. USDA Forest Service, Albany, CA

Bohlman GN, North M, Safford HD (2016) Shrub removal in reforested post-fire areas increases native plant species richness. For Ecol Manag 374:195–210

Bottero A, D'Amato AW, Palik BJ, Kern CC, Bradford JB, Scherer SS (2017) Influence of repeated prescribed fire on tree growth and morialily in Pinus resinosa forests, Northern Minnesota. For Sci 63:94–100

Carlson CH, Dobrowski SZ, Safford HD (2012) Variation in tree mortality and regeneration affect forest carbon recovery following fuel treatments and wildfire in the Lake Tahoe Basin, California, USA. Carbon Balance Manag 7:1–7

Christopoulou A, Fyllas NM, Andriopoulos P, Koutsias N, Dimitrakopoulos PG, Arianoutsou M (2014) Post-fire regeneration patterns of Pinus nigra in a recently burned area in Mount Taygetos, Southern Greece: the role of unburned forest patches. For Ecol Manag 327:148–156

Collins BM, Miller JD, Knapp EE, Sapsis DB (2019) A quantitative comparison of forest fires in central and northern California under early (1911–1924) and contemporary (2002–2015) fire suppression. Int J Wildland Fire 28:138–148

Coppoletta M, Merriam KE, Collins BM (2016) Post-fire vegetation and fuel development influences fire severity patterns in reburns. Ecol Appl 26:686–699

Crotteau JS, Ritchie MW, Varner JM (2013) A mixed-effects heterogeneous negative binomial model for postfire conifer regeneration in Northeastern California, USA. For Sci 60:275–287

Davis KT, Dobrowski SZ, Higuera PE, Holden ZA, Veblen TT, Rother MT, Parks SA, Sala A, Maneta MP (2019) Wildfires and climate change push low-levation forests across a critical threshold for tree regeneration. Proc Natl Acad Sci U S A 116:6193–6198

Dixon GE (2002) Essential FVS: a user's guide to the forest vegetation simulator. in F. M. S. Center, editor. US Department of Agriculture, Forest Service, Fort Collins, CO, USA

Dobrowski SZ, Swanson AK, Abatzoglou JT, Holden ZA, Safford HD, Schwartz MK, Gavin DG (2015) Forest structure and species traits mediate projected recruitment declines in western US tree species. Glob Ecol Biogeogr 24:917–927

Duguy B, Alloza JA, Baeza MJ, De la Riva J, Echeverría M, Ibarra P, Llovet J, Cabello FP, Rovira P, Vallejo RV (2012) Modelling the ecological vulnerability to forest fires in Mediterranean ecosystems using geographic information technologies. Environ Manag 50:1012–1026

Dunbar-Irwin M, Safford H (2016) Climatic and structural comparison of yellow pine and mixed-conifer forests in northern Baja California (Mexico) and the eastern Sierra Nevada (California, USA). For Ecol Manag 363:252–266

Finco M, Quayle B, Zhang Y, Lecker J, Megown KA, Brewer CK (2012) Monitoring trends and burn severity (MTBS): monitoring wildfire activity for the past quarter century using Landsat data. In: Morin RS, Liknes GC (eds) Moving from status to trends: Forest Inventory and Analysis (FIA) symposium, Gen. Tech. Rep. NRS-P-105. US Department of Agriculture, Forest Service, Northern Research Station, Baltimore, pp 222–228

Gobierno_de_España (2013) Guía técnica para la gestión de montes quemados: protocolos de actuación para la restauración de zonas quemadas con riesgo de desertificación. Ministerio de Agricultura, Alimentación y Medio Ambiente, Madrid

Haffey C, Sisk TD, Allen CD, Thode AE, Margolis EQ (2018) Limits to ponderosa pine regeneration following large high-severity forest fires in the United States Southwest. Fire Ecol 14:143–163

Hammett EJ, Ritchie MW, Berrill JP (2017) Resilience of California black oak experiencing frequent fire: regeneration following two large wildfires 12 years apart. Fire Ecol 13:91–103

Keeley JE, Bond WJ, Bradstock RA, Pausas JG, Rundel PW (2012) Fire in Mediterranean ecosystems: ecology, evolution and management. Cambridge University Press, Cambridge

Keeley JE, Safford HD (2016) Fire as an ecosystem process. In: Mooney H, Zavaleta E (eds) Ecosystems of California. University of California Press, Oakland, pp 27–45

Kitzmiller JH, Lunak G (2012) Growth of giant sequoia compared to ponderosa pine and other mixed-conifers in California plantations. West J Appl For 27:196–204

Landres PB, Morgan P, Swanson FJ (1999) Overview of the use of natural variability concepts in managing ecological systems. Ecol Appl 9:1179–1188

Long J, Lake FK, Lynn K, Viles C (2018) Chapter 11: Tribal ecocultural resources and engagement. In: Spies TA, Stine PA, Gravenmier R, Long JW, Reilly MJ (eds) Synthesis of science to inform land management within the Northwest Forest Plan area, Gen. Tech. Rep. PNW-GTR-966. U.S. Department of Agriculture, Forest Service, Pacific Northwest Research Station, Portland, pp 851–917

Long JW, Goode RW, Gutteriez RJ, Lackey JJ, Anderson MK (2017) Managing California black oak for tribal ecocultural restoration. J For 115:426–434

Lutz JA, Furniss TJ, Germain SJ, Becker KML, Blomdahl EM, Jeronimo SMA, Cansler CA, Freund JA, Swanson ME, Larson AJ (2017) Shrub communities, spatial patterns, and shrub-mediated mortality following reintroduced fire in Yosemite National Park, California, USA. Fire Ecol 13:104–126

Lyderson JM, Collins BM, Coppoletta M, Jaffe MR, Northrop H, Stephens SL (2019) Fuel dynamics and reburn severity following high-severity fire in a Sierra Nevada, USA, mixed conifer forest. Fire Ecol 15:43

Mallek C, Safford H, Viers J, Miller J (2013) Modern departures in fire severity and area vary by forest type, Sierra Nevada and southern Cascades, California. USA Ecosphere 4:1–28

Meigs GW, Donato DC, Campbell JL, Martin JG, Law BE (2009) Forest fire impacts on carbon uptake, storage, and emission: the role of burn severity in the Eastern Cascades, Oregon. Ecosystems 12:1246–1267

Meyer M, North M (2019) Natural range of variation of red fir and subalpine forests in the Sierra Nevada bioregion. PSW-GTR-263, Pacific Southwest Research Station

Meyer M, Steel ZL, North M, Wuenschel A, Ostoja SM (in Press) Reforestation decision support tool for tree mortality landscapes. Pacific Southwest Research Station, USDA Forest Service, Albany

Meyer MD, Long JW, Safford HD (eds) (2021) Postfire restoration framework for national forests in California. Gen. Tech. Rep. PSW-GTR-270. U.S. Department of Agriculture, Forest Service, Pacific Southwest Research Station, Albany, CA, 204 p

Miles SR, Goudey, CB (1997) Ecological subregions of California: section and subsection descriptions. In Pacific Southwest Region, editor. USDA Forest Service, San Francisco, CA

Minnich RA, Barbour MG, Burk JH, Fernau RF (1995) Sixty years of change in Californian conifer forests of the San Bernardino Mountains. Conserv Biol 9:902–914

Missoula_Fire_Sciences_Laboratory (2018) Fire regimes of California montane mixed-conifer communities: Information from the Pacific Southwest Research Station and LANDFIRE. In F. E. I. System, editor. U.S. Department of Agriculture, Forest Service, Missoula, MT, USA

Moreira F, Arianoutsou M, Corona P, De las Heras J (eds) (2011) Post-fire management and restoration of southern European forests. Springer Science & Business Media, Dordrecht

North MP, Stevens JT, Greene DF, Coppoletta M, Knapp EE, Latimer AM, Restaino CM, Tompkins RE, Welch KR, York RA, Young DJN, Axelson JN, Buckley TN, Estes BL, Hager RN, Long JW, Meyer MD, Ostoja SM, Safford HD, Shive KL, Tubbesing CL, Vice H, Walsh D, Werner CM, Wyrsch P (2019) Tamm review: reforestation for resilience in dry western US forests. For Ecol Manag 432:209–224

Ordóñez JL, Franco S, Retana J (2004) Limitation of the recruitment of Pinus nigra in a gradient of post-fire environmental conditions. Ecoscience 11:296–304

Ordóñez JL, Molowny-Horas R, Retana J (2006) A model of the recruitment of Pinus nigra from unburned edges after large wildfires. Ecol Model 197:405–417

Pellegrini AFA, Ahlstrom A, Hobbie SE, Reich PB, Nieradzik LP, Staver AC, Scharenbroch BC, Jumpponen A, Anderegg WRL, Randerson JT, Jackson RB (2018) Fire frequency drives decadal changes in soil carbon and nitrogen and ecosystem productivity. Nature 553:194–198

Politi PI, Arianoutsou M, Stamou GP (2009) Patterns of Abies cephalonica seedling recruitment in mount Aenos National Park, Cephalonia, Greece. For Ecol Manag 258:1129–1136

Safford H, Stevens JT (2017) Natural range of variation for yellow pine and mixed-conifer forests in the Sierra Nevada, southern Cascades, and Modoc and Inyo National Forests, California, USA. Gen. Tech. Rep. PSWGTR-256. U.S. Department of Agriculture, Forest Service, Pacific Southwest Research Station, Albany

Safford H, Van de Water K (2014) Using fire return interval departure (FRID) analysis to map spatial and temporal changes in fire frequency on National Forest Lands in California. Page 66 in P. S. R. Station, editor. USDA Forest Service

Safford HD, Vallejo VR (2019) Ecosystem management and ecological restoration in the Anthropocene: integrating global change, soils, and disturbance in boreal and Mediterranean forests. Dev Soil Sci 36:259–308

Serra-Diaz JM, Maxwell C, Lucash MS, Scheller RM, Laflower DM, Miller AD, Tepley AJ, Epstein HE, Anderson-Teixeira KJ, Thomposn JR (2018) Disequilibrium of fire-prone forests sets the stage for a rapid decline in conifer dominance during the 21st century. Sci Rep 8:1–12

Shive KL, Preisler HK, Welch KR, Safford HD, Butz RJ, O'Hara KL, Stephens SL (2018) From the stand scale to the landscape scale: predicting the spatial patterns of forest regeneration after disturbance. Ecol Appl 28:1626–1639

Steel ZL, Koontz MJ, Safford HD (2018) The changing landscape of wildfire: burn pattern trends and implications for California's yellow pine and mixed conifer forests. Landsc Ecol 33:1159–1176

Steel ZL, Safford HD, Viers JH (2015) The fire frequency-severity relationship and the legacy of fire suppression in California forests. Ecosphere 6

Stephens SL, Collins BM (2004) Fire regimes of mixed conifer forests in the north-central Sierra Nevada at multiple spatial scales. Northwest Science 78:12–23

Stephens SL, Fry DL (2005) Spatial distribution of regeneration patches in an old-growth Pinus Jeffrey-mixed conifer forest in northwestern Mexico. J Veg Sci 16:693–702

Stephens SL, Gill SJ (2005) Forest structure and mortality in an old-growth Jeffrey pine-mixed conifer forest in north-western Mexico. For Ecol Manag 205:15–28

Stephens SL, Lydersen JM, Collins BM, Fry DL, Meyer MD (2015) Historical and current landscape-scale ponderosa pine and mixed conifer forest structure in the Southern Sierra Nevada. Ecosphere 6

Stevens-Rumann CS, Morgan P (2019) Tree regeneration following wildfires in the western US: a review. Fire Ecol 15

Stevens-Rumann CS, Prichard SJ, Strand EK, Morgan P (2016) Prior wildfires influence burn severity of subsequent large fires. Can J For Res 46:1375–1385

Stevens JT, Collins BM, Miller JD, North MP, Stephens SL (2017) Changing spatial patterns of stand-replacing fire in California conifer forests. For Ecol Manag 406:28–36

Tepley AJ, Thompson JR, Epstein HE, Anderson-Teixeira KJ (2017) Vulnerability to forest loss through altered postfire recovery dynamics in a warming climate in the Klamath Mountains. Glob Chang Biol 23:4117–4132

Underwood EC, Safford HD, Molinari NA, Keeley JE (2018) Valuing chaparral ecological, socio-economic, and management perspectives preface. In: Underwood EC, Safford HD, Molinari NA, Keeley JE (eds) Valuing chaparral: ecological, socio-economic, and management perspectives, pp IX-XI

Vallejo VR, Alloza JA (2012) Post-fire management in the Mediterranean Basin. Isr J Ecol Evol 58:251–264

Van de Water KM, Safford HD (2011) A summary of fire frequency estimates for California vegetation before Euro-American settlement. Fire Ecol 7:26–58

Van Wagtendonk JW, Sugihara NG, Stephens SL, Thode AE, Shaffer KE, Fites-Kaufman JA (eds) (2018) Fire in California's ecosystems. University of California Press, Oakland

Welch KR, Safford HD, Young TP (2016) Predicting conifer establishment post wildfire in mixed conifer forests of the North American Mediterranean-climate zone. Ecosphere 7

Werner CM, Young DJN, Safford HD, Young TP (2019) Decreased snowpack and warmer temperatures reduce the negative effects of interspecific competitors on regenerating conifers. Oecologia 191:731–743

Westlind DJ, Kerns BK (2017) Long-term effects of burn season and frequency on ponderosa pine forest fuels and seedlings. Fire Ecol 13:42–61

White AM, Long JW (2019) Understanding ecological contexts for active reforestation following wildfires. New For 50:41–56

Wiens JA, Hayward GD, Safford HD, Giffen CM (2012) Historical ecological variation in conservation and natural resource management. Wiley-Blackwell, Oxford

Young DJN, Meyer M, Estes B, Gross S, Wuenschel A, Restaino C, Safford HD (2020) Forest recovery following extreme drought in California, USA: natural patterns and effects of predrought management. Ecol Appl 30

Young DJN, Werner CM, Welch KR, Young TP, Safford HD, Latimer AM (2019) Post-fire forest regeneration shows limited climate tracking and potential for drought-induced type conversion. Ecology 100

Zald HSJ, Gray AN, North M, Kern RA (2008) Initial tree regeneration responses to fire and thinning treatments in a Sierra Nevada mixed-conifer forest, USA. For Ecol Manag 256:168–179

Chapter 27
Remote Sensing Techniques for Monitoring Fire Damage and Recovery of Mediterranean Pine Forests: *Pinus pinaster* and *Pinus halepensis* as Case Studies

Alfonso Fernández-Manso, Carmen Quintano, Susana Suarez-Seoane, Elena Marcos, and Leonor Calvo

27.1 Maritime and Aleppo Pine Forests in a World of Increasing Forest Fires

Pine and mixed pine forests are highly representative ecosystems of the Mediterranean Basin, providing goods and services to society. Both *Pinus halepensis* Mill. (Aleppo pine) and *P. pinaster* Ait. (maritime pine) are native pines in the Mediterranean region and dominate the current forested Mediterranean landscape (Fig. 27.1). Aleppo pine is the most widely distributed and abundant among the Mediterranean pines, covering nearly 7 million ha in the western, eastern and southern Mediterranean Basin. By contrast, maritime pine is a widespread conifer distributed mainly in the western Mediterranean Basin, across a broad range of elevation, climate and soil, resulting in significant genetic variation (Fernandes and Rigolot 2007).

Both species are very frequently affected by forest fires in Southern Europe (da Ponte et al. 2019). They are highly flammable and dominate forest ecosystems with high horizontal and vertical fuel continuity (Fernández-García et al. 2019a, b). This

A. Fernández-Manso (✉)
Agrarian Science and Engineering Department, University of León, Ponferrada, Spain
e-mail: alfonso.manso@unileon.es

C. Quintano
Electronic Technology Department, University of Valladolid, Valladolid, Spain

Sustainable Forest Management Research Institute, University of Valladolid-INIA, Palencia, Spain

S. Suarez-Seoane · E. Marcos · L. Calvo
Area of Ecology, Faculty of Biological and Environmental Sciences, University of León, León, Spain

G. Ne'eman, Y. Osem (eds.), *Pines and Their Mixed Forest Ecosystems in the Mediterranean Basin*, Managing Forest Ecosystems 38,
https://doi.org/10.1007/978-3-030-63625-8_27

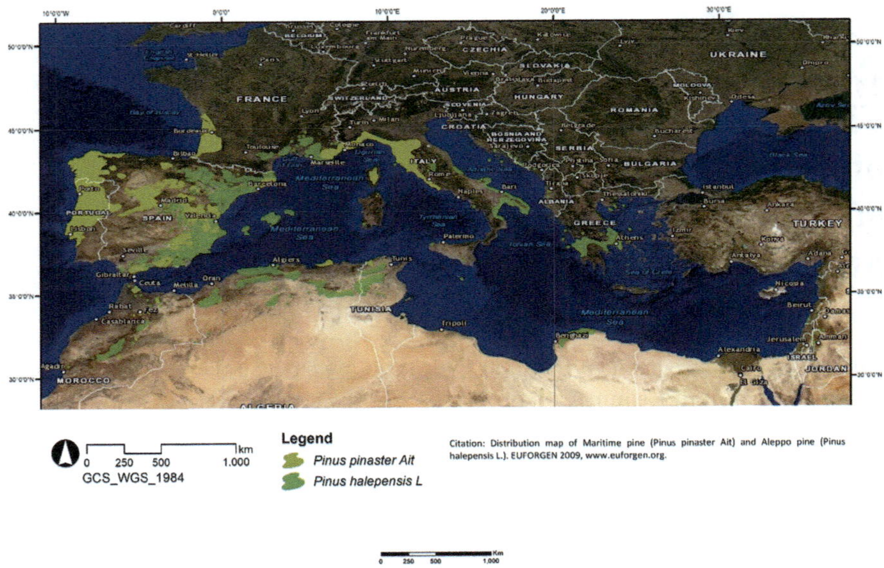

Fig. 27.1 Distribution map of *P. halepensis* and *P. pinaster* in the Western Mediterranean Basin. (Data obtained from Euforgen.org)

fuel profile facilitates the occurrence of crown fires that cause widespread mortality (Calvo et al. 2008). The response of these species to fire has been classified into two categories that, in general, are considered complementary: (i) individual survival mainly characterized by thick bark, thick protected buds, self-pruning, deep rooting and rapid growth; (ii) stand resilience characterized by the presence of a large canopy seed bank (serotiny) that ensures abundant post-fire seedling recruitment (de las Heras et al. 2012). These mechanisms, in general, ensure the resilience of Mediterranean pine forests after fire, although it is also determined by population traits such as large cone crops, massive regeneration after fire, rapid juvenile growth and a short juvenile period (Calvo et al. 2008). *Pinus halepensis* and *P. pinaster* have demonstrated, as an adaptation to fire, a high number of serotinous cones, mainly in those populations with high fire recurrence (Moya et al. 2018). The seeds stored in serotinous cones are protected from fire and are massively released after the disturbance, finding proper conditions for germination, seedling establishment and growth (Calvo et al. 2013). This mechanism ensures the regeneration of *P. pinaster* and *P. halepensis* forests, with very high maritime pine seedling recruitment after only one fire (Calvo et al. 2008). However, Fernández-García et al. (2019b) and García-Llamas et al. (2019a) discussed the effect of burn severity on its regeneration process, because high severity could cause significant seed mortality. Another parameter of the fire regime that could significantly reduce post-fire regeneration of both pine species is fire recurrence. Recurrence of fires at short intervals induces a reduction in the regeneration capacity because these pines need at least 15 or 20 years to develop a fully productive canopy seed bank (Pausas et al. 2008).

In general, fire frequency and burn severity directly affect pine regeneration, but additional indirect effects are observed through changes in the abundance and composition of the woody understory community (Fernández-García et al. 2018a). Frequent fires facilitate the encroachment of understory species, such as resprouter species (e.g. *Erica australis* L.) (Calvo et al. 1998; Fernández-García et al. 2018b), which compete with pines for light, space and nutrients (Calvo et al. 2008, 2013; Taboada et al. 2017). Although these two Mediterranean serotinous pine tree species are adapted to certain fire regimes, rural abandonment, which induces an increase in fuel loads and continuity (García-Llamas et al. 2019b), as well as predicted climate change, with warmer and drier summers, lead to increases in fire frequency and burn severity that can hinder pine regeneration (Fernández-García et al. 2019b).

In studies carried out after two large wildfires (in 2012 in a forest dominated by *P. pinaster* in northwest Spain (León province) and in a forest dominated by *P. halepensis* in eastern Spain (Valencia) with different fire regime conditions of burn severity and fire recurrence), we have demonstrated the significant effects of both drivers on the recruitment and development of pine seedlings, as well as on competition between pine seedlings and woody understory species (Fernández-García et al. 2018a, b). Fire frequency significantly decreased the density, cover and height of pine saplings in the *P. pinaster* and *P. halepensis* forests (Figs. 27.2 and

Fig. 27.2 *Pinus halepensis* seedling recruitment in the high burn severity and low recurrence scenarios

Fig. 27.3 Strong woody understory regeneration in the *P. pinaster* forest under high burn severity and high recurrence scenarios

27.3). At the same time, burn severity mainly reduced the density of pine saplings. The negative effects of inter-specific competition between cover of woody understory species and pine regeneration is more important in high fire frequency scenarios (Fernández-García et al. 2019b).

27.2 Evaluation of Fire Damage in Mediterranean Pine Forests by Remote Sensing Techniques

The effects of fire on ecosystems are controlled by fire regime parameters, amongst other factors. In this sense, burn severity can alter the resilience of vegetation and soil properties, both of which are critical for post-fire forest management. It has also been identified as one of the most critical factors determining the ecological effect of fire on ecosystems (Tanase et al. 2011). Consequently, the timely generation of reliable maps of burn severity that reflect induced changes in vegetation and soil properties is a high priority for supporting post-fire decision-making in both the short and long term (Miller et al. 2009).

Traditionally, burn severity assessment has been based on field methods such as the Composite Burn Index (CBI) and the GeoCBI Index (Key and Benson 2006; de Santis and Chuvieco 2009). However, field methods are often expensive and

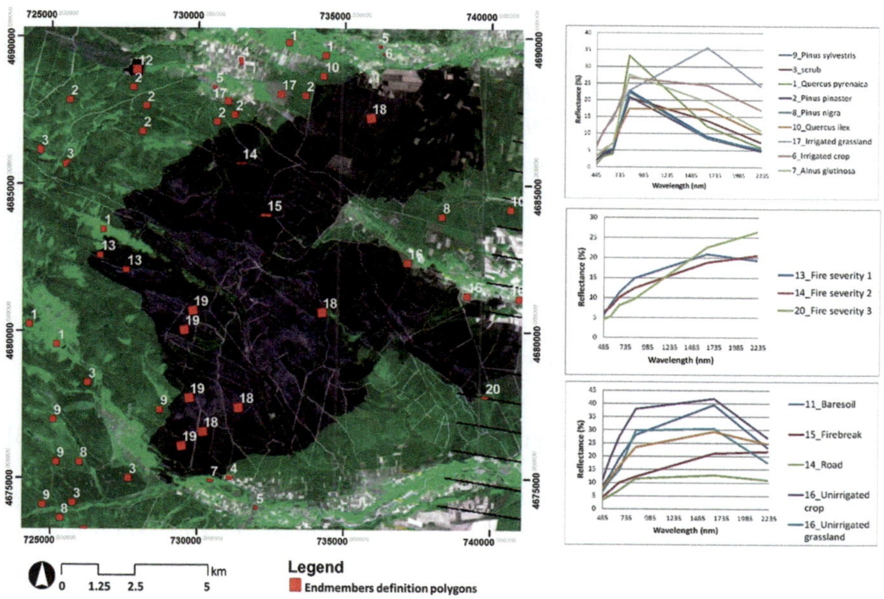

Fig. 27.4 Left: Field plots to measure burn severity after the Castrocontrigo fire. (León, Spain) over a Landsat 7 (ETM+) RGB:741 color composition. Right: Spectral signatures of vegetation (upper), char (center) and soil (lower) included in the spectral library to unmix the Landsat data

time-consuming, and provide limited spatial and temporal representation of the ecological effects of fire. Fire causes substantial spectral and thermal changes to the earth's surface, associated with the consumption of vegetation and exposure of soil and charred stems, which can be captured by remote sensing sensors. Following a fire event, a dramatic reduction in near infrared reflectance (NIR, 700–1300 nm) and an increase in shortwave infrared reflectance (SWIR, 1500–2300 nm), both associated with carbonization and elimination of vegetation, is the principal information recorded by pre- and post-fire sensors. As an example, Fig. 27.4 displays the spectral library of the Castrocontrigo fire (León, Spain) in a Landsat 7 image one month after fire. Spectra 13, 14 and 20 are representative of the burned surfaces, and contrast with spectrum 2, associated with *P. pinaster*, the dominant species in the area.

In recent years, spectral indices based on remotely sensed data have been developed to map and monitor the effects of forest fires. Among them, the Normalized Burn Ratio (NBR) (based on NIR and SWIR spectral bands) and, in particular, differenced NBR (dNBR), is the most widely used index (Key and Benson 2006). Relative dNBR (RdNBR) has subsequently been proposed to eliminate the influence of pre-fire vegetation. It hypothetically enables creation of categorical classifications using the same thresholds for fires that occur in similar types of vegetation without the need to acquire data calibration for each fire (Miller et al. 2009). More recently, a new index for estimating burn severity, the relativized burn ratio (RBR)

Table 27.1 Spectral indexes derived from Normalized Burn Ratio

Acronym	Spectral index	Equation	Reference
NBR	Normalized burn ratio	$\frac{NIR - SWIR}{NIR + SWIR}$	López-García and Caselles (1991)
dNBR	Differenced NBR	$prefireNBR - postfireNBR$	Key and Benson (2006)
RdNBR	Relative dNBR	$\frac{dNBR}{\sqrt{ABS(prefireNBR / 1000)}}$	Miller et al. (2009)
RBR	Relativized burn ratio	$\frac{dNBR}{prefireNBR + 1.001}$	Parks et al. (2014)

(Parks et al. 2014), has been proposed as an alternative to dNBR and RdNBR (Table 27.1).

Burn severity maps are usually obtained from these spectral indices (Arnett et al. 2015; Cardil et al. 2019; Fernández-García et al. 2018b; Harris et al. 2011; Lhermitte et al. 2011; McCarley et al. 2017; Quintano et al. 2018; Stambaugh et al. 2015, among others). In these maps, burn severity is represented by discrete thematic categories, generally distinguishing between unburned and low, moderate and high burn severity levels. Since the indices provide continuous data to classify severity into discrete categories, it is necessary to establish cut-off thresholds. The burn severity map of the Gátova fire (Fig. 27.5; Valencia, Spain) is based on the classification of a dNBR image (Landsat 8) using the thresholds proposed by Botella and Fernández-Manso (2017). Details 1–4 in Fig. 27.5 show the different levels of burn severity in a fire that occurred in Aleppo pine ecosystems in 2017.

From an ecological point of view, burn severity classification is closely linked to the time elapsed between fire occurrence and the measurement of fire damage. Thus, two different assessment intervals are distinguished: initial and extended, each one associated to different information content and functions. The initial assessment is the first opportunity to obtain an essentially complete ecological assessment of fire damage, since it is ideally performed immediately after the fire has been extinguished, and good quality data are also available. Elapsed time should range between fire occurrence and 8 weeks later (Key and Benson 2006). By contrast, the extended assessment is made during the first growing season after the fire. It enables assessment of the delayed survival of vegetation that has burned, but whose roots or stems remain viable and regenerate, and delayed mortality, where plants appear externally healthy immediately after the fire, but eventually die due to root damage. Therefore, the choice of assessment type depends on the objectives pursued.

New methodologies have been tested to improve burn severity estimates. Quintano et al. (2013) obtained a highly accurate estimate using Multiple Endmember Special Analysis (MESMA) fraction images instead of spectral indices. Their success is related to the characteristics of the short-term domain after the

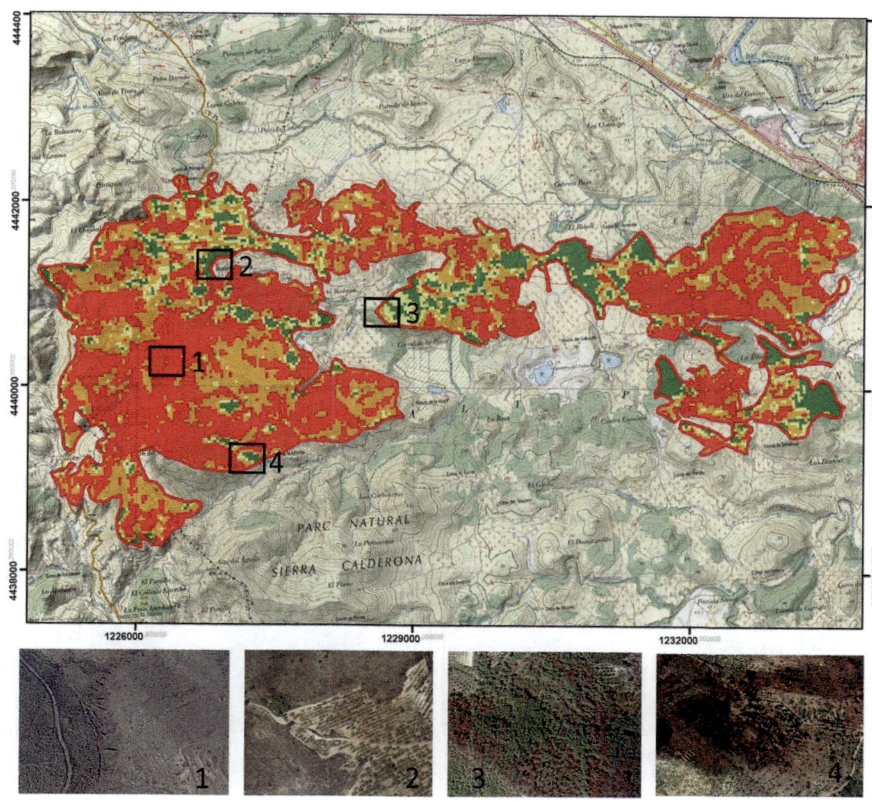

Fig. 27.5 Burn severity map of the Gátova fire (Valencia) (red: high burn severity; orange: moderate burn severity; yellow: low burn severity; green: unburned). Details 1–4 represent the burn severity in the locations indicated on the map

fire that generally produces a mixture of vegetation, ash and burnt soil, which is essentially a sub-pixel question in the spatial resolution of commonly used multispectral sensors. Fraction images have a physical meaning; they estimate the fraction of the different land covers present in a pixel, thus making them easier to interpret. Other research works on post-fire burn severity mapping have been also based on fraction images, in particular on char fraction (Dennison et al. 2006; Fernández-Manso et al. 2019; Quintano et al. 2013, 2017, 2019; Tane et al. 2018; Veraverbeke et al. 2012, 2014).

Perhaps the most novel approach in the study of burn severity is that which is based on variables related to energy balance. Post-fire changes in vegetation, soil and water balance corroborate the impact of forest fires on the energy balance. Fire induces modifications in vegetation structure and species composition and alters latent heat flux, closely related to evapotranspiration, and other variables in the energy balance equation (Quintano et al. 2015). Evapotranspiration (ET), land surface temperature (LST) and land surface albedo (LSA) can discriminate different

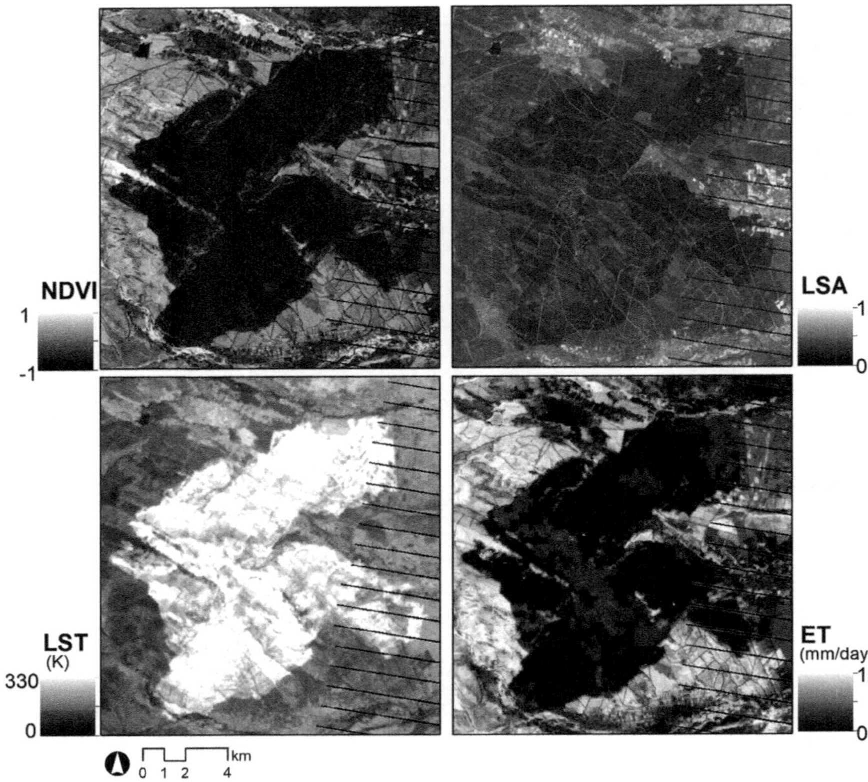

Fig. 27.6 Representation of the normalized difference vegetation index (NDVI), evapotranspiration (ET), land surface temperature (LST), and land surface albedo (LSA) for the Castrocontrigo fire. (León, Spain) from a Landsat 7 ETM+ acquired on 6 September 2012

levels of burn severity using the normalized difference vegetation index (NDVI) as a reference (see Fig. 27.6).

Burn severity mapping is not limited to optical sensors. Amongst other applications in forestry management, LiDAR sensors are extremely useful for assessing burn severity. They provide metrics for pre- and post-fire vegetation structure, which is modified to different degrees depending on burn severity. The combination of LiDAR and multispectral satellite data for burn severity assessment is an ongoing research topic. Recently, a maximum entropy model trained with EO-1 Hyperion MESMA fraction images and LiDAR-derived vegetation structure metrics was proposed to assess burn severity (Fernández-Manso et al. 2019). The data were acquired in Valencia (Spain) in a forest dominated by Aleppo pine. The obtained model has good performance as the area under curve (AUC) values proved (AUC values >0.85).

To date, the development of new remote sensing techniques has allowed more accurate estimation of burn severity and evaluation of the effect of fuel conditions on fire severity (García-Llamas et al. 2019a). However, a suitable methodology for estimating soil burn severity in large fires has yet to be proposed. For example, spectral indices (reflective, thermal and mixed) derived from Landsat 8 OLI/TIRS used to estimate burn severity in *Pinus* forests along the Mediterranean-Transition-Oceanic climatic gradient, showed a greater ability to determine ecosystem and vegetation burn severity than soil burn severity (Fernández-García et al. 2018b) and differences in index performance among the climatic regions. The use of hyperspectral imagery shows more potential for determining changes in soil properties after wildfire, particularly with respect to the carbon stock (Peón et al. 2017). However, the results must be validated with soil samples analyzed in a laboratory. For this reason, the use of field data combined with remote sensed assessments is essential for measuring and understanding the effects of fire in forest ecosystems (Cardil et al. 2019) and transferring this information to managers.

Field burn severity assessment can be carried out by means of visual indicators or indices, which generally reflect changes in vegetation, forest floor and mineral soil. One of the most commonly used is the CBI index (Key and Benson 2006), which was developed as an operational methodology for burn severity assessment and is well-adapted to estimating burn severity variations in forests. However, this index must be modified for use in different forest ecosystems since burn severity depends on soil and vegetation, among other factors. First of all, it is important to know whether the CBI index reflects the real impact on the ecosystem after a wildfire. Both Marcos et al. (2018) and Fernández-García et al. (2019c) reported a good match between changes in soil properties (biological, chemical and physical) and the CBI index in Mediterranean pine forests (Fig. 27.7).

Similarly, the understory CBI shows a high capacity to predict microbial community response to fire (Whitman et al. 2019). This approach would make it possible to better understand relationships between remote sensing techniques and field measurements of severity. For this reason, when working with CBI, it is necessary to select indicators that reflect changes in soil and vegetation, such as char depth, litter consumption or foliage consumption (Marcos et al. 2018; Fernández-García et al. 2019c) more accurately. From a management point of view, the immediate identification of fire-induced changes in soil properties that may result in an unacceptable risk to human safety or natural resources must be a priority objective immediately after fire, to facilitate formulation of post-fire management plans.

Fig. 27.7 Examples of soils affected by different soil burn severity values in Composite Burn Index (CBI) units: (**a**) Low soil burn severity (CBI 0.90) with blackened litter and no changes in mineral soil); (**b**) Moderate soil burn severity (CBI 2.15) with light fuel consumption, where charred remains are not completely recognizable; (**c**) High soil burn severity (CBI 3.00) where all ash is white and the soil has completely reddened

27.3 Remote Sensing Techniques for Monitoring Post-Fire Mediterranean Pine Regeneration

Post-fire regeneration is a heterogeneous process that varies spatially depending on fire regime parameters, pre-fire species composition, and environmental factors such as edaphic or climatic parameters (Taboada et al. 2017; Fernández-García et al. 2019b). In Mediterranean fire-prone pine forests affected by large fires, the assessment of local variations in both understory and forest canopy recovery has substantial implications for defining priority areas for management actions (Ruíz-Gallardo et al. 2004). Since the field-based monitoring of vegetation recovery in large burnt areas might be unfeasible in terms of data collection, forest managers require new tools aimed at collecting data efficiently while minimizing costs. In this sense, remote sensing techniques facilitate collection of environmental data at different spatial and spectral resolutions over large areas with little effort. The combination of this information with field data in spatially explicit modeling approaches has become an essential tool for assessing vegetation recovery at the landscape scale (Fernández-Guisuraga et al. 2019).

In recent decades, post-fire vegetation recovery in forest landscapes dominated by *P.* and *P. halepenis* has been successfully evaluated under different fire-regime scenarios of recurrence and severity using temporal series of satellite imagery collected at medium spatial resolution, such as Landsat MSS, TM, ETM+ or OLI (Díaz-Delgado et al. 2003; Viana-Soto et al. 2017). In this approach, the most frequently applied methodologies have been: (i) Time Series Analysis based on spectral indices, such as NDVI (Vicente-Serrano et al. 2011); and, (ii) Spectral Mixture Analysis (Fernández-Manso et al. 2016). Currently, a main challenge in fire ecology is to transfer these approaches to high and ultrahigh resolution time series of satellite data in order to evaluate post-fire regeneration in spatially heterogeneous landscapes (Hirschmugl et al. 2017).

High spatial resolution satellite imagery, such as that provided by Deimos-2, GeoEye-2, QuickBird or WorldView-2 on-board sensors, offers the possibility for detailed assessment of post-fire regeneration in terms of fractional vegetation cover, species richness or basal area of tree species, among other parameters. For instance, Fernández-Guisuraga et al. (2019) demonstrated in a pine forest landscape affected by a megafire in 2012 (León province, northwest of Spain) that spectral indices and textural information derived from WorldView-2 provided highly valuable information on post-fire structure of both the *P. pinaster* population (number and cover of pine seedlings) and the understory community (cover, height and richness of woody species, percentage of bare soil, necromass and leaves) that can be used as a basis for post-fire management. Indeed, WorldView-2 imagery facilitates evaluation of adaptive strategies of regeneration in the understory woody species; the fact that resprouters are the functional group of species demonstrating the highest cover and habitat suitability at high fire recurrence has implications for landscape dynamics.

Despite the general success of this approach, high resolution satellite imagery may have limited applicability in particularly heterogeneous and dynamic areas. In

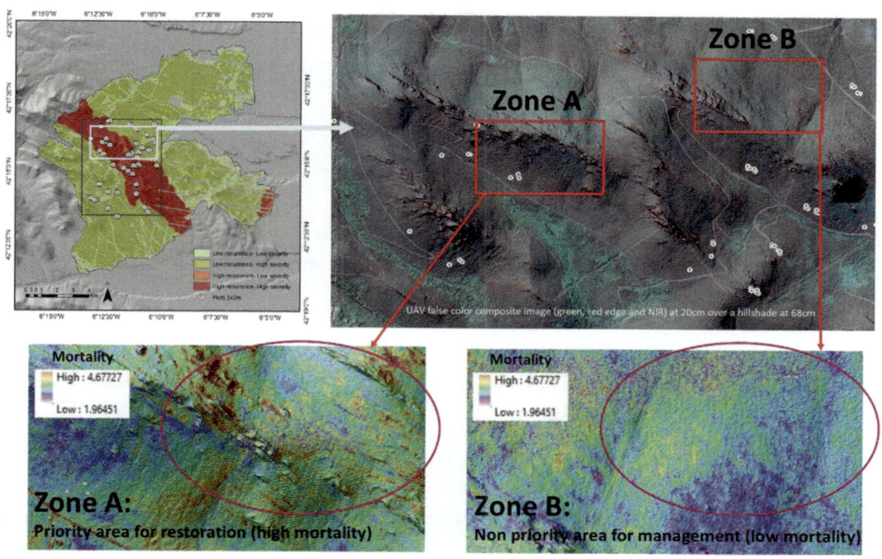

Fig. 27.8 Priority areas for restoration identified in Castrocontrigo (León, Spain) from a pine mortality model based on multspectral drone imagery at ultra-high spatial resolution

such a case, the use of unmanned aerial vehicles (UAVs) confers advantages such as the achievement of ultra-high spatial resolution (better than 20 cm) imagery on demand. In fact, Fernández-Guisuraga et al. (2018) found that multispectral orthomosaics collected by drones provided more information in terms of spatial variability than WorldView-2 satellite imagery, becoming a potential alternative for evaluating post-fire regeneration in highly heterogeneous burned areas dominated by *P. pinaster*. Furthermore, the combination of multispectral drone imagery with population data (pine mortality rates) of the dominant tree species in mechanistic/correlative models offers a suitable framework for landscape decision-makers, since it allows detection of vulnerable zones requiring restoration across large and heterogenous, disturbed areas, as well as the landscape structures supporting population recovery (Fig. 27.8).

27.4 Conclusions: Wildfire Risk Management for the Reduction of Fire Damage in Mediterranean Pine Forests

Remote sensing plays a fundamental role in risk assessment – accurately predicting the occurrence of forest fires and measuring their impacts on people and ecosystem goods and services (Vaillant et al. 2016). Risk is a function of threat and vulnerability. Vulnerability is limited by the exposure, sensitivity and adaptability of a system

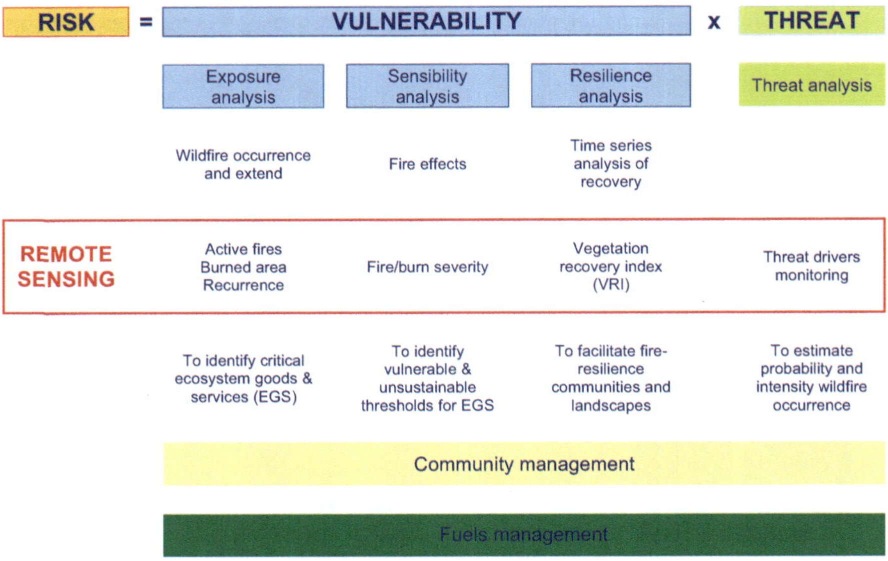

Fig. 27.9 Risk is a function of threat and vulnerability. Reducing risk involves managing community and fuels

(resilience). Risk management requires that the threat and/or vulnerability be addressed. In all these tasks, remote sensing is presented as a powerful tool (Fig. 27.9).

Analysis of forest fire occurrence and extent is essential in the study of vulnerability. Remote sensing offers spatially continuous detection and observation of fires with daily temporal resolution. A database (whose spatial and temporal resolution depends on the used sensor) of fire occurrence and extent can be created using these data (Schroeder et al. 2014). In this sense it is of vital importance to historically analyze the attributes of the fire regime in Mediterranean ecosystems (Fernández-García et al. 2018a).

Landscape sensitivity can be quantified in terms of the effects of forest fire on key ecosystem goods and services. The focal point of sensitivity quantification is to develop burn severity databases. A thorough assessment of ecological change after fires is the key to understanding, predicting and measuring fire effects. The study of burn severity in Mediterranean pine forests using remote sensing data and techniques has produced very accurate estimates at different spatial resolutions (Fernández-Manso et al. 2016).

In Mediterranean landscapes where forest fires are a stressful factor, it is essential to analyze ecosystem resilience. Remote sensing techniques have enabled the development of vegetation recovery rates (VRI). VRI time series trend modeling provides predictions of post-fire resilience of fire-affected vegetation cover for each level of burn severity. Fernández-Manso et al. (2016) have successfully developed a VRI for *P.* in this field.

From a fire risk management point of view, remote sensing based studies such as those by García-Llamas et al. (2019a, b) have identified the main drivers of burn severity in the Mediterranean Basin and provided indications on how it can be reduced. Efficient post-fire management strategies can be proposed based on these studies (for instance, modification of fuel structure or actions aimed at the community). In all these tasks, remote sensing is an essential tool for defining optimal forest management strategies in these systems to reduce fire damage.

Acknowledgments The results presented in this chapter were part of different projects financially supported by the Spanish Ministry of Economy and Competitiveness, the Spanish Ministry of Science, Innovation and Universities, and the European Regional Development Fund (ERDF), in the framework of the GESFIRE (AGL2013-48189-C2-1-R) and FIRESEVES (AGL2017-86075-C2-1-R) projects; and by the Regional Government of Castile and León within the framework of the FIRECYL (LE033U14) and SEFIRECYL (LE001P17) projects.

References

Arnett JTTR, Coops NC, Daniels LD, Falls RW (2015) Detecting forest damage after a low-severity fire using remote sensing at multiple scales. Int J Appl Earth Obs 35:239–246

Botella MA, Fernández-Manso A (2017) Study of post-fire severity in the Valencia region comparing the NBR, RdNBR and RBR indexes derived from Landsat 8 images. Spanish J Remote Sens Revista de Teledeteccion (RAET) 49:33–47

Calvo L, Tárrega R, Luis E (1998) Space-time distribution patterns of *Erica australis* L. subsp. *aragonensis* (Willk) after experimental burning, cutting and ploughing. Plant Ecol 137:1–12

Calvo L, Santalla S, Valbuena L et al (2008) Post-fire natural regeneration of *Pinus pinaster* forest in NW of Spain. Plant Ecol 197:81–90

Calvo L, Torres O, Valbuena L et al (2013) Recruitment and early growth of *Pinus pinaster* seedlings over five years after a wildfire in NW Spain. Forest Syst 22:582–586

Cardil A, Mola-Yudego B, Blázquez-Casado A et al (2019) Fire and burn severity assessment: calibration of Relative differenced Normalized Burn Ratio (RdNBR) with field data. J Environ Manag 235:342–349

da Ponte A, Costafreda-Aumedes S, Vega-Hidalgo C (2019) Lessons learned from arson wildfire incidence in reforestation and natural stands in Spain. Forests 10:229

de las Heras J, Moya D, Vega JA et al (2012) Post-fire management of serotinous pine forests. In: Moreira F, Arianoutsou M, Corona P, de las Heras J (eds) Post-fire management and restoration of Southern European forests, Managing forest ecosystems, vol 24. Springer, Dordrecht, pp 121–150

de Santis A, Chuvieco E (2009) GeoCBI: a modified version of the Composite Burn Index for the initial assessment of the short-term burn severity from remotely sensed data. Remote Sens Environ 113:554–562

Dennison PE, Charoensiri K, Roberts DA, Peterson SH, Green RO (2006) Wildfire temperature and land cover modeling using hyperspectral data. Remote Sens Environ 100:212–222

Díaz-Delgado R, Lloret F, Pons X (2003) Influence of fire severity on plant regeneration by means of remote sensing imagery. Int J Remote Sens 24:1751–1763

Fernandes P, Rigolot E (2007) The fire ecology and management of maritime pines (*Pinus pinaster* Ait.). For Ecol Manag 241:1–13

Fernández-García V, Quintano C, Taboada A et al (2018a) Remote sensing applied to the study of fire regime attributes and their influence on post-fire greenness recovery in pine ecosystems. Remote Sens 10:733

Fernández-García V, Santamarta M, Fernández-Manso A et al (2018b) Burn severity metrics in fire-prone pine ecosystems along a climatic gradient using Landsat imagery. Remote Sens Environ 206:205–217
Fernández-García V, Miesel J, Baeza J et al (2019a) Wildfire effects on soil properties in fire-prone pine ecosystems: indicators of burn severity legacy over the medium term after fire. Appl Soil Ecol 135:147–156
Fernández-García V, Fulé PZ, Marcos E et al (2019b) The role of fire frequency and severity on the regeneration of Mediterranean serotinous pines under different environmental conditions. For Ecol Manag 444:59–68
Fernández-García F, Marcos E, Fernández-Guisuraga JM et al (2019c) Impact of burn severity on soil properties in a *Pinus pinaster* ecosystem immediately after fire. Int J Wildland Fire 28:354–364
Fernández-Guisuraga JM, Sanz-Ablanedo E, Suárez-Seoane et al (2018) Using unmanned aerial vehicles in postfire vegetation survey campaigns through large and heterogeneous areas: opportunities and challenges. Sensors 18:586
Fernández-Guisuraga JM, Suárez-Seoane S, Calvo L (2019) Modeling *Pinus pinaster* forest structure after a large wildfire using remote sensing data at high spatial resolution. For Ecol Manag 446:257–271
Fernández-Manso A, Quintano C, Roberts DA (2016) Burn severity influence on post-fire vegetation cover resilience from Landsat MESMA fraction images time series in Mediterranean forest ecosystems. Remote Sens Env 184:112–123
Fernández-Manso A, Quintano C, Roberts DA (2019) Burn severity analysis in Mediterranean forests using maximum entropy model trained with EO-1 Hyperion and LiDAR data. ISPRS J Photogramm Remote Sens 155:102–118
García-Llamas P, Suárez-Seoane S, Taboada A et al (2019a) Assessment of the influence of biophysical properties related to fuel conditions on fire severity using remote sensing techniques: a case study on a large fire in NW Spain. Int J Wildland Fire 28(7):512–520
García-Llamas P, Suárez-Seoane S, Taboada A et al (2019b) Environmental drivers of fire severity in extreme fire events that affect Mediterranean pine forest ecosystems. For Ecol Manag 433:24–32
Harris S, Veraverbeke S, Hook S (2011) Evaluating spectral indices for assessing fire severity in chaparral scosystems (Southern California) using MODIS/ASTER (MASTER) airborne simulator data. Remote Sens 3:2403–2419
Hirschmugl M, Gallaun H, Dees M et al (2017) Methods for mapping forest disturbance and degradation from optical earth observation data: a review. Curr For Rep 3:32–45
Key CH, Benson NC (2006) Landscape Assessment (LA) sampling and analysis methods. USDA Forest Service General Technical Reports RMRS-GTR-164-CD
Lhermitte S, Verbesselt J, Verstraeten WW, Veraverbeke S, Coppin P (2011) Assessing intra-annual vegetation regrowth after fire using the pixel based regeneration index. ISPRS J Photogramm Remote Sens 66:17–27
López-García MJ, Caselles V (1991) Mapping burns and natural reforestation using thematic mapper data. Geocarto Int 1991(6):31–37
Marcos E, Fernández-García F, Fernández-Manso A et al (2018) Evaluation of composite burn index and land surface temperature for assessing soil burn severity in Mediterranean fire-prone pine ecosystems. Forests 9:494
McCarley TR, Kolden CA, Vaillant NM, Hudak AT, Smith AMS, Wing BM, Kellogg BS, Kreitler J (2017) Multi-temporal LiDAR and Landsat quantification of fire-induced changes to forest structure. Remote Sens Environ 191:419–432
Miller JD, Knapp EE, Key CH et al (2009) Calibration and validation of the relative differenced Normalized Burn Ratio (RdNBR) to three measures of fire severity in the Sierra Nevada and Klamath Mountains, California, USA. Remote Sens Environ 113:645–656

Moya D, González-De-Vega S, García-Orenes F et al (2018) Temporal characterisation of soil-plant natural recovery related to fire severity in burned *Pinus halepensis* Mill. forests. Sci Total Environ 640–641:42–51

Parks SA, Dillon GK, Miller C (2014) A new metric for quantifying burn severity: the relativized burn ratio. Remote Sens 6:1827–1844

Pausas JG, Llovet J, Rodrigo A et al (2008) Are wildfires a disaster in the Mediterranean basin? A review. Int J Wildland Fire 17:713–723

Peón J, Recondo C, Fernández S et al (2017) Prediction of topsoil organic carbon using airbone and satellite hyperspectral imagery. Remote Sens 9:1211

Quintano C, Fernández-Manso A, Roberts DA (2013) Multiple Endmember Spectral Mixture Analysis (MESMA) to map burn severity levels from Landsat images in Mediterranean countries. Remote Sens Environ 136:76–88

Quintano C, Fernández-Manso A, Calvo L et al (2015) Land surface temperature as potential indicator of burn severity in forest Mediterranean ecosystems. Int J Appl Earth Obs Geoinf 36:1–12

Quintano C, Fernández-Manso A, Roberts DA (2017) Burn severity mapping from Landsat MESMA fraction images and land surface temperature. Remote Sens Environ 190:83–95

Quintano C, Fernández-Manso A, Fernández-Manso O (2018) Combination of Landsat and Sentinel-2 MSI data for initial assessing of burn severity. Int J Appl Earth Observ Geoinform 64:221–225

Quintano C, Fernández-Manso A, Calvo L, Roberts DA (2019) Vegetation and soil fire damage analysis based on species distribution modeling trained with multispectral satellite data. Remote Sens 11:1832

Ruíz-Gallardo JR, Castaño S, Calera A (2004) Application of remote sensing and GIS to locate priority intervention areas after wildland fires in Mediterranean systems: a case study from Southeastern Spain. Int J Wild Fire 13:241–252

Schroeder W, Oliva P, Giglio L et al (2014) The new VIIRS 375m active fire detection data product: algorithm description and initial assessment. Remote Sens Environ 143:85–96

Stambaugh MC, Hammer LD, Godfrey R (2015) Performance of burn-severity metrics and classification in oak woodlands and grasslands. Remote Sens 7:10501–10522

Taboada A, Tárrega R, Marcos E et al (2017) Fire recurrence and emergency post-fire management influence seedling recruitment and growth by altering plant interactions in fire-prone ecosystems. For Ecol Manag 402:63–75

Tanase M, de la Riva J, Pérez-Cabello F (2011) Estimating burn severity at the regional level using optically based indices. Can J For Res 41:863–872

Tane Z, Roberts DA, Veraverbeke S, Casas A, Ramirez C, Ustin S (2018) Evaluating endmember and band selection techniques for multiple endmember spectral mixture analysis using post-fire imaging spectroscopy. Remote Sens 10:389

Vaillant NM, Kolden CA, Smith AMS (2016) Assessing landscape vulnerability to wildfire in the USA. Curr For Rep 2(3):201–213

Veraverbeke S, Hook SJ, Harris S (2012) Synergy of VSWIR (0.4–2.5μm) and MTIR (3.5–12.5μm) data for post-fire assessments. Remote Sens Environment 124:771–779

Veraverbeke S, Stavros EN, Hook SJ (2014) Assessing fire severity using imaging spectroscopy data from the Airborne Visible/Infrared Imaging Spectrometer (AVIRIS) and comparison with multispectral capabilities. Remote Sens Environ 154:153–163

Viana-Soto A, Aguado I, Martínez S (2017) Assessment of post-fire vegetation recovery using fire severity and geographical data in the Mediterranean region (Spain). Environments 4:90

Vicente-Serrano SM, Pérez-Cabello F, Lasanta T (2011) *Pinus halepensis* regeneration after a wildfire in a semiarid environment: assessment using multitemporal Landsat images. Int J Wildland Fire 20:195–208

Whitman T, Whitman E, Woolet J et al (2019) Soil bacterial and fungal response to wildfires in the Canadian boreal forest across a burn severity gradient. Soil Biol Biochem 138:107571

Part VI
Policy and Ecosystem Services of Mediterranean Pine and Mixed Forest Ecosystems

Asaf Karavani

Chapter 28
Mediterranean Forests and the United Nations Sustainable Development Goals

Nicolas Picard and Valentina Garavaglia

28.1 Introduction

By its resolution 71/285 of 27 April 2017, the United Nations General Assembly adopted the United Nations strategic plan for forests 2017–230 prepared by the United Nations Forum on Forests (UNFF). This strategic plan defined six voluntary and universal global forest goals (GFG) that highlight the role that forests and trees outside forests can play in sustainable development by providing economic, social, environmental and cultural benefits for present and future generations. The GFGs reflect at the forest sector level the more comprehensive and cross-sectoral commitments of the 2030 Agenda for Sustainable Development adopted by the United Nations General Assembly in September 2015 (DESA 2019).

The potential contribution of Mediterranean forests to achieving these global goals primarily relies on their tree species. Several pine species are biogeographical markers and emblematic of Mediterranean forests (Quézel et al. 1999). Many species and varieties of pines (mainly *Pinus brutia* Tenore (Turkish pine), *P. halepensis* Mill. (Aleppo pine), *P. nigra* (black pine), *P. pinaster* Ait. (maritime pine) and *P. pinea* L. (stone pine)) are effective pioneer trees that make a large contribution to forest restoration, thanks to their rapid juvenile growth rate and drought resistance. At the same time, pines have always been a major wood resource across the region and source of non-wood forest products (NWFP), such as resin (Bourlion et al. 2018b). The importance of pines is thus intimately linked to the importance of Mediterranean forests as a whole.

According to the Global Forest Resources Assessment of the Food and Agriculture Organization (FAO), in 2015 Mediterranean countries had a cumulative forest area of 88 million ha (10% of the land area of those countries and 2% of the world's

N. Picard (✉) · V. Garavaglia
Food and Agriculture Organization of the United Nations, Forestry Department, Rome, Italy
e-mail: nicolas.picard@fao.org

G. Ne'eman, Y. Osem (eds.), *Pines and Their Mixed Forest Ecosystems in the Mediterranean Basin*, Managing Forest Ecosystems 38,
https://doi.org/10.1007/978-3-030-63625-8_28

forests), with an additional 32 million ha of other wooded lands (FAO 2015). When considering only forests of the Mediterranean biogeographical type, the forest area was 25.5 million ha, which is 18% of the land area of the Mediterranean biome (Bourlion et al. 2018b). At the regional level, forest area increased slightly at a rate of 0.7% per year during 1990–2015, mainly because of forest expansion, and to a large extent, to the detriment of other wooded lands. During the same time interval, the carbon stock in the forests of Mediterranean countries increased at a rate of 1.6% per year, a rate superior to the rate of change of forest area, indicating that Mediterranean forests are becoming denser (Bourlion et al. 2018b). Mediterranean forests thus are a source of natural capital that can play a role in the achievement of the GFGs.

Nonetheless, this overall regional trend is in line with sub-regional differences and inequalities between countries. The regional trend is mostly driven by a few countries (France, Italy, Spain and Turkey) which have larger forest areas than other Mediterranean countries (Picard 2019). Forest expansion in these countries can be caused by reforestation programs (facilitated by the Common Agricultural Policy in the European Union) or can be the outcome of the natural process of forest recolonization after the abandonment of croplands. In comparison, countries of the southern and eastern Mediterranean have smaller forest areas. Active reforestation programs in some southern and eastern countries are sometimes partially offset by ongoing deforestation and forest degradation due to overgrazing and overexploitation of fuelwood.

The overexploitation of forest resources in some parts of the Mediterranean may increase with the growing human population, which is expected to reach 670 million inhabitants by 2050, compared to 537 million in 2015 (World Bank 2015). The increasing population is accompanied by a change in the spatial distribution of populations, either between urban and rural areas (rural exodus) or between countries (migration). These changes in turn modify societal expectations from forests. Climate change is another overarching driver of forest change. It is all the more important in the Mediterranean as the region is both more sensitive and will undergo more pronounced climate changes than other regions of the world, a situation that some authors have summarized by saying that the Mediterranean is a hotspot of climate change (Giorgi 2006).

Climate change and population growth are likely to interact, posing a major threat to the sustainability of Mediterranean forests. Yet, this issue is neither specific to the Mediterranean region, nor to forests. At the global level, the urgency to take action was recognized in the 2030 Agenda for Sustainable Development, which defined 17 Sustainable Development Goals (SDG). These goals address all sectors and all dimensions (social, economic and environmental) of sustainable development. The current chapter aims to examine the relevance of SDGs for Mediterranean pine and mixed forests, their interlinkages with regional and national commitments and policies and, conversely, how Mediterranean pine and mixed forests can contribute to achieving these goals.

28.2 Relevance of SDGs for Mediterranean Forests

Forests are recognized worldwide for their significant contribution to achieving the goals and targets in the 2030 Agenda for Sustainable Development. Recently, FAO presented substantial evidence from wide range of sources that this contribution encompass the entire 17 SDGs, directly or indirectly (FAO 2018). Nevertheless, forests are at the core of SDG 15 to "sustainably manage forests, combat desertification, halt and reverse land degradation, and halt biodiversity loss". Forest degradation is a concern in the Mediterranean region where overgrazing precludes regeneration and where unsustainable harvest of fuelwood reduces forest resources. The World Resources Institute's Atlas of forest and landscape restoration opportunities estimated that there are 78.4 million ha of degraded lands (including forests) that can be restored in the Mediterranean (WRI 2017), an estimate similar to the 80.0 million ha of degraded lands obtained by Martín-Ortega et al. (2017) for the Mediterranean. Based on the global map of tree restoration potential by Bastin et al. (2019), 28.0 million ha of non-forest lands could be restored as forest in the Mediterranean. These numbers are quite significant compared to the total forest area of the Mediterranean region. SDG 15 also refers to biodiversity, which is particularly relevant for the region. The Mediterranean Basin is a biodiversity hotspot (Myers et al. 2000) with over 25,000 plant species, 60% of which (including some species of pines) are endemic. Biodiversity is also reflected in the Mediterranean forest genetic resources that are of particular importance for the SDGs. Located between desert and temperate zones, the Mediterranean is the rear edge of the distribution range of major temperate pines and other tree species, with original genetic resources located in marginal or peripheral tree populations.

By being one of the main carbon stocks at the landscape level and thus directly contributing to the reduction of greenhouse gas emissions, and by playing a key role in the mitigation of, and adaptation to, climate change, forests are part of the natured-based solution to address SDG 13 to "take urgent action to combat climate change and its impacts". Due to slower growth rates and smaller tree stature than other types of forest, the density of aboveground biomass in Mediterranean forests is certainly less than in other forest types (IPCC 2019), thus limiting its per-hectare efficiency in climate mitigation, even if young Mediterranean pine forests can have levels of net ecosystem exchange similar to that of temperate forests (Rotenberg and Yakir 2010). Nevertheless, this limitation can be offset by the area that is available for forest expansion in the Mediterranean. In a country like France, a significant part of the forest biomass increase comes from the Mediterranean region, which has witnessed massive land use change from cropland to forests (Denardou et al. 2017). Adaptation to climate change will be important in the region and cannot be dissociated from mitigation. Uniform maximization of carbon storage, for instance, would also increase vulnerability to drought and the risk of wildfires, two parameters of key importance for adaption to climate change. Mediterranean forests will also be important for adaptation strategies at the landscape level by reducing the risk and damage from extreme weather events that are expected to intensify due to climate change (e.g. floods and dust storms).

By providing nutritious food for human populations (NWFP) and cattle (fodder), and generating decent incomes in rural areas while protecting the environment, Mediterranean forest can contribute to SDG 2 to "end hunger, achieve food security and improved nutrition and promote sustainable agriculture", in particular its target 2.3. Concerning the poorest populations that rely on forests to graze animals or to collect NWFP for informal markets, this role played by forests also relates to SDG 1 to "end poverty". One specificity of the Mediterranean forests is that the value of Mediterranean forests lies more in the non-wood goods and services that they provide than in the wood products (Croitoru 2007). In most cases, the incomes of people living in or close to forests do not depend on the forest but on different and competing land uses such as agriculture or grazing. Local villagers, however, depend on forests for fuelwood, grazing and fodder (Bourlion et al. 2018b). Game, mushrooms, honey, aromatic and medicinal plants, berries and nuts (including acorns) are Mediterranean NWFP that directly, even if marginally, contribute to food production in the region. Fodder for livestock also contributes to food production. The contribution of Mediterranean trees to food production becomes more important when considering trees outside forests, including orchards, agroforestry systems and fruit trees in urban or peri-urban environments (Salbitano et al. 2018). One NWFP that specifically relates to Mediterranean pines is the pine nut extracted from the cones of *P. pinea* L. The Mediterranean pine nut represents 10% of world pine nut production (average over 2017/18–2018/19 according to the International Nut and Dried Fruit database). Beyond food production, Mediterranean forests and trees contribute to soil erosion prevention, watershed protection (improving water quality), and pollinator abundance. These ecosystem services provided by forests contribute to food security, and specifically to target 2.4 of SDG 2.

Because green entrepreneurship and the development of local value chains based on the goods and services provided by forests can boost economic activity and job creation for youth in rural landscapes across the Mediterranean region, Mediterranean forests can also address SDG 8 to "promote inclusive and sustainable economic growth, employment and decent work for all" (in particular its target 8.3). An example of economic growth supported by Mediterranean pine forests is given by the revival in the last 5 years of the resin sector in Spain, based on the tapping of *Pinus spp*. (Soliño et al. 2018). The cork sector is another example of the economic potential of Mediterranean forests.

Urban and peri-urban forestry, by proposing nature-based solutions to improve the resilience of cities to climate change and improving the well-being of human populations in urbanized environments, also addresses SDG 11 to "make cities inclusive, safe, resilient and sustainable" (in particular its target 11.7). By impacting the quality of water, Mediterranean forests also contribute to target 6.6 of SDG 6 ("By 2020, protect and restore water-related ecosystems, including mountains, forests, wetlands, rivers, aquifers and lakes"). By providing fuelwood, Mediterranean forests contribute to target 7.2 of SDG 7 ("By 2030, increase substantially the share of renewable energy in the global energy mix"). Fuelwood can not only be a combustible harvested by local populations in rural areas, but also a material to produce bioenergy as a substitute for fossil fuel in industrial settings, such as the Gardanne power station in southern France.

The declination of forest-related SDGs into regional and national objectives and actions (that will be addressed in the next section) relies on international development, capacity building and exchange of knowledge within the Mediterranean. This regional collaboration in turn addresses SDG 10 to "reduce inequality within and among countries" and SDG 17 to "strengthen the means of implementation and revitalize the global partnership for sustainable development".

28.3 Connection Between SDGs and Other International Forest-Related Instruments

28.3.1 Global Level

Many SDGs are echoed in the commitments undersigned at the three Rio Conventions. Due to their high biodiversity, their vulnerability to degradation and desertification, and their role in the mitigation of, and adaptation to, climate change, Mediterranean forests are at the crossroad between the three Rio conventions. The Strategic Plan for Biodiversity 2011–2020 and the Aichi Targets adopted at the tenth Conference of the Parties (COP) of the Convention on Biological Diversity (CBD) in 2010 in Nagoya define targets on biodiversity that concur with the targets of SDG 15, and are highly relevant for Mediterranean forests. For instance, Aichi Target 15 to restore 15% of degraded ecosystems by 2020 echoes target 15.3 of SDG 15 ("By 2030, combat desertification, restore degraded land and soil, including land affected by desertification, drought and floods, and strive to achieve a land degradation-neutral world").

At the 12th COP of the United Nations Convention to Combat Desertification (UNCCD) in October 2015 in Ankara, country Parties endorsed the vision of SDG target 15.3 as a strong vehicle for driving UNCCD implementation, while at the same time contributing to other SDGs, including those relating to climate change mitigation and adaptation, biodiversity conservation, ecosystem restoration, food and water security, disaster risk reduction, and poverty. The Global Mechanism of the UNCCD has engaged in Land Degradation Neutrality (LDN), a target for responding to the challenge of sustainably intensifying the production of food, fuel and fibre to meet future demand without the further degradation of our finite land resource base. As of October 2019, 14 Mediterranean countries have committed to setting LDN targets. Further reinforcing these commitments on restoration and LDN, on 1 March 2019 the United Nations General Assembly declared 2021–2030 the UN Decade on Ecosystem Restoration with the aim to massively scale up the restoration of degraded ecosystems as a measure to fight the climate crisis and enhance food security, water supply and biodiversity.

The Paris Agreement adopted at the 21st COP of the United Nations Framework Convention on Climate Change (UNFCCC) in 2015 in Paris acknowledges the role of forests in mitigating and adapting to climate change. Specifically, article 5 of the Agreement encourages for "policy approaches and positive incentives for activities relating to reducing emissions from deforestation and forest degradation, and the role of conservation, sustainable management of forests and enhancement of forest carbon stocks in developing countries; and alternative policy approaches, such as joint mitigation and adaptation approaches for the integral and sustainable management of forests." Most (88%) Mediterranean countries have ratified the Paris Agreement and almost all (96%) have produced a Nationally Determined Contribution (NDC) or an intended NDC to specify their national efforts to reach climate goals. The role of the forest sector is explicitly mentioned in most of the NDCs of the Mediterranean countries (Picard 2019).

The United Nations Forest Instrument (UNFI – formerly "Non-legally binding instrument on all types of forests"), negotiated by the UNFF and adopted by the United Nations General Assembly in December 2007, highlighted the importance of sustainable forest management as a way to maintain and enhance the economic, social and environmental values of all types of forests, for the benefit of present and future generations. The UNFI defined four voluntary, non-binding global objectives on forests that are also highly relevant for the Mediterranean region as they address forest protection, restoration, afforestation and reforestation, forest degradation, forest-based economic, social and environmental benefits. The United Nations strategic plan for forests 2017–2030 has defined six voluntary Global Forest Goals that incorporate the global objectives on forests of the UNFI, support the objectives of the International Arrangement on Forests, and build on and contribute to the SDGs, the Aichi Biodiversity Targets, and the Paris Agreement on climate change.

28.3.2 Regional Level

Global strategies and SDGs are in turn reflected in a number of regional documents dealing with Mediterranean forests. A few are mentioned here to illustrate their consistency with global commitments. The Strategic Framework for Mediterranean Forests adopted by the FAO Committee on Mediterranean Forestry Questions–*Silva Mediterranea* in 2013, aims to provide a common policy direction for the integrated management of Mediterranean forest ecosystems, focusing on the need to develop goods and services, promote resilience, and strengthen both capacity and resources. Its nine strategic lines define some of the challenges contained in the SDGs with respect to Mediterranean forests. For instance, the restoration of degraded forest ecosystems (target 15.3 of SDG 15) is addressed in its sixth strategic line. These strategic lines were prioritized in the strategy that was adopted by *Silva Mediterranea* at its 23rd session in April 2019.

In 2017, the Agadir Commitment endorsed by ten Mediterranean countries at the Fifth Mediterranean Forest Week (Morocco, March 2017) recognized the operational dimension of Mediterranean forest and landscape restoration for the achievement of several SDGs. The Agadir Commitment aims to restore 8 millions ha of degraded Mediterranean ecosystems by 2030, thus contributing to the Bonn Challenge. In 2019, the Brummana Declaration for the role of Mediterranean forests to fulfil the NDCs adopted during the Sixth Mediterranean Forest Week (Lebanon, April 2019) acknowledged as a cross-cutting issue the importance of raising awareness and promoting good practices for Mediterranean forests, including their role in achieving the SDGs and combatting climate change.

The Mediterranean Forest Research Agenda 2009–2020, developed by the Mediterranean Regional Office of the European Forest Institute and its network, presents a common vision for the challenges of Mediterranean forests and the scientific priorities to address them. Four elements were identified in this strategic document: (1) climate change and land use change; (2) wildfire; (3) goods and services provided by forests; and (4) silviculture and multifunctional management of forests.

The Mediterranean Strategy for Sustainable Development 2016–2025, which was adopted by the Contracting Parties of the Barcelona Convention in 2016, is a guiding document for all stakeholders to translate the 2030 Agenda for Sustainable Development at the Mediterranean level, with particular focus on the green economy in the context of sustainable development and poverty eradication. This strategy highlighted the connection between forests and agriculture in the Mediterranean.

The Ministerial Declaration on Environment and Climate Change, adopted in 2014 by the Union for the Mediterranean, identified combatting desertification and drought, conserving biodiversity and improving resilience to climate change as priorities for the Mediterranean region.

28.3.3 National Level

Mediterranean countries have developed various forest-related strategies and policy documents, either to comply with commitments under the three Rio Conventions (such as National Biodiversity Strategies and Action Plans for the CBD, National Action Programme for the UNCCD, or NDCs for the UNFCCC), as readiness documents to access multilateral funds, or simply to define the national policy regarding the forest sector (national forest programs and strategies). Forest-related sectors such as bioenergy, green economy, biodiversity, and agriculture (including agroforestry) may have also issued national policy documents that include measures regarding forests.

A review of forest-related policy documents in Mediterranean countries was conducted by Buttoud et al. (2018). It showed that sustainable forest management is consistently prioritized in national forest policies in the Mediterranean. Priorities for forest products (non-wood and wood forest products alike) are also universally expressed in policy documents throughout the region, while prioritization of ecosystem services remains fragmented and is implemented only intermittently. The policies of a number of Mediterranean countries include pledges related to forest and landscape restoration and afforestation. It is worth noting in this regard that Mediterranean pines have long been used for land restoration in the Mediterranean basin and in other parts of the world, particularly since the nineteenth century (e.g. in the Algerian Green Dam). Mixed with more resilient oak species, the objectives have been to increase forest productivity, restore degraded land, protect watersheds and fix eroded coastal dune systems (Pausas et al. 2004).

Biodiversity is deeply rooted in forest policies and policy documents throughout the region, the primary focus being the biodiversity–climate change nexus. Regionally, climate change mitigation and adaptation policies and instruments are in their initial phase. The average year for the publication of National Forest Programmes in the Mediterranean region is 2009. As a result, NDCs and the Paris Agreement are not mentioned in most forest policies. It is anticipated that these existing forest policies will be revised in order to foster the role of forests in NDCs.

Overall, national forest policy documents are well aligned with regional and global objectives. Nevertheless, actions to be taken are often described in a qualitative way, with little or no quantitative targets. For instance, with a few exceptions, NDCs do not assign quantitative objectives to the land use, land-use change and forest (LULUCF) sector, nor clarify the share that the LULUCF sector will have in the reduction of greenhouse gas emissions (Picard 2019).

28.4 Forest-Based Solutions and Enabling Conditions to Address SDGs in the Mediterranean

The unicity of Mediterranean forests at the global level make them particularly relevant for developing specific forest-based solutions to global changes. Due to anthropogenic pressure, any forest-based solution in the Mediterranean will need to closely associate the three dimensions of sustainable forest management (social, economic, environmental). In particular, any measure that does not recognize the total economic value of forest goods and services is likely to fail in conserving the forest in a non-degraded status. This economic value includes the value of ecosystem services that are too often treated as externalities, with an environmental cost that becomes perceptible only when these services have become disservices.

The value of Mediterranean forests lies more in their non-wood products and ecosystem services rather than in their wood. Therefore, forest-based solutions in the Mediterranean should be developed based on endemic or specific NWFP that can reach niche markets (cork, carob, mastic, pine nut, thyme, rosemary, truffle, argan oil, cork oak acorns, etc.), as well as on ecosystem services (particularly for water, prevention of soil erosion, and recreation). However, regarding wood, Mediterranean forests are unlikely to be able to compete with other production forests on international markets.

Valorising NWFP in economic markets, combined with participatory approach, is a way to reinforce sustainable management of forests. An example is provided by the Maâmora forest in Morocco that is a state-owned cork oak forest producing cork at a value of 27.9 € ha^{-1} (Daly Hassen 2016). The forest is also used by local populations for grazing livestock (valued at 142.8 € ha^{-1}) and collecting acorns (valued at 133.3 € ha^{-1}) that enter informal value chains and end up being exported as food for animals. These activities originally generated substantial income but were either unsustainable (overgrazing) or illegal (acorn harvesting). Contracts were signed between the forestry administration and the associations that represent local communities, in order to better manage the acorn resource (by allowing associations to collect acorns provided that they respect technical rules set by the administration and contribute a share of the harvested acorns to reforestation activities), and alleviate grazing pressure (by granting financial compensation for temporarily setting aside areas to allow forest regeneration). Local associations were further involved as labor in forest activities such as thinning and clearing, surveillance, etc. These win-win contracts encouraged the communities to be more responsible for the forests and allowed the forest administration to be more efficient in its reforestation program at the Maâmora (Qarro 2016). Reorganizing value chains to ensure a fairer distribution of the value from the producer to the end-user, developing innovative practices for NWFP (e.g. for traceability or automation of harvesting), eco-labelling of NWFP and eco-marketing (including eco-tourism), adjusting regulations on NWFP (e.g. import-export rules) are other ways to promote NWFP (Bourlion et al. 2018a). All these improvements would contribute to the adaptation of local communities to climate change using forests (Fig. 28.1).

Biodiversity can also play a role in supporting economic activities in Mediterranean forests. The Barbary macaque in the Maghreb is an example of an emblematic animal species living in forests that has promoted tourism. Natural parks also attract tourists or local urban populations. This recreational use of forests also contributes to the adaptation of populations using forests. Nevertheless, biodiversity is expected to have a greater contribution to climate goals through its contribution to the adaptation of forests to climate change (Fig. 28.1). The variability of genetic material is indeed key to maintaining the capacity of forests to adapt to climate change. Given the speed of climate change, it is likely that natural evolutionary process and natural processes of tree species migration will not be enough to cope with these changes. *In situ* and *ex situ* conservation of reproductive materials will then be essential to implement adaptive management, including assisted migration when no other alternatives are available. Because mixed stands have a greater resilience to climate change and its secondary effects (pest and pathogen outbreaks, wildfire) than monospecific stands, biodiversity can also play a role in reducing forest degradation (Fig. 28.1).

Wood can play a role in climate mitigation, either by storing carbon in living biomass or in long-lived wood products, or as a substitute for other materials with a larger carbon footprint (Fig. 28.1). These different options have different, and even possibly conflicting, implications on forest management. For instance, maximizing the standing biomass would require letting trees attain their maximum size, while

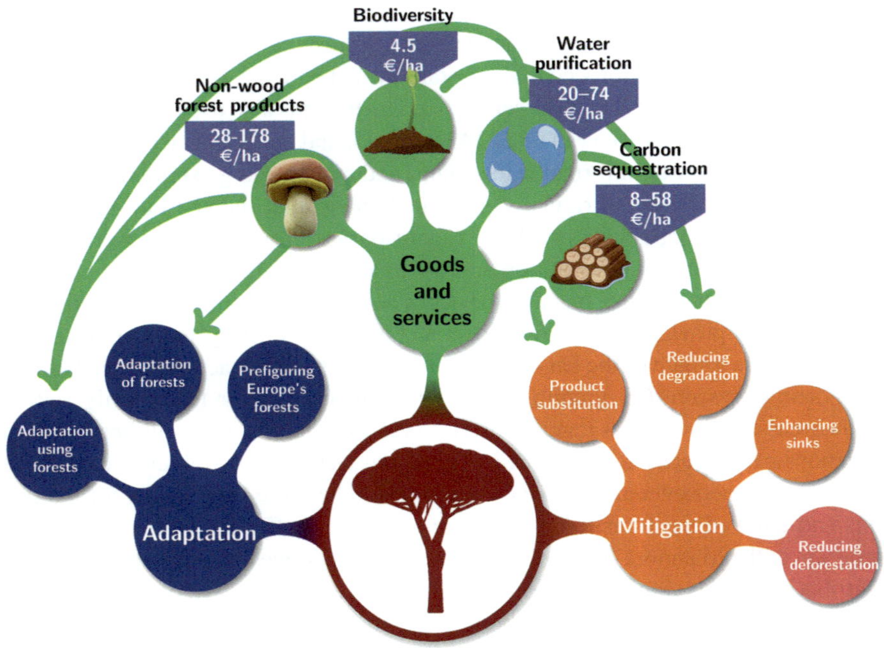

Fig. 28.1 Mind map on how to address climate-related goals by promoting the goods and services provided to the communities by Mediterranean forest ecosystems. Economic values are taken from Daly Hassen (2016) and are examples from specific sites in the Mediterranean; they are not representative of the whole region. Arrows indicate which good or service can contribute to adaptation or mitigation measures

conversely, optimizing the substitution of fossil fuel by wood for bioenergy production would require short felling cycles of small-sized trees. It is quite certain that neither of these options would be uniformly optimal. Uniform maximization of the standing biomass would for instance lead to higher vulnerability of forest stands to drought and wildfire (Barbati et al. 2018). Accordingly, different options should be streamlined at the landscape level.

Landscape approaches are key for forest-based solutions to address SDGs in the Mediterranean. An approach that seems particularly suited to the Mediterranean context is Forest and Landscape Restoration (FLR). FLR aims to restore an agreed optimal balance of the ecological, social and economic benefits of forests and trees within a broader pattern of land uses at the landscape level, by bringing people together to identify, negotiate and implement practices. FLR is particularly relevant for the Mediterranean because it addresses degradation, which is the main concern for the region, searches for trade-offs, and Mediterranean forests are intimately connected to many other sectors, and implements a participatory approach.

28.5 Monitoring and Reporting of SDGs in the Mediterranean

Monitoring and reporting of forest-based solutions will be required to ensure their effective contribution to forest-related goals at national and international levels. Different forest monitoring systems may coexist within a country, depending on the target quantity to monitor (such as forest attributes, forest fires, forest health, carbon emissions, etc.). Some Mediterranean countries have a permanent national forest inventory (NFI) system that allows them to monitor forest resources and guide national forest policies (Tomppo et al. 2010). NFI are the source of national forest statistics that are reported at the international level, such as the Global Forest Resources Assessment program of FAO. These data may also be used for carbon accounting in the forest sector for NDCs. Wildfires are an important issue in the Mediterranean and specific surveillance and monitoring systems have been set up to assess their impact. The European forest fire information system that also extends to non-European countries is an example of such a system (San-Miguel-Ayanz et al. 2012). Monitoring and early warning systems have also been set up to monitor forest health and receive alerts on pest and pathogen outbreaks (Saintonge et al. 2011). Regarding biodiversity, the IUCN (International Union for Conservation of Nature) Red List of Ecosystems, the IUCN Red List of Threatened Species or the European Red List of Habitats are different systems to monitor biodiversity in forests in Mediterranean countries.

These different monitoring systems raise the question of their articulation and synergies. Different forest monitoring systems are often implemented by different stakeholders within a country, with different spatial and temporal scales, which often makes their integration difficult. The issue of integrating forest monitoring systems becomes even more complicated when trying to scale up local monitoring systems (typically established to monitor the outcomes of a local project) to the national level. The nested approach is an example used in the REDD+ (reducing emissions from deforestation and forest degradation) mechanism to integrate carbon accounting at the national level from subnational carbon accounting systems (Lee et al. 2018). Even if no Mediterranean country is currently engaged in the REDD+ mechanism, this thinking may be of interest for the carbon accounting of Mediterranean countries with respect to their NDC. Scaling up data from subnational monitoring systems to the national level will be facilitated by the use of common monitoring tools. Some tools like Collect Earth (Bey et al. 2016), which are flexible, spatially explicit, easy to use and cost effective, can be used both at the subnational and national levels within integrated approaches.

The scaling of monitoring systems from national to regional or global levels is also important for SDG reporting. National data and statistics can be aggregated to obtain a regional view, but this requires the boundaries of the region to coincide with countries' borders. The Mediterranean region in the bioclimatic sense spans the Mediterranean countries. Therefore, monitoring forests in the Mediterranean bioclimatic region would require disaggregating national data to bioclimatic sub-regions, then re-aggregating them. Even with robust and well-documented systems like NFIs, such operations can be difficult unless national methodologies are aligned and harmonized to obtain consistent statistics and indicators across countries (Packalen et al. 2019).

The integration of forest monitoring systems for SDG reporting is also a question of consistency of indicators. A forest-related indicator developed at a local level would be of little use for SDG reporting if its scope did not match the scope of the SDG indicators. Three SDG indicators are directly related to forests (all of them for SDG 15): forest area as a proportion of total land area (indicator 15.1.1); progress towards sustainable forest management (indicator 15.2.1); and Mountain Green Cover index (indicator 15.4.2). Other SDG indicators will require forest information, such as the proportion of land that is degraded over total land area (indicator 15.3.1), which will necessitate monitoring of the trends in forest cover and forest carbon stocks; or the volume of production per unit labor by classes of farming/pastoral/forestry enterprise size (indicator 2.3.1).

The multiplication of forest-related targets at global, regional and national levels might result in a multiplication of indicators to be filled in, resulting in a reporting burden that few Mediterranean countries may be able to cope with. The alignment of global and regional forest goals with SDGs is already a safeguard against this risk. The global core set of forest-related indicators that is being developed by the Collaborative Partnership on Forests is a further effort to build synergies between forest monitoring systems.

28.6 Conclusions

Mediterranean forests can play a role in achieving SDGs by implementing forest-based solutions based on the goods and services provided by pine and mixed forest ecosystems. The specificity of these forests, namely that their non-wood forest products and services are more valuable than their wood, should be duly acknowledged when designing solutions that recognize the value of the forest. Moreover, these solutions should be designed at the landscape level to connect the forest sector to other key sectors, and be based on participatory approaches, especially when local populations strongly interact with the forest. Giving value to the Mediterranean forests will help conserve them for sustainable development, either by curbing their degradation where the overexploitation of forest resources reduces the natural capital, or by initiating active management where forests are abandoned. Forest-based solutions to achieve SDGs will require enabling conditions to be fulfilled, primarily national policies that are consistent with international and regional goals. They will also require forest monitoring systems to be established in a synergistic way.

References

Barbati A, Scarascia Mugnozza G, Ayan S et al (2018) Adaptation and mitigation. In: Bourlion N, Garavaglia V, Picard N (eds) State of mediterranean forests 2018. FAO, Rome/Plan Bleu, pp 128–146

Bastin JF, Finegold Y, Garcia C et al (2019) The global tree restoration potential. Science 365(6448):76–79

Bey A, Sánchez-Paus Díaz A, Maniatis D et al (2016) Collect Earth: land use and land cover assessment through augmented visual interpretation. Remote Sens 8(10):807

Bourlion N, Daly Hassen H, Gatto P et al (2018a) Valuing ecosystem services as part of the decision-making agenda. In: Bourlion N, Garavaglia V, Picard N (eds) State of mediterranean forests 2018. FAO, Rome/Plan Bleu, pp 216–228

Bourlion N, Garavaglia V, Picard N (eds) (2018b) State of Mediterranean forests. FAO, Rome/ Plan Bleu

Buttoud I, Schioppa A, Garavaglia V et al (2018) Trees outside forests in the Mediterranean region. In: Bourlion N, Garavaglia V, Picard N (eds) State of mediterranean forests 2018. FAO, Rome/ Plan Bleu, pp 180–194

Croitoru L (2007) How much are Mediterranean forests worth? Forest Policy Econ 9(5):536–545

Daly Hassen H (2016) Assessment of the socio-economic value of the goods and services provided by Mediterranean forest ecosystems: critical and comparative analysis of studies conducted in Algeria, Lebanon, Morocco, Tunisia and Turkey. Plan Bleu, Valbonne

Denardou A, Hervé JC, Dupouey JL et al (2017) L'expansion séculaire des forêts françaises est dominée par l'accroissement du stock sur pied et ne sature pas dans le temps. Rev For Fr 69(4–5):319–339

DESA (2019) Global forest goals and targets of the UN strategic plan for forests 2030. United Nations Department of Economic and Social Affairs, New York

FAO (2015) Global Forest resources assessment 2015: desk reference. FAO, Rome

FAO (2018) The state of the World's forests 2018 – Forest pathways to sustainable development. FAO, Rome

Giorgi F (2006) Climate change hot-spots. Geophys Res Lett 33(8):L08707

IPCC (2019) 2019 refinement to the 2006 IPCC guidelines for National Greenhouse gas inventories. Volume 4: Agriculture, forestry and other land use. Calvo Buendia E, Tanabe K, Kranjc A et al (eds). IPCC, Geneva

Lee D, Llopis P, Waterworth R et al (2018) Approaches to REDD+ nesting: lessons learned from country experiences. World Bank, Washington, DC

Martín-Ortega P, García-Montero L, Pascual C et al (2017) Global Drylands assessment using collect earth tools and opportunities for forest restoration. For Méd 38(3):259–266

Myers N, Mittermeier RA, Mittermeier CG et al (2000) Biodiversity hotspots for conservation priorities. Nature 403(6772):853–858

Packalen T, Lier M, Korhonen KT et al (2019) DIABOLO policy brief. Responding to European, national and regional challenges with harmonised forest information. Natural Resources Institute Finland (Luke), Helsinki

Pausas JG, Bladé C, Valdecantos A et al (2004) Pines and oaks in the restoration of Mediterranean landscapes of Spain: new perspectives for an old practice – a review. Plant Ecol 171:209–220

Picard N (2019) The role of the forest sector in the Mediterranean Nationally Determined Contributions. For Méd 40(3):225–236

Qarro M (2016) Améliorer la gouvernance des espaces boisés méditerranéens à travers la mise en œuvre de démarches participatives, forêt de la Maâmora. Maroc, Plan Bleu

Quézel P, Médail F, Loisel R et al (1999) Biodiversity and conservation of forest species in the Mediterranean basin. Unasylva 197(50):21–28

Rotenberg E, Yakir D (2010) Contribution of semi-arid forests to the climate system. Science 327(5964):451–454

Saintonge FX, Nageleisen LM, Assali F et al (2011) Santé des forêts marocaines: adaptation de la stratégie d'observation du Département de la santé des forêts (DSF) au contexte marocain. Rev For Fr 63(1):7–16

Salbitano F, Conigliaro M, Acil N et al (2018) Trees outside forests in the Mediterranean region. In: Bourlion N, Garavaglia V, Picard N (eds) State of mediterranean forests 2018. FAO, Rome/ Plan Bleu, pp 51–71

San-Miguel-Ayanz J, Schulte E, Schmuck G et al (2012) Chap 5: Comprehensive monitoring of wildfires in Europe: the European Forest Fire Information System (EFFIS). In: Tiefenbacher J (ed) Approaches to managing disaster. IntechOpen, London, pp 87–108

Soliño M, Yu T, Alía R et al (2018) Resin-tapped pine forests in Spain: ecological diversity and economic valuation. Sci Total Environ 625:1146–1155

Tomppo E, Gschwantner T, Lawrence M et al (eds) (2010) National forest inventories: pathways for common reporting. Springer, Geidelberg

World Bank (2015) Population estimates and projections. World Bank Open Data, Washington, DC. https://datacatalog.worldbank.org/dataset/population-estimates-and-projections. Accessed 31 October 2016

WRI (2017) Atlas of Forest and landscape restoration opportunities. World Resources Institute, Washington, DC. https://www.wri.org/resources/maps/atlas-forestand-landscape-restoration-opportunities. Accessed 11 May 2017

Chapter 29
Ecosystem Services Provided by Pine Forests

Iván Torres, José M. Moreno, César Morales-Molino, and Margarita Arianoutsou

29.1 Introduction

Pines are a key element in forests and other woodland areas across the Mediterranean region. Pines form many natural pure and mixed stands, but have also been planted in many areas inside and outside their native ranges, significantly increasing their original representation in Mediterranean landscapes. They have been widely used in reforestation projects for soil protection and hydrological purposes due to their rapid growth, tolerance to poor soils and summer drought, and easy cultivation in nurseries (Fady et al. 2003; Burylo et al. 2012; Enescu et al. 2016; Mauri et al. 2016). Furthermore, they provide economically valuable forest products such as timber and firewood, resin and pine nuts, giving justification for their maintenance, management and expansion.

The multi-purpose usefulness that promoted the expansion of pines by humans falls nowadays within the broad term of ecosystem services, conceived as the direct and indirect contributions of ecosystems to human well-being (Montes et al. 2011). Several classifications have been proposed for ecosystem services according to different criteria (Burkhard and Maes 2017), but the categorization used by the Millennium Ecosystem Assessment (2005) comprises provisioning (i.e., something that can be exploited), regulating (e.g., protection from erosion) and cultural services

I. Torres (✉) · J. M. Moreno
Departamento de Ciencias Ambientales, Universidad de Castilla-La Mancha, Toledo, Spain
e-mail: ivan.torres@uclm.es

C. Morales-Molino
Institute of Plant Sciences and Oeschger Centre for Climate Change Research, University of Bern, Bern, Switzerland

M. Arianoutsou
Department of Ecology and Systematics, National and Kapodistrian University of Athens, Athens, Greece

G. Ne'eman, Y. Osem (eds.), *Pines and Their Mixed Forest Ecosystems in the Mediterranean Basin*, Managing Forest Ecosystems 38,
https://doi.org/10.1007/978-3-030-63625-8_29

is widely used. The ecological processes that underpin these three groups of services are considered a fourth type in the Millennium Assessment, named supporting services, but more recent assessments omit this category (Montes et al. 2011). Pine forests provide all these groups of services, which depend on their persistence and proper functioning. However, these services may be put at risk by threats to pine forests such as wildfires, land-use change or diseases; the onset of global change has enhanced them and also generated other new threats.

This chapter addresses the ecosystem services provided by natural and planted pine forests across the Mediterranean Basin and addresses the main threats to pine forest maintenance and provisioning of ecosystem services.

29.2 Ecosystem Services Provided by Pine Forests

29.2.1 Provisioning Services

Pine forests and plantations provide multiple provisioning, regulating and cultural ecosystem services. Wood production has traditionally been the main service responsible for the management, plantation and resulting expansion of pine forests (Masiero et al. 2016; FAO and Plan Bleu 2018). In the Mediterranean region, conifers represent 41% of the total growing stock (FAO and Plan Bleu 2018). The main product from most pines in the Mediterranean region is timber, with prevalence of *Pinus sylvestris* L., *P. nigra* J.F.Arn and *P. pinaster* Ait. (Durrant et al. 2016; Enescu et al. 2016; Viñas et al. 2016a). Industrial timber is the most important forest product (Masiero et al. 2016). In Spain, more than half of the wood cut in 2017 corresponded to conifers (9.6 Mm^3 of cut wood), although approximately one third of this (3.6 Mm^3) corresponded to *P. radiata* D.Don plantations outside the Mediterranean region *sensu stricto* (MAPA 2017). Subsequently, *P. pinaster* was by far the most cut, with 3.4 Mm^3 of cut wood, which far exceeds other native species (e.g., next were *P. sylvestris* and *P. nigra* with 1.4 and 0.6 Mm^3, respectively) and all of the native broad-leaved species (e.g. *Populus spp.* accounted for 0.5 Mm^3 cut) (MAPA 2017). In Spain, the main destination of cut wood from most Mediterranean pines is timber, but the wood of *P. halepensis* Mill. is used for pellets, woodchips and briquettes (MAPA 2017). Other than timber, pines also provide biomass. In the southern Mediterranean, the primary forest product is firewood (Masiero et al. 2016) for which *P. halepensis* and *P. brutia* Ten. are used (Fady et al. 2003; Mauri et al. 2016). Biomass from pines is a resource with increasing potential to supply renewable energy demands (Perea-Moreno et al. 2017). All these activities require management actions such as shrub removal or thinning that improve stand structure and lower fire risk (García-Gonzalo et al. 2014; Madrigal et al. 2017), otherwise Mediterranean pinewoods are at risk of burning, as has happened extensively in recent decades (see Sect. 29.3.1).

In addition, pine forests provide a wide range of non-wood forest products (NWFP) (FAO and Plan Bleu 2018). In fact, NWFP account for the main use of forests in the Middle East and northern Africa (Croitoru 2007; Croitoru and Liagre 2013). Resin is one such forest product; it is obtained from *P. pinaster* in the western Mediterranean, although *P. halepensis* can also be tapped in the eastern Mediterranean (Calama et al. 2010; Spanos et al. 2010). Nowadays resin is a minor product, with an estimated value between €2.6 M and €3.2 M in Spain, Greece and Turkey (Masiero et al. 2016), but there has been an increase of resin production in Spain in recent years, from 1402 t in 2009 to 13,001 t in 2017 (MAPA 2017). Resin used to be the main product in many inland Spanish forests until it became unprofitable and exploitation of the forests was abandoned, resulting in improper management of pine stands and thus increasing fire risk, but also promoting a transition to *Quercus*-dominated or mixed *Quercus-Pinus* stands after fire (Torres et al. 2016).

Another very valuable NWFP that Mediterranean pine forests provide is pine nuts from *P. pinea* L. Production of pine nuts is very variable, because *P. pinea* is affected by climatic conditions and is also a masting species (Calama et al. 2011; Mutke et al. 2012; Pasalodos-Tato et al. 2016). Between 2003 and 2017, the average production in Spain was 6100 t/year (MAPA 2017), while the estimated annual pine nut production in the Mediterranean is estimated at around 30,000 t in shell, mostly in Portugal and Spain, but also in Italy and Turkey (Mutke et al. 2012). This high value has led to extensive planting of *P. pinea* in recent years in countries such as Portugal and Turkey (Küçüker and Baskent 2017; Valdiviesso et al. 2017).

Additionally, pine forests promote the growth and production of wild edible mushrooms with commercial value (e.g., *Boletus* and *Lactarius*, among others) to which forest management may be oriented (Tomao et al. 2017). As with other NWFPs, the estimated value of mushrooms may be equal or even greater than that of timber (Bonet et al. 2008; Calama et al. 2010; Rincón et al. this volume, Chapter 20).

Another NWFP produced by pines with an important contribution to the local economy is pine honey, produced by bees that collect the honeydew secreted by aphids sucking on *P. brutia* and *P. halepensis*. It represents about half the total honey production in Greece and Turkey (de-Miguel et al. 2014) and is the main source of income for nearly ten thousand families in Turkey (Croitoru and Liagre 2013).

29.2.2 Regulating Services

In addition to provisioning services, pine forests supply important regulation services. Forests are essential in preventing soil erosion (Anaya-Romero et al. 2016; Guerra et al. 2016; FAO and Plan Bleu 2018), and pines are particularly well suited to this since they are pioneer species that can withstand poor soils, steep slopes, strong radiation and other harsh conditions, and therefore have been used to reforest degraded landscapes (Fady et al. 2003; Burylo et al. 2012; Enescu et al. 2016; Mauri et al. 2016) and stabilize sand dunes (Cutini 2002; Pereira 2002; Viñas et al. 2016a, b). For example, most pine plantations in Spain have been established to regulate hydrological processes that carried important erosion and flood risk (Valbuena-Carabaña et al. 2010). Also, they have been used as pioneer species, which are more drought-tolerant and better adapted to poor soils, in order to restore and promote growth of other species. Pine forests, particularly those that have received management treatments (Rodríguez-Calcerrada et al. 2007; Prévosto et al. 2011), create suitable conditions for the recruitment of oak and other broad-leaved species in their understory (Sheffer 2012 and references therein; Van de Peer et al. 2018; Waitz and Sheffer this volume, Chapter 16), thus increasing the biodiversity of the ecosystem, which plays an important role in the maintenance and delivery of the goods and services that pine forests provide (Martínez-Jáuregui et al. 2019). Pine forests also harbor associated species with high conservation value. This is the case of sub-Mediterranean *P. nigra* forests, a priority habitat in the EU Habitats Directive, in which the largest European vulture, the black vulture (*Aegypius monachus* L.) nests in part of its range (Zaghi 2008).

Another important regulating service that pine forests provide is carbon sequestration. Pines are fast-growing species and store large amounts of carbon in their wood (Klein this volume, Chapter 7) and in the soil. In Italy, Mediterranean pines (*P. pinaster*, *P. halepensis*, *P. pinea*) accounted for 132.8 t carbon per ha in 2008–2009, which is higher than the carbon storage of most Mediterranean broad-leaved species (Gasparini and Di Cosmo 2015). In Spain, more than half of the total CO_2 stored in pine forests is due to *P. pinaster* and *P. sylvestris* forests (Montero et al. 2005; del Río et al. 2017), and pine forests accumulated over 535 Mt CO_2 in 1990 (Montero et al. 2005). The relevance of pine forests as a carbon pool becomes more prominent in more Mediterranean environments, with increased aridity, such as southern Spain, where conifer forests (mainly *P. pinaster*) account for the highest density of carbon (Anaya-Romero et al. 2016) and, during the twentieth century, pine plantations (mainly *P. halepensis*, *P. pinaster* and *P. sylvestris*) captured nearly as much carbon as oaks in a mountain ecosystem (Padilla et al. 2010).

29.2.3 Cultural Services

Pine forests provide a wide array of recreational and cultural services (see Orenstein this volume, Chapter 30). Hunting, as a highly regulated activity, is the most salient in the available statistics. Although no data exist for the specific contribution of pine forests to income from hunting, this activity accounts for 75% of the annual income from the category "other uses" (which also includes profit from water, recreational uses or aggregate extraction) of forest statistics in Spain (MAPA 2017). However, this evaluation is probably biased since hunting provides an easily accountable source of income, while carbon fixation and emission, recreation and other landscape values are omitted or undervalued (Campos et al. 2019). In fact, when these are accounted for, such as in the *P. sylvestris* forests in Sierra de Guadarrama, central Spain, very close to a large city such as Madrid, recreation accounted for half the value of timber, followed by conservation value (Caparrós et al. 2001). Recreational use is an increasing value of forests, including pine forests. For instance, there has been a pronounced increase in the forest area with recreational use as a main management goal since 2005 in south-eastern European countries (Forest Europe 2015). Finally, pine forests are used in a diverse range of scientific and educational activities.

29.3 Ecosystem Services Under Global Change

29.3.1 Fire and Ecosystem Services Provided by Pines

Wildfires can have positive and negative impacts on the ecosystem services provided by pine forests. In the short term, many of the services provided by pine forests are at risk of deterioration when they are affected by fire (Górriz-Mifsud et al. 2016). Fires obviously cause changes in esthetics that affect recreation and other cultural services. Also obvious are the potential losses of wood and wood products, but fires also affect the provisioning of other NWFPs. For instance, fire strongly decreases the production of mushrooms in pine forests (Martín-Pinto et al. 2006; Gassibe et al. 2011, 2014; Mediavilla et al. 2014). Regulating services can also be affected; many of the pine forests in Spain that come from plantations planted in the twentieth century are lacking adequate management and could easily be affected by fire (Valbuena-Carabaña et al. 2010). Therefore, the regulatory function that pine forests provide is highly vulnerable to fire, and soil erosion and runoff increase significantly after fire when pine forests and other Mediterranean vegetation burns (Cosandey et al. 2005; Mayor et al. 2007; Shakesby 2011; Wittenberg this volume, Chapter 23). Furthermore, a considerably large fraction of the carbon pool stored in pine forests is prone to mobilization during wildfire. Montealegre-Gracia et al. (2017) estimated CO_2 emissions ranging from 20 to 100 t/ha in a *P. halepensis* forest. In addition, common post-fire practices, such as salvage logging, seem to reduce carbon sequestration (Serrano-Ortiz et al. 2011). However, fires also produce pyrogenic organic matter that remains in the soil, acting as a potentially important carbon sink (Santín et al. 2015).

Pine forests are among the main vegetation types affected by wildfire, and will likely continue so since climate change will increase fire risk (Moreno et al. this volume, Chapter 21). Most pine species show adaptations to fire (Ne'eman and Arianoutsou this volume, Chapter 22), and the most Mediterranean pines (i.e., *P. halepensis*, *P. brutia*, *P. pinaster*) can regenerate even after crown fires thanks to their serotinous cones. In the case of pines with adaptations to frequent, low intensity surface fires, such as *P. nigra* (Tapias et al. 2004; Fulé et al. 2008), absence of fire can lead to fuel accumulation and crown fires after which the species, and the services provided by the pines, disappear and thus fire within a specific, historical fire regime may have beneficial results (Morales-Molino et al. 2017; Pausas and

Keeley 2019). However, increased frequency of fires with intervals shorter than the time needed to resupply the canopy seed bank and high fire severity can both constrain the regeneration of even the most fire-adapted *P. halepensis* and *P. brutia* (Kazanis and Arianoutsou 2004; Tessler et al. 2014; Kavgaci et al. 2016). For the more fire-sensitive mountain pines, *P. sylvestris* and *P. nigra*, a lack of regeneration after wildfires is a common phenomenon (Rodrigo et al. 2004; Vilà Cabrera et al. 2012; Christopoulou et al. 2014).

Because of the multiple services they provide, pines have been used in afforestation and reforestation projects throughout the Mediterranean Basin. However, due to their high susceptibility to burning, provision of such services may cease. The balance between planting of pines and pine stands being burned can illustrate the role of pines in forest fires. While most countries compile data about wildfires, these are usually not complete or accessible. To illustrate our point, we will focus on Spain, since fire statistics have been rather complete and available for over half a century. In a study of the fire history of Spain for the years 1974–1995, Moreno et al. (1998) showed that coniferous forests amounted to approximately 31% of the total burned area. Within the treed burned areas, the *Pinus* species that burned the most were *P. pinaster* (594 kha, 33%), *P. halepensis* (468 kha, 26%), *P. sylvestris* (162 kha, 9%), *P. pinea* and *P. nigra* (57 kha, 3% each). The burned pine trees were, on average, rather young, with a mean age for all pine species of 24.2 years, with little differences among species or over time in the last four decades (Table 29.1).

If we sum the entire burned area of *Pinus* species during the period of nearly 50 years of available statistics (1967–2015) (Fig. 29.1) and compare it with their plantation area from 1940 to 2013, we can roughly evaluate the role of pines in forest fires (Fig. 29.1). *Pinus pinaster* was the species most burned (826 kha) but was also the species most planted (912 kha), thus its balance was positive but only slightly. The next species in area burned was *P. halepensis*, with 709 kha burned, compared to 631 kha planted, thus it had a negative balance. The third-ranked species in area burned was *P. sylvestris* (192 kha), which had 659 kha planted, thus its balance was positive (70% remained unburned). In fact, many of these plantations were made in the wetter north-west of the country, and while this region is a hotspot for fires (Urbieta et al. 2019) many trees are still able to grow and be harvested

Table 29.1 Mean age at burning of five *Pinus* species based on fire reports for two periods (1974–1994 and 1974–2010)

	Mean age at burning	
Pinus species	1974–1994	1974–2010
P. halepensis	23.8	24.5
P. nigra	24.4	25.3
P. pinea	25.2	24.1
P. pinaster	23.2	24.2
P. sylvestris	21.4	23.0

Data from: Moreno et al. (1998), Vázquez de la Cueva (2016)

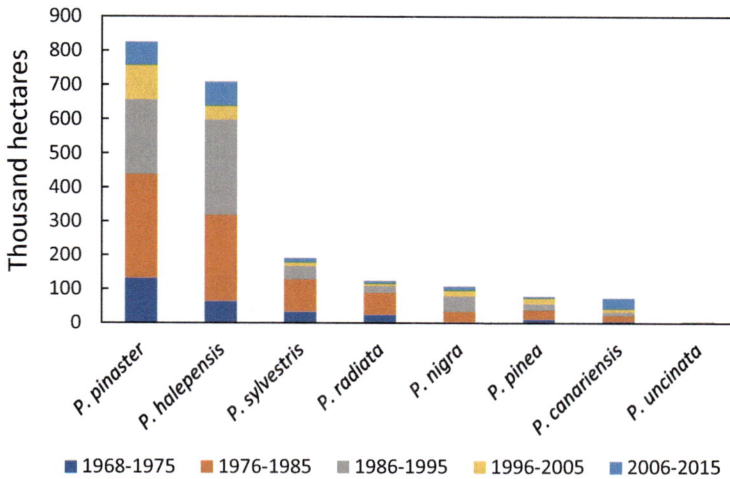

Fig. 29.1 Total area burned (kha) of the various species of *Pinus* growing in Spain (including the Canary Islands) from 1968 until 2015. Source: Estadística de incendios forestales, Ministerio para la Transición Ecológica y el Reto Demográfico, Government of Spain, https://www.mapa.gob.es/es/desarrollo-rural/estadisticas/Incendios_default.aspx)

(Schelhaas et al. 2018). *Pinus radiata* had 126 kha burned, and 254 kha planted. This species, which is planted in the wet, non-Mediterranean northern Spain, had a positive balance, but still half of the area planted burned. *Pinus nigra* had 110 kha burned of 441 kha planted, for a 75% positive balance, *P. pinea* had 79 kha burned from 337 kha planted (76% positive balance) and *P. canariensis* had 74 kha burned, from only 37 kha planted, for a negative balance of 50%. Overall, of 3299 kha of all the species that were planted, 2118 kha burned, that is, only 36% remained. This analysis, while rough, because there is no match between the areas planted and burned, shows that fires have very markedly affected *Pinus* species in Spain and that the probability of a pine plantation surviving in the long term was quite low overall during that period. The balance of the more typical Mediterranean species of the lowland areas (*P. halepensis*, *P. pinaster* and *P. pinea*) was nearly neutral (only 14% more area planted than burned).

29.3.2 Other Threats to Ecosystem Services

In addition to fire, pine forests are facing threats that may constrain their functionality and thus limit their ability to provide ecosystem services. For most pine species in Spain, increased aridity and competition with oaks severely limits their recruitment (Carnicer et al. 2014; Gazol et al. 2018). A decade-long drought episode in a *P. pinaster* forest in central Spain had negative impacts on growth and recruitment, and on tree survival during the decade following drought (Madrigal-González et al.

2017). For pines that have the limit of their distribution in the Mediterranean region, such as *P. sylvestris*, drought can limit growth and increase mortality at the most xeric sites (Gea-Izquierdo et al. 2014), and future predictions of growth for this and other species indicate a reduction in Spain and the rest of the Mediterranean (Gómez-Aparicio et al. 2011; Camarero et al. 2013; Sánchez-Salguero et al. 2017; Dorado-Liñán et al. 2019). Furthermore, the upward migration of *P. sylvestris* under climate change may be restricted by increased herbivory at the tree line during dry summers (Herrero et al. 2012).

Finally, increased warming induced by global change is affecting pines via increased pressure from pests and diseases – either new, introduced species, or existing species that become more aggressive with increased warming (Oliva this volume, Chapter 10). The case of the pine processionary moth (*Thaumetopoea pytiocampa* Denis & Schiffermüller, 1775) is illustrative of a pest with economic importance (Gatto et al. 2009) that is increasingly affecting pine stands due to global change. Warmer winter temperatures have been responsible for its expansion in recent decades (Hódar et al. 2003; Battisti et al. 2005). Furthermore, its expansion has been aided by human-mediated dispersal most likely by accidental introduction of pupae in the soil when transplanting large trees (Robinet et al. 2012), while the abundance of pine plantations throughout the Mediterranean also favors its expansion (Doblás-Miranda et al. 2017). The pine wood nematode (*Bursaphelenchus xylophilus* (Steiner & Buhrer) Nickle) is another example of a recently introduced pine pathogen with potential for expansion throughout the Mediterranean Basin (Naves et al. 2016).

29.4 Conclusions

Pine forests and plantations constitute a very important source of ecosystem services in the Mediterranean Basin. Regarding provisioning services, wood production for both timber and firewood are important but with different uses and intensities depending on the pine species and with different geographical patterns (i.e. firewood plays a more important role in the southern Mediterranean Basin). Non-wood forest products are equally important, but resin tapping has experienced a declining trend in recent decades while pine nuts, edible mushrooms and pine honey production are still important contributors to local economies, and to which management of pine stands may be oriented. Pine forests and plantations provide valuable regulating services that can explain the historical plantation of pines throughout the Mediterranean Basin for erosion control and hydrological regulation. Nowadays they still fulfill that role, while their relevance as habitats for biodiversity and as carbon sinks is recognized. Additionally, pine forests are providers of cultural services that when accounted for can result in similar values as those of some provisioning services. Many of these ecosystem services can be interrupted or heavily modified, at least in the short term, by risks related to global change.

Wildfire, drought and pests affect pine forests alone or in combination, and thus it is essential to properly evaluate services and risks to adapt their management.

Acknowledgments We (JMM, IT) acknowledge funding from the Government of Spain (project CGL2016-78357-R).

References

Anaya-Romero M, Muñoz-Rojas M, Ibáñez B et al (2016) Evaluation of forest ecosystem services in Mediterranean areas. A regional case study in South Spain. Ecosyst Serv 20:82–90

Battisti A, Stastny M, Netherer S et al (2005) Expansion of geographic range in the pine processionary moth caused by increased winter temperatures. Ecol Appl 15:2084–2096

Bonet JA, Pukkala T, Fischer CR et al (2008) Empirical models for predicting the production of wild mushrooms in Scots pine (*Pinus sylvestris* L.) forests in the Central Pyrenees. Ann For Sci 65:206

Burkhard B, Maes J (2017) Mapping ecosystem services. Advanced Books 1:e12837

Burylo M, Rey F, Bochet E et al (2012) Plant functional traits and species ability for sediment retention during concentrated flow erosion. Plant Soil 353:135–144

Calama R, Tome M, Sánchez-González M et al (2010) Modelling non-wood forest products in Europe: a review. For Syst 19:69–85

Calama R, Mutke S, Tomé J et al (2011) Modelling spatial and temporal variability in a zero-inflated variable: the case of stone pine (*Pinus pinea* L.) cone production. Ecol Model 222:606–618

Camarero JJ, Manzanedo RD, Sanchez-Salguero R et al (2013) Growth response to climate and drought change along an aridity gradient in the southernmost *Pinus nigra* relict forests. Ann For Sci 70:769–780

Campos P, Caparrós A, Oviedo JL et al (2019) Bridging the gap between national and ecosystem accounting application in Andalusian Forests, Spain. Ecol Econ 157:218–236

Caparrós A, Campos P, Montero G (2001) Applied multiple use forest accounting in the Guadarrama pinewoods (Spain). For Syst 10:91–108

Carnicer J, Coll M, Pons X et al (2014) Large-scale recruitment limitation in Mediterranean pines: the role of *Quercus ilex* and forest successional advance as key regional drivers. Glob Ecol Biogeogr 23:371–384

Christopoulou A, Fyllas NM, Andriopoulos P et al (2014) Post-fire regeneration patterns of *Pinus nigra* in a recently burned area in Mount Taygetos, Southern Greece: the role of unburned forest patches. For Ecol Manag 327:148–156

Cosandey C, AndréassianV MC et al (2005) The hydrological impact of the Mediterranean forest: a review of French research. J Hydrol 301:235–249

Croitoru L (2007) Valuing the non-timber forest products in the Mediterranean region. Ecol Econ 63:768–775

Croitoru L, Liagre L (2013) Contribution of forests to a green economy in the Middle East and North Africa region. Forêt Méditerranéenne 4:291–298

Cutini A (2002) Pines of silvicultural importance. CABI, Wallingford, UK, 608 pp

del Río M, Barbeito I, Bravo-Oviedo A et al (2017) Mediterranean pine forests: management effects on carbon stocks. In: Bravo F, LeMay V, Jandl R (eds) Managing forest ecosystems: the challenge of climate change. Springer, Cham, pp 301–327

de-Miguel S, Pukkala T, Yeşil A (2014) Integrating pine honeydew honey production into forest management optimization. Eur J For Res 133:423–432

Doblás-Miranda E, Alonso R, Arnan X et al (2017) A review of the combination among global change factors in forests, shrublands and pastures of the Mediterranean Region: beyond drought effects. Glob Planet Chang 148:42–54

Dorado-Liñán I, Piovesan G, Martínez-Sancho E et al (2019) Geographical adaptation prevails over species-specific determinism in trees' vulnerability to climate change at Mediterranean rear-edge forests. Glob Change Biol 25:1296–1314
Durrant TH, De Rigo D, Caudullo G (2016) *Pinus sylvestris* in Europe: distribution, habitat, usage and threats. European atlas of forest tree species. Publications Office of the European Union, Luxembourg, pp 132–133
Enescu CM, de Rigo D, Caudullo G et al (2016) *Pinus nigra* in Europe: distribution, habitat, usage and threats. European atlas of forest tree species. Publications Office of the European Union, Luxembourg, pp 126–127
Fady B, Semerci H, Vendramin GG (2003) EUFORGEN technical guidelines for genetic conservation and use for Aleppo pine (*Pinus halepensis*) and Brutia pine (*Pinus brutia*). Bioversity International
FAO, Plan Bleu (2018) State of Mediterranean forests 2018. Food and Agriculture Organization of the United Nations. Rome/Plan Bleu, Marseille
Forest Europe (2015) State of Europe's Forests 2015
Fulé PZ, Ribas M, Gutiérrez E et al (2008) Forest structure and fire history in an old *Pinus nigra* forest, eastern Spain. For Ecol Manag 255:1234–1242
García-Gonzalo J, Pukkala T, Borges JG (2014) Integrating fire risk in stand management scheduling. An application to Maritime pine stands in Portugal. Ann Oper Res 219:379–395
Gasparini P, Di Cosmo L (2015) Forest carbon in Italian forests: stocks, inherent variability and predictability using NFI data. For Ecol Manag 337:186–195
Gassibe PV, Fabero RF, Hernández-Rodríguez M et al (2011) Fungal community succession following wildfire in a Mediterranean vegetation type dominated by *Pinus pinaster* in Northwest Spain. For Ecol Manag 262:655–662
Gassibe PV, Fabero RF, Hernández-Rodríguez M et al (2014) Post-fire production of mushrooms in *Pinus pinaster* forests using classificatory models. J For Res 19:348–356
Gatto P, Zocca A, Battisti A et al (2009) Economic assessment of managing processionary moth in pine forests: a case-study in Portugal. J Environ Manag 90:683–691
Gazol A, Camarero JJ, Sangüesa-Barreda G et al (2018) Post-drought resilience after forest die-off: shifts in regeneration, composition, growth and productivity. Front Plant Sci 9:1546
Gea-Izquierdo G, Viguera B, Cabrera M et al (2014) Drought induced decline could portend widespread pine mortality at the xeric ecotone in managed Mediterranean pine-oak woodlands. For Ecol Manag 320:70–82
Gómez-Aparicio L, García-Valdés R, Ruiz-Benito P et al (2011) Disentangling the relative importance of climate, size and competition on tree growth in Iberian forests: implications for forest management under global change. Glob Change Biol 17:2400–2414
Górriz-Mifsud E, Varela E, Piqué M et al (2016) Demand and supply of ecosystem services in a Mediterranean forest: computing payment boundaries. Ecosyst Serv 17:53–63
Guerra CA, Maes J, Geijzendorffer I et al (2016) An assessment of soil erosion prevention by vegetation in Mediterranean Europe: current trends of ecosystem service provision. Ecol Indic 60:213–222
Herrero A, Zamora R, Castro J et al (2012) Limits of pine forest distribution at the treeline: herbivory matters. Plant Ecol 213:459–469
Hódar JA, Castro J, Zamora R (2003) Pine processionary caterpillar *Thaumetopoea pityocampa* as a new threat for relict Mediterranean Scots pine forests under climatic warming. Biol Conserv 110:123–129
Kavgaci A, Örtel E, Torres I et al (2016) Early postfire vegetation recovery of *Pinus brutia* forests: effects of fire severity, prefire stand age, and aspect. Turk J Agric For 40:723–736
Kazanis D, Arianoutsou M (2004) Factors determining low Mediterranean ecosystems resilience to fire: the case of *Pinus halepensis* forests. In: Arianoutsou M, Papanastasis VP (eds) Ecology, conservation and management of mediterranean climate ecosystems. Millpress. (electronic edition)
Klein T (this volume) Carbon allocation dynamics in Mediterranean pines under stress. In: Ne'eman G, Osem Y (eds) Pines and their mixed forest ecosystems in the Mediterranean Basin. Springer, pp 117–128

Küçüker DM, Baskent EZ (2017) State of stone pine (*Pinus pinea*) forests in Turkey and their economic importance for rural development. In: Mutke S, Carrasquinho I, Correia AC (eds) Mediterranean pine nuts from forests and plantations, 122. Zaragoza, CIHEAM, pp 111–117

Madrigal J, Fernández-Migueláñez I, Hernando C et al (2017) Does forest biomass harvesting for energy reduce fire hazard in the Mediterranean basin? A case study in the Caroig Massif (Eastern Spain). Eur J For Res 136:13–26

Madrigal-González J, Herrero A, Ruiz-Benito P et al (2017) Resilience to drought in a dry forest: insights from demographic rates. For Ecol Manag 389:167–175

MAPA (2017) Anuario de estadística forestal 2017. Ministerio de Agricultura, Pesca y Alimentación. https://www.mapa.gob.es/es/desarrollo-rural/estadisticas/forestal_anuario_2017.aspx. Accessed 10 Jan 2020

Martínez-Jáuregui M, White PC, Touza J et al (2019) Untangling perceptions around indicators for biodiversity conservation and ecosystem services. Ecosyst Serv 38:100952

Martín-Pinto P, Vaquerizo H, Peñalver F et al (2006) Early effects of a wildfire on the diversity and production of fungal communities in Mediterranean vegetation types dominated by *Cistus ladanifer* and *Pinus pinaster* in Spain. For Ecol Manag 225:296–305

Masiero M, Pettenella DM, Secco L (2016) From failure to value: economic valuation for a selected set of products and services from Mediterranean forests. For Syst 25:e051

Mauri A, Di Leo M, de Rigo D et al (2016) *Pinus halepensis* and *Pinus brutia* in Europe: distribution, habitat, usage and threats. European atlas of forest tree species. Publications Office of the European Union, Luxembourg, pp 122–123

Mayor AG, Bautista S, Llovet J et al (2007) Post-fire hydrological and erosional responses of a Mediterranean landscape: seven years of catchment-scale dynamics. Catena 71:68–75

Mediavilla O, Oria-de-Rueda JA, Martín-Pinto P (2014) Changes in sporocarp production and vegetation following wildfire in a Mediterranean forest ecosystem dominated by *Pinus nigra* in Northern Spain. For Ecol Manag 331:85–92

Millennium Ecosystem Assessment (2005) Ecosystems and human well-being: synthesis. Island Press, Washington, DC

Montealegre-Gracia AL, Lamelas-Gracia MT, García-Martín A et al (2017) Using low-density discrete Airborne Laser Scanning data to assess the potential carbon dioxide emission in case of a fire event in a Mediterranean pine forest. GISci Remote Sens 54:721–740

Montero G, Ruiz-Peinado R, Muñoz M (2005) Producción de biomasa y fijación de CO2 por los bosques españoles. INIA-Instituto Nacional de Investigación y Tecnología Agraria y Alimentaria

Montes C, Santos Martín F, Aguado M et al (2011) La evaluación de los ecosistemas del Milenio de España. Síntesis de resultados. Fundación Biodiversidad. Ministerio de Medio Ambiente, y Medio Rural y Marino

Morales-Molino C, Tinner W, García-Antón M et al (2017) The historical demise of *Pinus nigra* forests in the Northern Iberian Plateau (south-western Europe). J Ecol 105:634–646

Moreno JM, Vázquez A, Vélez R (1998) Recent fire history of forest fires in Spain. In: Moreno JM (ed) Large forest fires. Backhuys, Leiden, pp 159–185

Moreno JM, Morales-Molino C, Torres I, Arianoutsou M (this volume) Fire in Mediterranean pine forests, past, present and future. In: Ne'eman G, Osem Y (eds) Pines and their mixed forest ecosystems in the Mediterranean basin. Springer, pp 421–456

Mutke S, Calama R, González-Martínez SC et al (2012) Mediterranean stone pine: botany and horticulture. Hortic Rev 39:153–201

Naves P, Bonifácio L, de Sousa E (2016) The pine wood nematode and its local vectors in the Mediterranean Basin. In: Paine TD, Lieutier F (eds) Insects and diseases of Mediterranean forest systems. Springer, pp 329–378

Ne'eman G, Arianoutsou M (this volume) Mediterranean pines – adaptations to fire. In: Ne'eman G, Osem Y (eds) Pines and their mixed forest ecosystems in the Mediterranean basin. Springer, pp 457–480

Oliva J (this volume) Forest disease affecting pines in the Mediterranean Basin. In: Ne'eman G, Osem Y (eds) Pines and their mixed forest ecosystems in the Mediterranean basin. Springer, pp 183–198

Orenstein D (this volume) The cultural ecosystem services of Mediterranean pine forests. In: Ne'eman G, Osem Y (eds) Pines and their mixed forest ecosystems in the Mediterranean basin. Springer, pp 631–655

Padilla FM, Vidal B, Sánchez J et al (2010) Land-use changes and carbon sequestration through the twentieth century in a Mediterranean mountain ecosystem: implications for land management. J Environ Manag 91:2688–2695

Pasalodos-Tato M, Pukkala T, Calama R et al (2016) Optimal management of *Pinus pinea* stands when cone and timber production are considered. Eur J For Res 135:607–619

Pausas JG, Keeley JE (2019) Wildfires as an ecosystem service. Front Ecol Environ 17:289–295

Perea-Moreno AJ, Perea-Moreno MÁ, Hernandez-Escobedo Q et al (2017) Towards forest sustainability in Mediterranean countries using biomass as fuel for heating. J Clean Prod 156:624–634

Pereira JS (2002) *Pinus pinaster*. In: CABI (ed) Pines of silvicultural importance. CABI publishing, pp 316–328

Prévosto B, Monnier Y, Ripert C et al (2011) Can we use shelterwoods in Mediterranean pine forests to promote oak seedling development? For Ecol Manag 262:1426–1433

Rincón A, Pérez-Izquierdo L, de-Miguel S et al (this volume) Mycorhizae in Mediterranean pine and mixed forests. In: Ne'eman G, Osem Y (eds) Pines and their mixed forest ecosystems in the Mediterranean basin. Springer, pp 395–418

Robinet C, Imbert C-E, Rousselet J et al (2012) Human-mediated long-distance jumps of the pine processionary moth in Europe. Biol Invasions 14:1557–1569

Rodrigo A, Retana J, Pico FX (2004) Direct regeneration is not the only response of Mediterranean forest to large fires. Ecology 85:716–729

Rodríguez-Calcerrada J, Pardos JA, Gil L et al (2007) Summer field performance of *Quercus petraea* (Matt.) Liebl and *Quercus pyrenaica* Willd seedlings, planted in three sites with contrasting canopy cover. New For 33:67–80

Sánchez-Salguero R, Camarero JJ, Gutiérrez E et al (2017) Assessing forest vulnerability to climate warming using a process-based model of tree growth: bad prospects for rear-edges. Glob Change Biol 23:2705–2719

Santín C, Doerr SH, Preston CM et al (2015) Pyrogenic organic matter production from wildfires: a missing sink in the global carbon cycle. Glob Change Biol 21:1621–1633

Schelhaas M-J, Fridman J, Hengeveld G et al (2018) Actual European forest management by region, tree species and owner based on 714,000 re-measured trees in national forest inventories. PLoS One 13(11):e0207151

Serrano-Ortiz P, Marañón-Jiménez S, Reverter BR et al (2011) Post-fire salvage logging reduces carbon sequestration in Mediterranean coniferous forest. For Ecol Manag 262:2287–2296

Shakesby RA (2011) Post-wildfire soil erosion in the Mediterranean: review and future research directions. Earth-Sci Rev 105:71–100

Sheffer E (2012) A review of the development of Mediterranean pine–oak ecosystems after land abandonment and afforestation: are they novel ecosystems? Ann For Sci 69:429–443

Spanos K, Gaitanis D, Spanos I (2010) Resin production in natural Aleppo pine stands in northern Evia, Greece. Web Ecol 10:38–43

Tapias R, Climent J, Pardos JA et al (2004) Life histories of Mediterranean pines. Plant Ecol 171:53–68

Tessler N, Wittenberg L, Provizor E et al (2014) The influence of short-interval recurrent forest fires on the abundance of Aleppo pine (*Pinus halepensis* Mill.) on Mount Carmel, Israel. For Ecol Manag 324:109–116

Tomao A, Bonet JA, de Aragón JM et al (2017) Is silviculture able to enhance wild forest mushroom resources? Current knowledge and future perspectives. For Ecol Manag 402:102–114

Torres I, Pérez B, Quesada J et al (2016) Forest shifts induced by fire and management legacies in a *Pinus pinaster* woodland. For Ecol Manag 361:309–317

Urbieta IR, Franquesa M, Viedma O et al (2019) Fire activity and burned forest lands decreased during the last three decades in Spain. Ann For Sci 76:90

Valbuena-Carabaña M, de Heredia UL, Fuentes-Utrilla P et al (2010) Historical and recent changes in the Spanish forests: a socio-economic process. Rev Palaeobot Palyno 162:492–506

Valdiviesso T, Pimpao M, Trindade CS et al (2017) Reproductive phenology of *Pinus pinea*. In: Mutke S, Carrasquinho I, Correia AC (eds) Mediterranean pine nuts from forests and plantations, vol 122. CIHEAM, Zaragoza, pp 63–68

Van de Peer T, Mereu S, Verheyen K et al (2018) Tree seedling vitality improves with functional diversity in a Mediterranean common garden experiment. For Ecol Manag 409:614–633

Vázquez de la Cueva A (2016) Incendios forestales en la España peninsular (1974-2010) – Análisis temporal y espacial desde una perspectiva ecológica. Monografías INIA, Serie Forestal 29: 203

Vilà-Cabrera A, Rodrigo A, Martínez-Vilalta J et al (2012) Lack of regeneration and climatic vulnerability to fire of Scots pine may induce vegetation shifts at the southern edge of its distribution. J Biogeogr 39:488–496

Viñas RA, Caudullo G, Oliveira S et al (2016a) *Pinus pinaster* in Europe: distribution, habitat, usage and threats. European atlas of forest tree species; Publications Office of the European Union: Luxembourg. pp 128–129

Viñas RA, Caudullo G, Oliveira S et al (2016b) *Pinus pinea* in Europe: distribution, habitat, usage and threats. European atlas of forest tree species; European Commission: Brussels, Belgium, 204. pp 130–131

Waitz Y, Sheffer E (this volume) Dynamics of mixed pine–oak forests. In: Ne'eman G, Osem Y (eds) Pines and their mixed forest ecosystems in the Mediterranean basin. Springer, pp 345–362

Wittenberg L (this volume) Post-fire soil erosion – the Mediterranean perception. In: Ne'eman G, Osem Y (eds) Pines and their mixed forest ecosystems in the Mediterranean basin. Springer, pp 481–496

Zaghi D (2008) Management of Natura 2000 habitats. 9530 *(Sub)-Mediterranean pine forests with endemic black pines. European Commission

Chapter 30
The Cultural Ecosystem Services of Mediterranean Pine Forests

Daniel E. Orenstein

30.1 Introduction

30.1.1 Challenges in Defining the Cultural Significance of an Ecosystem

A tree, in the eyes of the humans who benefit from its existence, is far more than a biological entity. Likewise, a forest means far more to a culture than a collection of trees. Contemporary Mediterranean pine forests are the product of millennia of human–nature interactions in the Mediterranean Basin. They have been shaped by human activities, and in response, they have influenced cultural and historical developments from Spain and Portugal in the west, to Israel and Lebanon in the east. Through the products they provide, such as timber, resin, pine nuts, and mushrooms in their understory, to the refuge they provide visitors from densely populated cities and towns, they have contributed to shaping the lived experience in the region. Like most ecosystems, these forests provide cultural ecosystem services (CES) and catalyze relational values. But they are unique in the intense interaction they have facilitated with human civilizations since the Paleolithic Period. They are dynamic socio-ecological systems par excellence, which implies that the "basket" of cultural services changes and fluctuates in value, intensity, and spatial distribution over time.

In this chapter, I analyze and synthesize the CES provided by Mediterranean pine forests. CES are 'the non-material benefits people obtain from ecosystems' (MA 2005), and they include a range of services, including spiritual and religious values, knowledge and education, inspiration and aesthetic values, mediation of social relations, sense of place, recreation and tourism, mental and physical health,

D. E. Orenstein (✉)
Faculty of Architecture and Town Planning, Technion – Israel Institute of Technology, Haifa, Israel
e-mail: DanielO@ar.technion.ac.il

G. Ne'eman, Y. Osem (eds.), *Pines and Their Mixed Forest Ecosystems in the Mediterranean Basin*, Managing Forest Ecosystems 38,
https://doi.org/10.1007/978-3-030-63625-8_30

cultural and heritage values, and more (MA 2005; TEEB 2017; CICES 2018).[1] The conceptual framework of the International Platform on Biodiversity and Ecosystem Services (IPBES 2013) expanded upon these non-material benefits to include intrinsic values (the worth of nature itself) and the concept of relational values, or "preferences, principles and virtues about human–nature relationships" (Chan et al. 2018). In this chapter, when a pine forest contributes to one's identity formation, for example, I consider the forest to be producing or augmenting a relational value. While relational values are technically not "services", they are richly intertwined with cultural services, and so we consider them here.

CES, by definition, are the outcome of an intricate and dynamic interaction between human society and the natural environment (Tengberg et al. 2012; Fish et al. 2016). The forests' biophysical and ecological characteristics, and their provisioning and regulating services can be studied with a modicum of empirical, quantitative precision (if not completely); studying the cultural significance and benefits of forests, and their associated value(s), is arguably more complex and more fluid. CES assessment is subject to the dynamic changes of both ecosystems and society, varying from group to group and person to person, making their assessment highly context-dependent. CES can be studied and interpreted from multiple perspectives, focusing on different beneficiaries during different historical periods.

The temporal period of the assessment, the informants, the methodology of research, and even the political disposition or values orientation of the researcher will yield different pictures regarding which CES are provided, in what amounts and at what value. The type and value of CES is perceived and determined through the eye of the beholder. A pine forest will mean something different depending on whether you are farmer (on whose fields the trees are encroaching), a modern urban dweller (seeking to get away from the hectic urban lifestyle), a mushroom collector (who knows that their mushrooms flourish in the pine humus), a conservation biologist (who may scorn the perceived lack of biological diversity in the forest) or a forester (who appreciates the economic benefits of pine timber production). In Israel, your political disposition will likely impact your assessment regarding the cultural value of Mediterranean pine forests, most of which were planted as an act of nation building and to reinforce land claims.

Due to these complexities of time and space dynamics and the importance of context, assessing the CES of Mediterranean pine forests requires the use of diverse quantitative and qualitative research tools, drawn from multiple disciplines from the natural and social sciences to the humanities, to account for diverse perspectives. The fluidity of CES assessments makes a definitive and consensual valuation elusive, but also makes the story of human–pine forest relationships richer and more interesting.

---

[1] Cultural services are one of three types of ecosystem services commonly identified in the ecosystem service conceptual framework. The other two are provisioning services (tangible products, in the form of material or energy, obtained by humans from ecosystems) and regulating services (the contribution of ecosystem processes to moderating conditions of the biotic and abiotic environment for humans).

In the following chapter sections, I begin with a definition and overview of CES and expand upon the challenges of assessing them in dynamic socio-ecological systems and the demand for nuance in our assessments (Sect. 30.1.2). I then inventory the CES provided by Mediterranean pine and mixed-pine forests through a review of the recent academic literature (Sects. 30.2 and 30.3). Because the concept of ecosystem services (ES) is relatively new to the scientific community (the concept became mainstream with the publication of the Millennium Assessment in 2005), most of the literature explicitly discussing CES in Mediterranean pine forests is from the last decade. But there is also a tradition of researching human–nature interactions in the Mediterranean Basin prior to the proliferation of the ES conceptual framework, such as can be drawn from environmental psychology, anthropology, environmental history, landscape architecture and design, ecology and others. Here I focus on the first body of literature, which focuses explicitly on CES, but I include work from the latter group, particularly for historical perspectives.

Section 30.4 focuses on CES from Israel's pine forests, particularly their aesthetic value and preferences. I introduce the long-standing debate among scientists, environmental activists, foresters and the public regarding the ecological (and perhaps the aesthetic) role of pine forests in the region (planted and naturally-occurring), because that debate – I argue – shapes our perception of the cultural value of these forests and their CES. I conclude by considering both the need for a nuanced, contextualized approach to assessing CES in general, and in Mediterranean pine forests in particular, and consider how current and predicted environmental changes in the Mediterranean Basin may impact the future of pine forests and their CES.

A final caveat before beginning: For this chapter, pines will be treated as a collective genus (*Pinus*), and individual species will rarely be specified unless particularly relevant to the services they provide. While different species may provide different levels of services (e.g., shape and extent may affect shade quality, and some species provide higher quality pine nuts), most services are perceived by stakeholders in a generic sense that does not make fine distinctions between species.

30.1.2 Assessing Cultural Ecosystem Services: From the General to the Specific

According to the major ES assessment frameworks, there is a common repertoire of CES humans derive from ecosystems in general which are summarized in Table 30.1. These frameworks are also used to identify CES of Mediterranean pine forests (see below). However, a finely detailed and quantified assessment of CES, which includes identification, characterization, and valuation of services, is elusive for at least five reasons:

1. CES are, by definition, **intangible benefits**, which are often difficult to define and value monetarily, or in any quantitative terms. While there are economic tools for directly and indirectly assessing CES, some researchers note that those

Table 30.1 Categories of cultural ecosystem services according to comprehensive conceptual and assessment frameworks

MA 2005	Spiritual and religious values
	Knowledge and educational values
	Inspiration and aesthetic values
	Social relations
	Sense of place
	Cultural diversity, culture, and heritage values
	Recreation and tourism
UKNEA 2014	Cultural values
	Shaping identities
	Wellbeing, mental and physical health
	Obtaining skills and capabilities
IPBES 2016	Spiritual services
	Recreation
	Tourism
TEEB 2017	Recreation
	Mental and physical health
	Tourism
	Aesthetic appreciation and inspiration
	Spiritual experience and sense of place
CICES 2018	Scientific investigation
	Education and training
	Culture, heritage, and aesthetic experiences
	Symbolic and religious meaning
	Entertainment

From Teff-Seker and Orenstein (2019)

CES that are easily quantifiable receive a disproportionate emphasis in the literature (Milcu et al. 2013).

2. CES are often benefits **derived from a holistic system or landscape** rather than a single species or organism (consider, for example, spirituality, aesthetics, and recreation). As such, they cannot be linked to a particular species, and different species and assemblages of species could provide similar baskets of CES. This is especially relevant regarding Mediterranean pine forests, where we can delineate specific CES, but we cannot always attribute them specifically to pine trees (i.e., other tree species may be able to provide the same service).
3. CES are intimately **integrated with other ES types**, such as provisioning services (e.g., the cultural value of mushroom collection (Marini Govigli et al. 2019) or resin tapping (Soliño et al. 2018) in Mediterranean pine forests), and regulating services, (e.g., threat of fire and flood can also degrade the sense of psychological wellbeing provided by forests, and therefore assessed cultural values are inseparable from other services).

4. CES reflect the **complex interactions between two dynamic systems** – social and biophysical. As such, CES vary in time and between demographic groups and even individuals.
5. Measurements of CES value may be heavily influenced by factors unrelated to social or ecological characteristics, but rather be affected by **accessibility, physical proximity, or awareness** (Fleischer et al. 2014).

The fourth reason, linked to the characteristics of socio-ecological systems, demands further explanation. The Mediterranean pine forest is a socio-ecological system, par excellence (Naveh and Carmel 2004). In such a system, social (human) and bio-physical systems are interconnected, with changes in one impacting the other. Humans directly and indirectly alter biophysical systems, which are also impacted by global changes outside of the system, and their changes are perceived by humans via the change in ES provision (Fig. 30.1; Collins et al. 2011). As socio-ecological systems, the CES produced by ecosystems and/or perceived by humans are thus dynamic in time and space. As such, four characteristics of this socio-ecological system challenge our capacity to assess CES. These are:

1. **Temporal dynamics of ecosystems**: Ecosystems change due to natural and/or human-driven causes and this will affect the provision of ES. For example, Mediterranean forests were cleared throughout history for agricultural and grazing lands. Some of these lands have since returned to forest following rural

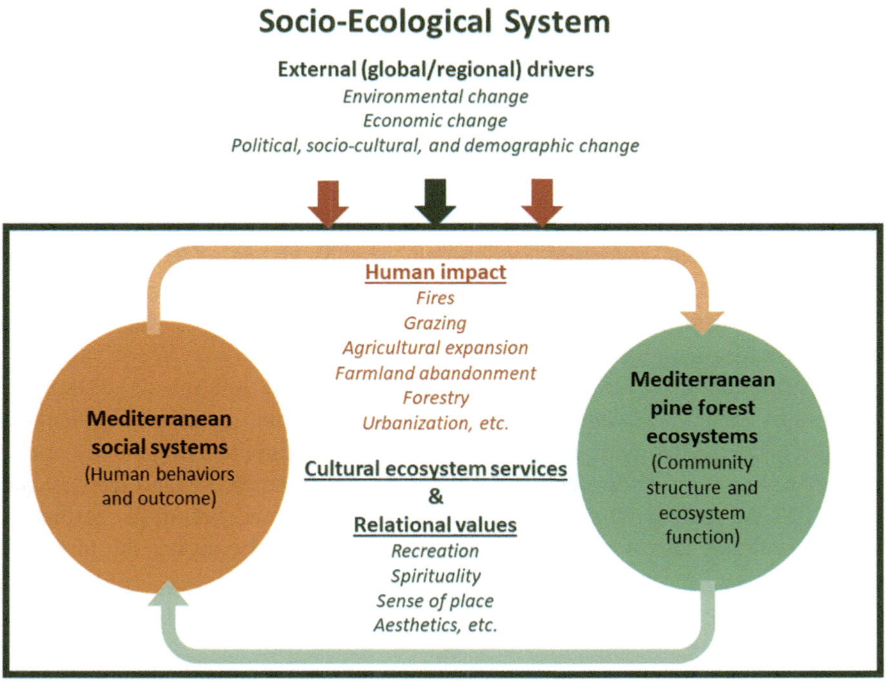

Fig. 30.1 Socio-ecological system. (Adapted from Collins et al. 2011)

abandonment. Each condition provides a unique set of CES to a specific population group. These changes may be welcomed by various segments of society, and not by others.

2. **Spatial dynamics of ecosystems**: As the perceived benefits from ecosystems reflect the interaction between people and nature, when moving from place to place across the Mediterranean Basin, we would expect the "basket" of services and the value of the benefit to also change according to social and ecological context.
3. **Temporal dynamics of social systems**: Societal values and behaviors change over time. "Nature" does not mean the same to today's citizens what it did for past generations. Major economic changes, such as movement from agrarian to urban industrial and post-industrial societies, change the way we value and use forests.
4. **Demographic dynamics of social systems**: Different demographic groups coexist with different value systems. Perceived benefits vary from group to group and between individuals at a single location or between locations. Differences in CES perceptions are observed between groups with varying occupations, level of formal education (and type of education), gender, income, nationality, ethnicity, etc. All of these can impact the type, value, and intensity of various CES received from pine forests.

To summarize, social and ecological systems are both dynamic, and they respond to one another in cyclical processes with feedbacks. Historical changes in society (agricultural and industrial revolutions, urbanization processes, developments in religion and politics) change the way humans perceive and interact with the natural environment. At the same time, the natural systems themselves are dynamic and change in response to human and non-human drivers, leading to a shifting "basket" of services. To further complicate the picture, at any given time in history, the perceptions of benefits from ecosystems differ among different demographic groups depending on, for example, where an individual works, what their underlying views are regarding human–nature interactions, etc. This is evident in CES assessment of forests in general (Holmgren and Scheffer 2017), and in Mediterranean pine forests in particular (Lopez-Santiago et al. 2014; Almeida et al. 2016; Maestre-Andrés et al. 2016; Martínez-Sastre et al. 2017; Negev et al. 2019).

For these reasons, CES assessments should be offered with the caveat that any given assessment is relevant for only the specific time, place, and social context in which it is conducted. Nonetheless, we can also find broad recurring themes across the recent literature with reference to Mediterranean pine forests and their associated CES. We review these CES below, focusing on (1) recreation; (2) aesthetics, and; (3) cultural identity and relational values. But first I consider the historical cultural role of pine forests in the Mediterranean Basin, and then explore broad, recurring CES appearing across the Mediterranean Basin.

30.2 Mediterranean Pine Forests and Their Cultural Significance in History

In this section we consider the historical significance of pine forests in the Mediterranean Basin. By doing so, we not only understand how the range of CES has changed throughout history, but we also begin to consider how modern perceptions regarding the "naturalness" of *Pinus* species in the region influence perceptions of CES received from a forest comprised of these trees.

Although pine forests (including mixed forests, where pines were a significant part) may have been a distinct component of Mediterranean ecosystems in the Paleolithic Period when hunter-gatherers reached the region (Peñuelas et al. 2017, studying the Iberian peninsula), several researchers suggest that pines were a relatively minor component of Mediterranean biota, and that they later proliferated due to human activities (Blondel and Aronson 1995).[2] Hunter-gatherers may have burned, or otherwise cleared, naturally occurring forests to ease hunting constraints and increase grasslands for targeted species, supporting a hypothesis that humans have an innate preference for savanna habitats, since humans had evolved in these ecosystems and had grown accustomed to them (Falk and Balling 2009).

Widespread deforestation of the Mediterranean Basin followed the arrival of humans, particularly during the Neolithic Period and the establishment of permanent settlements, resulting in greater landscape patchiness and desiccation of the region, where reduced vegetative cover resulted in greater water runoff, stream flow, and erosion (Blondel 2006). Following the first agricultural revolution and the rise of cities, the demand for timber and grazing lands led to rapid deforestation of the region, and along with it, crucial regulating ES, such as soil and moisture retention, were also degraded (Blondel 2006; Hughes 2011). Through a combination of natural climate change and human activities, the Mediterranean ecosystem was greatly altered, with one of the prominent results being the decline in prevalence of forests in general, and pine forests in particular. Some of the maquis we recognize today may be a direct result of historical deforestation and the proliferation of husbandry, agriculture, and urbanization.

In the Hebrew Bible, the pine makes four possible cameo appearances (in English translations), although the Hebrew word for pine (*oren*) appears only once, where a pine is planted and raised and eventually used for fuel for cooking and for warmth. While the pine was considered by some as common in Levant forests during the Biblical era, its general absence in the Bible is a quandary, and this fact – along with additional evidence – led Liphschitz and Biger (2001) to conclude that the region of Israel/Palestine was actually not home to *P. halapensis* Mill., which is so prolific in the region today due to afforestation activities (see below).

---

[2] In one comprehensive volume on Mediterranean-type ecosystems, there is only a single reference to naturally-occurring pine species in the Mediterranean Basin, with the remaining references to pines referring only to the negative ecological impacts of pine plantations in Australia and South Africa (Davis and Richardson 1995).

Further north, in Lebanon, pines (*P. brutia* Ten.) were a historical component of the cedar forests exploited by Sumerians, Babylonians, and Israelites, among others (Abu-Izzeddin 2000). Following three millennia of deforestation, Roman Emperor Hadrian implemented what Abu-Izzeddin considered to be the first nature conservation law – a command to preserve remaining pine and cedar forests in Lebanon as an Imperial Domain (Abu-Izzeddin 2000). The author goes on to describe nineteenth century re-forestation efforts that featured stone pines (*P. pinea* L.), valued not only for their wood and pine nuts, but also for their "sweet scent and cool shade" that attracted locals and tourists alike (a CES). Due to ongoing destruction of the forests of Lebanon into the twenty-first century, multiple efforts are underway to preserve the integrity of its remaining forested areas (Abu-Izzeddin 2000). The Romans are also credited with initiating pine afforestation for dune stabilization on the Antalya coastline of Turkey, and pine continues to be a target genus for afforestation by the Turkish Forest Service today (Çaliskan and Boydak 2017).

The Aleppo pine (*P. halepensis*) was a dominant tree on the forested landscape across the central and western Mediterranean (France, Spain, Morocco, Italy, and Algeria), and in Tunisia, according to one study, the tree had important social roles, alongside its environmental and economic importance (Jaouadi et al., 2019). In particular, its pine nut is described as a cultural icon due to its use, unique taste and role in local cuisine (Jaouadi et al. 2019).

Several studies on cultural landscapes and ES note the relatively recent arrival of pines in the Mediterranean Basin. Gragson et al. (2015) note that the increasing presence of pines in the western French Pyrénées was a phenomenon of the last 500 years. Martínez-Sastre et al. (2017) note in their analysis of the Sierra Morena mountain range in Spain that mixed pine forests (mostly *P. pinaster* Ait.) arrived with plantation programs in the 1960s and are currently being phased out in favor of native woody Mediterranean species of oaks and strawberry trees. Sacchelli (2017), assessing forest ES in a silver fir forest in central Italy, refers to artificial stands of black pine forest, which was not included in his ES assessment.

30.3 A Contemporary Survey of Cultural Ecosystem Services Around the Mediterranean Basin

Many CES drawn from the generic list (Table 30.1) can be intuitively attributed to Mediterranean pine and mixed-pine forests, and they are indeed noted in introductory sections for research on various aspects of Mediterranean ecosystems (e.g. MA 2005; Peñuelas et al. 2017; Gauquelin et al. 2018).

Using several research methods in a comprehensive ES assessment of a natural park (characterized as containing mixed oak-pine forests) in the Mediterranean north-east of Spain, Maestre-Andrés et al. (2016) found that when asked to identify ES offered by the park, stakeholders most often identified cultural services (alongside habitat services) in comparison to provisioning and regulating services.

Through these interviews, coupled with a broader survey-based inquiry, they identified the most highly valued ES to include spiritual experience, information for cognitive development, mental recreation, aesthetic information, physical recreation, gene pool protection (a habitat service) and distraction and leisure. Several of these were positively or negatively correlated with various demographic factors such as age, place of residence, educational level, and environmental values. Additional studies have focused on assessments of CES, either through open interviews with stakeholders, literature surveys, and/or questionnaires for the general population. I expand upon the most predominant assessment results including recreation, tranquility and relaxation, aesthetic benefits, and relational values.

30.3.1 Recreation, Tranquility, and Relaxation

One of the prominent global trends of the twentieth and twenty-first centuries is the increasing detachment of humans from the natural environment as the result of urbanization, which has physically distanced humans from natural areas, and changes behaviors (Soga and Gaston 2016). But with less access to natural open spaces and growing populations, there is also increased demand for open, natural areas for recreation and leisure use. Pine forested areas offer several qualities to fulfil this demand, including shade, refuge from noise and wind, and pleasant qualities such as sound, scents, and clean air. Multiple studies note that Mediterranean pine and mixed-pine forest provide recreational value (Maestre-Andrés et al. 2016; Jucker Riva et al. 2018) and tranquility and relaxation (Lopez-Santiago et al. 2014; Maestre-Andrés et al. 2016).

In Tuscany, Italy (Bernetti et al. 2019), pine forests are so popular with local tourists that they are subject to overload. A survey of Italian managers of National and Regional Parks revealed that the most relevant ES provided by forests was "improvement of tourism and recreational concerns" (Vizzarri et al. 2015). On Greek Ionian Islands, pine forest cover was positively associated to recreational services individually and in conjunction with other land cover types, such as olive groves (i.e., landscape mosaic; Lorilla et al. 2018).

According to Campos et al. (2019), the monetary value of recreational use of ecosystems, like regulating services and other cultural services, is often undervalued in standard national accounting, especially when recreational access is provided for free. These researchers used contingent valuation willingness-to-pay surveys to estimate the value of two CES, namely recreation and private amenity (expressed as the amount of money farmers would be willing to accept to give up their property) in the Andalusian Forests of Spain (where six *Pinus* species are present). They found that that these two CES account for 59% of the total value of the Andalusian forest ES, making recreation one of the most prominent ES provided by these forests. Since forest CES, and recreational use in particular, are prominent among Mediterranean pine and mixed-pine forests, it is logical that other researchers have found that forest CES are strongly and positively correlated to proximity to roads (Bernetti et al. 2019) and to major population centers (Roces-Díaz et al. 2018).

30.3.2 Relational Value

Like other ecosystems, Mediterranean pine forests contribute towards shaping the identity of those who live and work in them. Relational values are thus frequently noted as important with respect to these forests, and they can be defined in various ways, for example local identity and cultural heritage (Maestre-Andrés et al. 2016) or traditional cultural and emotional values (Peñuelas et al. 2017). We consider the forest to be contributing to relational values when some biological or landscape characteristic of the ecosystem contributes to an individual's or group's sense of attachment to the forest or contributes to defining the identity of the person or group. This is why provisioning or regulating services are often interlinked with cultural services and relational values. The act of collecting mushrooms, for example, is at once a provisioning service (something we eat), a cultural service (a recreational activity) and a relational value (it has defined my family identity for generations). When a pine forest helps retain soil and prevent erosion, it is providing a regulating service that also may increase one's sense of wellbeing with nature and security from environmental degradation.[3]

Soliño et al. (2018) note that in Spain's Castilla y León region, "the virtual disappearance of resin tapping in Spain [from *P. pinaster* forests] caused the abandonment of traditional forest activities and the subsequent losses of ecosystem forest services", and that "abandonment of this traditional forestry activity would lead to a loss of social welfare beyond the economic activity." In this way, the authors connect a provisioning service (resin) to relational values by noting that the implications of the decline of resin tapping resulted in the abandonment of rural settlement and the loss of cultural and natural heritage. Resin tapping was not only an economic activity, but a cornerstone supporting a way of life. In Cyprus, pine forests provide the ideal conditions for mushrooms and pine cones, which are both considered provisioning services (food and decoration, respectively; Ciftcioglu 2018). But as I and others suggest (e.g. Marini Govigli et al. 2019), activities such as hunting and mushroom collection overlap with CES and relational values, such as recreational activities and sense of place, respectively.

Scientists may, in subtle or not-so-subtle ways, introduce their own biases into their research and this may affect how pine forests are assessed with regard to relational values. Ecologists, for example, are particularly sensitive to land uses that reduce biodiversity. If they are studying processes that they know are causing damage to biodiversity, this may weaken their sense of attachment to a place.

[3] Alternatively, when a pine forest is viewed as a threat due to its flammability (a regulating disservice), it can degrade one's attachment to the forest (Depietri and Orenstein 2019).

Mediterranean pine forests, particularly pine plantations, are often, though not always (Pedley et al. 2019), assumed to have lower biodiversity potential than other natural ecosystems, and this affects opinions regarding the desirability of pine forests (see Rothschild 2019 for an example of opposition to pine afforestation in Israel). There may also be ambiguity regarding the "naturalness" of pine forests in the Mediterranean Basin, and many of the references to pine forests in the literature are to plantations and agro-forestry, which are implicitly less desirable than the "natural" forest or other vegetative structure.

30.3.3 Aesthetic Value: The Eye of the Beholder or Shape over Species?

CES assessments often focus on the aesthetic value of a landscape, which opens the topic to a long history of landscape aesthetics research predating the proliferation of the ES conceptual framework. There are two broad schools of thought regarding landscape aesthetics drivers and preferences, although both garner empirical support and they are not necessarily mutually exclusive. Some suggest that there are universal and consensual commonalities regarding landscape tastes, or *sensus communis* (Kant 1791, cited in Stamps 1997). A variation of this is the hypothesis from evolutionary psychology that savannas constitute the most preferred human landscape due to access to resources and the evolutionary advantage early humans had against both prey and predator when living in savannas (Falk and Balling 2009). On the other hand, landscape preferences have been shown to be distinct among different demographic groups (Zube and Pitt 1981; Buijs et al. 2009) and most landscape preference assessments are thus structured to investigate these differences.

Several studies have been conducted to assess the aesthetic value of Mediterranean pine forests, usually assessing the forests relative to other landscape alternatives, such as agricultural landscapes, shrublands, or forests with different structures or biological composition, such as the Montado of Portugal, which is dominated by cork oak. Results suggest that preferences for pine forests depend heavily on the landscapes to which the forest is being compared. Almeida et al. (2016), for example, measured landscape appreciation in a region of diverse landscape types in southern Portugal, including pine forests. Using photographs of 16 landscape types, they asked respondents to record their preferences. Pine forests ranked relatively low in preference in comparison to the other 15 possible landscapes, while mixed forests ranked slightly higher, but still much lower than Montado, olive groves, or vineyards. However, the authors showed that pine forests were preferred by individuals expressing an overall preference for productive landscapes (also including vineyards, irrigated agriculture, and intensive olive groves; Almeida et al. 2016). Pine forests were disliked by respondents with a preference for traditional agricultural landscapes. Lopez-Santiago et al. (2014) found that stakeholders in a transhumance landscape in Spain ranked pine forests higher in aesthetic value than croplands, which they attribute to a more general preference of people for green, forested landscape than to more arid ones.

Rosário et al. (2019) suggest that the Montado forest, with its sparser density and wide-canopy trees, is considered more aesthetic than forests dominated by the conical or columnar structures of some pines, which further supports the theory of preference for savanna landscapes over forested and other landscapes. Bernetti et al. (2019), analyzing big data in the form of Flickr social network photographs, showed that Mediterranean pine forest species (particularly the Italian umbrella pine, *P. pinea*, with their wide canopies) are among the more appreciated tree types in Tuscany, Italy, compared to lower-canopy oaks and broad-leaved species. Both studies suggest that dendrometry and forest density have greater importance than species when assessing tree aesthetics.

30.3.4 Cultural Services in Mediterranean Pine Forests Relative to Other Landscape Types

As with the measure of aesthetic value, cultural services in pine forests in general are also assessed relative to other land-cover types, such as rangeland, other forest types, or agricultural landscapes. In Spain, Lopez-Santiago et al. (2014) asked stakeholders to compare the perceived ES in agricultural and pine forest landscapes. CES in forests scored high among stakeholders relative to provisioning and regulating ES. Pine forests were perceived to provide four out of five cultural services, including aesthetic value, tourism, hunting and tranquility/relaxation. Furthermore, pine forests were perceived to provide more of these services than cropland. For cultural identity, cropland and pine forests were not significantly different.

In a second study in the Sierra Morena mountain range in Spain, the ES of pine forests were assessed using both social and biophysical measures and these forests were found to contribute less to traditional and local knowledge than alternative forest types (Martínez-Sastre et al. 2017). Nonetheless, the authors of this study also found that future land-use scenarios that included pine forests within a mosaic of multi-functional land uses were deemed optimal from a biophysical and social perspective (as opposed to landscapes dominated by olive plantations or livestock production). A cross-country comparison in the Mediterranean Basin found that stakeholders noted that various forms of pine forests supply CES (recreation or maintenance of traditional landscapes) more than alternative landscapes such as diversified shrubland or extensive grazing areas (Jucker Riva et al. 2018). In central Portugal, researchers concluded that forests (including pine forests) provided a higher amount of CES than both urban and agricultural landscape types (Leitão et al. 2019).

Perceptions of CES from different landscapes are often found to differ between demographic groups. In almost every empirical study of the CES provided by Mediterranean pine forests reviewed here, different demographic groups were shown to hold difference perceptions regarding forest CES. López-Santiago et al. (2014), for example, found that perceptions of ES differed between rural and urban inhabitants, herders and non-herders, younger and older respondents, genders, and groups with different formal educational achievement.

Differences in perceptions of CES provision can sometimes lead to conflict. For instance, conflicts arise between various groups regarding access to cultural services. In a natural park in Spain, conflicts arose between park managers, who sought to increase access to the park and its CES, and conservationists who advocated limiting public use (Maestre-Andrés et al. 2016).

The twentieth- and twenty-first-century proliferation of pine plantations is often referred to in the context of research on other Mediterranean cultural landscapes such as deciduous oak woodlands in Greece (Schaich et al. 2015), or transhumance cultural landscapes in Spain (Oteros-Rozas et al. 2014). While the plantations are often not the focal ecosystem of the research, references to pine plantations focus on their potentially negative impacts on biodiversity or CES, or their positive contribution to provisioning services (timber). On the other hand, Derak and Cortina (2014) compared the ES provided by *P. halepensis* plantations used for landscape restoration to ES provided by unrestored landscapes (grasslands, steppes, shrublands and abandoned agricultural fields) and found that, according to stakeholders, the pine forests enhanced the aesthetic value of the landscape.

Aesthetic perceptions are not only comparable between different landscape units, but also to landscape change over time. Human activity (forestry, agriculture, grazing, urban development) or inactivity (e.g. farmland abandonment), coupled with natural processes, such as forest fires and pest outbreaks, shape our familiarity and cultural affinity with certain landscapes (Holmgren and Scheffer 2017). The dynamic nature of Mediterranean landscapes leads to a dynamic character of cultural valuation of those sites. The phenomenon of how landscape flux impacts aesthetic perceptions of those landscapes differently in time and between demographic groups was described by López-Santiago et al. (2014), who compared aesthetic preferences between pine forests and agricultural landscapes among various stakeholders in Mediterranean Spain (Cuenca province).

The general picture that emerges is that Mediterranean pine forests provide the suite of CES common to many natural ecosystems, including recreational activities, rest and relaxation, activities associated to resource use (mushrooms, resin, timber, pine nuts), cultural landscapes and sense of place. The perception of their CES value is high relative to degraded landscapes, agricultural landscapes, and landscapes featuring sparser vegetation (e.g. urban areas), but low relative to other forest types and sometimes low relative to what are considered to be more "natural" landscapes. Within those generalities, there is also some diversity of opinion between different demographic groups – particularly depending on their economic attachment (or lack thereof) to the forest, their desire for recreational space, and their predispositions regarding the "naturalness" of the pine forest.

30.4 Cultural Ecosystem Services and Aesthetic Preferences in Israel's Pine Forests

Israel's pine forests, like those across the Mediterranean Basin, offer a familiar suite of CES. These include recreation, relaxation, mushroom collecting, physical activities, aesthetic landscapes, pensive activities, religious and spiritual meaning, relational values, and more (Ginsberg 2000, 2006; Naveh and Carmel 2004; Amir and Rechtman 2006; Tal 2013; Eid and Haller 2018; Negev et al. 2019). However, Israel's Mediterranean pine forests also have a tumultuous socio-ecological history.

30.4.1 The Politics and Ecology of Israel's Pine Forests

Pine forests are considered politically contentious in Israel because the story of forestry is tightly intertwined with the history and mythology of nation building. Until the beginning of the twentieth century, pine stands were reportedly rare in the region (Liphschitz and Biger 2001). Though not without its ecological and political critics, for the past century tree planting – particularly pine planting – has been considered a national mission (Amir and Rechtman 2006). One of Israel's most senior forestry experts, Dr. Gabriel Schiller, in the introduction to his self-published tome about afforestation, suggests that the history of forestry in Israel is analogous to the process of nation building itself. Range ecologist Prof. Avi Perevolotsky, who was chief scientist of Israel's Nature and Parks Authority, concurred with this view, writing in his introduction to Schiller's book "The act of forestry in Israel is not fundamentally different from the other aspects related to founding the state." (Schiller 2012). Thus, one's predisposition to the country in general may inform, or correlate with, one's predisposition to its forests (compare, for example, the narratives of Tal 2013 and Braverman 2009, for starkly contrasting views regarding afforestation in Israel).

The ecological debate focuses on the "naturalness" of the planted pine forests and their implications for biodiversity, ES, and ecosystem integrity. Critics suggest that the selected species, mostly pines, are not only foreign to the Israeli landscape, but that their presence is ecologically harmful (see Rotem et al. 2014 and Rothschild 2019 for a critical perspective, and Tal 2013 for an overview of the ecological critique). Recalling that CES are the product of the dynamic interaction between human society and the natural environment within socio-ecological systems, the relevance of these underlying ecological and political debates becomes evident. Political and ecological perspectives and values can, in fact, have a profound influence on perceptions of CES from Israel's pine forests.

30.4.2 Socio-ecological Dynamics and Israel's Pine Forests

The socio-ecological development of the landscape as co-produced by human and natural forces was well conceptualized in the context of Israel's Mediterranean ecosystem in the oeuvre of Professor Zeev Naveh and his colleagues (Naveh and Lieberman 1994; Naveh and Carmel 2004). Israelis often construct their individual, group, and national identities based on the aesthetic landscape of the country – real or imagined (Ben-David 1997; Naveh and Carmel 2004; Eid and Haller 2018). The dynamic nature of this interaction emphasizes the two aspects of change introduced previously, namely ecological and demographic, and their feedbacks (Fig. 30.1). Natural and human-driven processes, such as tree planting, agriculture, grazing, urbanization, and fire, change the landscape over time. These changes can affect the aesthetic value of the landscape (Misgav and Amir 2001; Tal 2013) – for better or for worse, depending on the beholder.

Concurrent with ecological changes, values also change over time, including those regarding the human–nature relationship. Such changes in modern Israel include, for example, a shift from the dominant desire for human control over nature, to an ideology of protection of nature, to a desire to co-exist with nature (Arieli 1997; Schwartz 2009; Orenstein and Silverman 2012; Tal 2013). These changes in value systems may also impact perspectives regarding the perceptions and valuation of CES of pine forests and how management agencies and the public at large assess the aesthetic landscape. Forest fire can impact relational values with the pine forest both positively and negatively (Fig. 30.2). Increasing exposure to fire risk can weaken one's identity with the forest (Depietri and Orenstein 2019), but citizen responses to forest fires in Israel's Carmel Forest in the past reflected an affirmation of stakeholders' identity with the forest, which has been measured by assessing donations to post-fire recovery programs (Shechter et al. 1998).

Fig. 30.2 Spontaneous regeneration of pines in the Carmel Forest following the 2010 fire; in focus group discussions, respondents appreciated the view, and few noted which trees were growing in the foreground. One ecologist appreciated the appearance of life after the fire, "despite the fact that they are pine trees". (Photo credit: Haim Zinger)

30.4.3 Landscape Perceptions and Aesthetic Preferences in Israel's Pine Forests

Research suggests that there is a broad consensus regarding the high aesthetic value of Mediterranean (and other natural) landscapes in Israel, though the preferred characteristics of this landscape can change over time and between groups. Of all of the various aspects of the natural environment, research has shown that landscape (Negev et al. 2019 for the mixed-pine Carmel Forest), scenery (Koniak et al. 2011 for Mediterranean landscapes) and landscape viewpoints (Becker and Choresh 2007 for pine forests), all synonymous with one another, are the most highly valued characteristic of open spaces. According to Becker and Choresh (2007), who applied a travel cost model to visitations in Biriya Forest, recreational visits to the forest (valued at 331 NIS per visit) were most positively affected by landscape view, followed by access to picnic facilities, walking paths, and access for automobiles. In their analysis of approximately 27,000 geotagged photos of landscapes in Israel uploaded to a photo sharing service over an 11-year time period, Lotan et al. (2018) found that planted (primarily pine) forests in close proximity to urban areas were one of two of the most commonly photographed natural environments in Israel.

Researchers of Israeli landscape preferences also focus on the differences in landscape aesthetic preferences among demographic groups, as will be discussed in the following sections.

30.4.4 Vegetation Composition and Structure Contributing to the Most Preferred Aesthetic Landscapes

Israel's Mediterranean landscapes are a patchwork of vegetative compositions and structures. They are a rich mosaic of maquis (*horesh*) and garrigue, combined with orchards, field crops, planted forests (Naveh and Carmel 2004), and, increasingly, urban development. In addition, the landscapes are grazed at various intensities by cattle, sheep, and goats. So, although planted forests are treated in this book as a unique ecosystem, from the perspective of the Israeli public, particularly when assessing landscape preferences, boundaries between landscape types are not so clearly delineated or they are perceived as a mosaic of landscapes (Fig. 30.3). As shown in Spain (Martínez-Sastre et al. 2017) and Greece (Lorilla et al. 2018), public aesthetic preferences may lean towards a mosaic of landscape types, including pine forests, rather than a single land-cover type.

Israel's Mediterranean forested ecosystems provide aesthetic benefits for Israelis and tourists alike (Ginsberg 2000; Amir and Rechtman 2006; Tal 2013; Negev et al. 2019). This aesthetic benefit co-occurs with other benefits and is valued in multiple ways, including health (vegetated landscapes promote health), economic (residences with green views are more highly valued than without; visitors are willing to

Fig. 30.3 A landscape mosaic at the southern edge of the Mediterranean zone in Israel, including olive groves and planted pine forests. (Photo credit: Daniel Orenstein)

pay for landscape and recreation), and social benefits (vegetated landscapes provide the backdrop for social gatherings) (Orenstein et al. 2015).

Vegetative composition and structure may impact the aesthetic value of Mediterranean landscapes in Israel. Misgav (2000) studied public aesthetic preferences for particular vegetation structure and composition, asking whether the public could distinguish between vegetation groups, and according to what physical properties. She identified 44 different vegetation compositions, which varied according to dominant species and general structure (e.g., forest, maquis, garrigue or scrub) and queried respondents regarding their aesthetic preferences. The most preferred landscapes were cultured landscapes, including planted and managed forests and olive groves, while scrub and garrigue were least preferred, giving a feeling of exposure, on the one hand, and blocking the field of vision, on the other.

Koniak et al. (2011), received quite different responses in their survey of landscape preferences in the Ramat HaNadiv nature park on Israel's northern Mediterranean coast. They surveyed hikers and picnickers in the park regarding their preferences for different landscapes, including garrigue, forested and dense shrub, and came to distinctly different conclusions. They found that hikers preferred, from most to least, open garrigue, dense shrub, and pine forest, concluding that hikers preferred "natural" landscapes. For picnickers, on the other hand, garrigue and pine forests were most highly preferred, while dense shrub was less preferred (Koniak et al. 2011). Among the most positively assessed elements of the landscape, scenery, flowers, native trees, and deer/gazelle were the most highly valued landscape features, while reptiles, insects and cattle were the lowest valued (although still positive). Using photographs and a paycard method, Divinski et al. (2018) concluded that planted pine forests were the least valued landscape according to respondents queried while riding a commuter train, who preferred woodland landscapes, followed by garrigue, grassland and only then, planted forests.

In their survey of 703 visitors and residents of the Carmel Forest, Raviv et al. (2020) reported that 89% of their respondents liked or very much liked maquis landscapes and 45% reported visiting them more frequently than any other local landscapes. The percentage of respondents liking or very much liking the pine forests was 70%, but only 12% reported that they visited pine forests more frequently than any other landscape. However (as reported by Itzhaki pers. comm.), when respondents specified the locations they visited, the top five places visited were either pine or mixed-pine forest areas. This might suggest that while the *idea* of pine forests is viewed somewhat with disfavor, in practice, pine forests remain the most popular areas for recreational visits either due to revealed (actual) preferences or due to the presence of recreational infrastructures within and around the pine forests (Raviv et al. 2020).

Zimroni et al. (2017), in their study of landscape aesthetic preferences, in which 70+ diverse stakeholders were queried in a focus-group setting about 16 different landscapes in the Carmel Forest (natural and built environments; pine forested and "natural" landscapes), which were projected on a high-definition large screen in a visualization theater. They documented contradicting opinions regarding aesthetic preferences. Results suggested that planted forests and olive orchards were highly valued aesthetically by most, though not all, respondents (Orenstein et al. 2015; Zimroni et al. 2017; Eizenberg et al. 2018). Some respondents, primarily ecologists and environmentalists who perceived the pine as a "foreign" or "invasive" plant, responded negatively to the presence of pine trees in the landscape. Upon identifying pine species in the landscape photographs, these respondents expressed their disdain for the landscape or the trees themselves, or expressed their approval, as one ecologist did, "despite the presence of pine trees".

Another major point of debate in the study by Zimroni et al. (2017) related to evidence of human presence in the landscape, such as roads, picnic tables, residential neighborhoods, or garbage bins. Here, too, ecologists and respondents associated to environmental organizations had a strong, negative opinion regarding such elements and their aesthetic impact. Many other respondents, on the other hand, saw these elements as positive, as they facilitate easy access to nature and a clean and orderly environment. In a survey of visitors at Biriya Forest, respondents were asked to choose their most and least favored landscapes between 16 different pine forest scenes. In some scenes, the forest floor was covered with woody debris that had fallen from the trees in a snowstorm, while the others were cleaned of debris. Results indicated that respondents (primarily picnickers in the forest) preferred a defined set of physical parameters, including green, orderly, and shady nature (Orenstein et al. 2018), echoing the findings of Misgav and Amir (2000, 2001).

30.4.5 Cultural Ecosystem Services in Israel's Pine Forests – Conflict and Compromise

The pine forests of Israel provide crucial CES for diverse populations. While they are criticized on ecological and political grounds, they also serve a much-needed function of providing recreational space, aesthetic landscapes, and spaces for reflection, meditation, research, and sporting activities. But, as we have shown here for the entire Mediterranean Basin, the type and value of CES provided by these forests are a subjective matter, depending on when and where they are being assessed and by and for whom. The value of CES, as we have seen with aesthetic preference, is a relative concept – depending on what other landscape options exist. In a country with rapidly depleting open spaces and a growing population, open spaces are increasingly in demand (Tal 2008). Pine forests are currently one of the major providers of CES in Israel. This has been supported in a number of studies using a variety of methodological tools, although it is also somewhat intuitive, considering the high demand for such landscapes for recreational and other intangible benefits (Tal 2008, 2013).

Where there is less consensus regarding the best vegetation constitution and structure of these landscapes, there remains a diversity of opinion regarding whether pine forests are the best alternative for providing CES. Factors affecting CES perceptions include the recreational goals of the respondent (Koniak et al. 2011; Negev et al. 2019), gender (Negev et al. 2019), type and level of formal education (Eizenberg et al. 2018; Negev et al. 2019), income (Becker and Choresh 2007), and/or ethnicity (Zimroni et al. 2017; Negev et al. 2019).

Misgav and Amir (2001) pointed out the potential conflict between the aesthetic value of cultured landscapes and the ecological value of restored "natural" landscapes, and recommended that land use managers take into account both objectives when designing landscapes and designate spatially explicit areas where cultural (aesthetic) value should be considered. Orenstein et al. (2018) made a similar recommendation, advocating a zoning approach to public space management, where high visitation pine forests are managed to facilitate intensive use, while pine planting and regeneration is discouraged in less visited areas or those that are more ecologically sensitive. These recommendations are commensurate with the new management doctrine adopted by Israel's forest service (Osem et al. 2012).

30.5 In Conclusion: A Holistic Assessment of CES in Mediterranean Pine Forests

Mediterranean pine forests provide a rich set of CES to contemporary societies across the Mediterranean Basin. But assessments of CES must be appraised with nuance and context because they will change from community to community, over time, and relative to what they are being compared. Two crucial management questions resulting from this review are (1) what is the basket of CES being provided by pine forests in specific places and for specific beneficiaries? and (2) is this the best way to provide these services with the least amount of negative ecological impacts?

While perceptions of Mediterranean pine forest CES depend largely on the landscape to which it is being compared and who is being asked, there are also some consistent themes that emerge from this review, including:

1. Mediterranean pine forests invoke relational values and a strong sense of place among some sectors of the societies who inhabit and utilize them.
2. Mediterranean pine forests provide large populations of stakeholders with recreational services, which are becoming increasingly important with the loss of open spaces near urban centers.
3. Mediterranean pine forests are green and productive – two qualities that rank highly in landscape preference surveys.
4. Many respondents across countries prefer trees with wide canopies and not densely distributed, which allow for both shade and a wide field of view. Only some *Pinus* species have wide canopies.

30.5.1 Changing Perceptions with Changing Climate

We conclude with a crucial third management question: (3) how will the provision of these CES be affected by climate change and other environmental changes in the Mediterranean Basin? The convergence of global and regional environmental challenges, from climate change to urbanization and population growth to biodiversity loss will negatively impact the provision of CES from Mediterranean pine forests. Studies suggest that the region will be increasingly drier (IPCC 2014), leading to a greater frequency and extent of forest fires (Piñol et al. 1998), desertification (IPCC 2014; Peñuelas et al. 2017), and loss of regulating services, including water and soil retention (Anaya-Romero et al. 2016). All of these will further degrade the provision of CES, including touristic and recreational value (Peñuelas et al. 2017), aesthetics, and relational values. Greater fire risk brings greater demand to mitigate the risk through forestry practices, including thinning pine stands (Depietri and Orenstein 2019), which, in the proximity of human settlement, may reduce the CES that the trees provide, or increase the demand for domestic grazing, which has been shown to reinvigorate relational values in pine forests (Eid and Haller 2018). Under conditions of predicted environmental change, some continue to recommend pine species as tools to counteract the impacts of desertification (Peñuelas et al. 2017), while others suggest that the continued use of pines for afforestation increases vulnerability to drought and fire (Jucker Riva et al. 2018; Depietri and Orenstein 2019). Somewhat ironically, nature may soon make our landscape management decision for us: climate trends may lead to a reduction in the cover of pine and mixed pine-oak forests and to greater dominance of oaks (Peñuelas et al. 2017). Whether oak forests, or whatever eventually replaces them, will provide CES equivalent to those provided by pines is a question for future researchers.

Acknowledgments I thank Hila Sagie, Alon Lotan, and Carly Golodets for their reviews of this chapter and for their important suggestions and improvements, and Ronit Cohen for her assistance with the literature review. This research was funded by an Israel Science Foundation grant (No. 1835/16).

References

Abu-Izzeddin F (2000) The protected areas project in Lebanon: conserving an ancient heritage. Parks 10:25–32

Almeida M, Loupa-Ramos I, Menezes H et al (2016) Urban population looking for rural landscapes: different appreciation patterns identified in Southern Europe. Land Use Policy 53:44–55. https://doi.org/10.1016/j.landusepol.2015.09.025

Amir S, Rechtman O (2006) The development of forest policy in Israel in the 20th century: implications for the future. Forest Policy Econ 8(1):35–51. https://doi.org/10.1016/j.forpol.2004.05.003

Anaya-Romero M, Muñoz-Rojas M, Ibáñez B, Marañón T (2016) Evaluation of forest ecosystem services in Mediterranean areas. A regional case study in South Spain. Ecosyst Serv 20:82–90. https://doi.org/10.1016/j.ecoser.2016.07.002

Arieli D (1997) Cultural construction of nature: the case of the Society for the Protection of Nature in Israel. Megamot 2:189–206. [in Hebrew]

Becker N, Choresh Y (2007) Economic valuation of nature recreation: estimating the Biria Forest value using the travel cost method. Ya'ar 9:27–34. [in Hebrew]

Ben-David O (1997) Tiyul (hike) as an act of consecration of space. In: Ben-Ari E, Bilu Y (eds) Grasping land: space and place in contemporary Israeli discourse and experience. SUNY Press, Albany, p 246

Bernetti I, Chirici G, Sacchelli S (2019) Big data and evaluation of cultural ecosystem services: an analysis based on geotagged photographs from social media in Tuscan forest (Italy). iForest 12:98–105. https://doi.org/10.3832/ifor2821-011

Blondel J (2006) The "design" of Mediterranean landscapes: a millennial story of humans and ecological systems during the historic period. Hum Ecol 34(5):713–729. https://doi.org/10.1007/s10745-006-9030-4

Blondel J, Aronson J (1995) Biodiversity and ecosystem function in the Mediterranean Basin: human and non-human determinants. In: Davis GW, Richardson DM (eds) Mediterranean-type ecosystems: the function of biodiversity. Springer, Berlin/Heidelberg, pp 43–119. https://doi.org/10.1007/978-3-642-78881-9_2

Braverman I (2009) Planted flags: trees, land, and law in Israel/Palestine. Cambridge University Press, New York

Buijs AE, Elands BHM, Langers F (2009) No wilderness for immigrants: cultural differences in images of nature and landscape preferences. Landscape Urban Plan 91(3):113–123. https://doi.org/10.1016/j.landurbplan.2008.12.003

Çaliskan S, Boydak M (2017) Afforestation of arid and semiarid ecosystems in Turkey. Turk J Agric For 41:317–330. https://doi.org/10.3906/tar-1702-39

Campos P, Caparrós A, Oviedo JL et al (2019) Bridging the gap between national and ecosystem accounting: application in Andalusian forests, Spain. Ecol Econ 157:218–236. https://doi.org/10.1016/j.ecolecon.2018.11.017

Chan KM, Gould RK, Pascual U (2018) Editorial overview: relational values: what are they, and what's the fuss about? Curr Opin Environ Sust 35:A1–A7. https://doi.org/10.1016/j.cosust.2018.11.003
CICES – Common International Classification of Ecosystem Services (2018) Version v5.1 spreadsheet and guidance document. Available at: https://cices.eu/resources
Ciftcioglu GC (2018) Revealing major terrestrial- and marine species-based provisioning ecosystem services provided by the socio-ecological production landscapes and seascapes of Lefke Region in North Cyprus. Environ Dev Sustain 20(1):197–221. https://doi.org/10.1007/s10668-016-9877-5
Collins SL, Carpenter SR, Swinton SM et al (2011) An integrated conceptual framework for long-term social-ecological research. Front Ecol Environ 9(6):351–357. https://doi.org/10.1890/100068
Davis GW, Richardson DM (eds) (1995) Mediterranean-type ecosystems: the function of biodiversity. Springer, Berlin/Heidelberg
Depietri Y, Orenstein DE (2019) Fire-regulating services and disservices with an application to the Haifa-Carmel region in Israel. Front Environ Sci 7:107. https://doi.org/10.3389/fenvs.2019.00107
Derak M, Cortina J (2014) Multi-criteria participative evaluation of *Pinus halepensis* plantations in a semiarid area of Southeast Spain. Ecol Indic 43:56–68. https://doi.org/10.1016/j.ecolind.2014.02.017
Divinski I, Becker N, Bar (Kutiel) P (2018) Opportunity costs of alternative management options in a protected nature park: the case of Ramat Hanadiv, Israel. Land Use Policy 71:494–504. https://doi.org/10.1016/j.landusepol.2017.11.015
Eid R, Haller T (2018) Burning forests, rising power: towards a constitutionality process in Mount Carmel Biosphere Reserve. Hum Ecol 46(1):41–50. https://doi.org/10.1007/s10745-018-9968-z
Eizenberg E, Orenstein DE, Zimroni H (2018) Back to the (visualization) laboratory: using focus groups to generate qualitative and quantitative results. J Plan Educ Res 38(3):345–358. https://doi.org/10.1177/0739456X17700252
Falk JH, Balling JD (2009) Evolutionary influence on human landscape preference. Environ Behav 42(4):479–493. https://doi.org/10.1177/0013916509341244
Fish R, Church A, Winter M (2016) Conceptualising cultural ecosystem services: a novel framework for research and critical engagement. Ecosyst Serv 21:208–217. https://doi.org/10.1016/j.ecoser.2016.09.002
Fleischer A, Shaler M, Konijnenkijk CC (2014) Factors affecting the pattern of use of community forests – a case study of three forests in Israel. Discussion paper no. 8.14. The Hebrew University of Jerusalem
Gauquelin T, Michon G, Joffre R et al (2018) Mediterranean forests, land use and climate change: a social-ecological perspective. Reg Environ Chang 18(3):623–636. https://doi.org/10.1007/s10113-016-0994-3
Ginsberg P (2000) Afforestation in Israel: a source of social goods and services. J Forest 98(3):32–36. https://doi.org/10.1093/jof/98.3.32
Ginsberg P (2006) Restoring biodiversity to pine afforestations in Israel. J Nat Conserv 14(3–4):207–216. https://doi.org/10.1016/j.jnc.2006.04.003
Gragson TL, Leigh DS, Coughlan MR (2015) Basque cultural landscapes of the western French Pyrenees. Il Capitale Culturale 12:565–596
Holmgren M, Scheffer M (2017) To tree or not to tree: cultural views from ancient Romans to modern ecologists. Ecosystems 20(1):62–68. https://doi.org/10.1007/s10021-016-0052-5
Hughes JD (2011) Ancient deforestation revisited. J Hist Biol 44(1):43–57. https://doi.org/10.1007/s10739-010-9247-3
IPBES (2013) Decision IPBES-2/4: conceptual framework for the intergovernmental science-policy platform on biodiversity and ecosystem services. Available at: https://ipbes.net/system/tdf/downloads/DecisionIPBES_2_4.pdf?file=1&type=node&id=14649

IPCC (2014) Climate Change 2014: impacts, adaptation, and vulnerability. Part A: global and sectoral aspects. Contribution of Working Group II to the fifth assessment report of the Intergovernmental Panel on Climate Change [Field CB, Barros DJ, Dokken KJ et al (eds)]. Cambridge University Press, Cambridge, 1132 pp

Jaouadi W, Naghmouchi S, Alsubeie M (2019) Should the silviculture of Aleppo pine (*Pinus halepensis* Mill.) stands in northern Africa be oriented towards wood or seed and cone production? Diagnosis and current potentiality. iForest 12:297–305. https://doi.org/10.3832/ifor2965-012

Jucker Riva M, Baeza J, Bautista S et al (2018) How does land management contribute to the resilience of Mediterranean forests and rangelands? A participatory assessment. Land Degrad Dev 29(10):3721–3735. https://doi.org/10.1002/ldr.3104

Koniak G, Noy-Meir I, Perevolotsky A (2011) Modelling dynamics of ecosystem services basket in Mediterranean landscapes: a tool for rational management. Landsc Ecol 26(1):109–124. https://doi.org/10.1007/s10980-010-9540-8

Leitão IA, Ferreira CSS, Ferreira AJD (2019) Assessing long-term changes in potential ecosystem services of a peri-urbanizing Mediterranean catchment. Sci Total Environ 660:993–1003. https://doi.org/10.1016/j.scitotenv.2019.01.088

Liphschitz N, Biger G (2001) Past distribution of Aleppo pine (*Pinus halepensis*) in the mountains of Israel (Palestine). Holocene 11(4):427–436. https://doi.org/10.1191/095968301678302869

Lopez-Santiago CA, Oteros-Rozas E, Martín-López B et al (2014) Using visual stimuli to explore the social perceptions of ecosystem services in cultural landscapes: the case of transhumance in Mediterranean Spain. Ecol Soc 19(2):27. https://doi.org/10.5751/es-06401-190227

Lorilla R, Poirazidis K, Kalogirou S et al (2018) Assessment of the spatial dynamics and interactions among multiple ecosystem services to promote effective policy making across Mediterranean island landscapes. Sustainability 10(9):3285. https://doi.org/10.3390/su10093285

Lotan A, Kost R, Mandelik Y et al (2018) National scale mapping of ecosystem services in Israel – genetic resources, pollination and cultural services. One Ecosyst 3:e25494. https://doi.org/10.3897/oneeco.3.e25494

MA (2005) Millennium ecosystem assessment. Island Press, Washington, DC

Maestre-Andrés S, Calvet-Mir L, van den Bergh JCJM (2016) Sociocultural valuation of ecosystem services to improve protected area management: a multi-method approach applied to Catalonia, Spain. Reg Environ Chang 16(3):717–731. https://doi.org/10.1007/s10113-015-0784-3

Marini Govigli V, Górriz-Mifsud E, Varela E (2019) Zonal travel cost approaches to assess recreational wild mushroom picking value: trade-offs between online and onsite data collection strategies. Forest Policy Econ 102:51–65. https://doi.org/10.1016/j.foropol.2019.02.003

Martínez-Sastre R, Ravera F, González JA et al (2017) Mediterranean landscapes under change: combining social multicriteria evaluation and the ecosystem services framework for land use planning. Land Use Policy 67:472–486. https://doi.org/10.1016/j.landusepol.2017.06.001

Milcu AI, Hanspach J, Abson D et al (2013) Cultural ecosystem services: a literature review and prospects for future research. Ecol Soc 18(3):44. https://doi.org/10.5751/es-05790-180344

Misgav A (2000) Visual preference of the public for vegetation groups in Israel. Landscape Urban Plan 48(3–4):143–159. https://doi.org/10.1016/S0169-2046(00)00038-4

Misgav A, Amir S (2001) Integration of visual quality considerations in development of Israeli vegetation management policy. Environ Manag 27(6):845–857. https://doi.org/10.1007/s0026702353

Naveh Z, Carmel Y (2004) The evolution of the cultural Mediterranean landscape in Israel as affected by fire, grazing, and human activities. In: Wasser SP (ed) Evolutionary theory and processes: modern horizons: papers in honour of Eviatar Nevo. Springer Netherlands, Dordrecht, pp 337–409. https://doi.org/10.1007/978-94-017-0443-4_18

Naveh Z, Lieberman AS (1994) Landscape ecology: theory and application, 2nd edn. Springer, New York

Negev M, Sagie H, Orenstein DE et al (2019) Using the ecosystem services framework for defining diverse human–nature relationships in a multi-ethnic biosphere reserve. Ecosyst Serv 39:100989. https://doi.org/10.1016/j.ecoser.2019.100989

Orenstein DE, Silverman E (2012) The future of the Israeli environmental movement: is a major paradigm shift under way? In: Orenstein DE, Tal A, Miller C (eds) Between ruin and restoration: an environmental history of Israel. University of Pittsburg Press, Pittsburg, pp 357–382

Orenstein DE, Zimroni H, Eizenberg E (2015) The immersive visualization theater: a new tool for ecosystem assessment and landscape planning. Comput Environ Urban Syst 54. https://doi.org/10.1016/j.compenvurbsys.2015.10.004

Orenstein DE, Lew M, Porat Y, Tzuk A (2018) Landscape value in the eyes of visitors to the Biriya and Meron Forests following treatments for snow damage. Technical Report. Keren Kayemet L'Israel, Haifa. [in Hebrew; Unpublished, available from author]

Osem Y, Brand D, Tauber I et al (2012) The bible of afforestation in Israel, policy and directives for forest planning and management. Keren Kayemet L'Israel, Jerusalem. [in Hebrew]

Oteros-Rozas E, Martín-López B, González JA et al (2014) Socio-cultural valuation of ecosystem services in a transhumance social-ecological network. Reg Environ Chang 14(4):1269–1289. https://doi.org/10.1007/s10113-013-0571-y

Pedley SM, Barbaro L, Guilherme JL et al (2019) Functional shifts in bird communities from semi-natural oak forests to conifer plantations are not consistent across Europe. PLoS One 14(7):e0220155. https://doi.org/10.1371/journal.pone.0220155

Peñuelas J, Sardans J, Filella I et al (2017) Impacts of global change on Mediterranean forests and their services. Forests 8(12):463. https://doi.org/10.3390/f8120463

Piñol J, Terradas J, Lloret F (1998) Climate warming, wildfire hazard, and wildfire occurrence in coastal Eastern Spain. Clim Chang 38:345–357. https://doi.org/10.1023/A:1005316632105

Raviv O, Zemah Shamir S, Izhaki I et al (2020) The socioeconomic value of multiple ecosystem types at a biosphere reserve as a baseline for one holistic conservation plan. Ecosyst Serv 41:101043. https://doi.org/10.1016/j.ecoser.2019.101043

Roces-Díaz JV, Vayreda J, Banqué-Casanovas M et al (2018) Assessing the distribution of forest ecosystem services in a highly populated Mediterranean region. Ecol Indic 93:986–997. https://doi.org/10.1016/j.ecolind.2018.05.076

Rosário IT, Rebelo R, Cardoso P et al (2019) Can geocaching be an indicator of cultural ecosystem services? The case of the montado savannah-like landscape. Ecol Indic 99:375–386. https://doi.org/10.1016/j.ecolind.2018.12.003

Rotem G, Bouskila A, Rothschild A (2014) Ecological effects of afforestation in the northern Negev. Society for the Protection of Nature in Israel, Tel Aviv, Israel

Rothschild A (2019) The need to stop afforestation in sensitive natural ecosystems in Israel, and conserve Israel's natural landscapes. Society for the Protection of Nature in Israel, Tel Aviv, Israel

Sacchelli S (2017) A decision support system for trade-off analysis and dynamic evaluation of forest ecosystem services. iForest 11:171–180

Schaich H, Kizos T, Schneider S et al (2015) Land change in eastern Mediterranean wood-pasture landscapes: the case of deciduous oak woodlands in Lesvos (Greece). Environ Manag 56(1):110–126. https://doi.org/10.1007/s00267-015-0496-y

Schiller G (2012) Growing forest trees and forests in Israel: a land on the desert's edge. Beit Dagan, Israel

Schwartz E (2009) Shifting paradigms of environmental approaches. Heschel Center for Environmental Leadership, Tel Aviv, Israel

Shechter M, Reises B, Zaitsev N (1998) Measuring passive use value: pledges, donations and cv responses in connection with an important natural resource. Environ Resour Econ 12(4):457–478. https://doi.org/10.1023/A:1008397411466

Soga M, Gaston KJ (2016) Extinction of experience: the loss of human–nature interactions. Front Ecol Environ 14(2):94–101. https://doi.org/10.1002/fee.1225

Soliño M, Yu T, Alía R et al (2018) Resin-tapped pine forests in Spain: ecological diversity and economic valuation. Sci Total Environ 625:1146–1155. https://doi.org/10.1016/j.scitotenv.2018.01.027

Stamps AE (1997) A paradigm for distinguishing significant from nonsignificant visual impacts: theory, implementation, case histories. Environ Impact Assess Rev 17(4):249–293. https://doi.org/10.1016/S0195-9255(97)00008-5

Tal A (2008) Space matters: historic drivers and turning points in Israel's open space protection policy. Israel Stud 13(1):119–151

Tal A (2013) All the trees of the forest: Israel's woodlands from the bible to the present. Yale University Press, New Haven

TEEB – The Economics of Ecosystems and Biodiversity (2017) http://www.teebweb.org/resources/glossary-of-terms/

Teff-Seker Y, Orenstein DE (2019) The "Desert Experience": evaluating the cultural ecosystem services of drylands through walking and focusing. People and Nature 1(2):234–248. https://doi.org/10.1002/pan3.28

Tengberg A, Fredholm S, Eliasson I et al (2012) Cultural ecosystem services provided by landscapes: assessment of heritage values and identity. Ecosyst Serv 2:14–26. https://doi.org/10.1016/j.ecoser.2012.07.006

Vizzarri M, Tognetti R, Marchetti M (2015) Forest ecosystem services: issues and challenges for biodiversity, conservation, and management in Italy. Forests 6(12):1810–1838. https://doi.org/10.3390/f6061810

Zimroni H, Eizenberg E, Orenstein DE (2017) Between cultures of natures and communities of knowledge: expert knowledge and local knowledge regarding perceptions of Mount Carmel forest. Isr Sociol 2:123–144. [in Hebrew]

Zube EH, Pitt DG (1981) Cross-cultural perceptions of scenic and heritage landscapes. Landsc Plan 8(1):69–87. https://doi.org/10.1016/0304-3924(81)90041-1

Part VII
Afforestation and Forest Management Policy

Yagil Osem

Chapter 31
Dynamics and Management of Western Mediterranean Pinewoods

Lluís Coll, Aitor Ameztegui, Rafael Calama, and Manuel E. Lucas-Borja

31.1 A Brief Historical Overview of Western Mediterranean Pinewoods

31.1.1 Origin and History of Anthropogenic Uses

Despite some debate on the role of pinewoods as native forests in the western Mediterranean Basin, recent paleoecological research has proved their long-term presence and importance in the area. Pines led the recovery of forest area after the Last Glacial Maximum, and in the early Holocene they were dominant throughout most of the Iberian Peninsula (Figueiral and Carcaillet 2005; Rubiales et al. 2010), southern France (Ali et al. 2008; Rius et al. 2012) and northern Italy (Magri et al. 2015). From the mid-Holocene the pines lost their prevalence to broad-leaved species, mainly deciduous and evergreen *Quercus* species (Rubiales et al. 2010). In the west and north of the Iberian Peninsula and western France, the substitution was mostly driven by the milder climatic conditions, which allowed the strong growth of

L. Coll (✉)
Department of Agriculture and Forest Engineering, University of Lleida, Lleida, Spain

Joint Research Unit CTFC-AGROTECNIO, Solsona, Spain
e-mail: lluis.coll@udl.cat

A. Ameztegui
Department of Agriculture and Forest Engineering, University of Lleida, Lleida, Spain

R. Calama
INIA-CIFOR, Madrid, Spain

iuFOR. Sustainable Forest Management Research Institute UVa-INIA, Madrid, Spain

M. E. Lucas-Borja
Higher Technical School of Agricultural and Forestry Engineering, Castilla-La Mancha University, Albacete, Spain

G. Ne'eman, Y. Osem (eds.), *Pines and Their Mixed Forest Ecosystems in the Mediterranean Basin*, Managing Forest Ecosystems 38,
https://doi.org/10.1007/978-3-030-63625-8_31

deciduous broad-leaved forests (Benito Garzón et al. 2007). In the easternmost interior areas of the Iberian Peninsula and southern France the demise of pines (obligate seeders) was more related to anthropogenic activities, including pastoralism and the recurrent use of fire, which put them at a competitive disadvantage against resprouting species such as evergreen *Quercus* species (Rodríguez-Sánchez et al. 2010; Morales-Molino et al. 2017).

Historically, exploitation of natural resources increased to cope with the requirements of increasingly complex civilizations (Valbuena-Carabaña et al. 2010). The expansion of agriculture, the incipient mining industry, shipbuilding activities, and war operations were all activities that resulted in a significant reduction in forest area. However, the role played in Spain by livestock herding (mainly sheep) should be considered. For centuries, the Spanish economy was deeply linked to the international trade of wool at the expense of a deep transformation of the woodlands, (see Valbuena-Carabaña (2010) for an extensive review of the socio-historical determinants of forests).

As a consequence, by the end of the nineteenth century most Western European landscapes had been drastically deforested. Forests were limited to the mountains and the poor quality soils, and it is estimated that they occupied only 15% of their potential range (Blondel 2006). The need to alleviate the prevailing deforestation and its consequences was evident across Western Europe, and most countries embarked on ambitious reforestation plans.

31.1.2 National Reforestation Plans and Natural Colonization of Abandoned Lands

The first wave of plantations, at the end of nineteenth century, promoted afforestations for hydrological purposes; these afforestations were used to regulate water flow, prevent landslides and stabilize coastal or eolian dunes (Valbuena-Carabaña et al. 2010). However, the largest afforestations of this period took place on private lands. In France, more than 600,000 ha of former swampy land and coastal dunes were planted with maritime pine (*Pinus pinaster* Ait.) between the eighteenth and the twentieth century (Métailié 1999). In Portugal, the area occupied by *P. pinaster* increased from only 5000 ha in 1879 to 430,000 ha in 1902 and reached 1,140,000 ha in 1934 (Figueiral 1995). Spain joined this reforestation wave a few decades later, but the extent of the plan carried out in Spain by far exceeded that of other European countries (Pemán and Serrada 2017). The dictatorship of General Franco assumed the ambitious plan prepared during the II Republic, and afforestation was also seen as a solution to the high unemployment resulting from the Civil War and the difficult years that followed (Valbuena-Carabaña et al. 2010). Between 1940 and 1975 almost 3 million ha were afforested in Spain, mostly on state-owned land, and 500,000 ha more were incorporated after the end of Franco's regime (1975–1984). Meanwhile, in France, the Fonds Forestier National afforested a significant part of the mountains of the Massif Central and the Languedoc, traditionally denuded of forests (Métailié 1999). All these reforestation plans shared the almost exclusive use of conifers, mostly pines, as the chosen species, due to their attributes of vitality, frugality, xerophilia and heliophilia (Vadell Giral et al. 2019). In Spain, *P. pinaster* and *P. sylvestris* L. increased their distribution 133 and ca. 100%, *Pinus halepensis* by 58%, and *Pinus nigra* by ca. 40% between 1940 and 1985 (Valbuena-Carabaña et al. 2010, and see Table 31.1). The most productive lands of northern Spain and Portugal were afforested mainly with exotic species such as *Eucalyptus* spp. (ca. 300,000 ha) or Monterrey pine (*P. radiata* D.Don, ca. 200,000 ha) (Vadell Giral et al. 2019).

These reforestation efforts have been joined in recent decades by important processes of natural reforestation, following the abandonment of traditional activities and the exodus from rural areas. This process has affected both central and southern France (Bodin et al. 2013) as well as important areas in the interior of the Iberian Peninsula, especially in mountainous areas (Ameztegui et al. 2010) (Fig. 31.1).

The former agricultural lands or pasture areas have been occupied mainly by pioneering species such as pines, contributing to the expansion of this genus in the landscape of western Mediterranean Europe. The advance of the forest along with depopulation has led some authors to speak of authentic 'green deserts' in the heart of the main countries of southwestern Europe.

31.2 The Dynamics of Western Mediterranean Pinewoods

31.2.1 Biotic and Abiotic Drivers of Natural Regeneration Processes

In view of the global changes, the distribution and persistence of current pinewoods in the Mediterranean landscapes relies on their capacity to successfully regenerate (Calama et al. 2017b). However, natural regeneration processes in these areas are slow and unpredictable since they are modulated by complex interactions between multiple factors operating at different spatio-temporal scales. Among them, canopy plant species characteristics, understory vegetation interactions, soil factors, pre- and post-dispersal seed predation, and microclimatic conditions (Modrý et al. 2004; Nathan and Ne'eman 2004; Calama et al. 2019). Although the success of natural forest regeneration depends on events occurring along the entire lifespan of the tree, early life stages such as seed emergence, seedling survival and early seedling growth are particularly critical due to their vulnerability to biotic and environmental constraints (Lucas-Borja et al. 2012; Prévosto et al. 2012). Drought and soil desiccation, for example, are primary constraints to seedling establishment in many forest areas characterized by long, dry summer periods (Herrera 1992). In such areas, recruitment processes are often restricted to sporadic rainfall periods or wet

Table 31.1 Different examples of silvicultural models for the main western Mediterranean pine species in Spain

Species	*P. pinea*	*P. sylvestris*	*P. nigra*	*P. pinaster*	*P. halepensis*
Total area (x 1000 ha)[a]	400	1184	625	1373	1926
Afforested area (1879–1997, x 1000 ha)[b]	254	516	364	731	518
Mean annual rainfall (mm)	430–800	600–1200	600–1200	400–800	300–700
Mean annual temperature (°C)	11–16	6–12	9–12	9–15	12–16
Altitude (m)	0–800	800–2000	800–2000	600–1300	0–1000
Main management objective	Cone production	Veneer wood	Quality saw-wood	Resin	Timber
Structure	Semiregular	Semiregular	Semiregular	Even-aged	Even-aged
Rotation length	140 years	140 years	150–160 years	100 years	60–70 years
Final density	125 stems ha^{-1}	250 stems ha^{-1}	350 stems ha^{-1}	200 stems ha^{-1}	400 stems ha^{-1}
Regeneration system	Shelterwood system	Selection group system	Shelterwood system	Shelterwood system	Clearcutting with seed trees

Regeneration period	30 years	40 years	30–40 years	20 years	10 years
Initial operations	Pre-commercial thinning at 20 years	Pre-commercial thinning and cleaning at 20 years	Pre-commercial thinning at 15 years	Pre-commercial thinning at 20 years	Mechanical tending at 10 years
Thinning	Two low thinnings at 35–60 years	3–4 selective crop thinnings every 15 years	4 low – mixed thinnings (25, 45, 60, 80 years)	Two mixed thinnings at 25–40 years	Mixed at 25 +, low thinning at 40 years
Other interventions	2 stem prunings up to 5 m + 3 crown prunings	2–3 stem prunings up to 6 m on crop trees	2–3 stem prunings up to 6 m on crop trees	2 stem prunings up to 4 m	Stem pruning up to 1.5 m + shrub cleaning
Timber production	1.5 m^3 ha^{-1} $year^{-1}$	8 m^3 ha^{-1} $year^{-1}$	5–6 m^3 ha^{-1} $year^{-1}$	2 m^3 ha^{-1} $year^{-1}$	5 m^3 ha^{-1} $year^{-1}$
Other products & services	Cones	Mushrooms Landscape	Soil protection	Resin	Higher resistance to fire
Source (silviculture)	Calama et al. (2017a)	Montero et al. (2008)	Beltrán et al. (2012) Alejano et al. (2008)	Mutke et al. (2013)	Beltrán et al. (2011) Del Río et al. (2008)

[a]Source: Montero and Serrada (2013)
[b]Source: Vadell et al. (2019)

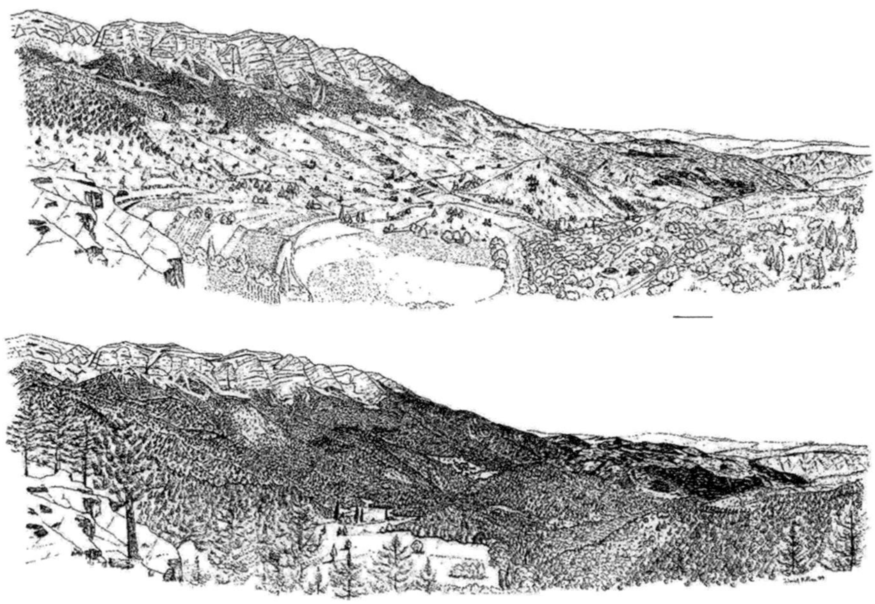

Fig. 31.1 Landscape changes (including forest encroachment into open areas and densification of existing forest stands) in the northern slopes of the Cadi mountain range (Lleida, NE Spain) between 1900 (above) and 1999 (below). (Reproduced with permission of: David Molina Gallart)

microsites, resulting in pulses of recruitment rather than continuous and regular seedling establishment (Prévosto et al. 2015).

31.2.1.1 Seed Dispersal

Pine seeds are dispersed across a wide variety of sites covering a range of abiotic and biotic conditions. Dominant winds or the actions of dispersers shape the spatial distribution of the seedlings. For example, initial recruitment can be seriously limited when large amounts of seeds are deposited in unsuitable places for recruitment or where predators have higher activity and access (Alcántara et al. 2000). Seed dispersal rates may differ between years since pine species appear to produce small crops much more frequently than large crops, a process is known as masting (Kelly 1994). In effect, while some Mediterranean pine species produce good seed crops annually (e.g. *P. halepensis*, Pardos et al. 2005) others produce one or two good crops every 4–6 years (e.g. *P. pinea* L., Mutke et al. 2005). Mountain pines in the Mediterranean region (e.g. *P. sylvestris*) produce a large crop every 3 years and irregular intervals have been identified for *P. nigra* Arn. ssp. *salzmannii*, depending on the site (Lucas-Borja and Vacchiano 2018).

31.2.1.2 Seed Predation

Seed predation can limit population recruitment by not only reducing seed availability (Schupp 1995; Lucas-Borja et al. 2010) but also changing the spatial distribution of seeds in the initial seed rain (Schupp 1995). This implies that the availability of safe sites for seeds depends on the interaction between seed rain distribution, habitat structure, and the preferences of post-dispersal seed predators. Ants, birds and rodents have been described as the main post-dispersal seed predators (Hulme and Hunt 1999; Ordóñez and Retana 2004). Predation by these groups of animals may be very variable, depending on seed characteristics, the phenological cycles of the plants and the animals and the feeding habits of predators, or habitat characteristics (Hulme 1998). For example, overall predation rates of *P. nigra* Arn. ssp. *salzmannii* seeds by the above mentioned three groups of predators have been found to be quite high for relict and typical forest locations during masting years, being lower in years of high seed production (Lucas-Borja et al. 2010).

31.2.1.3 Seed Emergence and Seedling Survival

After seed fall, seed emergence represents a risky transition from the seed stage (the most tolerant to environmental conditions) to the seedling one (the weakest and most vulnerable). Under adequate moisture and temperature, Mediterranean pine seeds will emerge readily following release from the cone (Richardson 2000). Emergence under field conditions can also be temporally variable, as some microhabitats provide appropriate conditions for emergence at different times (Nilsson et al. 2000). In Mediterranean areas, seed emergence is unlikely to occur during the dry summer, being often restricted to short periods in the wetter spring or autumn (Quilichini and Debussche 2000). During the early stages, seedling survival is primarily controlled by abiotic factors such as light, bedrock type, soil moisture and temperature, but other biotic factors, such as stand density, herbaceous competition and/or seedling herbivory, may intervene at this stage (Castro et al. 1999; Lucas-Borja et al. 2011a). Changes in stand density not only modify light conditions in the soil but also soil nutrients, soil microbes and soil moisture, which may affect the performance of the seedlings and the outcome of the regeneration (Balandier et al. 2006). A number of studies have been conducted within different Western Mediterranean pine ecosystems, examining the influence of the canopy cover on the success of natural regeneration, including *P. sylvestris* (Barbeito et al. 2011), *P. nigra* Arn ssp. *salzmannii* (del Cerro Barja et al. 2009; Lucas-Borja et al. 2011b; Tíscar Oliver and Lucas-Borja 2010) and *P. pinaster* (Rodríguez-García et al. 2015). Results are general, rather than species-specific. For example, higher basal area intervals have been found to promote seed emergence and seedling survival in *P. nigra* Arn ssp. *salzmannii*, whereas lower basal area intervals seem to be preferred for *P. pinaster* (Lucas-Borja et al. 2011b; Rodríguez-García et al. 2015).

31.2.1.4 Initial Seedling Growth

Initial growth of pine seedlings is enhanced under low canopy cover, in the absence of light limitations (Martín-Alcón and Coll 2016). Although these conditions do also favor the development of the ground vegetation, in general Mediterranean pines are highly tolerant of soil resource depletion by the understory layer. This is in part due to their high morphological and physiological plasticity that allows them, for example, to increase water use efficiency under competition (Picon-Cochard et al. 2006). In addition, pine seedlings demonstrate rapid root growth and remarkable soil penetration capacity (del Cerro Barja et al. 2009), that enable them to avoid competition from the neighboring ground vegetation by rapidly reaching deeper soil horizons. However, in stands with a very well-developed shrub layer, competition for light between seedlings and shrubs can limit the growth of the seedlings and delay the progressive advance of recruitment towards more advanced stages (Martín-Alcón and Coll 2016).

31.2.2 Stand Development and Natural Diversification Processes

Once the regeneration has been consolidated, and in the absence of natural disturbances or forest management, young Mediterranean pine stands (either natural or planted) develop and enter a stage of intense intraspecific competition (i.e. "stem exclusion stage" *sensu* Oliver and Larson (1996)) that can last for a prolonged period of time. This is the case in numerous pine plantations established during the last century in Mediterranean Western Europe in the context of large national reforestation programs (see Sect. 31.1). At that time, the use of pine species was intended to facilitate the establishment of shade-tolerant broad-leaved species and the progressive diversification of the stands (Nocentini and Coll 2012, Waitz and Sheffer this volume, Chapter 16) but the generalized lack of post-plantation operations led these stands to intense canopy closure and prolonged stem exclusion stages, negatively affecting understory plant diversity and regeneration (Gómez-Aparicio et al. 2009). Moreover, the diversification of Mediterranean pinewoods is driven not only by stand characteristics (i.e. tree density, canopy cover) but by site quality (e.g. soil characteristics) a number of biotic factors operating at different spatial scales, such as the availability of seed sources in the vicinity of the stands (González-Moreno et al. 2011) (Fig. 31.2), the abundance and behavior of seed dispersers (Gómez 2003, Zamora et al. 2010) and the activity of post-dispersal seed predators (González-Rodríguez and Villar 2012). When appropriate conditions are met for the dispersion and establishment of late-successional broad-leaved species into the pinewoods, the development of the seedling towards more advanced developmental stages mostly depends on microsite factors, of which browsing pressure and the

Fig. 31.2 Forest legacies such as this old-growth coppice oak tree (Solsona, NE Spain) are essential for the diversification of natural pinewoods coming from secondary succession processes. (Photo: Lluís Coll)

availability of light for the young plants are the most important (Gavinet et al. 2015, Martín-Alcón et al. 2015).

We note that the presence of broad-leaved resprouters (such as *Quercus* species, among others) in the understory of non-serotinous Mediterranean pinewoods (e.g. *P. nigra, P. sylvestris*) is essential for the resilience of these systems to catastrophic disturbances, as their resprouting ability allows the rapid recover of forest cover after the occurrence of these events (Puerta-Piñero et al. 2012, Martín-Alcón and Coll 2016). Thus in view of the expected increase in fire frequency and intensity in the Mediterranean region (Moriondo et al. 2006), promoting the progressive diversification of pine stands (both naturally or artificially) is advocated as an appropriate strategy to enhance forest resilience (Pausas et al. 2004, Gavinet et al. 2016, Martín-Alcón et al. 2016) (Fig. 31.3).

Indeed, we must remember that the vulnerability of Mediterranean pinewoods to fire (reviewed in Moreno this volume, Chapter 21) is not constant but changes along successional dynamics. For example, in an analysis conducted on Spanish black pine (*Pinus nigra*) stands, Sánchez-Pinillos et al. (2019) found that progressive diversification of the pinewoods with shade-tolerant species results in an increase of fine fuel loads and high vertical continuity (ladder fuels), leading to an increase in stand susceptibility to crown fires. In contrast, pine forests that remained

Fig. 31.3 Post-fire recovery of a burnt European black pine (*Pinus nigra*) forest via oak resprouts, near Solsona (Lleida, NE Spain). (Photo: Aitor Ameztegui)

monospecific showed more constant fine fuel loads and a certain decrease in ladder fuels with stand maturation.

In addition to fire, drought is expected to significantly modulate the future dynamics of western Mediterranean pinewoods (see Chap. 6). Pure pine stands, for example, may face important regeneration problems associated to the predicted increasing occurrence of extreme drought periods (see Sect. 31.2.1). The future dynamics of current mixed pine-oak forests (a system widely distributed in Western Europe) is also uncertain. At present, pine-oak forests produce rather stable communities that have been explained by complementary mechanisms associated to contrasting light- and water-use strategies (Zavala et al. 2000, Zavala and Zea 2004, Moreno-Gutiérrez et al. 2012), leading the species to occupy different hydrological niches during drought periods (see Comas et al. 2015 or del Castillo et al. 2016). However, the way the mechanisms underlying the coexistence of both species will be altered by the forecasted climate changes remains an open question.

31.3 The Management of Western Mediterranean Pinewoods

31.3.1 Management Objectives and Silvicultural Models

Western Mediterranean pinewoods are clear paradigms of multifunctionality, providing a wide array of goods (timber, fuelwood, pine nuts, resin, mushrooms, livestock) and services (CO_2 fixation, blue water, soil protection, recreation and biodiversity), the latter being unfortunately poorly recognized or valued (Croitoru 2007). While many of the abovementioned services are compatible with each other, joint optimization of each individual benefit is not always possible (Pasalodos-Tato et al. 2016), thus silvicultural prescriptions may rely on the primary objectives of the stand, the ecology of the subject species and the site characteristics and constraints (Table 31.1).

The characteristics of the stands that can be modified through silvicultural actions are:

- *Composition:* Although traditional silvicultural prescriptions in western Mediterranean pinewoods have focused on maintaining pure stands oriented to maximize a single product (e.g. timber, pine nuts, resin) recent studies have shown increases in biomass productivity and higher resilience to drought (de Dios García et al. 2015, Riofrío et al. 2016) when a mixture of species is maintained at the stand level.
- *Structure*: Typical management of western Mediterranean pinewoods has favored even-aged structures, derived from either intensive regeneration cuttings (by shelterwood and clearcutting systems) or large national reforestation programs (see Sect. 31.1). Although these even-aged structures have been shown to maximize biomass production for some species (Del Río et al. 2017a) or to be less vulnerable to wildfires (Beltrán et al. 2011), irregular and multi-aged structures have been proposed for western Mediterranean pinewoods as being optimal in terms of the provision of some services, such as cone production, soil protection against erosion or landscape and biodiversity (Calama et al. 2008). In addition, for shade tolerant species, such as *P. nigra*, uneven-aged forestry was proposed as an alternative to promote wood quality and attain natural regeneration (Alejano et al. 2008)
- *Stand stocking*: Contrary to Central-European and Boreal forest management systems that aim to maximize timber production by maintaining high-density stands and delaying thinning interventions, low-density stands have been proposed to maximize individual growth in water-limited environments such as the Mediterranean region (Calama et al. 2019). Maintaining low stand densities also favors non-wood forest products such as pine cones (Calama et al. 2016) and mushrooms (de Miguel et al. 2014) and optimizes blue water provision (Ameztegui et al. 2017).
- *Rotation length*: Maximum timber yield is attained in even-aged western Mediterranean pinewoods at young ages, ranging from 40 to 50 years in *P. pinea*

(Yagüe 1995) or *P. halepensis* (Ciancio 1986), to approximately 75 years in *P. sylvestris* (Montero et al. 2008) and to more than 100 years in *P. nigra* (Gómez-Loranca 1996). However, as management is not commonly focused on maximizing biomass production, typical rotations are much longer, reaching 150–160 years in *P. sylvestris* or *P. nigra* forests oriented to high quality (veneer or saw) wood production. In the case of *P. pinea* forests managed for cone production, trees have demonstrated increasing production of cones up to ages exceeding 140–150 years, albeit resulting in significant loss of timber quality due to rot (García-Güemes and Montero, 1998). The rotation length in *P. pinaster* aims to integrate the cycle of resin production (40 years), the period required for reaching the minimum diameter for tapping (around 40 years) and a 20-year period for attaining natural regeneration, resulting in a 100-year cycle (Serrano, 1994). Finally, in *P. halepensis* forests typical rotations are of approximately 60–70 years (Del Río et al. 2008). These rotations are reduced when minimizing vulnerability to fire is a management objective, or extended if the main aim is protecting the soil against erosion (Beltrán et al. 2011).

31.3.2 Silvicultural Interventions

Forest managers can attain the desired forest state defined by its composition, structure, stocking density and rotation length by planning different interventions in space and time, mainly: (*i*) regeneration cuttings, (*ii*) tending and thinning operations, and (*iii*) other treatments.

31.3.2.1 Regeneration Cuttings

While strip clear-cutting and clear-cutting with reserves of seed-tree systems are still successfully applied to naturally regenerate *P. pinaster* and *P. sylvestris* forests in the mountains of Central Spain, current trends focus on the use of more gradual regeneration cuttings (Calama et al. 2017b) and to an extension of the regeneration periods, in order to ensure the existence of the optimal conditions for regenerating forests. Group regeneration systems, where shelterwood fellings are gradually applied over small patches, are nowadays commonly applied in *P. nigra* and *P. sylvestris* forests, over extended periods of 30–40 years, resulting in typical semiregular structures (Lucas-Borja et al. 2016). In the case of *P. pinea* and *P. pinaster*, traditional shelterwood systems have been modified by the application of a single preparatory-seeding felling in order to attain the establishment of natural regeneration under the cover of a number of mature trees, which are released after 5–10 years by means of gradual and low-intensity secondary fellings (Manso et al. 2014). For *P. halepensis*, the management proposals for natural regeneration should aim to mimic small perturbations (Prévosto et al. 2012), resulting in an intensive shelterwood method (with a heavy seeding cutting and a single secondary liberatory felling) or clear-cutting with reserves (Beltrán et al. 2011). In any case, we note that regeneration cuttings for *P. sylvestris, P. nigra* and *P. halepensis* are commonly accompanied by the release of understory vegetation, soil removal, scarification, or prescribed burning in order to facilitate seed germination. Grazing must be excluded in the areas under regeneration.

31.3.2.2 Tending and Thinning

This set of operations includes all the intermediate cuttings used to modify the growth, quality, vigor, composition, or structure of a forest stand after its establishment with no objective for obtaining regeneration. In western Mediterranean pinewoods these interventions do not always increase total volume production of the stands (see Ruiz-Peinado et al. 2017; Del Río et al. 2017b) but enhance individual tree growth and vigor, prevent processes of decay following dry periods, and decrease the vulnerability of the stands to wildfire. Moreover, thinning treatments in Mediterranean pinewoods can facilitate other ecosystem services such as mushrooms (de Miguel et al. 2014) or blue-water provision (Ameztegui et al. 2017). In most Mediterranean forests, thinnings from below are recommended to remove dominated and less vigorous trees or co-dominant ones with undesirable shape. To achieve the desired results, these thinnings may be heavy and applied from the early stages, particularly in species managed for non-wood forest production such as *P. pinea* (cones) and *P. pinaster* (resin). In these forests, early and intense low thinnings should aim to attain the final density (i.e 150–200 stems ha^{-1}) in the middle of the rotation cycle. Thus, crown expansion of *P. pinea* is optimized and cone production is maximized (Moreno-Fernández et al. 2013), whereas in the case of *P. pinaster* the required time for attaining minimum tapping diameter is reduced (Mutke et al. 2013).

When the main objective is obtaining good quality timber (e.g. in *P. nigra* and *P. sylvestris* stands), the use of early and heavy thinning may not be adequate since it promotes tapered stems, large-sized knots along the stem axis, lower wood density, eccentric stem-cross sections and crowns with uneven lateral growth (Del Río et al. 2017b). In these cases, delayed and progressive mixed thinnings (Montero et al. 2008, Alejano et al. 2008) or, in the best sites, selective thinnings over 300–400 selected crop trees accompanied by stem pruning to promote wood quality are recommended.

Finally, naturally regenerated *P. halepensis* stands with very high densities (over 50,000 stems ha^{-1}) may require early and heavy application of pre-commercial thinnings in their initial stages (Del Río et al. 2008, Beltrán et al. 2011), followed by low or mixed thinning.

31.3.2.3 Other Prescriptions

Tree pruning is applied in western Mediterranean pinewoods in response to different objectives; in particular fire-prone areas, stem pruning reaching −1.5–2 m height is applied at young stand ages as a fire prevention measure, in order to interrupt vertical fuel continuity (Serrada et al. 2008). This operation can be repeated in concert with subsequent thinning operations. In forests managed for high-quality timber production, stem pruning up to 6 m is commonly applied over selected crop trees, to improve wood quality and prevent live knot formation (Montero et al. 2008). Finally, in *P. pinea* forests managed for cone production, particular prunings are practiced to obtain a clean stem that allows the use of vibrating-shakers for cone harvesting. In addition, inner crown prunings are applied to promote the formation of an umbrella-shaped crown.

31.3.3 From Multifunctionality to Adaptation: New Challenges in Managing Western Mediterranean Pinewoods

Most climate change scenarios for western Mediterranean pinewoods predict more frequent and severe processes of decay and dieback (Gea-Izquierdo et al. 2019), a lack of natural regeneration (Calama et al. 2017b), significant reductions of growth and yield (Bravo-Oviedo et al. 2010), a decrease in the provision of some non-wood forest products such as cones (Calama et al. 2019), more frequent and intense forest fires (Turco et al. 2018), and processes of local extinction of some pine species (Pardos et al. 2015). In this context, management of western Mediterranean pinewoods should rely on enhancing the adaptive capacity of the stands to climate change (Lindner and Calama 2013). This may in part be accomplished by promoting mixtures of pine species (e.g. *P. sylvestris* – *P. pinaster* in Mediterranean mountains) or pine – broad-leaved species (e.g. *P. nigra* – *Q. humilis* Mill.), applying early and selective crop thinnings, implementing gradual and less intense continuous cover regeneration fellings, favoring natural local processes of species substitution (e.g. *P. pinaster* being substituted by *P. pinea* in inner Spain, see Vergarechea et al. 2019) and controlling the understory vegetation by means of mechanical cleaning, prescribed burning or grazing activity (Duane et al. 2019).

In any case, the application of these treatments should be unavoidably accompanied by the establishment of experiments and monitoring programs aiming at disentangling their mid- and long-term effects on different components of the ecosystem (Vilà-Cabrera et al. 2018).

References

Alcántara JM, Rey PJ, Sánchez-Lafuente AM et al (2000) Early effects of rodent post-dispersal seed predation on the outcome of the plant–seed disperser interaction. Oikos 88:362–370

Alejano R, González JM, Serrada R (2008) Selvicultura de *Pinus nigra* Arn, subsp. *salzmannii* (Dunal) Franco. In: Serrada R, Reque J, Montero G (eds) Compendio de selvicultura aplicada en España. INIA-FUCOVASA, Madrid, pp 312–356

Ali AA, Roiron P, Chabal L et al (2008) Holocene hydrological and vegetation changes in southern France inferred by the study of an alluvial travertine system (Saint-Guilhem-le-Désert, Hérault). Comptes Rendus Geosci 340:356–366

Ameztegui A, Brotons L, Coll L (2010) Land-use changes as major drivers of mountain pine (*Pinus uncinata* Ram.) expansion in the Pyrenees. Glob Ecol Biogeogr 19:632–641

Ameztegui A, Cabon A, de Cáceres M et al (2017) Managing stand density to enhance the adaptability of Scots pine stands to climate change: a modelling approach. Ecol Model 356:141–150

Balandier P, Sonohat G, Sinoquet H et al (2006) Characterisation, prediction and relationships between different wavebands of solar radiation transmitted in the understorey of even-aged oak (*Quercus petraea*, *Q. robur*) stands. Trees 20:363

Barbeito I, LeMay V, Calama R et al (2011) Regeneration of Mediterranean *Pinus sylvestris* under two alternative shelterwood systems within a multiscale framework. Can J For Res 41:341–351

Beltrán M, Piqué M, Vericat P et al (2011) Models de gestió per als boscos de pi blanc (*Pinus halepensis* L.): producció de fusta i prevenció d'incendis forestals. Sèrie: Orientacions de Gestió Forestal Sostenible per a Catalunya (ORGEST). Centre de la Propietat Forestal. Departament d'Agricultura, Ramaderia, Pesca, Alimentació i Medi Natural. Generalitat de Catalunya

Beltrán M, Vericat P, Piqué M et al (2012) Models de gestió per als boscos de pinassa (*Pinus nigra* Arn.): producció de fusta i prevenció d'incendis forestals. Sèrie: Orientacions de gestió forestal sostenible per a Catalunya (ORGEST). Centre de la Propietat Forestal. Departament d'Agricultura, Ramaderia, Pesca, Alimentació i Medi Natural. Generalitat de Catalunya

Benito Garzón M, Sánchez de Dios R, Sáinz Ollero H (2007) Predictive modelling of tree species distributions on the Iberian Peninsula during the Last Glacial Maximum and Mid-Holocene. Ecography 30:120–134

Blondel J (2006) The 'design' of Mediterranean landscapes: a millennial story of humans and ecological systems during the historic period. Hum Ecol 34:713–729

Bodin J, Badeau V, Bruno E et al (2013) Shifts of forest species along an elevational gradient in Southeast France: climate change or stand maturation? J Veg Sci 24:269–283

Bravo-Oviedo A, Gallardo C, Del Río M et al (2010) Regional changes of *Pinus pinaster* site index in Spain using a climate-based dominant height model. Can J For Res 40:2036–2048

Calama R, Barbeito I, Pardos M et al (2008) Adapting a model for even-aged *Pinus pinea* L. stands for complex multi-aged structure. For Ecol Manag 256:1390–1399

Calama R, Gordo J, Madrigal G et al (2016) Enhanced tools for predicting annual stone pine (*Pinus pinea* L.) cone production at tree and forest scale in Inner Spain. For Syst 25:079

Calama R, Montero G, Vázquez-Piqué J et al (2017a) Indicadores de sostenibilidad en la gestión forestal actual de las masas de *Pinus pinea* L. 7° Congreso Forestal Español, Plasencia

Calama R, Manso R, Lucas-Borja ME et al (2017b) Natural regeneration in Iberian pines: a review of dynamic processes and proposals for management. For Syst 26

Calama R, Conde M, de Dios-García J et al (2019) Linking climate, annual growth and competition in a Mediterranean forest: *Pinus pinea* in the Spanish Northern Plateau. Agr For Met 264:309–321

Castro J, Gómez JM, García D et al (1999) Seed predation and dispersal in relict Scots pine forests in southern Spain. Plant Ecol 145:115–123

Ciancio O (1986) Sylviculture du Pin d'Alep. Options Mèditerranées. Series CIHEAM 86(1):25–28

Comas C, del Castillo J, Voltas J et al (2015) Point processes statistics of stable isotopes: analysing water uptake patterns in a mixed stand of Aleppo pine and holm oak. For Syst 24:e009

Croitoru L (2007) How much are Mediterranean forests worth? For Policy Econ 9:536–545

De Dios GJ, Pardos M, Calama R (2015) Interannual variability in competitive effects in mixed and monospecific forests of Mediterranean stone pine. For Ecol Manag 358:230–239

de Miguel S, Bonet JA, Pukkala T et al (2014) Impact of forest management intensity on landscape-level mushroom productivity: a regional model-based scenario analysis. For Ecol Manag 330:218–227

del Castillo J, Comas C, Voltas J et al (2016) Dynamics of competition over water in a mixed oak-pine Mediterranean forest: spatio-temporal and physiological components. For Ecol Manag 382:214–224

del Cerro BA, Lucas-Borja ME, Martínez Garcia E et al (2009) Influence of stand density and soil treatment on the Spanish black pine (*Pinus nigra* Arn. ssp. *salzmannii*) regeneration in Spain. Invest Agric Sist Rec For 18

Del Río M, Calama R, Montero G (2008) Selvicultura de *Pinus halepensis* L. In: Serrada R, Reque J, Montero G (eds) Compendio de Selvicultura Aplicada en España. INIA-FUCOVASA, Madrid, pp 289–311

Del Río M, Barbeito I, Bravo-Oviedo A et al (2017a) Mediterranean pine forests: management effects on carbon stocks. In: Bravo F et al (eds) Managing forest ecosystems: the challenge of climate change, 2nd edn. Kluwer Academic Publishers, pp 301–327

Del Río M, Bravo-Oviedo A, Pretzsch H et al (2017b) A review of thinning effects on scots pine stands: from growth and yield to new challenges under global change. For Syst 26:2

Duane A, Aquilué N, Canelles Q et al (2019) Adapting prescribed burns to future climate change in Mediterranean landscape. Sci Tot Environ 677:68–83

Figueiral I (1995) Charcoal analysis and the history of *Pinus pinaster* (cluster pine) in Portugal. Rev Palaeobot Palynol 89:441–454

Figueiral I, Carcaillet C (2005) A review of Late Pleistocene and Holocene biogeography of highland Mediterranean pines (*Pinus* type *sylvestris*) in Portugal, based on wood charcoal. Quat Sci Rev 24:2466–2476

García-Güemes C, Montero G (1998) Influencia de ciertas variables selvícolas en la pudrición provocada por *Phellinus pini* sobre *Pinus pinea*. Inv Agr Sist Rec For 7:203–217

Gavinet J, Vilagrosa A, Chirino E et al (2015) Hardwood seedling establishment below Aleppo pine depends on thinning intensity in two Mediterranean sites. Ann For Sci 72:999–1008

Gavinet J, Prévosto B, Fernandez C (2016) Do shrubs facilitate oak seedling establishment in Mediterranean pine forest understory? For Ecol Manag 381:289–296

Gea-Izquierdo G, Férriz M, García-Garrido S et al (2019) Synergistic abiotic and biotic stressors explain widespread decline of *Pinus pinaster* in a mixed forest. Sci Tot Environ 685:963–975

Gómez JM (2003) Spatial patterns in long-distance dispersal of *Quercus ilex* acorns by jays in a heterogeneous landscape. Ecography 26:573–584

Gómez-Aparicio L, Zavala MA, Bonet F et al (2009) Are pine plantations valid tools for restoring Mediterranean forests? An assessment along gradients of climatic conditions, stand density and distance to seed sources. Ecol Appl 19:2124–2141

Gómez-Loranca JA (1996) *Pinus nigra* Arn. en el Sistema Ibérico: tablas de crecimiento y producción. Monografías INIA 93. Madrid, Spain

González-Moreno P, Quero JL, Poorter L et al (2011) Is spatial structure the key to promote plant diversity in Mediterranean forest plantations? Basic Appl Ecol 12:251–259

González-Rodríguez V, Villar R (2012) Post-dispersal seed removal in four Mediterranean oaks: species and microhabitat selection differ depending on large herbivore activity. Ecol Res 27:587–594

Herrera CM (1992) Individual flowering time and maternal fecundity in a summer-flowering mediterranean shrub: making the right prediction for the wrong reason. Acta Oecol 13:13–24

Hulme PE (1998) Post-dispersal seed predation: consequences for plant demography and evolution. Persp Plant Ecol Evol Syst 1:32–46

Hulme PE, Hunt MK (1999) Rodent post-dispersal seed predation in deciduous woodland: predator response to absolute and relative abundance of prey. J Anim Ecol 68:417–428

Kelly D (1994) The evolutionary ecology of mast seeding. Trends Ecol Evol 9:465–470

Lindner M, Calama R (2013) Climate change and the need for adaptation in Mediterranean forests. In: Lucas-Borja ME (ed) Mediterranean forest management under the new context of climate change. Nova Science Publishers, New York

Lucas-Borja ME, Vacchiano G (2018) Interactions between climate, growth and seed production in Spanish black pine (*Pinus nigra* Arn. ssp. *salzmannii*) forests in Cuenca Mountains (Spain). New For 49:399–414

Lucas-Borja ME, Silva-Santos P, Fonseca TF et al (2010) Modelling Spanish black pine postdispersal seed predation in central-eastern Spain | Modelización de predación post-dispersión de semillas de pino laricio en el Centro-este de España. For Syst 19(3):393–403

Lucas-Borja ME, Fonseca T, Parresol BR et al (2011a) Modelling Spanish black pine seedling emergence: establishing management strategies for endangered forest areas. For Ecol Manag 262:195–202

Lucas-Borja ME, Fonseca TF, Silva-Santos P (2011b) Natural regeneration of *Pinus nigra* Arn spp *salzmannii* forest in Cuenca Mountains (Spain): a problem for sustainable forest management. In: Boehm DA (ed) Forestry: research, ecology and policies. Nova Science, New York, pp 37–52

Lucas-Borja ME, Fonseca TF, Lousada JL et al (2012) Natural regeneration of Spanish black pine [*Pinus nigra* Arn. ssp. *salzmannii* (Dunal) Franco] at contrasting altitudes in a Mediterranean mountain area. Ecol Res 27:913–921

Lucas-Borja ME, Candel-Pérez D, García-Morote FA et al (2016) *Pinus nigra* Arn. ssp. *salzmannii* seedling recruitment is affected by stand basal area, shrub cover and climate interactions. Ann For Sci 73:649–656

Magri D, Agrillo E, Di Rita F et al (2015) Holocene dynamics of tree taxa populations in Italy. Rev Palaeobot Palynol 218:267–284

Manso R, Pukkala T, Pardos M et al (2014) Modelling *Pinus pinea* forest management to attain natural regeneration under present and future climatic scenarios. Can J For Res 44:250–262

Martín-Alcón S, Coll L (2016) Unraveling the relative importance of factors driving post-fire regeneration trajectories in non-serotinous *Pinus nigra* forests. For Ecol Manag 361:13–22

Martín-Alcón S, Coll L, Salekin S (2015) Stand-level drivers of tree-species diversification in Mediterranean pine forests after abandonment of traditional practices. For Ecol Manag 353:107–117

Martín-Alcón S, Coll L, Ameztegui A (2016) Diversifying sub-Mediterranean pinewoods with oak species in a context of assisted migration: responses to local climate and light environment. Appl Veg Sci 19:254–267

Métailié J-P (1999) Los bosques de Francia en los siglos XIX y XX: Bosques nuevos y nuevas investigaciones. Hist Agrar 18:33–55

Modrý M, Hubený D, Rejšek K (2004) Differential response of naturally regenerated European shade tolerant tree species to soil type and light availability. For Ecol Manag 188:185–195

Montero G, Serrada R (2013) La situación de los bosques y el sector forestal en España – ISFE 2013. Sociedad Española de Ciencias Forestales. Lourizán (Pontevedra), Spain. Available online: http://secforestales.org/content/informe-isfe

Montero G, del Río M, Roig S et al (2008) Selvicultura de *Pinus sylvestris* L. In: Serrada R, Reque J, Montero G (eds) Compendio de selvicultura aplicada en España. INIA-FUCOVASA, Madrid, pp 502–534

Morales-Molino C, Tinner W, García-Antón M et al (2017) The historical demise of *Pinus nigra* forests in the Northern Iberian Plateau (South-Western Europe). J Ecol 105:634–646

Moreno J (this volume) Fire in Mediterranean pine forests: past, present and future. In: Ne'eman G, Osem Y (eds) Pines and their mixed forest ecosystems in the Mediterranean Basin. Springer, pp 421–456

Moreno-Fernández D, Cañellas I, Calama R et al (2013) Thinning increases cone production of stone pine (*Pinus pinea* L.) stands in the Northern Plateau (Spain). Ann For Sci 70:761–768

Moreno-Gutiérrez C, Dawson TE, Nicolás E et al (2012) Isotopes reveal contrasting water use strategies among coexisting plant species in a Mediterranean ecosystem. New Phytol 196:489–496

Moriondo M, Good P, Durao R et al (2006) Potential impact of climate change on fire risk in the Mediterranean area. Clim Res 31:85–95

Mutke S, Gordo J, Gil L (2005) Variability of Mediterranean stone pine cone production: yield loss as response to climate change. Agric For Meteor 132:263–272

Mutke S, Garcia del Barrio JM, Martínez-Jauregui M et al (2013) Bases para las buenas prácticas en la gestión del aprovechamiento resinero. INIA-SUSTFOREST, Madrid

Nathan R, Ne'eman G (2004) Spatiotemporal dynamics of recruitment in Aleppo pine (*Pinus halepensis* Miller). Plant Ecol 171:123–137

Nilsson M-C, Zackrisson O, Sterner O et al (2000) Characterisation of the differential interference effects of two boreal dwarf shrub species. Oecologia 123:122–128

Nocentini S, Coll L (2012) Mediterranean forests: human use and complex adaptive systems. In: Messier C, Puettmann KJ, Coates KD (eds) Managing forests as complex adaptive systems. Building resilience to the challenge of global change. The Earthscan Forest Library (series). Routledge, New York, pp 214–243

Oliver CD, Larson BC (1996) Forest stand dynamics. Wiley, New York

Ordóñez JL, Retana J (2004) Early reduction of post-fire recruitment of *Pinus nigra* by post-dispersal seed predation in different time-since-fire habitats. Ecography 27:449–458

Pardos M, Ruiz del Castillo J, Cañellas I et al (2005) Ecophysiology of natural regeneration of forest stands in Spain. For Syst 14(3):434–445

Pardos M, Calama R, Maroschek M et al (2015) A model-based analysis of climate change vulnerability of *Pinus pinea* stands under multi-objective management in the Northern Plateau of Spain. Ann For Sci 72:1009–1021

Pasalodos-Tato M, Pukkala T, Calama R et al (2016) Optimal management of *Pinus pinea* stands when cone and timber production are considered. Eur J For Res 135:607–619

Pausas J, Bladé C, Valdecantos A et al (2004) Pines and oaks in the restoration of Mediterranean landscapes of Spain: new perspectives for an old practice—a review. Plant Ecol 171:209–220

Pemán J, Serrada R (2017) El plan general de repobblación de España de 1939. In: La restauración forestal de España: 75 años de una ilusión. Ministerio de Agricultura y Pesca, Alimentación y Medio Ambiente, Madrid, Spain

Picon-Cochard C, Coll L, Balandier P (2006) The role of below-ground competition during early stages of secondary succession: the case of 3-year-old scots pine (*Pinus sylvestris* L.) seedlings in an abandoned grassland. Oecologia 148:373–383

Prévosto B, Amandier L, Quesney T et al (2012) Regenerating mature Aleppo pine stands in fire-free conditions: site preparation treatments matter. For Ecol Manag 282:70–77

Prévosto B, Gavinet J, Ripert C et al (2015) Identification of windows of emergence and seedling establishment in a pine Mediterranean forest under controlled disturbances. Basic App Ecol 16:36–45

Puerta-Piñero C, Espelta JM, Sánchez-Humanes B et al (2012) History matters: previous land use changes determine post-fire vegetation recovery in forested Mediterranean landscapes. For Ecol Manag 279:121–127

Quilichini A, Debussche M (2000) Seed dispersal and germination patterns in a rare Mediterranean island endemic (*Anchusa crispa* Viv. Boraginaceae). Acta Oecol 21:303–313

Richardson DM (2000) Ecology and biogeography of *Pinus*. Cambridge University Press, Cambridge

Riofrío J, Del Río M, Bravo F (2016) Mixing effects on growth efficiency in mixed pine forests. Forestry 90:381–392

Rius D, Vannière B, Galop D (2012) Holocene history of fire, vegetation and land use from the Central Pyrenees (France). Quat Res 77:54–64

Rodríguez-García A, Martín JA, López R et al (2015) Influence of climate variables on resin yield and secretory structures in tapped *Pinus pinaster* Ait. in central Spain. Agric For Meteor 202:83–93

Rodríguez-Sánchez F, Hampe A, Jordano P et al (2010) Past tree range dynamics in the Iberian Peninsula inferred through phylogeography and palaeodistribution modelling: a review. Rev Palaeobot Palynol 162:507–521

Rubiales JM, García-Amorena I, Hernández L et al (2010) Late Quaternary dynamics of pinewoods in the Iberian Mountains. Rev Palaeobot Palynol 162:476–491

Ruiz-Peinado R, Bravo-Oviedo A, López-Senespleda E et al (2017) Forest management and carbon sequestration in the Mediterranean region: a review. For Syst 26(2):eR04S

Sánchez-Pinillos M, De Cáceres M, Ameztegui A et al (2019) Temporal dimension of forest vulnerability to fire along successional trajectories. J Env Manage 248:109201

Schupp EW (1995) Seed-seedling conflicts, habitat choice, and patterns of plant recruitment. Am J Bot 82:399–409

Serrada R, Aroca JK, Roig S (2008) Selvicultura preventiva de incendios. In: Serrada R, Reque J, Montero G (eds) Compendio de selvicultura aplicada en España. FUCOVASA, Madrid, pp 950–979

Serrano M (1994) Métodos de ordenación de pinares en resinación. In: Madrigal A (ed) Ordenación de Montes Arbolados. ICONA, Madrid

Tíscar Oliver PA, Lucas-Borja ME (2010) Seed mass variation, germination time and seedling performance in a population of *Pinus nigra* subsp. *salzamannii* | Variación del peso y del tiempo de emergencia de las semillas, y crecimiento de las plántulas en una población. For Syst 19(3):344–353

Turco M, Rosa-Cánovas JJ, Bedia J et al (2018) Exacerbated fires in Mediterranean Europe due to anthropogenic warming projected with non-stationary climate-fire models. Nat Commun 9(1):3821

Vadell Giral E, de Miguel MS, Pemán García J (2019) La repoblación forestal en España: las especies utilizadas desde 1877 a partir de las cartografías forestales. Hist Agrar Rev Agric E Hist Rural 77:107–136

Valbuena-Carabaña M, de Heredia UL, Fuentes-Utrilla P et al (2010) Historical and recent changes in the Spanish forests: a socio-economic process. Rev Palaeobot Palynol, Iberian Floras through Time: Land of Diversity and Survival 162:492–506

Vergarechea M, Calama R, Fortin M et al (2019) Climate-mediated regeneration occurrence in Mediterranean pine forests: a modeling approach. For Ecol Manag 446:10–19

Vilà-Cabrera A, Coll L, Martínez-Vilalta J et al (2018) Forest management for adaptation to climate change in the Mediterranean basin: a synthesis of evidence. For Ecol Manag 407:16–22

Waitz Y, Sheffer E (this volume) Dynamics of mixed pine-oak forests. In: Ne'eman G, Osem Y (eds) Pines and their mixed forest ecosystems in the Mediterranean Basin. Springer, pp 345–362

Yagüe S (1995) Ordenación de pinares de pino piñonero (*Pinus pinea* L.) en la provincia de Ávila. Cuadernos SECF 1:221–236

Zamora R, Hodar JA, Matias L et al (2010) Positive adjacency effects mediated by seed disperser birds in pine plantations. Ecol Appl 20:1053–1060

Zavala MA, Zea E (2004) Mechanisms maintaining biodiversity in Mediterranean pine-oak forests: insights from a spatial simulation model. Plant Ecol 171:197–207

Zavala MA, Espelta JM, Retana J (2000) Constraints and tradeoffs in Mediterranean plant communities: the case of holm oak (*Quercus ilex* L.) – Aleppo pine (*Pinus halepensis* Mill.) forests. Bot Rev 66:119–149

Chapter 32
From First Generation of Pine Monocultures to Mixed-Forest Ecosystems: Biotic and Abiotic Determinants of Pine Forests' Dynamics in Mediterranean Israel

Yagil Osem and Yossi Moshe

32.1 Introduction

Pine plantations were established throughout the Mediterranean zone of Israel during the last century as part of a comprehensive afforestation effort (Osem et al. 2008). These forests were planted as even-aged monocultures based mainly on the local *Pinus halepensis* Mill. and to a lesser extent, the exotic *P. brutia* Ten., with unified silvicultural methodologies (e.g. site preparation, planting density and thinning regime) applied throughout a wide geo-climatic range. Establishing even-aged, monospecific plantations based on a few coniferous pioneer species proved efficient in achieving the historical goals and targets of afforestation. However, it resulted in the creation of forest ecosystems characterized by a short life cycle and limited diversity, complexity and ecological integrity (Noy-Meir 1989), and high sensitivity to pest outbreaks, fire, drought and extreme climatic events (snowfall and strong winds, Boneh and Silverstone 1999). However, as these forest systems matured, they underwent a series of vegetation processes depending on a variety of natural- and human-related drivers (Osem et al. 2012, 2017b). Here, we review two key processes that constitute the forest dynamics of pine plantations in Israel and discuss the major factors and interactions that drive them. Based on our understandings of these processes, we propose silvicultural guidelines for gradually turning these simply structured manmade forests into complex, sustainable mixed-forest ecosystems.

Y. Osem (✉)
Department of Natural Resources – Institute of Plant Sciences, Agricultural Research Organization – Volcani Center, Rishon LeZion, Israel
e-mail: yagil@agri.gov.il

Y. Moshe
Department of Natural Resources, Institute of Plant Sciences, Volcani Center, Israel
e-mail: yosi@agri.gov.il

G. Ne'eman, Y. Osem (eds.), *Pines and Their Mixed Forest Ecosystems in the Mediterranean Basin*, Managing Forest Ecosystems 38,
https://doi.org/10.1007/978-3-030-63625-8_32

Two key parallel processes that occur continuously, in the understory of pine plantations in Mediterranean Israel affect the future of their forest structure: (1) pine regeneration – the spontaneous recruitment and growth of pines and (2) Native broadleafs reestablishment – the spontaneous recruitment, resprouting and growth of native broadleaved tree species. While these two processes interact with one another, they are also bound to external regulators, mainly: (1) aridity – abiotic habitat factors determining the amount of available water for plants (e.g. rainfall, potential evapotranspiration, soil type and aspect); (2) pine overstory cover – the extent of shading by overstory pine canopies and consequent availability of light, water and nutrients for understory regeneration and growth; and (3) grazing – the consumption by livestock of vegetative material in the forest understory. Despite their critical importance as regulators of forest dynamics, these three factors have not yet been discussed, in an integrative way, as key components of a silvicultural management strategy for developing diverse and sustainable mixed-forest ecosystems in water-limited environments.

32.2 Regulators

32.2.1 Aridity

According to the wetness index (WI = annual precipitation/annual potential evapotranspiration) for climatic aridity, pine forests in Israel are distributed along an aridity range of WI = 0.13–0.50 (i.e., semiarid to dry-subhumid climate, Perelman 2018). This is mainly due to the high spatial variation in mean annual rainfall, in the range of 260–850 mm (Wilson 2016), falling mostly during winter (December–March, Fig. 32.1), with some minor additional variation in potential evapotranspiration

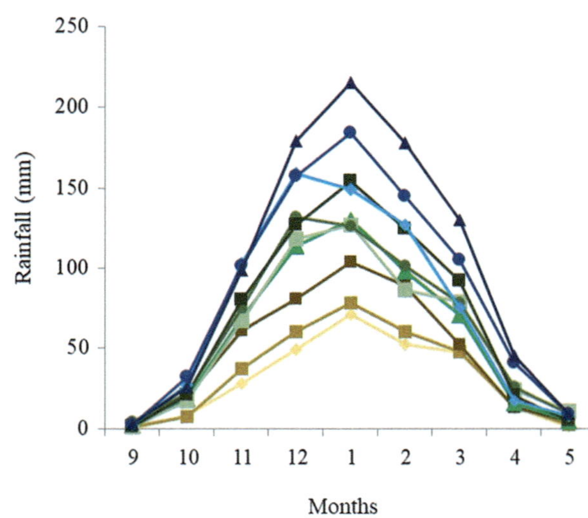

Fig. 32.1 Average monthly rainfall (mm) in 10 pine forest areas in Israel. ◆ Yatir; ■ Lahav; ■ Shaharia; ■ Eshtaol; ▲ Givat-Hamoreh; ● Nazereth; ■ HaKedoshim; ◆ Yaaranim; ● Maalot; ▲ Sasa. (Osem et al. 2009)

ranging between 1750 and 1980 mm/year. The development and function of pine forests in water-limited habitats in Israel is known to depend considerably on the amount and distribution of rainfall (Osem et al. 2009, 2012). In addition to climate, habitat aridity is also known to be influenced by edaphic and topographic conditions such as bedrock type, topographic slope and aspect and soil type and depth.

32.2.2 Pine Overstory Cover

The overstory cover of mature trees in *P. halepensis* forests in Israel, quantified by leaf area index (LAI), ranges from 1 to 3.5 m m^{-2} (Perelman 2018). This range corresponds to a stand density of 40–880 trees ha^{-1}, stand stem basal area (BA) of 2.5–28 m^2 ha^{-1} and overstory canopy cover (CC) of 30–100%. Overstory LAI is highly variable across the forests, influenced by forest management (e.g., thinning), as well as by a variety of natural disturbances (e.g., drought, wind-throw, snow). It was found to be linearly and significantly related to the level of aridity (WI), with mean LAI values ranging from 1.5 to 2.5 throughout the climatic gradient of the pine forests' distribution range in Israel (Perelman 2018). Tree density and overstory cover of mature *P. brutia* stands in Israel were found to be higher (by ~30% and 15% respectively) than that of *P. halepensis* stands, and not as closely related to aridity, but more influenced by bedrock type (higher on soft than on hard calcareous bedrock, Wilson 2016).

32.2.3 Grazing

Grazing by domesticated livestock, cattle, sheep and goats is very common in Israel's pine forests. Based on regional foresters' knowledge it was estimated that 60% of the forests in the central forest region of Israel and 95% in the northern and southern regions are under some kind of livestock grazing, (Zeidan S et al. pers. comm.). The proportion of forest area grazed by cattle vs. sheep and goats is also variable among the regions. In the north, cattle dominate (80%) the grazed forested area, while in the south and center sheep-goat herds dominate (95% of the grazed forested area). Grazing is considered by local foresters to be the primary tool for fire hazard reduction, seeking maximum herbaceous biomass removal (Ne'eman et al. 1997; Osem et al. 2017b). Grazing in Israeli forest systems is highly variable and detailed information on actual grazing regimes is lacking. Cattle grazing in pine forests typically occurs from mid-winter to late spring or summer (4–6 months) with grazing pressures ranging from 1 to 7 ha per cattle head, depending on forage availability, accessibility and other factors (Seligman et al. 2016). Sheep and goat grazing in the pine forests is usually allowed when the herbaceous vegetation reaches 70–80 g m^{-2} dry biomass and continue for several months until the remaining dry herbage is not more than 75 g m^{-2} (Seligman et al. 2016). According to an assessment of carrying capacity of planted pine forests in central Israel, the herbaceous vegetation can support ca. 100–200 sheep grazing days per ha while the woody vegetation can support as much as 200–400 goat grazing days per ha (Evlagon et al. 2012). However, the extent of forage exploitation in these areas is highly variable, ranging from less than 5% in some forests to more than 95% in others (Evlagon et al. 2012).

32.3 Pine Regeneration

Pine regeneration is the process through which the next conifer forest generation establishes in the understory of mature forests and is essential for pine forest sustainability. This process involves seed production and dispersal, seed germination, seedling survival and growth, tree maturation and production. All these processes are influenced by the abovementioned regulators.

32.3.1 Aridity

In a study that compared seed rain and seedling emergence within *P. halepensis* plantations, across three ecoregions differing in rainfall (600, 430 and 280 mm year^{-1}, WI = 0.33, 0.23 and 0.13), large differences were found in seed rain (~250, 150, 25 seeds m^{-2} year^{-1}), and in emerged seedling density (~13, 2, 0.3 seedlings m^{-2}), corresponding to the variation in aridity level (Osem et al. 2013, Fig. 32.2). Seed rain

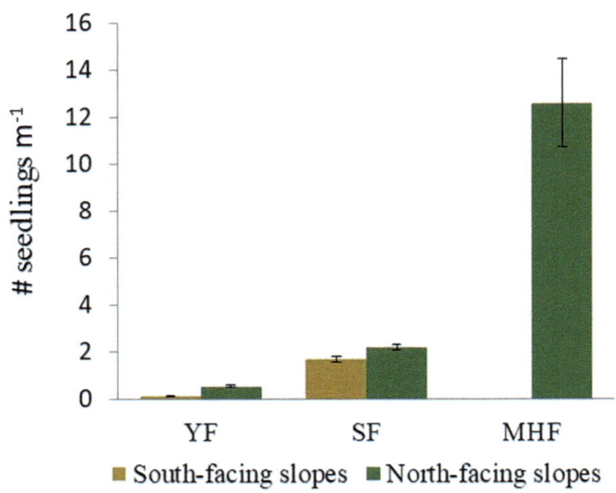

Fig. 32.2 Average density of emerged pine seedlings on north- and south-facing slopes, in late winter (March) in Israel: Mishmar-Ha'Emek Forest 1972–1975 (MHF), Shaharia Forest 2008–2009 (SF) and Yatir Forest 2005–2007 (YF). South-facing slopes were not measured in MHF. Error bars indicate standard error (SE). (Osem et al. 2013)

and seedling emergence were higher on north- vs. south-facing slopes, highlighting the importance of water availability for the natural regeneration potential of pine forests in Israel. A study on the regeneration of a *P. brutia* plantation in a relatively humid ecoregion (600 mm annual rainfall, WI = 0.33) found seed rain of ~154 seeds m^{-2} $year^{-1}$ and emerged seedling density of ~8 seedlings m^{-2}, which corresponded with those found for *P. halepensis* under similar rainfall amounts (Osem et al. 2015).

Regarding the density of established pine recruits (i.e., advanced regeneration) in the understory of mature (>30 year old) *P. halepensis* plantations, along the aridity gradient in Israel, Osem et al. (2009) found a threshold of 400 mm annual rainfall (WI = 0.2) below which pine regeneration is very limited. Thus, under such conditions the ability of natural regeneration as required for sustainable pine forest cover is lacking. However, above 400 mm annual rainfall, a wide range of recruit densities was found (0–1565 recruit ha^{-1}) with no clear relationship to annual rainfall amount nor to topographic aspect (Fig. 32.3).

Previously, in his work, Schiller (1978) highlighted the important effect of bedrock type on natural regeneration in three *P. halepensis* forests in northern and central Israel. Establishment rates were higher on soft (chalk and marl) than on hard (limestone and dolomite) calcareous bedrock types. This effect was attributed to higher soil/rock water holding capacity resulting in better water conditions and to lower competition by annual herbs in habitats on soft bedrock types. A comprehensive survey of advanced regeneration conducted over the long-term monitoring plot setup of Israel's mature pine forests, provided further support for the great importance of bedrock type, with this factor alone contributing 78% of the explained variance in *P. halepensis* regeneration (Wilson 2016). In this study, advanced regeneration was also found to be positively related to rainfall amount, but this relationship was prevalent only on soft bedrock and was relatively weak. According to this

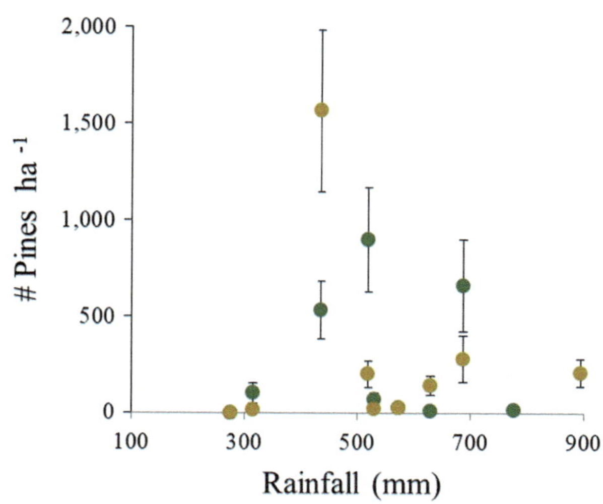

Fig. 32.3 Regeneration density of Aleppo pine trees in the forest understory along the rainfall gradient in Israel. ● North-facing slopes; ● South-facing slopes. (Osem et al. 2009)

analysis, natural regeneration of *P. halepensis* appears above an annual rainfall threshold of 350 mm (WI = 0.18) on soft calcareous bedrock. In contrast, very low regeneration was found in mature *P. brutia* forests throughout a similar long-term monitoring plot setup, with no clear relationship to rainfall amount or bedrock type.

Support for the importance of herbs in limiting pine regeneration was gained from a study looking at advanced pine regeneration in a semiarid *P. halepensis* forest (Shaharia Forest) located on the dry margins of the natural regeneration ability of *P. halepensis*. In this study natural regeneration occurred only in some of the more rainy years, and seedling survival was negatively and significantly related to the biomass of the herbaceous vegetation (Osem et al. 2013). Furthermore, manipulation of herbaceous biomass by ploughing or mowing improved pine seedling survival (Fig. 32.4).

32.3.2 Overstory Cover

The first study on the effect of overstory cover on *P. halepensis* recruitment in Israel, by Schiller (1978), was a silvicultural thinning experiment on mature stands (43-year-old, Mishmar-HaEmek Forest) growing in typically favourable sites for pine regeneration (600 mm annual rainfall, WI = 0.33, north-facing slope and soft chalk bedrock). Thinning forest stands of ca. 300 trees ha^{-1} (canopy cover = 66%) to 200, 100 and 0 trees ha^{-1} (CC = 53%, 37% and 0%, respectively) resulted in much lower seed rain and germinating seedling density but increased seedling survival and growth. Overall, the densities of established recruits 6 years after thinning were 3200, 6700, 1300 and 0 recruits ha^{-1} in the 0, 100, 200 and 300 (non-thinned control) trees ha^{-1} treatments, respectively (Table 32.1).

Osem et al. (2017a) conducted a similar experiment in *P. halepensis* stands growing in less favorable sites (540 mm annual rainfall, WI = 0.30, west-facing slopes and hard lime bedrock). In this study, mature forest stands (40 year old, HaKedoshim

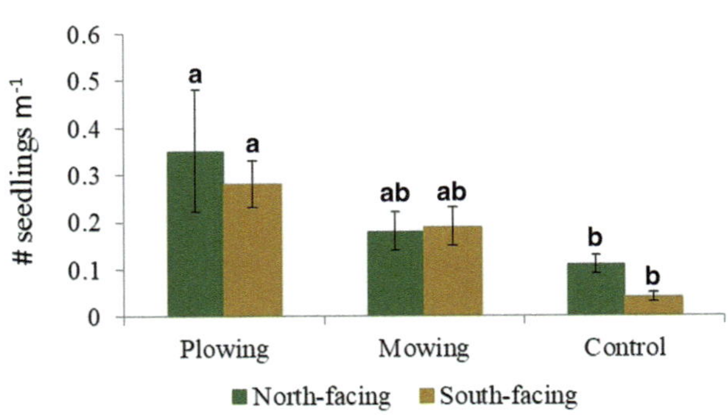

Fig. 32.4 Pine seedling density in mid-spring (May) as affected by site preparation treatments on south- and north-facing slopes in Shaharia Forest 2009. Different letters above columns indicate significant differences (P < 0.05). Error bars indicate standard error (SE). (Osem et al. 2013)

Table 32.1 One-way ANOVAs and post-hoc comparisons on the effect of thinning treatments on winter seedling density, seedling survival rate and recruit density in Mishmar Ha'Emek Forest

Treatment canopy cover	One-way ANOVA F	P	0 trees ha^{-1} 0%	100 trees ha^{-1} 36.9%	200 trees ha^{-1} 53.2%	300 trees ha^{-1} 66%
Annual seed rain (no. seeds m^{-2})	321.42	<0.0001	26.5 ± 1.94 C	92.10 ± 5.48 B	106.35 ± 7.3 B	249.8 ± 4.77 A
Winter seedling density (no. seedlings m^{-2})	16.26	<0.0001	1.56 ± 0.22 C	8.68 ± 1.69 BC	9.32 ± 1.7 B	19.96 ± 2.96 A
Seedling survival (%)	70.90	<0.001	24.3 ± 1.3 B	33.8 ± 1.2 A	15.6 ± 1.5 C	11.2 ± 0.5 D
Recruit density (no. recruits m^{-2})	17.84	<0.0001	0.6 ± 0.07 C	2.6 ± 0.14 A	2.0 ± 0.17 AB	1.7 ± 0.32 B
Density of 4–6-year-old recruits (no. recruits m^{-2})	136.3	<0.0001	0.32 ± 0.03 B	0.67 ± 0.04 A	0.13 ± 0.01 C	0.0 ± 0 D
Dry weight of 0.5-year-old seedlings (g)	5.7	0.004	1.51 ± 0.36 A	1.47 ± 0.35 A	0.9 ± 0.22 AB	0.11 ± 0.03 B

Means and standard errors are presented. Different capital letters indicate significant differences ($P < 0.05$). (Osem et al. 2013)

Forest) of ~600 trees ha^{-1} (LAI = 2.0; CC = 88%) were thinned to 300, 100 and 0 trees ha^{-1} (LAI = 1.5, 1.0 and 0; CC = 61%, 48% and 0%, respectively). The responses of pine recruitment and growth were similar to those described by Schiller (1978) but with pine recruit densities being lower by one level of magnitude i.e., 250, 180, 80 and 50 recruit ha^{-1} in the 0, 100, 300 and 550 (non-thinned control) trees ha^{-1} treatments, respectively.

Osem et al. (2015) examined the effect of overstory thinning on natural regeneration in *P. brutia* stands (50-year-old, Mount Horshan Forest) with 600 mm annual rainfall (WI = 0.33) and soft chalk and marl bedrock. Pine seed rain, seedling germination and survival (recruitment) were compared between non-thinned stands of ~230 trees ha^{-1} (CC = 95%) and thinned stands with ~100 trees ha^{-1} (CC = 55%). Similarly to the results of Schiller (1978), overstory thinning decreased seed rain (40%) and winter seedling emergence (66%) but increased seedling survival and resulting recruit density by more than threefold. Moreover, in the non-thinned plots most of the recruits were not taller than 20 cm, while in the thinned plots most of them were taller than 20 cm, with many reaching heights of over 1 m and even 2 m. Finally, when plotting recruit height vs. age it became evident that in non-thinned plots the amount of time needed for recruits to reach a certain height was about twice that needed in the thinned plots. We note that in this study, recorded pine recruit densities were extremely high, with 31,000 and 9000 recruits ha^{-1} in the thinned and non-thinned stands, respectively. Such recruitment levels are not in line with the common observation (as described above) according to which natural regeneration in *P. brutia* forests of Israel is quite limited compared to that of *P. halepensis*. This result may indicate exceptionally favorable conditions for pine regeneration prevailing in the specific studied area (e.g., a combination of high annual rainfall and soft calcareous bedrock).

Ashkenazi (2016) examined pine regeneration in *P. brutia* stands (50-year-old, Begin Park Forests, Judean Mountains) after they were thinned from ~250 trees ha^{-1} (CC = 78%, control) to 130 (CC = 62%, moderate thinning) and 50 (CC = 12%, heavy thinning) trees ha^{-1}. Seven years after thinning, pine recruit density was 430 and 700 recruits ha^{-1} in the heavy and moderate thinning treatments, respectively, while in the non-thinned control plots it was zero. This study was conducted in sites with 560 mm annual rainfall (WI = 0.31) and complex soft-hard calcareous bedrock formation. While the reported recruitment levels are much lower than those described in Mount Horshan Forest, they highlight the capacity of *P. brutia* stands to regenerate naturally in a variety of habitat types in Israel following overstory thinning.

32.3.3 Grazing

Livestock grazing is widely recognized as a major regulator of vegetation structure in Mediterranean ecosystems as well as a silvicultural tool for fire fuel load control. However, the effects of grazing on vegetation dynamics in the conifer forests of Israel have been hardly studied. In their study, Osem et al. (2015) specifically addressed the effect of cattle grazing on pine recruitment in mature (50-year-old) *P. brutia* stands (Mount Horshan Forest). Density of newly emerged seedlings, during the winter, was not found to be affected by grazing. However, the density of established pine recruits was significantly reduced by 25% under grazing in non-thinned plots (overstory CC = 95%) and by as much as 85% in thinned ones (CC = 55%). Moreover, in the thinned plots, with no cattle grazing (exclosures), high densities of recruits reaching 101–200 cm in height (~6000 recruit ha^{-1}) and 201–400 cm in height (~2000 recruit ha^{-1}) were found, while under grazing the existence of recruits taller than 100 cm was supressed completely (Fig. 32.5). Thus, grazing reduced pine recruits' survival and growth; this effect was particularly remarkable in thinned plots with low overstory cover. In contrast, grazing had a significant, positive effect on water availability for young recruits, as measured by predawn shoot water potential of 6-month-old seedlings; this probably indicates reduced water consumption by the competing understory vegetation in grazed areas (Osem et al. 2015, 2017b).

Cohen (2019) examined the effect of heavy goat grazing pressure on post-fire *P. halepensis* recruits (5000–10,000 recruits ha^{-1}) in the Mount Carmel region (600 mm annual rainfall, complex limestone, chalk and marl bedrock). The herd was introduced into the regenerating forest area 5 years after the fire (grazing regime: 2000 grazing days per ha during the spring) for three consecutive years (2016–2018). In this study, grazing did not affect pine recruit density, that is, it did not increase recruit mortality, but it restricted recruit growth as well as cone production with respect to the ungrazed control.

32.4 Broad-Leafs Reestablishment

Regeneration of Broad-leaved tree species in the understories of conifer plantations may take place either by seed germination, from external or internal seed sources, or by resprouting. Assessing this process is more complicated as it involves a variety of tree species differing in their ecological requirements.

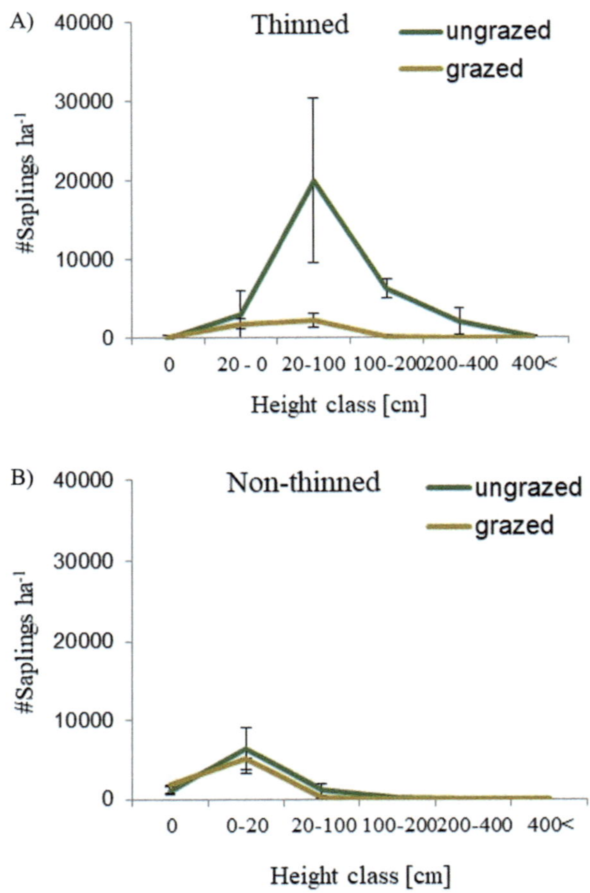

Fig. 32.5 Density of *Pinus brutia* saplings, by height classes, as affected by cattle grazing in (**a**) thinned and (**b**) non-thinned plots in Mount Horshan Forest, Israel, November 2009. Bars indicate SE, N = 20. (Osem et al. 2015)

32.4.1 Aridity

Osem et al. (2009) examined the natural recruitment of broad-leaved trees in *P. halepensis* forests distributed along the aridity gradient of Israel (280–900 mm annual rainfall), on north and south-facing slopes. This survey focused on ungrazed, mature stands (40–50 year old) with overstory tree densities corresponding to accepted thinning practices (300–350 trees ha^{-1}). Broadleaved trees appeared in the conifer forest understories in areas of at least 450 and 500 mm annual rainfall on north- and south-facing slopes, respectively. Above these thresholds, the density of naturally regenerating broad-leaved trees increased sharply with increasing rainfall amount reaching up to ~2000 and 8000 trees ha^{-1} in the most humid forests (>700 mm annual rainfall) on south- and north-facing slopes, respectively (significant rainfall×aspect interaction, Fig. 32.6). No effect of climatic aridity on the size of the regenerating trees was found; average tree height was typically in the range 1–2 m. One variable that was found to be the most strongly related to rainfall amount ($R^2 > 0.8$) was the species richness of the regenerating broad-leaved trees, which reached ~5 coexisting species per 200 m^2 in the most humid forests (Fig. 32.7). The effect of topographic aspect on tree species richness was minor. Overall, 12 species of broad-leaved trees were recorded throughout the climatic range. Among the broad-leaved tree species regenerating in the forest understory, the most dominant was common oak (*Quercus calliprinos* Webb, 40–90% of total broad-leaved tree density) which appeared from 500 mm annual rainfall and increased linearly with increasing rainfall amount up to ~1000 and 2000 trees ha^{-1} in the most humid forests, on north- and south-facing slopes, respectively. In several sites of the most humid forests, the dominance of *Q. calliprinos* was overtaken by other species, namely, *Arbutus andrachne* L., *Q. boissieri* Reut. and *Laurus nobilis* L.. These species were restricted to forest sites where average annual rainfall exceeded 600 mm. In contrast, *Ceratonia siliqua* L. was found to be relatively important in forests receiving 450–600 mm annual rainfall, but decreased in importance at higher rainfall levels. One other important tree species was *Pistacia palaestina* Boiss., which was quite common from 500 mm annual rainfall and above.

Wilson (2016) conducted a more extensive understory survey over the long-term monitoring plot setup of Israel's mature pine forests. He examined broad-leaved tree reestablishment in *P. halepensis* (48 stands) and *P. brutia* (32 stands) forests, along the rainfall gradient (250–850 mm annual rainfall) and across bedrock type (hard vs. soft calcareous bedrock), topographic aspect (north vs. south) and elevation (70–820 m). Broad-leafs reestablishment was highly variable throughout the climatic range with *Q. calliprinos* being the most dominant tree species. The extent

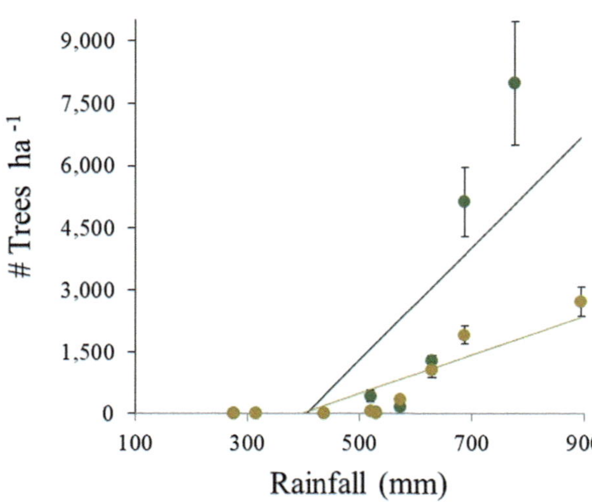

Fig. 32.6 Total density of regenerating trees in the forest understory on north-facing slopes (—— ●, R^2=0.6, P=0.01) and south-facing slopes (—— ●, R^2=0.77, P=0.002) along the rainfall gradient in Israel. (Osem et al. 2009)

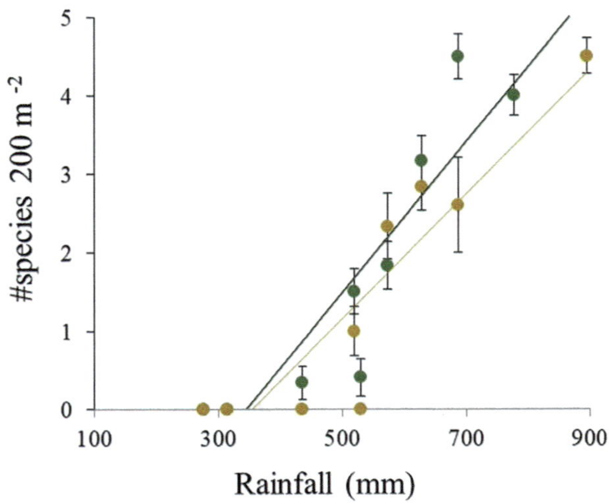

Fig. 32.7 Species richness of regenerating trees in the forest understory on north-facing slopes (——●, R^2=0.82, P=0.0009) and south-facing slopes (——●, R^2=0.83, P=0.0007) along the rainfall gradient in Israel. (Osem et al. 2009)

of broad-leafs reestablishment was positively and linearly related to annual rainfall amount in both *P. halepensis* and *P. brutia* forests. There were no significant effects of forest type (*P. halepensis* vs. *P. brutia*), topographic aspect, bedrock type or elevation level.

32.4.2 Overstory Cover

Osem et al. (2015) examined the effect of overstory thinning on broad-leafs reestablishment in *P. brutia* stands (50 year old, Mount Horshan Forest) with 600 mm annual rainfall and soft chalk bedrock. The density and size of regenerating trees in the forest understory were compared between non-thinned stands of ~230 trees ha^{-1} (CC = 95%) and thinned stands of ~100 trees ha^{-1} (CC = 55%). Native broad-leaved tree species including *Q. ithaburensis* Decne., *Q. calliprinos*, *P. palaestina*, *Rhamnus alaternus* L., *Styrax officinalis* L., *Phillyrea latifolia* L. and *C. siliqua* were found throughout the forest understory, with recruit densities of 1000–2400 trees ha^{-1}. Recruit density of *Q. ithaburensis*, the most dominant of the regenerating broad-leaved tree species in this forest, was nearly four-fold higher in the non-thinned than in the thinned forest plots. No significant effect of overstory thinning was found on the recruit density of the other broad-leaved tree species. Moreover, no effect of overstory thinning was found on the height distribution, canopy cover or species richness of broad-leaf recruits.

Ashkenazi (2016) looked at broad-leafs reestablishment in *P. brutia* stands (50 year old, Begin Park Forests, Judean Mountains) with 560 mm annual rainfall and complex soft-hard calcareous bedrock formation. In this study trees were thinned from ~250 trees ha^{-1} (CC = 78%, control) to 130 (CC = 62%, moderate thinning) and 50 (CC = 12%, heavy thinning) trees ha^{-1}. Seven years after thinning the density of regenerating broadleaved tree species was similar among treatments with an average of ~1250 trees ha^{-1}. Regenerating tree species were mainly *Q. calliprinos and P. palaestina* that were equally abundant, with an additional minor contribution of *Crataegus aronia* (L.) DC.

In another study conducted by Osem et al. (2017a), broad-leafs reestablishment was examined in *P. halepensis* stands with 540 mm annual rainfall on west-facing slopes and hard calcareous bedrock (lime). In this study, mature forest stands (40 year old, HaKedoshim Forest) of ~550 trees ha^{-1} (LAI = 2.0; CC = 88%) were thinned to 300, 100 and 0 trees ha^{-1} (LAI = 1.5, 1.0 and 0; CC = 61%, 48% and 0%, respectively). Naturally regenerating broadleaved tree species included *Q. calliprinos*, *P. palaestina*, *C. siliqua*, *P. latifolia* and *Olea europaea* L.. Among these, *Q. calliprinos* was highly dominant and the only species responding to overstory thinning. While prior to thinning the density of regenerating oaks was similar among the forest plots with an average of ~400 trees ha^{-1}, 7 years after thinning it continued to increase in the non-thinned and moderately thinned plots (550 and 300 trees ha^{-1}) but slightly decreased in the heavily thinned and clear-cut plots (100 and 0 trees ha^{-1}). In contrast, the growth rate of oak recruits was positively affected by overstory thinning (i.e., relative growth rate increased three-fold following complete removal of overstory cover).

Cooper et al. (2014) specifically examined the effect of overstory cover on the performance of young *Q. ithaburensis* trees growing in the understory of a mature *P. brutia* forest (Metzer Forest, 45 year old, ~600 mm annual rainfall, soft to semi-hard chalk bedrock). This study compared carbon assimilation, water use and growth rate of artificially sown oaks (of identical age) along an overstory cover gradient (irradiance range of 20–100%). Ten years after the creation of the overstory gradient, the size of the oaks was strongly affected by overstory cover, with oaks growing under no cover (irradiance = 100%) reaching more than 10 times the height and 300 times the stem basal area of those growing under dense overstory cover (irradiance≈20%, Fig. 32.8). Correspondingly, irradiance level had a positive effect not only on carbon assimilation rate but also on water use efficiency.

32.4.3 Grazing

Osem et al. (2015) studied the effects of cattle grazing on broad-leafs reestablishment in mature (50 year.old) *P. brutia* stands (Mount Horshan Forest). Recruits of native broad-leaved tree species including *Q. ithaburensis*, *Q. calliprinos*, *P. palaestina*, *R. alaternus*, *S. officinalis*, *P. latifolia* and *C. siliqua* were found throughout the forest understory. Recruit density of *Q. ithaburensis*, the most abundant of these species, was nearly three-fold higher in ungrazed than in grazed areas. However, no significant effect of grazing was found on the recruit density of any of the other

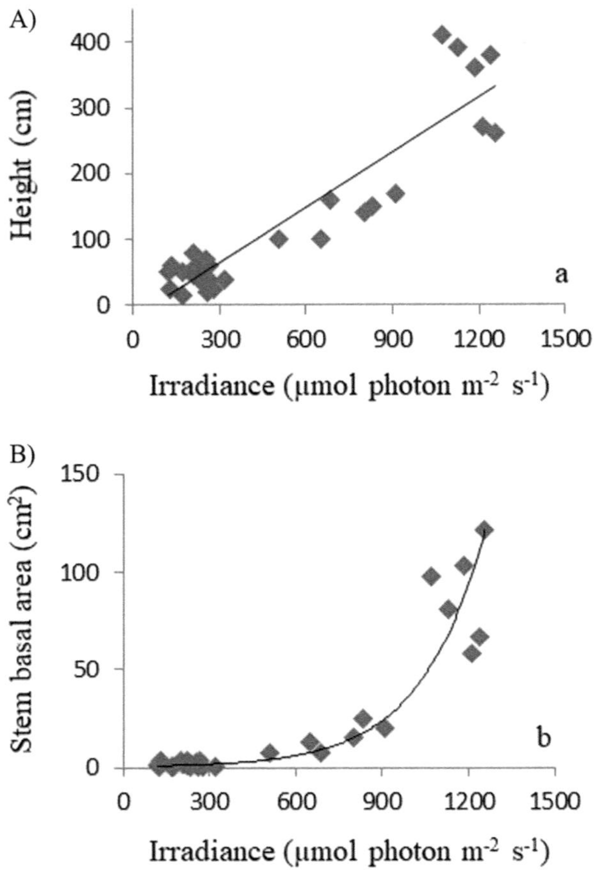

Fig. 32.8 Tree height (**a**, y = 0.30x–19.58, R^2 = 0.85, **P < 0.0001**) and stem basal area (**b**, y = $0.39e^{0.005x}$, R^2 = 0.81, **P < 0.0001**) of 20-year-old oaks (*Quercus ithaburensis*) as a function of daily average irradiance in Metzer Forest, Israel, 2009. (Cooper et al. 2014)

broad-leaved tree species. When examining total recruit density by height class it became evident that the density of larger recruits (20 cm to 4 m in height) was higher in the ungrazed than in the grazed areas, while the density of smaller recruits (<20 cm) was higher in grazed than in ungrazed areas (Fig. 32.9). This resulted in the total cover of broad-leaved tree recruits being more than three-fold higher in ungrazed than in grazed areas. Furthermore, the species richness of the broad-leaved tree recruits was higher in ungrazed than in grazed areas.

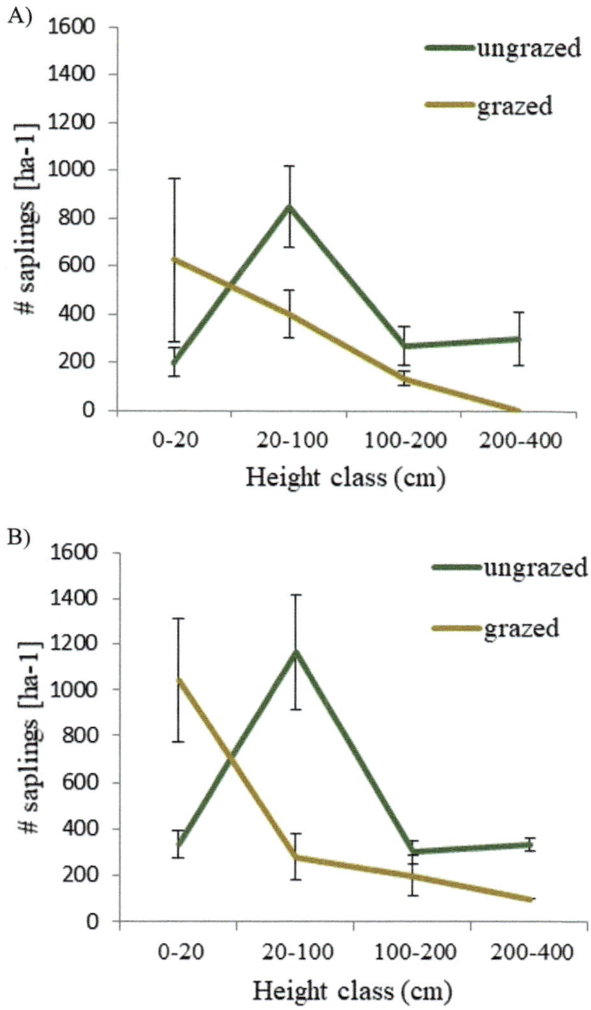

Fig. 32.9 Density of broadleaved tree species saplings by height classes as affected by cattle grazing in thinned (**a**) and non-thinned (**b**) plots. Mount Horshan Forest, Israel, November 2009. Bars indicate SE, N = 20. (Osem et al. 2015)

32.5 Integrative Discussion

The ways by which aridity, overstory cover and livestock grazing regulate tree regeneration in the understory of Israel's pine forests may be summarized as follows:

Natural pine recruitment in the understories of Israel's pine forests (i.e., advance regeneration) occurs in *P. halepensis* stands in forests receiving more than ~350–400 mm annual rainfall depending on bedrock formation (i.e., lower threshold on soft- than on hard-calcareous bedrock). Above these thresholds, the relationship between aridity and pine recruitment is weak and observed mainly on soft bedrock formations. This difficulty is most likely related to a negative feedback by competing annual herbs that normally increase at higher rainfall amounts and on hard vs. soft bedrock formations. In addition, pine recruitment is negatively affected by overstory cover and livestock grazing which consistently hamper seedling survival. Moreover, aridity, overstory cover and livestock grazing all restrict the growth (size gain) of pine recruits.

Currently, the extent of advance regeneration occurring in *P. brutia* stands is very limited. This may be partly explained by the fact that mature *P. brutia* stands in Israel are considerably denser with higher overstory cover than *P. halepensis* stands (Wilson 2016). In addition, *P. brutia* seedlings are probably more sensitive to drought. Nevertheless, this pine species has proven its capacity to regenerate naturally under some habitat conditions following overstory thinning. More research is needed to determine the environmental thresholds for natural regeneration of *P. brutia* in Israel.

broad-leafs recruitment appears in the understory of Israel's pine forests at locations exceeding 450–500 mm of annual rainfall with a linear increase in tree recruit density and species diversity with increasing rainfall. This effect of climatic aridity is accompanied by an additional positive effect of north- vs. south-facing slopes, further illustrating the importance of water availability for broad-leafs recruitment. In contrast to pine recruitment, broad-leafs recruitment is positively influenced by overstory cover, while the effect of livestock grazing on this process appears to be minor with the exception of *Q. ithaburensis* which appears to be rather sensitive to grazing. When considering the growth of broad-leaf recruits, the effects of aridity, overstory cover and grazing are all consistently negative as observed for pine recruits.

In Mediterranean climate regions which are undergoing climate change (i.e., higher temperatures, reduced precipitation), a lack of natural regeneration is considered to be one of the main limitations threatening the sustainability of forest ecosystems (Benavides et al. 2013; Manso et al. 2014; Benavides et al. 2016; Vergarechea et al. 2019). This threat is further exacerbated by other human-related factors such as inappropriate forest management practices (Martin-Alcon et al. 2015; Ruano et al. 2015; Simões et al. 2016) and overgrazing (Plieninger et al. 2011; Lempesi et al. 2017). Climatic aridity leading to water scarcity, particularly summer drought during the early seedling establishment phase, is frequently viewed as the main bottleneck for natural regeneration in Mediterranean forests (Vizcaino-Palomar et al. 2014; Benavides et al. 2016; Principe et al. 2019; Vergarechea et al. 2019). This constraint may be either strengthened or mitigated by local conditions such as topography and soil conditions (González-Martínez and Bravo 2001; Principe et al. 2019), light regime (Prevosto et al. 2012; Adili et al. 2013; Calama et al. 2013, 2015; Benavides et al. 2013, 2016; Martin-Alcon et al. 2015; Vergarechea et al. 2019) and the characteristics of the neighboring vegetation (Benavides et al. 2013, 2016; Simões et al. 2016; Vergarechea et al. 2019). The effect of overstory cover on natural regeneration in Mediterranean forests varies across forest types and tree species. While overstory shading can have positive (Calama et al. 2013, 2015,

Martin-Alcon et al. 2015 – with regard to broad-leafs, Vergarechea et al. 2019) or negative (Martin-Alcon et al. 2015 – with regard to pines) effects on early seedling survival, it usually has negative effects on the growth rate of recruits (Adili et al. 2013; Benavides et al. 2013, 2016; Martin-Alcon et al. 2015; Calama et al. 2015). In addition to its direct effect on light availability for recruits, overstory cover may influence natural regeneration through indirect effects on the neighboring understory vegetation which may compete with (Prevosto et al. 2012; Benavides et al. 2013, 2016; Vergarechea et al. 2019) or facilitate (Benavides et al. 2013, 2016) tree recruits. In contrast to overstory cover, the effects of livestock grazing on natural regeneration in Mediterranean forests appear to be quite consistent, with commonly negative effects on recruit survival and growth observed across various forest types (González-Martínez and Bravo 2001; Cierjacks and Hensen 2004; Alias et al. 2010; Plieninger et al. 2011; Lempesi et al. 2017). However, this outcome is usually associated with intense grazing regimes, while moderate and carefully managed livestock grazing may facilitate natural regeneration in water-limited Mediterranean forests through the reduction of competing vegetation (Tyler et al. 2008).

32.6 Management Guidelines for Developing Mixed Forests

Considering these conclusions, I present management guidelines for turning simply structured monoculture pine forests, in Mediterranean Israel, into sustainable mixed-forest ecosystems.

32.6.1 Aridity

- In forest habitats with <350 mm annual rainfall (WI < 0.17), neither pine forest renewal nor broad-leafs establishment should rely on natural regeneration processes.
- In forest habitats with >400 mm annual rainfall (WI ≥ 0.21), on soft calcareous bedrock, natural pine regeneration will most probably suffice for forest renewal. In such habitats, management of pine regeneration should focus on early age thinning to avoid the development of excessively dense forest stands. The target density of regenerating pine stands should decrease with increasing aridity. For example, recent management guidelines in Israel recommend the final density of *P. halepensis* stands to be set before the age of 30 year at 300, 250 and 200 trees ha^{-1} for annual rainfall levels of 250–350, 351–500 and >500 mm, respectively.
- In habitats with >400 mm annual rainfall, on hard calcareous bedrock as well as on various other soft-hard bedrock complexes, the extent of natural pine regeneration is highly unpredictable. Advance regeneration assessments may assist in reducing this uncertainty in such habitats, and monitoring successes following regeneration treatments (e.g., regeneration thinning) is necessary.

- When pine recruitment appears insufficient, herb control may be considered for promoting natural regeneration, although effective implementation of such a practice has not yet been developed and demonstrated in Israel's pine forests.
- Broad-leafs reestablishment should be expected in habitats with >450 mm annual rainfall (WI $\geq$ 0.23), with the density and species diversity of broadleaved tree recruits increasing with higher water availability.

32.6.2 Overstory Cover

- High pine overstory cover ($\geq$50%, LAI $\geq$ 2) promotes the recruitment of native broad-leaved trees.
- Low overstory cover ($\leq$20%, LAI $\leq$ 1) is required to allow successful pine recruitment and growth.
- Overstory thinning for pine regeneration should limit overstory cover to the range 0–20% cover, or LAI = 0–1, depending on target stand structure: 0–5% coverage for single and 6–20% coverage for multi-age structure.
- Decreased overstory cover will enhance the growth rate of both pine and native broad-leaved trees.
- Selective thinning may be applied in regenerated mixed stands focusing on regulating tree cover according to habitat aridity, creating variable age structures and promoting tree species diversity. This requires further study.

32.6.3 Grazing

- Grazing exclusion is required to allow survival of young pine recruits and enhance the growth of both pine and broad-leaf recruits.
- Grazing should be excluded for a sufficient time period (approximately 10 year), allowing tree recruits to reach a minimal height of 3 m.
- Reintroduction of grazing following the satisfactory development of tree recruits would facilitate further tree development through herbaceous and shrub biomass removal and reduced competition, fire hazard reduction and pruning effect.
- More research is required to examine the various grazing regimes and their effects on forest dynamics.

Using these guidelines for prescribing overstory thinning and livestock grazing regimes according to site water conditions should facilitate the conversion of first generation, even-aged pine monocultures into mixed multi-aged forests of variable stand structure, and promote the diversity and complexity of forest ecosystems in Mediterranean Israel.

32.6.4 Concluding Remarks

The diversification of forest systems is broadly considered to be a desirable course (Schuler et al. 2017; Sánchez de Dios et al. 2019) promoting biodiversity (Felton et al. 2016; Pukkala 2018; Wang et al. 2019), productivity (Pretzsch et al. 2012, de-Dios-Garcia et al. 2015), and multiple ecosystem services (Felton et al. 2016; Schuler et al. 2017; Pukkala 2018), as well as forest health (Felton et al. 2016; Aoki et al. 2018), resistance (Pretzsch et al. 2013, de-Dios-Garcia et al. 2015), and resilience (Pukkala 2018). Nonetheless, currently, very little is known about these processes in the recently evolving mixed forests of Mediterranean Israel. Furthermore, silvicultural strategies to optimize ecosystem services and promote the resistance and resilience of these semi-natural ecosystems, in the face of climate change, are yet to be developed.

Acknowledgements The studies reported in this review were all supported by the forest service of Israel – KKL. I thank Ira Mor, Michael Sprintsin, Chanoch Zoref and Avi Perevolotsky for critical reading of the MS.

References

Adili B, El Aouni MH, Balandier P (2013) Unravelling the influence of light, litter and understorey vegetation on *Pinus pinea* natural regeneration. Forestry 86:297–307

Alias S, Bianchi L, Calamini G et al (2010) Shrub facilitation of *Quercus ilex* and *Quercus pubescens* regeneration in a wooded pasture in Central Sardinia (Italy). IFOREST 3:16–22

Aoki CF, Cook M, Dunn J et al (2018) Old pests in new places: effects of stand structure and forest type on susceptibility to a bark beetle on the edge of its native range. For Ecol Manag 419:206–219

Ashkenazi M (2016) Fuelbreak treaments and their effect vegetation structure and fire behaviour in Mediterranean coniferous forests. MSc thesis submitted to the Hebrew University of Jerusalem, Istrael

Benavides R, Rabasa SG, Granda E et al (2013) Direct and indirect effects of climate on demography and early growth of *Pinus sylvestris* at the rear edge: changing roles of biotic and abiotic factors. PLoS One 8:e59824

Benavides R, Escudero A, Coll L et al (2016) Recruitment patterns of four tree species along elevation gradients in Mediterranean mountains: not only climate matters. For Ecol Manag 360:287–296

Boneh O, Silverstone A (1999) Forest renewal in the northern region. Scientific report submitted to the Forestry Division, Keren Kayemeth Le'Israel (KKL), Eshtaol, Israel. (in Hebrew)

Calama R, Puertolas J, Madrigal G et al (2013) Modeling the environmental response of leaf net photosynthesis in *Pinus pinea* L. natural regeneration. Ecol Model 251:9–21

Calama R, Puertolas J, Manso R et al (2015) Defining the optimal regeneration niche for *Pinus pinea* L. through physiology-based models for seedling survival and carbon assimilation. Trees-Struct Funct 29:1761–1771

Cierjacks A, Hensen I (2004) Variation of stand structure and regeneration of Mediterranean holm oak along a grazing intensity gradient. Plant Ecol 173:215–223

Cohen N (2019) Effect of goat grazing and pine seedling removal on the plant community, in general, and pine seedlings, in particular, following the Carmel fire. MSc thesis submitted to the Hebrew University of Jerusalem, Israel

Cooper A, Shapira O, Zaidan S et al (2014) Oak restoration in water-limited pine plantations: interactive effects of overstory light interception and water availability on understory oak performance. Eur J For Res 133:661–670. https://doi.org/10.1007/s10342-014-0794-6

de Dios-Garcia J, Pardos M, Calama R (2015) Interannual variability in competitive effects in mixed and monospecific forests of Mediterranean stone pine. For Ecol Manag 358:230–239

Evlagon D, Kommisarchik S, Gurevich B et al (2012) Estimating normative grazing capacity of planted Mediterranean forests in a fire-prone environment. Agric Ecosyst Environ 155:133–141

Felton A, Nilsson U, Sonesson J et al (2016) Replacing monocultures with mixed-species stands: ecosystem service implications of two production forest alternatives in Sweden. Ambio 45:s124–s139

González-Martínez SC, Bravo F (2001) Density and population structure of the natural regeneration of Scots pine (*Pinus sylvestris* L.) in the High Ebro Basin (Northern Spain). Ann For Sci 58:277–288

Lempesi A, Eleftheriadou A, Delivasi Z et al (2017) Effects of grazing intensity on the regeneration of woody species in an oak woodland. Not Bot Horti Agrobot Cluj Napoca 45:597–601

Manso R, Pukkala T, Pardos M et al (2014) Modelling *Pinus pinea* forest management to attain natural regeneration under present and future climatic scenarios. Can J For Res 44:250–262

Martin-Alcon S, Coll L, Salekin S (2015) Stand-level drivers of tree-species diversification in Mediterranean pine forests after abandonment of traditional practices. For Ecol Manag 353:107–117

Ne'eman G, Perevolotsky A, Schiller G (1997) The management implications of the Mt. Carmel research project. Int J Wildl Fire 17:343–350

Noy-Meir I (1989) An ecological viewpoint on afforestation in Israel: past and future. Allg Forst Zeitsch 24–26:614–618

Osem Y, Ginsberg P, Tauber I et al (2008) Sustainable management of Mediterranean planted coniferous forests: an Israeli definition. J For 106:38–46

Osem Y, Zangy E, Bney-Moshe E et al (2009) The potential of transforming simple structured pine plantations into mixed Mediterranean forests through natural regeneration along a rainfall gradient. For Ecol Manag 259:14–23. https://doi.org/10.1016/j.foreco.2009.09.034

Osem Y, Zangy E, Bney-Moshe E et al (2012) Understory woody vegetation in manmade Mediterranean pine forests: variation in community structure along a rainfall gradient. Eur J For Res 131:693–704. https://doi.org/10.1007/s10342-011-0542-0

Osem Y, Yavlovich H, Zecharia N et al (2013) Fire-free natural regeneration in water limited *Pinus halepensis* forests: a silvicultural approach. Eur J For Res 132:679–690

Osem Y, Fogel T, Moshe Y et al (2015) Managing cattle grazing and overstorey cover for the conversion of pine monocultures into mixed Mediterranean woodlands. Appl Veg Sci 18:261–271. https://doi.org/10.1111/avsc.12152

Osem Y, Zangi E, Ashkenazi M et al (2017a) Long term ecological research site for studying the management of planted forests in Israel: the effects of overstory and understory silvicultural treatments on forest vitality and health, natural regeneration and biodiversity in *Pinus halepensis* forest. Scientific report submitted to the Forestry Division, Keren Kayemeth Le'Israel (KKL), Eshtaol, Israel. (in Hebrew)

Osem Y, Fogel T, Moshe Y et al (2017b) Understory structure and function following release from cattle grazing and overstory thinning in Mediterranean conifer plantations. Ann For Sci 74:22. https://doi.org/10.1007/s13595-017-0622-5

Perelman Y (2018) Leaf area organization in *Pinus halepensis* forests as related to abiotic environmental factors. MSc thesis submitted to the Hebrew University of Jerusalem, Israel

Plieninger T, Schaich H, Kizos T (2011) Land-use legacies in the forest structure of silvopastoral oak woodlands in the Eastern Mediterranean. Reg Environ Chang 11:603–615

Pretzsch H, Dieler J, Seifert T et al (2012) Climate effects on productivity and resource-use efficiency of Norway spruce (*Picea abies* [L.] Karst.) and European beech (*Fagus sylvatica* [L.]) in stands with different spatial mixing patterns. Trees-Struct Funct 26:1343–1360

Pretzsch H, Schutze G, Uhl E (2013) Resistance of European tree species to drought stress in mixed versus pure forests: evidence of stress release by inter-specific facilitation. Plant Biol 15:483–495

Prevosto B, Amandier L, Quesney T et al (2012) Regenerating mature Aleppo pine stands in fire-free conditions: site preparation treatments matter. For Ecol Manag 282:70–77

Principe A, Matos P, Sarris D et al (2019) In Mediterranean drylands microclimate affects more tree seedlings than adult trees. Ecol Indic 106:105476

Pukkala T (2018) Effect of species composition on ecosystem services in European boreal forest. J For Res 29:261–272

Ruano I, Manso R, Fortin M et al (2015) Extreme climate conditions limit seed availability to successfully attain natural regeneration of *Pinus pinaster* in sandy areas of Central Spain. Can J For Res 45:1795–1802

Sánchez de Dios R, Velazquez JC, Ollero HS (2019) Classification and mapping of Spanish Mediterranean mixed forests. IFOREST 12:480–487

Schiller G (1978) The effect of environmental factors on the natural regeneration of a *Pinus halepensis* forest. PhD thesis submited to the University of Tel-Aviv, Israel

Schuler LJ, Bugmann H, Snell RS (2017) From monocultures to mixed-species forests: is tree diversity key for providing ecosystem services at the landscape scale? Landsc Ecol 32:1499–1516

Seligman N, Unger E, Henkin Z et al (eds) (2016) About vegetation, animals and people: rangeland management in Israel. Nekudat Chen, Jerusalem

Simões MP, Belo AF, Fernandes M et al (2016) Regeneration patterns of *Quercus suber* according to montado management systems. Agrofor Syst 90:107–115

Tyler CM, Davis FW, Mahall BE (2008) The relative importance of factors affecting age-specific seedling survival of two co-occurring oak species in southern California. For Ecol Manag 255:3063–3074

Vergarechea M, Calama R, Fortin M et al (2019) Climate-mediated regeneration occurrence in Mediterranean pine forests: a modeling approach. For Ecol Manag 446:10–19

Vizcaino-Palomar N, Revuelta-Eugercios B, Zavala MA et al (2014) The role of population origin and microenvironment in seedling emergence and early survival in Mediterranean maritime pine (*Pinus pinaster* Aiton). PLoS One 9:e109132

Wang XY, Hua FY, Wang L et al (2019) The biodiversity benefit of native forests and mixed-species plantations over monoculture plantations. Divers Distrib 25:1721–1735

Wilson T (2016) Principal abiotic factors influencing the structure and function of mature pine forests in Israel. MSc thesis submited to Texas A&M University

Chapter 33
Forest Policy in Israel – From Creation of Forest Plantations to Sustainable Management of Forest Ecosystems

Yagil Osem, Yahel Porat, Chanoch Zoref, and Omri Bonneh

33.1 Introduction

The afforestation enterprise in the Mediterranean region of Israel, initiated at the turn of the twentieth century, has led to the establishment of nearly 100,000 ha of planted forests, covering major parts of the landscape that were unsuitable for cultivation. This amounts to nearly 4% of Israel's total area and ca. 8% of its Mediterranean region, and represents less than 0.02 ha per capita (KKL 2007). The major objectives of afforestation in Israel in its early stages were "protecting public land", "providing employment" and "improving the landscape to encourage settlement" (Amir and Rechtman 2006). Wood production was also regarded as an expected benefit (Liphschitz and Biger 2004) and provided a basis for the management strategy of the planted forests (planting densities, thinning regimes, rotation age, etc.). In light of these objectives, the Israeli Forest Service (KKL) concentrated on planting a few coniferous species, characterized by high establishment and growth rates under a wide variety of edaphic and climatic conditions, and requiring minimal care. A concept of dense, monospecific, even-aged pine forests proved efficient in achieving the historic goals of afforestation in Israel. However, it resulted in the creation of simply structured forest ecosystems characterized by a short life cycle and limited

Y. Osem (✉)
Department of Natural Resources – Institute of Plant Sciences, Agricultural Research Organization – Volcani Center, Rishon LeZion, Israel
e-mail: yagil@agri.gov.il

Y. Porat
Forestry Division, Keren Kayemeth LeIsrael (KKL), Jerusalem, Israel

C. Zoref
Mountain Region, Keren Kayemth LeIsrael (KKL), Jerusalem, Israel

O. Bonneh
Northern Region, Keren Kayemeth LeIsrael (KKL), Jerusalem, Israel

G. Ne'eman, Y. Osem (eds.), *Pines and Their Mixed Forest Ecosystems in the Mediterranean Basin*, Managing Forest Ecosystems 38,
https://doi.org/10.1007/978-3-030-63625-8_33

ecological stability (Noy-Meir 1989). Now, as the potential stock of land for new afforestation in the Mediterranean region of Israel is nearly exhausted, and as many planted forest stands approach the end of their life expectancy, the focus of forest managers is turning towards the sustainable management of established forest ecosystems. The definition of sustainable forest management (SFM) for conifer forests in Israel is unique, since they are man-made systems and there are hardly any natural or historical reference forests to serve as benchmarks. This definition must determine the future roles and management objectives of these forests while considering Israel's current ecological, social and economic constraints (Osem et al. 2008).

33.2 Forest Management Policy in Israel from the 1920s to the 1990s

Afforestation began in Israel nearly 100 years ago; however, out of the entire afforested area (100,000 ha), nearly 85% was established from the 1950s. During the first few decades, afforestation in Israel was characterized by dense conifer plantations based mainly on the native Aleppo pine (*Pinus halepensis* Mill.), a pioneer species. Criticism regarding this afforestation policy arose from its early beginning (Gindel 1952; Liphschitz and Biger 2004). During the early 1970s extensive tree mortality of mature Aleppo pine stands in the Shaar HaGay Forest that was attributed mainly to the Israeli pine bast scale (*Matsucoccus josephi* Bodenheimer et Harpaz), attracted significant public attention and increased the controversy regarding Israel's afforestation policy. This event resulted in the recommendation to replace Aleppo pine by other conifer species (Orshan et al. 1975). Since the 1970s, the exotic Turkish pine (*Pinus brutia* Tenn.), native to the north-eastern Mediterranean (including Lebanon but not Israel), has gradually become the leading planted pine species accompanied by several other Mediterranean conifers such as stone pine (*Pinus pinea* L.), Mediterranean cypress (*Cupressus sempervirens* L.) (Bonneh 2000), and to a lesser extent, Canary Island pine (*P. canariensis* C. Smith.). During the 1980s and 1990s, mortality of maturing Aleppo pine stands, caused by the Israeli pine bast scale, expanded over many forest stands in Israel. This, together with increasing criticism by local "green organizations" regarding the non-natural, commercial-like, approach to afforestation (see review by Perevolotsky and Sheffer 2009), has led to a gradual transformation, which began prior to any formal decision, from pine monocultures to mixed conifer-broadleaved plantations using native broad-leaved species (Fig. 33.1).

In 1990, a new management policy for Israel's planted forests (hereafter, the 1990 policy) was declared (KKL 1990). The new policy was designed to overcome the following major deficiencies that were identified in Israel: (a) low forest diversity due to the extensive planting of Aleppo pine monocultures, (b) overstocking due to high planting densities and failure to keep to the thinning plans (Fig. 33.2) and, (c) low forest resilience to various disturbances attributed to a lack of tree

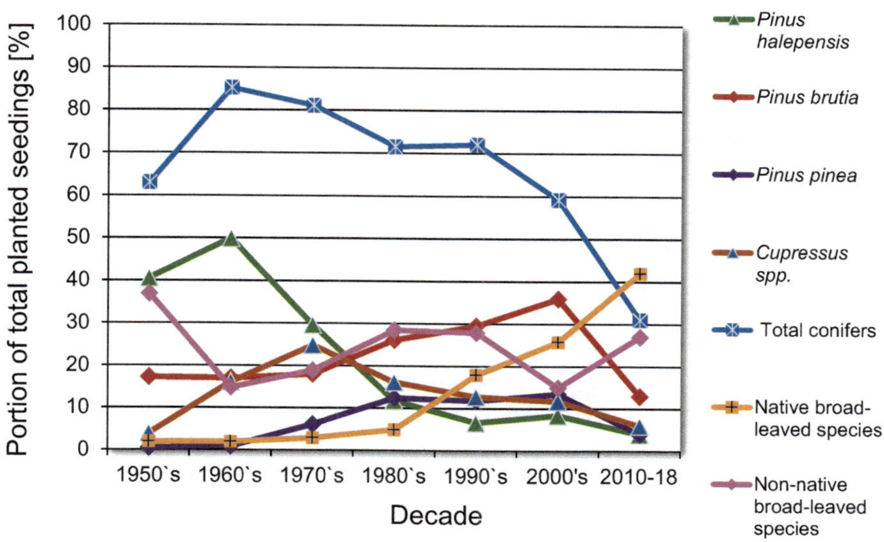

Fig. 33.1 Changes along the decades of afforestation in the planting proportions of native and non-native, coniferous and broad-leaved species

genetic fitness (Schiller and Gruneald 1987), overstocked stands and forest homogeneity (Osem et al. 2008).

The following new guidelines were defined by the 1990 policy:

A. Implementation of tree improvement programs (particularly for Aleppo pine).
B. Planting mixed forests based on a variety of coniferous and native broad-leaved species.
C. Selecting forest species for planting according to local site conditions.
D. Reducing pine planting densities (from 2500–3000 to 1300–1600 seedlings per ha).
E. Reducing pine stocking rates throughout the forest rotation period.
F. Reducing the number of thinning cycles to three (instead of six) throughout the rotation period.

33.2.1 Practical Implications of the 1990 Forest Management Policy

The 1990 policy was the beginning of a process in which significant changes in the practice of afforestation and forest management were implemented. This process developed gradually through the following three decades. During this time, criticism of afforestation and forest management by local green organizations continued contributing to the accelerated adoption of more environmentally friendly

Fig. 33.2 Overstocking in mature pine (*Pinus halepensis*) plantations. HaKedoshim Forest, Judean Mountains 2009

silvicultural practices by KKL (see verdict by the Israeli Supreme Court 288/00, 2001). Several examples of these changes are described below.

33.2.1.1 Site Preparation

The "agricultural approach", in which the optimization of conditions for the planted seedlings is the ultimate goal of site preparation, was gradually replaced by an approach aiming to reduce the ecological impacts of site preparation operations (Bonneh 2000). These changes included:

A. Replacement of strip till using bulldozers equipped with subsoilers by spot preparation of planting pits using excavators.

B. Abandoning the practices of pre-planting shrub elimination by broadcast burning or aerial herbicide spraying over the entire planting site, and replacing them by ground herbicide application around planting pits only.
C. Retention of logging residues and slash on the planting site.

33.2.1.2 Planting

Since 1990, planting of native broad-leaved species has increased gradually from 5 to 25% (Bonneh 2000). During the 2000s this increase continued, and broad-leaved species now comprise 50% of the total planted seedlings (Bonneh 2016; Fig. 33.1). Broad-leaved seedlings are planted as separate stands of various sizes (Fig. 33.3) or within the conifer matrix as single trees or small groups. The use of these various planting patterns creates a complex patch mosaic. Wherever grazing cannot be prevented effectively by fencing the entire planting area, individual fencing is installed around each seedling (Bonneh et al. 2015). In this mode of planting, 3–4-year-old, 10–25 L containerized seedlings are used and irrigation systems are installed. Another mode of planting, which has been employed since 2000, involves transplanting of mature oak (*Quercus calliprinos* Webb., *Q. ithaburensis* Decne.) and carob (*Ceratonia siliqua* L.) trees (20–50 years old) into afforestation sites and

Fig. 33.3 Planting native broad-leaved trees in afforestation sites. Refa'im Valley in the Judean Mountains 2008

within existing conifer forest areas. Most of these transplanted trees come from infrastructure development sites.

33.2.1.3 Thinning and Pruning in Pine Forests

Since 1990, heavy thinnings in young conifer stands (8–12 years old) and timely thinnings in mature forests have become increasingly significant. This change was perceived as essential for promoting the resilience of the pine forests, their biodiversity and regeneration capacity (Osem et al. 2015, 2017; Bonneh 2017).

33.2.1.4 Natural Regeneration of Broad-Leaved Species

In many of the conifer forests in Israel, it became evident that native broad-leaved species can be recruited through natural regeneration processes, rather than by planting. Native broad-leaved species, such as the common oak (*Q. caliprinos*) were shown to be fairly shade tolerant and able to regenerate and establish naturally in the pine forest understory (Goldring 1977; Osem et al. 2009). Following this understanding, a practice of selective overstory pine thinning was adopted, in order to release broad-leaved trees growing in the understory and enhance their growth (Osem et al. 2015; Bonneh 2016; Fig. 33.4). Recently, direct oak acorn seeding was also recommended as a useful method to disseminate and establish oaks in planted conifer forests (Herr et al. 2011).

33.2.1.5 Post-Disturbance Forest Regeneration

Planted pine stands in Israel are occasionally exposed to various disturbances, of which wildfire, insect outbreaks, and droughts are the most common. These disturbances often cause various degrees of tree mortality. A post-disturbance selective forest renewal strategy, which preserves trees that survived the disturbance, coupled with planting or nurturing of the natural regeneration of coniferous and native broad-leaved species, provides an opportunity to transform pure even-aged stands into mixed uneven-aged forests (Bonneh 2015, 2016; Fig. 33.5). In the past, following wildfire, salvage cutting of burned pine trees were implemented soon after the fire (Bonneh 1996). Since the early 1990s, a different strategy has been gradually implemented in which foresters are instructed to wait at least 1 year before conducting any post fire cuttings, in order to examine the extent of tree mortality and the amount of natural regeneration.

Fig. 33.4 Development of mixed conifer-broadleaved forest through release thinning of the planted pine (*Pinus helepensis*) overstory and natural regeneration of native broad-leaved trees (*Quercus caliprinos*). Bar-Giora Forest, Judean Mountains 2010

33.2.2 Conclusions

The 1990 policy aimed at gradually transforming Israel's first generation forests from simply structured and susceptible conifer monocultures to more diverse, resilient forests. The main practical implications of this policy which are still being implemented in the present are:

A. Conversion of conifer monocultures into mixed conifer-broadleaved forests by planting broad-leaved seedlings in various patterns within the conifer matrix to create diverse and complex forest structures.
B. Transplanting of mature native broad-leaved trees to diversify forest structure while rescuing these trees.
C. Avoiding thinning delay and increasing thinning intensities of conifer forests to maintain the health and vitality of the conifer overstory and promote natural regeneration and plant diversity in the understory of these forests.
D. Using forest disturbances such as wildfire, insect outbreaks, and drought events as opportunities to enhance forest structural complexity.

Fig. 33.5 Mixed two-aged planted forest with scattered 80-year-old Aleppo pine and Mediterranean cypress trees that survived a fire and various conifers and native broad-leaved species that were planted to restore the burned forest. Balfour Forest, Lower Galilee, 2014

E. Using livestock grazing (goats, sheep, beef cattle) in a differential manner as a means to control understory encroachment, reduce fire hazard and diversify forest structure.

33.3 A New Forest Management Policy: From Practical Modifications to the Development of a Strategy

In 1995, a new "National Master Plan (NOP 22) for Forests and Afforestation" was approved by the Israeli Government (KKL 1995). The plan defined six different forest formations with only 42% of the forest area in Israel designated for planted conifer forests, whereas the rest of the area was designated for various native broad-leaved forests and other wooded land formations. In 2007, the board of directors of KKL adopted a policy of sustainable forest development (KKL 2004). This policy emphasized the variety of ecosystem services to be provided by forests to the local public and to future generations. Furthermore, the new policy called for a gradual transformation of the pioneer pine forest generation to a second generation of mixed uneven-aged forests with this process relying as much as possible on processes of natural regeneration and succession. In light of the directives contained within NOP

22 and the Policy of Sustainable Development adopted by KKL, an updated policy for forest management in Israel was developed (Osem et al. 2014).

33.3.1 The New Forest Management Policy of Israel (2014): Goals, Principles and Implementation

The new forest management policy of Israel (IFMP) recognized the need to implement changes in forest management according to the latest developments in the fields of forestry, ecology, social and economic sciences (e.g., Puettmann et al. 2009; O'Hara 2014). The IFMP furnished an obligating professional basis for managing Israel's forests in a goal-oriented and sustainable fashion. It defined and outlined the forestry goals in Israel, the various forest types, and the planning and management principles to be employed in them.

33.3.1.1 Forestry Goals

According to the IFMP, the primary goal of forest management in Israel is to provide a variety of ecosystem services to the public, recognizing that human existence and wellbeing depend on biological diversity and the services provided by ecosystems.

Subsidiary goals contained within this statement were defined as follows:

A. Recreation and outdoor activity services
B. Landscape beauty and diversification
C. Provision of supporting and regulating ecosystem services (e.g., carbon sequestration, primary productivity)
D. Supporting Israel's unique biological diversity
E. Provision of soil and water conservation services (preventing soil erosion, increasing water infiltration)
F. Provision of a variety of economic benefits to society (wood products, pasture for livestock, tourism)
G. Protection of open landscapes
H. Protection of Israel's native tree species and their reintroduction into the landscape
I. Protection and restoration of natural heritage landscapes
J. Ecological restoration of damaged sites and ecosystems
K. Creation of buffer zones around communities to mitigate the effects of environmental hazards such as noise, air pollution, visual blemishes to the landscape, and to protect against wildfire
L. Strengthening the public's identification with the forest and nature, and educating people to protect them

33.3.1.2 Principles of Forest Management in Israel

Ten principles for the implementation of forest planning and management were defined by the IFMP as follows:

A. *Goal-oriented, adaptive management*: the planning and execution of management operations according to a pre-determined set of management goals (Baskerville 1986) and a systematic approach for improving these operations by learning from management outcomes (Wang 2004; Allen et al. 2011).
B. *Multiple use*: managing the forests to attain a diverse set of ecological, social and economic goals (Cubbage et al. 2007; Scmitüsen and Rojas-Briales 2012).
C. *Habitat suitability*: determination of forest management goals, forest structure and species composition as a function of specific habitat conditions and location within surrounding landscapes and ecosystems (Reubens et al. 2011).
D. *Natural processes*: reliance upon local species and natural processes (e.g., natural regeneration) as much as possible (Kuuluvainen 2009; Björse and Bradshaw 1998).
E. *Rational intervention*: setting the level of silvicultural intervention to correspond with forest management objectives and attainment of these objectives through the lowest possible level of intervention (Götmark 2013).
F. *Vitality, tolerance and stability*: shaping and managing forests to become vital and healthy systems able to withstand stresses and unexpected hazards.
G. *Diversity, complexity and patchiness*: preservation and encouragement of a variety of landscape types, habitats, vegetation types, species and genotypes (Puettmann et al. 2009).
H. *Continuity*: managing the forest as an interconnected landscape element within its surroundings over time, and maintaining continuous vegetative cover (Honnay et al. 2005; Peura et al. 2018).
I. *Protection and stewardship of natural heritage assets*
J. *Environmental protection*: minimizing the environmental impacts of forest establishment and management activities as much as possible.

33.3.1.3 Forest Planning and Management

According to the IFMP, the long-term forest planning process is based on the division of forestlands into landscape units of different designated land-uses (land-use units). In each unit the desired vegetation formation is defined based on its land-use designation, existing vegetation formation, projected vegetation dynamics and specific environmental conditions. The long-term forest plans are fully coordinated with approved national statutory plans (e.g., NOP 22). The primary forest land-use designations are:

A. *Multiple-use forest*: these forestlands are designated to preserve a continuous open landscape; express the landscape diversity of Israel; and provide a broad range of ecosystem services. This is the proposed designation for the majority of

Israel's forestlands. Management of these lands will be of an *extensive* nature, based primarily on natural regeneration, succession and adaptation. The multiple-use forests will be multi-aged, structurally complex and patchy, and maintain a variety of vegetation formations.

B. *Recreation*: these forestlands are designated for intensive recreational activities and outdoor experiences.
C. *Heritage assets and unique landscape types*: these forestlands are designated for the preservation and enhancement of heritage and landscapes assets of high cultural value.
D. *Firebreaks*: these forestlands are managed to deter and/or stop the advance of forest fires across the landscape.
E. *Natural assets and unique habitat types*: these forestlands are designated for the preservation and enhancement of natural assets of high ecological importance such as unique or endangered habitats, communities or species.
F. *Community forest*: these forestlands are located next to existing communities and primarily used by their local populace.
G. *Research*: these forestlands are designated for research purposes to advance forest management techniques.

In order to optimize the management process, four types of forest management plans have been designated. Each type varies according to its resolution and time scale:

A. *Master Plan*: deals with an individual forest or group of adjacent forests. This plan defines the chief management goals of the general forestland and subdivides the landscape unit into smaller parcels; each one is further defined for its designated land-use and desired vegetation type. This plan is valid for 25 years.
B. *Multi-Year Work Plan*: presents a long-term work program to attain the goals contained within the master plan. This plan is valid for 10 years.
C. *Annual Work Plan*: presents comprehensive work plans based on those detailed in the multi-year work plan.
D. *Detailed Operations Plan*: describes the operation of management tools in each specific site designated for treatment in the annual work plan.

33.3.1.4 Employment of Management Tools

According to the IFMP, the main goal of forest management operations is to direct natural processes occurring within the forest ecosystem towards the realization of planning objectives. In general, the majority of management activities should be applied to planted forests requiring intervention in order to achieve the desired goals and vegetation structure. Other vegetation types within the forestlands which are native (woodland, shrubland, dwarf shrubland, etc.) should be managed at a lower level of intervention, with the main goals of protecting them and directing their utilization in such a way as to maintain their integrity. Similarly, the management of planted forests strives to create a sustainable forest ecosystem that will require less intensive and more self-sustaining forms of management over time. The IFMP presents the principal set of tools available to the forester, including planting, thinning and pruning; agro-technical techniques including mowing, plowing, and weed control; as well as grazing and prescribed burning.

33.3.2 *Major Changes Generated by the New IFMP*

Despite the recognition that forests growing under the ecological conditions of Israel are incapable of producing commercial wood, the management of pine forests has continued for many years to follow classic commercial silvicultural guidelines which concentrated on maximizing stand stocking and optimizing stand production. The IFMP reflects significant conceptual changes in the approach of forest management in Israel with the aim of fundamentally modifying the practice of forest planning and treatment. This sub-chapter summarizes these changes based on four major trends, which represent the essence of the conceptual change.

33.3.2.1 From Forest Establishment to Forest Management

Forest Establishment From its very beginning, forestry in Israel has concentrated on the creation of new forests as part of its effort to rehabilitate degraded habitats and shape the landscape to encourage human development. Material and professional resources were mostly directed towards forest planting with its ensemble of related practices, including nursery production, site preparation, planting techniques, irrigation and herb control.

Forest Management Currently, land reserves for new forest planting in the Mediterranean region of Israel are almost completely exhausted, while the amount of established forests has constantly accumulated. As part of this trend, the focus of forestry is now turning towards forest management operations such as thinning, fire prevention, forest renewal, livestock grazing management and more, to achieve a variety of management goals while nurturing the forests' health and vitality and promoting their longevity, resilience and regeneration.

33.3.2.2 From Forest Condition to Forest Goals

Forest Condition As part of the continuous effort to establish new forests, forest management tended to focus on the condition of the forest trees, and silvicultural treatments were performed, almost entirely, on the basis of stand condition i.e., tree density, tree age, tree size and health, using a commercial forestry approach and indices. Forest management units were based on planting plots of unified tree age and species (stand). As a result, forest areas differing in their geographic location, management goals, and other characteristics such as their accessibility and environmental context, were treated uniformly with no clear definition of priorities, nor any deliberate divergence in intervention regime. Management plans were, basically, "yearly work plans" which responded to stand condition.

Forest Goals A growing understanding of the need to set clear objectives and targets as part of effective forest management (Baskerville 1986) together with emerging practical difficulties in applying a unified, intensive, silvicultural intervention regime across the entire forest area, have led to the embracement of a different approach. Forestlands are divided into "land-use designation units" (e.g., recreation, natural assets, firebreaks etc.), representing different forest management objectives. The desired forest structure, silvicultural intervention regime and treatment priority are all defined, primarily, according to the land-use designation, while taking into account forest condition as well as other practical considerations such as accessibility for treatment and other logistic constraints. According to this approach, forest stands of similar condition (density, tree age etc.) may be bound to entirely different management plans leading them towards different structures and functions. In order to fulfill the management objectives of forest land-use designation units, strategic master plans (25 years) and multi-year work plans (10 years) are required before yearly work plans can be set.

33.3.2.3 From Trees to Ecosystems

Trees As part of the classic commercial silvicultural approach that was implemented, forests were considered to be "tree stands", which can be characterized according to tree density, age and size. Planting density, thinning regime and "end of rotation" (clear-cut) timing were determined according to yield charts and inventory surveys. Vegetation components other than the planted trees were considered to be undesired competitors while various insects and fungi, which effect tree performance, were regarded as pests. Furthermore, disturbances diverting the forest stand from its "normal" development course were seen as undesired events typically requiring silvicultural intervention.

Ecosystems The growing awareness of the essential role of various biotic and abiotic factors in the function of forests as complex, adaptive and resilient ecosystems has led to the embracement of a management approach which perceives the forest, whether natural or man-made, as an ecosystem (Führer 2000; Leech et al. 2009). This approach is based on the understanding that the variety of organisms and interactions occurring in the forest system are essential for its capacity to provide ecosystem services. In order to promote forest biodiversity and complexity, forest managers are required to shape the forests as complex landscape mosaics incorporating a diversity of vegetation formations, trees and other plant species of various ages, sizes and spatial configurations. Vegetation formations with sparse tree cover or no trees at all, are now accepted as a desired component of the forest landscape mosaic. Various elements such as dead wood and litter are perceived to be an integral component of the system. Species, communities and habitats which are recognized as uniquely valuable are defined as natural assets receiving particular management attention. A fundamental principle in this ecosystem-based management approach is the desire for temporal and spatial continuity of the forest ecosystem (Honnay et al. 2005; Peura et al. 2018). Various disturbances, which take place at various spatial scales, are perceived to be integral components of the forest ecosystem dynamics; forest managers should recognize their importance in shaping forest structure and function and even emulate some of them actively (Seymour et al. 2002) in order to promote structural complexity (Bengtsson et al. 2000).

33.3.2.4 From Maximal Control to Rational Intervention

Maximal Control managing forest stands according to yield tables has led to the definition of an optimal stand structure according to tree age and habitat conditions (site index). In order to achieve this optimal course of stand development in each forest stand, intensive intervention was required throughout the entire forest area. Any divergence from this previously determined course due to disturbances or spontaneous processes such as tree mortality, natural regeneration and others, were considered to be undesired outcomes.

Rational Intervention The recognition of the importance of natural processes for ecosystem functionality and the necessity of self-organization processes for ecosystem adaptability and resilience (Thompson et al. 2009), has led to the concept of rational intervention. While in some parts of the forest area, an intensive management regime is necessary according to the forest land-use designation and management objectives (e.g., firebreaks, intensive recreation areas), in most of the forest area natural processes should be preferred.

33.3.3 Conclusions

The IFMP presents a new approach to managing Israel's forests as complex adaptive ecosystems which provide a variety of benefits to human society, namely, ecosystem services. Forest mangers intervene in the ecosystem by implementing a differential management regime according to various forest land-use designations in order to optimize a variety of ecosystem services over the entire managed forest area. The level of intervention is managed with the aim of minimizing interference with natural processes from an understanding that the capability of the forest to provide the variety of ecosystem services depends on these processes.

33.4 Long-Term Forest Planning for Implementing the New IFMP

33.4.1 HaKedoshim Forest as a Model

In parallel to the approval of the IFMP it was decided to prepare an example of a master plan and multi-year work plan (hereafter, long-term forest plans). HaKedoshim Forest (Forest of the Martyrs) was chosen as a 'model forest' for this purpose. Below are the steps and components of such a plan that demonstrate the applicability of the IFMP.

33.4.1.1 Defining the Current State of Forest Landscape Units

According to IFMP, in order to define the designation and desired state of a landscape unit, the "current state" must be defined first. In other words, it is necessary to define the existing vegetation formation in the field and understand the trends of vegetation development. In addition, it is important to identify the important natural and cultural assets within each landscape unit and map human activities in the present and future, as well as understand the emerging trends. Therefore, in the first stage, information is collected across a broad area surrounding the forest (i.e., reference area) for a range of aspects. A major part of this stage includes mapping all of the vegetation formations existing in the forest area (Fig. 33.6). Based on this map and on the habitat conditions (rainfall, rock, aspect) within each landscape unit, a simple model was developed and is used to define the predicted vegetation formation for the next 10 years if no human intervention or significant disturbance were to occur (i.e., projected vegetation formation).

33.4.1.2 Defining Land-Use Designations for Landscape Units in the Forest

Following the preliminary study of the landscape, and based on a range of considerations that take into account the current state and future needs, land-use designation units are determined (Fig. 33.7). In the HaKedoshim Forest model, 66% of the forest was designated as multi-purpose forest managed to provide a range of ecosystem services, with a focus on regulating ecosystem services and biodiversity conservation. The rest of the forest area was designated for conservation of natural assets and unique habitats, as well as unique cultural heritage sites and landscape formations, firebreaks, leisure and recreation areas and a field site for ecological monitoring and research (Figs. 33.7, 33.8a, 33.8b, 33.8c and 33.8d).

33.4.1.3 Multi-Year Work Plan

The long-term management plan is intended to shape or preserve the desired vegetation formation in accordance with the land-use designation and the current vegetation state.

As part of the multi-year work plan, the land-use designation units are divided into sub-units in accordance with their current vegetation state (management units). For each management unit, the vegetation formation that best suits the land-use designation is defined, while considering the current and projected vegetation states (Fig. 33.9). Subsequently, the management tools required to obtain the desired vegetation formation are defined and the implementation of management activity is prioritized. Ranking forest management units according to treatment priority allows forest managers to direct their management efforts more efficiently, based on management objectives and a broad understanding of a range of considerations. According to the principle of "rational intervention", the desired state of any particular management unit and the proposed management tools can be completely different, depending on the land-use designation.

33.4.2 Pioneer Long-Term Forest Plans – Prominent Trends

Following the HaKedoshim Forest model, a national-scale process to prepare long-term forest plans was initiated. Analysis of the first six plans (covering ca. 2000 ha each) that were prepared during the years 2013–2016, allows a first examination of the aims and prominent trends in the current and desired state of Israel's forests, in light of IFMP principles.

Fig. 33.6 A map of current forest vegetation formations in the HaKedoshim Forest master plan

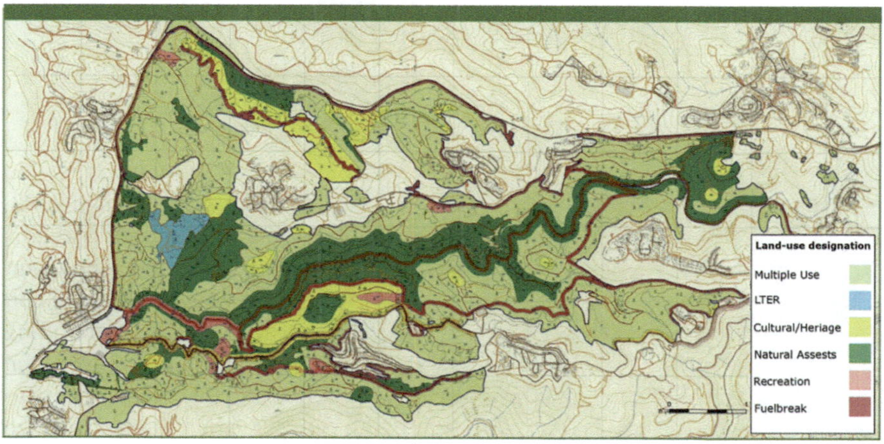

Fig. 33.7 Map of forest land-use designation units in the HaKedoshim Forest master plan

Fig. 33.8a Forestland designated for conservation of unique natural assets and habitats according to the forest master plan. A unique population of Gilboa Iris (*Iris haynei* Baker) in Mount Gilboa Forest, 2019

Fig. 33.8b Forestland designated for conservation of unique cultural landscape heritage according to the forest master plan. Reconstruction of ancient agriculture near Sataf Spring, Sataf Forest, Judean Mountains, 2019

33.4.2.1 The Importance of Forests for Provision of Cultural Services, Regulatory Services and Conservation of Unique Natural Assets

The distribution of land-use designations in the six forests for which long-term forest plans were prepared (e.g., Fig. 33.10) emphasizes, on one hand, the great importance attributed to multi-purpose forests (nearly 40% of the forestlands) and the variety of regulatory services they provide (e.g., infiltration of runoff to groundwater, prevention of erosion and floods, carbon assimilation, provision of pollinators and biological control). On the other hand, approximately 40% of the forest landscape was designated for conservation of heritage values and unique landscape formations, provision of leisure and recreation services and community forests. This fact emphasizes the great importance of the forests for the provision of diverse cultural services. Emphasis was also placed on conservation of unique natural assets and habitats (15%) while the remaining area was designated for construction of firebreaks (ca. 5%).

Fig. 33.8c Forestland designated as a firebreak according to the forest master plan. The Kisalon River Valley in the Judean Mountains, 2017

33.4.2.2 An Increase in the Structural Complexity of the Forests

Examination of the distribution of existing vegetation formations (Fig. 33.11), in the forests for which plans were prepared, shows that the planted conifer forest is still the dominant vegetation formation within the forests (48% of the landscape). The rest of the forestland is divided between natural vegetation formations (e.g., woodlands and shrublands, 26%) and planted areas of eucalyptus forests, mixed conifer-broadleaved forests, orchards and others. In addition, the structural complexity of the existing conifer forests is relatively high and includes a complex understory layer, often comprising young conifer trees and native woodland trees, shrubs and vines. Following these processes of natural regeneration and the planning and management decisions taken in the long-term forest plans, a marked increase in landscape complexity of the forests can be discerned. For example, when comparing the distributions of the current vs. desired vegetation formations (Fig. 33.12), a decrease in the proportion of the simply structured conifer forest (from 48% to 29% of the landscape) and a marked increase in the area of mixed conifer-broad-leaved forests (from 6% to 14% of the landscape) can be observed. Nevertheless, the proportion of natural woodland and shrubland areas is maintained.

Fig. 33.8d Forestland designated for leisure and recreation according to the forest master plan. Picnic area in Eshtaol Forest, Judean Plain, 2017

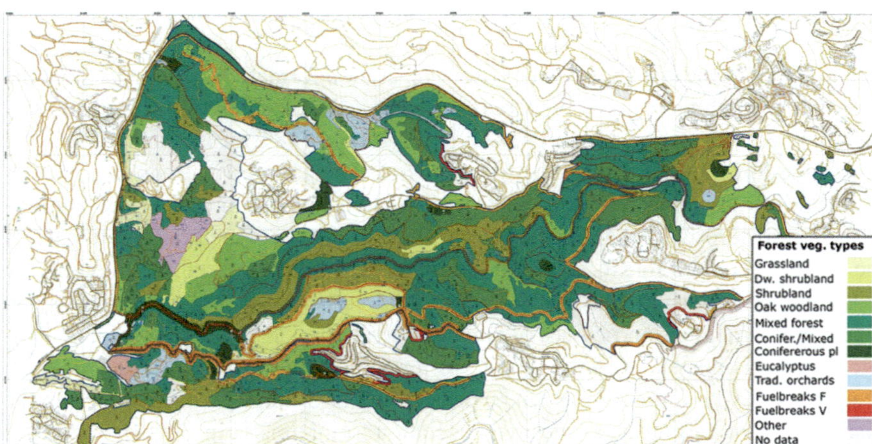

Fig. 33.9 Map of management units with the desired forest vegetation formations in the HaKedoshim Forest master plan. The desired vegetation formation is derived from the landscape designation, the existing vegetation formation and the natural process predicted to take place within it

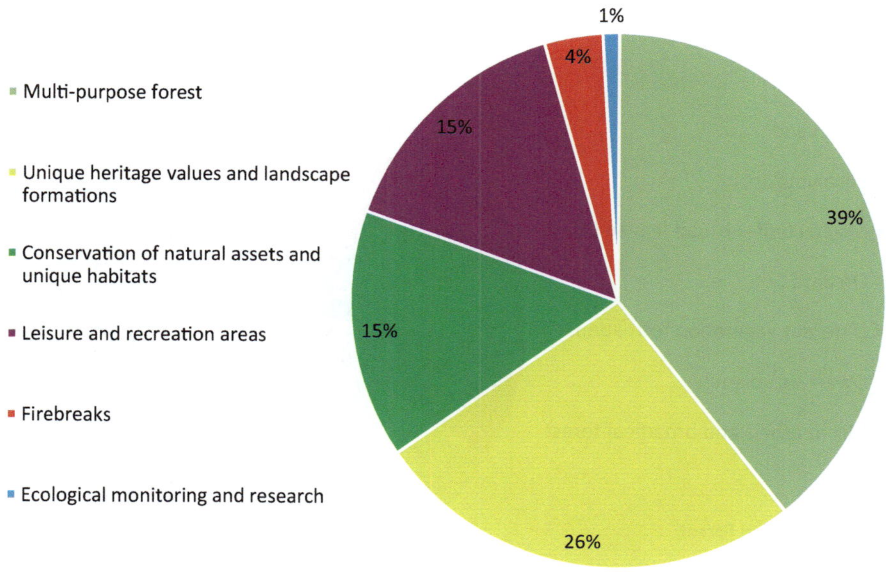

Fig. 33.10 Distribution of landscape designation of the first six forest plans

33.4.2.3 Variation among Forests of Different Regions

While the forest vegetation formations in the more humid, mountainous and hilly areas of northern and central Israel comprise mainly planted conifer forests, or natural woodlands and shrublands, the vegetation formations in the forests of the agricultural plains and along dry river courses of the drier northern Negev comprise mainly eucalypt forests, mixed forests of *Acacia* and broad-leaved species and thickets of common reed (*Phragmites australis* (Cav.) Trin. Ex Steud.). This high variation of current vegetation formations together with the diversity in the planning context of forests in Israel, create high variability in the distribution of planned land-use designations among the different forests. For example, in some of the forests of northern and central Israel which are dominated by pine plantations, ca. 60% of the forest area was designated as multi-purpose forest while in other parts of these regions dominated by natural woodlands and shrublands, approximately 90% of the landscape was designated for landscape conservation as cultural heritage sites. In forest of the drier northern Negev, however, approximately half of the forest area was designated for provision of leisure and recreation services while the other half was mostly designated for conservation of natural assets and unique habitats such as dry river banks. These trends illustrate the diversity of forest conditions in Israel and the requirement of a well-tailored long-term plan for each forest.

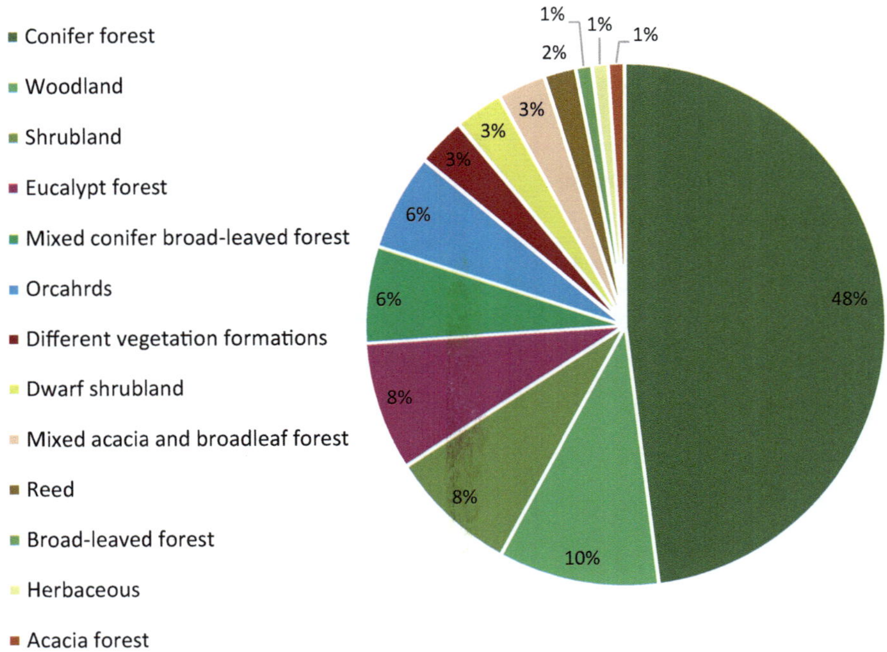

Fig. 33.11 Distribution of current vegetation formations in the first six long-term forest plans

33.4.2.4 The Importance of Long-Term Forest Plans for Making Strategic Management Decisions

An interesting example showing the effect of long-term forest plans on management decisions is the planning of Biriya Forest following an intense snowstorm in December 2013. This storm caused significant damage to approximately half the conifer forest area. Following the completion of a long-term plan for this forest ca. 60% of the area was designated as multi-purpose forest; thus, a forest rehabilitation process based on natural regeneration leading to the spontaneous development of a mixed conifer-broadleaved forest formation was chosen accordingly. In contrast, 17% of the area was designated for providing leisure and recreation services; thus, a forest rehabilitation plan based on planting a range of tree species including cedar and native woodland species was chosen. Another 9% were defined as firebreaks requiring heavy thinning and sanitation with no planting. While multi-purpose land-use units received the lowest priority for silvicultural intervention, firebreaks were prioritized as first for treatment. Following this planning process, it became clear that the main challenge in rehabilitating Biriya Forest was not the establishment of new plantings but rather the application of thinning treatments at various intensities in order to create firebreaks in some of the forested area and encourage natural regeneration and mixed forest development in most of the remaining area.

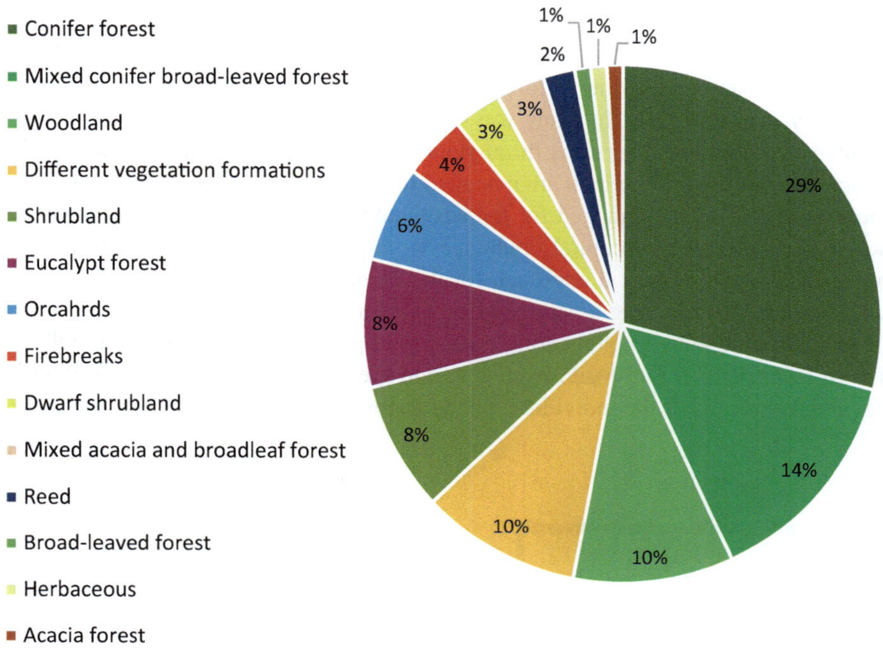

Fig. 33.12 Distribution of desired vegetation formations in the first six long-term forest plans

33.4.3 Current Status and Future Challenges

As of 2019, master plans and multi-year work plans have been completed for 12 forests covering ca. 30,000 ha of forestland (ca. 20% of the total forest area managed by KKL). As part of this process, thematic plans are being prepared for complex forest management issues that require additional in-depth planning and management detail, such as leisure and recreation in the forest, protection from fire, and conservation and rehabilitation of biodiversity and unique natural assets and habitats. Preparation of these plans allows the forest manager to focus on creating the desired forest state while managing the other unique infrastructure and assets existing in the forest. Most of the forest managers in Israel have already been exposed to the new IFMP when formulating the long-term plans and are beginning to implement it in practice. The Ministry of Agriculture and Rural Development, which regulates forest management in Israel, has adopted the new IFMP and encourages the completion of long-term forest plans for all the forestlands in Israel. Annual work plans submitted by KKL for approval by the Ministry of Agriculture are required to be based on the long-term forest plans. The long-term plans are transparent to the Israeli public through their publication on the KKL website. The KKL is facing two major future challenges: the completion of long-term forest plans for the entire forested area of Israel and implementation of these plans as part of routine forest management while considering the variety of relevant stakeholders and coordinating field activity with them.

33.5 General Conclusions

Forestry in Israel, including afforestation and forest management, has changed markedly over the years. In many ways these changes are similar to those implemented by other forest services throughout the world. The transition from simply structured forest stands, which are managed for a limited set of goals, into complex multi-purpose forests managed for the sustainable provision of diverse ecosystem services is a major silvicutural challenge worldwide. The new IFMP provides a novel, real-world, practical approach for the planning and management of forests with the aim of providing a variety of ecosystem services and balancing between them. This is done by implementing a structured process bridging between holistic sets of goals defined for broad forest areas and specific work plans which are prioritized and executed at the individual management-unit scale (Fig. 33.13). We argue

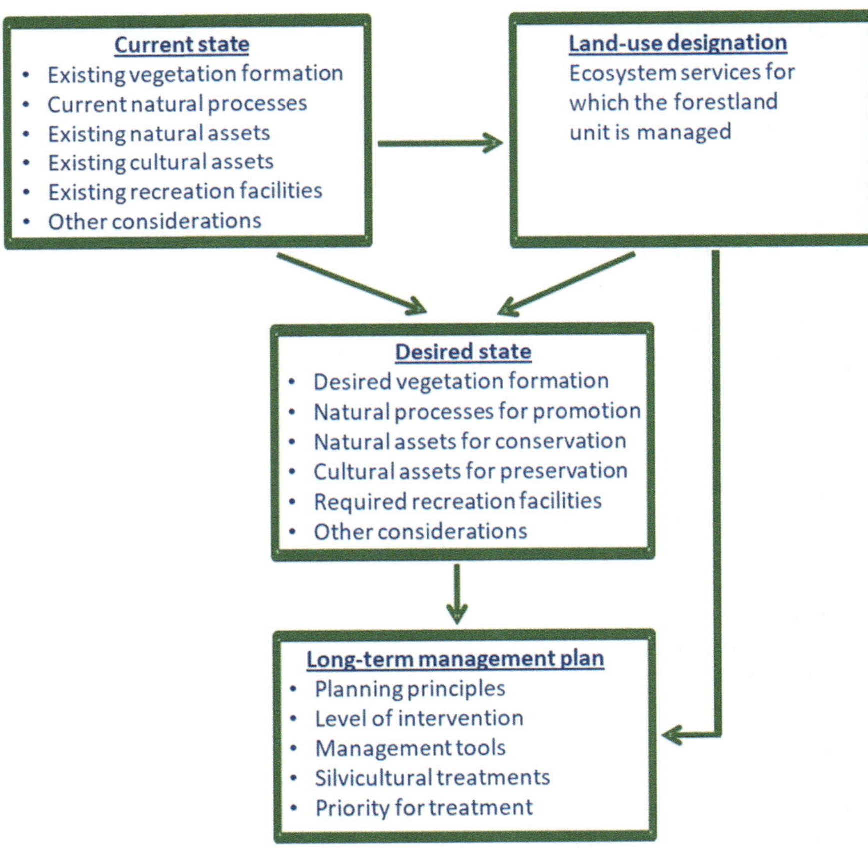

Fig. 33.13 Conceptual diagram of the long-term planning process for forest management units according to the Forest Management Policy of Israel (IFMP)

that this approach is highly relevant for a variety of forest types managed according to various forestry goals throughout the world.

References

Allen CR, Fontaine JJ, Pope KL et al (2011) Adaptive management for a turbulent future. J Environ Manag 92:1339–1345

Amir S, Rechtman O (2006) The development of forest policy in Israel in 20th century: implications for the future. Forest Policy Econ 8:35–51

Baskerville G (1986) Understanding forest management. Forest Chron 62:339–347

Bengtsson J, Nilson SG, Franc A et al (2000) Biodiversity, disturbances, ecosystem function and management of European forests. For Ecol Manag 132:39–50

Björse G, Bradshaw R (1998) 2000 years of forest dynamics in southern Sweden: suggestions for forest management. For Ecol Manag 104(1–3):15–26

Bonneh O (1996) Regeneration and rehabilitation of burned forests by KKL Forest Department. Ecol Environ 3:21–23. (in Hebrew with English abstract)

Bonneh O (2000) Management of planted pine forests in Israel: past, present and future. In: Ne'eman G, Trabaud L (eds) Ecology, biogeography and management of *Pinus halepensis* and *P. brutia* forest ecosystems in the Mediterranean Basin. Backhuys Publishers, Leiden, pp 377–390

Bonneh O (2015) Evolution of Israeli forestry from pure even-aged pine plantations to sustainable mixed uneven-aged forests. Part B: practical implications of the new policy on the forest planting stage. Forest, Journal of Forests, Woodlands and Environment 15:5–20. (in Hebrew with English abstract)

Bonneh O (2016) Overview of mixed forests in Israel. COST action FP1206 EuMIXFOR country report. [Online publication, www.mixedforests.eu]

Bonneh O (2017) Evolution of Israeli forestry from pure even-aged pine plantations to sustainable mixed uneven-aged forests. Part C: practical implications for forest tending. Forest, Journal of Forests, Woodlands and Environment 17:5–21. (in Hebrew with English abstract)

Bonneh O, Herr N, Har'el R et al (2015) The influence of four types of individual-seedlings protection fences on the survival and development of four native broadleaf species in planting sites that were subjected to cattle grazing. Forest, Journal of Forests, Woodlands and Environment 15:40–47. (in Hebrew with English abstract)

Cubbage F, Harou P, Sills E (2007) Policy instruments to enhance multifunctional forest management. Forest Policy Econ 9:833–851

Führer E (2000) Forest functions, ecosystem stability and management. For Ecol Manag 132:29–38

Gindel I (1952) Forests and afforestation in Israel. Shoham M Print Ltd, Tel-Aviv. (in Hebrew)

Goldring Y (1977) Regeneration of the natural flora in a planted pine forests. MA, The Hebrew University of Jerusalem

Götmark F (2013) Habitat management alternatives for conservation forests in the temperate zone: review synthesis, and implications. For Ecol Manag 306:292–307

Herr N, Moshe Y, Madmoni A (2011) Tabor oak (*Quercus ithaburensis*) acorn seeding as a method to establish a forest in the Ramat Menashe Biosphere Region. Forest, Journal of Forests, Woodlands and Environment 12:28–36. (in Hebrew with English abstract)

Honnay O, Jacquemyn H, Bossuyt B et al (2005) Forest fragmentation effects patch occupancy and population viability of herbaceous plant species. New Phytol 166:723–736

KKL (2004) A policy for sustainable development. KKL, Eshtaol. (in Hebrew)

KKL (2007) Green information: forest statistics. The Department of Forest Management, and GIS, the Forestry Department, Keren Kayemeth LeIsrael-Jewish National Fund, 16 p

KKL (Keren Kayemeth LeIsrael) (1990) New program for managing the planted forests in Israel. KKL, Eshtaol. (in Hebrew)

KKL, Ministry of Interior, Israel Land Administration (1995) National Master Plan NOP 22: forests and afforestation. KKL, Jerusalem. (in Hebrew)

Kuuluvainen T (2009) Forest management and biodiversity conservation based on natural ecosystem dynamics in Northern Europe: the complexity challenge. Ambio 38:309–315

Leech S, Wiensczyk A, Turner J (2009) Ecosystem management: a practitioners' guide. J Ecosyst Manage 10:1–12

Liphschitz N, Biger G (2004) Green dress for a country. Afforestation in Eretz-Israel, the first hundred years 1850–1950. KKL and Ariel Publishing House, Jerusalem. (in Hebrew and English)

Noy-Meir I (1989) An ecological view point on afforestation in Israel: past and future. Algemeine Forst Zeitschrift 24–26:614–618

O'Hara K (2014) Multiaged silviculture: managing for complex forest stand structures. Oxford University Press

Orshan G, Bar-Yosef M, Gil Y et al (1975) Working group report on Aleppo pine die-back in Sha'ar Hagay forest. La-Yaaran 25(3–4):33–35. (in Hebrew with English abstract)

Osem Y, Ginsberg P, Tauber I et al (2008) Sustainable management of Mediterranean planted coniferous forests: an Israeli definition. J For 106:38–46

Osem Y, Zangy E, Bney-Moshe E et al (2009) The potential of transforming simple structured pine plantations into mixed Mediterranean forests through natural regeneration along a rainfall gradient. For Ecol Manag 259:14–23

Osem Y, Brand D, Tauber I et al (2014) The forest management policy of Israel Forestry Department, KKL, Israel (Hebrew with English abstract) http://www.kkl.org.il/files/HEBREW_FILES/yieur/forest_management_new.pdf

Osem Y, Fogel T, Moshe Y et al (2015) Managing cattle grazing and overstory cover for the conversion of pine monocultures into mixed Mediterranean woodlands. Appl Veg Sci 18:261–271. https://doi.org/10.1111/avsc.12152

Osem Y, Fogel T, Moshe Y et al (2017) Understory structure and function following release from cattle grazing and overstory thinning in Mediterranean conifer plantations. Ann For Sci 74(1):22. https://doi.org/10.1007/s13595-017-0622-5

Perevolotsky A, Sheffer E (2009) Forest management in Israel: the ecological alternative. Isr J Plant Sci 57:35–48

Peura M, Burgas D, Eyvindson K et al (2018) Continuous cover forestry is accost-efficient tool to increase multifunctionality of boreal production forests in Fennoscandia. Biol Conserv 217:104–122

Puettmann KJ, Coates KD, Messier C (2009) A critique of silviculture – managing for complexity. Island Press, Washington/Covelo/London

Reubens B, Moeremans C, Poesen J et al (2011) Tree species selection for land rehabilitation in Ethiopia: from fragmented knowledge to an integrated multi-criteria decision approach. Agrofor Syst 82:303–330

Schiller G, Gruneald C (1987) Resin monoterpenes in range-wide provenances trials of *Pinus halepensis* Mill. in Israel. Silvae Genet 36:109–114

Scmitüsen F, Rojas-Briales E (2012) From sustainable wood production to multifunctional forest management – 300 years of applied sustainability in forestry. Working Papers International Series, Forest Policy and Forest Economics, Institute for Environmental Decisions, Department of Environmental System Sciences, ETH, Zurich

Seymour RS, White AS, deMaynadier PG (2002) Natural disturbance regimes in northeastern North America – evaluating silvicultural systems usin natural scales and frequencies. For Ecol Manag 155:357–367

Thompson I, Mackey B, McNulty S et al (2009) Forest resilience, biodiversity, and climate change. A synthesis of the biodiversity/resilience/stability relationship in forest ecosystems. Secretariat of the Convention on Biological Diversity, Montreal. Technical series no. 43, 67 p

Wang S (2004) One hundred faces of sustainable forest management. Forest Policy Econ 6:205–213

Chapter 34
Water-Based Forest Management of Mediterranean Pine Forests

Ana María Sabater, Eduardo Vicente, Luna Morcillo, Antonio del Campo, Elisabeth Karlstad Larsen, Hassane Moutahir, Francisco Pastor, José Luís Palau, Juan Bellot, and Alberto Vilagrosa

34.1 Introduction

Considerable concern has emerged in recent years regarding large-scale, climate-induced reductions in forest growth and survival. The amount and frequency of drought events, which occur jointly with warmer temperatures, have been linked to forest tree mortality and decline processes worldwide (Allen et al. 2010; Grossiord et al. 2017). Forests provide many environmental benefits, particularly those related to the water cycle, such as enhancing groundwater recharge, reducing soil erosion, and even promoting summer storms (Derak and Cortina 2014; Millán 2014; Ellison

A. M. Sabater · H. Moutahir · A. Vilagrosa (✉)
Mediterranean Center for Environmental Studies (CEAM Foundation), Joint Research Unit University of Alicante-CEAM, Alicante, Spain

Department of Ecology, University of Alicante, Alicante, Spain
e-mail: a.vilagrosa@ua.es

E. Vicente · J. Bellot
Department of Ecology, University of Alicante, Alicante, Spain

The Multidisciplinary Institute for Environmental Studies (IMEM), University of Alicante, Alicante, Spain

L. Morcillo · F. Pastor · J. L. Palau
Mediterranean Center for Environmental Studies (CEAM Foundation), Joint Research Unit University of Alicante-CEAM, Alicante, Spain

A. del Campo
Re-ForeST (IIAMA), Universitat Politècnica de València, Valencia, Spain

E. K. Larsen
Mediterranean Center for Environmental Studies (CEAM Foundation), Joint Research Unit University of Alicante-CEAM, Alicante, Spain

Royal Horticultural Society Science and Collections, Wisley, UK

G. Ne'eman, Y. Osem (eds.), *Pines and Their Mixed Forest Ecosystems in the Mediterranean Basin*, Managing Forest Ecosystems 38,
https://doi.org/10.1007/978-3-030-63625-8_34

et al. 2017). Adaptive forest management can maintain these environmental services by promoting benefits for populations. This chapter provides in-depth knowledge about Mediterranean pine forest management with respect to the water balance and ecohydrological processes.

34.2 Ecophysiology and Functional Characteristics of Mediterranean Pines Related to Water Use in Forests

In plant science, water stress is usually caused by reduced water availability (i.e., drought) and/or excessive evaporative demand, leading to soil water deficiency (Gilbert and Medina 2016). Mediterranean forests are strongly limited by a long and intense drought period that coincides with the hottest summer period. Climate projections and field simulations suggest that this situation will become even more extreme in the near future (IPCC 2014).

Plant drought responses and tolerance to water shortage are critical for their adaptation and success in most environmental ecosystems, especially in water-limited environments (Larcher 1995; Grossiord et al. 2017). Functional drought resistance strategies that species develop to face many stress factors have been the subject of many studies (Levitt 1980; Vilagrosa et al. 2003). In general, plant resistance to drought relies on two hypothetical, contrasting physiological approaches: drought avoidance and drought tolerance (Levitt 1980; Jones 1992). The underlying mechanism of either approach involves stomatal response and regulation, namely isohydric and anisohydric, respectively (Tardieu and Simonneau 1998; McDowell et al. 2008). Isohydric plants reduce stomatal conductance as soil and atmospheric conditions dry by maintaining the leaf water potential relatively constant regardless of drought intensity. In contrast, anisohydric species allow the leaf water potential to decline as the soil water potential declines with drought (West et al. 2012).

Pinus spp. and other conifers have been considered generally isohydric, drought-avoiding species, compared to other Mediterranean species such as *Quercus* spp. or *Juniperus* spp. (Chirino et al. 2011; Klein et al. 2013b; see also Resco de Dios this volume, Chap. 6; Rubio-Cuadrado et al. this volume, Chap. 9). Pine trees withstand drought by: (i) the presence of structures to control water loss through their leaves, such as a cutinized epidermis, high stomatal control and fibrous hypodermal layers; (ii) the adaptive restriction of their transpiring surface via shedding of old needles and production of shorter new needles; (iii) maintaining osmotic control capacity during unfavorable dry periods; (iv) hydraulic adjustments at the xylem level; (v) penetrating of their roots into moderately deep soil layers (Klein et al. 2011; Matías et al. 2017; Vicente et al. 2018; Andivia et al. 2019; Cherif et al. 2019). Considering these capacities, several authors have ranked different Mediterranean pines according to their drought resistance: *P. halepensis* Mill. is the most resistant species, followed by *P. pinea* L., while *P. brutia* Ten. is slightly more sensitive to unfavorable ecological conditions such as high exposure or low soil water availability. While pines from other geographical distributions, such as *P. sylvestris* L., *P. nigra* J.F.Arn

Table 34.1 Main functional characteristics in *Pinus halepensis, P. brutia* and other closely related pines (*P. nigra, P. sylvestris, P. pinea*, etc.) and sclerophyllous Mediterranean species (*Quercus* spp., *Pistacia* spp., *Rhamnus* spp., etc.)

Characteristic	*P. halepensis*	*P. brutia*	Closely related pines	Co-existing sclerophyllous species	References
Serotiny	Yes	Yes	Not always	no	de la Heras et al. (2012), Tavsanoglu and Pausas (2018)
Gs (mol m^{-2} s^{-1})	0.27	0.3–04	0.2–0.4	0.15–0.49	Awada et al. (2003), Vilagrosa et al. (2003, 2005)
A (mmol m^{-2} s^{-1})	20–25	13.4–17	8–15	7–14	Awada et al. (2003), Vilagrosa et al. (2003), Fkiri et al. (2020), Vallejo et al. (2003)
WUE (mmol CO^2 mmol H_2O)	2.1–4.8	8–10	1.5–11	1.9–6.5	Vilagrosa et al. (2003) Fkiri et al. (2020), Gratani and Varone (2004), Cregg and Zhang (2001)
SLA (cm^2 g^{-1})	54	35–40	30–50	70–150	Awada et al. (2003), Vilagrosa et al. (2003), Gulías et al. (2003), Pardos et al. (2009) Anestis et al. (2018)
Ψ_{tlp} (MPa)	−2.2–−2.6	−2.1–−3.2	−1.7-−3	−3.2-−3.9	Vilagrosa et al. (2005), Dirik (2000), Grunwald and Schiller (1988), Corcuera et al. (2002), Fernández et al. (1999)
π_o (MPa)	−1.7	1.5-−1.9		−2.6-−2.8	Vilagrosa et al. (2005) Dirik (2000)
PLC_{50} (MPa)	−4.2	−3.7	−3-−5	−4.5-−7.0	Vilagrosa et al. (2003); Delzon et al. (2010)
R/S ratio	Low	Low	Low	High	Tavsanoglu and Pausas (2018)
Rooting depth	Moderate	Moderate	Shallow-Moderate	Deep	Tavsanoglu and Pausas (2018)

Gs stomatal conductance, *A* CO_2 assimilation, *WUE* water use efficiency, *SLA* specific leaf area, Ψ_{tlp} leaf water potential at the turgor-loss point, π_o osmotic potential at full turgor, $PLC_{50:}$ water potential at 50% loss hydraulic conductivity, *R/S ratio* root/shoot ratio

and *P. uncinata* Mill. ex Mirbel are less adapted to water scarcity (Matías et al. 2017). Interspecific comparisons made with other co-occurring species, such as *Quercus coccifera* L., *Q. ilex* L., *Q. pubescens* Willd. or *Pistacia lentiscus* L., indicate that pines are more sensitive to water deficit (Baquedano and Castillo 2007; Poyatos et al. 2008; Chirino et al. 2011; Table 34.1).

34.3 Ecohydrology and Water Balance in Pine Forests

The ecophysiological characteristics of plants determine different water use patterns (Vilagrosa et al. 2014; Vicente et al. 2018). Therefore, species composition, abundance and tree density determine water balance and other hydrological processes in ecosystems. In terms of water balance components, the amount of water intercepted and used by the vegetation to maintain its functioning (i.e. transpiration) constitutes the green water fraction. The available surplus remaining in the ground (i.e. runoff, groundwater) is the blue water fraction. Therefore water balance components can be simplified into green and blue water fluxes (Falkenmark and Rockström 2006). Blue water can be directly used by society and/or feed ecosystems (i.e. rivers, hydrological cycles). Disentangling how these fluxes are driven by biotic and abiotic factors is essential for management policies that aim to guarantee both forest well-being and water supply for society.

Within the green water component, actual evapotranspiration (ETa) fluxes play a major role. Worldwide ETa can range from 40 to 70% of total precipitation in rainy forests (Van der Ent et al. 2010; Jasechko et al. 2013). These values can be higher for Mediterranean pine forests, reaching up to 90% or even 100%, depending on climate conditions (Ungar et al. 2013). Similar results have been reported for other dry and semi-arid Mediterranean ecosystems, e.g., ETa of sclerophyllous holm oak can reach 85% of total precipitation (Gracia et al. 1999). In some cases, ETa may be even higher when trees use the water stored in deeper soil layers.

The ETa term comprises transpiration, interception and soil evaporation. Transpiration (T) is the water flux that most contributes to ETa and can represent 65–90% of terrestrial ETa (Jasechko et al. 2013; Zhang et al. 2016). In Mediterranean ecosystems, these values can increase to up to 75% of overall forest ETa annually (Lawrence et al. 2007). Several studies that have focused on Mediterranean pine forests suggest that their contribution to ETa can reach values of ~50% (Table 34.1). For example, Raz-Yaseef et al. (2010) estimated T/ETa percentages at ~45–55% in *P. halepensis* forests with precipitation regimes of 224–377 mm year^{-1}. Eliades et al. (2018) estimated 37–59% for precipitation values of ~359–507 mm year^{-1} for *P. brutia*. Other pine species such as *P. sylvestris* have presented similar results with values of 45–65% (Poyatos et al. 2005). Therefore, transpiration at the plot scale is a key factor that determines the ecosystem-water balance in drylands or water-limited regions (Ungar et al. 2013). Furthermore, from comparisons among forests we know that Mediterranean pines perform more efficient water use in drier environments than in wetter ones (Klein et al. 2013a; Manrique-Alba et al. 2015; Rohatyn et al. 2018; Table 34.2), indicating that they undergo physiological adjustments when faced with aridity.

Dry sub-humid and semi-arid areas are also strongly dependent on precipitation patterns and seasonality, and not only on the total annual precipitation (see Sect. 34.3). In this context, Raz-Yaseef et al. (2012) established an effective precipitation threshold in events above 30 mm to water reaches the main location of the plants rooting systems. This highlights the importance of high intensity and depth events, such as pulse-storms, for water balance and tree water use to maintain their functionality.

Table 34.2 Annual values of transpiration (T), actual evapotranspiration (ETa) and precipitation (P) in different pine forests

Species	P (mm year^{-1})	T (mm year^{-1})	ETa (mm year^{-1})	T/P (%)	ETa/P (%)	T/ETa (%)	References
P. halepensis	120*	64.8*	114*	52*	95*	54*	Larsen, pers. comm.
P. halepensis	224	111	217 ± 12	50	97 ± 5	52 ± 3	Raz-Yaseef et al. (2010)
	231	134	248 ± 13	58	107 ± 5	54 ± 3	
	308	115	250 ± 9	37	81 ± 3	46 ± 2	
	377	156	325 ± 18	41	86 ± 5	48 ± 3	
P. halepensis	263	150	267 ± 3	57	101 ± 1	56 ± 1	Ungar et al. (2013)
P. halepensis	279	–	262 ± 5	–	94 ± 2	–	Helman et al. (2017a)
P. halepensis	279	–	244	–	88	–	Helman et al. (2017b)
	820	–	586	–	72	–	
P. halepensis	285	–	267	–	94	–	Rohatyn et al. (2018)
	727	–	583	–	80	–	
P. halepensis	394	–	161.5	–	41	–	Manrique-Alba et al. (2015)
	406	–	163	–	40	–	
	404	–	168.6	–	42	–	
	894	–	326.6	–	37	–	
P. halepensis	743	154	643	21	87	24	del Campo et al. (2014)
P. brutia	295	–	269	–	91	–	Helman et al. (2017b)
	696	–	536	–	77	–	
P. brutia	359	115	305	32	85	38	Eliades et al. (2018)
	507	309	542	61	107	57	
P. pinea	434	–	465	–	107	–	Helman et al. (2017b)
	605	–	373	–	62	–	

Nine-month values are marked by an asterisk (*). Values calculated using different methodologies are shown as the average ± standard error

34.4 Role of Mediterranean Pine Forests in Soil Water and Deep Drainage

As mentioned in previous sections, soil water deficit is crucial for plant functioning and survival. However, what determines the amount of water in the soil? Mediterranean forest soils are subjected to high evaporation and low precipitation, which usually lead to low soil water content and a year-round xeric regime in semi-arid dry-lands (United Nations Environment Aridity Index; AI = 0.5–0.2; Middleton and Thomas 1992). However, in dry sub-humid Mediterranean climates (AI = 0.65–0.5), the soil might also be subjected to marked seasonality with a precipitation season followed by a drought period.

In the driest Western Mediterranean Basin areas, deep drainage depends mostly on convective storm events that produce enough precipitation to promote a wetting front beyond the rooting zone. Under these conditions, the pine forest structure (i.e. tree cover, density) has a minor, but direct, impact on net precipitation (NetP, i.e. gross rainfall minus interception) and, hence, on percolation rates (proportion of NetP that moves downward beyond the rhizosphere) given the low interception of high-intensity events. Pine cover plays, however, an important indirect role in limiting soil water erosion and improving soil infiltration conditions. In wetter conditions, on the other hand, a percolation season may take place during wet/cool periods and hence drainage can be associated with less intense precipitation events (del Campo et al. 2019). Under these conditions, high pine forest cover and/or stocking densities can have an opposing effect regarding percolation either by reducing the NetP in light precipitation events (i.e. reducing percolation) or by reducing soil evaporation and, hence, keeping higher antecedent soil moisture levels (i.e. increasing the percolation chances for a given NetP event). Studies conducted in Aleppo pine plantations demonstrate that during wet periods, the dominant process is the former, e.g., percolation was reported to diminish with tree cover increasing from 22 to 84% to rates from 42% down to 13% of the gross precipitation (del Campo et al. 2014). Furthermore, climate and vegetation relationships, and topography and soil properties, together with their horizontal and vertical heterogeneity, further influence infiltration and percolation rates for a particular forest soil (Beven and Germann 2013).

Further insights into the factors affecting belowground hydrological processes in Mediterranean pine forests are presented in Fig. 34.1 for both semi-arid (CAL) and dry sub-humid (HU) ecosystems (del Campo et al. 2019). As expected, precipitation features (mostly depth, not shown) are the most important predictors of variation in soil moisture recharge ($\Delta\theta$) and deep drainage (D) rates, regardless of site. Evaporative conditions (meteorological conditions during precipitation) in semi-arid climates (CAL) affect NetP and have relatively more impact for both variables ($\Delta\theta$ and D) than in dry sub-humid ones (HU). Likewise, antecedent soil moisture (ASM) conditions (a state variable that defines the magnitude of soil water replenishment before percolation occurs) is of greater relative importance (RI) in dry sub-humid ecosystems (HU), where both colder climate and marked wet season promote higher and more dynamic soil water contents. However in CAL, the soil is mostly dry all year round, and only a few scattered precipitation events (storms of high intensity and marked depth) can produce significant $\Delta\theta$ and further D (thus explaining the relatively lower importance of ASM). Further evidence of this performance is observed when the RI of precipitation convectiveness (β_t, a metric derived from event intensity) is isolated from the set of precipitation event features, and its impact on $\Delta\theta$ and D, which is notably higher in CAL.

The role of ASM and precipitation characteristics on the magnitude of percolation and water distribution in the soil profile during the infiltration process has been well established (Calvo-Cases et al. 2003). Other studies conducted in dryland pine forests have reached similar results (Bellot and Chirino 2013; García-Prats et al.

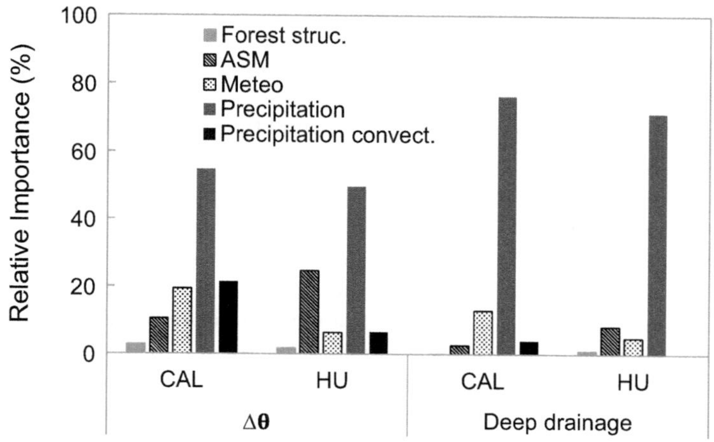

Fig. 34.1 Relative importance (RI) of different predictors on the processes of soil water replenishment (Δθ) and deep drainage in two distinct Mediterranean forests: semi-arid (CAL) and dry sub-humid (HU). The predictors are forest structure (plant cover, leaf area index and tree density), antecedent soil moisture (ASM), meteorological conditions during the precipitation event (Meteo) and precipitation event characteristics (mm, duration and intensity). Precipitation convectiveness (proportion of an event above a threshold intensity of 50 mm/h) was further isolated from the precipitation set to demonstrate its individual RI. The RI values were derived from boosted regression trees, as detailed in del Campo et al. (2019)

2016). In fact, the impact of vegetation structure, comparing to other explaining variables, is modest although it is also related to ASM.

34.5 Vegetation Management and Forest Ecohydrology

Widespread Mediterranean pine forests are the consequence of extensive reforestation programs conducted during the last century, principally in Spain and Turkey (FAO 2010). Several pine species, such as *P. halepensis, P. pinaster and P. pinea,* were out-planted in large zones of scrub, alpha grass steppes and shrublands. These actions addressed two key objectives: i) to increase vegetation cover for soil protection against erosion; ii) to increase plant biomass production. The consequence was large areas forested with pines in high densities (i.e., 1000–2500 trees ha^{-1}). Increased vegetation cover has a direct impact on precipitation partitioning and, consequently, on the distribution of the water reaching the soil surface (see Sect. 34.2). These factors condition these forests' ecohydrological responses and productivity (Porporato et al. 2001). In addition, it has been recently pointed out that pine stands with large basal areas or those densely colonized by other species can be more vulnerable to drought episodes, and there is much concern about *Pinus* spp. persisting under the most limiting conditions (McDowell et al. 2008; García de la

Serrana et al. 2015; Morcillo et al. 2019). In this context, adaptive management is needed to ensure a balance between blue and green waters in forest ecosystems, especially in the context of global change (Birot et al. 2011).

Stand thinning, as a vegetation management tool, can increase water availability for the remaining trees improving their growth and functionality (Gracia et al. 1999; Olivar et al. 2014; Helluy et al. 2020). When thinning is applied at high intensity, trade-offs take place between tree and stand regarding water use and net primary production. Tsamir et al. (2019) estimated that a reduction of 67% in tree density (from 300 trees/ha to 100 trees/ha) in a *P.halepensis* stand, increased tree transpiration by a 14%, while carbon assimilation and tree growth increased notably, around a 103%. However, at stand level, transpiration and net primary production were significantly reduced (36–47% and 3–35%, respectively). In the same way, Calev et al. (2016) showed that after removing 48% and 68% of stand basal area in a *P. halepensis* forest, individual tree basal area increased in a 43% and 143%. This experiment also addressed that thinning has the potential to reduce the effects of drought stress in pine forests, decreasing tree mortality in a 60% and 96%. Also, thinning keep shoot water potentials less negative during intense drought conditions, becoming more pronounced over time. In the same line, Manrique-Alba et al. (2020) demonstrated that a 50% reduction of tree basal area consistently decreased the whole stand water use efficiency and its drought sensitivity, even 20 years after the treatment was applied. Consequently, these effects, made the remnant trees more resilient to harsh climatic conditions and less dependent to climate fluctuations.

In terms of water balance, thinning experiments conducted in the Western Mediterranean Basin report that higher tree density and larger basal area in pine forests increase water interception and decrease runoff and water infiltration. Consequently, the green/blue water ratio increases. Some experimental studies report decreased interception with values in the range 17–63% after removing 26–76% of the tree basal area (Chirino et al. 2015; Molina and del Campo 2012). In south-eastern Spain, simulated thinning increments of 20% pine cover in two different forests reduced NetP by about 7.3% in the semi-arid area, and by 6.4% in the sub-humid one (Moghli et al. 2019; Fig. 34.2). In holm oak forests, del Campo et al. (2018) reported a 60% decrease in interception after removing 41% of the basal area. Similar results were observed in north-eastern Spain, where surface water flow increased in the short term after removing 20% of the tree basal area (principally *Q. ilex* and *Arbutus unedo* L.) (Bellot and Sabater 2018). Hence, modification of the vegetation can be managed through water-oriented forest treatments to improve soil water storage and groundwater recharge (Manrique-Alba et al. 2015; García-Prats et al. 2016).

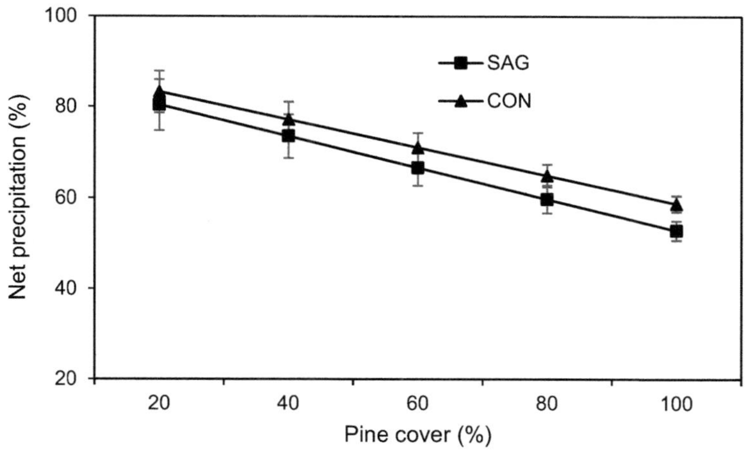

Fig. 34.2 Changes in net precipitation (% of total precipitation) when increasing the stand pine cover (%) in two study areas in SE Spain. A semi-arid area (Sierra de las Águilas; SAG; P = 299 mm year^{-1}) and a sub-humid area (Confrides; CON; P = 651 mm year^{-1}). Simulations were performed by the HYDROBAL® ecohydrological model over the 1953–2012 period

34.6 From Plant to Atmosphere and From Atmosphere to Precipitation: The Role of Pine Forests in the Water Cycle

Actual evapotranspiration from vegetation contributes to the total water that reaches air and atmospheric flows. In dry sub-humid and semi-arid areas, this contribution can be critical for producing precipitation (Wilcox et al. 2003; Eliades et al. 2018). As each vegetation type has a distinctive impact on the soil-vegetation-atmosphere water exchange, any large-scale land cover changes or forest management options, e.g., replacing forest with food crops or decreasing forest tree density, can modify precipitation patterns and local weather regimes (Spracklen et al. 2012; Millán 2014).

In the Western Mediterranean Basin, sea-land breezes are the dominant daily wind regime in summer (Millán et al. 2005; Kallos et al. 2013). This establishes a daily cycle of moisture being transported from sea to inland (Fig. 34.3). Cloudiness and potential summer storms can develop when sea breezes encounter orographic obstacles, and instabilities emerge in upper atmospheric layers (Fig. 34.3a; Millán 2014). The same study found that these summer storms can account for up to ~16% of yearly precipitation in some regions in eastern Spain. However, across the moisture path from sea and coastal areas to inland areas, some advected marine air interacts with the free atmosphere. Consequently, it loses part of its moisture content by mixing with upper and drier air masses (vertical entrainment of air masses at the upper layers of the troposphere) as shown in the study by Benetó et al. (2018). In addition, ecosystem impacts, such as deforestation, wildfires, human-rural interfaces or cities development, alter the amount of potential forest ETa that reaches the

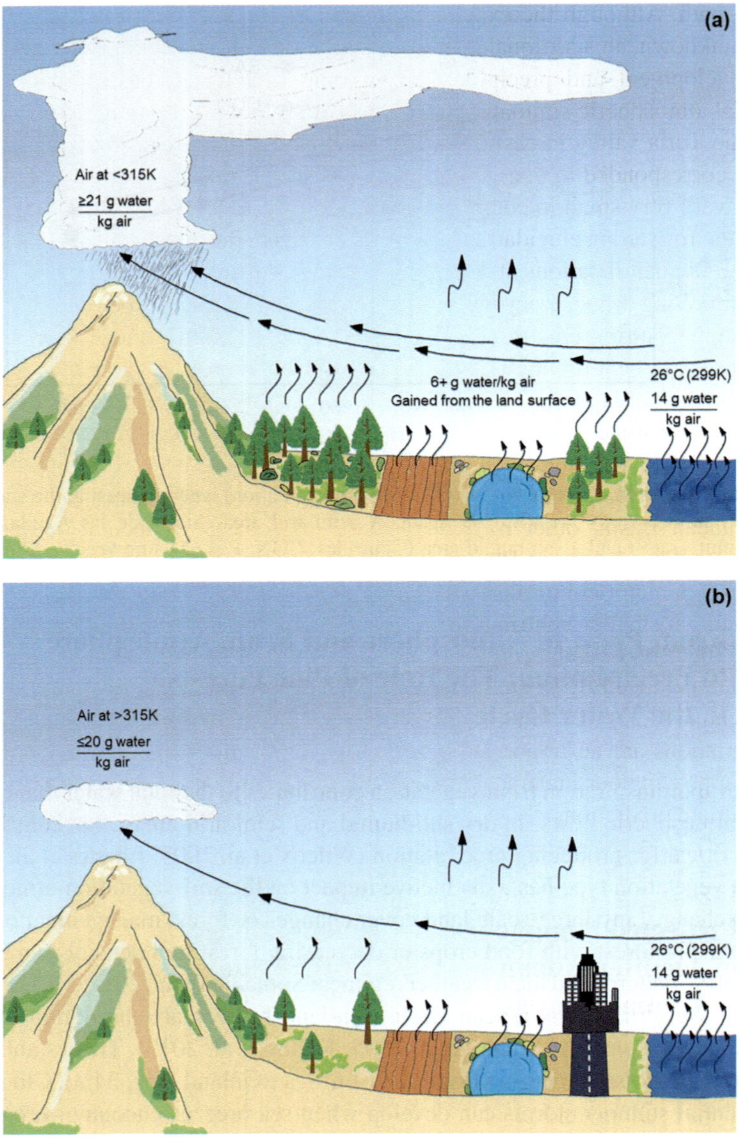

Fig. 34.3 Schematic description of different conditions determining summer storm formation: (**a**) Summer storm formation requires combined sea–land breezes to enter in order to produce the necessary water vapor content required to counterbalance its own warming along its path aboveground to keep the layer for cloud formation (CCL, Cloud Condensation Level) and storm triggering as low as possible. According to Millán (2008, 2014), in the summer months the CCL in inland parts of eastern Spain is ~2000 m a.s.l.. Under these specific conditions, $\geq$ 21 g water/kg air is necessary, with an average temperature of about 26 °C, to trigger summer storms. When air moves inland from the coast, the initial climatic water vapor is 14 g water/kg air (Millán 2014). Thus an additional contribution by evapotranspiration (7 g water/kg air) is required. (**b**) No summer storm. Water vapor content is not enough to strike this counterbalance, and the CCL rises higher than the mountain's boundary, preventing storms from forming. Adapted from Millán (2014)

moisture path. Although the exact contribution of forest ETa to total precipitable water is unknown, an additional moisture supply to sea evaporation is necessary for cloud development and precipitation triggering in inland areas (Millán 2014). Numerical atmospheric simulations of a mass of air moving along a Mediterranean valley (the Turia valley in eastern Spain) showed that the greatest contribution of moisture corresponded to marine advection, while surface and vegetation moisture supplies were very small in comparison (Benetó et al. 2018; Fig. 34.3a).

One step forward in elucidating the role of forests in the water cycle is to produce rigorous ETa quantifications at both local and regional scales, which will be essential for improving the accuracy of weather forecasts and climate modeling. Although the moisture contribution from vegetation or land is small compared to marine advection, it can be a tipping point that determines the onset of cloudiness with or without precipitation (Fig. 34.3). The moisture supply from vegetation has been shown to demonstrate similar relative importance for advection from dry soils in the southern US Great Plains (Santanello et al. 2013). In western (Lebeaupin et al. 2006; Pastor et al. 2015) and eastern (Katsafados et al. 2011) parts of the Mediterranean Basin, other numerical simulations have revealed a relationship between sea temperature and storm formation, which supports the role of marine advection as a major contributor of water and energy, with the vegetation moisture supply being a smaller contributor. Other studies argue that it is the soil moisture anomaly effect on boundary-layer stability, rather than the total amount of evaporated water, that needs to be studied to understand how soil moisture affects precipitation (Santanello et al. 2009; Seneviratne et al. 2010). Therefore, accurate forest ETa estimations are an important aspect of modeling the soil-water balance, and also for improving our understanding of the soil moisture–climate relationship and future climate predictions.

34.7 Projected Changes in the Water Balance of Mediterranean Forests in the Context of Global Change

In the future, the Mediterranean region is projected to become much drier and hotter than it is today (Giorgi and Lionello 2008; IPCC 2014; Vicente-Serrano et al. 2020). Future warming in this region is expected to exceed global rates by 25%, with projected decreases in precipitation of up to 30% and temperature increases of 2–4 °C in the 2080s for Southern Europe (Cramer et al. 2018). These changes are expected to have negative impacts on all the forest water balance components that affect the green/blue water ratio (Wullschleger and Hanson. 2006). Indeed, aquifer recharge (AR) may be the most affected, particularly in semi-arid areas, including the Mediterranean, where decreases of more than 70% are expected (Döll 2009; Ouhamdouch et al. 2019). The projected water balance in two Mediterranean pine forests in south-eastern Spain under different climate scenarios shows a significant

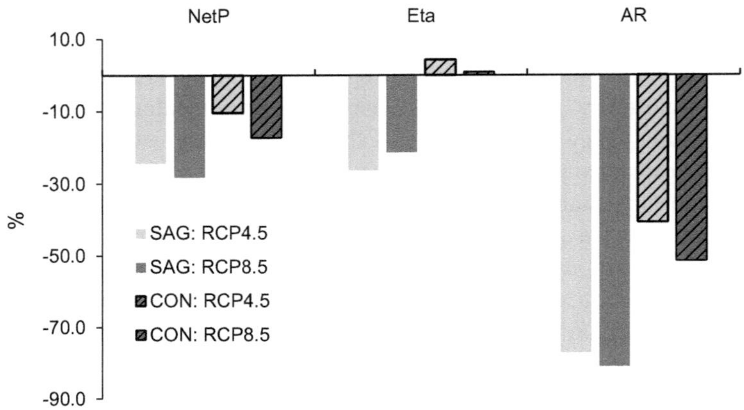

Fig. 34.4 Relative changes (%) in net precipitation (NetP), actual evapotranspiration (Eta) and aquifer recharge (AR) during the projected period 2040–2099 in relation to 1953–2012 for two climate scenarios (RCP4.5 and RCP8.5) in a semi-arid area (Sierra de Las Águilas; SAG, P = 299 mm year^{-1}) and a sub-humid area (Confrides; CON, P = 651) in SE Spain. This figure was created with data from Moghli et al. (2019)

decrease in AR, with reductions exceeding 77% in the semi-arid area (Fig. 34.4). This decrease, due principally to reduced NetP is more marked in the semi-arid area than in the sub-humid area (Fig. 34.4), and in forest pines compared to non-forested areas (Moutahir 2016; Moghli et al. 2019). Moghli et al. (2019) pointed out that the decrease in ETa is minor, and it even increases in the sub-humid area because of the relative soil water availability in this area and rising temperatures (Fig. 34.4). Increasing the water that reaches the soil by increasing the NetP through reducing vegetation density may be a solution to improve water balance and to mitigate climate change effects and die-off processes in forests (Wyatt et al. 2015). Several observed and projected studies around the world have reported water yield increase after vegetation density reduction (Harper et al. 2019; Schenk et al. 2020). Wyatt (2013) reported that water yield increased by 0–50% when basal area was reduced by 5–100% in coniferous forest worldwide. Previous studies for the Four Forest Restoration Initiative predicted that a 20–40% reduction in forest basal area was required for detectable increases in surface water yield (Neary et al. 2008; O'Donnel et al. 2016). In the Mediterranean area, similar findings were reported for the *Q. ilex* and *P. halepensis* forests indicating that water balance can be improved by management practices (Vicente et al. 2018; Manrique-Alba et al., 2020). Under projected climate conditions, Sun et al. (2015) reported that water yield increased 3–13% when the leaf area index was reduced 20–80% for the conterminous United States (CONUS). In this same work, the authors concluded that forest thinning in the dry regions has the added benefit of reducing tree water stress under climate change while increasing watershed water yield. However, it is important to mention that this positive effect of the vegetation density reduction on water balance is assumed to quickly decline after treatment due to the vegetation regrowth and forest understory and their associated increased evapotranspiration (Wyatt et al. 2015).

Predicted changes in structure and species composition could also alter forest water balance, as observed in a review of the effects of the woody plant encroachment global phenomenon (Acharya et al. 2018). At Mediterranean scale simulations performed under different climate change and vegetation cover change scenarios, including stand structure and species changes, have shown that the projected reduction in NetP can be totally or partly compensated for by reducing pine cover by 20% at sub-humid and semi-arid sites, respectively (Moghli et al. 2019). In a small watershed in south-eastern Spain (3 ha), using a distributed hydrological model and different restoration scenarios, Moutahir et al. (2019) showed that restoration with pine trees had higher effect on water balance than a mix of Mediterranean shrub species in aquifer recharge under both wet and dry conditions (−6% vs −2.3% in the wet year and − 2.2% vs −0.6% in the dry year).

The importance of monitoring forests' water balance have been remarked in each section. However, obtaining accurate estimations of each component of the water balance is complex. For estimations of tree transpiration, sap flow measurements have been presented as an accurate method. Also, tree transpiration can be up-scaled to stand transpiration by estimating stand basal area of conducting sapwood of each tree. In Mediterranean pine forests, several studies monitored transpiration with an hourly or half-hourly resolution (Schiller and Cohen, 1995; Fernandes et al. 2016; Eliades et al. 2018), but the main limitation of sap flow measurements is that only accounts for transpiration of the individuals measured. For correct estimations of shrubs transpiration and evaporation from the soil other type measurements or estimations are needed (i.e. gas exchange by infrared portable devices at plant species level are useful systems). For estimations of total evapotranspiration and carbon fixation, eddy covariance measurements are employed in many forest ecosystems worldwide (Baldocchi 2003). Eddy covariance can be employed in areas covering up to millions of square meters (Burba and Anderson 2010), but requires a homogenous surface and is expensive for small-scale research projects. Furthermore, the use of eco-hydrological modelling tools such as Hydrobal model (Bellot and Chirino 2013) and the Soil and Water Assessment Tool (SWAT, Arnold et al. 2012) can be applied in larger-scale field studies. SWAT approach requires an extensive list of parameters including vegetation characteristics and local climatic variables e.g. precipitation, soil water content and temperature. However, the benefit is that it can also estimate Mediterranean forest and shrubs ecosystem and is not limited to single species-measurements. Finally, remote-sensing-based models could be also applied to evapotranspiration estimation in a large-scales (from regional to global scale), but often the algorithms are complex and an extra-effort to data validation with in-situ measurements is required.

34.8 Concluding Remarks

Mediterranean ecosystems, particularly pine forests, are limited by water availability. Pines species are typically isohydric species capable of quickly reducing their water fluxes as the soil dries out. These characteristics, together with vegetation structure and functionality, will condition the water balance and hydrological processes in this type of forest. Pine forests can use ~50% of the total precipitation. Percolation and deep drainage in dry sub-humid ecosystems are related to antecedent soil moisture, and closely related to precipitation features (intensity and depth) in both semi-arid and dry sub-humid ecosystems. If conveniently applied, adaptive forest management practices can be a good procedure for coping with future increments in aridity in order to minimize the risk of tree drought stress and forest decline. Further knowledge on soil-plant-atmosphere water fluxes is needed to gain a better understanding of the role that forests play in the formation of summer storms, which may represent an important amount of water-cycle inputs in some ecosystems. Global climate change is likely to alter rainfall and temperature regimes, increasing the frequency and intensity of extreme events, and shifting plant functional types, which could modify interception, water infiltration, evapotranspiration, subsurface flow, groundwater recharge, and climatic feedbacks. However, these effects are largely uncertain and further research is warranted.

Acknowledgments We thank Pau Benetó and Aymen Moghli for data contribution, and Vicente Quiles for his illustration contribution in Fig. 34.3. AM Sabater is supported by European Social Fund and Generalitat Valenciana (GVA) with a PhD contract (ACIF –2017/9830). LM thanks to Spanish MICINN (PTA2019-018094). This chapter is partially funded by the Spanish Ministry of Science, Innovation and Universities through Research Projects "VERSUS"(CGL2015-67466-R), INERTIA (PID2019-111332RB-C22), BLUEWATER (PID2019-111332RB-C21), CEHYRFO-MED (CGL2017-86839-C3-2-R), RESILIENT-FORESTS (LIFE17 CCA/ES/000063) and SilvAdapt.net (RED2018-102719-T), and by the Conselleria de Cultura (Generalitat Valenciana, SPAIN) through Research Project "IMAGINA–PROMETEOII/2019/110". The CEAM Foundation is supported by the Generalitat Valenciana (Spain).

References

Acharya BS, Kharel G, Zou CB et al (2018) Woody plant encroachment impacts on groundwater recharge: a review. Water 10(10):1466

Allen CD, Macalady AK, Chenchouni H et al (2010) A global overview of drought and heat-induced tree mortality reveals emerging climate change risks for forests. For Ecol Manag 259(4):660–684

Andivia E, Zuccarini P, Grau B et al (2019) Rooting big and deep rapidly: the ecological roots of pine species distribution in southern Europe. Trees 33(1):293–303

Anéstis I, Kamaláki V, Koutsogiannáki IE (2018) Structure and functional characteristics clusters of *Pinus brutia* and *Pinus nigra* along altitude in D. Lesvos. Thesis. Environment Deparment, Aegean University, Greece

Arnold JG, Moriasi DN, Gassman PW et al (2012) SWAT: model use, calibration, and validation. Trans ASABE 55(4):1491–1508

Awada T, Radoglu K, Fotelli MN et al (2003) Ecophysiology of tree seedlings of three Mediterranean pine species in contrasting light regimes. Tree Physiol 23(1):33–41

Baldocchi DD (2003) Assessing the eddy covariance technique for evaluating carbon dioxide exchange rates of ecosystems: past, present and future. Glob Change Biol 9(4):479–492

Baquedano FJ, Castillo FJ (2007) Drought tolerance in the Mediterranean species Quercus coccifera, Quercus *ilex, Pinus halepensis* and *Juniperus phoenicea.* Photosynthetica 45(2):229–238

Bellot J, Chirino E (2013) Hydrobal: an eco-hydrological modelling approach for assessing water balances in different vegetation types in semi-arid areas. Ecol Model 266:30–41

Bellot J, Sabater AM (2018) Canvis en el cabal i l'erosió per efecte de l'aclarida parcial de la Conca de la Teula. Resultats preliminars. Actes de les cinquenes jornades sobre el bosc de Poblet i les Muntanyes de Prades. Paratge Natural d'Interès Nacional de Poblet. Departament de Territori i Sostenibilitat, Generalitat de Catalunya, L'Espluga de Francolí

Benetó P, Palau JL, Pastor OF (2018) Comparison of two land surface models in WRF during convective summer storms over the Túria Basin River. In: 16th Plinius conference on Mediterranean Risks, Montpellier (France), 9–11 October 2018

Beven K, Germann P (2013) Macropores and water flow in soils revisited. Water Resour Res 49(6):3071–3092

Birot Y, Gracia C, Palahi M (eds) (2011) Water for forests and people in the Mediterranean region – a challenging balance. What science can tell us 1. European Forest Institute

Burba, Anderson D (2010) Efficacy of using eddy covariance method for gas and energy flux measurements in disciplines and applications beyond micrometeorology

Calev A, Zoref C, Tzukerman M, Moshe Y et al (2016) High-intensity thinning treatments in mature *Pinus halepensis* plantations experiencing prolonged drought. Eur J For Res 135(3):551–563

Calvo-Cases A, Boix-Fayos C, Imeson AC (2003) Runoff generation, sediment movement and soil water behaviour on calcareous (limestone) slopes of some Mediterranean environments in southeast Spain. Geomorphology 50(1–3):269–291

Cherif S, Ezzine O, Khouja ML et al (2019) Hydraulic traits performances of three pine species in Tunisia. J Agric Sci 11(13)

Chirino E, Bellot J, Sánchez JR (2011) Daily sap flow rate as an indicator of drought avoidance mechanisms in five Mediterranean perennial species in semi-arid southeastern Spain. Trees 25(4):593–606

Chirino E, Osorio MH, Granados ME et al (2015) Balance hídrico del suelo en pinares con diferente densidad de arbolado. Efectos sobre el establecimiento de brinzales de especies rebrotadoras bajo el dosel Cuadernos de la Sociedad Española de Ciencias Forestales 41:291–304

Corcuera L, Camarero JJ, Gil-Pelegrín E (2002) Functional groups in *Quercus* species derived from the analysis of pressure-volume curves. Trees 16(7):465–472

Cramer W, Guiot J, Fader M et al (2018) Climate change and interconnected risks to sustainable development in the Mediterranean. Nat Clim Chang 8(11):972–980

Cregg BM, Zhang JW (2001) Physiology and morphology of *Pinus sylvestris* from diverse sources under cyclic drought stress. For Ecol Manag 154:131–139

de la Heras MD, Vega JA et al (2012) Post-fire management of serotinous pine forests. In: Moreira F, Arianoutsou M, Corona P et al (eds) Post-fire management and restoration of southern European forests, Managing forest ecosystems, vol 24. Springer, pp 121–150. https://doi.org/10.1007/978-94-007-2208-8_6

del Campo AD, Fernandes TJ, Molina AJ (2014) Hydrology-oriented (adaptive) silviculture in a semiarid pine plantation: How much can be modified the water cycle through forest management? Eur J For Res 133(5):879–894

del Campo AD, González-Sanchis M, Lidón A et al (2018) Rainfall partitioning after thinning in two low-biomass semiarid forests: impact of meteorological variables and forest structure on the effectiveness of water-oriented treatments. J Hydrol 565:74–86

del Campo AD, González-Sanchis M, Molina AJ et al (2019) Effectiveness of water-oriented thinning in two semiarid forests: the redistribution of increased net rainfall into soil water, drainage and runoff. For Ecol Manag 438:163–175

Delzon S, Douthe C, Sala A (2010) Mechanism of water-stress induced cavitation in conifers: bordered pit structure and function support the hypothesis of seal capillary-seeding. Plant Cell Environ 33(12):2101–2111

Derak M, Cortina J (2014) Multi-criteria participative evaluation of *Pinus halepensis* plantations in a semiarid area of southeast Spain. Ecol Indic 43:56–68

Dirik H (2000) Analysis of Pressure-Volume (PV) curves within dry season of Calbarian pine (*Pinus brutia* Ten.) provenances from different bioclimatic zones. Istanbul Üniversitesi Orman Fakültesi Dergisi A 50:100–103

Döll P (2009) Vulnerability to the impact of climate change on renewable groundwater resources: a global-scale assessment. Environ Res Lett 4(3):035006

Eliades M, Bruggeman A, Lubczynski MW et al (2018) The water balance components of Mediterranean pine trees on a steep mountain slope during two hydrologically contrasting years. J Hydrol 562:712–724

Ellison D, Morris CE, Locatellu B et al (2017) Trees, forests and water: Cool insigths for a hot world. Glob Environ Change 43:51–61

Falkenmark M, Rockström J (2006) The new blue and green water paradigm: breaking new ground for water resources planning and management. J Water Resour Plan Manag 132(3):129–132

FAO (Food and Agriculture Organization) (2010) Global forest resources assessment: main report. Food and Agriculture Organization of the United Nations, Rome

Fernandes TJ, Del Campo AD, Herrera R et al (2016) Simulnaeous assessment, through sap flow and stable isotopes, of water use efficiently (WUE) in thinned pines shows improvement in growth, tree-climate sensivity and WUE, but not in WUEi. For Ecol Manag 361:298–308

Fernández M, Gil L, Pardos JA (1999) Response of *Pinus pinaster* Ait. provenances at early age to water supply. I. Water relation parameters. Ann For Sci 56(2):179–187

Fkiri S, Rzigui T, Elkhorchani A et al (2020) Tree growth and leaf gas exchange variability of three Mediterranean Pinus spp. growing in a common garden in Northeastern Tunisia. Euro-Mediterr J Environ Integr 5(1):2

García de la Serrana R, Vilagrosa A, Alloza JA (2015) Pine mortality in southeast Spain after an extreme dry and warm year: interactions among drought stress, carbohydrates and bark beetle attack. Trees 29(6):1791–1804

García-Prats A, del Campo AD, Pulido-Velazquez M (2016) A hydroeconomic modelling framework for optimal integrated management of forest and water. Water Resour Res 52(10):8277–8294

Gilbert ME, Medina V (2016) Drought adaptation mechanisms should guide experimental design. Trends Plant Sci 21(8):639–647

Giorgi F, Lionello P (2008) Climate change projections for the Mediterranean region. Glob Planet Change 63(2–3):90–104

Gracia CA, Sabaté S, Martínez JM et al (1999) Functional responses to thinning. In: Roda F, Retana J, Gracia CA et al (eds) Ecology of Mediterranean evergreen oak forests. Springer, Berlin, pp 329–338

Gratani L, Varone L (2004) Adaptive photosynthetic strategies of the Mediterranean maquis species according to their origin. Photosynthetica 42(2):551–558

Grossiord C, Sevanto S, Borrego I et al (2017) Tree water dynamics in a drying and warming world. Plant Cell Environ 40(9):1861–1873

Grunwald C, Schiller G (1988) Needle xylem water potential and water saturation deficit in provenances of *Pinus halepensis* Mill. and *P. brutia* Ten. For Mediterr 10(2):407–414

Gulías J, Flexas J, Mus M et al (2003) Relationship between maximum leaf photosynthesis, nitrogen content and specific leaf area in Balearic endemic and non-endemic Mediterranean species. Ann Bot 92(2):215–222

Harper R, Smettem KRJ, Ruprecht JK et al (2019) Forest-water interactions in the changing environment of south-western Australia. Ann For Sci 76(4):95

Helluy M, Prévosto B, Cailleret M et al (2020) Competition and water stress indices as predictors of *Pinus halepensis* Mill. radial growth under drought. For Ecol Manage 460:117877

Helman D, Lensky IM, Osem Y et al (2017a) A biophysical approach using water deficit factor for daily estimations of evapotranspiration and CO_2 uptake in Mediterranean environments. Biogeosciences 14:3909–3926
Helman D, Lensky IM, Yakir D et al (2017b) Forests growing under dry conditions have higher hydrological resilience to drought than do more humid forests. Glob Chang Biol 23(7):2801–2817
IPCC (Intergovernmental Panel on Climate Change) (2014) Climate change 2014: synthesis report. Contribution of working groups I. II and III to the fifth assessment report of the Intergovernmental Panel on Climate Change. IPCC, Geneva, Switzerland
Jasechko S, Sharp ZD, Gibson JJ et al (2013) Terrestrial water fluxes dominated by transpiration. Nature 496(7445):347–350
Jones HG (1992) Plants and microclimate. A quantitative approach to environmental plant physiology, 3rd edn. Cambridge University Press, Cambridge
Kallos G, Mitsakou C, Alastuey A et al (2013) Mechanisms of climate variability, air quality and impacts of atmospheric constituents in the Mediterranean Region. In: Regional assessment of climate change in the Mediterranean. Springer, Dordrecht, pp 119–156
Katsafados P, Mavromatidis E, Papadopoulos A et al (2011) Numerical simulation of a deep Mediterranean storm and its sensitivity on sea surface temperature. Nat Hazards Earth Syst Sci 11(5):1233–1246
Klein T, Cohen S, Yakir D (2011) Hydraulic adjustments underlying drought resistance of *Pinus halepensis*. Tree Physiol 31(6):637–648
Klein T, Di Matteo G, Rotenberg E et al (2013a) Differential ecophysiological response of a major Mediterranean pine species across a climatic gradient. Tree Physiol 33(1):26–36
Klein T, Shpringer I, Fikler B et al (2013b) Relationships between stomatal regulation, water-use, and water-use efficiency of two coexisting key Mediterranean tree species. For Ecol Manag 302:34–42
Larcher W (1995) Physiological plant ecology. Ecophysiology and stress physiology of functional groups. Springer, Berlin/Heidelberg
Lawrence DM, Thornton PE, Oleson KW et al (2007) The partitioning of evapotranspiration into transpiration, soil evaporation, and canopy evaporation in a GCM: Impacts on land–atmosphere interaction. J Hydrometeorol 8(4):862–880
Lebeaupin C, Ducrocq V, Giordani H (2006) Sensitivity of torrential rain events to the sea surface temperature based on high-resolution numerical forecasts. J Geophys Res 111(D12)
Levitt J (1980) Responses of plants to environmental stresses: water, radiation, salt and other stresses. Academic, New York
Manrique-Alba ÀM, Ruiz-Yanetti S, Chirino E et al (2015) El balance hídrico en el suelo de 4 pinares (*Pinus halepensis*) de repoblación de la provincia de Alicante. Cuadernos de la Sociedad Española de Ciencias Forestales 41:183–194
Manrique-Alba À, Beguería S, Molina AJ et al (2020) Long-term thinning effects on tree growth, drought response and water use efficiency at two Aleppo pine plantations in Spain. Sci Total Environ 138536
Matías L, Castro J, Villar-Salvador P et al (2017) Differential impact of hotter drought on seedling performance of five ecologically distinct pine species. Plant Ecol 218(2):201–212
McDowell N, Pockman WT, Allen CD (2008) Mechanisms of plant survival and mortality during drought: why do some plants survive while others succumb to drought? New Phytol 178(4):719–739
Middleton N, Thomas DSG (1992) World atlas of desertification. United Nations Environmental Programme, Edward Arnold, London, pp 15–45
Millán MM (2008) Drought in the Mediterranean and summer floods in the UK and Central and Eastern Europe: what global climate models cannot see regarding the hydrological cycles in Europe and why. Internal Gammeltoft-RACCM-CIRCE report produced for the European Commission

Millán MM (2014) Extreme hydrometeorological events and climate change predictions in Europe. J Hydrol 518:206–224
Millán MM, Estrela MJ, Miró J (2005) Rainfall components: variability and spatial distribution in a Mediterranean area (Valencia Region). J Clim 18(14):2682–2705
Moghli A, Moutahir H, Bellot J (2019) How can vegetation cover changes affect water balance in the context of climate change? In: del Barrio G, Lázaro R (ed) Geoecology and desertification – from physical to human factors. Proceedings of the International symposium in memory of Prof Juan Puigdefabregas, Almeria, pp 20–22. ISBN: 978-84-09-12447-3
Molina AJ, del Campo AD (2012) The effects of experimental thinning on throughfall and stemflow: a contribution towards hydrology-oriented silviculture in Aleppo pine plantations. For Ecol Manag 269:206–213
Morcillo L, Gallego D, González E et al (2019) Forest decline triggered by phloem parasitism-related biotic factors in Aleppo pine (Pinus halepensis). Forests 10(8):608
Moutahir H (2016) Likely effects of climate change on water resources and vegetation growth period in the Province of Alicante, Southeastern Spain. Thesis, University of Alicante, Alicante, Spain, p 206
Moutahir H, Touhami I, Moghli et al (2019) Water balance estimation under wildfire and restoration scenarios in semiarid areas: effects on aquifer recharge. In: Advances in sustainable and environmental hydrology, hydrogeology, hydrochemistry and water resources. Springer, Cham, pp 351–353
Neary DG, Koestner KA, Poff B (2008) Water yield from harvesting and thinning southwestern mountain forests: Historical experience from US Forest Service research. In: World Environmental and Water Resources Congress 2008, Ahupua'A, pp 1–8
O'Donnell FC, Lopez SM, Springer AE (2016) Understanding forest restoration effects on water balance: the 4FRI paired watershed study B.E. Ralston (eds) In: Proceedings of the 12th Biennial conference of research on the Colorado River Plateau. Geological Survey Scientific Investigations, Reston, US, pp 79–87
Olivar J, Bogino S, Rathgeber C et al (2014) Thinning has a positive effect on growth dynamics and growth–climate relationships in Aleppo pine (*Pinus halepensis*) trees of different crown classes. Ann For Sci 71(3):395–404
Ouhamdouch S, Bahir M, Ouazar D et al (2019) Evaluation of climate change impact on groundwater from semi-arid environment (Essaouira Basin, Morocco) using integrated approaches. Environ Earth Sci 78(15):449
Pardos M, Calama R, Climent J (2009) Difference in the cuticular transpiration and sclerophylly in juvenile and adult pine needles relates to the species-specific rates of development. Trees 23(3):501–508
Pastor F, Valiente JA, Estrela MJ (2015) Sea surface temperature and torrential rains in the Valencia region: modelling the role of recharge areas. Nat Hazards Earth Syst Sci 3(2)
Porporato A, Laio F, Ridolfi L et al (2001) Plants in water-controlled ecosystems: active role in hydrologic processes and response to water stress. III. Vegetation water stress. Adv Water Res 24:725–744
Poyatos R, Llorens P, Gallart F (2005) Transpiration of montane Pinus sylvestris L. and Quercus pubescens Willd. forest stands measured with sap flow sensors in NE Spain. Hydrol Earth Syst Sci 2(3):1011–1046
Poyatos R, Llorens P, Piñol J et al (2008) Response of Scots pine (Pinus sylvestris L.) and pubescent oak (Quercus pubescens Willd) to soil and atmospheric water deficits under Mediterranean mountain climate. Ann For Sci 65(3):1
Raz-Yaseef N, Yakir D, Rotenberg E et al (2010) Ecohydrology of a semi-arid forest: partitioning among water balance components and its implications for predicted precipitation changes. Ecohydrol Ecosyst Land Water Process Interact Ecohydrogeomorphol 3(2):143–154
Raz-Yaseef N, Yakir D, Schiller G et al (2012) Dynamics of evapotranspiration partitioning in a semi-arid forest as affected by temporal rainfall patterns. Agric For Meteorol 157:77–85

Resco de Dios V (this volume) What is a Mediterranean pine? Physiological trade-offs under stress and perturbation. In: Ne'eman G, Osem Y (eds). Springer, Pines and their mixed forest ecosystems in the Mediterranean Basin, pp 103–115

Rohatyn S, Rotenberg E, Ramati E et al (2018) Differential impacts of land use and precipitation on "ecosystem water yield". Water Resour Res 54(8):5457–5470

Rubio-Cuadrado Á, López R, Rodríguez-Calcerrada J et al (this volume) Stress and tree mortality in Mediterranean pine forests: anthropogenic influences. In: Ne'eman G, Osem Y (eds) . Springer, Pines and their mixed forest ecosystems in the Mediterranean Basin, pp 141–181

Santanello JA Jr, Peters-Lidard CD, Kumar SV et al (2009) A modeling and observational framework for diagnosing local land–atmosphere coupling on diurnal time scales. J Hydrometeorol 10(3):577–599

Santanello JA Jr, Peters-Lidard CD, Kennedy A et al (2013) Diagnosing the nature of land–atmosphere coupling: a case study of dry/wet extremes in the US Southern Great Plains. J Hydrometeorol 14(1):3–24

Schenk ER, O'Donnell F, Springer AE et al (2020) The impacts of tree stand thinning on groundwater recharge in aridland forests. Ecol Eng 145:105701

Schiller G, Cohen Y (1995) Water regime of a pine forest under a Mediterranean climate. Agri For Meteorol 74(3–4):181–193

Seneviratne SI, Corti T, Davin EL et al (2010) Investigating soil moisture–climate interactions in a changing climate: a review. Earth-Sci Rev 99(3–4):125–161

Spracklen DV, Arnold SR, Taylor CM (2012) Observations of increased tropical rainfall preceded by air passage over forests. Nature 489(7415):282–285

Sun G, Caldwell PV, McNulty SG (2015) Modelling the potential role of forest thinning in maintaining water supplies under a changing climate across the conterminous United States. Hydrol Process 29(24):5016–5030

Tardieu F, Simonneau T (1998) Variability among species of stomatal control under fluctuating soil water status and evaporative demand: modelling isohydric and anisohydric behaviours. J Exp Bot 49(90001):419–432

Tavsanoglu C, Pausas JG (2018) A functional trait database for Mediterranean Basin plants. Sci Data 5:180135

Tsamir M, Gottlieb S, Preisler Y et al (2019) Stand density effects on carbon and water fluxes in a semi-arid forest, from leaf to stand-scale. For Ecol Manag 453:117573

Ungar ED, Rotenberg E, Raz-Yaseef N et al (2013) Transpiration and annual water balance of Aleppo pine in a semiarid region: implications for forest management. For Ecol Manag 298:39–51

Vallejo VR, Cortina J, Vilagrosa A et al (2003) Problemas y perspectivas de la utilización de leñosas autóctonas en la restauración forestal. In: Rey Benayas JM, Espigares Pinilla T, Nicolau Ibarra JM (eds) Restauración de Ecosistemas en Ambientes Mediterráneos. Posibilidades y Limitaciones. Servicio de Publicaciones de la Universidad de Alcalá. Universidad de Alcalá, Alcalá de Henares, pp 11–42. ISBN: 84-8138-549-2

Van der Ent RJ, Savenije HH, Schaefli B et al (2010) Origin and fate of atmospheric moisture over continents. Water Resour Res 46(9). https://doi.org/10.1029/2010wr009127

Vicente E, Vilagrosa A, Ruiz-Yanetti S et al (2018) Water balance of Mediterranean Quercus ilex L. and Pinus halepensis Mill. forests in semiarid climates: a review in a climate change context. Forests 9(7):426

Vicente-Serrano SM, Quiring SM, Peña-Gallardo M et al (2020) A review of environmental droughts: increased risk under global warming? Eart-Sci Rev 201:102953. https://doi.org/10.1016/j.earscirev.2019.102953

Vilagrosa A, Bellot J, Vallejo VR et al (2003) Cavitation, stomatal conductance, and leaf dieback in seedlings of two co-occurring Mediterranean shrubs during an intense drought. J Exp Bot 54(390):2015–2024

Vilagrosa A, Cortina J, Rubio E et al (2005) The role of ecophysiology in reforestation. For Syst 14(3):446–461

Vilagrosa A, Hernández EI, Luis VC et al (2014) Physiological differences explain the co-existence of different regeneration strategies in Mediterranean ecosystems. New Phytol 201(4):1277–1288
West AG, Dawson TE, February EC et al (2012) Diverse functional responses to drought in a Mediterranean-type shrubland in South Africa. New Phytol 195(2):396–407
Wilcox BP, Breshears DD, Allen CD (2003) Ecohydrology of a resource-conserving semiarid woodland: effects of scale and disturbance. Ecol Monogr 73(2):223–239
Wullschleger SD, Hanson PJ (2006) Sensitivity of canopy transpiration to altered precipitation in an upland oak forest: evidence from a long-term field manipulation study. Glob Chang Biol 12(1):97–109
Wyatt CJW (2013) Estimating aquifer response following forest restoration and climate change along the Mogollon Rim, Arizona. Thesis. Northern Arizona University, Flagstaff, Arizona, p 100
Wyatt CJ, O'Donnell FC, Springer AE (2015) Semi-arid aquifer responses to forest restoration treatments and climate change. Groundwater 53(2):207–216
Zhang Y, Peña-Arancibia JL, McVicar TR et al (2016) Multi-decadal trends in global terrestrial evapotranspiration and its components. Sci Rep 6:19124

GPSR Compliance

The European Union's (EU) General Product Safety Regulation (GPSR) is a set of rules that requires consumer products to be safe and our obligations to ensure this.

If you have any concerns about our products, you can contact us on ProductSafety@springernature.com

In case Publisher is established outside the EU, the EU authorized representative is:

Springer Nature Customer Service Center GmbH
Europaplatz 3
69115 Heidelberg, Germany

Batch number: 10732064

Printed by Printforce, the Netherlands